Wireless Web Development

RAY RISCHPATER

Wireless Web Development
Copyright ©2000 by Ray Rischpater

ISBN: 1-893115-20-8

Printed and bound in the United States of America
10 9 8 7 6 5 4 3 2 1

Technical Reviewer: Charles Stearns
Developmental Editor: Elizabeth d'Anjou
Copy Editor: Katharine Dvorak
Corporate Irritant and Layout Specialist: Susan Glinert/Bookmakers
Indexer: Nancy Guenther
Project Manager: Grace Wong
Cover Design: Derek Yee Design

Distributed to the book trade worldwide by Springer-Verlag New York, Inc.
175 Fifth Avenue, New York, NY 10010 USA.
In the United States, phone 1-800-SPRINGER; orders@springer-ny.com
www.springer-ny.com

For information on translations, please contact Apress directly:
901 Grayson Street, Berkeley, CA 94710-2617
www.apress.com

Dedication

THIS BOOK IS DEDICATED TO the memory of Meredith C. White and Raymond W. White, who could only imagine the possibilities of a world with the wireless Web, and to the future of Jarod Raymond Rischpater, who can only imagine the world without the wireless Web.

Acknowledgments

I AM INDEBTED TO the staff at Apress for their contribution throughout this book. Nancy Delfavero tackled this project with enthusiasm, support, understanding, and clear feedback on my initial chapters. Leading the charge from the rear, Elizabeth d'Anjou did the same, providing invaluable input on all parts of this book. Elizabeth deserves special thanks for her hard work and professionalism during the final weeks of this project, which seemed all too short for the work at hand. Gary Cornell and Grace Wong, both of whom deserve my sincere thanks, organized their efforts at Apress.

My technical editor, Charles Stearns, gave innumerable helpful suggestions during the development of the examples in this book. Equally important, he tolerated the late-night phone calls and weekend visits where I rambled incessantly about this or that detail of an example or section. I should note, of course, that despite Charles' efforts (and those of my editors), the responsibility for any remaining errors in this book falls squarely on my shoulders.

I'd like to thank members of the staff at AllPen Software, Inc., (now a part of Spyglass, Inc.) for their support and input during the development of the chapters on wireless HTML and synchronizing browsers. As a group, we at AllPen pioneered many aspects of wireless browsers and wireless content development; this book reflects many of the lessons we've learned. Similarly, I must thank the staff at AstroApps Technologies, Inc., with whom I have been able to discuss many of the issues around dynamic content development and content development for constrained devices in general.

Finally, this book would not have been possible without the love, support, and devotion of my family. My wife, Rachel, supported this project *from* its infancy, while my son, Jarod, supported it *in* his infancy. Rachel's love and enthusiasm for the project buoyed me on the days when I found writing difficult or impossible, and I owe her a world of gratitude for her gift of the freedom to write and to raise Jarod.

Contents at a Glance

Contents

Introduction

JUST AS THE DUST FROM THE Internet boom settles, the next land grab is beginning. In conjunction with advancements in wireless technology, the World Wide Web has made the dream of handheld devices for information access a reality.

Wireless data exchange is poised to be the Internet craze of the next decade. More than 200 million people currently subscribe to a wireless service, and this number is expected to reach one billion within the next five years. Around the world, providers are now rolling out third-generation cellular systems that bring digital voice and high-speed data services to subscribers. Whether you've already developed a successful presence, or you're just setting out to stake your claim, you cannot afford to overlook this exciting part of the Web.

Who Should Read This Book

If you're interested in developing a Web site for use with wireless devices in a mobile environment, then this book is definitely for you. If you're a software developer, Webmaster, or intranet administrator, you can use the techniques and tools described in this book to deliver information to your customers, wherever they may be.

In this book I discuss not only the use of traditional Web technologies for the wireless market, but also give you the knowledge you need to author content for two emerging environments: Handheld Device Markup Language (HDML), and the new Wireless Application Protocol (WAP). WAP provides lightweight devices that have very limited resources with the ability to browse content from servers in a manner analogous to the way desktop browsers access servers on the Web.

Throughout this book I make the assumption that you have some familiarity with traditional Web development, including knowledge in areas such as the following:

- **Client-Server Interaction:** Understanding how a Web browser and a server interact will help you understand how technologies such as the Wireless Application Protocol can be used with an existing Web server to reach new customers.

- **HyperText Markup Language (HTML):** Familiarity with the HyperText Markup Language is a good idea if you're going to be developing pages for wireless Web browsers.

- **Common Gateway Interface (CGI)**: As you'll soon find out, Web sites for wireless access are even more likely to be dynamic than their traditional kin. CGI's are an important tool for bringing your information to your viewers.

If, like me, you wondered as a kid how they fed the guy in the radio, or you visualized electrons rolling around inside your computer like little peas, never fear. The foundation of this book lies in understanding the development of Web content, not the mysteries of radio transmissions or how mobile devices actually work.

Moreover, I assume you're relatively new to the subject of wireless Web development. Throughout this book, as you learn how to construct Web sites for wireless users, you'll gain important insights into the differences between wireless and traditional Web content development.

A word of warning: If your intended platforms are high-speed local-area wireless networks and relatively robust laptops, much of this book may not apply to you. Wireless Ethernet coupled with traditional computers in the guise of laptops need little special attention from Web developers. (Some of the discussion in Chapters 4, 5, 6, and 7 may be of use to you, however.) But, if you're working with wireless local-area networks and significantly more constrained devices (say, a Microsoft Windows CE or Palm Computing Platform device), this book is for you.

What You Will Find in This Book

This book intersperses comments and ruminations about the nature of developing for wireless clients with examples using the latest technologies for the World Wide Web, including the Wireless Application Protocol (WAP) and Handheld Device Markup Language (HDML). It's written to be both a sourcebook of general ideas you can call upon when developing your site and a cookbook of techniques and tricks for making the most of wireless presentation.

Chapter 1, "A Wireless Data Primer," provides a primer for those new to wireless data and mobile devices.

Chapter 2, "The Wireless Landscape," introduces the wireless world, including opportunities, key features, and pitfalls.

Chapter 3, "The Wireless User Interface," discusses the fundamentals of producing wireless World Wide Web content for your subscribers.

Chapter 4, "The World Wide Web without Wires," introduces the use of existing Web protocols for wireless Web content.

Chapter 5, "Hypertext Markup Language the Wireless Way," examines the use of HTML in wireless Web content and shows you a subset of HTML appropriate for marking up wireless content.

Chapter 6, "Look, Mom—No Radio! Web Synchronization," looks at the exciting niche of content obtained by synchronization with Web servers.

Chapter 7, "Server-Side Content Management Made Easy," offers server-side techniques for enabling wireless devices, including applications of Apache's server-parsed directives and PHP: Hypertext Processor.

Chapter 8, "The Wireless Application Protocol," introduces the Wireless Application Protocol (WAP), and shows you how to mark up content using the Wireless Markup Language (WML) for screen phones and similar devices.

Chapter 9, "Dynamic Content with WMLScript," shows you how you can add dynamic behavior to your WAP content using WMLScript, a light-weight scripting language for screen phones.

Chapter 10, "The Handheld Device Markup Language," compares WAP with Handheld Device Markup Language (HDML), the predecessor to WAP. Originally designed for screen phones, HDML is a powerful alternative to WAP for some applications.

Chapter 11, "Custom Applications: When a Browser Won't Work," discusses when and why you would choose to write a custom application using the wireless protocols instead of tailoring content for a browser.

Chapter 12, "Other Technologies," wraps up things up with an overview of other technologies, including the Palm Computing Platform Palm VII and server-assisted wireless browsing that can affect Web development for wireless devices.

Appendix A, "Resources for Wireless Web Developers," provides a list of resources, including browsers for testing and software kits for developing your wireless Web content.

Appendix B, "The Unified Modeling Language for Web Developers," reviews the Unified Modeling Language (UML), the visual language for modeling relationships within computer systems that I use in many of the figures in this book.

How You Can Use this Book

Of course, you're free to dip into this book wherever you wish, and skip through it as you please. But if you're new to the wireless frontier, I recommend you plan on starting with the first two chapters. Those with a technical streak will enjoy Chapter 1; those with an interest in marketing and business may find Chapter 2 a better place to start in this book.

You should definitely read Chapter 3, as in it I introduce many of the constraints that define the wireless Web. You'll face these constraints when developing your content, and I refer back to them throughout my discussion of HTML, WAP, and HDML.

With an understanding of the capabilities of the wireless Web under your belt, you should feel free to roam through the remaining chapters at whim. Readers interested in traditional Web technologies applied to wireless markets will want to focus on Chapters 4, 5, 6, and 7; Chapter 7 will be especially interesting to many of you, because it discusses synchronizing existing Web content to mobile devices. Readers developing for the wireless phone market will want to read Chapters 8, 9, and 10, which discuss the technologies available for these platforms. Chapters 10 and 11 provide answers for readers interested in creating custom wireless applications, or investigating other technologies, respectively.

If you've already spent time working with wireless content or mobile devices, you may want to use this book to fill in any gaps in your knowledge, or as a reference. Flip right to any chapter that interests you and dive in. Go right ahead!

A Word on Presentation

As with other technical books, it helps to make a distinction between what's meant for people to read and what's meant for computers to read.

Any text in this book that looks `like this` is either a tag in one of the Web markup languages, or a variable or statement in some computer language being discussed. Whole listings of code or markup languages are set in the same style, as shown here:

```
<HTML>
<META HandheldFriendly=TRUE>
<TITLE>Hello world!</TITLE>
<BODY><P>Hello world!</P></BODY>
</HTML>
```

It is widely held that a picture is worth a thousand words. I've tried to use illustrations in this book for two purposes: to show you the results of marking up content using one of the wireless Web markup languages (HTML, HDML, or WML), or to describe the behavior of a system in some way. To represent the operation of a system, I use the Unified Modeling Language (UML). UML provides a powerful way to represent different aspects of systems in a compact notation in a clear and intuitive manner. If you're new to UML, you may want to read Appendix B first for an introduction.

Resources

Throughout the book, I discuss developer tools and resources for HTML, WAP, and HDML authors. For the latest information regarding these and other developer resources and where you can find them, see Appendix A.

Looking Ahead

The development of the wireless Web is in its infancy. This is the time for new play-ers and new ideas. I'm pleased to be sharing how to develop for the wireless Web with you, and look forward to seeing the content and solutions you develop.

A Wireless Data Primer

To DEVELOP GOOD WIRELESS CONTENT, you need to understand something about the wireless marketplace. Over the last ten years a dizzying array of technologies, services, and companies contributed to today's wireless landscape. Unlike traditional computing, where one-year-old technology can be pitifully obsolete, wireless technology can persist for years after its development. Many of today's most successful wireless networks, for example, were developed over ten years ago. Knowledge of these networks reveals some of the strengths and weaknesses of the wireless Web.

In this chapter, I walk you through a brief history of the wireless data industry, and then review some of the technologies behind today's wireless networks. (Those readers eager to get to the market opportunities are invited to skip this chapter and move along to Chapter 2; if you're ready to start developing, skip ahead to Chapter 3. Either way, refer back to this chapter when you're curious about a particular technology or term.) Because many of today's wireless data communications are coupled with voice services, I present the development of wireless data through the deployment of voice networks. Understanding a wireless voice network will help you to better understand a wireless data network.

A Bit of Wireless Data History

The history of the consumer side of wireless data is closely tied to the history of wireless telephony. By the early 1940s, several thousand police squad cars equipped with two-way mobile radios established the importance of wireless communications. Making use of point-to-point communications from a central location, those early systems became an essential part of police communications. In 1946, the first public mobile telephone systems were introduced in 25 United States cities, providing businesspeople and wealthy individuals the ability to make calls to destinations 30 miles or more away, while in transit.

Typically these mobile systems were not connected to the public telephone system. Rather, coverage over a metropolitan area was provided by a central service that used preassigned channels to encompass mobile radios, which greatly limited the number of potential subscribers. Thus, the early mobile network expanded slowly due to limitations in available frequency and technological shortcomings. During the 1950s and 1960s, telephone companies around the world worked to address these limitations by developing the theories behind cellular radio technology.

The First Cellular Systems

When AT&T proposed a cellular telephone network to the Federal Communications Commission (FCC) in the late 1960s, the principles of cellular communication were known, but the technology to implement a cellular telephone network wouldn't be available until more than a decade later.

The first cellular system was deployed in 1979 by the Japanese firm Nippon Telephone and Telegraph Company. It was followed in 1981 by a cellular system that covered much of Europe. In 1983, the FCC finally allocated enough radio *spectrum* (that is, frequencies) to support the first domestic cellular telephone network. This network, called the Advanced Mobile Phone System (AMPS), remains in place today, providing what is typically called "analog cellular" service to customers via cellular providers.

Industry Partnerships

Over the past 15 years we've seen a series of revolutionary movements in both wireless telephony and wireless data. In the 1980s, International Business Machines and Motorola joined forces to form ARDIS, a separate entity responsible for providing wireless data to mobile workers. Originally targeted at providing wireless messaging for IBM and Motorola field service technicians, the ARDIS network was rapidly adopted by outside clients such as Otis Elevator.

The ARDIS system provided nationwide service and remains one of the country's largest wireless networks with coverage in more than 11,000 cities. One characteristic of ARDIS is its *deep-building coverage*: antennas and power are calibrated to support mobile users who are deep inside buildings and well shielded from radio energy. Mobile workers can remain on the wireless network from within elevator shafts, basements, and other hard-to-penetrate areas.

In 1986, Ericsson developed the Mobitex protocol, which is the foundation of the RAM Mobile Data network. This network was initially established in Europe and the United States to provide wireless data to vertical markets, such as field technicians and service engineers. BellSouth acquired RAM Mobile Data Limited in June 1998 and deployed 13 networks worldwide based on the Mobitex standard. Both the ARDIS and RAM networks provide users in the United States with dedicated data networks, while the RAM network provides service in the United Kingdom, Australia, the Netherlands, Belgium, Singapore, and other countries.

Recent Advances

In the early 1990s, advancements in cellular voice technology enabled existing cellular networks to be used for data transmission by employing Cellular Digital

Packet Data (CDPD). CDPD works with existing first- and second-generation AMPS cellular systems, enabling simple and inexpensive installation in areas already covered by AMPS. More recently, the advent of digital cellular networks, including Global Service for Mobile (GSM) in Europe along with the various personal communication services (PCS) networks in the United States, has provided an even broader range of services for wireless data.

By the early part of the next decade, there will be as many wireless subscribers as there are wired service subscribers. In many developing countries, for example, wireless networks are being deployed faster than their traditional wired counterparts because the cost of installation, service, and upgrades is significantly lower than that of wired networks. With the cost of wireless installation rapidly falling as standards converge and the volume of the installed base increases, costs to providers and consumers have similarly dropped. As a result, the rate at which the public has adopted wireless services is on a par with the rate at which the videocassette recorder became a household item. In fact, wireless services have caught on much faster than other significant product developments, including the automobile and the telephone.

A Bit of Technical Talk

If you're interested in the nuts and bolts of wireless data technologies, you'll probably find the material in this section fascinating. If you're not, you may regard this section as optional reading. (Then again, if you need to separate fact from fiction when talking to the providers of specific services, the following information may come in handy.)

The Basic Concept behind Cellular Radio

The whole concept of cellular radio communications is quite simple. Rather than using a single high-power station with multiple channels to cover a large area, many smaller stations are distributed across the same area. Each station uses a different subset of the same channels formerly used by the high-power station. This way, adjacent stations (each of which covers an area called a *cell)* do not interfere with each other, much as radio stations in different parts of the country can use the same frequency without getting in each other's way. Figure 1-1 shows a hypothetical cellular network.

Cellular network users use *wireless access terminals* to access the wireless network. These devices may be wireless telephone handsets (that is, "cell phones"), screen phones, or wireless terminals, depending on the network and product. All of these devices are small, mobile, and are intended to be carried around by the subscriber. They communicate with *base stations*, which are fixed stations responsible

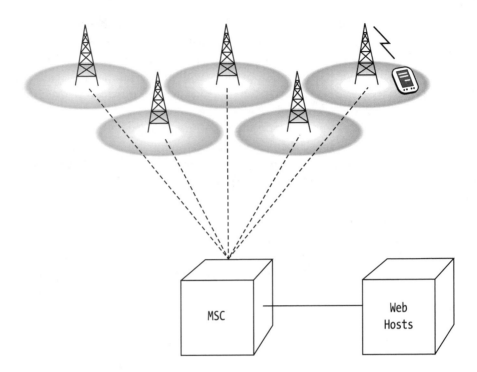

Figure 1-1. A simplified cellular network

for coordinating all wireless access terminals within a specific cell. In turn, the base stations coordinate with a *mobile switching center* (MSC), which is responsible for coordinating activities between base stations.

In a cellular system, channels are reused only when sufficient geographic distance exists between them to prevent interference among shared frequencies. The challenge here lies in designing and deploying a cellular system that can manage and maintain these frequencies, and track the callers who are in contact with each cell station as they move between cells.

A cellular base station moderates the reuse of frequency on behalf of the wireless terminals in its cell in accordance with geography, system use, and radio propagation. As the wireless terminal physically moves between multiple cells (typically when carried by a user in motion, such as a pedestrian or driver), it receives instructions regarding the frequencies to be used for communications. Cellular radio works as it does because of the increased investment by providers in the system's individual stations and the growing complexity of the wireless terminals.

Packet-Switched Radio Networks

Traditional telephony networks are *circuit switched*. With circuit switching, the network establishes a point-to-point link between two nodes for the duration of the communication of a message. The analog phone system also worked this way, but with the advent of digital networks, this is no longer the case with a large number of calls.

The Internet and the wireless networks I discuss in this book generally operate as *packet-switched* networks. Over a packet-switched network, the message originator breaks up a message into many small pieces, or *packets*. Each packet may take a separate route to its destination. When the packets reach their destination, they are reassembled to form the message sent. The originator and receiver must negotiate the details about the ordering of packets, what to do about lost or corrupt packets, and similar details.

The BellSouth RAM network, ARDIS, and other similar networks are examples of *wireless* packet-switched networks, whereas the Internet is an example of a *wired* packet-switched network. On the BellSouth RAM network, messages are broken into packets and exchanged using the Mobitex protocol, which was originally developed by Ericsson (mentioned earlier in this chapter). ARDIS, on the other hand, uses the MDC4800 and RD-LAP protocols developed by Motorola.

Most content developers don't have to worry about "raw" wireless protocols because their interaction will be with the Internet protocols or the *Wireless Application Protocol* (WAP), a standard protocol for the distribution of wireless data independent from a specific network. Many packet-switched networks now support Transmission Control Protocol/Internet Protocol (TCP/IP) directly through gateways that move IP traffic from the Internet to a selected wireless network. On the client-side device, a customized TCP/IP stack is installed that uses the same wireless interface and protocols as the underlying network (called the *bearer* or *bearing* network) for TCP/IP traffic.

Frequency Division Multiple Access

Central to the notion of wireless communications is the need to share frequency between multiple users. One of the simplest schemes for accomplishing this is *Frequency Division Multiple Access* (FDMA). Used by AMPS, this system works by assigning specific frequencies to a user for the duration of a communication. During the course of a session, an assigned frequency is only used by its assignee. FDMA is generally used with analog systems or simple digital systems, such as local-area or medium-area cordless phones that may use digital encoding schemes to carry voice traffic over assigned frequency channels.

Time Division Multiple Access

Time Division Multiple Access (TDMA) is a mechanism used by packet radio networks. TDMA assigns each user a specific time slot over a unit of time called a *frame*. Time slots may be reserved by the protocol in order to coordinate new users within a defined area (for example, to facilitate roaming between adjacent cells). The system is essentially a "buffer-and-burst" system, in which units buffer data to be exchanged until it's their turn to transmit.

The CDPD system discussed earlier in this chapter uses a blend of FDMA and TDMA to carry IP-based traffic over the AMPS network. Callers on the network are assigned a fixed pair of frequencies (one for the downstream channel from the cell to the wireless terminal, and one for the upstream channel from the wireless terminal to the cell) for the duration of a call within a specific cell. Cells typically have unused frequencies ready to assign to users to satisfy demand during peak operating times. In addition, unused gaps of time on the frequency between two calls are commonplace. The CDPD system employs TDMA to allocate use of these gaps on a particular frequency among multiple users with small packets of data, and uses FDMA to allocate the use of frequencies among voice users.

Pure TDMA systems are also found. The United States Digital Cellular (USDC) system uses TDMA, as does the GSM in Europe. In these systems, voice traffic is digitized and then broken into packets for telephony. The digital nature of these networks gives them the ability to carry pages and short messages, such as the Short Messaging System (SMS) available under GSM, in which messages are carried as digital content between packets of voice data. These networks are well suited to supporting the WAP.

Frequency Hopped Multiple Access and Code Division Multiple Access

Other, more complex systems also exist for sharing multiple users. *Frequency Hopped Multiple Access* (FHMA) and *Code Division Multiple Access* (CDMA) are applications of *spread-spectrum technology*, which are designed to share a wide allocation of spectrum among multiple users.

FHMA

FHMA uses many small chunks of frequency over time, each lasting only a split second. Each user on an FHMA network is given a unique code that defines a particular pseudo-random set of frequencies. Although frequency use appears to be random, a receiver that knows the unique code of the transmitter can calculate the frequencies on which to listen for it. The user's terminal transmits on each frequency in the set for a small amount of time, which makes a user's signal appear to "hop" from frequency to frequency for the duration of a message.

The receiving station uses that same sequence to tune its receiver and reconstruct the message. The hops may be faster than the signal rate, in which case the system is called a *fast-frequency hopper*. Or, the hops may be slower than the signal rate, in which case the system is a *slow-frequency hopper*. The receiver selects the hopping sequence appropriate for the given user. FHMA is popular among local-area and medium-area networks, such as those employed by wireless PCMCIA (Personal Computer Memory Card International Association) cards.

CDMA

By comparison, CDMA uses one relatively large frequency region continuously by multiplying the original signal with a wide bandwidth signal (called a spreading function). Like FHMA, CDMA is based on a pseudo-random sequence generated from a unique code, which determines the spreading function to be used. (Conceptually, a CDMA system can be compared to an extremely fast FHMA system in which the hop rate is of a magnitude many times higher than the signal rate.) Reception in a CDMA system requires monitoring the resultant high-bandwidth signal and correlating the signal with the sequence of the transmitter.

Qualcomm uses CDMA in its successful digital cellular products, providing the backbone to one of the PCS networks found in many areas. Like FHMA and TDMA, CDMA can be used to exchange packet data for wireless data. The network protocols devised by Qualcomm support data and voice traffic, which is carried by digitizing and compressing analog data and then breaking the data into packets, which are then sent via CDMA to their destination.

Putting It All Together: How a Cellular System Works

Whereas the cellular system concept is fairly simple to grasp, the actual system for supporting a cellular network is staggeringly complex. Figure 1-2 shows an example of such a system. The network must provide a means for the setup and handoff of users between cells, as well as authorization, traffic routing, and billing.

Initial Information Exchange

When a user enters a region serviced by a cell station, his or her wireless terminal announces its presence and requests service from the mobile switching center. Typically, the wireless terminal listens for the strongest cell station. That station then gives the terminal instructions on how to announce its presence to the cell

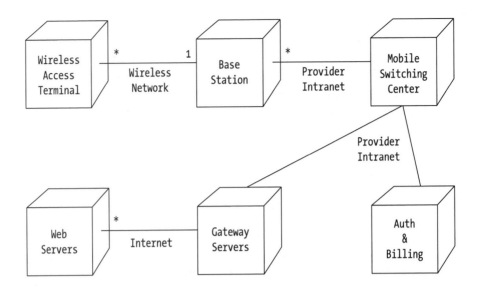

Figure 1-2. Deployment view for a hypothetical cellular network

station from which it will seek service. During this exchange, the wireless terminal is authenticated, billing information is processed, and instructions for operating within the cell (such as frequency, code, or time constraints for multiple access) are provided to the terminal.

This dialog, collectively referred to as *registration,* must occur whenever a terminal roams from cell to cell. Registration provides hosts on the network with the information needed to route data to a specific wireless terminal.

Competing networks support so-called "roaming" features. Therefore, if a terminal roams outside of its native service area, details on subscribed services, such as voice calling preferences for three-way conversations or call forwarding, may still be obtained.

Movement Monitoring and Intersystem Handoff

As the terminal moves within the cell, the base station continually monitors the terminal's signal strength and passes the measurements to the MSC. When appropriate, the MSC instructs the base station and unit to perform a *handoff,* causing the unit to begin using a new base station. This may be done entirely through the action of the base station monitoring the unit, or with the assistance of the terminal if the unit is capable of monitoring adjacent cell activities during its normal operation. Less frequently, an MSC may initiate an intersystem handoff—enabling the terminal to "roam"—in which it hands the terminal to an adjacent MSC. This is done in the event the terminal moves into an area covered by another provider, or when the terminal moves out of the MSC's coverage area.

Data Exchange

While the terminal's movement and signal strength is constantly monitored, the data itself is being exchanged between the wireless terminal and base station. This data is carried to the MSC, where it is brought to the Internet through one or more gateways responsible for translating the wireless protocols to the appropriate Web protocols.

At the MSC, gateway servers are responsible for translating the various protocols used by the wireless network and the Internet on behalf of hosts on both networks. For wireless access by WAP, the protocols of the bearer network may be entirely different than those used by the Internet. In addition, the WAP gateway provides an application-level translation from the WAP to Web protocols, and vice versa.

Summary

The wireless data marketplace has exploded after a nascent period that has lasted for the past ten years. Wireless networks moved from being special-purpose tools for a few mobile workers and wealthy consumers to being on the brink of widespread adoption by average consumers.

Most commercial wireless networks are based on existing cellular networks, any one of which use one of a number of schemes to share frequencies between mobile wireless access terminals over a wide area. These networks divide a geographic region into cells, each of which has a station responsible for bridging the gap between wireless terminals and wired network.

This path from wireless network to wired Internet involves hops from a cell's base station, through one or more switching centers, through a gateway server, and then to an Internet destination. Packets from the Internet follow a similar route in reverse when finding their way to a wireless access terminal.

CHAPTER 2
The Wireless Landscape

THE NATURE OF INFORMATION ACCESS is undergoing a quiet revolution. As the World Wide Web has become available to consumers around the world via their home computers, a similar type of network is emerging that eliminates the need for a computer or wires to access the Web. As this network emerges, more and more users are accessing Internet content through wide-area radio networks, such as those commercially deployed by the major telecommunications companies.

This new *wireless network* consists of millions of small, portable devices, readily available whenever and wherever a user needs information. Handheld computers, smart phones, and similar gadgets are becoming increasingly available with wireless connectivity options.

An Explosive Growth Industry

As more and more companies employ the Web to integrate various sources of information with their partners' and consumers' information systems, the Web's value as a reference tool has risen exponentially. With today's rapidly growing wireless infrastructure, is becoming more and more possible to access information from the Web from virtually any source at any time.

The growth of wireless Web access can be traced to the scores of companies who invest big bucks in this emerging technology hoping to get a piece of the revenues from this growing market. These companies provide the infrastructure, hardware, and software necessary to make wireless Web access happen. They recognize the commercial potential of acquiring millions of wireless data subscribers over the next several years.

The cellular phone is a perfect example of this business development. The growing public demand for cellular phones over the past 15 years has resulted in one the most successful new industries of this century. Hundreds of millions of wireless handsets have been sold to date. It is estimated that by 2005 more than one billion wireless handsets will be in use. Many, if not all, of those handsets will be equipped to handle both voice and data.

Uses of the Wireless Web

Handset owners will use wireless devices whenever and wherever computers are impractical. For instance, a laptop may be usable in some instances where a desktop computer isn't, but access to the Internet might not be available. For workers who need a computer when they're on the road or out in the field, handheld wireless computers are available to meet a variety of needs at a relatively low cost. There are also wireless handset users who won't even own a desktop computer.

All of these people will turn to the wireless network for up-to-date information and services. In fact, the trend has already begun. Some pager subscribers receive customized news updates throughout the day, and other users are turning to phones with Internet access.

This interaction has a number of useful applications. For example, a family on vacation in an unfamiliar town can get instant travel directions. With these directions, they can find a hotel, reserve a room, and find a restaurant for dinner, all without needing to refer to a printed map or ask for directions. A student can use a wireless terminal to look up the meaning of an unfamiliar word during a classroom lecture, without interrupting the lecture. Consumers can comparison shop across the Internet while within a store. As the wireless Web grows, interaction with the Web will become more immediate and more tailored to the user.

Then there is wireless *pay-as-you-use* service. With two-way data networks, some phone services can accept a credit card number entered directly into the handset to pre-pay for a number of days of data use or minutes of voice use. Users can buy a shrink-wrapped phone and subscribe to a service as they need it. This gives subscribers unprecedented budget flexibility while lowering paperwork costs for service providers burdened with contracts administration overhead. In addition, some phones will feature an automated recharge capability right from the handset.

Opportunities for Wireless Development

Opportunities for developers of wireless content exist in several areas, including:

- Partnerships between existing content providers and wireless Web developers to offer existing content to subscribers,

- Partnerships between existing wireless service providers and content providers to make new content available to subscribers,

- Offerings of new wireless Web content.

In many cases, the opportunities will include a mix of all three, either initially or after consolidation rolls together content and service providers. The purchase of BigBook by GTE, a union of content and service providers on the traditional Web, will be a model operation for many wireless content providers.

Content Provider Partnerships

Existing providers are working to move to the wireless Web. Consider the following:

- Yahoo! Anywhere gives users of the Qualcomm pdQ phones access to Yahoo! Instant Messaging wherever they are.

- MapQuest To Go! provides turn-by-turn directions to users of AvantGo software and the Palm VII.

- The Internet Movie Database provides access to information on the cast, crew, and reviews of movies for both Web and wireless Web viewers.

In most cases, content providers offer these services as they position themselves for the coming generation of wireless subscribers. In other cases, such as the Internet Movie Database, the initial work was done outside the content provider, and only later was the content provider brought into the mix.

The business models behind these partnerships vary. In many cases, wireless Web developers will be able to use their skills as contractors or employees to participate in the building of wireless sites. Less often, developers may be able to share in the ongoing subscriber revenue stream from wireless services.

Service Provider Partnerships

In the short run, wireless service providers will be seeking content to enhance the value of their wireless services. During this time, some service providers may be willing to help finance not only the start-up but also the ongoing maintenance costs incurred by a wireless content provider, especially for new and novel content. This relationship may enable enterprises with limited funding to break into the wireless marketplace.

Over the long run, partnerships with wireless service providers will amount to endorsements, and the business enterprise of the wireless content provider will benefit from the increased usage volume from referrals.

Although service partnerships are helpful, they are by no means required. A look at the Palm.Net wireless service from Palm, Inc., is instructive. The service launched with wireless access to ABC News, ESPN, E*Trade, and other services. Today the service has access to even more providers, including Excite, My Documents To Go, and Starbucks Coffee, among others.

New Wireless Content

The most lucrative opportunity for wireless Web developers is in the creation and deployment of new wireless content. A developer is free to obtain content, process it, and make it useful to the mobile user as he or she sees fit.

Opportunities abound in enterprise and business applications, enabling corporations to provide greater access to increasing numbers of mobile users. A similar, untapped resource also exists in the consumer marketplace. Horoscopes, jokes, product reviews, travel reviews — anything a consumer would want can be provided wirelessly.

While some of these opportunities may require third-party partnerships with existing content providers, others will not. Imagine a consumer reporting site dedicated to providing purchase point information about thousands of products in different categories, organized by category and Universal Product Code. Just such a service is in the making: it's called BarPoint.com. Here, content editors aggregate content from distributors, reviewers, and wireless consumers around the country to create an entirely new business. Other organizations will choose to buy content and repackage it, as Excite has done with its presentation of stock quotes from Standard & Poor's ComStock. In either case, businesses are able to provide content that wasn't previously available, just as the revolution of geographic mapping data now available on the wired Web made new kinds of content available to wired Web users.

Fundamentals of Wireless Development

Once you've staked a claim in the wireless world, it's time to get down to business. Two fundamental principles demand a new approach to developing Web content for the wireless market. These principles are:

- Consumers become subscribers.

- Subscribers are mobile, which makes the data appear to be mobile.

Understanding these principles and knowing how they interrelate is key to designing truly valuable wireless sites.

Subscribers versus Consumers

In the wireless marketplace, device users are *subscribers*. In addition to renting or purchasing the device they use to access a network, they also pay—on a regular basis—for access to the service that provides the data. Subscribers provide a reoccurring revenue stream, but only as long as they find the services they purchase are worth what they are paying. This is in sharp contrast to consumers, who pay once for a product or service of limited duration.

A wireless subscriber tends to be more aware of the cost of information access than the average desktop user. This awareness affects the relationship between the subscriber and the content provider. Subscribers become resentful if they end up paying for something they don't want, and many subscribers will quickly catch on to a content provider's misuse or abuse of subscriber-paid bandwidth and take their business elsewhere. Such transgressions may be minor, such as adding extra images, or they may be more serious, such as excessive advertising paid for by the subscriber.

Subscribers consider themselves active participants in an ongoing business transaction, rather than just passive consumers. This shift in the customer-provider relationship forces a different kind of business model on site developers interested in providing wireless content. Traditional Web models, such as advertising and purchasing partnerships, will take a back seat to strategies involving pay-per-use and revenue sharing between service providers and content providers. These new models may be simple (for example, a subscriber pays a fee per transaction), or they may be complex (a service provider invests heavily in a content provider in order to jump-start a service, and then the content provider shares revenues with the service provider to repay the initial investment and generate additional revenue for the service provider).

When setting out to create a wireless Web site, you should ask yourself the following questions:

- What are you providing to your subscribers?

- What are you asking for in return?

- What implicit costs are you asking your subscribers to bear?

- Is what you're providing equivalent in value to the sum of the implicit and explicit costs?

- What kind of long-term relationship with your subscribers are you seeking?

The answers to these questions are a litmus test as to the true utility of your service to your subscribers. Moreover, they will help you explore how you can continue to add value to your content in the future.

Subscriber Mobility and Data Mobility

Wireless subscribers are *mobile* subscribers. As such, they interact differently with their devices and the information their devices display than do users of desktop computers or users of WebTV-like Internet terminals. Mobile subscribers expect instant access to information. In their minds, the information is with *them,* and not on some distant server accessed over a network using thousands of radio towers and zillions of miles of cable. This creates a perception of data mobility—users feel as if they're carrying a metropolitan phone book, travel guide, map, newspaper, and the activities of New York Stock Exchange clipped to their belt.

In Chapter 3, you see how the need for data mobility affects when and how users interact with their mobile devices and how this need can affect your site design. This, in turn, requires you to pay special attention to the following factors:

- **Ease of use:** Access to desired information must be fast and hassle-free for users. In many cases, your viewers will be performing other tasks, such as driving or talking, while working with your content. More important, many people viewing your content may never have even used a computer. Your great uncle, the plumber who bailed out your bathroom last Thanksgiving, or your grade school math teacher (now retired) may be one of your staunchest subscribers.

- **Information access:** Subscribers will be using your site to access critical information as part of a task or activity they're trying to complete; they're not engaging in idle recreation. Therefore, your site must be organized well to enable rapid access to any desired information. Not only does this organization help provide an easy-to-use site, it also meets your viewer's needs to accomplish his or her tasks quickly and efficiently.

- **Viewer expectations:** Your subscribers will expect your site to act in a manner similar to that of the rest of their wireless device. In fact, they will expect your site to be part of the device. Most users won't recognize the boundaries between your content and their hardware, especially on a device that has little differentiation in its interface. For example, lists, buttons, and other controls should be crafted with the target device in mind. The viewer is using a phone or handheld computing device, not a cumbersome personal computer. Interaction between the user and your site must be intuitive and the desired results must be able to be obtained quickly.

Understanding data mobility and your subscriber's environment will help you establish content in a manner best suited to its use. To foster this understanding, ask yourself these questions:

- In what sort of surroundings will your content be viewed?

- How much attention will your subscribers give to your content?

- What sort of mood will your subscribers be in when they view your content? What about when they're done viewing it?

- Will your subscribers perceive the information you're providing as an integral part of the device?

- Will the information reflect their lifestyles? If not, why?

The Computing Capacity Gap: Handheld versus Desktop

Devices used by mobile subscribers have vastly different capabilities than their desktop kin. Factors such as power consumption, physical size, and affordability create a vast disparity between these two types of devices.

The oft-quoted Moore's Law (named after Intel co-founder Gordon Moore), which says that computing capacities double every 18 months, applies equally to both mobile devices and desktop units. Moore's original prediction was that chip capacity would double every two years. In reality, the pace is even faster. Take a look at Figure 2-1, which shows functionality per unit cost for both desktop and handheld devices. Over time, it becomes possible to buy significantly more computing capacity for the same unit cost.

However, although the computing capacity of individual components and entire systems may be increasing at an exponential rate, a feature gap exists between the different platforms. And this gap, as Figure 2-1 shows, won't close up anytime soon. The same advances in technology that created the mobile-device market—increased miniaturization, increased computational power, and decreased power consumption—will also improve their desktop counterparts, further widening the gap in information-processing capabilities. Throughout this book, I focus on how these differences affect how you need to develop content for the desktop. Chapter 3 explores one side of this discrepancy—network bandwidth—in great detail.

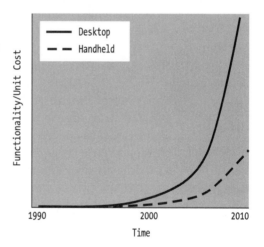

Figure 2-1. Functionality per unit cost over time

Because of the relatively limited processing capacities (for now, anyway) of mobile devices, wireless Web content must be simple, use relatively little memory and bandwidth, and require a small amount of processing by the device's central processing unit (CPU) prior to the content being displayed.

Wireless Platform Options

The first choice you need to make as you prepare to take the wireless plunge is picking the platform your subscribers will use. Being knowledgeable about the available platform choices is key to designing a wireless-ready Web site.

You have two major platforms, and one smaller—but nevertheless relevant—third platform, to consider: the existing Web (I discuss developing for this platform in Chapters 4 through 6), the growing Wireless Application Protocol (WAP) (see Chapters 8 and 9), and the Handheld Device Markup Language (HDML) (see Chapter 10), a foundation of WAP that has re-emerged as an open standard. These choices aren't mutually exclusive; unless you're developing for a specific platform, you'll want to plan on supporting all platforms to make your content available to the widest possible audience.

The World Wide Web

If you haven't been trapped under a heavy object for the past five years, you are well aware of the sweeping influence of the World Wide Web. The establishment of a worldwide network based on open data standards has fueled an explosion of information exchange. Millions of computers are already on the Web, with many thousands added each month.

The Client-Server Model

The Web uses a client-server model to provide data to viewers (see Figure 2-2). On behalf of its user, the Web client makes a request to the Web server for content. The Web server then obtains the content, either from its file system or by dynamically creating it on the fly, and returns the content to the Web client. The Web client then displays the resulting content

In Figure 2-2, the two large boxes represent computers. Lines between boxes indicate a network connection and the stars at each end of the line indicate that one or more boxes can be on each side of the line. (See Appendix B for further information on Unified Modeling Language [UML] diagrams for wireless Web deployment.) The smallest boxes represent software modules, hardware, firmware, libraries, components, and such that do things inside the computer.

The protocol for making the request to the Web server, and several possible schemes for formatting the content, have been standardized so that anyone can write the software necessary for implementing this process. These standards are also used for wireless devices. Many of today's handheld computing devices can run Web browsers capable of presenting Web pages from virtually any site on the

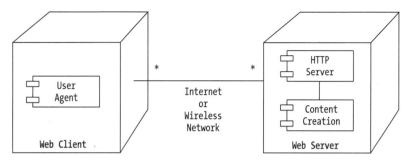

Figure 2-2. The Web deployment model

Internet. These products, such as Personal Digital Assistants (PDAs), tend to be at the high end of the mobile computing spectrum.

The strengths of the client-server approach are obvious, although the implementation can be fraught with problems. On the up side, content developed for mobile devices can be created using any of the multitudes of authoring tools now available. The open standards enable developers to author content for many devices with just a single investment in technology.

Formatting Limitations

Using existing client-server technologies and content may be tempting, but unfortunately, the rewards are few. Many handheld devices lack the capacity to fully render a rich Web page, especially if the page contains a good deal of images or multimedia. Even when a device can render a full Web page, physical constraints such as screen size can alter page layouts because the Hypertext Markup Language (HTML) dynamic layout policy applied to limited screen sizes can result in bizarre-looking Web pages. Layout, color, display contrast, and other factors can make a page that looks beautiful on a desktop device unreadable on a handheld one. (In Chapter 3, I discuss in detail how bandwidth limits the kind of content you can deliver to wireless devices; in Chapter 5, I introduce you to a subset of HTML that can be used with almost any wireless device.)

In addition, whereas the basic formats for the Web—including HyperText Transfer Protocol (HTTP), HTML, and simple image formats such as GIF and JPEG—are available on mobile devices, many of the more sophisticated aspects of the Web are not. Most mobile devices, for example, don't support all the features of HTML 4.0 and advanced formats (such as Adobe's Portable Document Format) aren't available at all. Where support for scripting exists, its functionality is a far cry from that found on a desktop. Multimedia, such as sound sequences or Shockwave-style animation, are not likely to be available on a device with a paltry 8MB of RAM.

In general, content developed for conventional Web access when viewed on wireless devices will look lackluster by comparison. The opposite also holds true: Content meant for display on wireless devices, when viewed through a desktop browser, will lack much of the glamour and glitz needed to attract and hold a Web surfer.

There is hope, however. Developing for wireless devices using Web standards isn't impossible, nor even that difficult once you understand a few basic principles. (I discuss those principles in Chapters 4 and 5.) Supporting wireless Web clients can lead to a better Web site design for all viewers. Lightweight content optimized for wireless devices looks clean and uncluttered, and loads faster than gee-whiz pages with lots of flashy graphics, sound, and interactive scripts. This, in turn, can result in a more productive user experience, even for desktop viewers.

The Wireless Application Protocol

In June, 1997, Ericsson, Motorola, Nokia, and Phone.com (formerly Unwired Planet) formed the Wireless Application Protocol (WAP) Forum. As manufacturers of cellular phones and cellular infrastructure, Ericsson, Motorola, and Nokia sought a single standard for wireless data; Phone.com brought a proposed standard to the table that provided a starting point for everyone. The WAP Forum aimed to do the following:

- Integrate Web content and advanced data services to be used on wireless phones and other wireless terminals.

- Create a global wireless protocol specification that works across all wireless technologies.

- Enable the creation of content and applications that scale across a wide range of wireless networks and device types.

- Embrace and extend existing standards and technology wherever appropriate.

As a result of the forum's efforts, the WAP specifications are recognized by more than 80 companies, and WAP is now becoming the standard mechanism for integrating the Web with wireless devices.

WAP leverages the data distribution model adopted by the Web (see Figure 2-3). WAP clients running on wireless devices use the lowest levels of the protocols behind WAP between the client user agent and the WAP gateway to encode requests and responses. It does this in a way that's most efficient for wireless devices. In turn, the WAP gateway uses HTTP to interact with the Web server. The Web server provides its content in the Wireless Markup Language (WML), although it may use any of the standard Web technologies (including CGIs, content databases, server side includes, and so on) to generate the content to be served.

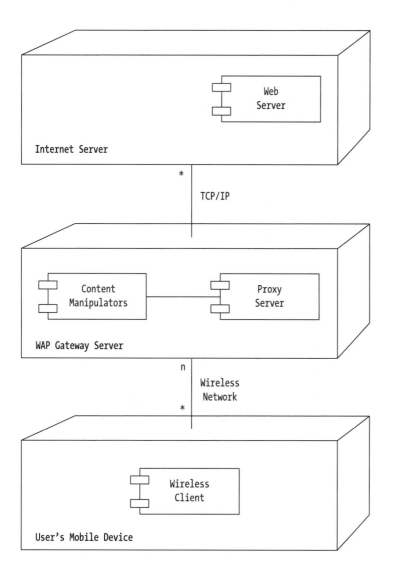

Figure 2-3. The WAP deployment model

Producing WML content is a fairly simple process. The content is authored using WML, either by hand or with content-editing tools designed for WML development. In turn, this content is hosted by a Web server, which is neither the same server used for desktop browsers nor a secondary one. Mobile devices make their requests of the server via the WAP gateway, requesting WML documents instead of HTML documents. (In Chapter 8, you learn how to create WML content; Chapter 9 shows you how to make your content more interactive using WMLScript, the scripting environment of the WAP.)

Hypertext Device Markup Language

The Handheld Device Markup Language (HDML) preceded WAP, and is in fact the foundation for much of what constitutes the WML. Created by Unwired Planet (now Phone.com), it was a significant addition to the WAP Forum's activities and is now maintained by Phone.com as a separate open standard competing with the WAP standard.

As of this writing, the marketplace is split—devices shipping within the United States appear to favor HDML whereas devices shipping within Europe and Asia are more likely to use WAP. This trend can be traced to the early days of WAP. The first wireless Web projects done in the United States were based on previous work by Unwired Planet before the WAP Forum released its specifications. In Europe, on the other hand, manufacturers who are also leading members of the WAP Forum largely dominate the market, and WAP appears in the majority of networks being deployed by European operators. In the Pacific, Japan has accepted the WAP standard more quickly than HDML as Japanese wireless companies such as NTT Docomo seek to interconnect the wireless network with the Internet.

This trend probably won't continue for long. Phone.com is working to make more and more HDML-specific features a part of the WAP specification, and many new devices yet to be introduced will comply with WAP and not Phone.com specifications.

The data distribution model of HDML is fundamentally identical to that of the WAP model. A Web server delivers content in HDML over HTTP, and the content is carried to a gateway server (called an *UP.Link server* in HDML parlance), where it is provided to wireless clients in response to a request.

Writing HDML is very similar to writing WML (or even HTML). In Chapter 10, you learn how to produce HDML content for wireless screen phones.

Comparing WAP and HDML

What does HDML bring to the table that WAP doesn't? Some good reasons exist for supporting HDML as well as WAP in your site, including:

- Coordination with operators who have existing partnerships with content providers already using HDML or UP.Link servers.

- Better support for customers using HDML phones, such as the Alcatel OneTouch.

- Support for client-initiated prefetch actions, which enables the client to fetch content for caching.

Of the three reasons I list, the first one is probably the most important. If you're planning on partnering with an operator, be sure you've adopted the standard the operator deploys. Similarly, if you know that your users are already biased toward one standard or the other, be sure you recognize that bias at the outset.

Both WAP and HDML provide interoperability between devices from different manufacturers, and enable competing vendors to share data across various networks from a variety of sources. This serves the same purpose as the older Web standards—fostering interoperability.

In addition, WAP and HDML were designed to make the most of existing standards and concepts in order to leverage existing technologies and services and ease the way for new content developers. As a result, deploying a wireless solution using WAP or HDML is easier than it would be with an earlier solution, such as those found on any of the wireless networks (for example, BellSouth's RAM network).

Hardware Options

The emphasis on interoperability from organizations that are creating the standards, such as the WAP Forum and the World Wide Web Consortium (W3C), hasn't limited the creativity of hardware manufacturers building wireless access terminals. If anything, the wide diversity in hardware belies the clear delineation of a few well-adopted standards for the management of wireless data.

In some cases, hardware used for wireless access has been modified to operate in a way that differs from its Web-oriented purpose. This is especially true of hand-held computers, which are sporting wireless modems or links to cellular phones in increasing numbers. Some of these devices have integrated wireless technology direct from the manufacturer, while others employ expansion cards that provide wireless access on one of the wireless networks.

Some hardware is custom tailored for wireless access. For instance, so-called "smart phones" have multi-line liquid crystal displays (LCDs) and software to support wireless data access. Other hardware items, such as the Research In Motion pager, a two-way pager capable of interfacing with a user's e-mail account, are single-purpose devices.

Wireless terminals capable of interfacing with the Web fall into one of four general categories, which I describe later in this chapter. The size and price range of the hardware in each of these categories dictates which features each device can offer.

Over the past five years, handheld device manufacturers have been releasing equipment with growing data functionality while making many attempts to endow these devices with wireless access. In some instances, wireless access has been tightly integrated with the device through the use of a voice network, creating a sort of cellular phone on steroids or a PDA that can hear and speak, a.k.a., a "super phone."

More recently, manufacturers have tried to skirt the issues of high price and low adoption rate of super phones by releasing simpler, lower-cost screen phones, which are watered-down super phones based on bare-bones operating systems and software. However, as with evolving technologies, there's always crossover between product families. Recently, software companies including Spyglass and others have worked to create WAP browsers that can run on platforms that are arguably PDAs, leading one to wonder if a PDA running a WAP browser is a screen phone or a PDA.

I'll steer clear of the semantic arguments and discuss the three broad categories of devices. Knowing which devices your subscribers are using can help guide you in applying a particular technology. (You must also have an understanding of your subscriber's environment in order to tailor your content to his or her needs. In later chapters of this book, I cover the various operating platforms, user constraints, and device limitations and how they figure into your planning and creation of content.)

Super Phones

Super phones offer a computing platform based on a PDA combined with cellular phone hardware. A number of these devices have been around for a while, ever since PDA manufacturers began working with telephony providers to integrate the power of handheld computing with wireless access technology. Super phones are epitomized by the Nokia 9000 "clamshell" phone running the GeoWorks operating system; another example of a super phone product is the Qualcomm pdQ, a Code Division Multiple Access digital phone integrated with a Palm Computing Platform device.

Many of these devices currently offer an integrated Web browser, enabling users to view Web pages wirelessly. None of these devices presently support WAP or HDML, but that's certain to change quickly.

Super phones represent the first steps toward the creation of the *smart phone*, which I discuss in the next section. As their functionality continues to improve, smart phones will increasingly incorporate the features of super phones. Predominantly two kinds of manufacturers develop super phones: phone handset vendors seeking additional differentiation between products, and PDA manufacturers seeking new markets via tighter integration between data and voice services. Consequently, the differences between PDAs and super phones are sure to blur over time.

Some super phone features—such as tight integration between data and voice, compact size, portability, and the availability of PDA-like features—that make super phones desirable to some consumers also slow their deployment. The relatively powerful computing hardware also increases their cost, directing sales away from the devices in a price-sensitive market. Nevertheless, super phones constitute a hybrid that performs well for specific applications involving intranets maintained for a particular set of users.

Smart Phones

Smart phones are the newest of all devices in the wireless marketplace. An extension of the existing cell-phone product concept, a smart phone combines the best features of a digital cellular phone with the ability to retrieve data from the Internet using either WAP or HDML. These phones are often bundled with additional applications, such as simple messaging, phone books, and games. Increasingly, the applications themselves are being developed in WML or HDML and run on the phone, rather than in the phone's embedded programming environment.

Smart phones are likely to become the common denominator for transmission of wireless data in the next few years. They are the least expensive of all wireless terminals and are simple enough for the general population to use. As such, potential smart phone users represent an enormous untapped market for wireless applications.

Personal Digital Assistants

Often regarded by the uninitiated as a glorified daily planner or electronic organizer, the PDA has been awaiting wireless Web access for a number of years while manufacturers sought advanced connectivity to increase demand. By 1995, products such as the Motorola Marco and Motorola Envoy that used the ARDIS two-way paging network offered wireless connectivity.

Today, most PDAs are clamshell or palm-sized electronic tablets that accept snap-on modules or PC cards that enable the devices to interface with wireless service from digital networks. PDAs are increasingly used in conjunction with wireless networks and for browsing Web content that is downloaded while the device rests in a cradle connecting the PDA to a PC. Web browsers of one sort or another are available for nearly all PDAs.

The PDA marketplace is split into two camps: the Palm Computing Platform on one side, and everybody else on the other. The lion's share of the PDA market is owned by Palm, Inc., and Symbian, who together have shipped more than eight million handheld devices as of this writing. While Palm is a well-known name domestically, few in the United States are familiar with the Symbian name (formerly Psion). The brand is popular in the United Kingdom and other parts of Europe where Symbian sells more devices than all of the manufacturers of Microsoft Windows CE units combined worldwide. Nearly all PDA manufacturers now recognize the importance of integrating telephony with their devices, although the added expense of incorporating wireless data makes an integrated wireless PDA a rarity.

Contrary to some analysts' projections, the PDA marketplace is not likely to disappear anytime soon. The demand for cheap personalized computing devices that can carry data is not likely to dry up, and the cost of wireless hardware and services will not drop quickly enough for some segments of the market. Although wireless operation is becoming more economical (consider the Palm Computing Platform Palm VII wireless connected organizer), some folks will balk at spending money for a device with integrated wireless connectivity and service. These factors combine to preserve the PDA market niche, especially for price-conscious consumers. Nevertheless, as an increasing number of PDA manufacturers realize the importance of wireless connectivity, more and more PDAs will sport integrated wireless technology, blurring the distinction between smart phones, super phones, and PDAs altogether.

Laptops

Unlike the other categories of wireless terminals I discuss here, the computing capabilities of a laptop device (or notebook computer) are generally comparable to those of their desktop counterparts. Laptop devices are constrained primarily by the wireless network, and not by their weight, battery life, or software. And, unlike other wireless terminals, laptops are usually operated while the user is stationary.

Laptop users are able to run applications identical to those found on desktop computers, the traditional means for Web access. Many laptops have memory, disks, and processors roughly comparable to that of desktop units and are used in much the same way. Laptop and desktop users employ essentially the same software and technologies to access the Web; however, functionality can be compromised if a laptop user explicitly selects a low-bandwidth wireless network.

Over the next several years, many laptops are likely start including wireless interfaces that are integrated into the units, added as expansion options, or provided through cellular phones. When that time comes, a laptop user will have the option of accessing the Web through traditional means, or wirelessly through either conventional Web technology or through the use of a WAP browser running on a laptop. Therefore, if you're predominantly targeting laptop users, you may still need to support WAP or HDML and recognize the limitations of a bandwidth-constrained connection.

What Makes the Wireless Market Unique?

At first blush, the wireless market seems to be largely an extension of the Web. Consumer growth by the millions, rapidly plummeting access costs, and increased integration among competing information sources will all be experienced within the wireless marketplace. Despite these similarities, however, many factors contribute to make the wireless data market significantly different than the Web market.

Centralized Distribution

The current investment in infrastructure, services, and support by the major wireless providers is but one clue that the distribution of wireless data will differ quite a bit from Web access. The early days of Web access saw small providers catering to local communities and specific classes of users, and eventually consolidating into larger organizations eager to provide consumers with Web access. By comparison, the wireless environment has relatively few providers catering to nationwide markets. The high cost of deploying a wireless network makes the idea of startup wireless Web service providers, such as the new Internet Service Providers (ISPs) that have sprung up over the past few years, difficult to imagine.

For the most part, the distribution of wireless services will be managed by existing large corporations in the telephony and data marketplaces. This changes the nature of opportunities for content providers because their distribution partners will be fewer. Nevertheless, after witnessing the growth of the Internet, these content and service providers are keenly aware that content drives the growth of the network. These distribution partners are already eager to obtain additional content services. For instance, consider the plethora of wireless solutions programs currently offered by major network providers such as France Telecom, AT&T, and BellSouth.

The interoperating standards adopted by wireless data providers, content providers, and the manufacturers of wireless terminals will make centralized distributors either an important factor for content providers, or almost nonessential. For some content providers, partnering with a specific type of network provider will speed deployment; for many others, the interoperation of multiple standards across networks means that little or no differentiation exists among network providers.

Bandwidth Limitations

Bandwidth is at a premium in any network, but it has become a watchword for wireless networks. To use a plumbing analogy, *bandwidth* is the amount of content that can be passed down a pipe. The "diameter" of the pipe determines how much content can flow at any given time, while the speed at which the content is flowing determines its *latency,* the length of time it takes the content to flow a certain distance. (I describe latency a bit further on in this chapter; see Chapter 3 for more information on latency and user expectations.)

The bandwidth of today's third-generation cellular networks is just beginning to approach that of traditional land-based modems. Many wireless subscribers may currently be using systems that provide throughput of a magnitude much less than that of their existing modem Web access. (I discuss the technical implications of these limitations here; in Chapter 3, I discuss in detail the way bandwidth constraints affect content and market issues.)

How Bandwidth Affects Throughput

By necessity, wide-area wireless networks have a relatively low *throughput* (the data transfer rate). A channel's capacity is a function of the amount of information it carries. In addition, the wider the channel, the more energy that is required to maintain an equal signal strength at the reception point. Therefore, higher bit rates result in more bits sent per unit of time, which means that additional frequency spectrum is required and more power is needed to travel the same distance compared to a lower bit rate. This phenomenon is a factor not only for base stations, but also in the design of every wireless terminal.

How Bandwidth Affects Latency

Latency is a measure of the amount of time it takes for a request to make a roundtrip from the client and back. Wireless networks suffer from greater latencies than wired networks for two reasons. First, the low bandwidth of wireless networks results in longer transit times for data traveling through the network. In addition, the increased complexity of a wireless network (relative to a wired network such as the Web) raises latency because data must traverse additional nodes to reach the final destination.

A wireless network's throughput constraints are usually visible to the user. For example, high throughput on the Web results in fast download times for images, text, and large files. Latency, on the other hand, is generally observable only by the computers managing the data. In a networked computer game, for example, latency is more of an issue than throughput because the importance of real-time data surpasses that of the volume of data being exchanged. Consequently, throughput drives specifics of content design, while latency drives the specifics of software and protocol design.

Subscriber Demands for Ease of Use

Many wireless service subscribers are not computer literate. Wireless subscribers will range from a teen using her parent's cell phone to a retiree inexperienced with the latest computer developments. The large number of wireless subscribers—in some areas outpacing the growing population with access to a computer—suggests that many wireless subscribers will not have the computing skills of traditional Web users.

These novice subscribers expect a wireless terminal to operate as if it were a telephone or pager—not a computer. They expect the devices to be entirely reliable, easy to use, and provide instantaneous response. Subscribers may further assume that these devices will operate like a traditional telephone, with no extensions to the interface. The user's unstated assumption is that these are consumer devices designed for a single purpose. (I discuss these issues further in Chapter 3.)

Efforts to meet these consumer expectations have been largely unsuccessful, with a few notable exceptions. Although a pager or cell phone is significantly more reliable than a desktop computer (when was the last time you saw someone reboot a cell phone?), the user-friendliness of these simple devices is still debatable. Few users are able to employ the advanced features offered by these phones and their service plans; fewer still will be able to navigate the content on these devices unless some significant improvements are made. Therefore, content providers as well as manufacturers need to keep simplicity in mind when designing the device-to-human interface.

Today's wireless devices are a union of hardware, software, and content forms. Once manufacturers have figured out how to make these devices small *and* easy to use, and the operating system designers have figured out how to create a human-friendly interface that anyone can quickly grasp, any remaining complexities that have to be made simple for the user are within the province of the content provider. Content will increasingly be the ingredient that differentiates these devices, and for many users, content will determine what these devices actually are.

Wireless Usage Patterns

Using a computer requires a fair amount of preparation. You need a computer table, an ergonomically correct chair, a monitor, a keyboard, a mouse, and possibly other peripherals. The system must be booted, applications launched, and a network connection made. Using a computer involves work, planning, and goal setting.

The same cannot be said of wireless devices. Their portability, compact size, and relative unobtrusiveness all contribute to our finding them in use nearly everywhere. Users pluck them from bags or pockets to retrieve messages, check the weather, or examine the latest stock prices. Students use them to perform simple research queries, such as "What is the meaning of this word?" Wireless devices are especially useful to mobile workers who can call up the information residing on their company networks from the palm-sized devices in their hands.

These factors also serve to decrease the amount of time per access a user spends with these devices as compared with the time they can expect to spend on a desktop computer. Typically, mobile device users are engaged in more than one activity. They may be waiting in a checkout line while using their devices, or they may be actively involved in conversation, driving, walking, or some other activity. Their primary focus is not on the device or the software it uses, but rather on the information being sought or their primary activity. This decreases the per-instance usage. The device, therefore, becomes a per-query or per-idea tool, rather than a general-purpose computing device that can monopolize a user's attention.

Price Sensitivity

Market studies show that products and services must overcome certain price barriers before they can gain widespread acceptance by the average consumer. Although this price barrier varies by geographic location, broad market statistics demand relatively low costs for the wireless products and services.

In terms of product acceptance, wireless developers currently face price barriers of $99, $150, and $300, depending on the functionality of the device. Customer expectations and economics dictate these price limits. A simple wireless device, such as a base-model cell phone, should be priced at or below $99 for the greatest acceptance; more feature-filled units should remain priced at or under $300. (A $900 price point also exists, but the expected sales volumes at that price are still too low to support most wireless applications.) As prices fall, consumer demand increases, but the increase is nonlinear—huge increases in demand are seen at each price point. Most consumers are immune to the justifications of higher prices, and will not spend outside of a price barrier on the basis of features or capabilities.

Market surveys on wireless services show a similar cost breakdown, with fee-per-access barriers at $29, $49, and $99 per month, depending on the extent of the service provided.

It's up to the content provider to deliver the most for each dollar subscribers pay for their access. Subscribers are seeking information, not sophisticated scripting, crazy formatting, or complex logos and graphical bullets. Content providers will probably find that the simplest sites are often the most popular, partly because simplicity equals low access cost for many viewers. Moreover, simple formatting guarantees the highest likelihood of fast access for the consumer. Thus, simple formatting itself becomes important when considering a customer's concern over pricing.

Potential for Market Growth

As I mentioned in Chapter 1, the adoption rate for wireless telephony devices has matched or exceeded that of other successful consumer products to date. This level of acceptance can also be expected for wireless data as the availability of wireless services grows and compelling content becomes available. Today's numbers on the actual volume of wireless data usage typically exceed the predicted numbers from previous years. Although the rate of adoption may grow slowly at first, exponential growth likely will be experienced over the next several years.

In many developing countries, the cost of deploying a wireless network is significantly lower than the cost of installing a copper wire network. Some countries are laying down only minimum wiring for their data infrastructure, preferring instead to spend their limited finances on the latest in wireless technology.

During the next decade, some young consumers' initial experience with a telephone may well be a phone call made on a wireless handset. Around the world, growing numbers of people are using wireless phones as their primary means of communication; some are even replacing their existing telephone service with a wireless handset. These factors, along with the expanding integration of wireless data with wireless telephony services, serve to speed the deployment of wireless data worldwide.

Data Mobility

Today's computer users think of their data as residing somewhere in the conceptual space that surrounds their PC. Many users may not recognize that their information could be on a hard drive, floppy disk, network server, or Internet server. Although the information they seek may be on a computer halfway across the world, the user's interaction with this information occurs at a few physical locations using just a few computers at most—a laptop for business, a desktop computer at work, and another at home. The information appears to be available to them locally while they use their computer, rather than in another physical location (or stored on a hard drive, CD-ROM, or other media). In a wireless context, however, users' perceptions are different.

A number of factors make information content, rather than the hardware, software, or service, the most important element of the wireless network mix, such as:

- The user's increasing view of a wireless access terminal as an information source.

- The blurring of the distinction between manufacturers and service and content providers, as consumers focus increasingly on content and less on service brand.

- The desire for instant access to information.

These factors point to a single trend: the user's attention will shift from the device where the information is obtained to the mobility they're experiencing while accessing the information itself. This gives the user the perception that the data itself is mobile. I call this trend *data mobility.* Data mobility encompasses all of the factors I've discussed, from the blurring of brand distinction to the mobility of the user, and centers around the notion that the user will expect data to be with them, regardless of the location of the source of the data.

Summary

In this chapter, I discussed opportunities for wireless Web developers in both abstract and concrete terms. Whether you seek relationships with content providers, service providers, or both, your work will likely lie in the aggregation of existing content and repackaging that content for wireless devices. If you're lucky, you'll be able to stand the world on its ear with a novel wireless application using data that's not presently available on either the wired or wireless Web.

When you go about defining this content, you need to consider how the fundamentals of wireless development affect your content. Remember that wireless users are subscribers, not consumers; these subscribers are mobile, and that they expect their data to be mobile, too.

Equally important, you should think about how your subscribers access your content, considering both the wireless networks they'll use and the devices they'll carry. Some wireless networks (such as the wireless Web using HTML) are well-suited to devices that most closely resemble computers, such as PDAs; other networks are primarily the domain of telephones, with small screens and limited capabilities at low cost that can reach a large audience.

The wireless market isn't about bringing content identical to what's already on wired networks to mobile users. Limitations of bandwidth, a demand for ease-of-use, and how wireless devices are used all make the wireless market a new field in which all can make a mark.

CHAPTER 3

The Wireless User Interface

DEVELOPING CONTENT FOR THE WIRELESS WEB is somewhat like developing content for the wired Web during its early days. Although the tools for wireless Web authoring are better than the early Web authoring tools, the constraints are similar. The network connecting users to wireless subscribers is slow, the devices lack sophistication, and most end users are inexperienced in accessing information over a new medium.

Data mobility—the user's belief that their data resides with them, regardless of its origin or means of transport—revolves around bandwidth—the rate of information transfer. Bandwidth is a scarce commodity over today's wireless networks, and relative to wired networks, will remain so for the near future.

While the term is usually reserved for technical discussions of data rates, bandwidth applies equally to user interfaces. Your users will access your content in many different situations, and their attention won't be dedicated to your content alone. A user on the go has less "mental bandwidth" (or capacity for absorbing and processing content) than a user seated at a desk. A wireless user also faces some "physical bandwidth" constraints. It's hard to keep your eye on the road and on the screen of a handheld device while you're driving at 40 miles per hour down a highway off ramp.

As a result, your content must be crafted to respect both the physical bandwidth limitations of wireless devices—slow data rates, small screens, and limited input—and the mental bandwidth limitations of your subscribers. In this chapter, I discuss users' interface expectations and how you can work to meet those expectations. Central to your efforts is the design of effective information presentation and collection—topics I explore in detail.

Meeting User Expectations

Users will have high expectations of wireless Web sites. Unfailing reliability, speedy operation, and ease of use will be of paramount importance.

Latency

To a user, any service delay can be an annoyance. Fortunately, most users will understand that the fulfillment of a wireless query takes time and that some amount of delay is inevitable, if only because they've become accustomed to delays across all Internet services.

Users also recognize the convenience factor attached to wireless services, and are willing to trade some speed for this convenience. Most users want access instantly, but will typically wait up to ten seconds for it. The number of users willing to wait for longer periods, however, quickly diminishes. In short, faster is *always* better.

Traditional Web users have learned to become patient with sluggish content access over modem lines, especially those lacking high-speed connections. They deal with the delay by rationalizing ("Well, it *is* downloading a big picture."), killing time (waving the cursor around the screen, sipping some coffee, balancing a checkbook), or avoiding a site altogether ("I don't really go there much; it's too slow."). Wireless users tend to skip the first two coping steps and move straight to avoidance.

Developers can't do much to directly decrease latency (the delay between the request and the response or between the exchange of individual data packets) on a wireless network or to improve network performance. They can, however, unwittingly *increase* latency, stretching the time it takes to access a page to several minutes, and thereby make a wireless Web page virtually unusable to most viewers.

Some browsers will refuse to render any content until a large percentage of the page is available, blurring the distinction between *throughput* (which I discuss next) and latency in the mind of the user. Users cannot distinguish between how long it takes the request and response to occur, the page to draw, and how long it takes the entire page to appear.

Because of this, you should design small pages and format them so that a minimum of memory and computing horsepower are necessary to render them fully. "Small" is a relative term here, dependent on the network and the device responsible for rendering the content. From a user's perspective, single screens are ideal, and are a good compromise if you want to achieve decent performance over a variety of networks. In general, individual documents should be no bigger than 1KB or 2KB; images should be as small as possible while retaining their meaning. Text should be concise, clear, and formatted simply. (I explore formatting in more depth later in this chapter.)

Throughput

Given the relatively large number of users competing for limited wireless resources, any user's traffic occupies a small part of the total channel. Consequently, a handheld device remains idle for a length of time between the transmission and reception of data. *Apparent throughput* (how much data appears to be exchanged, as opposed to actual *throughput*, which can be measured by looking at the average amount of data exchanged over time) on a wireless network is largely a function of *per-packet latency.* If wireless terminals were equipped with lights such as those found on modems, users would be able to see a request go out and a response come back as two discrete operations. Although individual packets coming and going won't be visible to the user (most browsers do not render on a per-packet basis), the impact of these delays due to slow throughput will be.

Implementation issues force most wireless networks to have staggeringly long inter-packet times relative to wired networks, often extending to tenths of a second. Although wireless network stacks deal with these delays efficiently, the applications that use these stacks do not readily manage them. Most Web browsers, for example, do not render certain images until set segments are loaded. Other browsers, in the interest of conserving memory and computational efficiency, do not perform text layout until large portions of the page have been loaded. Thus, a high-throughput wireless network with relatively high latency may appear to be a low-throughput, high-latency network to the client application and users.

Controlling page length by formatting appropriately, presenting information piecemeal rather than in one big chunk, and using images judiciously can help advance the perception of high throughput.

Designing the User Interface

User interface design for a Web page may seem like a superfluous concern. After all, isn't the user interface for a Web page largely supplied by the browser's user interface? The answer is Yes and No.

As any experienced Web developer knows, you can do a lot within a markup language to establish a look and feel for a particular site. Images, fonts, the application of text styles, word choice, and colors all have an impact on how a site is perceived. These factors contribute to a specific user interface that can complement the site, or can make a site much more cumbersome to use than it should be.

Generally, each wireless hardware platform has a browser released by the platform vendor and at least two competing browsers written by third-party developers. Many browsers on wireless devices are essentially full-screen applications, with few menus, icons, or other adornments. Vendors must keep the interface of the browser simple in order to make the best possible use of limited screen space.

Similarly, wireless content demands simple page designs. Therefore, Web sites viewed on a wireless device play a larger role in defining the interface than would sites viewed on a full-size computing device. An extreme case of this occurs with the Wireless Application Protocol (WAP), where the browser on many screen phones has no interface of its own, and on many of these devices the browser dominates the entire display.

Fonts

Although the HyperText Markup Language (HTML) provides the ability to dictate which among a number of fonts the browser should use to draw a page, the Wireless Markup Language (WML) does not. Moreover, HTML browsers on many handheld devices don't have the same rich set of fonts that is available on desktop computers. Your best bet is to stick with the default fonts selected by the device and user.

Most platforms pick a default font that users will find readable in various settings. Some Personal Digital Assistant (PDA) browsers even enable the user to pick a default-viewing font by picking a size from a list of small, medium, or large, with the browser selecting an appropriate font to match. Most users expect the browser to explicitly follow their font selection, rather than letting the content choose its own font, especially when their selection is appropriate to the operating environment.

Even using fonts by name on platforms with multiple fonts can cause problems. Fonts with the same name may vary from device to device, and the same font used on multiple devices may lead to drastically different output depending on the font directives. This can negate the desired effect of a site's specific look and feel.

Scrolling

On a desktop, most of the time users don't even think about the scrolling operation, especially if they're viewing large screens with maximized windows because the scrolling action happens comparatively rarely. Not so on a wireless device, where scrolling is a necessary evil.

On wireless devices in which a 400-pixel display is virtually unheard of, scrolling becomes much more frequent. Many devices have less than a 200-pixel vertical display, less any space used by menu bars or other controls. By creating short pages, you'll keep the need to scroll at a minimum.

User Input

Mobile users don't want to bother with inputting a lot of information at most sites, given that inputting text can be slow, tedious, and awkward on most small devices. Users expect to be able to input text rapidly and with minimal delays.

A common mistake is to prompt the user with an input line when a more specific means of obtaining user input, such as selecting an item from a choice, is equally appropriate. For example, it's easier for a user to pick a state from a list than enter its name. The judicious use of specific input types, in addition to controls such as lists and radio buttons, can significantly decrease the amount of input the user has to perform. Some browsers, such as the Palm VII Web Clipping Application, support specific input types for sophisticated data such as time and dates, invoking operating-system–specific dialog boxes for quickly entering information.

Creating User-Friendly Content

There's more to creating user-friendly content than just trial and error. What follows are some general guidelines for creating content suited to your subscriber's needs. In Chapter 5, I look closely at these issues for HTML developers, while in Chapters 8 and 10 I discuss these issues in detail for WML and Handheld Device Markup Language (HDML) developers, respectively.

Using Images Wisely

The old adage about a picture being worth a thousand words doesn't always apply to the wireless Web when you consider the cost of delivering the images. Plus, many "pictures" on the Web are not illustrative — they're used merely to divide a page into sections or to add some color or background patterns, or to just provide a particular look and feel for a site. Many Web sites have adopted the trick of using "empty" images as spacers while others use *image maps* (graphics with embedded hyperlinks that enable users to click a region of the image to navigate to a new page) with text in special fonts to ensure a distinctive appearance.

Virtually none of these embellishments works for wireless content. Many handheld devices have grayscale screens, and even color displays do not provide the same contrast quality as desktop displays and laptop screens. Users also view content displayed on handhelds in a variety of lighting, which can add to the contrast problem. More important than the visual problem, however, is the simple fact that images take precious time to load and display.

Therefore, your content should employ images sparingly. Confine their use to those times when you need to drive home an important point or when an image is a more efficient use of screen space than text. Some other guidelines to keep in mind include the following:

- Try to avoid logos as images for branding and advertising, because what might look appropriate on a desktop is often a messy smudge on a handheld. Complicated logos, or those that rely on color to promote recognition, rarely scale well to the screens used in these devices. In addition, a logo provides little value to the user and consumes wireless usage and loading time.

- Images should be high contrast, preferably rendered in black and white (or in colors that appear black and white when viewed on monochrome devices). Line drawings without gray shading are ideal, as are simple block drawings.

- Images should be as small as possible for the content they need to display. Generally, the largest images should be no more than 50% of the device's target display area; selectable images (used as links, for example) should be in the neighborhood of 16 × 16 pixels in size. (Touchable on-screen items are an exception to the "as small as possible" rule. Most users have difficulty selecting items on a touch screen that are less than 16 pixels wide.)

- Where icons are appropriate, make sure they are simple and clear. Avoid rectangular icons, which users can easily misinterpret at a glance as characters on a display.

Figure 3-1. A selection of graphics suitable for a monochrome device

Figure 3-1 shows a sample of icons suitable for use with a black-and-white client device. For color devices, you could add bold colors, but sparingly.

These guidelines also apply to image maps. While some mobile Web browsers support image maps, WML devices do not, and navigating multiple image maps will be prohibitively tedious and expensive for the average user. Although an attractive feature on a desktop device, image maps quickly become a nuisance on a mobile one. One exception is a user interface metaphor that makes navigating with images necessary, such as when maps are displayed. But even when navigation through an image map is appropriate, using images for panning, scrolling, and zoom selection around the border of the image is not.

Some Web browsers enable users to select whether images are downloaded, but a wireless content provider who gives the user this choice is looking for trouble. A page designed around graphics inevitably falls short when viewed as a text-only page. Nothing is more aggravating to a viewer than discovering much of the content of a page is buried in graphical content. Almost as irritating for this user is to have to sift through dozens of little browser-supplied graphic icons representing images that weren't downloaded, indicating that if he or she downloads these images, the page might actually make sense. It is far better to design a spartan site with few graphics to begin with, and then let the user decide whether the graphics you've selected are worth the download time.

When reworking an existing page for wireless access, there's an irresistible temptation to recycle existing graphics, especially icons and diagrams. *Don't do it!* Most graphics simply don't look that good on a handheld device after they've been dithered and scaled, even by an artist. More important, don't be taken in by the siren's song of automated translation. Two strategies have appeared on the market: software packages you can run over the images from your site, or a proxy server designed to provide mobile access to desktop images for handheld devices. These technologies were designed to be applicable to a variety of sites and images, but your content will inevitably suffer unless you invest the time to create graphics that are truly appropriate for it.

Writing Concisely

Brevity is key. Not only will subscribers appreciate a concise presentation of the information they need, but keeping everything short and sweet also ensures that a minimum of content needs to be delivered to the viewer.

Sentences should be simple and vocabulary clear. Avoid run-on sentences, subordinate clauses, and the other sorts of things your English teacher didn't like. Never use a long word when a short word will do. Examine newspaper or news magazine writing for examples of clear, succinct prose; it is a style that will be familiar to your readers.

Formatting for Readability

You should place the most relevant information at the top of a document—many readers don't want to scroll to get what they need. This is especially important for screen phones in which the display area is limited to a few lines. This generally precludes the use of banners and large headings for pages, and poses an additional challenge for those trying to use advertising as a revenue source.

Keeping It Short

Pages should be brief. Short pages not only reduce the need to scroll, but several small pages may also load more quickly than a single long page. Small pages also require less memory for the browser to format and display and are more likely to work on a variety of devices.

How short is short? To some extent, that depends on the platform. The Palm II wireless device, for example, works best with pages that contain a few hundred characters or less. This relatively small amount of information displays on a single screen without any scrolling, and equally important, the proprietary wireless scheme used by Palm, Inc.'s browser can compress pages that small into one or two wireless packets, significantly reducing the time a page takes to load (and decreasing user cost).

By extension, screen phones will accommodate even less information at one time, sometimes as little as a hundred characters. On the other hand, super phones and PDAs may perform adequately with pages containing around a thousand characters, depending on screen size, formatting, and the nature of the information displayed.

Making Use of Tables

Tables are a good medium for demonstrating numeric data or ensuring text alignment, but should be employed sparingly. Tables should be used for creating tabular material and not for general purpose formatting. Presenting a weather forecast, a stock quote, or summarizing important information are all excellent uses of tables; using a table to create two columns of text or to flow a paragraph around an image is not only an abuse of the table markup commands, but generally won't portray the effect you're seeking.

Keep tables simple and use as few columns as possible by ensuring that the table only presents pertinent information, and by breaking tables up if they stretch more than two or three columns. On a narrow display, a wide table may end up badly formatted or incomprehensible.

Most important, avoid nesting tables; not all wireless browsers support nested tables, and even those that do often render them poorly.

Letting the Browser Do Its Job

Don't second-guess the browser. The host of mobile devices makes writing device-independent content an even more important consideration than when you're writing for the Web. Keep the material simple, concentrate on providing the information, and let the browser do its job in interpreting markup tags in a way that's appropriate for the device. Attempting to coax a particular layout or alignment using browser tags not only needlessly increases the size of a page, it's a waste of time for the user, and guaranteed to fail more often than work.

Other Tips

You can do a number of other things (just as you would in print) to promote readability, such as the following:

- Group related items together on the screen.

- Avoid using more fonts and typefaces than are necessary. In general, enable the viewer to select fonts, rather than use font directives in formats that enable you to change the font. Use bold face and italic text sparingly.

- Don't abuse the markup language. Pages with complex markup tend to look cluttered and not useful.

- Use color sparingly, and remember that many users will not have access to a color screen.

- Use numbered or bulleted lists to provide a series of steps, rather than individual paragraphs. Lists enable you to tightly group related material while keeping listed items clearly separate.

Choosing Input Methods

Many kinds of wireless content will require textual or numeric user input. For some sites, such as a dictionary or thesaurus site in which the user enters a word and (Presto!) a definition appears, input is mandatory. Others sites, such as weather or travel guides, may require input at some entry points (for example, when looking for a review of a specific restaurant) but not at others (such as a list of the restaurants in a region). Sites requiring user input present a special challenge for the wireless designer.

Many devices have numeric keypads, requiring a tremendous amount of finger gymnastics to spell even a short name. So, wherever possible, use buttons and pick lists in place of free text fields.

Menus

Menus—often called *pick lists*—are an excellent alternative to keystrokes when you know ahead of time what the choices are. Keep the number of choices short; most users on the go are able to keep only six or seven items in their short-term memory at one time. Not only will short pick-list menus linger longer in a user's memory, but a short list eliminates the need for scrolling. Keep in mind that most screens display no more than ten to twelve lines.

Text Input

Some amount of text entry is inevitable. For example, consider the large market for travel and navigation sites. Input regarding current location and destination provides interesting opportunities for the content developer. But some platforms over the next few years will provide or are already providing geolocation services, making it unnecessary for the user to enter his or her current position.

When designing a site that includes a positioning mechanism, ask yourself whether information on a user's current location is required. For example, a navigation aid may want the cross-streets nearest to the user, while a weather forecaster may only need a city name or zip code.

A zip code is often the best compromise for location information. It is short, easy to enter, and usually available even when the user is not familiar with his or her current locale (hotel stationery, for instance, can supply a zip code). Unfortunately, this method may not work if the traveler doesn't have the zip codes of the travel destination handy. A good compromise is to enable the user to enter either a zip code or a city name, as shown in Figure 3-2. As the figure depicts, you should do your best to keep the required input to a minimum; this is no time to force the user to enter unnecessary information.

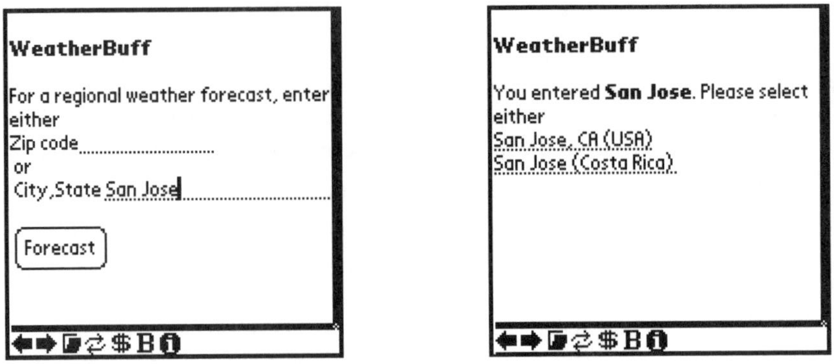

Figure 3-2. Optimizing user input for location finding

Check Boxes

Check boxes work well when the user must choose from a number of options. Few Web sites presently use free-form input when check boxes will do the job. (Who wants to spend extra time writing those back-end CGI scripts anyway?) This design consideration is even more important with wireless devices. Check boxes tend to be easier to use than multiple-pick lists, especially on a device that may lack a Shift key (which is used to indicate multiple selections).

Buttons

Buttons provide an opportunity for the wireless content developer to get creative and emulate a device's native interface. Many handheld wireless platforms present a software developer with a steep learning curve or without a software development kit (SDK) at all. (See Chapter 9 for a discussion of writing native applications for wireless platforms that provide SDKs.)

On wireless platforms with an SDK, writing a dedicated wireless application can become prohibitively expensive—especially if your goal is to support more than one platform. Buttons on Web pages provide one mechanism for making a wireless site *look* more like an application than a Web site. Coupled with WAP and technologies such as WMLScript, buttons provide a Web site with a look and feel that's more native to the platform.

Placement of controls presents an interesting challenge for many content developers. In general, layout languages will not allow you to provide exact placement information for text or controls. However, if you're targeting specific platforms, you may find that you can obtain some control over the screen layout through careful editing of content. If so, consider the target device carefully and be sure to preview everything on it!

Touch Screens

Most devices employ a touch screen that the user operates with a fingertip or a stylus. Remember that, unlike with a mouse or graphics tablet, the user's interaction with the device will obscure a large percentage of the display. Therefore, buttons should be located underneath any relevant text, enabling the user to see the information while making a selection. Also, most users cannot accurately select areas smaller than 16 pixels square. Some browsers may size buttons intelligently, but others may not.

Previewing Content

The key to meeting user expectations is envisioning your content from their perspective. This is not a new idea; site developers have been previewing their Web sites for years. Remember, however, that the range of wireless devices on the market is very wide. If you're developing a site aimed at a few different kinds of devices, it's well worth the expense to obtain those devices and begin regularly previewing content on all of them.

Simulating the Viewer's Environment

Whenever possible, try to access your content in the same way your subscribers will. There's little point in measuring the performance of your site on a desktop browser with your Ethernet Local Area Network (LAN) if you'll be deploying a Cellular Digital Packet Data (CDPD) wireless solution using a Windows CE handheld computer.

In some cases, accurately simulating your subscriber's environment will be difficult or impossible. When developing general sites, for instance, it may not be possible to assemble the full assortment of devices your clients will be using. In other cases, you may find that the network you plan to deploy over isn't available in your area. The availability of CDPD is a case in point, in which the mix of strong metropolitan coverage and weak rural coverage can cause problems for developers.

Ironically, resurrecting old technology provides a good approximation of how the latest in wireless technology might work. To emulate most handheld Web browsers, you can use an early Web browser (Netscape 2.0 works well) and a slow LAN connection, such as the 9.6KB or 14.4KB modem you couldn't give away at your last garage sale. (Or, rather than dumpster-diving for that precious modem, you could adjust your computer's baud rate on the appropriate serial port.) Using older software helps when you examine sites for dependence on scripting, nonstandard tags, and certain image formats, while the slower modem simulates your user's connection speeds.

Examining Display Parameters

When using your browser to preview your work, set your display to no more than 256 colors or grayscale if applicable. You may want also to deliberately pick a poor contrast for your monitor to simulate the displays present on many devices. Adjusting the brightness and contrast controls on your monitor to simulate poor screen contrast may seem like a step backward for content development, but it's for the greater good of your content. Also, be sure to scale your browser window to approximate the size of the target screen.

Table 3-1 lists some common display parameters for wireless clients. (*Note:* Most browser applications take up 30 to 40 pixels of space along the top and bottom of the display for status and other information; this margin is not reflected in the numbers shown in Table 3-1.)

DEVICE	SCREEN SIZE	SCREEN COLORS
Screen Phone*	12 characters × 4 lines	Monochrome
Palm Computing Platform	160 × 160 pixels	4 shades of gray or 256 colors
Handspring Visor	160 × 160 pixels	4 shades of gray
Microsoft P/PC	320 × 320 pixels	16 shades of gray or 256 colors
Microsoft H/PC	640 × 240 pixels	256 colors
Microsoft H/PC Pro**	640 × 480 pixels	256 colors

* Minimal functionality as defined by the WAP Forum. Note that the soft buttons at the bottom of the display consume one of these four lines.

** Some H/PC Pro devices now sport displays with SVGA (800 × 600) resolution and 256 colors.

Table 3-1: Typical Display Parameters for Wireless Clients

Figure 3-3 shows a desktop browser previewing a page intended for wireless access. To preview the content, I used Netscape Communicator 4.6 to get an idea of a site's appearance. This enabled me to create a page on the desktop with some idea of how it was going to be laid out on Palm, Inc.'s Palm III.

On the Palm III, I viewed the content with ProxiNet, a popular Web browser available to users of the Palm Computing Platform. Note that ProxiNet's layout is actually better than the Netscape view given the window size. Considerations such as these — actual font choices, the layout of a page, and available characters — demonstrate the importance of viewing your content with the target hardware and software.

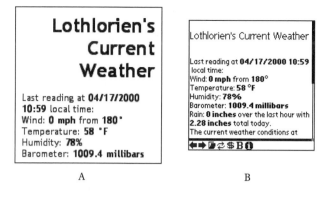

Figure 3-3. A sample page rendered with (A) Netscape and (B) ProxiNet

Summary

As you've doubtless gathered by now, the watchword for successful wireless Web development is *brevity.* Keeping pages short will help both objective and subjective perceptions of your content. The page will not only be easier for the viewer to read thanks to its brevity and uncluttered appearance, but also will be more pleasing to look at.

Use images sparingly if at all, because they cannot be displayed by all platforms and consume a large amount of bandwidth. In no case should you use image maps, because when you do, viewers without image display cannot navigate the links on your site.

You'll want to create your pages with a minimum of type fonts and styles so that the pages appear correct on the widest number of wireless access terminals. This approach, along with brief content, will also help minimize scrolling—a source of frustration for many users because of the small screens on most devices. When formatting your content, use markup commands sparingly, and preview often on the kinds of devices your viewers will use.

Input poses challenges for many users. You should avoid prompting users for input unless absolutely necessary; whenever possible, give users a choice of items rather than making them input names, addresses, or other information.

CHAPTER 4

The World Wide Web without Wires

AFTER ALL MY WARNINGS in previous chapters about the constraints wireless devices suffer, you may be ready abandon any notion of using the World Wide Web to distribute your content. After all, a simple Web page can have 20 or 30 different files (images, sounds, and HTML), each taking several kilobytes and filling an 800×800-pixel window. This isn't the sort of content you would want to bring to a handheld device; even if the content displayed well, the amount of time it would take to download this much data could stretch into minutes.

Why Use the Web at All?

So why use the Web at all, especially when there are other protocols such as the Wireless Application Protocol (WAP) and the Handheld Device Markup Language (HDML) being developed to distribute content to wireless devices more efficiently? (Chapter 2 looks at each of these in a comparative light.) There are several reasons, all of which have to do with the fact that the Web is ubiquitous these days. Almost all wireless devices can access the Web, and developers already have a wide range of tools at hand for creating Web-based content. Furthermore, there are obvious advantages, in terms of integrating content with existing environments and services, to adapting content that has already been created for the Web.

Ubiquity of Web Browsers

Probably the most obvious reason to use the Web for wireless development is the ubiquity of applications that use Web protocols. Nearly every handheld computer ships with a Web browser of some kind, and for those that don't there are third-party Web browsers readily available. On most devices, including the hugely popular Palm Computing Platform and Microsoft Windows CE devices, users can choose one of many browser solutions tailored to meet their needs in terms of size, performance, and features. The over eight million handheld devices presently in use (as counted by sources such as International Data Corporation), most of which enable Web access, form an established user base for any Web content developed for use with these devices.

Ready Availability of Development Tools

The world of Web development offers a plethora of tools for creating content, from the HyperText Markup Language (HTML), which can be created in a simple text editor, to What-You-See-Is-What-You-Get (WYSIWYG) tools designed for the maintenance of entire Web sites. Back-end development is supported by over a dozen commercial applications and hundreds of open-source tools written in the Perl, C, C++, and Java languages.

By comparison, tools for WAP are still on the cutting edge. This isn't necessarily a bad thing, but it can make development challenging. New tools have bugs that need to be worked out; frequent releases of these tools can mean you spend time updating environments rather than authoring content; and the rapid changes that are a reality of new tools may well require you to redo much of your work from scratch.

Easy Integration with Existing Systems

While WAP is designed to work well with existing Web-based services, there's no argument that Web services undoubtedly make integration with the Web easier, because WAP involves the use of new technologies to deploy its services. The Web, on the other hand, is a tried and true medium for content exchange. Extending existing content services to wireless units using the Web requires only new content— no new markup languages to learn, no new distribution tools or servers to get.

This ease of integration—and the widespread acceptance of Web-based technology—is especially important if you plan to develop for major businesses (the *enterprise* market). Information technology (IT) leaders in enterprise computing can be paranoid when new technology is introduced; their need for security and stability make them resistant to the kinds of changes that are necessary to support WAP applications on an intranet. Installing new back-end scripts for database management and, potentially, new servers, incurs the kinds of fiscal and risk-related costs most IT managers work to avoid.

When *Not* to Use the Web

When all these factors are taken into account, it often makes more sense to use the Web rather than WAP or HDML for wireless Web development, despite the drawbacks. In any particular case, your final decision should be predicated on knowledge of your clients. If your subscribers will largely be viewing your content with handheld computers, such as Personal Digital Assistants (PDAs) or super phones, then using the Web is the obvious choice. These devices have capable browsers; with some fine-tuning of the HTML and graphics, Web content displayed on them can shine. On the other hand, if you're targeting screen phones, you'll definitely want to work with WAP and HDML. These devices generally don't have full-fledged Web browsers, and can't display traditional HTML content. (See Chapter 1 for a discussion of the range of handheld devices.)

In some cases, it may make sense to provide your content in both formats so it can reach the widest possible audience. A Web site providing travel information to mobile workers, for example, should cater to the users of both handheld computers and screen phones.

A Review of Web Standards

The power of the World Wide Web lies in its open, documented standards for the sharing of information. These standards describe how data is formatted and exchanged between different computers, enabling virtually any computer to access the information on any server connected to the Web.

Before I discuss how to use the Web to distribute content to wireless devices, a brief review of the Web standards and their use in wireless Web development is in order. The HyperText Transfer Protocol standard is widely used to exchange data between computers, while HTML, the Graphical Interchange Format, and the Joint Photographers Expert Group format are commonly used to specify data formats for kinds of content. There are also standards for security between hosts and for specifying scripting behavior, both of which I touch on briefly.

HTTP

The HyperText Transfer Protocol (HTTP) is the backbone of the Web. While it is designed to support the exchange of virtually any content, it is largely used for the transfer of text and images.

A unique Uniform Resource Locator (URL) identifies each unit of content, or object (for example, `http://www.apress.com/CatalogMain.htm`). A URL is a concatenation of the protocol used (`http://`), the name of the host serving the document (such as `www.apress.com`), and the path and file name of the document being served (`/Catalog/` is the directory, and `CatalogMain.htm` the document being requested). URLs serve as arguments to *methods*, which are plain text keywords instructing the HTTP server to do something on the client's behalf. An object can be any aggregate stream of bytes—an HTML page, an image, a sound, or an application-specific blob of data. Common objects include Web pages (written in HTML), responses to forms from the client (commonly called *form data*), and image data.

The two methods—operations invoked by HTTP clients in a request—most commonly used by HTTP are GET and POST. As you can probably guess, HTTP clients use GET to obtain the object associated with a given URL, and use POST to submit an object to the server. Thus, clients obtain regular Web content using GET, and clients submit form content using a POST request. For historical reasons, there's also a way to use GET to present form data, because the POST method was added to the protocol later. Form data can also be posted to a server using HTTP/0.9, using an extension of the URL.

Clients and servers can modify a message using a *header*, which precedes an object. Headers are used to identify the client and server, and enable the client or server to give additional instructions. For example, a *keep-alive* header tells the server to maintain a connection over multiple transactions. Enabling keep-alive is vital for good operation on wireless networks; it allows the client to reuse a single network socket to request multiple objects, saving the overhead of renegotiating a connection to the server for each one. (Newer servers, notably those implementing HTTP/1.1, will automatically behave this way when interacting with HTTP/1.1 clients and don't explicitly require the keep-alive header.)

As a Web developer, you probably won't make much use of HTTP itself, as it operates behind the scenes to provide the medium of exchange between client and server. However, you should understand the following:

- What keep-alive is (and make sure your server is new enough to support it— or use HTTP/1.1 servers exclusively).

- The difference between the GET and POST methods for returning form data—that is, with GET, the form data is embedded in the URL, whereas with POST, data is returned in the object body. (The use of forms in a wireless context will be covered in more detail later in this chapter.)

HTML

While HTTP describes how hosts exchange data, the HTML describes how data is presented to the viewer. It is a *markup language*, a set of instructions for formatting a document.

The term "markup language" has pre-computer roots. In publishing, editors "mark up" manuscript to indicate how text should be formatted by the compositor. Computer typesetting applications use a similar concept: typesetting commands embedded into the text of an electronic manuscript to produce a formatted document are dubbed "markup commands." The tradition continues on the Web: using HTML, authors can place markup commands in text files and the client will interpret these commands to determine how to present the data. In this section, I discuss the general aspects of HTML. In Chapter 5, I look closely at which HTML markup commands are appropriate for wireless Web developers.

It's important to realize that markup implies a set of *guidelines*, rather than hard-and-fast rules about how a document should appear. For example, the <B> directive hints to the browser that content so marked should stand out in some way. The fact that browsers choose to use a bold font is a largely historical one. Good browsers, especially good micro-browsers on wireless devices, use this flexibility to their advantage, and weigh markup requests with device constraints. Consequently, some markup tags may go ignored on some browsers because their devices have no way to support the markup requested.

An HTML document consists of American Standard Code for Information Interchange (ASCII) text interspersed with *tags* that instruct the client as to how the viewer should format the text. Tags come in pairs surrounding the text they apply to, and are set off using angle brackets (< and >). The first tag of the pair is generally a description of the operation to be performed, while the ending tag is the same term preceded by a /.

Here is a simple HTML file:

```
<HTML>
<HEAD>
   <TITLE>Hello, World!</TITLE>
</HEAD>
<BODY>
<P>Hello, world!</P>
</BODY>
</HTML>
```

Every HTML document begins and ends with the <HTML> tag, indicating that this document is marked up with HTML. A document consists of an HTML *heading*, enclosed within starting and ending <HEAD> tags, and a *body*, enclosed within starting and ending <BODY> tags. (The HTML heading should not be confused with the HTTP header; it's an integral part of the HTML document, rather than part of HTTP. The HTML heading is used by clients to set the title of a window, the window size, and so forth; the body contains the actual Web content to be displayed.

Our sample uses the <TITLE> tag to specify a window title, and the <P> tag to indicate the beginning and end of a paragraph, which in this case says simply, "Hello, world!" HTML is versatile: there are tags for selecting fonts, sizes, and styles; tags to create bullet and auto-numbered lists; tags that can include images and sounds with a document; tags for tables; and tags for constructing pages consisting of several frames, called *framesets*.

If you're wholly unfamiliar with HTML, I suggest you take a break from reading this chapter and learn a little more about it. There are several good ways to do this:

- Visit the World Wide Web Consortium (W3C) Web page about HTML, at http://www.w3.org/MarkUp/Guide/.

- Search the Web (a Yahoo! search works wonders) for introductory HTML pages and read a few.

- Get your hands on a free WYSIWYG editor for HTML, such as Netscape Communicator. Write some pages in it, then use its View ➤ Source command to see the HTML it generated. You can edit the source by hand (using any text editor), and preview it with a Web browser to see what your changes do. Experimentation is the best way to learn how tags behave.

In the next chapter I walk you through many of the commonly used HTML tags. If you're a quick study, that may be all you need to get started marking up HTML content.

GIF

The Graphical Interchange Format (GIF), made popular in the 1980s by the Compuserve network, is the workhorse of Internet graphic formats. GIF is the mainstay for transferring both simple and complex images on the Web.

GIF is good for crisp line art and general-purpose images. It supports 8 bits of color data, making 255 colors (and a transparency "color" that can be used to make the background of an image transparent) available to the artist. GIF first encodes the image as scan lines and then compresses one or many scan lines into fixed-size packets, making it ideally suited to the compression of line drawings and drawings with solid color.

JPEG

The Joint Photographers Expert Group (JPEG) format was formed to provide high-quality compression for highly detailed images such as photographs and satellite imagery. (This format, by the way, is actually technically called JFIF by the Joint Photographers Expert Group, which stands for JPEG File Interchange Format, but in common parlance so many people refer to it as JPEG that I do the same here.) The format supports higher color depths than does GIF (24 bits of color information, providing up to 16.8 million colors) and can support multiple images within a JPEG file. Unlike GIF, JPEG compression works using *color reduction*, which takes advantage of a peculiarity of the human eye: people are more likely to perceive differences in brightness than differences in hue. JPEG records the brightness and hue of any given point on an image separately, then reduces similar hues in adjacent pixels to the same hue and re-encodes the image using the reduced data. Brightness information is preserved, while similar hues are averaged, yielding a reduction in image size. This kind of compression is often referred to as "lossy," because the resulting image contains less information than the original.

The JPEG format enables the artist to choose the amount of color reduction that occurs. At greater levels of compression, the color reduction becomes more noticeable to the viewer, taking the form of artifacts that appear as blocky boxes of slightly off-color pixels. In images that consist of many solid horizontal and vertical intersections, these artifacts are only marginally observable, because in many cases the artifacts will fall along the lines formed by the intersections. On the other hand, in complex images with a variety of shapes and lines at nonperpendicular angles, these artifacts can be distracting at best and disruptive at worst. Pictures with these abrupt transitions that show artifacts when compressed by JPEG are best stored as GIF files.

The JPEG format can dramatically reduce the space an image occupies with little noticeable difference to end users. Its algorithm works best with images that contain gradual transitions between colors (such as bitmapped images from photographs), as this is where the opportunity exists to reduce the amount of data needed to record the image. For images with a relatively small number of solid colors (such as line diagrams or colored cartoons), JPEG fares more poorly—the algorithm can't find hues that can be easily combined, and any combining it does may affect the image's quality significantly.

Other Image Formats

Other formats for transferring images on the Web abound. The traditional operating-system image formats—PICT, BMP, and XBM—are still supported by many Web browsers, although some browsers will only display formats native to their platforms. (The Windows CE Microsoft Pocket Internet Explorer, for example, can draw BMP images, but not XBM or PICT images.) In general, it's best to avoid these three formats, because they are platform-dependent and not necessarily well suited to the compression of images.

A new format, the Portable Network Graphic (PNG) format is being adopted by a growing number of Web browsers for desktops. PNG provides patent-independent compression (avoiding the potential headaches of Unisys' claims over GIF, for example), and offers some attractive features, including variable transparency, cross-platform control of image brightness, and two-dimensional interlacing. (*Two-dimensional interlacing* encodes an image in successive layers of detail so that as the image is drawn, higher detail becomes apparent as the client obtains more data. This gives a user on a low-bandwidth connection the perception of a quick-loading image because there is something to watch sooner.) However, most of the browsers on wireless devices do not yet support PNG, so content developers targeting the wireless market should avoid it for now.

Scripting

Scripting languages—Java, JavaScript, VBScript, or the proprietary scripting environments found in some smaller browsers—enables developers to create interactive content that runs on the client. Web browsers can download and run programs written in these languages. These programs appear as part of one or more content pages. While not protocols *per se*, their use is closely linked to the Web protocols mentioned thus far.

While desktop Web browsers support both JavaScript and Java, you'll be hard-pressed to find a browser for a handheld device with similar features. Some software vendors—notably Spyglass, Inc.—offer browsers that support one or another of the standard scripting environments, but few hardware manufacturers have actually deployed browsers with this level of sophistication due to the hardware costs of running a browser with scripting support.

The reason for this is simple: memory. All scripting implementations consume enormous amounts of memory (in handheld terms). While language providers such as Sun Microsystems are working to produce smaller scripting implementations, browser and hardware vendors aren't likely to adopt these immediately.

When they become available, scripting languages for the wireless Web are likely to be significantly simpler than those currently in use. This simplicity will reduce the burden on wireless access terminals, but increase the burden on content providers: those interested in providing material for both the conventional Web and its wireless counterpart will have to develop their content in two distinct versions, each using different scripting dialects.

Security

Security is as important on the wireless Web as it is on its wired counterpart. Electronic commerce, transfer of corporate data, and other private transactions may need to be secured from prying eyes.

Fortunately, wireless networks are actually safer than wired networks for several reasons. Wireless networks often offer some form of encryption. And while seldom truly secure, the encryption schemes wireless networks use safely obscure data from most prying eyes. Some wireless schemes, notably Code Division Multiple Access (CDMA), are devilishly hard to eavesdrop. In addition, the interface between wireless clients and the network—except when the data is actually in the air—is usually part of a closed telecommunications network, making interception more difficult.

As discussed in Chapter 1, wireless networks generally use some sort of frequency sharing—distributing data from a single source across multiple channels. Any attempt to eavesdrop on transactions being carried out this way would require capturing and reassembling very large amounts of data. On spread-spectrum networks, inside information about how a signal is spread might even be required in order to reconstitute the data from a particular source. This obfuscation does not guarantee security by any means, but it discourages many attacks by would-be hackers.

In addition, many wireless networks have been designed with the need for higher security in mind. Most all-digital wireless networks are less than five years old, and their architects have had supporting electronic commerce as a high priority right from the drawing-board stage. Therefore, wireless networks usually include active security measures such as encryption and authentication.

Nonetheless, security remains an issue, especially where the wireless network meets the wired one. To ensure that data does not become vulnerable once it arrives on the wired network, there are established security technologies available on both browsers and servers, such as Secure Sockets Layer (SSL), using standard encryption algorithms. The wide availability of these options makes them a good choice for wireless Web applications using HTML and HTTP—and is yet another point in favor of using traditional Web protocols. A chain is only as strong as its weakest link; this is as true in data security as anywhere else.

Composition Tools for HTML

One major advantage of producing content in HTML is the variety of authoring environments available. Many developers now use HTML in much the same way as they use word processors, taking advantage of the vast array of templates, style sheets, and WYSIWYG editing aids now available. These aids make elaborate HTML painless to create.

Using these aids for wireless development, however, is fraught with danger. These tools can create complex HTML for even the simplest looking of pages, and embed a lot of content that is unnecessary for clients. These features bloat the size of HTML files considerably, which leads to longer download times, higher access costs, and possibly, irritated customers.

Keeping It Simple…

Despite the pantheon of sophisticated authoring tools available, the best tool for developing Web content for wireless devices is still a simple text editor. Working with raw HTML can be somewhat error-prone, but the benefits are tremendous. For example:

- It helps you avoid the temptation of overly complex formatting.

- It avoids the space-hogging, editor-specific extra content that some advanced editing tools insert into HTML documents, which takes up valuable bandwidth but is ignored by browsers (see following Netscape example).

- It enables you to fine-tune a file's layout and performance by hand.

It is a truism about the nature of people—in this case, Web authors—that, if given an opportunity to do something in a complex manner, they will choose to do so. Sophisticated HTML editors make creating frames, nested tables, spacer images, and such almost trivially easy; and these are exactly the things you want to *avoid* when creating content for wireless devices. Even if you exercise restraint with these tools, you may find yourself spending hours "just trying" something you think will look good on a device, poring over numerous iterations of an embellishment your content didn't really need in the first place. By using a text editor and writing the HTML by hand, you can lessen significantly the temptation to use all the bells and whistles at your disposal, simply because it will be a little more work to do so.

Many HTML editors save information for their own purposes—but not useful to end browsers—in the form of *HTML comments* or *meta tags*. HTML comments, of course, were originally meant for the people writing the HTML, while meta tags are meant to contain browser directives not expressible using the standard markup language. In many cases, neither is appropriate for a client's consumption. This practice has become accepted as these tools have saturated the marketplace. After all, there appears to be little harm in downloading this extra content meant for HTML editors to clients, and embedding it in the HTML document enables remote clients to edit it with the same tools. This capability can be useful when people write content in a distributed environment.

However, space is at a premium in all aspects of wireless Web content, and these editor-specific tags can occupy a significant amount of space in an HTML document, especially if the document is short. Consider, for example, the following sample of HTML as presented by Netscape Communicator, a popular free HTML composer:

```
<!doctype html public "-//w3c//dtd html 4.0 transitional//en">
<html>
<head>
   <meta http-equiv="Content-Type" content="text/html; charset=iso-8859-1">
   <meta name="GENERATOR" content="Mozilla/4.7 (Macintosh; I; PPC) [Netscape]">
   <title>samp.html</title>
</head>
<body>

<h1>
Welcome!</h1>

<table WIDTH="100%" >
<tr>
<td><img SRC="image.jpg" height=26 width=26></td>
```

```
<td>The weather here today is sunny, with a temperature of 68&deg; F.
<br>Expect similar conditions during your visit.</td>
</tr>
</table>

</body>
</html>
```

The header for this document is 181 bytes long, but 154 of those bytes are not necessary for a client. The first meta tag in the header (`<meta http-equiv="Content-Type" content="text/html; charset=iso-8859-1">`) indicates the content type and character set, which is information that will be provided by a well-configured HTTP server anyway. The second one (`<meta name="GENERATOR" content="Mozilla/ 4.7 (Macintosh; I; PPC) [Netscape]"`) specifies the HTML editor that was used to create the document (Netscape Communicator, in this case)—information that is of no interest to a wireless client. Thus, 85% of this header is useless to the client, and serves only to waste transmission time and memory space on the device.

The same page can be expressed in (hand-written) HTML as follows:

```
<html>
<head><title>samp.html</title></head>
<body>
<h1>Welcome!</h1>
<table WIDTH="100%">
<tr>
<td><img SRC="image.jpg"></td>
<td>The weather here today is sunny, with a temperature of 68&deg; F.
<br>Expect similar conditions during your visit.</td></tr>
</table>
</body>
</html>
```

This version of the content occupies a mere 411 bytes of space, as opposed to the original content's 913—a savings of over 50%. This sort of penny-pinching may be a waste of time when the material is intended for today's desktop browsers, which can carry even the long form of this document in a single packet. But it will realize a significant savings over wireless networks.

...the Easy Way

Of course, if you're generating hundreds of pages for a large site, the thought of creating or revising them all by hand can be a bit daunting. For one thing, marking up documents with HTML is tedious, as simple errors such as unmatched tags tend to creep in after a long day's work.

Luckily, help is available. Before the age of WYSIWYG editors for HTML, several programs were written that would examine an HTML file and point out errors. Some of these programs went so far as to check hyperlinks, enabling you to validate an entire site. Others focused on checking the grammar of a specific HTML file.

The best known of these tools is Tidy, written by Dave Raggett, and available from the W3C at http://www.w3.org/People/Raggett/tidy/. Tidy examines HTML for mismatched tags, open tags with no matching close tags, and other problems. Unlike other programs that only validate HTML, Tidy attempts to fix errors as it detects them, making less work for the content author.

Here's our previous HTML example run through Tidy:

```
MacTidy (vers 19th October 1999) Parsing "Kona:Desktop Folder:MacTidy
1.0b6:test.html"
line 13 column 1 - Warning: <table> lacks "summary" attribute
line 15 column 5 - Warning: <img> lacks "alt" attribute


"Kona:Desktop Folder:MacTidy 1.0b6:test.html" appears to be HTML 3.2
2 warnings/errors were found!

<!DOCTYPE html PUBLIC "-//w3c//dtd html 4.0 transitional//en">
<html>
<head>
<meta http-equiv="Content-Type" content=
"text/html; charset=iso-8859-1">
<meta name="GENERATOR" content=
"Mozilla/4.7 (Macintosh; I; PPC) [Netscape]">
<title>samp.html</title>
</head>
<body>
<h1>Welcome!</h1>

<table width="100%">
<tr>
<td><img src="image.jpg" height="26" width="26"></td>
<td>The weather here today is sunny, with a temperature of 68&deg;
F.<br>
Expect similar conditions during your visit.</td>
</tr>
</table>
</body>
</html>
```

Because we didn't make any mistakes, it didn't have much to say. The tool begins by identifying the file being examined (useful if you're verifying an entire site), and then prints any errors or warnings it has regarding the file's contents. Once this information is printed, the tool prints the proposed file with corrections.

This input file had no errors, but two suggestions. One is pertinent only to the latest version of HTML; the <TABLE> tag accepts a SUMMARY attribute that is used by some browsers to provide a summary linking to the table itself. The other warning is one we should heed, our image tag lacks an ALT label. HTML uses the ALT attribute when a browser cannot display an image, something that may well happen on the wireless Web. I should have written

```
<td><img src="image.jpg" alt="sun" height="26" width="26"></td>
```

so that browsers choosing not to display images would include the word "sun" near the image indicator.

If we introduce errors and try again, Tidy will tell us about them. In the following example, I "accidentally" mismatched the closing <h1> tag:

```
<html>
<head><title>samp.html</title></head>
<body>
<h1>Welcome!</h3>

<table>
<tr>
<td><img src="image.jpg"></td>
<td>The weather here today is sunny, with a temperature of 68&deg;
F.<br>
Expect similar conditions during your visit.</td>
</tr>
</table>
</body>
</html>
```

Running this imperfect sample through Tidy produced the following result:

```
MacTidy (vers 19th October 1999) Parsing "Kona:Documents:Writing:Books:WWD:Exam-
ples:4:tidy-broken.html"
line 4 column 13 - Warning: replacing unexpected </h3> by </h1>
line 6 column 1 - Warning: <table> lacks "summary" attribute
line 8 column 5 - Warning: <img> lacks "alt" attribute

"Kona:Documents:Writing:Books:WWD:Examples:4:tidy-broken.html" appears to be HTML 3.2
```

```
3 warnings/errors were found!

<!DOCTYPE html PUBLIC "-//W3C//DTD HTML 3.2//EN">
<html>
<head>
<title>samp.html</title>
</head>
<body>
<h1>Welcome!</h1>

<table>
<tr>
<td><img src="image.jpg"></td>
<td>The weather here today is sunny, with a temperature of 68&deg;
F.<br>
Expect similar conditions during your visit.</td>
</tr>
</table>
</body>
</html>
```

Once you've verified the HTML in a document, you can reduce the amount of space it takes up. The Perl HTML::Clean module by Paul Linder (available from `http://people.itu.int/~lindner/`, or the Comprehensive Perl Archive Network at `http://www.cpan.org`) can eliminate white space, carriage returns, and other debris that unnecessarily increase the length of an HTML file. (It can also extract meta tags from documents that have been authored using tools.) The resulting HTML is significantly more terse than it was written, and takes less time to traverse the wireless network.

While Linder's package comes with a robust tool to condense HTML, a simpler Perl program such as the one that follows uses his module as well:

```
#!/usr/bin/perl
use HTML::Clean;

$h = new HTML::Clean( $ARGV[0] );
$h->compat();
$h->strip();
$data = $h->data();

print $$data;
```

To run this program on an HTML file, enter the file's name on the command line. The resulting condensed HTML will be printed. Using this program, the HTML example we've been working with would be condensed as follows:

```
<!doctype html public "-//w3c//dtd html 4.0 transitional//
en"><html><head><title>samp.html</title></head><body><h1>
Welcome!</h1><table width="100%"><tr><td><img src="image.jpg" height=26 width=26
alt=""></td><td>The weather here today is sunny, with a temperature of 68&deg; F.
<br>Expect similar conditions during your visit.</td></tr></table></body></html>
```

Condensed HTML isn't pretty—it's downright hard to read, and editing it by hand in this state would likely introduce errors. Yet this file occupies a scant 360 bytes—a 63% savings over the original content.

While automated tools aren't perfect, and new pages should always be inspected by a person to ensure proper formatting and good aesthetics, tools such as those described here can go a long way in helping you produce quality, error-free content, with the least possible effort. Moreover, tools such as HTML::Clean can be used to help make HTML produced on a fancy editor smaller, too. Running this tool over all of the HTML produced for a site prior to distributing it can make a world of difference to your wireless content.

Presentation Tips

The basic rules of wireless Web authoring are straightforward: keep it simple and keep it small. Yet it can be surprisingly difficult to master the strategies for presenting information in this manner. In Web content designed for the desktop or print, a single "page" of information commonly consists of a half-page to a page of print. If you think of this page as coming from your college notebook, think of the wireless Web equivalent as a crib sheet—a mere 3×5-inch card on which you must fit all your information. Unfortunately, in this context, you don't have the option of writing small and carrying a magnifying glass to your final exam!

Text

Perhaps the most important fact to remember about presenting text concisely is the value of summaries. Most users of mobile devices don't want copious amounts of data; they want just the facts. Summaries can help them quickly extract what they're looking for from the sea of data behind your site.

One approach is to offer the summary *first*, with a wireless link to the full content. Thus, a user is able to read the headline, skim the summary, and then decide if it's worth the extra time and bandwidth to download the longer version.

Some news sites are already doing this for the AvantGo HTML wireless browser (which you'll encounter in Chapter 6) enabling readers to pick the depth of information they need. Figure 4-1 shows a fictitious example of using content summaries to optimize wireless content.

Figure 4-1. Optimizing wireless content by using a summary (left) and a full story (right)

More often, however, only summaries are available to wireless viewers. The summary itself can include a URL (not a hyperlink) to a page with the full story for the viewer who wants to access it from a desktop via a separate link. This tactic is especially appealing to content providers who derive revenue from desktop viewers through subscriptions or advertising.

It's especially important to make your headlines and summaries relevant and useful because wireless users rely so heavily on them. Many writers, aiming to draw the reader into a story, are tempted to make headlines sensational or even absurd; this tactic may be acceptable in print, but it is rarely appropriate for wireless content. Following the link from a dubious headline can cost the viewer precious time—and hence money—downloading a story of no interest. Naturally, this makes for a resentful customer. Keep headlines and summaries to the point!

Tables

For use on mobile browsers, it is generally best to create tables that are long and narrow, as opposed to wide and short. Although the market appears evenly split between portrait-oriented and landscape-oriented screens, they often scroll vertically and only rarely horizontally; some narrow-screen platforms don't even support horizontal scrolling. A long, skinny table is therefore more likely to be accessible to all users than a short, fat one.

Another pitfall to avoid is nesting tables. Although HTML supports nesting tables, the results on handhelds are indeterminate at best and catastrophic at worst. Some clients refuse to draw nested tables; others attempt it—and manage so poorly that the meaning of a table is obscured or changed without warning to the user.

In any case, there is usually a more appropriate way to represent information than nesting tables. If you find yourself tempted to nest a table, stop and examine what content you're trying to bring to the reader. Often, you may realize that your purpose is in fact to control the screen layout (for example, using a table to create a page with paragraphs of text in multiple columns). It's always better to let the client perform the screen layout, as it's more familiar with the characteristics of the device than you will be.

Forms

Handling form input for wireless devices presents challenges both in interface design and in handling the resulting data. The two major factors to keep in mind are:

- Provide end users with an easy-to-use interface.

- Keep the quantity of form data that results from the responses as small as possible for its return trip over the wireless network.

Fortunately, these goals are not mutually exclusive. Most mobile users want to enter as little information as possible when using an online form. In most cases, they will be at least partially distracted. They may be walking, driving, or otherwise preoccupied—often with at least a part of their body in motion, so inputting may require difficult feats of both mental and physical balance.

Therefore, you should design forms so that they require the minimum possible amount of input. For example, if you need to find out the user's location, ask for a zip code or telephone number with area code, rather than a full address. Often, this information can be used to pinpoint a location closely enough to provide information such as restaurant recommendations or movie listings.

Radio buttons, pop-up menus, and pick lists are the easiest input methods for mobile users. Most mobile devices offer a touch screen or rocker key for rapid scrolling; many have both. These tools make managing lists easier than inputting text, as it requires fewer motions to manipulate a rocker button or touch an item on a display than to navigate a keyboard. Items best left to lists include the following:

- Countries or states.

- Color or size choices.

- Categories of data, such as "Simple" or "Detailed" stock quotes.

- Choices of operations, such as "Accept" or "Cancel" an order.

Unfortunately, the most common scheme for reducing the inputting burden placed on users isn't well supported on the wireless Web. *Cookies* are bits of information from specific HTTP headers that are stored on the client between transactions which enables servers to maintain the illusion of a session without the need for server complexities. This information might be used for various purposes: to manage the state of a transaction, to authenticate a user, to keep track of preferences, and so on. Not all wireless HTML browsers support cookies, although the number is growing gradually. Moreover, the convenience of cookies can pose a problem for handheld devices; as more and more sites adopt them, they tend to bloat the "hidden" cost of Web access, as they are exchanged with the browser but aren't directly seen by the subscriber.

Some handheld devices support another technique for reducing input and managing transactions, however. This mechanism uses a hidden identifier within forms that is generated uniquely by the client as a transaction starts, and then carried throughout a transaction's lifetime via a hidden variable within the form. For example, the HTML

```
<INPUT TYPE="hidden" NAME="client" VALUE="%deviceid%">
```

creates a hidden field on Palm Computing Platform devices running the Palm Web Clipping Application that servers can access by checking the value of the `client` variable when examining form results. This resulting identifier can be carried from form to form between client and server, enabling the server to track state in an internal database keyed by the value of the `client` variable. You should only do this when necessary to maintain state however, to keep the client from needlessly downloading irrelevant content.

Unfortunately, the way to get the unique identifier for a device is not standard, reducing this trick's usefulness for content aimed at a broad viewer base. As the use of unique identifiers for these devices spreads, one can hope that a standard will emerge and that unique identifiers can be used in concert with server-side data storage to emulate the behavior of cookies.

Thus, forms work best when interaction can be kept to a single request-response sequence ("Give me the quote for this stock.", "Where will I have dinner tonight?", or "What wine goes best with salmon?"). Doing so minimizes both the amount of information a user must enter and the complexity of the back-end service necessary to track clients, their requests, and results.

Images

Probably the feature of any Web page that makes the biggest impression on viewers is its use of images. While you'll want to use images sparingly in your wireless-oriented content for the reasons discussed in Chapter 3 (hardware, memory, and network performance constraints all play a role in the decision), you may not want to abandon the use of images completely.

Choosing Proper Images

In some cases, images can provide more meaning in less space than can words. Consider Figure 4-2, inspired by the Web site Weather Underground. Both versions of this information take only a handful of bytes; in fact, depending on the format chosen, the amount of memory required could actually be *smaller* for the image than for the text. Which version provides the viewer with the most information with the least effort? While either may be appropriate for a deskbound user, the graphic is a better choice for a wireless user. The graphic carries the same message as the text, uses a similar amount of memory and network resources, yet is easier to read and takes up less screen space than the text.

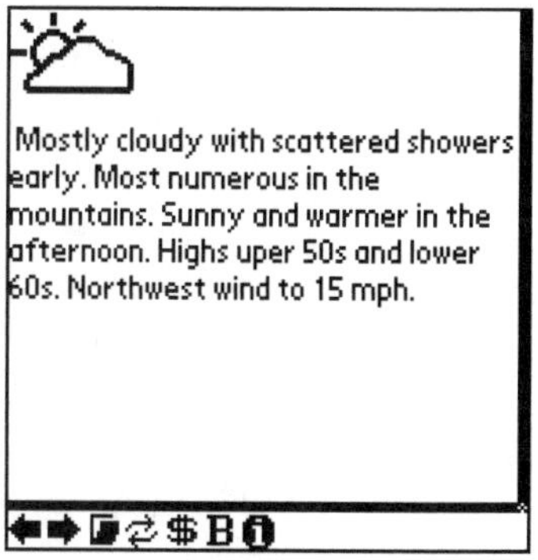

Figure 4-2. Contrasting graphics with text

Keep graphics simple; you should stick to either grayscale or just a few bold colors. Images shouldn't make up more than a small fraction of your site's content, or they can easily dominate both screen space and bandwidth. Eschew complexity.

Selecting a format for your images is as important as creating the images themselves. The right image format can minimize download time and avoid wasting bandwidth by downloading data that isn't applicable to your content.

While graphics rule on the standard Web, they have a much less prominent role in the wireless world. Often, the cost—in download time, screen space, and, ultimately, dollars—of transferring an image to a wireless device is simply too high to be worthwhile to the viewer. Although these limitations are decreasing as the capabilities of both the wireless network and the devices that use it evolve, the wireless Web is likely to stay several steps behind the wired one in terms of graphics for some time yet. There's no point in losing your first wave of subscribers over the inappropriate use of graphics.

Avoiding Bullets and Spacers

Graphic bullets and spacers are two examples of items to avoid. HTML can produce nicely formatted bulleted lists using the `<UL>` tag, with each item delineated
by the `<LI>` tag. Some authors, however, create special graphic bullets—three-
dimensional spheres, miniature logos, spinning icons, and so forth—for use in
lists. While some users may find the best of these efforts tasteful, the practice is
deplored by even by many desktop Web users. Using images this way can dramatically increase the amount of time it takes for even a desktop page to load; the
equivalent performance hit for a handheld device is phenomenal.

Using image spacers poses a similar problem. HTML was designed as a
device-independent markup language: it enables the client to format content in
whatever way is best given the particular platform and user. Many Web designers
take exception to this loss of control, however, and seek an identical appearance
for their content across different devices and viewers. A common trick used to
achieve this goal is to create small, transparent or solid-color images of a specific
size and insert them as shims or spacers between the main graphic images on the
page. These spacer images, however, cause problems for most mobile clients. They
increase the time it takes to load a page and defeat the browser's algorithms for
optimizing the data for small-screen display. Why force a user who is paying for
your content by the minute—or by the byte—to download empty images?

Optimizing Logos

A marketing department may deem the use of a corporate logo essential even on
wireless sites. If you must include a logo in your content, take the time to create a
version of it that is optimized for presentation on low-definition displays. Make it
as small as you can and only use it on the first page of your content.

This lesson is one I've presented before, but I can't stress it enough: use only
images that are essential to your content. Any image on a page aimed at wireless
devices should add distinct value and carry more information than an equiva-
lently sized body of text on the display.

Choosing a Format: GIF or JPEG?

So, when you do use an image, and which format is best? The answer is painfully
simple: try and see. Simple, because it's easy to try both GIF and JPEG. Painful,
however, because most content providers believe it will take too much time to test
various possibilities to determine the outcome. Many people, it seems, would rather
just pick a rule of thumb—and be wrong—than actually verify the best choice.
Remember how valuable small, high-quality content is to wireless subscribers.
Taking the time to use the best image compression can bring your users tremendous

savings and significant performance enhancement. This, in turn, can help to create a loyal subscriber base.

Various tools exist to experiment with compression rates and the kinds of image compression available. High-end content development shops will most often have access to software packages such as Adobe Photoshop, while others may be using the Gnome Image Toolkit. One good tool, which is free for all to use, is cjpeg, provided by the Joint Photographers Expert Group as a reference implementation of a JPEG encoder. Using a tool such as Photoshop or cjpeg, you can select the "quality" level at which to encode your image. (As you might expect, higher levels of quality result in less compression—and larger data files.)

Only a little work is necessary to find the best compression scheme and quality for a particular image. As different images may appear different when compressed (JPEG's lossy compression, for example, may alter alternating solid colors), choosing the correct compression scheme is important.

Figure 4-3 shows six versions of a test image: (A) uncompressed, (B) compressed using GIF, and (C-F) compressed using JPEG at varying levels.

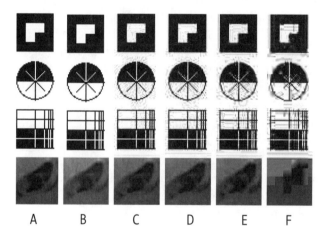

A   B   C   D   E   F

Figure 4-3. A test image compressed with GIF and varying levels of JPEG compression

Table 4-1 shows the difference in file sizes produced by the different compression methods. GIF provides a clear version of both the line and bitmap parts of the test image, but its compression ratio (ratio of uncompressed to compressed file size) is not even 3:1. The same level of quality—see image C in Figure 4-3—can be achieved with a high-quality JPEG compression, which has a compression of 8.5:1. Depending on your taste, the line diagram degrades by about image D, and clearly the quality of image E is inappropriate for most content. GIF remains the best choice for line-only artwork, and compression quality for the bitmap image in this example is probably best left around 50, yielding a 14:1 compression of the image size.

TEST	SCHEME	QUALITY (CJPEG)	SIZE	RATIO
A	None	N/A	32079	–
B	GIF	N/A	11,384	2.8:1
C	JPEG	90	3790	8.5:1
D	JPEG	50	2205	14.5:1
E	JPEG	25	1651	19.4:1
F	JPEG	12	1253	25:6:1
G	JPEG	1	779	40:1

Table 4-1: Compression Schemes, Quality, and Space Savings

This same experiment would have different results with a different image. For example, line drawings, especially those with long runs of horizontal lines, typically show 4:1 or better compression with GIF (with no loss in image quality, because GIF does not discard data during compression), but there may be image degradation at the same compression ratios using JPEG. Because JPEG decompression involves additional computation steps, a smaller image may actually take *more* time to display than a larger GIF image.

As you can see, the factors determining the results for any particular image are complicated—hence the simple rule, "try and see." On the other hand, when working with pictures or graphics with smooth color transitions, you can often achieve 15:1 compression using the JPEG format and a moderately low-quality setting. At this level, blocks of continuous color begin to suffer from degradation, which may or may not be noticeable.

To recap, when preparing images for the wireless Web, first ensure that every image is relevant and necessary, then make the images as small as they can be and remain useful. Use few, contrasting colors, or keep to black and white. Finally, compress the data as much as possible while keeping quality at an acceptable level—test both GIF and JPEG (you may want to try several levels of JPEG quality) to determine the best choice for each image.

Summary

For many applications, existing Web standards such as HTTP, HTML, GIF, and JPEG are well suited to wireless development. Despite relatively large size requirements, the widespread adoption of these protocols among handheld devices and wireless terminals offer many advantages for the developer.

These standards must be used with care, however. HTML, for example, has a number of tags not appropriate for wireless development, and just because wireless Web browsers *can* display images doesn't mean you should use them. In general, the features of these standards should be balanced with the need for succinct content and respect for the user's perception of data mobility.

HTML provides a good markup language for wireless Web content when you follow certain rules. Either your HTML should be written by hand or you should use tools such as Tidy to remove the extraneous tags introduced by many editors. The pages you create should be brief, and descriptive links should be used to connect related content, enabling your users to pick and choose the material they want to view. You can use forms to enable users to select from different choices or to obtain information about one of a group of things, but remember, they will be using the browser in settings when their attention is often elsewhere.

In general, avoid using images. When you find it necessary to use an image, be sure that it conveys more meaning than a like-sized block of text would. You should take pains to optimize your images for wireless viewers, both by experimenting with how the images are compressed and by viewing the results on wireless devices. You'll find you need to make tradeoffs in the image quality, size, and download times.

In short, when using these protocols, you should pay close attention to the amount of bandwidth your content requires as you produce it. To ensure high-quality presentation of your content as well as reasonable delivery speeds across the wide spectrum of devices in use today, your best bet is to try your content frequently on the devices for which are intended.

CHAPTER 5

Hypertext Markup Language the Wireless Way

As DISCUSSED IN CHAPTER 4, the HyperText Markup Language (HTML) has significant advantages as a medium for wireless access terminals. However, you must approach its use with the constraints of handheld devices in mind. If you are used to developing in HTML for desktop browsers, you will have to start thinking a little differently, making sure you stick to HTML tags that are appropriate for the wireless world. In this chapter, I first review the different versions of HTML in light of their suitability for use with the wireless Web. I then walk through a tag-by-tag discussion of using HTML in your development of wireless content.

> **NOTE** *All of the examples on this chapter were generated using Qualcomm's pdqSuite 1.0, AvantGo's AvantGo 3.1, and Puma Technology's ProxiNet 3.02a for the Palm Computing Platform. You'll see similar results using Microsoft's Pocket Internet Explorer for Windows CE, although there are marginal differences between its representation of HTML and those shown here (just as there are differences between browsers on the Palm Computing Platform). Of course, you should always test your content on target devices before you make assumptions about its appearance!*

Picking a Version of HTML

The most important thing to consider when beginning to work with HTML for wireless browsers is which version of HTML you'll use. When developing content for the World Wide Web, you'll usually want to use the most up-to-date version, with all of its features. But if the market for your content is the world of handheld devices, you can't be sure the latest versions of HTML will be supported on the platforms your customers use. As a result, you'll want to use earlier versions.

The use of older HTML is due to the way in which wireless Web browsing crept up on the marketplace. Many wireless browsers currently in use were written with cores several years old. Others evolved from prototypes (created for academic research, to satisfy someone's individual curiosity, or as corporate proof-of-concept demonstrations) that later were developed into software products as the demand for portable Web access grew. The humble origins of these browser applications make them incompatible with many of the latest features of HTML.

As long as hardware manufacturers are providing handheld devices with pre-installed browsers, this disparity between wireless browsers and wired ones is likely to continue. Hardware vendors today tend to have a significant amount of software savvy, but few take the time and expense to track the rapid changes in Web standards hammered out by the World Wide Web Consortium (W3C) and industry giants such as Netscape and Microsoft. Because hardware sales success is measured by the number of units sold, not by the products' compliance with W3C standards, there simply is no compelling business reason for them to do so. Smaller software vendors may provide more up-to-date browser applications for these devices, but the reality is that most users will have access only to those that are installed during manufacturing.

So, what versions of HTML *are* recommended for wireless content development? Well, nearly all wireless Web browsers support HTML 2.0, and most support a subset of the features of HTML 3.2. Some support certain features of HTML 4.0, but virtually none meet all of the W3C's requirements for implementing HTML 4.0 in all of its glory. Therefore, most developers of wireless content have found it best to stick with working in HTML 2.0 or, for some markets, version 3.2.

Determining exactly which version is appropriate requires knowledge of your user base. Keep in mind the adage "less is often more." If you know which wireless browsers will access your service, a quick look at the specifications provided by the browser vendor will often tell you which HTML version to use. If you're creating a site for general public access, however, you'll need to use the lowest common denominator of features between HTML 2.0 and 3.2.

Marking up the Document Heading

I mentioned in Chapter 4 that HTML documents are divided into two sections, a *heading* and a *body*. The heading is an optional segment containing meta-information about the document. If included, it is delineated by the tags <HEAD> and </HEAD>. The actual document to be displayed by the browser is then marked with the <BODY> tag.

Table 5-1 lists the HTML 4.0 heading tags that are of use in developing for wireless browsers; I discuss each one in more detail in the sections to follow.

TAG	INTRODUCED IN	PURPOSE
`<TITLE>`	HTML 2.0	Specifies the document's title, which may appear in a document's window bar, offline cache, or other location.
`<META>*`	HTML 2.0	Provides meta information not supported by HTML markup commands for the client.
`<BASE>*`	HTML 2.0	Specifies the base URL from which relative URLs in the body should be derived.

*These tags are empty tags and do not have a corresponding closing tag.

Table 5-1: HTML Tags for Wireless Web Document Headings

Specifying the Document Title

Probably the most common heading tag is `<TITLE>`, which, obviously enough, specifies the title of the document. Most wireless browsers honor this tag, but not necessarily in the way you might expect. For example, browsers on the PalmOS platform show as much of the title as will fit at the top of the display; Microsoft Pocket Internet Explorer shows the title in the task bar rather than in the browser window.

Wireless browsers can use the `<TITLE>` tag to identify a document in a list of bookmarks, for example, as most browsers provide a default title when a new bookmark is created. The browser derives this title from the document's `<TITLE>` tag.

Generally, your documents should include a title that briefly describes its content. Appropriate titles include the following:

- HTML 4.0 Specifications

- Where to Go on the Kona Coast

- Jarod Rischpater's Home Page

You should avoid titles such as:

- Home (too short—home for what?)

- Sandy's Stuff (while this is cute, most reader won't remember Sandy or the stuff on the page or know what's on the page from the title)

- Rachel and Ray's Not So Totally Exhaustive Survey of Coffee Establishments in and around Boulder Creek (too long)

Providing Meta Information to a Client

The <META> tag is an empty tag that enables you to establish arbitrary name/value pairs, called *entities*, for specific browsers. You can use these tags to specify browser-dependent behavior, such as how and when the client should reload a document, or to assert that a specific document is in fact formatted for a particular device. Table 5-2 outlines common <META> entities in wireless Web publishing and how they are used.

NAME	VALUE	PURPOSE
HTTP-EQUIV	varies	Specifies HTTP header information (not to be confused with the HTML heading) to browsers in the HTML document, rather than the HTTP headers.
PalmComputingPlatform	true	Indicates that the body of this document is appropriate for Palm Web Clipping applications.
HandheldFriendly	true	Indicates that the body of this document is appropriate for the AvantGo browser.

Table 5-2: HTML Meta Values for Wireless Web Document Headings

For wireless Web developers, the most important meta entities are those indicating that the document's content is appropriate for small devices. Two common meta entities that serve this purpose are PalmComputingPlatform and HandheldFriendly. The meta PalmComputingPlatform indicates suitability for Palm's Web Clipping application platform (see Chapter 12); and HandheldFriendly is used by the AvantGo browser (see Chapter 6). You'll find many general-purpose wireless Web pages have the following in their headers:

```
<META NAME="HandheldFriendly" CONTENT="true">
<META NAME="PalmComputingPlatform" CONTENT="true">
```

Both of these tags let the appropriate browsers know that the document was designed for handheld use (because each browser is different, a different meta tag must be used). In the absence of the tags—or if these values were equal to false—these browsers will assume the content was targeted for desktop devices and might display radically different results. For example, the Palm VII browser discards the end of long pages lacking the PalmComputingPlatform meta tag—probably not what a user would expect. And, when a browser encounters a meta entity it does not recognize, it ignores the entity entirely.

Another useful <META> entity is HTTP-EQUIV, which replaces a HyperText Transfer Protocol (HTTP) header with the value specified. Content authors use this meta entity when they want to express control over a document's behavior in ways that are only specified by the HTTP protocol. For example, HTTP-EQUIV can be used to force the browser to refresh a page periodically or to expire a page in the cache. This ability can be handy for pages serving information that changes often, such as stock quotes or the weather; users generally will want this kind of information to be up to the minute. Not all browsers, however, support HTTP-EQUIV, and some that do only support some uses of it.

A line such as

```
<META HTTP-EQUIV="expires"
CONTENT="Sun, 31 Dec 2005 11:59:00 GMT">
```

instructs the browser that this Web page should be discarded from the cache on New Year's Eve 2005.

The line

```
<META HTTP-EQUIV="Refresh" CONTENT="300">
```

indicates that the Web page should be reloaded every five minutes (300 seconds). Be very careful when using the HTTP-EQUIV = Refresh, however, as your content is dictating the behavior of your customer's browser, and may be incurring wireless fees subscribers aren't aware of as the browser goes off and fetches pages periodically. (Incidentally, the content loaded as a result of the refresh must also have this tag, or the refresh won't happen again. This is because the page loaded by the browser is different—the browser does not retain memory of this directive across pages.)

The strength of a <META> tag in ordinary Web development is that it can be used with any name/value pair. But this fact can be a weakness in wireless Web development, as it means a <META> tag can consume a great deal of space. Many WYSIWYG (What-You-See-Is-What-You-Get) Web page editors use <META> tags to specify their identities, as well as a whole host of related information that serves no purpose for wireless clients (recall the first example in "Keeping It Simple…", in Chapter 4). This kind of entity can take precious time to download and occupy space best left for more important material. If your desktop-oriented Web content uses meta tags this way, remove them before posting it for wireless use!

Specifying a Base URL

HTML headings use the <BASE> tag to specify the base Uniform Resource Locator (URL) from which other URLs in a particular document are derived. If your content refers to other content on the same server, it should *always* specify a <BASE>

tag and use relative URLs to keep your document body smaller. Consider the difference in the following two versions of the same page:

```
<HTML>
<HEAD>
    <BASE HREF="http://www.colors.org/pCp/index.html">
    <TITLE>People Color Picker</TITLE>
</HEAD>
<BODY>
<H1 ALIGN=right>People Color Picker</H1>
<P>Choose a color below to learn about people who have that color as their favorite
color:</P>
<MENU>
    <LI><A HREF="chart.html">Chartreuse</A></LI>
    <LI><A HREF="magen.html">Magenta</A></LI>
    <LI><A HREF="cyan.html">Cyan</A></LI>
</MENU>
</BODY>
</HTML>
```

This HTML document, which specifies a base URL in the heading, uses 798 bytes, while the following, which does not, occupies 823 bytes:

```
<HTML>
<HEAD>
    <TITLE>People Color Picker</TITLE>
</HEAD>
<BODY>
<H1 ALIGN=right>People Color Picker</H1>
<P>Choose a color below to learn about people who have that color as their favorite
color:</P>
<MENU>
    <LI><A HREF="http://www.colors.org/pCp/chart.html">Chartreuse</A></LI>
    <LI><A HREF="http://www.colors.org/pCp/magen.html">Magenta</A></LI>
    <LI><A HREF="http://www.colors.org/pCp/cyan.html">Cyan</A></LI>
</MENU>
</BODY>
</HTML>
```

While this example doesn't provide a marked increase in performance (the savings is a mere 5% or so), savings become dramatic when a page has many links to long domain names. With 20 links, the savings will reach nearly 50% over all of the links in the document.

There's another reason to use the <BASE> tag, too. Keeping a document's links in one place as much as possible makes moving around Web pages easier. When www.colors.org decides to move their "People Color Picker" site to a new server, they'll only have to edit the <BASE> reference if they use the first example. If they use the second, they'll have to change each link in the entire document.

Tags to Avoid

While the empty tags <ISINDEX>, <STYLE>, <SCRIPT>, and <LINK> are valid HTML, these tags aren't generally supported by wireless browsers. Most wireless browsers should ignore these tags, but including them may cause problems, and will always waste space. As virtually no Web browsers on handheld computers support these three features, there's little point in including them in your content.

Marking up the Document Body

Most HTML tags are used for marking up the document itself. HTML is a *structural* markup language, meaning that its tags are used to specify how a document is structured, not how it is to be rendered. An author specifies the various structural elements of a document (sections, paragraphs, lists, tables, and so on), and each browser uses this information to determine how best to render the content for its particular display.

Creating a Section Head

Documents may be separated into sections and subsections, down to six levels, delineated with the section heading tags <H1>, <H2>, <H3>, <H4>, <H5>, <H6>. Section heads should always be marked with these tags rather than with specific typography (such as *bold* or *italic*, which can be created as described under "Specifying a Text Style," later in this chapter). This leaves client applications free to present them in whatever format works best, and even to construct navigation aids, such as a table of contents. (NetHopper, for the now-defunct Apple Newton platform, did this.) Although the order and occurrence of headings is not constrained by the definition of HTML, you should avoid skipping levels (for example, from <H1> to <H3>), as this can cause problems in some representations.

Some wireless browsers may use the same formatting for several levels of section headings. This can be confusing, but often it is the only option, especially on platforms with a limited number of fonts. Figure 5-1 shows an example of using the same formatting for several heading levels on the Palm Computing Platform using pdqSuite, ProxiNet, and AvantGo. Most can present at least the first level of section heading fairly clearly; many can also define a second level fairly well, at least using the bold face of the standard font. Subsequent levels (<H3> through <H6>) are generally not well distinguished from the body text in wireless browsers, however, as you can see in the figure.

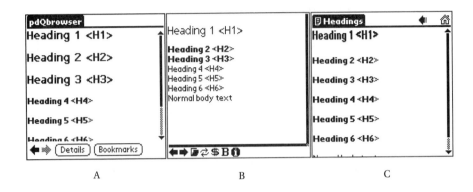

Figure 5-1. Section headings displayed in (A) pdqSuite, (B) ProxiNet, and (C) AvantGo

Formatting Blocks of Text

HTML provides three kinds of tags you can use when marking blocks of text. These tags enable you to indicate whether a block of text is a paragraph, preformatted text such as a computer listing, or a quotation from another source.

The tags for use with blocks of text are summarized in Table 5-3. Figure 5-2 shows how block quotes are displayed using pdqSuite, ProxiNet, and AvantGo. As you can see, each has its quirks.

TAG	PURPOSE	NOTES
<P>	Marks a paragraph.	
<PRE>	Preformatted text.	Generally set in fixed-width font if one is available.
<BLOCKQUOTE>	Long quotations of other material.	May not be differentiated from body text on all browsers.

Table 5-3: HTML Tags for Wireless Web Text Blocks

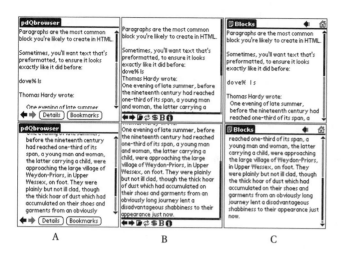

Figure 5-2. Blocks of text displayed in (A) pdqSuite, (B) ProxiNet, and (C) AvantGo

Paragraphs

Within a section, documents are composed of blocks of text. The most common block of text is a *paragraph*. Paragraphs in HTML correspond to the paragraphs you're reading in this book and are indicated with the <P> tag, like this:

```
<P>
Within a section, documents are composed of blocks of text. The most common block
of text is a <EM>paragraph</EM>; paragraphs in HTML correspond to the paragraphs
you're reading in this book.
</P>
```

Neither the opening nor closing tags need to be on their own line, but doing so can help to make things more readable.

Prior to HTML 2.0, paragraphs weren't marked as blocks with "start paragraph" and "end paragraph" tags, but separated with a single <P> tag. However, when the W3C enhanced HTML and developed the standard for HTML 2.0, it was necessary to maintain the begin/end paradigm for the <P> to adhere to the SGML specifications (of which HTML is a subset). Most browsers will support either style, but the start/end style is now seen more frequently, and you should use it for all new pages.

Preformatted Text

Another kind of text block, called a *preformatted* region, is set in a fixed-width font (like this). Blocks of preformatted text are demarcated with the <PRE> tag.

These can be used to depict tab-delimited tables, or other material in which character spacing is important, such as in the following code listing:

```
<PRE>
Hi    72
Low   63
</PRE>
```

However, not all browsers on wireless devices support fixed-width fonts, and because many people find them hard to read, they are generally a poor choice for rendering most kinds of content.

Quotations

The *block quote* format is used for citing a large body of text. It most often appears when long runs of text are quoted from another source—such as one article quoting another. This format enables readers to differentiate between regular text and quoted text (on desktop browsers, block quotes are often preceded and followed by blank lines and slightly indented). The tag used to specify a block quote is <BLOCKQUOTE>.

Making a List

Lists make up a significant portion of the content on the wireless Web because they're often more concise than paragraphs. In fact, after summaries, they're your best tool for reducing fluff and getting to the substance of your content. Put as much of your material as you can into lists, as they best reflect how your users are likely to approach your content. Examples of material suitable for list format include:

- Directions for navigating.

- Steps in any ordered task, such as preparing a recipe or using a software feature.

- A series of items of equivalent or related priority, such as the various stocks in a portfolio.

HTML offers several ways to create lists, most of which are well supported by the browsers on handheld devices. Each item within a list is marked with the <LI> tag, while the whole list is set off with one of several tags that indicate the type of list.
You can nest lists as deeply as you choose, whether they are all of the same kind or of two or more different kinds. (An outline is an example of nested ordered lists.) Nested lists can be used to create outlines or to make points in successively

greater detail. The results of nesting lists vary considerably, however. Most desktop browsers set off lists from surrounding paragraphs with indentation, and set off a list nested within a list by indenting further. Unfortunately, this scheme works poorly on the narrow screens of handheld devices, where repeated indentation quickly uses up valuable screen space, leaving little for the list items themselves.

Figure 5-3 shows several types of lists rendered as a succession of screens in each browser. Note the readability problems, discussed previously, with both definition lists and nested lists.

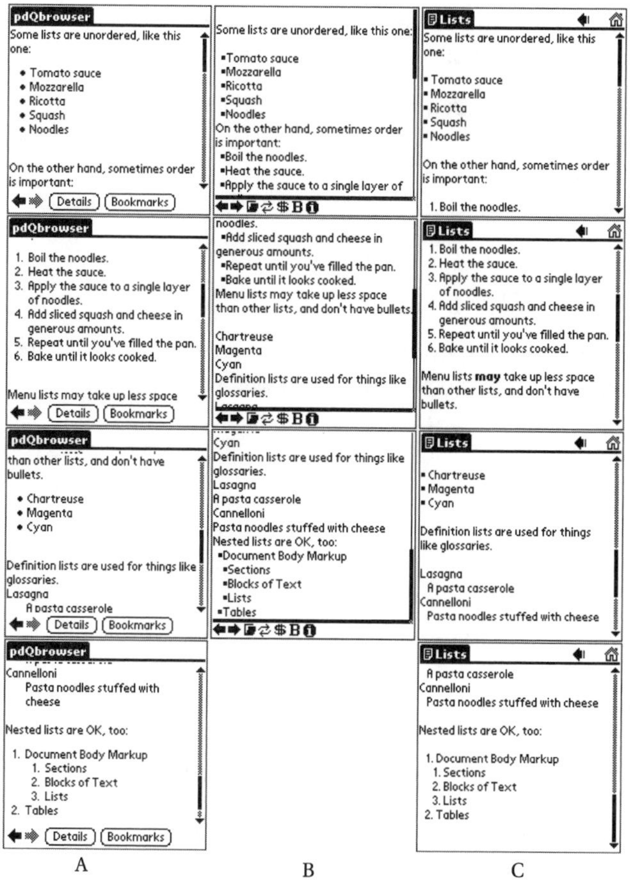

Figure 5-3. Lists displayed in (A) pdqSuite, (B) ProxiNet, and (C) AvantGo

Table 5-4 summarizes the various list tags available in HTML. We'll take a closer look at each one in the sections to follow.

TAG	PURPOSE	NOTES
<UL>	Unordered list	
<OL>	Ordered list	Elements are numbered or lettered in sequence.
<DIR>	Directory list	Not often used.
<MENU>	Menu list	Good for presenting a list of choices, where bullets aren't necessary.
<DL>	Definition list	Used for glossaries or definitions of terms.

Table 5-4: HTML Tags for Wireless Web Lists

Unordered Lists

In many cases, the order of items isn't important. This is often the case with bulleted lists, where items are listed for clarity and all have equal weight. For example, here's a list of colors in no particular order:

```
These wonderful aprons are available in the following colors:
<UL>
    <LI>Umber</LI>
    <LI>Ochre</LI>
    <LI>Sienna</LI>
</UL>
```

You can use the <UL> tag to create unordered lists, which will be drawn with bullets, dashes, squares, or similar indicators at the beginning of each list item.

Ordered Lists

When the order of items in your list is crucial, use the *ordered* list tag <OL>. The browser will generate a number or letter as the leading element for each list item, just as it generated bullets for the unordered list. For example:

```
To make a lasagna:
<OL>
  <LI>Boil the noodles.</LI>
```

```
    <LI>Heat the sauce.</LI>
    <LI>Apply the sauce to a single layer of noodles.</LI>
    <LI>Add sliced squash and cheese in generous amounts.</LI>
    <LI>Repeat until you've filled the pan.</LI>
    <LI>Bake until it looks cooked.</LI>
</OL>
```

Menus

Another option is the *menu* list, which you create with the `<MENU>` tag. On screen, a menu list looks a lot like an unordered list without bullets. This means it can be a simple way to create separate lines without using the empty `<BR>` tag (described later in this chapter in "Raw Typographic Text Styles").

The `<MENU>` tag traditionally has not been used much in HTML, but it is likely to get a new lease on life in the wireless Web world, where it can be a handy way to present a menu of choices that tie many short one-screen pages together. (Recall the People Color Picker example in the section "Specifying a Base URL.") Menu lists are also appropriate for lists of headlines, subjects, topics, and, well, software menus.

Definitions

Definitions are specialized content often presented in a list. In a definition list, each item begins with the text to be defined, which is followed by the definition. This format, often used in glossaries or in dictionaries, can be created using the `<DL>` tag. Unlike the other kinds of HTML lists, however, the definition list does not use the `<LI>` tag to separate items; rather, special tags denote the item being defined—`<DT>`—and the definition—`<DD>`—as shown here:

```
<DL>
    <DT>Lasagna</DT>
    <DD>A pasta casserole</DD>
    <DT>Cannelloni</DT>
    <DD>Pasta noodles stuffed with cheese</DD>
</DL>
```

The default layout of a definition list—even in a desktop browser—is rather simplistic. There's usually little to connect the item being defined with its definition; in fact, they often appear on separate lines. For that reason, many people prefer to present definitions by kludging together a similar format using paragraphs, line breaks, and bold tags, such as in the following example:

```
<P>
<B>Lasagna</B> A pasta casserole</BR>
<B>Cannelloni</B>Pasta noodles stuffed with cheese
</P>
```

Of course, this kind of specificity nullifies both the purposes of HTML and the definition list. HTML tags are supposed to specify the structure, not the style, of a document, while the <DL> tag should allow flexibility in how the definition parts are displayed. Some browsers might choose to display definitions in a separate view or space, for example. But in the real world, no browsers take advantage of this flexibility, so it's hard to argue against the method I outlined. It produces a result that not only looks better to most people, but also requires less scrolling and is considerably easier to read.

Specifying a Text Style

Many developers who are tempted to play with type styles in HTML often are disappointed to find out that its typesetting capabilities are reminiscent of the earliest word processors.

Typesetting for the wireless Web is even more restrictive; many mobile devices support only a few fonts, and often only a few styles of a particular font. Because of their limited memory and low performance levels, many of these devices use bitmapped fonts that look good only at specific sizes. Devices that use *scalable fonts* (fonts with a compact representation that scale well to various sizes) may provide a better appearance, but there's little standardization among the names and appearances of fonts and faces on handheld devices.

Idiomatic Text Styles

Most typographic decisions are best left to the browser, which is presumably tuned to make the appropriate typographic decisions on a particular platform based on structural tags such as section headings, text blocks, and so on. When additional differentiation is required, it's generally best to use HTML's *idiomatic* text styles, which indicate the intent (again, the structure) of the formatting rather than specifying a particular representation. This enables the browser to pick the most appropriate presentation for a particular environment.

Idiomatic tags are listed in Table 5-5. As you can see, these tags cover most situations where you would choose varying type styles and typefaces in printed text.

Using these idioms rather than typographic tags frees you to think about content rather than the specifics of representation.

TAG	PURPOSE	NOTES
`<CITE>`	Rendering a bibliography citation	May be underlined, italicized, or neither.
`<CODE>`	Computer print	Fixed-width font if available.
`<EM>`	Emphasized text	Traditionally rendered as italics.
`<KBD>`	Material to be entered at a keyboard	Fixed-width font if available.
`<SAMP>`	A sequence of literal characters	Fixed-width font if available.
`<STRONG>`	Important material	Usually rendered as boldface.
`<VAR>`	A mathematical variable	Usually italicized.

Table 5-5: HTML Tags for Wireless Web Idiomatic Typography

Raw Typographic Styles

If you feel you just have to have control over the particular appearance of a document, you can use *typographic* tags, shown in Table 5-6. However, you'll quickly discover that not all the tags look the way you expect on most handheld clients. Differences in supported fonts, typefaces, and screen quality make working with raw typography dicey unless your site is tailored to a specific device, or unless you're creating content independently for each kind of device. Your best bet is to stick with the default font, using the bold tag sparingly to emphasize specific information to the reader.

TAG	PURPOSE	NOTES
`<B>`	Boldface	Most likely supported on any device.
` *`	Line break	Inserts a line break at the current position.
`<HR>*`	Horizontal line	Inserts a horizontal rule at the current position.
`<I>`	Italic	Occasionally supported.
`<U>`	Underlined	Easily confused with hyperlinks.

Table 5-6

(continued)

Table 5-6 (continued)

TAG	PURPOSE	NOTES
<TT>	Teletype	Fixed-width font if available.
<FONT>	Enables the selection of a specific font, or modifies font size.	Introduced informally after HTML 3.2; codified in HTML 4.0.

* Denotes an empty tag.

Table 5-6: HTML Tags for Wireless Web Raw Typography

The <B> tag surrounds text to be set in bold. Almost all mobile browsers support boldface type, and this can be a good way to differentiate content when raw typography is appropriate.

Italic text may be indicated using the <I> tag. Generally, italics are used for emphasis weaker than what boldface implies. About half of the mobile browsers support italic text, although many of those that do have wretched italic fonts. These factors make using italics a poor choice on almost all platforms.

You can underline text on some browsers using the <U> tag, but I strongly discourage doing this. Underlines are used to mark hyperlinks on browsers; on grayscale screens, no colors will be available to tell the user the difference between an underlined title and a hyperlink. (In fact, I discourage using this tag in desktop content, too, for the same reason.)

If a fixed-width font is required by your content, you can request it using the <TT> tag. (The *TT* is short for *TeleType*.) Unlike the <PRE> tag, the <TT> tag sets the text in fixed-width font, but treats all white space as single spaces, and does not preserve other formatting. Since fixed-width fonts are hard to read, this tag should be used sparingly if at all.

One raw typographic tag you should definitely avoid is the <FONT> tag. Introduced between HTML 3.2 and 4.0, it is widely used by desktop browsers to select both the size and typeface of a font. Wireless Web browsers rarely support this tag, however. Even when it is supported, its output isn't well defined, as different platforms may have different fonts installed. In addition, the appearance of scaled fonts cannot be predicted with the same kind of accuracy on a handheld as on a desktop display.

Raw typographic tags also exist for separating text blocks. These are less problematic than those for specifying fonts and faces, and can be useful in providing separations where headings may not be appropriate. The empty <BR> tag inserts a single line break without starting a new paragraph; the empty <HR> tag inserts a horizontal line.

Aligning Text

It is also possible to specify an alignment for text blocks in HTML using the `ALIGN` attribute with the paragraph tag. The values for this attribute are summarized in Table 5-7.

ATTRIBUTE	VALUES	HTML VERSION	PURPOSE
ALIGN	left	HTML 3.2 and above	Specifies left alignment of a block.
ALIGN	center	HTML 3.2 and above	Specifies center alignment of a block.
ALIGN	right	HTML 3.2 and above	Specifies right alignment of a block.

Table 5-7: HTML Atributes for Wireless Web Block Alignment

Although not widely supported, this attribute is worth using when you want to set off a section's content. It can be used to align text at the left, right, or center, as shown in the following example:

```
<P ALIGN=left>Left</P>
<P ALIGN=center>Center</P>
<P ALIGN=right>Right</P>
```

Some browsers may not support the `ALIGN` attribute, but do support the older `<CENTER>` tag:

```
<P>Left</P>
<CENTER>
<P ALIGN=center>Center</P>
</CENTER>
```

For centering text, whether you use `ALIGN` or `<CENTER>` depends on what is most important for your users. `ALIGN` will create a smaller HTML file; for backwards compatibility, the `<CENTER>` tag is more appropriate. It's worth trying both on the browsers your subscribers are likely to use to see which is handled better.

Figure 5-4 illustrates examples of both idiomatic and raw typography, along with text alignment. Note that very few of the tags in either set are well supported by either browser. A few other browsers (Microsoft's Pocket Internet Explorer and Foilage Software's iBrowser both for Windows CE) do better, although these are generally less prevalent on the market today.

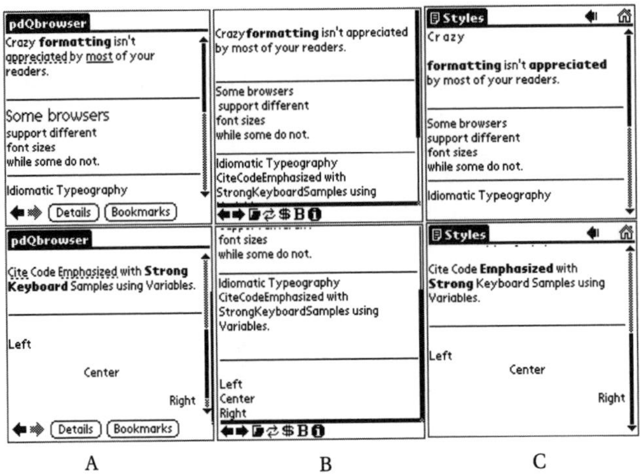

Figure 5-4. Raw typography and text alignment displayed in (A) pdqSuite, (B) ProxiNet, and (C) AvantGo

Adding Hyperlinks and Images

Of course, tags are available for linking documents and adding images to content. In general, I recommend using only the most basic tags for linking documents and for inserting images; these tags and their attributes are listed in Table 5-8. The tags for more specialized features such as audio are inappropriate for most wireless browsers.

TAG	ATTRIBUTE	PURPOSE
`<A>`		Specifies an anchor
	`HREF`	Specifies a hyperlink.
	`NAME`	Specifies a name for other hyperlinks.
`<IMG>`		Specifies an image to be included.
`SRC`	`SRC`	Specifies the source of the image.
`ALT`	`ALT`	Specifies an alternate title for the image.
`WIDTH`	`WIDTH`	Specifies the width the image should occupy.
`HEIGHT`	`HEIGHT`	Specifies the height the image should occupy.

Table 5-8: HTML Tags for Wireless Web Hyperlinks and Multimedia

The anchor tag (`<A>`) creates a link between two documents or between two parts of a single document. The client will render the argument of this tag in a way that indicates a hyperlink is available—usually as underlined text.

The anchor tag can have various attributes; in general, the only attributes appropriate for wireless development are the required HREF and NAME attributes. You'll use the NAME attribute to name a link, which you can then reference using the anchor tag HREF attributes, as in the following example:

```
<HTML>
<HEAD>
    <TITLE>Anchors &amp; Images</TITLE>
    <META NAME="PalmComputingPlatform" CONTENT="true"></HEAD>
<BODY>
<H1 ALIGN=right>
<A NAME="Top"></A>Anchors and Images</A>
</H1>
<P><IMG SRC="lost.GIF" ALT="Signs"></P>
<P><A HREF="#top">Back to the top</A></P>
</BODY>
</HTML>
```

This example, which is shown in Figure 5-5, uses the HTML `<A NAME=Top></A>` to place a marker before the section heading at the top of the page. When the link labeled "Back to the Top" (HTML `<A HREF="#top">Back to the top</A>`) is selected, the browser will scroll back to the position of the mark at the top of the page.

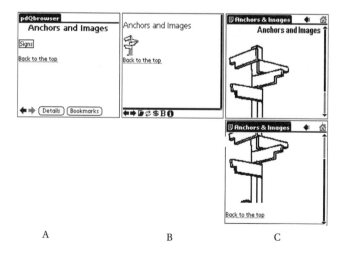

A   B   C

Figure 5-5. Anchors, links, and the IMG tag displayed in (A) pdqSuite, (B) ProxiNet, and (C) AvantGo

This example also shows the simplest <IMG> tag, which is used to include an image for display. (See Chapters 3 and 4 for detailed discussions on the mechanics of using images on the wireless Web.) This tag also has a number of possible attributes that you *will* generally want to use to create robust content.

The first of these, the ALT attribute, provides a title for the image, which will be displayed in the event that the image is not loaded, or used to describe the image on browsers supporting text-to-speech. (While no browsers presently support text-to-speech, it's probable that this will change in the near future as interactive car computers come of age.) For example, the HTML

```
<P><IMG SRC="lost.GIF" ALT="Signs"></P>
```

produces the content shown in Figure 5-5(A) on a browser that does not provide image support.

Other attributes used are the WIDTH and HEIGHT tags, which may be used to specify the size of an image. Most wireless browsers that support the <IMG> tag also support the scaling of images using these attributes, although this feature is easily abused. Many content providers use these attributes to create *thumbnail* images—smaller versions of a larger image—on pages that offer a choice of images. This practice forces clients to download a large image, which is then subsequently scaled to the device's display; users must thus expend the time and memory necessary to download all of the large images, rather than just those they choose to see. Needless to say, this can be especially onerous given the limited bandwidth of most handheld devices. In fact, the very notion of a page of images is a bad idea on the wireless Web.

If some degradation in image quality can be tolerated, however, the reverse—using these attributes to enlarge a smaller image—works well. In general, images will become grainy and distorted as they are scaled up, but, depending on your content, this approach can work well when both small and large versions of the same image are to be displayed.

On the other hand, the ISMAP attribute, which is used to specify an image map, is not appropriate for wireless devices. Few handheld devices support image maps, and those that do may not display downloaded images depending on preferences selected by the user. In general, it is inappropriate to use images for navigation; text-based lists of navigation locations load more quickly, display more reliably, render on a wider selection of devices, and are easier for a mobile user to use.

Creating a Table

Tables can be thought of as blocks of text within the body of an HTML document. The use of tables in a wireless Web page, however, deserves some additional attention. Tables are an important part of wireless Web presentation because they often provide information more concisely than paragraphs or lists. Many, but not all, wireless Web browsers support the table tags that were introduced in HTML 3.2 (see Table 5-9).

TAG	INTRODUCED IN	PURPOSE
<TABLE>	HTML 3.2	Marks the beginning and end of a table.
<CAPTION>	HTML 3.2	Marks a caption for a table.
<TR>	HTML 3.2	Marks a table row, consisting of multiple table cells.
<TH>	HTML 3.2	Marks a table heading cell within a row.
<TD>	HTML 3.2	Marks a table cell within a row.

Table 5-9: HTML Tags for Wireless Web Tables

Tables on wireless browsers can be problematic, however. The small screen provided by many devices—notably the Palm and Microsoft palm-sized PC plat-forms—make tables with more than a few columns difficult to render. Some Web designers try to use tables as a means to control the layout of a page, which can defeat the optimal rendering of many handheld browsers. Other designers nest tables for particular effects, a practice that can consume inordinate amounts of memory on a handheld device where rendering sophisticated tables is often too difficult for the client. In these cases, the page could take too long to download and then be ugly or unreadable on the client, doubly disappointing your subscribers.

Tables consist of rows of cells. The start and end of a table are delimited by the <TABLE> tag. The <TD> tag surrounds each ordinary cell, while the <TH> tag indicates a table header cell; these basic table tags are illustrated in the following example:

```
<TABLE BORDER=1>
   <CAPTION>Times</CAPTION>
   <TR>
      <TH><P>Start</P></TH>
      <TH><P>Stop</P></TH>
   </TR>
   <TR>
      <TD><P>5:52</P></TD>
      <TD><P>5:57</P></TD>
   </TR>
   <TR>
      <TD><P>6:23</P></TD>
      <TD><P>6:37</P></TD>
   </TR>
</TABLE>
```

This HTML creates a three-row table with two columns. (See Figure 5-6 for an example of how it might look.) The table elements are to be drawn with a single-width border. The table's caption, "Times", should appear somewhere near the table. The first row—the first <TR>...</TR> segment—is used to label the columns of the table, while all of the other table cells (between the <TD> and </TD> tags) simply contain times.

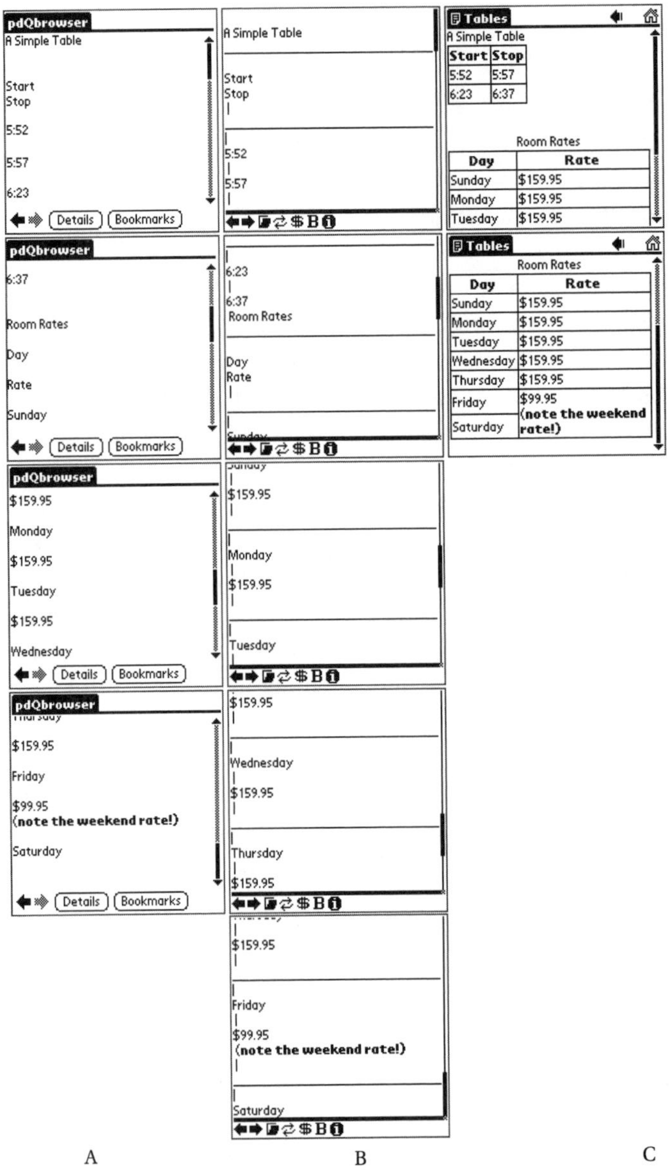

Figure 5-6. Table displayed in (A) pdqSuite, (B) ProxiNet, and (C) AvantGo

Browsers may make the column heads marked with <TH> stand out in any variety of ways, such as with bold type or a larger font, if available. Many, however, do not clearly differentiate between header cells and normal cells, and this is probably why many wireless Web developers don't bother using the <TH> tag.

The <CAPTION> tag, not surprisingly, enables you to provide a caption for a table. This contextual tag, like the <TH> tag, enables browsers to make appropriate typographic decisions best suited to a specific device. Many developers, however, try to control the appearance of a table caption; all too often you'll see something similar to the following:

```
Times<BR>
<TABLE BORDER=1>
  <TR>
      <TH><P>Start</P></TH>
      <TH><P>Stop</P></TH>
  </TR>
  <TR>
      <TD><P>5:52</P></TD>
      <TD><P>5:57</P></TD>
  </TR>
</TABLE>
```

This is bad form. I encourage you to use <CAPTION> instead, because some browsers may use table captions for special purposes. For example, on a device with a very small screen, a table may be given its own page, and the caption may be used as a link between the containing document and the table.

In some circumstances, you may want a single cell in a table to occupy multiple rows or columns. The ROWSPAN and COLSPAN attributes of a cell enable it to stretch across multiple cells in a row (using ROWSPAN) or a column (with COLSPAN). For example, in the table outlined here, the weekend rate occupies both the Friday and Saturday cells:

```
…
  <TR>
    <TD>
       <P>Friday</P>
    </TD>
    <TD ROWSPAN=2>
       <P>$99.95<BR>
       (<B>note the weekend rate!)</B></P>
    </TD>
```

```
    </TR>
    <TR>
      <TD>
        <P>Saturday</P>
      </TD>
    </TR>
...
```

The related NOWRAP attribute instructs the browser not to wrap the contents of a specific cell. Avoid using this tag, as it can cause inordinately long cells, making table layout unwieldy.

Table elements can use the ALIGN attribute (introduced earlier when we discussed the paragraph tag) to align the contents of a particular cell or heading. When the ALIGN attribute is used for a table, it instructs the browser to center the table, not the contents of the cells within the table.

The WIDTH and HEIGHT tags are used by table elements to indicate a desired size for the element. Not all browsers support this, and they are considered advisory at best—many browsers will work to fit the width and height of a cell as best they can, but use these values as starting points only.

Other HTML 4.0 tags, including <THEAD>, <TBODY>, and <TFOOT> are not available on most wireless Web browsers, and consequently should be avoided.

These examples and the material in this section should help bring home several points I've mentioned already:

- Keep tables simple.

- Don't nest tables.

- Opt for long, skinny tables over wide, short ones.

- Don't use tables to override a browser's page layout choices.

Table 5-10 reviews the attributes you're likely to use when marking up your tables.

ATTRIBUTE	APPLIES TO	VALUES	INTRODUCED IN	PURPOSE	NOTES
ALIGN	`<TABLE>`, `<TH>`, `<TD>`, `<CAPTION>`	`left`, `center`, `right`	HTML 3.2	Specifies alignment.	Deprecated in HTML 4.0.
WIDTH	`<TABLE>`, `<TH>`, `<TD>`	`integer`	HTML 3.2	Specifies width of item.	Specify in pixels or percent. Avoid specifying widths wherever possible to allow optimal rendering of tables on different platforms.
HEIGHT	`<TR>`	`integer`	HTML 3.2	Specifies height of an item.	Specify in pixels or percent.
BORDER	`<TABLE>`	`integer`	HTML 3.2	Specifies border size.	Used as a relative measurement (often indicates width of border in pixels).
NOWRAP	`<TD>`, `<TH>`		HTML 3.2	Specifies that a table entry should not be wrapped.	Not often supported, and inappropriate for most handhelds.
ROWSPAN	`<TD>`, `<TH>`	`integer`	HTML 3.2	Specifies how many rows this cell should span.	
COLSPAN	`<TD>`, `<TH>`	`integer`	HTML 3.2	Specifies how many columns this cell should span.	Use with caution, as multiple column spans can break on small devices.

Table 5-10: HTML Attributes for Wireless Web Tables

Creating a Form

Supporting interactive queries is at the heart of mobile Web design. All other factors being equal, user expectations tend toward greater customization and greater interactivity. The ability to target content—to provide a subscriber with only the content he or she desires—is often what differentiates one site from many other similar ones on the wireless Web.

Form tags (see Table 5-11) provide developers working in HTML with the means to provide simple input areas where users can enter information. This information can be used to customize the content for the particular user.

TAG	ATTRIBUTE	PURPOSE	NOTES
<FORM>		Marks the form region.	
	ACTION	Specifies the destination URL.	
	METHOD	Indicates the HTTP method to use.	Must be either GET or POST.
<INPUT>*		Specifies a form control.	
	TYPE	Specifies the type of the control.	See Table 5-12.
	NAME	Specifies the name of the field for use with the CGI.	
	VALUE	Value to be returned to the CGI.	
	CHECKED	Is item selected	Include to select the given item.
	SIZE	Specifies the size of the input item.	
	MAXLENGTH	Specifies the maximum length of the input item.	
	SRC	Specifies a URL for an image on a button.	Avoid using in wireless pages.
	ALIGN	Used to specify the image alignment on buttons.	Avoid using in wireless pages.
<SELECT>		Specifies a list.	
	NAME	Specifies the name of the field for use with the CGI.	

Table 5-11   *(continued)*

Table 5-11 (continued)

	SIZE	Specifies the number of items to display in the list.	
	MULTPLE	If present, indicates how many items may be selected.	When omitted, list supports only one selection at a time.
<OPTION>*		Specifies a list element within a <SELECT> region.	
	SELECTED	Is item selected by default.	If present, item is selected.
	VALUE	Value to be returned to the CGI.	
<TEXTAREA>*			
	NAME	Specifies the name of the field for use with the CGI.	
	ROWS	Specifies the number of rows in region.	
	COLS	Specifies number of characters per line in region.	

*Denotes an empty tag.

Table 5-11: HTML Tags for Wireless Web Forms

Defining the Form

Forms themselves are identified using the <FORM> tag. For each <FORM> tag, you must specify two attributes: an action and a method. The ACTION attribute indicates the URL of the server and program where the form is to be sent, while the METHOD indicates how the form should be sent — using an HTTP POST or GET request method. The implementation of the Common Gateway Interface (CGI) scripts on the server determines what HTTP method should be used. As you develop CGI scripts, bear in mind that for wireless development it's generally more appropriate to use the newer POST method than the older GET method.

Short forms can use the GET method, which has been supported since the early days of the Web; longer forms may call for the POST method, which supports longer form bodies. Form data is returned as a series of name/object pairs. The syntax of the response depends on which method was used: POST returns the pairs as a separate object body; GET returns the pairs concatenated to the end of the destination URL. For example, a form using the GET method would return a single URL in this example:

```
GET http://www.lothlorien.com/cgi-bin/locate.php3?callsign=kf6gpe HTTP/1.0
```

while the corresponding URL requested by a POST method would look like this:

```
GET http://www.lothlorien.com/cgi-bin/locate.php3 HTTP/1.0
callsign=kf6gpe
```

Note that only the URL of the CGI script is passed in the request; everything else is the form data. In practice, you'll never see this level of detail, although you may need to know which method is being used depending on how you implement your CGI script. Most server-side scripting environments (see Chapter 7) handle this detail for you.

Collecting User Input

User input to a form is supported by four kinds of tags. Most form objects are specified with the empty `<INPUT>` tag, which creates an input line (or button, depending on its attributes). `<SELECT>` and the empty `<OPTION>` tag are used to create a pop-up or menu list of items on a form; the actual appearance of the list is up to the client to determine.

Within the `<SELECT>` region, items are demarcated with `<OPTION>` tags. Note that this tag is in fact empty; there isn't an `</OPTION>` tag. In addition, each option must use its VALUE attribute to specify a value to return to the Web server.

You can create multi-line form elements that enable the user to enter sentences or paragraphs using the empty `<TEXTAREA>` tag. As I've mentioned before, you should avoid requiring lengthy text input from mobile users because their devices may not have keyboards. On such a device, entering even a small amount of text can quickly become frustrating.

All form objects should be given a NAME attribute, which specifies the name of the input item for the CGI.

The `<INPUT>` element uses the TYPE attribute to select the kind of input item to create—an input line, check box or radio boxes, a password input line, or a button. (Table 5-12 lists the possible values of this attribute.) A single line of text input can be indicated using the value TEXT for the TYPE attribute. For a similar short text input object, but one where you do not want the field to echo characters as they're input, use the value PASSWORD for the TYPE attribute instead.

TYPE	PURPOSE	NOTES
TEXT	Creates a single text input line.	
PASSWORD	Creates a single text input line.	Input is not echoed to the user.
CHECKBOX	Creates a checkbox element.	
RADIO	Creates a single radio button.	Radio buttons can be grouped by using the same NAME attribute value.
SUBMIT	Creates a submit button.	Multiple submit buttons can be created with unique NAME attributes.
RESET	Creates a reset button.	The reset button resets the form's values to its default.
HIDDEN	Creates a hidden name/value pair.	Used by dynamic forms to cache content between multiple requests and responses.
FILE	Requests the user to select a file to upload to the server.	Not appropriate for use with wireless browsers.
IMAGE	Creates a button with an image.	Not appropriate for use with wireless browsers.

Table 5-12: HTML Attributes for Wireless Web Input with <INPUT>

The CHECKBOX value for TYPE can be used to provide multiple possible selections, and the RADIO value for TYPE to offer two or more mutually exclusive selections. These elements should be used to select optional behaviors, not items from an array of choices. Selecting a radio button or a checkbox indicates a characteristic of something ("the blue t-shirt", for example), not an actual action ("buy a t-shirt"). Both support grouping using the NAME attribute; a group of RADIO buttons with the same name will only enable one element within that group to be selected. Name grouping for checkboxes isn't available.

An <INPUT> tag can be used to handle form actions using the SUBMIT and RESET TYPE values. Each provides a button labeled with the NAME attribute indicated for the tag. When pressed, the SUBMIT button packages the form's contents and submits the results to the URL indicated in the form's ACTION via the HTTP method indicated by the form's METHOD value. The RESET button, on the other hand, sets all the fields to supplied default values (or empties the fields if no defaults are provided) and displays the empty form to the user. Multiple inputs with SUBMIT values for the TYPE attribute, each with a distinct name, are permissible, enabling a single form to be sent to any of several URLs depending on the desired action.

The following form, shown in Figure 5-7 on various browsers, illustrates the concepts behind the different form input tags and their appearance for a hypothetical page offering weather reports:

```
<HTML>
<HEAD>
    <TITLE>North American Weather</TITLE>
</HEAD>
<BODY>
<H1 ALIGN=right>North American Weather</H1>
<P>Enter the postal code and country:</P>
<FORM ACTION="http://www.wirelesswx.com/cgi-bin/getwx.cgi" METHOD="POST">
    <P>Postal Code:
        <INPUT TYPE="text" NAME="Code" VALUE="" SIZE=10
                                            MAXLENGTH=10><BR>

        Country:
        <SELECT NAME="Country">
        <OPTION VALUE="C">Canada
        <OPTION VALUE="U">United States of America
        <OPTION VALUE="M">Mexico
        </SELECT><BR>
        <INPUT TYPE="checkbox" NAME="cond" VALUE="cond">
        Current conditions<BR>
        <INPUT TYPE=checkbox NAME="fore" VALUE="fore">
        Forecast</P>
    <P>Units:
        <INPUT TYPE="radio" NAME="unit" VALUE="m" CHECKED>Metric
        <INPUT TYPE="radio" NAME="unit" VALUE="e">English
    </P>
    <P>
        <INPUT TYPE="submit" NAME="Weather" VALUE="Weather">
        <INPUT TYPE="reset" VALUE="Clear">
    </P>
</FORM>
</BODY>
</HTML>
```

As this example demonstrates, not all form elements need to be visible. The <INPUT> HIDDEN TYPE value enables additional name/value pairs (indicated with the NAME and VALUE attributes) that do not appear on the form to be returned to the server. In a situation where multiple forms are dynamically generated by back-end scripts, this capability can be useful for carrying context between forms, such as order numbers or a user's authentication. HIDDEN elements are not completely secure, but they can be used to replace some of the functionality offered by cookies.

The FILE attribute, if used in conjunction with the POST method, enables some clients to submit files in response to a form. This isn't necessarily a good idea for content intended for wireless clients; not all wireless devices support file systems, and sending an entire file might be prohibitively expensive for the subscriber.

While it's possible to create buttons with images on them using the IMAGE attribute, don't. Wireless clients may not support the display of images or a subscriber may have turned off image downloads. In either case, the user is faced with a series of similar buttons that have no labels.

Ideally, most forms for mobile clients should be even simpler than the weather information form shown in Figure 5-7. Note the use of a <SELECT> list to enter a country (as opposed to free-input or checkboxes) and of a checkbox to indicate actions to be taken by the server.

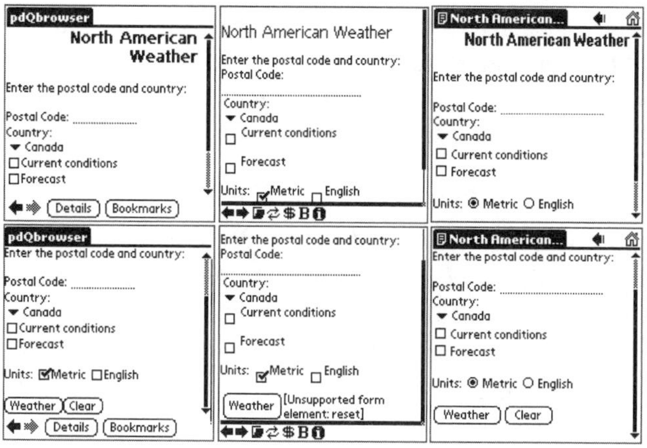

Figure 5-7. A form displayed in (A) pdqSuite, (B) ProxiNet, and (C) AvantGo

Using Other HTML Tags

The constraints of mobile devices eliminate a great deal of the glamour of Web content development. Frames, image maps, sounds, scripts, and other such gimmicks simply are not appropriate for a device the size of a calculator or notepad. Even if these handheld items evolve to the point where they do have the computational horsepower, memory, and wireless bandwidth to support such features, most users will not appreciate navigating through such junk in search of their content. Remember that wireless users are mobile for a reason — they are probably in a hurry.

When developing wireless content, stick to the subset of HTML I outlined in this chapter. I have already mentioned several pitfalls to avoid when using these elements, including nesting tables and using images, or worse, image maps for navigation (*all* links should be marked with text). Some whole categories of features, however, should be avoided altogether; specifically, the following:

- **Frames**: Displaying frames properly on a screen can occupy many hundreds of pixels on one side—generally far more than are available on a mobile device. For the kinds of information that are traditionally placed in frames, try creating multiple pages with a common navigation element.

- **Sounds**: While some browsers do support audio, the bandwidth an audio file requires is excessive for many wireless users. And in any case, how many users really want their wireless Web devices to start talking to them in public?

- **Scripting**: While not formally a part of HTML, the scripting technologies presently available—Java, JavaScript, and VBScript—are widely used in pages aimed at desktop Web viewers. Scripting isn't even available on many mobile Web clients, however, and those clients that do provide support for scripting may not adhere to accepted specifications for functionality. Waiting until mobile-friendly scripting technologies such as Sun's Java KVM are adopted is your best bet.

As time goes on, many of these features may become more widely supported and/or more readily available within the capabilities of a handheld device. You may find that using them becomes more and more tempting. Remember, however, that subscribers won't upgrade wireless terminals in the same way they do software. Legacy browsers unable to handle these features well are likely to still be in common use several years after their manufacture.

Summary

HTML is a powerful tool for content development on both the World Wide Web and the wireless Web. Not all HTML commands are appropriate for wireless browsers, however. By using a carefully selected subset of HTML, you can create concise Web pages that use reasonable quantities of bandwidth and address the special needs of wireless subscribers. Remember these general tips as you create your HTML content:

- Keep content simple.

- Allow the browser to select the best methods of presentation for a given combination of structural element and device.

- Keep tables small as much as possible; if they must be large, make them long rather than wide.

- Avoid the use of frames, images, and bandwidth-intensive content.

Be sure to declare a document heading with the document's title and any meta tags your subscriber's browsers will need. Often, you'll want to be sure to include browser-specific meta tags such as `PalmComputingPlatform` and `HandheldFriendly` to ensure that these pages display correctly on specific browsers.

Use HTML's idiomatic tags whenever possible. That means you'll want to use the section head tags, block tags, and idiomatic tags such as `<STRONG>` to call out important content. Avoid raw typography such as `<B>`, `<I>`, and the `<FONT>` directive. The same guidelines apply to lists, too; HTML provides a rich set of tags for creating ordered and unordered lists, as well as definition lists for glossaries and lists of choices. Use images judiciously, always in conjunction with a text label using the `<IMG>` tag's `ALT` attribute. Other kinds of multimedia—sounds or movies—are always inappropriate for wireless-oriented content.

Tables should be laid out simply, using the table caption, table head, and cell tags. Preferably, these tables should be long and skinny, or better yet, short and skinny. The table tags new to HTML 4.0 aren't widely supported, so be sure you use only the older HTML 3.2 tags we've discussed.

Form input enables you to target your content to your viewers. Be sure to do so, keeping in mind how they're likely to use the browser. Using the `<SELECT>` and `<OPTION>` tags is preferable to asking the user to enter text with the `<TEXTAREA>` tag. When getting form input, you'll find it easier to manage data submitted with the HTTP POST request, which you can specify using the `<FORM>` tag's `METHOD` attribute.

Above all, when in doubt, keep your markup simple.

Look, Mom—No Radio!
Web Synchronization

A WEB BROWSER WITH NEITHER a constant network connection nor a wireless connection may seem like a useless item. After all, how could it obtain information from the Web—let alone update this information regularly?

The operation of the *synchronized browser* answers these questions. As it turns out, mobile users may find that they use a synchronized browser *more* than they use a wireless one.

Introduction to Web Synchronization Concepts

A synchronized browser is just what the name implies—a Web browser that synchronizes an image of Web content with a remote server. In its simplest form, a synchronized browser simply takes a snapshot of a remote server's document, by traversing (or "spidering," as some call it) all links from a predetermined page. Each document is stored—and the entire site can be viewed—in the synchronized browser from the browser's local store. Links between documents are preserved; the user is essentially unable to tell whether the browser is loading content from the device's local storage or from the remote origin server.

It is true that synchronized browsers don't simply copy content. Rather, they compare the local copy with the copy on the origin server so that the client only needs to download the material that has changed since the last synchronization. In addition, more sophisticated synchronized browsers can perform simple compression and content optimization operations on content before transferring the content to the device, making it possible to browse a wider variety of content and store more material for offline viewing.

Most of the constraints of the wireless Web apply equally to synchronized Web access, such as the following:

- Client devices have limited presentation abilities (memory, sound, display, and user input).

- The connection pipe between device and Web is relatively slow (most devices synchronize at rates between 19.2kbps and 112kbps).

- Users generally view the content in a variety of settings with less than their full attention paid to the device or its content.

Synchronized access is generally available only to the users of handheld computers such as the Personal Digital Assistants (PDAs) or super phones described in Chapter 2. These devices sport the synchronization technologies required for the process, and are capable of running the powerful third-party applications needed to implement it, such as AvantGo (available for both Microsoft Windows CE and the Palm Computing Platform) and Microsoft Mobile Channels, a proprietary Web synchronization application for Microsoft Windows CE.

AvantGo supports both connected and synchronized browsing, enabling users to access the Web wirelessly or refer to synchronized content. By contrast, Microsoft Mobile Channels only enables offline browsing of synchronized content stored in *channels*, similar to the desktop channel format developed by Microsoft for Internet Explorer. Later in this chapter, I show you in detail how to write content for both of these browsers.

Channels

Synchronized browsers rely on the notion of a *channel*, which is a group of related pages. Users select a channel using their desktop Web browser from a Web portal, an individual Web site, or other Web source. The handheld platform's desktop synchronization interface—typically a technology such as the Palm Computing Platform HotSync or Microsoft Windows CE Services—uses the information in a channel in conjunction with a desktop-computer–side application to synchronize Web pages between the channel's origin servers and the handheld device. At any time, the user can update the handheld's copy of the Web site by synchronizing his or her handheld device. During synchronization, the handheld device uses the desktop synchronization software to interrogate the servers of each channel and determine what pages of content have changed (or been added or deleted) since the last synchronization. Once a list of changes has been built, the handheld downloads the new or changed pages, and deletes pages no longer relevant to the channel.

The user views the device-resident content using a special browser capable of presenting content from the synchronized database. This synchronized browser may or may not be capable of connected Web access or of storing form queries for retrieval during a subsequent synchronization.

The format of a channel varies from browser to browser; some browsers use a HyperText Markup Language (HTML) file to represent a channel, while others use other formats, such as an eXtensible Markup Language (XML) document describing the channel. The simplest approach—used by AvantGo—is to use a given HTML page as a channel, and define the channel's contents as all of the links from that given page out to a specific *link depth*. The channel consists of this page, all pages referred to by the given page, and so on out to the link depth. In contrast, Microsoft Mobile Channels uses an XML document to specifically name each Web page in a channel.

Appropriate Content

What kind of content is appropriate for synchronized browsers? More than you might think. In fact, the technology behind synchronized browsers can solve problems that are not well managed by either wireless or wired Web access.

Most users, if asked, will say they seek up-to-date information. But, of course, what they mean by "up-to-date" is relative. Newspapers, for example, are considered an up-to-date form of information for many purposes, yet they're only available once or twice per day. While the information users are seeking with their wireless Web devices does change, for many kinds of content the rate of change is less than the frequency with which most users synchronize their handheld devices—hourly or daily. News items obtained this way will be no more out of date than printed newspapers, and very likely less. Directions to a destination are not generally subject to change on a daily or weekly basis (road hazards, on the other hand, may change often, and thus are probably not a good fit of information type for synchronized browsing). Weather reports are generally good for a day or so. A special-interest publication, such as a science journal or hobby magazine, is also a good fit, enabling users to convert small amounts of down time into relaxing leisure time as they catch up on personal interests using their portable devices. Especially if their devices can store relatively large amounts of content (up to a megabyte or so for some devices), many users find their purposes satisfied by the "relatively up-to-date" content provided by synchronized Web browsers.

Appropriate Format

Synchronized browsers are a specialized kind of Web browser, and they use Web protocols to read content. During synchronization, content is collected from origin servers on the Web using the HyperText Transfer Protocol (HTTP), and the content obtained is marked up using HTML or stored as images in the Graphical Interchange Format (GIF) or Joint Photographers Expert Group (JPEG) format. The synchronized browser displays content in HTML, along with GIF and JPEG images. Developers should keep the same concerns in mind when creating content for synchronized browsers as they do when developing for other mobile Web-based devices: relatively little memory, small screen size, and limited processor speed. (Chapters 3, 4, and 5 discuss in detail ways developers can address these concerns.)

AvantGo

AvantGo, Inc.'s browser is the leading application for synchronizing Web content. Available for both Microsoft Windows CE and the Palm Computing Platform, it has an estimated half-million regular users, with an installed base of over a million users as of this writing.

The AvantGo Technology

While AvantGo sells their back-end server as a software package for Fortune 1000 companies, they also operate a server farm with the same software for consumers. Consumers may freely download and use the AvantGo client application and synchronize content from these servers. In addition, AvantGo maintains a vast directory of third-party content, providing a portal for synchronized Web content.

Because most readers will use this commercial service—and because licensees of AvantGo's enterprise products receive support from AvantGo—I focus primarily on developing for the AvantGo subscriber service in the following discussion. (The concepts for deploying a synchronized Web site with the Fortune 1000 server product are largely similar.)

The deployment view of the AvantGo technology is shown in Figure 6-1. The first component is the AvantGo server. This server runs applications that manage the data delivered to the mobile devices. In addition, this server provides administration for users and groups, as well as hosting the Mobile Application Link server responsible for the actual data delivery. Developers need not concern themselves with the details of the system, however, unless they are creating a vertical application with a specific server—in which case they will need to understand the architecture of the AvantGo server, and will work directly with AvantGo for installation, configuration, and support of the server product.

The second component of the AvantGo technology is the user's desktop computer. AvantGo installs a synchronization module on the desktop. This module is used by the handheld platform's synchronization software to obtain Web pages and deliver them to the handheld device during synchronization. Users can also perform synchronization via a traditional modem or wireless networks, enabling mobile users to synchronize Web data when away from their desktop computers.

The final component, shown in Figure 6-1, is the AvantGo browser. The browser obtains its information—Web pages stored in a compressed form—from the client side of the Mobile Application Link. Using this application, users can view content, create forms for submission during the next synchronization, and perform administrative functions such as the removal of channels from the device.

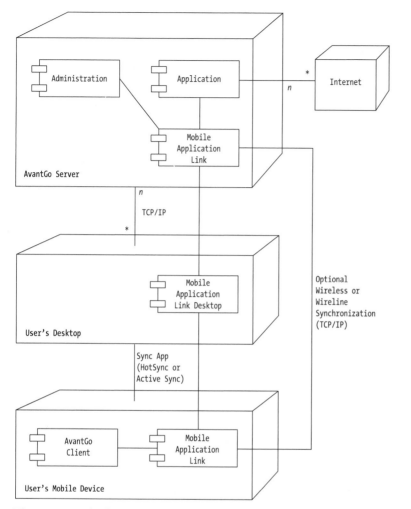

Figure 6-1. A deployment view of the AvantGo synchronized Web technology

The AvantGo Service

The AvantGo service provides hundreds of channels for mobile consumers. AvantGo works with content providers to develop content for leisure, business, news, and other channels consumers find interesting. This service can be obtained at AvantGo's Web site http://www.avantgo.com). There are no subscription fees—use of AvantGo and the account are free—although accounts are required to store users' preferences and subscriptions. (AvantGo's business model presumably draws income from the sales of its Fortune 1000 server products and other activities.)

At the AvantGo Web site, users can download the necessary applications, register for an account, and manage their list of channels to create a portal of synchronized content for themselves.

As a subscriber, you first select channels of interest on the AvantGo service site. Once you have done this, when you synchronize your mobile device, your desktop computer uses the Mobile Application Link to access your AvantGo.com account, retrieve the list of subscribed channels, and then download your channels' contents to your device.

In addition to subscribing to channels, the AvantGo service enables users to designate a particular Uniform Resource Locator (URL) as a channel, so that any Web site can be synchronized with their handheld devices. While this approach is not appropriate for many sites, AvantGo's servers and client software do a tolerable job of reducing HTML targeted for a desktop so that the content is usable on mobile devices.

Developing Content for the AvantGo Service

Creating content for the AvantGo service is very similar to creating wireless Web pages with HTML. You'll want to spend a bit of extra time thinking about how your content is organized, but once you've done that, all you need to do is mark up your content and create a channel.

Content Organization

AvantGo's channels treat content as a tree rooted on a central page. When synchronizing, AvantGo begins at an indicated page and collects all pages referenced by it up to a specified link depth. The resulting snapshot of the Web site contains the index page and content to support all links from that page and subsequent pages out to a specific depth, as illustrated in Figure 6-2.

This method of organizing content—as a tree rather than a web—is slightly different from what most developers are used to. When creating content intended for use with AvantGo, it's important to recognize that any items that are more links away from your root page than the selected link depth won't be transferred to the device. Moreover, you have to be careful when linking your content to other sites; an inadvertent link can cause AvantGo to attempt to download whole sections of a Web site that may be only appropriate for desktop viewers.

AvantGo content developers have the option of enabling the synchronization to either accept or avoid content from servers other than the one indicated in the root page. By choosing to avoid content from other servers, you can maintain greater control over what content is transferred to the device, ensuring it is suitable for presentation.

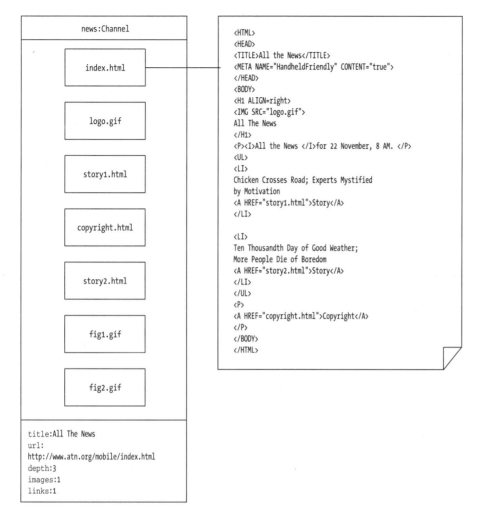

```
news:Channel

        index.html

        logo.gif

        story1.html

        copyright.html

        story2.html

        fig1.gif

        fig2.gif

title:All The News
url:
http://www.atn.org/mobile/index.html
depth:3
images:1
links:1
```

```
<HTML>
<HEAD>
<TITLE>All the News</TITLE>
<META NAME="HandheldFriendly" CONTENT="true">
</HEAD>
<BODY>
<H1 ALIGN=right>
<IMG SRC="logo.gif">
All The News
</H1>
<P><I>All the News </I>for 22 November, 8 AM. </P>
<UL>
<LI>
Chicken Crosses Road; Experts Mystified
by Motivation
<A HREF="story1.html">Story</A>
</LI>

<LI>
Ten Thousandth Day of Good Weather;
More People Die of Boredom
<A HREF="story2.html">Story</A>
</LI>
</UL>
<P>
<A HREF="copyright.html">Copyright</A>
</P>
</BODY>
</HTML>
```

Figure 6-2. A channel seen as an aggregation of documents (static view)

Markup

Most wireless-accessible Web sites are already marked up to suit AvantGo, which uses traditional HTML and GIF, and JPEG images. However, it does have some specific requirements.

AvantGo will attempt to display any HTML, although it does best with simple HTML targeted at small devices. HTML intended primarily for AvantGo users can be marked with a special meta tag. The following tag instructs AvantGo to trust that the author of the HTML has used only markup supported by AvantGo and respected the memory constraints of the mobile devices that will be displaying the content:

```
<META NAME="HandheldFriendly" CONTENT="true">
```

AvantGo operates on the assumption that any HTML not marked with this tag was initially targeted at desktop viewers, and will make corrections designed to improve its appearance on handheld devices, such as omitting tables or truncating documents.

You can optimize your content for AvantGo in several ways. Keep document titles short, with the first several characters unique; AvantGo will clip the title to fit a small device's screen (even if the meta tag is used, because some screens simply don't have the space for long titles). Use only monochrome images, or no more than four shades of gray for the best display results on all devices. Make sure any tables are kept simple, with only a few rows and columns, and remember that they will only be displayed if your page uses the `HandheldFriendly` meta tag.

Channel Creation

If your desktop Web site offers a version that works as a channel for AvantGo, you will want to advertise this fact. Most such sites do this with a small image linking the user to a *channel subscription*, which AvantGo represents as a special URL pointing back to the subscriber's AvantGo account.

Within this URL, AvantGo indicates the location of the AvantGo subscription server, along with the information describing your channel. This URL is actually an HTTP GET form object, with the following fields: `title`, `url`, `max`, `depth`, `images`, and `links`. The meaning of most of these fields is self-explanatory; for details, see Table 6-1.

Following is a sample channel subscription for the example shown in the next section:

```
http://avantgo.com/mydevice/autoadd.html?title=APRS/Find&url=
http://www.lothlorien.com/~dove/WCA/aprs-find/AvantGo/avantgo-aprs-find-index.html
&max=25&depth=2&images=1&links=1
```

FIELD	PURPOSE
title	The channel title.
url	The URL of the channel's root document.
max	The anticipated maximum size (kilobytes).
depth	The number of links to traverse from the root when harvesting pages.
images	Set to 1 if the channel should include images.
links	Set to 1 if the channel should include documents from hosts other than the host serving the root.

Table 6-1: Fields Used to Define an AvantGo Channel

An AvantGo Example

The use and capabilities of AvantGo are best illustrated in an example. Figure 6-3 shows an example of a site that lists locations of amateur radio stations. It uses AvantGo to display data collected from the APRServe server and the Automatic Position Reporting System (APRS) network. (See the sidebar that follows for information on the APRS network and APRServe server.) For now, we'll assume the existence of a back-end server Common Gateway Interface (CGI) script that returns a page detailing the position of a station when given its call sign. (I describe the construction of this back end in Chapter 7.)

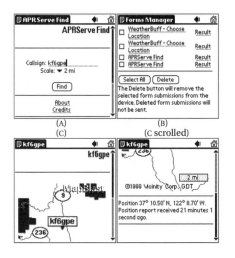

Figure 6-3. The example channel shown using AvantGo on a Palm Computing Platform device (A) queuing a form submission, and in (B) and (C), displaying the form transaction results

---

What Is Automatic Position Reporting System?

The Automatic Position Reporting System, or APRS, is a system developed by Bob Bruninga, WB4APR, that uses amateur ("ham") radio to transmit position reports, telemetry, and messages between users. Amateur radio operators routinely use it in public service, during emergencies, and just for fun. The APRS network covers the continental United States and much of the world via a far-flung assortment of amateur radio stations and Internet gateways that carry APRS traffic from one area to another.

Stations within the APRS network are referred to by their call signs. Each one is a unique identifier indicating a specific amateur radio operator.

For more information about APRS, visit one of these sites:

- APRS: Automatic Position Reporting System (`www.aprs.net`)

- Bob Bruniga's APRS Page (`web.usna.navy.mil/~bruninga/aprs.html`)
- Tuscon Amateur Packet Radio (`www.tapr.org`)

For more information about amateur radio, visit the Amateur Radio Relay League's Web site at `www.arrl.org`.

The example channel will consist of three HTML pages:

- The form that enables the user to enter a call sign (this is the root page).

- An acknowledgment page crediting those who have contributed to the channel.

- A page with background information about amateur radio and APRS.

This organization is typical of most AvantGo channels: static information such as credits, copyrights, and other information is included within the channel, while dynamic forms point to Web scripts elsewhere to obtain the information requested by the user. In many cases, if the channel is constructed to answer a specific question ("Where is that person?" "How do I get there?" "What is the value of my portfolio?"), the form where this question is entered should be the root page for the channel. The content in our example channel is likely to be useful to viewers in the time between device synchronization with the Web site. While some stations may move from place to place, many APRS stations operate from fixed locations such as homes, fire departments, or shelters during an emergency.

In our example, entering a call sign and pressing "Find" queues a form submission for the next AvantGo synchronization (see Figure 6-3A). (Users with a modem or a direct network link can use AvantGo to obtain the information immediately.) After synchronizing, the user can use the AvantGo Forms Manager to view the result of the form transaction (see Figures 6-3B and C and see the resulting information.

The HTML for the root page is straightforward:

```
<HTML>
<HEAD>
  <TITLE>APRServe Find</TITLE>
 <META NAME="HandheldFriendly" CONTENT="true">
</HEAD>
<BODY>
<H1 ALIGN=right>APRServe Find</H1>
<CENTER>
<HR>
<FORM ACTION="http://www.lothlorien.com/cgi-bin/locate-station.php3" METHOD=POST>
  <P>
```

```
Callsign: <INPUT TYPE=text NAME=CALLSIGN VALUE=""
            SIZE=10 MAXLENGTH=10><BR>
Scale:
<SELECT NAME=SCALE>
  <OPTION VALUE=50000>500 ft
  <OPTION VALUE=100000>1 mi
  <OPTION VALUE=200000>2 mi
  <OPTION VALUE=800000>5 mi
  <OPTION VALUE=3200000>50 mi
</SELECT>
</P><P>
<INPUT TYPE=hidden NAME=version VALUE=2>
<INPUT TYPE=submit NAME=Submit VALUE="Find">
</P>
</FORM>
<HR>
<P>
<A HREF="about.html">About</A><BR>
<A HREF="credits.html">Credits</A>
</P>
<HR>
<P>
&copy; 2000 Ray Rischpater, KF6GPE. Comments may be sent to the
<A HREF="mailto:dove@lothlorien.com">author</A><BR>
APRS is a registered trademark of APRS Software and Bob Bruninga,
WB4APR.
</P></CENTER>
</BODY>
</HTML>
```

The only way this page differs from a standard Web page in HTML is the initial meta tag indicating it is handheld-friendly. It's a good habit to include this meta tag in *all* pages destined for AvantGo browsers, even those, like this one, that contain no elements that AvantGo would need to adapt.

This root page points to three other Web pages:

- The Web CGI responsible for handling form submissions.

- The page about.html containing information about the channel.

- The page credits.html containing information about APRS and amateur radio.

In turn, each of these pages points back to the main page. Here's a snippet from the credits page:

```
<HTML>
<HEAD>
  <TITLE>APRServe Find</TITLE>
  <META NAME="HandheldFriendly" CONTENT="true">
</HEAD>
<BODY>
<H1 ALIGN=right>APRServe Credits</H1>
...
<CENTER>
<HR>
<A HREF="avantgo-aprs-find-index.html">Find</A><BR>
<A HREF="about.html">About</A><BR>
<HR>
</CENTER>
...
<BODY>
</HTML>
```

The final component of this example is the channel subscription URL, which enables users to access their AvantGo subscription manager and subscribe to the channel with a single click on a hyperlink. The following URL specifies our channel:

```
http://avantgo.com/mydevice/autoadd.html?title=APRS/Find&url=
http://www.lothlorien.com/~dove/WCA/aprs-find/AvantGo/avantgo-aprs-find-index.html
&max=25&depth=2&images=1&links=1
```

This indicates that the channel's root URL is `http://www.lothlorien.com/~dove/WCA/aprs-find/AvantGo/avantgo-aprs-find-index.html` and that its title is "APRS/Find". The channel is expected to occupy no more than 25KB on a user's device and is two links deep. Images and offsite links are allowed (to permit the display of the map data provided by the channel).

There are two ways to create this URL. The easiest is to use AvantGo's "custom channels" link for subscribers and use the channel creation wizard. Once the channel is created, it can be exported as a URL. The more difficult—but faster—method is to manually write the URL, being careful to include all the necessary fields.

As illustrated in the example, AvantGo requires very little in the way of support from content developers. Creating an AvantGo channel is as easy as constructing some HTML that is appropriate for a mobile device, posting it to the Web, and creating a channel subscription URL to make subscribing easy for your viewers.

Microsoft Mobile Channels

With the release of Microsoft Windows CE for the palm-size PC, Microsoft made possible the distribution of Web content via a technology called Mobile Channels. Based on Microsoft's Active Channels for Internet Explorer 4.0, Microsoft Mobile Channels provides the ability to browse Web content offline on a variety of Windows CE devices. While Microsoft Active Channels is technology for enabling the deferred download and automated update of material, Microsoft Mobile Channels aims to be a true synchronized browser.

While the same fundamental concepts of Web synchronization apply to Microsoft Mobile Channels and AvantGo, developing content for Microsoft Mobile Channels is radically different than writing for AvantGo. For one thing, developers using Mobile Channels are responsible for gathering lists of the URLs referenced by pages in their channels. For another, pages meant to be read via Mobile Channels must be written specifically for it. Finally, the support for Mobile Channels resides entirely on the user's handheld device and target desktop. No additional Web services or server middleware is required. Figure 6-4 illustrates the deployment view of a Microsoft Mobile Channels site.

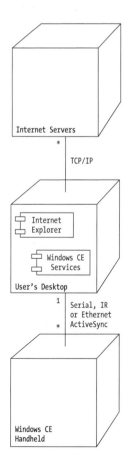

Figure 6-4. Deployment view of a Microsoft Mobile Channels site

Microsoft Mobile Channels Architecture

As shown in Figure 6-4, a Mobile Channel contains the following three components:

- The Channel Viewer on the mobile device.

- A medium for synchronizing handheld and desktop (Windows CE Services).

- Microsoft Internet Explorer, for maintaining the channel subscription.

These components are all readily available to Windows CE users. The mobile channel itself consists of the following two parts:

- Web content (HTML, GIF, JPEG, and BMP files) formatted for the Windows CE device.

- A Channel Definition File.

The *Channel Definition File* (CDF) is responsible for aggregating the content and describing the channel. (See Figure 6-5 for an example of a CDF to manage a news channel.) Like the root page of an AvantGo Web site, it points to the content referenced by the channel; it also describes the channel itself, including the title, summary, and any graphics, such as a logo used to identify a channel. This news channel has two stories, along with accompanying graphics and a logo for the news vendor.

Developing a Microsoft Mobile Channel

There are three steps involved in developing a Microsoft Mobile Channel: determining how the channel should be laid out, marking up the contents, and creating a CDF.

Content Organization

The organization of content in a mobile channel is as flexible as in a conventional Web site. Documents can contain links to other documents to any desired depth, but all links should point to other documents residing on the Windows CE device (in other words, within the channel). Mobile Channels will enable you to insert a pointer to a document outside of the channel, but selecting such a link causes an error in current versions of the Channel Viewer.

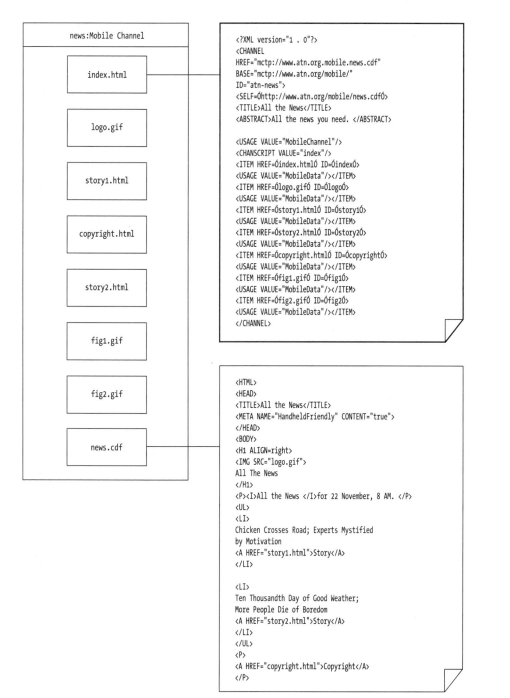

```
news:Mobile Channel

index.html

logo.gif

story1.html

copyright.html

story2.html

fig1.gif

fig2.gif

news.cdf
```

```
<?XML version="1 . 0"?>
<CHANNEL
HREF="mctp://www.atn.org.mobile.news.cdf"
BASE="mctp://www.atn.org/mobile/"
ID="atn-news">
<SELF=Óhttp://www.atn.org/mobile/news.cdfÓ>
<TITLE>All the News</TITLE>
<ABSTRACT>All the news you need. </ABSTRACT>

<USAGE VALUE="MobileChannel"/>
<CHANSCRIPT VALUE="index"/>
<ITEM HREF=Óindex.html0 ID=ÓindexÓ>
<USAGE VALUE="MobileData"/></ITEM>
<ITEM HREF=Ólogo.gifÓ ID=ÓlogoÓ>
<USAGE VALUE="MobileData"/></ITEM>
<ITEM HREF=Óstory1.htmlÓ ID=Óstory1Ó>
<USAGE VALUE="MobileData"/></ITEM>
<ITEM HREF=Óstory2.htmlÓ ID=Óstory2Ó>
<USAGE VALUE="MobileData"/></ITEM>
<ITEM HREF=Ócopyright.html0 ID=ÓcopyrightÓ>
<USAGE VALUE="MobileData"/></ITEM>
<ITEM HREF=Ófig1.gif0 ID=Ófig1Ó>
<USAGE VALUE="MobileData"/></ITEM>
<ITEM HREF=Ófig2.gifÓ ID=Ófig2Ó>
<USAGE VALUE="MobileData"/></ITEM>
</CHANNEL>
```

```
<HTML>
<HEAD>
<TITLE>All the News</TITLE>
<META NAME="HandheldFriendly" CONTENT="true">
</HEAD>
<BODY>
<H1 ALIGN=right>
<IMG SRC="logo.gif">
All The News
</H1>
<P><I>All the News </I>for 22 November, 8 AM. </P>
<UL>
<LI>
Chicken Crosses Road; Experts Mystified
by Motivation
<A HREF="story1.html">Story</A>
</LI>

<LI>
Ten Thousandth Day of Good Weather;
More People Die of Boredom
<A HREF="story2.html">Story</A>
</LI>
</UL>
<P>
<A HREF="copyright.html">Copyright</A>
</P>
```

Figure 6-5. A Microsoft Mobile Channel Definition File

A document's Web URL does not reference a document within a channel. Rather, each document is assigned an *item ID* during the process of creating the channel, which itself is denoted using a *channel ID*. Documents are referenced by these identifiers with URLs that are in the form: `mctp://channel-id/item-id`. The leading protocol indicator, `mctp://`, states that the document is to be fetched from the device's Microsoft Mobile Channel database; the channel ID and item ID indicate the specific object referenced.

In general, an index page linked to independent content is an appropriate method of organization for Mobile Channels as well as for AvantGo content. Not only are subscribers generally familiar with this model, but also maintaining unidirectional links (from the index page to the individual items) is easier than maintaining a graph of links between index and changing content.

Content Markup

Most Windows CE devices sport robust multimedia capabilities (relative to other handheld computers), and the Microsoft Channel Viewer takes advantage of these capabilities. While these devices still have less memory and smaller screens than a desktop PC, many handheld Windows CE have color screens and multiple fonts, allowing for sophisticated output.

The Channel Viewer accepts HTML content, but requires the Mobile Channels Transport Protocol (`mctp`) rather than the HTTP. Therefore, material designed for the Web requires some editing before it can be included in a channel. As a result, many developers choose to generate versions of content that are specific to the mobile channel rather than attempt to share content between a mobile channel and other forms of Web distribution.

When designing content for Windows CE, all of the considerations discussed in Chapter 4 apply. Above all, keep it simple, and keep pages small in data terms (no more than a few kilobytes). Do not use frames, which are not supported by the Windows CE channel viewer. As all Windows CE devices presently available are capable of grayscale image display, and many sport 8- or 16-bit color displays, you may feel a strong temptation to take advantage of these capabilities. If your content is targeted solely at Windows CE devices that use Mobile Channels, this may be appropriate, but be careful not to lose sight of the fact that space and bandwidth are limited.

The Channel Definition File

The bulk of the work involved in creating a mobile channel is writing the CDF. As shown previously in Figure 6-5, this file (the top file) contains all of the information about a channel, including its title, description, and contents. The format is an extension of the Microsoft Active Channel format, which uses XML.

Writing the Channel Definition File

Fortunately, only a nodding acquaintance with XML is required to write a channel definition. Here I provide the briefest of overviews of the language here—for my complete nickel tour, see Chapter 12.

About XML

The eXtensible Markup Language was created to provide Web developers with a metalanguage for describing arbitrary kinds of data—one that could be extended as necessary. Like HTML, XML uses tags to define operations to be performed on content. Unlike in HTML, the meaning of XML tags depends entirely on context. XML provides a means for defining a tag and indicating the style in which it should be rendered. The channel definition format does not use this feature, however.

XML Tags for Mobile Channel Creation

Windows CE CDFs are restricted to a finite set of XML tags understood by Microsoft Internet Explorer. Table 6-2 summarizes these XML tags.

TAG	ARGUMENTS	PURPOSE
ABSTRACT		Gives a synopsis of a channel or item.
	BASE	Provides a base path for all URLs referenced within the channel definition.
CHANNEL		Defines a channel.
CHANSCRIPT		Specifies the script page to be used to render the display of a channel.
	DAY	Specifies a unit of time in days.
EARLIESTTIME		Used within a SCHEDULE tag, indicates the earliest time a download should commence. Ignored.
	HOUR	Specifies a unit of time in hours.

Table 6-2   *(continued)*

Table 6-2 (continued)

TAG	ARGUMENTS	PURPOSE
	HREF	Provides the MCTP URL for the channel. Specifies the URL for the data item's source and for the logo's source.
	ID	Gives a unique identifier for the channel. Specifies the MCTP identifier for the item.
ITEM		Specifies a Mobile Channel data item.
LOGO		Used to provide Microsoft Internet Explorer with logo information for the channel. Logos may be GIF or JPEG files referenced by URL.
	MINUTE	Specifies a unit of time in minutes.
SCHEDULE*		Specifies the download schedule for a channel.
SELF		Gives the URL of the channel definition itself.
	STYLE	Indicates the style of the logo. Should be Image-Wide, Image, or Icon.
TITLE		Gives the title of a channel or item.
USAGE		Indicates how a channel should be used.
	VALUE	Indicates that a channel is a Mobile Channel or that an item is destined for a Mobile Channel.
	version	Indicates the version of XML used.
XML		Indicates that the document is an XML file.

*This tag is ignored by the Microsoft Mobile Channel Viewer.

Table 6-2: XML Tags for Defining a Microsoft Mobile Channel

As in HTML, tags in XML are delineated using angle brackets (< and >) and surround an optional block of text to be marked. Most, but not all, XML tags are used in pairs. XML tags follow the same convention as in HTML of indicating end tags with a slash (/). The XML tags that aren't paired, called *empty tags*, are closed with a slash and angle bracket (/>).

You must begin a channel definition with the declaration that it complies with XML 1.0, using the special tag <?xml version="1.0"?>. After this declaration, you must define the channel item itself is with the <CHANNEL> tag. The <CHANNEL> tag itself declares a channel's URL, its base URL (to be used when resolving references), and a unique identifier, as shown in the following example:

```
<CHANNEL
 HREF="mctp://host/folder/name.cdf"
 BASE="http://host/folder/"
 ID="unique-identifier">
…
</CHANNEL>
```

The HREF attribute of the <CHANNEL> tag defines the location of the channel using its full path and the MCTP. The BASE attribute simply indicates the relative path to files for the mobile channel on the Web server. The identifier specified in the ID attribute must be unique across all channels on a device; the Microsoft Mobile Channel Viewer uses this identifier to organize its cache and store channel items.

The channel itself is described within the <CHANNEL> tag. Although the items can come in any order, your channels will be more readable if you begin them with descriptive tags such as <SELF> and <TITLE>, and leave off listing the items in the channel until the end. For example:

```
<CHANNEL …>
  <SELF HREF="http://host/folder/name.cdf"/>
  <TITLE>A short title</TITLE>
  <ABSTRACT>A brief synopsis of the channel</ABSTRACT>

  <USAGE VALUE="MobileChannel"/>
  <CHANSCRIPT VALUE="index"/>

  <ITEM…></ITEM>

  …
</CHANNEL>
```

All channels *must* include a <USAGE/> tag with the value MobileChannel to indicate that they are Mobile Channels. (Forgetting to include this tag results in a valid channel subscription for Microsoft Internet Explorer, but no channel being synchronized to the device — a rather puzzling affair at best to debug.) You'll use the <CHANSCRIPT> tag to indicate the item responsible for rendering the channel. (Think of it as the channel's root page.)

Mobile Channels may specify an update schedule, just as a desktop's Microsoft Active Channels channel may (only Microsoft Internet Explorer honors the download schedule, however). A device will always attempt to synchronize its entire list of channels against the desktop. Consequently, specifying a schedule may help your site's server load characteristics, but does not generally serve to speed the process of device synchronization greatly. This is because the synchronization will copy the entire site from the Microsoft Internet Explorer cache (or your server) each time synchronization occurs. A download schedule can look something like this:

```
<SCHEDULE>
  <INTERVALTIME HOUR="6"/>
  <EARLIESTTIME HOUR="1"/>
  <LATESTTIME HOUR="23"/>
</SCHEDULE>
```

This schedule indicates that the channel may be updated by the client between 1:00 a.m. and 11:00 p.m. local time, but that updates should not occur more frequently than every six hours.

You can use the empty tag `<LOGO/>` to associate a logo with a channel. Windows CE recognizes three kinds of logos, each with a distinct purpose, and you can specify one of each:

- **Icon logo**: Displayed on the navigation bar within the Mobile Channel Viewer on the Windows CE device. It should be 16 × 16 pixels. Your subscribers will most likely associate this icon with your content. This logo is specified with the `ID` attribute equal to `LOGOIC` and a style equal to `Icon`.

- **Image logo**: Shown on the main screen of the Mobile Channel Viewer when users are browsing the list of channels downloaded on a device. It should be 80 × 24 pixels. This logo is specified with the `ID` attribute equal to `LOGO` and a style equal to `Image`.

- **Image-wide logo**: Shown on the desktop computer's list of channels. It should be 194 × 32 pixels. This logo is specified with the `ID` attribute equal to `LOGO-WIDE` and a style equal to `Image-Wide`.

These images may be color or grayscale, and furnished as GIF or JPEG files. For example:

```
<LOGO STYLE="Image" HREF="images/logo.gif" ID="LOGO"/>
<LOGO STYLE="Image-Wide" HREF="images/logo-wide.gif"
      ID="LOGO-WIDE"/>
<LOGO STYLE="Icon" HREF="images/icon.gif" ID="LOGOIC"/>
```

The actual items in the channel are defined in XML using paired <ITEM> tags, specifying the source URL and identifier within the channel for each item. For example:

```
<ITEM HREF="index.html" ID="index">
  <USAGE VALUE="MobileData"/>
</ITEM>
```

This item, assuming it's in the channel and has the ID unique-identifier, would be referred to by the MCTP reference, mctp://unique-identifier/index. Within the constraints of memory and performance considerations, a channel may contain any number of <ITEM> declarations.

The Microsoft Active Channels definition includes many more tags, but because they are ignored by the Microsoft Mobile Channels browser, I did not list them here.

A Mobile Channel Example

Figure 6-6 shows a sample Mobile Channel patterned after the AvantGo APRS/Find example shown in the previous section. This example assumes that the user indicates the stations sought using a desktop Web browser.

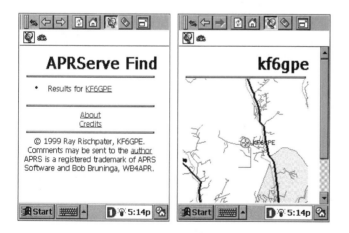

Figure 6-6. Microsoft Mobile Channels example

The example channel consists of the following:

- An index page, with links to each station being tracked.

- Location pages for each station being tracked.

- A page about the Mobile Channel.

- A page about APRS and amateur radio.

The index page, which shows links to the remaining content in the channel, is shown in detail as follows:

```
<HTML>
<HEAD>
  <TITLE>APRServe Find</TITLE>
</HEAD>
<BODY>
<H1 ALIGN=right>APRServe Find</H1>
<P>
<HR>
</P>
<UL>
  <LI>Results for
    <A HREF="mctp://aprs-find/APRS-kf6gpe">KF6GPE</A></LI>
</UL>
<CENTER>
<HR>
<A HREF="mctp://aprs-find/APRS-about">About</A><BR>
<A HREF="mctp://aprs-find/APRS-credits">Credits</A>
<HR>
&copy; 2000 Ray Rischpater, KF6GPE. Comments may be sent to the
author.<BR>
APRS is a registered trademark of APRS Software and Bob Bruninga,
WB4APR.</CENTER>
</BODY>
</HTML>
```

When converting documents for insertion into a Mobile Channel, remember to reference all hyperlinks in the channel with the Mobile Channels protocol, `mctp://`. (Forgetting to change the protocol declaration is a common mistake.) These references are resolved in the CDF, which also contains the other information necessary to create the channel. The CDF for the example shown in Figure 6-6 is as follows:

```
<?XML version="1.0"?>
<CHANNEL HREF="mctp://www.lothlorien.com/~dove/WCA/aprs-find/MobileChannels/
aprs.cdf"
  BASE="http://www.lothlorien.com/~dove/WCA/aprs-find/MobileChannels/"
  ID="aprs-find">
<SELF
```

```
  HREF="http://www.lothlorien.com/~dove/WCA/aprs-find/MobileChannels/aprs.cdf"/>

<TITLE>APRS/Find</TITLE>
<ABSTRACT>
Locate other APRS operators using the APRServe MapBlast! interface.
</ABSTRACT>
<USAGE VALUE="MobileChannel"/>
<CHANSCRIPT VALUE="APRS-INDEX"/>

<ITEM HREF="mc-index.html" ID="APRS-INDEX">
  <USAGE VALUE="MobileData"/>
</ITEM>
<ITEM HREF="http://www.lothlorien.com/cgi-bin/mc-aprs-map.pl?CALLSIGN=kf6gpe"
ID="APRS-kf6gpe">
  <USAGE VALUE="MobileData"/>
</ITEM>
<ITEM HREF="http://cyan.census.gov/cgi-bin/mapper/map.gif?lon=-
122.145000&lat=37.175000&iwd=200&iht=200&wid=0.035789523809383&ht=0.03578952380938
3&mark=-122.145000,37.175000,cross,KF6GPE;&on=places,majroads,streets,railroad"
  ID="APRS-kf6gpe-map">
  <USAGE VALUE="MobileData"/>
</ITEM>
<ITEM HREF="about.html" ID="APRS-about">
  <USAGE VALUE="MobileData"/>

</ITEM>
<ITEM HREF="credits.html" ID="APRS-credits">
  <USAGE VALUE="MobileData"/>
</ITEM>
</CHANNEL>
```

Each document in the channel has a corresponding <ITEM> entry in the CDF. The unique identifier of each item, specified with an ID argument, is used within the HTML content to refer between channel items.

Users obtain this CDF from a link on a Web page using the following HTML snippet:

```
<A HREF="aprs.cdf">APRS/Find Microsoft Mobile Channel</A>
```

This HTML enables a channel user to download the channel file to a desktop computer, which then adds the channel to the user's subscription list maintained by Microsoft Windows CE Services.

Summary

Synchronizing browsers provides a powerful alternative to wireless browsers for bringing information to handheld devices. Their use often overlaps with wireless access, as many kinds of information can be kept current via occasional synchronization with the Web. This synchronization enables users to have access to relatively up-to-date content without being burdened with fees to access this information. Sites for synchronizing browsers are small, with a central index page providing links to the rest of the content.

The two major synchronized browser products currently available are AvantGo and Microsoft's Mobile Channels for Windows CE. The fundamental approach to developing for both is the same, although there are practical differences.

You can make virtually any wireless HTML site available to AvantGo users with only a little extra work and planning. You must ensure that the content is organized in a tree rooted at an index page. All content must be within a few links of the root and include tags indicating AvantGo support. This small effort is usually worthwhile, given how widely this technology is being adopted.

Microsoft's Mobile Channels for Microsoft Windows CE is best used in situations in which the content will only be accessed with Windows CE devices, such as in intranet environments with a controlled user base. Available on many of the most popular Windows CE devices, the Microsoft Mobile Channel architecture uses a Channel Definition File, rather than HTML links, to determine what constitutes a channel. To develop content for this environment, you must define the channel carefully in a complex file, and edit all the content specifically for Windows CE using XML tags.

While XML is an endlessly extensible language, the subset of XML tags necessary for this task is small and well defined. These tags are used by Microsoft Mobile Channels to specify the pages to be included in synchronization, describe the synchronized content, and provide scheduling information for downloads.

Server-Side Content Management Made Easy

MOST, IF NOT ALL, mobile users seek access to dynamic content and immediate information. With few exceptions, people can wait for access to static content until they return to their homes or offices. However, even static content—maps, travel directions, or other material—can often be made more useful by adding dynamic information; for example, in the case of travel directions, up-to-date information on traffic, road hazards, and so on.

Even when content is truly static, a dynamic content strategy such as server-side scripting can generalize the content of a site for different classes of devices. Separating the format from the data when authoring lets you create one set of information for many kinds of clients. Figures 7-1 and 7-2 show the same content created by server-side scripts for Netscape and the Palm VII, respectively. You'll see how to do this using PHP: Hypertext Preprocessor (**PHP**) later in this chapter.

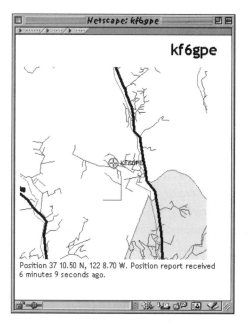

Figure 7-1. Server-side script result viewed with Netscape

Figure 7-2. Server-side script result viewed with Palm VII Clipper

In this chapter, we look closely at two server-side scripting technologies that you can use to create content for different kinds of devices. I focus on the aspects of these technologies that pertain to mobile devices.

What Is Server-Side Scripting?

Server-side scripting (also known as "active server" technology) has been around for a few years. Informally, any Common Gateway Interface (CGI) script that returns Web content is often called *server-side scripting*, but the term technically refers only to technologies characterized by the following:

- The use of a simple scripting language with conditional constructs that can generate content for clients.

- The placement of these scripts within or around HTML.

- The ability of these scripts to insert content within the HTML that contains them.

In a server-side scripting environment, the Web server invokes server scripts in response to client requests to create responses. These scripts can perform database operations, select formatting, manipulate content, and integrate content with data.

The two main server-side environments currently available are PHP, a free, open-source environment available for UNIX, Linux, and Windows; and Microsoft's Active Server Pages (ASP), for Windows-based servers only.

Server-side scripting provides several benefits for developers of mobile data. First and foremost, it provides a framework for separating data from its presentation. This division leaves content developers free to divide their work between information and layout or to create many layouts for a single data set. While other technologies (notably style sheets) exist that support content-format separation, they require client support that is not likely to be available on mobile platforms for some time.

Equally important is the support for conditional logic that a server-side scripting environment provides. Server-side scripts are able to examine what kind of client is making a request and prepare a document that is formatted optimally for that client. In many cases, this operation is as simple as the server selecting from a set of template scripts the one that is appropriate for a particular client. In other applications, the server selects an appropriate format template and merges it with the origin data.

Finally, server-side processing frees the client to perform its most important task: presenting data to the user. Although most clients can make the determination to ignore tags that the client cannot render, you shouldn't rely on the client to do this. By forcing the client to make this determination, these tags may waste bandwidth by sending tags to a client that are then ignored.

While server-side scripting was developed for providing HyperText Markup Language (HTML) content, there is nothing to stop an active server from providing content to Wireless Application Protocol (WAP) or Handheld Device Markup Language (HDML) clients. The Web-based deployment model for both WAP and HDML enables the same HyperText Transfer Protocol (HTTP) server to be used to deliver content in both HTML and a screen-phone format such as WAP. While scripting libraries capable of formatting content for these devices may not be available, server scripts can create these pages using the raw tags available in either WAP or HDML. In turn, the same Web server can serve these pages and make them available to the client via the network operator's gateway.

Using Server-Side Scripting in Mobile Applications

To use server-side scripting for mobile data, you'll need to prepare your site as follows:

- Select an appropriate server-side scripting technology.

- Separate your site's content into data and formatting information.

- Identify the interfaces to your data sets and formatting sets.

- Develop the data set.

- Develop the format set.

I examine each of these steps in detail below.

Selecting a Server-Side Scripting Solution

When selecting an appropriate server-side scripting solution, you'll want to consider a number of factors, including your personal experience, your server's infrastructure, your business model, and any existing restrictions on your information technology. Most technologies are functionally equivalent from the perspective of mobile data, so they are distinguished primarily by how well their feature sets fit with your skills and other requirements of site design. However, other constraints exist.

Organizations that are able to make only a minimal initial investment of funds would be wise to look at one of the open-source active server solutions, such as PHP or Java servlets. These technologies run on a variety of computing platforms and are freely available for modification and use. Although there is little commercial support for them, in knowledgeable hands, these systems can be installed and configured at little or no cost.

In other cases, platform constraints or organizational politics may dictate the choice of a server scripting solution. Organizations with existing Information technology appropriations will likely use the solution selected by their IT staff. Or, for example, a company focused on demonstrating the use of Microsoft technologies for mobile data will probably select Microsoft ASP, regardless of the strengths or weaknesses of it or other choices. Similarly, an organization with a heavy investment in a particular technology should make the most of that investment and pursue its service using that technology.

Separating Content into Data and Format Sets

Central to the use of server-side techniques for mobile data is the separation of content into data and format. *Data* refers to the intellectual property your site delivers. *Format* refers to the look and feel, such as the templates into which your data will be inserted.

Determining the difference between data and format can be more difficult than one would first think. Is your beautiful logo part of a page's format, for example, or is it data? Or is it simply a resource that spans formatted pages?

In general, content will change more often than information used to define your site's look and feel. Data often changes over time and needs to be updated to maintain accuracy. It must therefore be easy to access and change. Often, special tools are used to access or change information, such as a database front end, or scripts for entering and validating up-to-date data.

Format sets, on the other hand, are largely static. Once a site has been created, you will update format sets infrequently, often only to add support for a new kind of device in use.

Another possibility is to define data and format according to who maintains them. Content specialists are responsible for maintaining data—the specialist(s) may be simply yourself, or may be a group of content editors. By contrast, format information is largely the domain of graphic artists, Web designers, software script developers, and similar resources—those responsible for the look and feel of a site. Again, for small organizations or small sites, there may not be different staff members responsible for each of these tasks, but the people who fulfill them will switch roles if they perform different tasks.

Closely related to the idea of a format set is that of a *resource* set, which consists of common format elements used across multiple formats. For example, your site's logo might be part of a set of elements used by many templates—one for rich HTML, one for simple HTML, one for screen phones, and so on. Advertising content might also belong to a resource set, presented in different formats for different devices. (On the other hand, the advertising service might also divide data and format as described here when providing content to you in multiple formats for multiple devices.) Whether it is worthwhile to create a resource set largely depends on the number of elements that are common to several of your formats. If there are only a few, it's sufficient to keep the elements that comprise the resource set in their own directory, and handle them separately as you would any other site. If you run a larger service with complex formats that have many shared elements, you may choose to enforce a more rigorous partitioning scheme, perhaps storing resource content on secondary servers.

Similarly, how you store your data, format, and resource sets will depend largely on the scope of your service. For a small site, you may choose to keep data in text files and formats for various pages and devices in different directories. For a larger site or one with many sources of input, you may find it more practical to keep data—or even both content and format sets—in a relational database. Or you may want to use a mixture of strategies, mixing and matching data files and database access.

For example, the creators of a site providing financial market data would probably consider the financial data—stock market indexes, stock quotes, financial news, and other information—to be their data set. Their format set would contain the scripts, HTML, Wireless Markup Language (WML), and HDML for the various devices they support. An organization of this size would probably define a resource set of information such as its logo, organization name, trademarks, and disclaimers. The data set would be stored in a variety of locations, no doubt including networked access to commercial data feeds and databases for storing and retrieving news stories.

On the other hand, a smaller site, such as one responsible for keeping the calendar and other information for a local club, would have a much simpler division. The site's data set would likely be a set of text files containing meeting times and other special events for the calendar, and perhaps a simple contact list. The format set would contain the scripts and format for this data, probably organized for only the devices currently in use by club and prospective club members.

Identifying Interfaces

Identifying interfaces to your format and data sets is a function of both your data storage choices and of the scripting environment you will use to provide the data to clients. It is largely a process of software design because you will use these interfaces entirely within the scripts produced to manage and present your content. Broad categories of interfaces include simple text-file parsers that select information from a file, Open Database Connectivity (OBDC) database interfaces that can access back-end databases via Structured Query Language (SQL), and network interfaces that interact with other information servers.

Many simple sites choose to store their data in text files, using simple tools in Perl, tkl, C, or another language to search and sort them. These sites use single-purpose tools, crafted for the site's data, and include means to add data, remove or edit data, and search for data by a relevant key such as filing date, topic, or name. Server scripts to access the data sets for presentation with format elements use the same tools used to edit and manipulate the data set for content editors.

More sophisticated sites use relational databases, accessed by tools created for content editors. These sites use integration tools such as OBDC to make requests of the database on behalf of clients. In turn, the resulting information is merged with format elements and returned to the client. Most server-side scripting environments provide explicit tools for access to databases, or can be extended to do so using third-party products.

Large-scale sites, especially those interacting with legacy information sources such as financial or billing systems, may not provide traditional network interfaces. Such systems often require special network-based tools in order to gain access to their information. Server-side scripts are required to use these tools to make requests on behalf of clients. While this process is similar conceptually to database integration, it is more complex, because it calls for additional software rather than ready-made tools for integration.

A special class of these systems uses the eXtensible Markup Language (XML) for data exchange. Systems providing XML representation for their data often make this data available over HTTP; integrating with one of these systems simply involves the selection of an HTTP and XML layer in software. While much work remains to be done in establishing XML interfaces for many data sources, XML will likely replace legacy interfaces to data, and may in some settings extend or replace the need for OBDC database connectivity in Web applications.

Creating Data Sets

Once you have defined your interfaces, you can build and populate your data and format sets. This ongoing process occurs in phases throughout the lifetime of a service.

The first phase of this step emphasizes the tools and infrastructure needed to store the data set. This part of the task may be as simple as putting together some Perl scripts and denoting a directory on the server that will store data, or it may involve the procurement and installation of an enterprise database and user interface tools to maintain this database.

The next step is to populate the initial data and format sets, and is likely to involve teams of content specialists. This is often the "ramp up" phase, in which a large volume of data is inserted into a system for the first time.

Once a system is running, considerably fewer resources are required for the day-to-day operation and maintenance of its data set. Depending on the kind of data involved, the ongoing responsibilities may consist only of validation of new data, asking questions such as "Is it valid?" and "Does it meet our quality criteria?" Other kinds of data sets, however, may involve additional work, especially those involving the collection of news or other editorial information.

Creating Format Sets

The creation of format sets is an activity that largely occurs before the launch of a service and only infrequently during a service's active lifetime. It involves the preparation of format information and scripts that present aggregated content to mobile devices.

There are three steps to developing a format set. In the first, you'll make decisions regarding the elements of the set, looking from the perspective of the kinds of devices to be supported. In the second, you'll develop markup content for these devices, which is often thought of as *style sheets*, but not to be confused with the style sheets used by Cascading Style Sheets (CSS) or other Web technologies. Finally, you'll produce scripts that integrate this information with the data set. Often, an iterative process in which scripts are developed and placed within the markup content may blur the second and third steps together.

For many creators of new sites, determining which devices to support is a challenge. The goal, of course, is to provide the highest fidelity of content to each device with the minimum amount of effort. For some installations, specific device choices are driven by market concerns—a service targeted for screen phones will obviously invest heavily in the production of WAP and HDML content, for example. But even in these cases, the general question of what to support remains.

It is useful to think of clients as falling into four categories: high-definition HTML, low-definition HTML, WAP, and HDML. High-definition HTML will be viewed by laptops using a wireless service and by high-end portable computers such as Windows CE clamshells with robust Web browsers. Low-definition HTML targets simple Personal Digital Assistant (PDA) devices such as the Palm Computing Platform, palm-size PCs running Microsoft Windows CE, and similar devices. WAP and HDML will be targeted at the screen-phone browsers supporting these technologies.

Each of these categories has different format and testing requirements. The high-definition HTML category, for example, would support simple images and several paragraphs of text per page. The low-definition HTML category, on the other hand, calls for the sparing use of images, and for content in a format where abstracts can be collected and in-depth information made available only when requested. The categories intended for screen-phone support use the briefest possible presentation, no images, and might not provide access to the full text of content available to other device categories.

Once you have established the fundamental categories for your site, you need to draw up correspondences between the actual client identifications and the categories in which they reside. All browsers (except the very first versions of Mosaic) return an HTTP User-Agent header, specifying the name of the browser and its version number. For example, AvantGo returns the string "Mozilla/3.0 (compatible; AvantGo 3.2)". Your scripts can obtain the value of this header, use it to determine the kind of browser requesting the content, and then select the appropriate format to use.

Server-Side Parsing with Apache

The Apache Web HTTP server has been the most popular Web server on the Internet since April 1996. A recent site survey by Netcraft found that over 57% of the Web sites on the Internet use Apache, making it more widely used than all other Web servers combined.

This Apache server is available for most UNIX variants, including Solaris and Linux, as well as Microsoft Windows, and is available from many Web hosting companies. It is an open-source project; its source code available for examination and modification to suit your purposes. As with many other open-source projects, there are many customizations and enhancements available, and of course, you can extend the software if necessary.

The Apache server supports *modules* that extend its functionality. One module commonly used to support simple server-side scripts is the mod_include module—also called the "server parsing module"—which permits server parsing of HTML files. This module (available by default with the base installation) lets you use simple directives in your HTML source to include dynamic content created by external scripts.

The server-parsing module scans source files for server directives that perform simple tasks, such as replacing a directive with the output from an executed program, or obtaining the current file's size or last modified date. Simple if-then conditional operations are supported, enabling a variable to be tested and different content returned based on the value of a variable.

Server parsing with Apache is appropriate for simple sites where dynamic content is limited to the selection of markup. These sites can use server parsing to support desktop and synchronized browsers, or to differentiate between a high-definition and a low-definition HTML client. While the intrepid developer could choose to use server parsing for larger files, solutions such as PHP are a better choice because they are designed to support large aggregates of dynamic content.

Using Server Parsing

By default, files with the .shtml suffix are parsed by the Apache server. (Site administrators can change this suffix or make additional suffixes server-parsed, so you should check your local configuration before you begin.) In other words, the

Apache server scans them and the server replaces any directives with the content that the directives create. In case of errors, vague messages appear in the resulting HTML content; the Apache HTTP error log keeps track of the details.

A *server directive* is a single word instructing the server to perform a task; it may also contain options modifying how the task should be processed. Server directives may occur anywhere in the source content. They are contained within an HTML comment, and always start with a hash symbol (#). For example, the following code samples are both valid server directives:

```
<!--#set var="dog" value="husky"-->
<!--#echo var="LAST_MODIFIED"-->
```

By encapsulating server directives in a comment, you can use HTML editors and validation tools to compose your content without receiving spurious error messages.

When the Apache server processes content, it replaces server-parsed directives with the value of the directive's evaluation. For example, the directive

```
<!--#echo var="LAST_MODIFIED"-->
```

is replaced by the value of the variable LAST_MODIFIED, which happens to be the date on which the current file was last modified.

Server Directives

The Apache server recognizes twelve server directives in server-parsed files, which are listed in Table 7-1.

DIRECTIVE	OPTION	PURPOSE
config		Controls various aspects of server parsing.
	errmsg	Sets the error message to be returned for parsing errors.
	sizefmt	Sets the size format to be used when displaying sizes. Specified as bytes or abbrev.
	timefmt	Sets the time format to be used when displaying times.
echo	var	Prints the variable specified by the var option.
else		Specifies the failure clause for a conditional expression.
elif	expr	Specifies an additional conditional clause for a conditional expression.

Table 7-1   *(continued)*

Table 7-1 (continued)

DIRECTIVE	OPTION	PURPOSE
endif		Terminates a conditional expression.
exec	cmd	Executes the command specified by cmd.
	cgi	Executes the CGI given by cgi.
fsize	file	Prints the size of the file specified.
flastmod	file	Prints the last modified time of the file specified.
if	expr	Begins a conditional expression.
include	file	Includes the specified file at the current location for parsing.
printenv		Prints all environment variables.
set	var, value	Sets the variable specified by var to the value indicated.

Table 7-1: Apache Server-Parsed Directives

Output Format Configuration

The config directive is used to control how the server presents the results of server-parsed data within the current document. Most often, it is used with the errmsg option to specify what error message is inserted in the content if a directive fails. For example, if there's an error elsewhere in a directive in a page, the directive

```
<!--#config errmsg="Foiled again!" -->
```

causes the message "Foiled again!" to be inserted in your content any time an error occurs. By default, the error message displayed is the rather bland "[an error occurred while processing this directive]".

Reporting File Size

The fsize directive reports the size of a file. The sizefmt option determines whether its output is an exact number (when sizefmt is bytes) or rounded to the nearest kilobyte (when sizefmt is abbrev). For example, the directives

```
<!--#config sizefmt="bytes" -->
This file is <!--#fsize file="index.html"--> bytes.
```

could return content similar to

```
This file is 892 bytes.
```

Using `config sizefmt="abbrev"`, the same file would result in the content

```
This file is 1k bytes.
```

Tracking File Modifications

The `flastmod` directive reports the time that a file was last modified. This example prints the last modification date of the file `index.html`:

```
This file was last modified on <!--#flastmod file="index.html"-->.
```

The format of the output from `flastmod` depends on the `timefmt` option of the `config` directive. Its argument is a C `printf` style format string, consisting of text and field format directives, which are represented by a % symbol followed by a letter. For example, the server directive

```
<!--#config timefmt="%A %B %C"-->
This file was last modified on <!--#flastmod file="index.html"-->
```

results in output with the following date style:

```
This file was last modified on 17 August 2000.
```

The possible field format flags for the `timefmt` option are shown in Table 7-2. A time string may be built of any number of field format flags.

FORMAT	PURPOSE
%a	Day name (abbreviated).
%A	Day name (full).
%b	Month name (abbreviated).
%B	Month name (full).
%c	Preferred representation of date and time for current local and platform.
%d	Day of the month as a decimal number.
%H	Hour of the day (24-hour clock).

Table 7-2   *(continued)*

Table 7-2 (continued)

FORMAT	PURPOSE
%I	Hour of the day (12-hour clock).
%j	Day of the year.
%m	Month number.
%M	Minutes past the hour.
%p	Either "a.m." or "p.m.", depending on whether the time is before or after noon.
%S	Seconds past the minute.
%U	Number of the current week in the year (the first Sunday starts week 1; days before the first Sunday are in week 0).
%W	Number of the current week in the year (starting with the first Monday of the year).
%w	Day of week as a decimal number (Sunday = 0).
%x	Preferred representation of the date without the time for the current locale and platform.
%X	Preferred representation of the time without the date for the current locale and platform.
%y	Current century (last two digits).
%Y	Current century (all four digits).
%Z	Time zone designation or abbreviation.
%%	Inserts a percent sign.

Table 7-2: Time Format Field Flags

Including Other Content

The `include` directive specifies a file to be included in the content. The server simply replaces this directive with the specified file's contents. For example, if the file `test` contains the message "Hello World", the directive

```
<P><!--#include file="test" --></P>
```

is converted to the following HTML:

```
<P>Hello World></P>.
```

This directive can be used to build up a larger file from primitives such as a header, dynamic content, and footer.

Executing External Scripts

Similar to include, but more powerful, is the exec directive, which executes the script at a specified location. Thus, the directive

```
<P><!--#exec cmd="date" --></P>
```

is converted to the following HTML:

```
<P>Fri Feb 25 15:50:46 PST 2000</P>.
```

The use of include and exec files can pose security risks for Web sites, however, as Web authors and others may have unrestricted access to these files and commands. The constraint that only commands in the Apache's CGI directories can be executed using the exec directive provides some protection. Moreover, the Apache server must have access (via UNIX access controls) to the file that is included or executed.

On many servers, the exec function is disabled. If this is the case, only files and commands available in Web and CGI directories can be executed, and the directive must include the virtual option:

```
<P><!--#exec virtual="cgi-bin/date.cgi" --></P>
```

Before using the include and exec directives, confirm how your server's environment has been configured.

Using Variables

The Apache server script parser supports both user-defined and server-defined variables that contain strings. These variables can be evaluated and their results included in content for the client or tested by server directives to control execution of other server directives. The server defines the variables DATE_GMT, DATE_LOCAL, DOCUMENT_NAME, DOCUMENT_URI, LAST_MODIFIED, and HTTP_USER_AGENT. The first two of these are set to the current time relative to Greenwich Mean Time (GMT) and the local time zone, respectively. The next two are set to the document's filename and the URL path of the document requested by the user. The LAST_MODIFIED variable is set to the last modification date of the document requested by the user. The variable HTTP_USER_AGENT is set to the value of the HTTP_USER_AGENT header passed by the client during a request. (I'll explain more about this a bit later in this chapter.)

The echo directive converts a variable to HTML:

```
This page was last updated on <!--#echo var="LAST_MODIFIED" -->.
```

resulting in a message showing the last updated date for the current file. This directive is handy for footers of frequently modified pages.

The printenv directive prints all currently set variables, which makes it useful in debugging.

Making Decisions

Variables have a second use: conditional flow control. Apache provides conditional evaluation through the if, else, and endif directives. These directives evaluate the expression given by the if directive's expr option. If the expression is true, the content between the if and else directives is evaluated and returned to the client; otherwise, the content between the else and endif directives is evaluated and returned, as shown in the following example:

```
<!--#if expr="$DATE_LOCAL=\"Mon Dec 25 00:00:00 PST 2000\""-->
 <B>Ho ho ho!</B>
<!--#else-->
  <!--#echo var="DATE_LOCAL" -->
<!--#endif-->
```

This rather fruitless snippet prints "Ho ho ho!" in bold if the page is viewed on exactly Christmas morning at midnight; otherwise, it prints the current date.

Expressions may contain the comparison operations = and !=, along with the ordinal comparisons <= and >=. Logical aggregation of operations is supported using the C-style ! (not), && (and) and || (or) operations. Anything not recognized as a variable or operator is treated as a string for the purposes of comparisons. Regular expressions can be specified using // instead of quotes, allowing partial matching to occur. Thus, the following example would print a Yuletide message any time on Christmas day:

```
<!--#if expr="$DATE_LOCAL=/Dec 25/"-->
 <B>Ho Ho Ho!</B>
<!--#else-->
  <!--#echo var="DATE_LOCAL" -->
<!--#endif-->
```

Multiple comparisons can be performed using elif directives. Optionally, one or more elif directives may lie between an if and an else directive, each with its own comparison. The contents between the first elif resulting in true and the next if, elif, else, or endif block are returned to the client. The following example uses the HTTP_USER_AGENT to differentiate between AvantGo, Palm VII, and other browsers:

```
<!--#if expr="$HTTP_USER_AGENT = /AvantGo/" -->
  You're viewing this site with AvantGo.
<!--#elif expr="$HTTP_USER_AGENT = /Elaine/" -->
  You must be using a device with the Palm Clipping Technology.
<!--#else-->
  Hmm. Looks like you're using a no-name browser like Netscape, Microsoft Internet
Explorer, or something else.
<!--#endif-->
```

Client-Specific Content with Apache Server Directives

One of the best uses of server-parsed pages is managing content for multiple client devices. When content is static and only one or two different kinds of devices are supported, server parsing can be significantly easier to implement than a full active server solution.

The server-parsing environment defines the variable HTTP_USER_AGENT. This variable can be used to identify the client and return client-specific content. During a transaction, the HTTP client sends a User-Agent header with a string identifying the client. This string is placed in the HTTP_USER_AGENT variable by the server before the script is parsed.

I wrote the HTML that follows for a Web site with both desktop and mobile users. In it, I test the HTTP_USER_AGENT variable to return a Web page formatted for either AvantGo or a desktop browser.

```
<HTML>
<HEAD>
   <BASE HREF="http://www.lothlorien.com/~slvares/">
   <TITLE>San Lorenzo Valley ARES ARES</TITLE>
</HEAD>
<BODY BGCOLOR="#FFFFFF">
<!--We provide different content if we're being viewed by
AvantGo.-->

<!--#if expr="$HTTP_USER_AGENT = /AvantGo/" -->
<META NAME="HandheldFriendly" content="True">

<H1 ALIGN=right>San Lorenzo Valley ARES</H1>
<P>
Welcome to the San Lorenzo Valley ARES page, providing training
and reference information for the ARES volunteers in and around the San Lorenzo
```

```
            Valley in California, USA.
            </P>
            <H2>Current events:</H2>
            <P><!--#include file="events.html"--></P>
            <CENTER>
            <P>
            <HR>
            <A HREF="calendar.html">Calendar</A> |
            <A HREF="meetnets.html">Meetings&amp; Nets</A>
            </P>
            <P>
            <HR>
            &copy; 1997-2000 SLVARES. This page was downloaded on
            <!--#echo var="DATE_LOCAL" -->.</CENTER>
            </P>
            <P>
            Visit http://www.lothlorien.com/~slvares/avantgo.html with your
            desktop browser to obtain a list of SLV ARES resources available
            to AvantGo users or see additional information about the SLV ARES
            organization.</P>

            <!-- Content was requested by a conventional browser -->
            <!--#else -->
            <H1 ALIGN=right>San Lorenzo Valley ARES</H1>
            <CENTER>
            <HR>
                    <P>Welcome to the San Lorenzo Valley ARES page,
                    providing training and reference information for the ARES
                    volunteers in and around the San Lorenzo Valley in
                    California, USA.</P>

                    <P>See what's new <A HREF="whatsnew.html">here</A>, or
                    <A HREF="whatsnew.html#form">register</A> to get email
                    whenever these pages change.</P>
            <P>
            <HR>
            <A HREF="#ce">Current Events</A> -
            <A HREF="emergency.html#background">Background</A> -
            <A HREF="emergency#training">Training</A> -
            <A HREF="calendar.html">Calendar</A> -
            <A HREF="emergency.html#section">Our Section</A> -
            <A HREF="emergency.html#weather">Weather</A> -
            <A HREF="emergency.html#earthquakes">Earthquakes</A> -
            <A HREF="emergency.html#general">General</A>
```

```
</P>

<HR>
</CENTER>

<A NAME="ce">
<H2>Current events:</H2>
<P><!--#include file="events.html"--></P>

<CENTER>
<HR>
  <TABLE BORDER=0 WIDTH="100%">
  <TR>
     <TD>
     <P>
     This page is available in a simplified form for handheld
     computers running Palm Computing's PalmOS or Microsoft
     Windows CE. Here are details, or simply subscribe to the
     AvantGo channel by clicking this button.
     </P>
   </TD>
  <TD>
    <CENTER>
    <A HREF="http://avantgo.com/mydevice/
autoadd.html?title=San%20Lorenzo%20Valley%20ARES&url=http://www.lothlorien.com/
~slvares/
&max=100&depth=3&images=1&links=0&refresh=daily&hours=2&dflags=127&hour=0&quar-
ter=00&s=00">
       <IMG SRC="avantgo-channel.gif"
       ALT="Subscribe to AvantGo Channel">
       </A>
       </CENTER>
    </TD>
  </TR>
</TABLE>
<HR>

<P>
<FONT SIZE="-1">
&copy; 2000 SLVARES. Comments may be sent to the
<A HREF="mailto:slvares-webmaster@lothlorien.com">
SLVARES webmaster</A>.<BR>
This page was last updated on
<!--#echo var="LAST_MODIFIED" -->.
</FONT></CENTER>
```

```
</P>
<!--#endif -->
</BODY>
</HTML>
```

Figure 7-3 shows how this page would appear in Netscape and AvantGo.

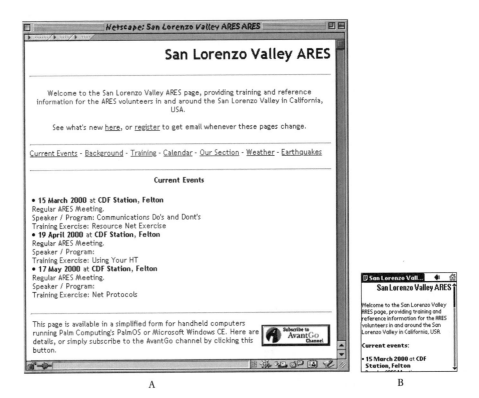

Figure 7-3. Results of server parsing for client identification of (A) Netscape and (B) AvantGo.

The page actually uses several server script directives. The first is the conditional statement evaluating the HTTP_USER_AGENT variable, which uses the server directive if-else-endif expression to return content for either the AvantGo or a conventional browser. The server directive include is used in both versions of the page to include the upcoming events calendar (sparing the administrator the need to edit the main page when the calendar changes). The echo directive displays the last modification date for the page.

I chose a short page with a link structure appropriate for synchronizing a browser for the AvantGo page (see Chapter 6), while the conventional page, formatted for a desktop browser, contains related links that are of little use to mobile users.

Note that using a regular expression when testing HTTP_USER_AGENT ensures that the server will return the AvantGo page for *all* versions of AvantGo. If I'd written the

following, it would have returned AvantGo pages only to AvantGo clients with
version 3.2:

```
<!--#if expr="$HTTP_USER_AGENT =
            "Mozilla/3.0 (compatible; AvantGo 3.2)" -->
```

I could make this page support multiple browsers using the if-elif-else-endif
construction, as follows:

```
<!-- HTML header here -->
...
<!-- Support for AvantGo -->
<!--#if expr="$HTTP_USER_AGENT = /AvantGo/" -->
  <META NAME="HandheldFriendly" content="True">
  ...
<!--Support for PalmVII -->
<!--#elif expr="$HTTP_USER_AGENT = /Elaine/" -->
  <META NAME=" PalmComputingPlatform" content="True">
  ...
<!--Everybody else -->
<!--#else-->
   ...<!--endif-->
<!-- HTML trailer here -->
```

As you can imagine, however, this approach grows cumbersome quickly. It's
more efficient to use the include directive in conjunction with long conditionals:

```
<!-- HTML header here -->
...
<!-- Support for AvantGo -->
<!--#if expr="$HTTP_USER_AGENT = /AvantGo/" -->
  <META NAME="HandheldFriendly" content="True">
  <!--#include file="avantgo/index.html"-->
<!--Support for PalmVII -->
<!--#elif expr="$HTTP_USER_AGENT = /Elaine/" -->
  <META NAME="PalmComputingPlatform" content="True">
  <!--#include file="palm/index.html"-->
<!--Everybody else -->
<!--#else-->
  <!--#include file="other/index.html"-->
<!--#endif-->
<!-- HTML trailer here -->
```

Of course, you can use a conditional to contain only a portion of a document, too. The following snippet sends a background image to all clients except Windows CE devices (presumably because Windows CE users are known to use a low-bandwidth connection):

```
...
<!--Don't send our background to Windows CE handhelds! -->
<!--#if expr="$HTTP_USER_AGENT = /Windows CE/" -->
  <body>
<!--#else-->
  <body background="images/swiss-cheese.gif">
<!--#endif-->
...
```

For simple sites with problems such as these, server parsing with Apache provides a solution that's efficient, quick to implement, and easy to understand later when adding new content.

Server-side scripts grow increasingly inefficient as the size of your site grows, however. You'll need to do more work to separate your content, and you'll need to resort to scripts executed with the exec directive for dynamic behavior. A combination of languages such as Perl, tkl, and C is used to write these scripts. These scripts are costly for the developers because they take programming skills to create, and for services because they require additional computing resources each time a script is executed. Under most implementations, the server uses a separate process for each script executed, which can quickly consume server resources.

PHP-Powered Wireless Web Sites

A better solution for larger sites is an active scripting technology such as PHP: Hypertext Preprocessor. PHP is a true scripting language, with looping constructions, a rich library of functions, and database interfaces. PHP can be embedded within a document, or it can generate documents through the execution of scripts. Scripts generally execute within the process and memory space of the Web server, although you can invoke system commands using two functions that create a new process and run any command on the system. Best of all, it's a free open-source project.

PHP (maintained at http://www.php.net/) was started by Rasmus Lerdorf and blossomed with the efforts of several contributors. It's a robust, stable, active scripting environment, used by over a hundred thousand Web sites, many of them commercial services. Many Internet Service Providers (ISPs) make PHP available to subscribers, enabling you to create a service quickly while leaving the task of installing PHP and maintaining servers and the Internet connection to an expert.

Introduction to PHP

As a thorough introduction to PHP would (and does) occupy an entire book, I only scratch the surface in the following discussion. I show you how PHP is integrated with a Web page, as well as the basic syntax of PHP commands, functions, and flow-control statements. With this knowledge, you will be able to understand the examples in the following section, and should be able to create simple PHP pages. To learn more about PHP, consult the online documentation at the PHP Web site or your local bookstore.

Usage

PHP scripts exist between special delimiters within a content file (such as an HTML file, although, as you'll see, they can also be used with HDML, WML, and potentially any other markup language served by HTTP servers). PHP scripts begin with the special tag `<?php` and end with the tag `?>`. (You may encounter scripts enclosed simply by the tags `<?` and `?>`, but this abbreviated form is discouraged in wireless Web development because the tag `<?php` is XML-compliant and more likely to work with future server-side wireless applications.) The content is saved as a file with a specific suffix, usually `.php` or `.php3`, which tells the server the PHP interpreter should process the file. As the server serves the file to the client, the PHP interpreter is used by the server to interpret the scripts. For example, the following page produces the output shown in Figure 7-4:

```
<HTML>
<HEAD>
<TITLE>Hello</TITLE>
</HEAD><BODY>
<?php
  print("Hello world. Today is ");
  print(Date("l d F Y"));
  print(".\n");
?>
</BODY>
</HTML>
```

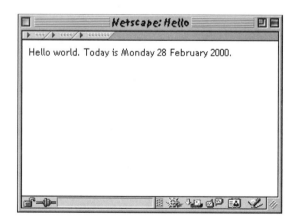

Figure 7-4. A simple PHP script's results

This page consists of some introductory HTML and a single PHP script. The script prints a series of three strings to the client: two are static, and the PHP function Date creates the third, which returns a string containing the date in the requested format.

I can rewrite this script so the PHP is responsible for generating *all* the required content, as shown here:

```php
<?php
  print("<HTML>\n");
  print("<HEAD>\n");
  print("<TITLE>Hello</TITLE>\n");
  print("</HEAD><BODY>\n");
  print("Hello world. Today is ");
  print(Date("l d F Y"));
  print(".\n");
  print("</BODY>\n");
  print("</HTML>\n");
?>
```

In practice, this approach creates pages that are both easier to maintain and easier to extend to support different clients. It can, however, be somewhat baffling to the newcomer, because the entire page's content is written in PHP. In the sections that follow, I follow the convention that PHP mixed with other Web content is *always* bracketed by the <?php and ?> tags; I leave these tags off for PHP script snippets not meant to stand alone, reminding you that they're only examples, not complete scripts. It's important to remember that you can use PHP to create content for other markup languages, too. Doing this requires an extra step: the PHP script must inform the client of the type of content returned using an HTTP header. PHP provides a function to modify this header, as shown here:

```php
<?php
  // Specify our content type
  header("Content-type: text/x-hdml");
  // and produce an HDML deck.
  print("<HDML VERSION=3.0>\n");
  print("<DISPLAY MARKABLE=\"TRUE\" TITLE=\"Hello\" NAME=\"hello\">\n");
  print("Hello world. Today is ");
  print(Date("l d F Y"));
  print(".\n");
  print("</DISPLAY></HDML>\n");
?>
```

The first and third lines of this script show one way to include a comment in a script. PHP supports C++-style comments. The characters // can be used to mark the rest of a line as a comment, and anything between /* and */ is treated as a comment.

The second line instructs the Web server to return the HTTP header Content-type to the server, specifying that the content in this document will be HDML content written in text. By using the appropriate content type declaration, you can have your PHP scripts return HTML, images, HDML, or WML. (This instruction *must* occur before any output from your content or PHP script, however, or the Web server will assume that your content is HTML.)

Syntax

As you've already seen, a PHP script is expressed by a set of comments, keywords, and function invocations. These expressions, which are separated from one another by a ;, as in most other languages patterned after Pascal and C, are comprised of identifiers and operators. An *identifier* is a variable or function name; an *operator* is a one- or two-character keyword that manipulates identifiers—examples include arithmetic operators (+, − ,/, and *, along with . [the "dot" or string concatenation operator]). Statements are grouped into a *block* using curly braces, { and }.

Variables

Variables contain data that may change during the evaluation of a script. They are named using a sequence of letters, numbers, and underscores, subject to the following rules:

- The first character of a variable identifier must be a dollar sign: $.

- The second character of a variable identifier must be a letter or an underscore: _ .

- All other characters of the identifier must be letters, numbers, or underscores.

PHP is case-sensitive; thus, the identifiers $myHome and $MYHOME are different. As you'll see when I show you how functions are named, as a side effect of the fact that all variable names begin with a $, is that you can use the same name for both a variable and a function. It's a good idea to avoid doing this, however, because you can easily get confused between the two.

Variables may contain strings, integers, or floating-point numbers. PHP is loosely typed; a variable may contain any data type. Type conversions occur automatically when necessary, so that expressions such as the following:

```
$x = "4"; $result = $x * $x; // $x is converted to an integer
print( $result ); // $result gets converted to a string before
                  // output
```

handle the conversion between $x as a string and a number transparently.

PHP dynamically creates variables as they're encountered. There's no need to declare a variable as you must in many compiled languages.

Variables can also hold arrays. An array can be indexed either by an integer or by the names of items within the array (these are known as *associative arrays* in other programming languages). The `array` function enables you create and initialize an array in a single step, as shown here:

```
$pets=array( "dog"=>"Sake",
             "cat"=>"Kashka",
             "bird"=>"Tau" );
$days=array( 0=>"Sunday",
             1=>"Monday",
             2=>"Tuesday",
             3=>"Wednesday",
             4=>"Thursday",
             5=>"Friday",
             6=>"Saturday");
```

To access an array, you specify the index using the `[]` operator:

```
$today=$days[4]; print( $pets["dog"] . "never did get the hang of  " .
       $today . ".\n" );
```

Functions

Functions are defined using the `function` statement. Following the `function` statement, you specify the function's name and any arguments first and then the body of the function as a block.

```
function doHeading( $heading )
{
  print("<H1>" . $heading . "</H1>" );
}
```

Variables used within a function have scope only through the function in which they are used. A function can return a result value to the function's caller using the `return` statement, as shown here:

```
function square ($x )
{
  $result = $x * $x;
  return $result;
}
```

When a function must access a variable declared outside a script—called a global variable, because its scope extends throughout a script—it must be explicitly declared using the global statement. This helps prevent accidental use or modification of global variables:

```
$PI = 3.1415926536;

function area( $radius )
{
  global $PI; // use this file's $PI.
  return $PI * square( $radius );
}
```

Functions can declare default values for variables, too. Later we will look at the following example.

```
function h1( $head, $align="center" )
{
  print("<H1 ALIGN=$align>" . $head . "</H1>\n");
}
```

This function produces an HTML header tag with the alignment specified by the caller of the function; by default, $align is set to "center" and the header is centered on the page.

Flow Control

PHP provides conditional flow control in the form of the if-else, if-elseif-else, and switch-case statements. The syntax of these statements strongly resembles that of other programming languages: they accept an expression to evaluate and blocks to execute if the expression is true or false. A sample if statement looks like this:

```
if ( $anchor )
{
  print("<A NAME=\"$anchor\">\n");
}
```

This statement prints a string containing the anchor tag and the anchor name if and only if $anchor is not an empty variable.

The if statement may be followed by an optional else clause, or by one or more elseif clauses to support multiple logic paths; for example:

```
if ( strstr( $userAgent, "Windows CE" ) ||
     strstr( $userAgent, "WinCE" ) )
{
    $UserAgentType = "WinCE";
    $UserAgentClass = "html-lo";
}
else if ( strstr( $userAgent, "AvantGo" ) )
{
    $UserAgentType = "AvantGo";
    $UserAgentClass = "html-lo";
}
else
{
    $UserAgentType = "unknown";
    $UserAgentClass = "html-hi";
}
```

Sometimes, it's clearer to use the switch statement. The switch statement takes a set of case statements and code to execute when one of the case conditions matches. Execution then continues through the remaining cases:

```
switch( $shape )
{
  case "triangle":
    $kind="pointy";
   break;
  case "square":
    $kind="rectangular";
    break;
  case "circle":
  case "oval":
    $kind = "round";
     break;
  default:
    $kind="unknown";
 }
```

Here, if $shape is either "circle" or "oval", $kind will be set to "round". The break statement causes execution to continue outside the current block, so in the case that $shape is equal to "triangle", $kind is set to "pointy", and execution continues outside the switch statement.

The switch statement can support a block that will be executed if no other conditional matches, called the *default*, which is declared using the default statement.

In the previous example, $kind is set to "unknown" if $shape doesn't match one of the strings "triangle", "square", "circle", or "oval".

Looping

Loop statements in PHP come in three flavors: while, do-while, and for. The while statement executes a block only while its test condition is true. The script that follows, for example, executes the block each time it loops back to it as long as $n is greater than 0:

```
$n = $initial;
while ($n > 0)
{
  $result = $result * $n;
  $n = $n - 1;
}
```

Note that if $n is 0 when the while statement is reached, the block is skipped altogether. In contrast, the do-while statement *always* executes its block the first time it is encountered, then applies the condition. Thus, the following example will not necessarily yield the same results as the last one:

```
$n = $initial;
do
{
  $result = $result * $n;
  $n = $n - 1;
}
while ($n > 0);
```

In this case, the operation of multiplying $result by $n will always be executed at least once, before the comparison with $n takes place.

In many cases, the for statement is both more convenient and easier to read than do or do while. It is functionally equivalent to the while statement, but divides a loop into four constructions:

- The initial conditions for the loop.

- The statements to be executed each time the loop is executed.

- A conditional statement (loop execution continues if it returns true).

- The block to be executed through the loop.

For example, our first loop could be rewritten using the `for` statement:

```
for ( $n = $initial; $n > 0; $n = $n - 1 )
{
  $result = $result * $n;
}
```

Here $n is set to $initial once. The loop block (the statement $result = $result * $n) is executed each time $n is found to be greater than 0; after the loop block, $n is decremented. Once $n is equal to 0, the loop stops, and the instructions immediately after the loop are executed.

The `break` statement we encountered in the `switch-case` statement is especially handy when working with loops. For example, the section of code shown here examines the array of headers that are returned by `getallheaders`, stopping when the `User-Agent` header is found:

```
$headers = getallheaders();
 for ( reset($headers);
       $index = key( $headers );
       $value = next( $headers ) )
  {
   if ( strcasecmp( $index, "user-agent" ) == 0 )
   {
     $userAgent = $value;
     break;
   }
  }
```

In this routine, the PHP `getallheaders` function retrieves the HTTP headers the client and server exchanged in the associative array $headers. The `for` loop initializes an iteration across all array elements using the `reset` function and then loops through each value in the array. On each pass, the `for` statement sets $index to the name of the current entry, with the `key` function and $value set to its contents using the `next` function. On each loop, the `next` function also sets the key and next functions to the next element of the array. When the end of the array is reached, both key and next return an empty string; this string is evaluated as `false`, and the loop terminates. On each pass through the loop, $index is compared to the string `"user-agent"`, and if the two are equal (irrespective of case), the value of that array element is saved in the variable $userAgent, and the loop ends.

Importing Files

PHP provides literally hundreds of functions for developing Web content. I've already mentioned several, including `print`, `reset`, `key`, and `next`—there isn't nearly room to cover them all. But I do want to introduce one more function you'll find useful, `include`, which imports the contents of a file into the current script. The `include` function is most often used to share common functions across multiple scripts. The examples in the next section make use of a number of functions that are used in multiple PHP scripts, all of which are saved in the file `nameUserAgent.php`. This means that that the statement

```
include( "nameUserAgent.php" );
```

within any script allows access to all the functions within that file. The `include` function thus permits you to keep common functions in one location; if a function changes, you need to edit only one file.

A PHP-Based Wireless Web Site

In previous chapters, I presented several examples of code I created involving APRS/Find, a wireless Web application that obtained an APRS station's position given its radio call sign. This information is obtained and returned in a format optimized for the current client device by a set of PHP scripts. PHP allows the service to return content formatted for a host of different devices, including screen phones, using the same script.

Script Organization

I divided these back-end scripts into several functional groups. One script, `nameUserAgent.php3`, is responsible for identifying the current client based on HTTP headers. The `format.php3` and `aprs-image.php3` scripts use this information to define a series of functions which the `locate.php3` script uses to build a response page. The `locate.php3` script also accepts the form data posted by a wireless Web client. It then invokes a command to determine the requested station's position and uses the other scripts to build a response.

    The `nameUserAgent.php3` script defines the `nameUserAgent` function, which other scripts use to determine what kind of client the request originated from. It uses a table of comparisons to establish two pieces of information about the client: its *type* and its *class*. I define a *type* as a loose identification of a device by its manufacturer or form factor, such as Palm, WinCE (Microsoft Windows CE), or phone. I think of *class* as the kind of content a client can accept. A client's class is named by one of three strings: "`html-hi`", which indicates desktop-fidelity HTML; "`html-lo`", which indicates a wireless browser that accepts HTML content; and "hdml", which indicates HDML content.

Format scripts select a specific set of functions for formatting content based on the class of the device. These are stored in a file called xxx-format.php3 (that is, either html-hi-format.php3, html-lo-format.php3, or hdml-format.php3). Each of these files defines the functions shown in Table 7-3. In addition, the locate.php3 script uses the indication of a browser's class to select the format scripts that will generate the map shown in the response using the xxx-aprs-image.php3 scripts (html-hi-aprs-image.php3, html-lo-aprs-image.php3, or hdml-aprs-image.php3).

FUNCTION	ARGUMENTS	PURPOSE
docBegin	$title	Begins a document with the given title.
docEnd		Ends a document.
paraBegin	$tag	Begins a paragraph and anchors it with the indicated tag.
paraEnd		Ends the current paragraph.
h1	$heading, $align	Creates a level 1 heading with the given text and alignment.
hr		Draws a horizontal rule.
br		Creates a line break within a paragraph.
bold	$text	Outputs the specified text in boldface if supported.

Table 7-3: Functions Used to Generate Result Data

Response Generation

The heart of the response generation occurs in the locate.php3 script. The server invokes this script in response to a client's form POST request. Forms using this script provide two variables in their POST request: callsign and scale. The former contains a call sign of the desired station, while the latter contains an optional map scale for the resulting map. The values of these variables are available within the PHP scripts as $callsign and $scale.

A simple form that interacts with the locate.php3 script looks like this:

```
<HTML>
<HEAD>
    <TITLE>APRS/Find</TITLE>
</HEAD>
<BODY>
<H1 ALIGN=right>APRS/Find</H1>
```

```
<P>

<HR>
<FORM ACTION="http://masamune.lothlorien.com:8080/~dove/aprs-find/locate.php3"
METHOD=POST>
    <P>Callsign: <INPUT TYPE=text NAME=callsign VALUE="" SIZE=10 MAXLENGTH=10><BR>
    Scale:<BR>
    <SELECT NAME=scale value="">
        <OPTION VALUE=100000>1 mi
        <OPTION VALUE=200000>2 mi
        <OPTION VALUE=500000>8 mi
        <OPTION VALUE=3200000>50 mi
    </SELECT> <INPUT TYPE=hidden NAME=version VALUE=2 size=1 maxlength=1><BR>
    <INPUT TYPE=submit NAME=Submit VALUE="Map" size=0 maxlength=0>
</FORM><A HREF="about.html">About</A> - <A HREF="credits.html">Credits</A>
<HR>
</P>
</BODY>
</HTML>
```

A closer look at locate.php3 reveals how the script selects the output format and the data obtained and presented. The first few lines are responsible for determining the client's class and type:

```
<?php
  // Determine our user agent.
  include( "nameUserAgent.php3" );
  nameUserAgent();
  // Load our format scripts for this user agent.
  $uac = userAgentClass();
  $uat = userAgentType();
  include ( "${uac}-format.php3" );
  include ( "${uac}-aprs-image.php3" );
 //the meat of the script goes here
?>
```

This portion of the script includes the nameUserAgent.php3 script, which I discuss next. Suffice to say, this script defines the functions nameUserAgent, userAgentClass, and userAgentType, which together identify the client by its HTTP headers. The locate.php3 script uses this information to include one of the xxx.format.php3 files and one of the xxx-aprs-image.php3 files. Together, these two files define the functions responsible for formatting the output generated by the locate.php3 script. I examine them more closely in the sections about data presentation that follow.

Once the determination of format scripts is complete, the `locate.php3` script uses a binary provided by the host system to locate the station by its call sign, as shown here:

```
...
// Find the given station
  $result = exec(
    "/home/httpd/cgi-bin/locate-station.pl $callsign",
    $info );
  $call = $info[0];
  $lat = $info[1];
  $lon = $info[2];
  $prettyLat = $info[5];
  $prettyLon = $info[6];
  $last = $info[7];
  $course = $info[4];
  $status = $info[3];
...
```

The `locate-station` Perl program contacts a remote server, passes on the call sign, and parses the response data. The `locate.php3` script interprets the response from this command, which is presented as a series of lines. The PHP function `exec` executes its command argument as if it were entered in a command shell. This command's output is returned in the array `$info`. The last line of the command's result—the last line of the `$info` array—is also returned by the `exec` function. The `locate.php3` script parses the response from `locate-staton` into a series of variables. These variables describe the station's call sign; position as both floating-point numbers and human-readable strings and status information about the station, including the time its position was last received; its course, if moving; and its most recent status message reported.

With the result data in hand, the `locate.php3` script uses our format functions to generate the following response:

```
...
  // Present the results
  docBegin( $call );
  h1( $call, "right" );
  paraBegin();

  if ( $result == "FOUND" )
  {
    map( $lat, $lon, $call, $scale, $uat );
    br();
```

```
     print( "Position $prettyLat, $prettyLon. " );
     print( "$last" );
     br();
     if ( $status )
     {
       print( "$status" );
       br();
     }
     if ( $course )
     {
       print( "$course" );
       br();
     }
   }
   else
   {
     print( "The server encountered an error because $status\n" );
   }
   paraEnd();
   docEnd();
?>
```

The work done by this portion of the script is straightforward. The script uses the format functions to output a document header using the station's call sign as a title. Then it creates a header, followed by the beginning of a new paragraph.

The script outputs either a position report or an error message. This output is sandwiched between the markup language's document heading tags and closing tags. The position report contains a map, generated by the map function defined in the xxx-aprs-image.php3 file included at the beginning of the script, and the station's latitude, longitude, course, status, and last reporting time. Once the script reports the dynamic content, it ends the open paragraph and document and then terminates.

Client Identification with PHP

As mentioned previously, the nameUserAgent.php3 file contains the functions that perform client identification. This file defines three global variables used to track the client's identity, along with functions to obtain the values of these variables. Although it's not necessary, I chose to use access functions to return the value of these variables, rather than referring to them directly outside of the file. This is a good example of *data abstraction*, as it hides the representation of the data from entities that don't need that level of detail. By obscuring the nature of these variables,

I minimize the opportunity for other scripts to accidentally manipulate or destroy them. You can see an example of data abstraction in the userAgentType function:

```php
<?php
$UserAgentType = "unknown";

function userAgentType( )
{
  global $UserAgentType;
  return $UserAgentType;
}
?>
```

The nameUserAgent function sets the various global variables used by the data abstraction layer:

```php
<?php
function nameUserAgent( )
{
  global $UserAgentType;
  global $UserAgentClass;
  $userAgent = "";
  $userAccepts = "";

  // Find the User-Agent header and get its value.
  $headers = getallheaders();

  for ( reset($headers);
        $index = key( $headers );
        $value = next( $headers ) )
  {
    if ( strcasecmp( $index, "user-agent" ) == 0 )
    {
      $userAgent = $value;
    }
    if ( strcasecmp( $index, "http-accept" ) == 0 )
    {
      $userAccepts = $value;
    }
  }
  // Set the client name appropriately.
  if ( strstr( $userAgent, "Windows CE" ) ||
       strstr( $userAgent, "WinCE" ) )
```

```php
  {
    $UserAgentType = "WinCE";
    $UserAgentClass = "html-lo";
  }
  elseif ( strstr( $userAgent, "AvantGo" ) )
  {
    $UserAgentType = "AvantGo";
    $UserAgentClass = "html-lo";
  }
  elseif ( strstr( $userAgent, "Elaine" ) )
  {
    $UserAgentType = "Palm";
    $UserAgentClass = "html-lo";
  }
  elseif ( strstr( $userAgent, "UPG1" ) )
  {
    $UserAgentType = "phone";
    $UserAgentClass = "hdml";
  }
  elseif ( strstr( $userAgent, "Mozilla" ) &&
           !strstr( $userAgent, "compatible;" ) )
  {
    $UserAgentType = "desktop";
    $UserAgentClass = "html-hi";
  }

  return $UserAgentType;
}
?>
```

This script performs two basic tasks: it obtains information from the HTTP headers and uses that information to make an educated guess about the kind of browser making the request.

This script uses the PHP function `getallheaders` to obtain all headers exchanged between client and server. It then searches for the `User-Agent` and `Http-Accept` headers, used by convention to specify the browser's identity and the kind of content it will accept. With this information, a set of comparisons are used to identify first the client type and class. Special gymnastics are required to identify Phone.com HDML clients, for example, because the user-agent reported by the Phone.com Up.Link server reports only that it is an Up.Link server. Instead, the Up.Link server uses the `Http-Accept` header to indicate the content the client can accept. (Some clients in this example can only accept HTML, and the only question to resolve is whether they can handle high-fidelity HTML content.) Doing this enables us to differentiate between HDML and WAP clients, should the need arise in the future.

I obtained the `User-Agent` and `Http-Accept` header values used in this function from two sources. Many mobile browser vendors make their headers available to content developers as a matter of course in developer documentation. For some browsers, however, this information may be difficult to collect, either because the browser's vendor has not provided the information or because the vendor no longer exists. An easy way to obtain the information is to examine the log file of an HTTP server after the client downloads a page. Even easier, however, is to use this simple PHP script, which prints the headers back to the client in response to a request.

The following script simply loops over all headers exchanged between client and server, and prints them as successive paragraphs in an HTML document to be displayed by the client:

```php
<?php
  $headers = getallheaders();
  print ("<HTML><BODY>");

    for ( reset($headers);
          $index = key( $headers );
          $value = next( $headers ) )
    {
    print ("<P>");
    print ( $index );
    print ("=<BR></P><P>");
    print ($value);
    print ("</P>");
    }
print ("</BODY></HTML >");
?>
```

Returning to the implementation of `nameUserAgent`, we note it is obviously incomplete because it omits several popular browsers including Internet Explorer, Opera (both for the desktop), pdqSuite for the Palm Computing Platform, and several others. To keep from overwhelming low-fidelity browsers, I used a copy of the format script for the `html-lo` class of devices as the default format script for unknown browsers.

Presentation for HTML Browsers

By separating the instructions for how the data is presented—its format—from the data itself, PHP allows you to easily add new format modules on a per-browser or per-class basis. In this example, there are only a few basic differences among the versions for different browsers because the content is so simple. In fact, only the `docBegin` function and the map presentation differ between the `html-hi` and `html-lo` browser classes.

The `html-hi-format.php3` file for this example is instructive because it demonstrates one way to separate data from formatting. This file defines functions (again, see Table 7-3) that provide abstractions for the HTML tags used by `locate.php3`. These abstractions, in turn, can be modified to produce different markup tags to suit different browsers as necessary.

The `html-hi-format.php3` file looks like this:

```php
<?php
function docBegin( $title )
{
  // Specify our content type
  header("Content-type: text/html");

  // emit heading tags.
  print("<HTML>\n");
  print("<HEAD>\n");
  print("<TITLE>$title</TITLE>\n");
  print("</HEAD>\n");
  print("<BODY>\n");
}

function docEnd( )
{
  print("</BODY>\n");
  print("</HTML>\n");
}

function paraBegin( $anchor = "" )
{
  if ( $anchor )
  {
    print("<A NAME=\"$anchor\">\n");
  }
  print("<P>");
}

function paraEnd( )
{
  print("</P>");
}

function h1( $head, $align="center" )
{
  print("<H1 ALIGN=$align>$head</H1>\n");
```

```
}

function bold( $head )
{
  print("<B>$head</B>\n");
}

function hr( )
{
  print("<HR>\n");
}

function br( )
{
  print("<BR>\n");
}
?>
```

One pair of functions that shows how the format abstraction layer works is the docBegin and docEnd functions, which are used to generate the tags that open and close a document. The docBegin function specifies the content type of the document sent to the browser, along with the heading of the document. The HTML implementations specify an HTML document, and use the HTML tags <HTML>, <HEAD>, <TITLE>, and <BODY> to perform this task. After specifying the content type, the docBegin function outputs a bare-bones HTML header, the document title, and the opening <BODY> tag. Scripts use subsequent calls to functions in this file, along with the print function to build up the response document. When the document creation is complete, the calling script uses the docEnd function to close the document (in the case of an HTML device, the output is the closing </BODY> tag).

Similarly, the aprs-image.php3 script files define the map function for various browser classes. This function is responsible for using a supplied latitude, longitude, scale, and legend to generate an <IMG SRC> statement containing a URL, which in turn obtains the corresponding map from a map server. The html-hi-aprs-image.php3 script defines the map function as follows:

```
<?php
$imgsize = 300;

function map( $lat, $lon, $legend, $scale, $type )
{
  global $imgsize;
  $size = ( $scale * 0.0000137671428571 - 0.606057142857 ) / 60;

  if ( strpos( $ legend, "-" ) )
```

```
{
  $call = strtoupper(
    substr( $legend, 0, strpos( $legend, "-" ) ) );
}
else
{
  $call = strtoupper( $legend );
}

print("<IMG SRC=\"");

print("http://tiger.census.gov/cgi-bin/mapgen?lon=$lon&lat=$lat&" .
      "iwd=$imgsize&iht=$imgsize&wid=$size&ht=$size&" .
      "mark=$lon,$lat,cross,$call;&on=places,majroads,streets,railroad");
  print("\" ALT=\"Map around $lat $lon\">\n");
}
?>
```

This function builds a URL in the format required by the database from which the maps are obtained (see the "The Tiger Map Database"). It specifies the map's center latitude and longitude as well as the image size, scale, legend, and themes to be displayed. The function derives the latitude, longitude, and scale information from its arguments. It then converts the map scale, which is provided by $scale from the form input, into decimal degrees and processes the legend to make a more attractive label on the map. The size of the image is controlled by the global variable $imgsize, which was defined as a variable at the top of the file to make it easy to revise if the script needs changing.

The Tiger Map Database

One of the handiest resources on the Web for prototyping new services is the United States Census Bureau Tiger Map Database service at http://tiger.census.gov. The Tiger Map Database offers free geographic and thematic maps as GIF images based on 1990 Census data for the entire United States.

While not appropriate for commercial systems—the Census Bureau does not guarantee the server's reliability, and the system does not support commercial deployment per se—the server provides excellent maps for rapid prototyping and demonstration uses. An experiment in Web-based mapping, the system has been replaced by the American FactFinder, a more sophisticated system. While the American FactFinder contains more up-to-date content and is a more reliable service, the Tiger Map database remains a great resource for online maps for prototyping, and will likely be available for years to come.
(continued)

The Tiger Map Database (continued)

URLs supplied to the map server can specify a map's size in degrees, the resulting image size in pixels, along with the specifics of the location to display. Annotations can be added at specific points, as was done with the APRS/Find wireless application. In addition, the server can overlay thematic data such as population, age distribution, and other information on generated maps.

If you flip back to Figure 7-1, you can see the server's response for Netscape, a high-fidelity browser. Figure 7-2 shows how the same data looks on a Palm Computing Platform Palm VII, which I classified as a low-fidelity browser. While the information contained is the same, the map format is radically different.

The `html-lo-aprs-image.php3` script, which `locate.php3` includes when the client is a low-fidelity HTML browser, requests a significantly smaller map. The script looks like this:

```php
<?php
$imgsize = 140;
$imgsizePalm = 80;
function map( $lat, $lon, $legend, $scale, $type )
{
  global $imgsize;
  global $imgsizePalm;
  $size = scale( $scale );
  $legend = safelegend( $legend );

  print("<CENTER>\n");
  print("<IMG SRC=\"");
  // Handle the Palm platform (presumably clipper) carefully.
  // Maps larger than about 80 pixels give it grief.
  if ( $type == "Palm" )
  {
    print("http://tiger.census.gov/cgi-bin/mapgen?lon=$lon&lat=$lat&" .
    " iwd=$imgsizePalm&iht=$imgsize&wid=$size&ht=$size&" .
    "mark=$lon,$lat,cross,$call;&on=places,majroads,streets,railroad"
    );
  }
  else
  {
print("http://tiger.census.gov/cgi-bin/mapgen?lon=$lon&lat=$lat&" .
        "iwd=$imgsize&iht=$imgsize&wid=$size&ht=$size&" .
        "mark=$lon,$lat,cross,$call;&on=places,majroads,streets,railroad");
  }
```

```
  print("\" ALT=\"Map around $lat $lon\">\n");
  print("</CENTER>\n");
}
?>
```

This map function centers the resulting (smaller) map; furthermore, it uses information about the browser's type (obtained by locate.php3 and passed to map) to handle the Palm Computing Platform Clipper application slightly differently. Experiments with the Palm VII shows that the browser occasionally misbehaves with maps more than 80 pixels wide; the script avoids this problem by checking the browser type and acting accordingly.

Presentation for HDML Clients

The organization of HDML is similar to that of HTML, but not identical. HDML, discussed in Chapter 10, is organized around the fundamental metaphor of a deck of cards rather than a page. An HDML card is a small chunk of information, typically a paragraph or less. The screen phones that HDML is designed to work with typically display the information on a single screen.

Establishing a correspondence with the abstraction functions shown earlier in Table 7-3 poses more of a challenge for HDML than for HTML. Primarily, this is a result of HDML's different interface metaphors. In the abstraction layer, each paragraph of a document is represented on a card, ignoring requests for boldface, alignment, and similar options. While HDML can support some of these features (notably text alignment), doing so leads to better presentation on an HDML device. The hdml-format.php3 file defines the format abstraction layer with the following functions:

```php
<?php
$HdmlHeader = "";

function docBegin( $title )
{

  // Specify our content type
  header("Content-type: text/x-hdml");

  // emit header tags.
  print("<HDML VERSION=3.0>\n");
}

function docEnd( )
{
```

```
      print("</HDML>\n");
    }

    function paraBegin( $anchor = "" )
    {
      global $HdmlHeader;

      if ( $HdmlHeader )
      {
        print("<DISPLAY TITLE=\"$HdmlHeader\" " .
                  "MARKABLE=TRUE NAME=\"$anchor\">\n");
        print("$HdmlHeader\n");
        br();
      }
      else
      {
        print("<DISPLAY NAME=\"$anchor\">");
      }
      $HdmlHeader = "";
    }

    function paraEnd( )
    {
      print("</DISPLAY>\n");
    }

    function h1( $head, $align="center" )
    {
      global $HdmlHeader;

      $HdmlHeader = strtoupper($head);
    }

    function bold( $head )
    {
      print("$head");
    }

    function hr( )
    {
      br();
    }

    function br( )
```

```
{
  print("<BR>\n");
}
?>
```

This file uses a bit of skullduggery to create document headings. HDML does not support document headings, as it is a language optimized to display documents of less than 1KB on a 4-line display. Nonetheless, the `locate.php3` script calls for labeling a document, and it seems likely that other pages might use headings in a similar way.

The `h1` function stashes aside the requested heading in the global variable `$HdmlHeading` and returns. In turn, the `paraBegin` function is responsible for declaring a card that will contain one paragraph. This card is named and titled using the heading information set aside in `$HdmlHeading`, if one is available, and this heading is used as the first line on the card before the card's contents.

Although HDML supports the display of simple images, I decided that it was prudent to omit map display for screen phone users. Maps are usable in grayscale but they don't display well in monochrome, and the information density of a map is far higher than the fidelity of a screen phone's display. Consequently, the `hdml-aprs-image.php3` file defines an empty `map` function:

```
<?php
function map( $lat, $lon, $legend, $scale, $type )
{
}
?>
```

The result is such that when the request comes from a screen phone, the output of the `locate.php3` script is limited to the station's call sign, position, and other information, as shown in Figure 7-5.

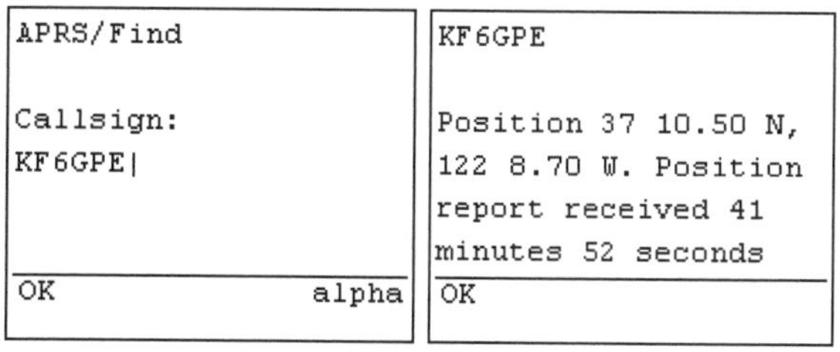

The form used by an HDML browser is so different from the HTML form that I chose to write it from scratch, rather than extend my abstraction layer to support the creation of forms for both platforms. The deck containing the form looks like this:

```
APRS/Find            KF6GPE

Callsign:            Position 37 10.50 N,
KF6GPE|              122 8.70 W. Position
                     report received 41
                     minutes 52 seconds
OK          alpha    OK
```

Figure 7-5. APRS/Find display on an HDML phone

```
<HDML VERSION=3.0>
<ENTRY MARKABLE="TRUE" NAME="home" KEY="callsign" FORMAT="*M">
  <ACTION TYPE="ACCEPT" TASK="GO" DEST="#result">
  <ACTION TYPE="HELP" TASK="GO" DEST="#about">
  APRS/Find
```

```
  <BR>
  Callsign:
</ENTRY>

<NODISPLAY NAME="result">
 <ACTION TYPE="ACCEPT" TASK="GO"
  DEST="http://masamune.lothlorien.com:8080/~dove/aprs-find/locate.php3"
  METHOD="POST" POSTDATA="callsign=$callsign&scale=0&version=2">
</NODISPLAY>

<DISPLAY NAME="about">
  <ACTION TYPE="ACCEPT" TASK="GO" DEST="#home">
  <!--- About and credit info. -->
</DISPLAY>
</HDML>
```

The form in this deck consists of two cards: the <ENTRY> card and the <NODISPLAY> card. The third card is responsible for showing the user trademark and credit information about APRS and the servers behind APRS/Find.

Summary

Authors of sites that target several different devices have two options. They can present information in the lowest common denominator format or prepare content for different classes of devices.

Combining server-scripting technologies with careful organization of your site's contents makes content preparation for multiple kinds of devices a much easier task. Authors can use active server technologies such as PHP: Hypertext Preprocessor or Microsoft Active Server Pages to build large sites, while small sites can often be supported with a simple technology such as Apache's server-parsed directives.

To use any scripting technology effectively, it is crucial to divide your content into data and format information. By separating what your site presents—its data—from how it is presented, you can use scripts to assume the responsibility of formatting data.

These scripts will use cues from the browser, such as the HTTP headers User-Agent and Http-Accept, to determine the nature of the device originating the request. Armed with this information, the scripts can select format tags most appropriate for the client device.

The Apache server-side parsing module provides some primitive server directives that you can use with simple sites to differentiate among different client types. Its chief strength is its simplicity; with only a handful of directives to learn, you can master it in an afternoon, or use its ability to run external scripts to enhance your site. This simplicity is also its drawback; it's difficult to organize a complex site using just markup content and Apache server directives.

PHP, on the other hand, provides content developers with a robust scripting language—actually, an entire programming environment. Using PHP, you can write scripts to output content as well as interact with system-level scripts and other resources. PHP provides constructions for functions and flow control, making it easy to organize complex scripts in large sites. The example of such a site in this chapter demonstrates how to use functions to encapsulate the functional parts of a script and return content targeted to a number of different kinds of devices.

CHAPTER 8

The Wireless Application Protocol

THE WIRELESS APPLICATION PROTOCOL (WAP) is an emerging standard for the markup and presentation of wireless content targeted explicitly at small mobile handheld devices such as cell phones, pagers, and wireless access terminals. At its heart is the Wireless Markup Language (WML), which dictates how online content should be marked up for presentation on wireless access terminals.

In this chapter, I show you how to use WAP to make your content available to the Internet phones deployed today. You learn how WAP brings content to these devices, and how to create and mark up content using WML.

In Chapter 9, you'll see how to extend WML's power with WMLScript, the lightweight scripting environment that's a part of the WAP standard. If you're already familiar with the basics of WML, you may want to skip straight to Chapter 9 to learn how scripting can enhance your wireless content.

The WAP Standard

The WAP standard, which includes WML, is maintained by the Wireless Application Protocol Forum (WAP Forum), a multinational consortium of wireless infrastructure providers that includes Nokia, Phone.com, Ericsson, Motorola, and others. More recently, computing vendors such as Palm, Microsoft, Sony, Spyglass, and Symbian have joined the WAP Forum.

At the lowest level, the WAP stack uses a *bearer network* such as Cellular Digital Packet Data (CDPD) or the Short Messaging System (SMS) to carry data. Like TCP/IP (Transmission Control Protocol/Internet Protocol), the higher level protocols are bearer-independent, enabling WAP to operate on a variety of wireless networks. The bearer network carries data in packets called *datagrams*, using a protocol such as TCP/IP's User Datagram Protocol (UDP) or a network-native protocol. WAP uses the Wireless Datagram Protocol, which abstracts the bearer network's characteristics, such as TCP/IP's UDP.

Above the datagram layer are the Wireless Transport Layer Security (WTLS) and Wireless Transaction Protocol (WTP), which provide the facilities necessary for secure transactions between hosts. WTLS is based on the Transport Security Layer (TSL)—formerly the Secure Sockets Layer (SSL)—in that it provides a means for hosts to negotiate a secure protocol for the exchange of data. WTLS provides assurances for data integrity, privacy, and authentication, in conjunction with the WTP, which is responsible for establishing a reliable two-way transaction service.

Protocols such as the Wireless Session Protocol (WSP), which provides a consistent interface for session-oriented services, use the WTP. The WSP provides both a connection-oriented service and a connectionless service based on WDP with or without security options. In turn, WML uses WSP as the bearer of data from servers to the handheld.

The deployment view for WAP services strongly resembles that of existing HyperText Markup Language (HTML) services, as shown in Figure 8-1. Both dedicated WAP servers and existing Web servers can serve WAP content in WML. WML content served by HyperText Transfer Protocol (HTTP) servers on the Internet is translated into a tightly compressed binary format suitable for wireless transmission and gated to wireless networks that implement WAP using a WAP gateway. Typically, wireless service providers make WAP gateways available to their subscribers transparently, so subscribers can access WML content served on the Internet. This gives them access to both dedicated WAP servers and traditional Web servers offering WAP content. Subscribers need not even know whether a particular server they access is a dedicated WAP server or a Web server offering WAP content.

Why Use WML?

While there are many similarities between the WAP architecture and the protocol architecture for the World Wide Web, there are also several important differences. In many ways, WAP has taken the best features of the infrastructure for the Web and enhanced their operation for mobile devices. WAP, therefore, has many advantages for the wireless content developer.

WAP Is Efficient

WAP is more efficient than traditional Web-based protocols. Explicitly designed for wireless use, WAP enables clients and servers to exchange the most data using the least amount of bandwidth.

Content marked up using WML generally occupies fewer bytes than content marked up in HTML. Not only is WML content more compact because of the encoding performed prior to exchange with the client, but also WML markup itself generally results in pages that are more compact. WML's markup fosters concise writing because its constructs remind authors of the simplicity of the viewing device. In addition, the tags defined in WML provide efficiencies not available in HTML. Similarly, the WAP protocols below WML, including the WTP and WSP, are more efficient than TCP/IP or HTTP.

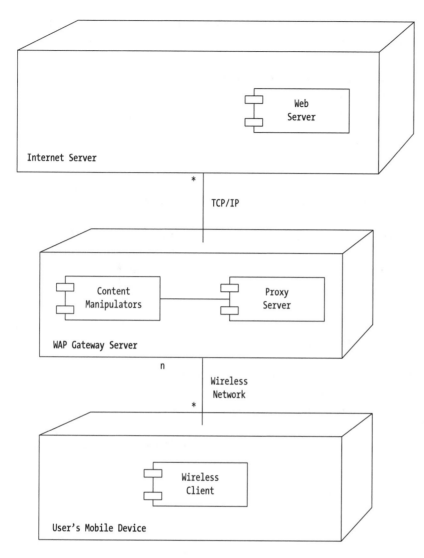

Figure 8-1. The deployment view for WAP services

WAP Provides for Constrained Clients

WAP was designed from the outset for clients with minimal computing capabilities. Unlike HTML, which has been adapted with varying degrees of success to
operate on low-capability devices, WAP has been optimized to provide the greatest
flexibility with minimal resources.

WAP's use of gateway servers at points throughout the wireless network enables
much of the processing required by Web browsers to be performed by servers,
rather than by individual clients. While these gateway servers are an additional

link in the chain between origin server and client that require computing resources, maintenance, stability, and robust operation, their health is generally assured by the companies that operate them.

Of course, in order to make the most of this feature, it's important to be sure that your content is well-formed—that there are no errors in your WAP content that will cause problems for either the gateway or handheld clients.

WAP Is Predictable

The WAP Forum keeps tight control over both the WAP standard and deployment of WAP-compatible devices. If a browser is compatible with a specific version of WAP, you know it will work with content authored to the specification. This is in sharp contrast to the wireless HTML browsers available, where the multitude of browser vendors and compromises has yielded an environment where it's not clear which tags will be available on which device.

WAP Provides Dynamic Behavior

Unlike in HTML, responses to behaviors can be coded in WML. WML provides a simple event mechanism that allows different content to be displayed. User actions, such as pressing a key, can be tied to scripts that cause changes in content. The WML browser also provides timers that can load a different page or trigger the change of variables when the indicated time elapses. These provide greater flexibility than the static content that HTML can deliver.

Why Not Use WML?

WML is not the right answer for every situation, however. Currently, WAP is not well-suited to viewing by Personal Digital Assistants (PDAs) or laptops, and there may be intellectual property restrictions on the use of WAP.

If Clients Are Laptops or PDAs

As of this writing, most WAP browsers are available only for dedicated devices. Those implementations that do run on laptops or PDAs tend to be proof-of-concept demonstrations of implementations eventually targeted for screen phones or other hardware. Moreover, the users of laptops or PDAs are likely to have user interface expectations that WAP applications cannot meet.

If a service is targeted exclusively for these devices, it is easier to deploy it using Web technologies. Not only will the browser be more reliable when running on the target device, but also the user's overall experience will more closely match his or her expectations.

If Intellectual Property May Be an Issue

Whereas open protocols are the norm on the Web, WAP is based on the pooled resources of the members of the WAP Forum. In many cases, WAP Forum members have extended licenses to share intellectual property (software, hardware, algorithms, or patents) with other WAP Forum members or outside parties. There is no guarantee, however, that one or more holders of intellectual property will not require the payment of licensing fees or other consideration for your use of these WAP technologies.

These restrictions may apply not just to the development of WAP clients and servers, but also to WAP content. If you're exploring the commercial use of WAP in your service, it's best to check with the WAP Forum partners regarding intellectual property restrictions and licensing before you set out.

WML User Interface Design

Content written with WML for WAP browsers is often referred to as a *WAP application*. This expression is partially marketing hyperbole but partially factual; content targeted for a WAP browser is, arguably, an application of WAP technology—hence, a WAP application.

Pedantic vocabulary arguments aside, designing WML content is similar in concept to designing Web content, with a few notable differences.

Cards and Decks

The fundamental principle behind the organization of WML documents is the metaphor of cards and decks. A *card* is a grouping of one or more user interface elements, such as text, lists from which the user can select an item, or lines for user input. A card usually represents a single screen on a WAP terminal, although there's no guarantee that a given card as constructed will fit on a given terminal's display. A *deck* is a single file containing WML content, consisting of one or more cards.

The user navigates to a card, reviews its contents, may enter requested information and/or make choices, then moves on to another card. The instructions embedded within a card may invoke services on an origin server as needed. The user's device obtains decks from origin servers on the Internet as they are needed.

The card/deck metaphor is a constant reminder of one of the significant differences between WML and HTML: the constraints of the target device. HTML developers typically think in terms of pages, or at least screens, consisting of color content several hundred pixels on a side. WML application developers have no such luxury—many devices offering WML browsers have monochrome screens and only a hundred or so pixels on a side. Memory is similarly constrained, and the bandwidth available for communication between device and network is scant. Organizing content into the small parcels of information contained in cards reminds developers of the limited viewing capabilities at the reader's disposal.

Emerging Paradigms

The standards advanced by the WAP Forum for WML are exceptionally general. They dictate intentions, not implementations. Thus, a WML specification indicates that a device must handle particular operations, but the specifics of *how* these operations behave are left up to device vendors.

For example, the WML specification requires several unassigned interface elements, which application developers can assign to particular tasks. It does not specify how these elements should appear or behave; many manufacturers have chosen to provide them in the form of soft keys located just below the liquid crystal display (LCD) screen, so that the LCD can show content-assigned titles above each key. However, it would be equally appropriate to offer these as physical keys along the side of a unit, or as voice commands, or as any other imaginable interface that meets the intention of the WML specifications.

Specific paradigms are becoming prevalent for wireless screen phones. Most have similar constraints as well as similar user interfaces, which generally include the following:

- A numeric keypad.

- Text input via shifting or multiple key presses.

- Several special-function keys, such as SEND and CLEAR.

- A directional cursor control using a rocker keypad.

- Soft keys that can be customized by WAP content to any WML functions.

As these devices become more and more prevalent, users and manufacturers are creating a de facto interface standard for them. For example, almost all of these devices have special-function soft keys assigned to the WML operations ACCEPT and OPTIONS. Developers should follow the convention of assigning the most frequently performed tasks to the ACCEPT function key and reserving the OPTIONS function key for less common uses, such as clearing a form.

User Input

Data entry on handheld devices is tedious and error-prone. Most do not have a full keyboard, so several actions are required to select a single character. Your WML content should therefore avoid the need for data input wherever possible. Any time you expect the user to enter a specific data type, take advantage of WML's capability to validate input on the device to catch mistakes before they are sent to your content servers. (See "Interaction" later in this chapter.) Content providers are adopting several appearance standards for different kinds of input; for example, email addresses and keywords are entered using only lower case. Descriptions, names, addresses, and free-form text are entered using mixed case.

Option lists, which enable the viewer to select an item from a list and view additional material, are a good alternative to free-form paragraphs. However, these lists need to be carefully constructed. A long list will require excessive scrolling, and likely confuse most viewers. Studies have shown that most people can manage between five and nine items in their short-term memory. A good compromise is to keep most option lists to the size of an average display, and be sure that none is more than nine items long.

Content for WAP browsers should be brief, clear, and succinct. Devices using WAP are not suited to lengthy reading sessions. Most users will be in an environment where viewing more than a few lines of text is difficult or impossible, as they will be occupied with other matters. They would find scrolling long cards of text, manipulating many cards, and interacting at length with complex decks to be a hindrance to their activities. Some vendors advise breaking long content into multiple cards to simplify scrolling, but the best solution is to avoid lengthy text from the beginning.

Images

The use of images on handheld devices presents several problems. Client devices are suited to display only the simplest of images, such as low-resolution icons. Moreover, images are only available in WAP applications as separate Uniform Resource Locators (URLs), making a new network request necessary each time an image is displayed. Because such network requests can drastically increase the amount of time it takes a card to display, images should be used sparingly, if at all, and only in situations in which they take less space than a text alternative. For example, an icon indicating weather conditions would be appropriate, while an icon denoting the direction of a stock has no advantages over a simple character, such as "+" or "–".

Never use an image to give information that is not available from its text attribute or from the surrounding text, because some devices may not display images at all. Remember also that if you want to be sure an image will appear correctly on all platforms, you need to be sure the image is no larger than the smallest target device's display.

Your First WAP Application

WAP is an eXtensible Markup Language (XML)–based markup language. Although understanding XML can help you write well-formed WML, there's no need to study XML just to work with WML. (The curious reader can skip ahead to the XML section of Chapter 12 before continuing with this chapter.)

No book about any computer technology would be complete without a good complement of "Hello World" examples. Here is one, with its output shown in Figure 8-2:

```
<?xml version="1.0"?>
<!DOCTYPE wml PUBLIC "-//WAPFORUM//DTD WML 1.1//EN" "http://www.wapforum.org/DTD/
wml_1.1.xml">

<wml>
    <card  title="Hello world">
    <p>
  Hello world!
    </p>
    </card>
</wml>
```

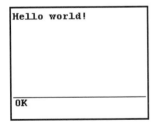

Figure 8-2. Output of the Hello World! XML sample

The first two lines of the previous code simply indicate that this document is an XML document and that the deck uses XML meeting the grammar specification for WML 1.1.

The WML itself is contained between the <wml> and </wml> tags. This WML document itself consists of a single *card*, or aggregation of interface items. Within this card is a single paragraph, consisting of the text "Hello World" for the browser to display.

Along the bottom of the screen are the labels for the phone's two soft keys. A WML browser must support at least two such soft keys or equivalent flexible interface elements named ACCEPT and OPTIONS. Because our WML document does not specifically reassign these keys the WML browser assigns default action. In Figure 8-2, the left one is presently labeled OK, while the right key is unlabeled. (At present, the default action for the left key—the ACCEPT key—is to perform the same action as the PREV key, bringing the viewer to the previously-viewed page.)

Viewing WML Content

While you will generally mark up WML content by hand, viewing it requires a WAP browser. You have several options for doing this:

- A WML browser in a simulator or other environment as provided by Phone.com, Nokia, or other vendors.

- A native-device browser provided by a software developer for testing, which uses a network connection on a PDA or similar device.

- An actual screen phone, such as an end consumer would use.

Of course, you should *always* test your content on a production device of some sort; if you plan to develop any WML content at all, a WAP-compliant screen phone is one of the most important investments you can make. While developing content, the best way to work is to use one of the simulators available through the developer programs offered by WAP Forum members such as Phone.com and Nokia, then perform final testing with a real screen phone. These simulators provide a WAP-compliant browser running on a desktop computer, enabling you to use your desktop environment to simulate a WAP terminal. Figure 8-3 shows Phone.com's simulator in action; it allows you to preview content and interact with the phone's buttons.

For the intrepid, several vendors have released WAP browsers for PDA devices in one form or another. As of this writing, Spyglass, Inc. has an agreement with Nokia to distribute their WAP browser, which presently runs on Microsoft Windows CE. The Edge Consultants offer WAPMan, a WAP browser for the Palm Computing Platform. These tools may work well for some developers, but aren't likely to be as robust or as useful in debugging as the simulators intended for content developers.

WML Syntax

WML documents follow the same rules as XML for organization and syntax. If you have written HTML by hand, the rules are similar enough that you will pick up the syntax almost instantly.

Basic Syntax

WML documents consist of character data interspersed with tags, which are enclosed by < and > characters. These tags define the structure of a document. On the next few pages you'll encounter the following tags: `<wml>`; `<card>`; `<p>` (paragraph); and `<b>` (bold).

Figure 8-3. The Phone.com UP.SDK WML browser

Unlike HTML tags, WML tags are case-sensitive; thus, a browser interprets `<wml>` and `<WML>` as two different tags. Like HTML tags, WML tags actually have two parts—an opening tag and a closing tag—that enclose the affected information between them. For example, the beginning of a WML document is marked with `<wml>`, while the ending is marked with the tag `</wml>`. These opening and closing tags contain the entire contents of the deck.

Some tags are *empty*; that is, they do not contain data. These tags do not require closing tags. While HTML makes no distinction between tags and empty tags such as `<br>`, WML does. Empty tags in WML are written using `<` and `/>`. For example, the line break tag is written as `<BR/>`.

Attributes

Tags may contain *attributes*, which describe their characteristics. Attributes are simply written as name-value pairs separated by an equal sign (=); the attribute is placed after the tag name but before the closing of the tag. For example, the paragraph tag accepts the `mode` attribute to indicate how to wrap text. This causes the text in a paragraph to be kept on one line and displayed marquee-style:

```
<p mode="nowrap">
```

Comments are supported in WML, just as in XML. Within a WML document, the `<!-` and `-->` characters delimit a comment. Comments contain information intended for content authors; they do not affect the browser's behavior and are not displayed by clients.

Variables

WML enables content developers to declare variables within a WML deck. A browser will substitute a variable's name for its value in WML. Variable names can follow any of the following syntax options:

- `$variable`

- `$(variable)`

- `$(variable:conversion)`

WML interprets any text after the $ as part of the variable's name, unless the entire variable name is in parentheses. (There are times where you'll want a variable to snuggle up close to the next word, for example, when you're constructing a URL from several variables, like this: `http://$(hostname)/index.wml`.) Variables are always evaluated before other markup is performed. Variables can be set either by user elements using WML form elements or by the `<setvar>` tag. The scope of a variable is generally global across the browser; there are no ways to hide variables.

Events and Tasks

WML defines several events that are triggered when the browser changes state, such as when it loads a new deck. WML decks can use these events to trigger actions called *tasks*. WML uses the `<onevent>` tag for this purpose, as well as some of the attributes of some specific tags that define items that can generate events.

WML has defined tags to create tasks that tell the browser to refresh a page, jump to a specific URL, jump to the previous URL stored in the browser's history, and so on.

Reserved Characters

WML reserves the <, >, ', ", &, and $ characters; to use one of these in any text contained in a WML deck, you'll need to use the corresponding character tags shown in Table 8-1.

CHARACTER	TAG	PURPOSE
<	&lt;	"less than" symbol
>	&gt;	"greater than" symbol
'	&apos;	apostrophe
"	&quot;	quotation marks
&	&amp;	ampersand
$	$$	dollar sign

Table 8-1: Reserved WML Characters and Their Corresponding Character Entities

Marking up Documents with WML

Like HTML, WML provides *contextual* tags for marking up content. When you use these tags to organize your content, the WAP browser can then select the optimum formatting that captures your document's organization.

WML tags can be organized into four broad categories:

- Organizational tags and their attributes organize a page into cards and decks.

- Navigation tags and their attributes specify how navigation occurs between cards.

- Interaction tags and their attributes specify the interactive components of a card, such as events, selection lines, and input forms.

- Format tags specify how the content on a card will appear.

Document Organization

A WML deck consists of an *XML prologue,* a *heading,* and one or more cards, all contained by the <wml> tag. I discuss each of these items in more detail next. The deck is contained within a single WML document, and written as character data, generally, in conventional ASCII (American Standard Code for Information Interchange), although any proper subset of the Unicode character set will suffice because WML is an XML derivative. Most developers won't need to use any encoding except ASCII, however, unless they're developing content for languages other than English.

By convention, WML files are named with the suffix .wml, and bear the MIME type text/vnd.wap.wml. Gateways and browsers use a file's suffix and MIME type to determine how its content should be handled.

The Document Prologue

Because WML is expressed in XML, all WML documents must begin with an XML prologue. The XML prologue for a WML deck indicates the version of XML syntax and Document Type Definition (DTD) used within the file containing a deck. This information is used by XML-compliant gateways and other applications to determine how a particular document should be interpreted. An XML prologue looks like this:

```
<?xml version="1.0"?>
<!DOCTYPE wml PUBLIC "-//WAPFORUM//DTD WML 1.1//EN" "http://www.wapforum.org/DTD/
wml_1.1.xml">
```

The first line simply says that this file follows the syntax used by XML 1.0. The second line identifies the document as a WML document, meeting the WML 1.1 specification (a DTD) made available at the indicated URL. It generally suffices to simply copy and paste these two lines into any WML document you create. If you're writing for another version of WML, such as 1.0 or 1.2, you'll need to take care to specify the appropriate DTD. Gateway servers will use this DTD to validate your content before passing it off to wireless clients for viewing.

The Document Heading

After the prologue comes an optional *heading,* which is used to give information about the deck's contents. This header contains access control and meta information that pertains to all cards within the deck. Most often, this kind of meta information relates to caching and bookmark behavior. The WML heading tags are outlined in Table 8-2; I take a closer look at each one next.

TAG	ATTRIBUTES	PURPOSE
`<access>`	`domain, path`	Specifies access control information for a deck.
`<card>`	varied	Delineates a card element within a deck.
`<head>`		Specifies the heading of a deck.
`<meta>`	`name, http-equiv, user-agent`	Indicates the kind of `<meta>` tag being provided.
	`content`	Provides the value for a `<meta>` tag.
	`scheme, forua`	Specifies particular details of how the `<meta>` tag should be handled.
`<wml>`		Specifies a WML deck.

Table 8-2: WML Tags Used to Organize a Deck

WML provides a simple form of access control that enables a deck to restrict who can link to its cards. This feature is useful in situations where all content should come from specific sources, such applications that provide access to account or billing information. The `<access>` tag limits access based on the domain and path information of the link. The `domain` attribute of this tag indicates the domain of other decks that can access cards in the deck, while the `path` attribute specifies the URL root of the other decks permitted access within the given domain. In this example, only those decks served from the `wap.apress.com` domain under the root directory can link to the cards in this deck:

```
<access domain="wap.apress.com" path="/"/>
```

The `<meta>` tag specifies additional meta information for a deck. Like the HTML tag, it is used as a catch-all for various functions that aren't represented in tags elsewhere. Many of these behaviors and meanings are browser-specific; others determine the behavior of a gateway between the origin server and wireless terminal. The `<meta>` tag uses one of several property attributes to indicate the kind of meta information being given, and the content attribute to indicate the value of the indicated meta information. Presently, the meta attributes supported in WML are as follows:

- `name`: Specifies a meta value. (Some meta name values are used by specific browsers to perform operations not supported in WML, such as alter caching behavior.)

- `http-equiv`: Specifies a meta value to be treated as equivalent to an HTTP response header value.

- `user-agent`: Specifies a meta value for a specific WAP client. This value indicates that only one vendor's WAP client will understand the meta contents; the meta tag is only passed to that vendor's client when accessing content through a gateway.

The `<meta>` tag includes two attributes that indicate how its contents should be handled by gateway servers and the client. The `scheme` attribute indicates a particular form or structure to be used to interpret the property value, enabling the content of a meta tag to be interpreted by a specified component of a particular browser. The `forua` attribute indicates whether the meta tag's contents should be sent to the browser, or only interpreted by an intermediate gateway and discarded. When `forua` is equal to `true`, the `<meta>` tag is passed to the client; if it is `false`, any intermediary gateways must strip the value before sending the final content to the client.

The dizzying array of meta tag attributes is best clarified with an example. As meta tags are generally used to control a specific browser's behavior in ways not supported by WML, let's consider the problem of restricting how long a page should be cached by a browser. The Phone.com WML browser provides a set of meta tags based on HTTP headers to manage its cache. The following example shows a meta tag to restrict a document's time in the cache to an hour. We explicitly specify an `http-equiv` meta with the value `Cache-Control`, and direct gateways to forward this meta tag to clients by including the `forua=true` attribute:

```
<meta http-equiv="Cache-Control" content="max-age=3600" forua="true"/>
```

Because many meta values are browser-specific, selecting an appropriate set of general-purpose meta tags to perform an operation not supported by WML can be difficult.

WML Cards

Each card contains a bite-sized piece of information suitable for presentation to the user on a single screen. While you think of your WML documents in terms of decks, your users perceive content in terms of screens, each of which is declared as a card in your deck. Table 8-3 summarizes the attributes of the `<card>` tag that you'll use to control card behavior. (You'll find other WML attributes for the `<card>` tag in the WML specifications, but these aren't supported by most of today's browsers.)

ATTRIBUTE	PURPOSE
id	Specifies a card's name for navigation.
ordered	When true, a card's input elements must be navigated in order.
onenterbackward	Specifies a URL to present when the card is entered using a `<prev>` task.
onenterforward	Specifies a URL to present when the card is entered using a `<go>` task.
ontimer	Specifies a URL to present when the card is entered using a `<timer>` task.
title	Specifies a card's title for bookmarks or elsewhere.

Table 8-3: Attributes of the `<card>` *Tag*

Anything you want the user to see in a deck has to be on a card. This includes interface elements, images, and styled text. In addition, a card may contain actions that a user can perform while viewing it. WML provides several attributes of cards that control their behavior.

The `title` and `id` attributes are self-explanatory. With the `title` attribute, a developer can assign a human-readable name to a card, and with the `id` attribute he or she can assign it a navigation reference. The `id` attribute is similar to the anchor name tag `<A NAME="...">` in HTML, which specifies a location within a document.

The `ordered` attribute controls how multiple input elements on a card will be displayed. A WAP browser may be unable to display a set of items on a single card because of screen or memory constraints and may choose to place the items on separate "virtual" cards. If the `ordered` attribute is `true` (the default), the user will only be able to navigate through the elements in the order they are specified. If the `ordered` value is `false`, the browser has the option of creating an index of the items and enabling the user to select items from the index in any order. Thus, a set of several input lines can be presented sequentially or can be navigated in an arbitrary order.

Three attributes—the `onenterforward`, `onenterbackward`, and `ontimer` attributes—are used as shorthand to declare a card that contains an `<onevent>` tag binding an event to a URL. You'll use these when you want to override what a particular card shows depending on how the user accessed the card. Using these tags, you can specify the URL of a card to show instead of the current card.

The browser will load the URL specified by the `onenterforward` attribute, and a card is viewed by the user using a `<go>` task. On the other hand, if the user views the card using a timer event or `<prev>` task, the card's contents are shown.

Likewise, the browser will load the URL specified by an `onenterbackward` attribute when the user navigates to a card with that attribute using a `<prev>` task or the PREV button, but show the card's contents when it is viewed by other means.

Finally, the `ontimer` attribute loads the specified card when a card's timer expires.

Navigation

Navigation between cards in a deck or between decks is based on *links*. Within a deck, any card may have an `id` attribute, which makes it possible for other locations to link to it. Links themselves are referenced by URLs, using the traditional format.

In addition to using URLs, WML has adopted the HTML standard of naming locations within a resource. A WML card is specified by the document URL, followed by a hash symbol (#), followed by the card's `id` attribute value. If no card identifier is specified, a URL names an entire deck. If the URL does not indicate a card, and the context calls for a specific card, the URL refers to the first card of the deck. WML supports relative URLs, too. The base URL of a WML deck is the one that identifies the deck. Relative URLs are evaluated in relation to the base URL.

In addition to supporting URLs, all WAP browsers are required to keep a history of the pages viewed by the user in a stack, sorted in the order they were viewed. This history can be accessed by the `<prev>` tag, which pops the most recently visited card from the history and returns the viewer to it, as if the browser's BACK button had been clicked.

Several WML tags are provided for defining how navigation occurs. Table 8-4 shows these tags, each of which is worth a closer look.

TAG	ATTRIBUTES	PURPOSE
`<a>`	`href`	Shorthand for the tags `<anchor><go href="url"/></anchor>`. Must contain the `href` attribute.
`<anchor>`	`title`	Anchors a task to a region of formatted text. Must contain both a task and character data. The optional `title` may be used by the browser for help, a tool tip, the label of a soft key, or other purposes.
`<go>`	`href, sendreferer, method, accept-charset`	Specifies a task element instructing the browser to go to the URL indicated by the `href` attribute.

Table 8-4   *(continued)*

Table 8-4 (continued)

TAG	ATTRIBUTES	PURPOSE
`<option>`	onpick, name	Identifies a particular selectable item within a `<select>` tag and labels the action key with the indicated name attribute. If the item is selected, the browser navigates to the URL indicated by the onpick attribute.
`<select>`	key, default, ikey, idefault, title	Establishes a list of option tags.

Table 8-4: WML Tags Used in Navigation

Menus

The `<select>` tag presents a menu of items from which the user can select (see Figure 8-4). While this tag has historically been used only in forms, I also recommend you use it whenever you present a list of links, because users find it easy and convenient.

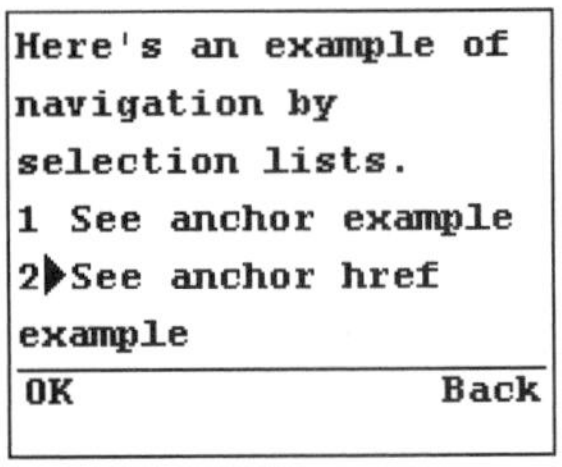

```
Here's an example of
navigation by
selection lists.
1 See anchor example
2▶See anchor href
example
OK              Back
```

Figure 8-4. The `<select>` tag in use

The `<select>` tag contains a series of `<option>` tags, each containing the information pertinent to one of the possible options. An `<option>` tag has a label assigned by the author, which is shown in the main browser screen. In addition, it can have a name attribute, which indicates a label that the browser displays above the ACCEPT soft key when the user has scrolled to that item. If an option is to operate as a navigation item, it should also have an onpick attribute that specifies the URL for the browser to go to when the user picks an item, for example:

```
<card id="selection" title="Selection Example">
  <p>
    <select>
      <option onpick="#anchor">See anchor example</option>
            <option onpick="#href">See anchor href example</option>
    </select>
  </p>
</card>
```

The `<select>` tag is well suited for the traditional "index" card found on most sites, which provides links to a variety of content on separate cards (often in other decks) from a central resource. Use it also to present a list of topics, such as a Web portal's "News/Stocks/Weather" choices.

Hyperlinks and Tasks

Traditional hyperlinks are available within WML, although you may find you use them less frequently. The creators of HTML introduced hyperlinks to provide supplemental information for specific topics within a body of text. However, the mechanics of using a screen phone make reading large bodies of text—and consequently, following trains of hyperlinks—cumbersome at best.

Hyperlinks are supported using the `<go>` tag, which defines a navigation task, and the `<anchor>` tag, which binds this task to a region of text. For example, the WML that follows creates the card shown in Figure 8-5:

```
<card id="anchor" title="Anchor Example">
  <p>Here's an example of navigation by anchors. You can see the
    <anchor><go href="#selection"/>selection list
    example</anchor>, the <anchor><go href="#href"/>anchor href
    example</anchor>, the <anchor><go href="#task"/>event/task
    example</anchor>, or the <anchor><go href="#key"/>key/task
    example</anchor>.
  </p>
</card>
```

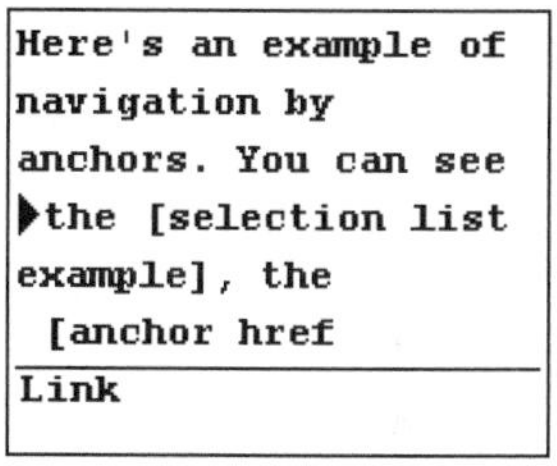

Here's an example of navigation by anchors. You can see ▶the [selection list example], the [anchor href Link

Figure 8-5. Hyperlinks in a WML document

The rendering and selection of hyperlinks varies from device to device. Many devices use a directional cursor key that enables the user to select items by moving a cursor between selections. The `<anchor>` tag may provide an optional title attribute used to identify the item within the browser interface. How this information appears differs from browser to browser: the Phone.com browser uses it to label a soft key, while other browsers may choose to show the information in a supplemental display line or provide other feedback.

The `<go>` tag has attributes that specify the kind of navigation event that should take place. The mandatory href attribute indicates the destination URL. Other attributes, which are generally only used when defining a form navigation task, are discussed in the next section.

You can use the `<anchor>` tag to anchor *any* task to a region of text. Thus, the following WML can create a hyperlink that brings the user to the previously viewed page:

```
<anchor><prev/>Go back from whence you came!</anchor>
```

This example is equivalent to the `<prev>` tag with the following content:

```
<prev>Go back from whence you came!</prev>
```

As in most languages, there's more than one way to say the same thing! WML provides shorthand for the <anchor>/<go> tag pair: the <a> tag, which combines the two, shown here:

```
<a href="http://some.host.com/content.wml">A link</a>
```

It may be used wherever an <anchor>/<go> tag pair would be used, provided the default actions for both the <anchor> and <go> tags is appropriate.

Navigation tasks can be used with other events as well, such as *timers*. A timer event is generated when the delay specified by a <timer> task in a deck has elapsed. For example, this card uses a timer and the card's ontimer attribute to force the browser to navigate to the card named "key" in the current deck after the timer expires. The timer is set to elapse after five seconds (timer values are expressed in tenths of a second):

```
<wml>
 <card id="task" title="Task Example" ontimer="#key">
  <timer value="50" name="time"/>
  <p>In five seconds, this card will take you to the key/task example.</p>
</card>
<card id="key" title="Key Example" ontimer="#task">
  <p>It didn't take that long, did it? <p>
</card>
</wml>
```

WML Navigation Example

Combining the samples in the previous section, I created an example to illustrate the variety of ways navigation can be coded in WML. Look for navigation using selection lists, hyperlinks, hyperlinks using the <a> tag, and two kinds of navigation tasks.

```
<?xml version="1.0"?>
<!DOCTYPE wml PUBLIC "-//WAPFORUM//DTD WML 1.1//EN"
                              "http://www.wapforum.org/DTD/wml_1.1.xml">

<wml>
  <card id="selection" title="Selection Example">
    <do  type="options" label="Back">
      <prev/>
    </do>
    <p>Here's an example of navigation by selection lists.</p>
    <p>
```

```
      <select>
        <option onpick="#anchor">See anchor example</option>
        <option onpick="#href">See anchor href example</option>
        <option onpick="#task">See event/task example</option>
        <option onpick="#key">See key/task example</option>
      </select>
    </p>
</card>

<card id="anchor" title="Anchor Example">
  <do  type="options" label="Back">
    <prev/>
  </do>
  <p>Here's an example of navigation by anchors. You can see the
    <anchor><go href="#selection"/>selection list example</anchor>,
    the <anchor><go href="#anchor"/>anchor href example</anchor>,
    the <anchor><go href="#task"/>event/task example</anchor>, or
    the <anchor><go href="#key"/>key/task example</anchor>.
  </p>
</card>

<card id="href" title="HREF Example">
  <do  type="options" label="Back">
    <prev/>
  </do>
  <p>Here's an example of navigation by anchors.</p>
  <p>
    <a href="#selection">See selection list example</a><br/>
    <a href="#href">See anchor href example</a><br/>
    <a href="#task">See event/task example</a><br/>
    <a href="#key">See key/task example</a><br/>
  </p>
</card>

<card id="task" title="Task Example" ontimer="#key">
  <timer value="50" name="time1"/>
  <do  type="options" label="Back">
    <prev/>
  </do>

  <p>In five seconds, this card will take you to the key/task
  example.</p>
</card>
```

```
<card id="key" title="Key Example" ontimer="#task">
  <timer value="100" name="time2"/>
  <do  type="options" label="Back">
    <prev/>
  </do>
  <do  type="accept" label="Selection">
    <go href="#selection"/>
  </do>
  <do  type="options" label="Anchor">
    <go href="#anchor"/>
  </do>

  <p>In ten seconds, this card will take you to the event/task
  example.</p>
  <p>You can use one of the soft keys to go elsewhere.</p>
  </card>
</wml>
```

This example also shows the use of the <do> tag to bind a task to an interface element. As used here, the tag is similar to the <anchor> tag: the task to be bound is wrapped within the <do> tag, which specifies the details of the interface element to be bound. While the <anchor> tag binds tasks to text, the <do> tag binds any interface element to a task.

Interaction

WML is designed to be an interactive markup language. It has several constructs that enables interactivity for simple operations such as substituting variables or modifying dynamic interface elements to be executed entirely on the client side, without time-consuming wireless queries.

One way WML accomplishes this is through *tasks*—operations the browser performs in response to *events*. The WML standard defines several kinds of events, such as the user viewing a new card or a time interval elapsing. In addition, there are tasks, such as navigation, that are not restricted to a single card. You can connect these in different ways in your WML by specifying an event and a task to be performed when the event occurs using the <onevent> tag.

In addition to events and tasks, WML makes selection objects and text fields available to enable decks to accept user input. Your deck can use this input locally to create dynamic scripts or to interact with remote services, such as stock quote services, weather reports, and so on.

Events

WML presently defines four events that occur when the browser performs a specific function:

1. The browser generates an onenterforward event when a user navigates to a card using an element that is tied to a <go> tag.

2. The browser generates an onenterbackward event when a user navigates to a card using a <prev/> tag. (This also occurs when the user presses the PREV key.)

3. The browser generates an onpick event if a user selects or deselects an item generated by an <option> tag.

4. The browser generates an ontimer event when a specified <timer> element expires.

Developers often use the onenterforward and onenterbackward events to manipulate the chain of cards in a form to prevent the misuse of history and to guarantee valid form data during submissions. For example, many decks use the following snippet:

```
<card …>
  <onevent  type="onenterbackward">
    <prev/>
  </onevent>
  <!--card content here-->
</card>
```

This bit of WML prevents the given page from being visited as a result of a PREV key press or <prev/> operation. If one is requested by the user, the browser will instead navigate to the card preceding this one. A variation, which forces the viewer to the beginning card in a deck if the PREV key pressed, is:

```
<card id="home" …>
  <!–home card content here-->
</card>
<card>
  <onevent  type="onenterbackward">
    <go href="#home">
  </onevent>
  <!--card content here-->
</card>
```

Less often, a content provider may use a similar trick to prevent *forward* navigation to a card, using the onenterforward event.

Both the `onenterforward` and `onenterbackward` events can also be used, as shown in the following example, to refresh a page's variables, which is important when their values may have changed:

```
<card>
  <onevent  type="onenterbackward">
    <refresh><setvar name="ticker" value=""/></refresh>
  </onevent>
  <onevent  type="onenterforward">
    <refresh><setvar name="ticker" value=""/></refresh>
  </onevent>
  <!--card content here-->
</card>
```

The browser generates the `onpick` event by the `<select>` tag in response to the selection of an `<option>` item. This tag is used in conjunction with variables to prepare form data or provide an interactive feel, such as in the following deck:

```
<?xml version="1.0"?>
<!DOCTYPE wml PUBLIC "-//WAPFORUM//DTD WML 1.1//EN"
                            "http://www.wapforum.org/DTD/wml_1.1.xml">
<wml>
  <card  id="c1" title="Volume">
  <onevent type="onpick">
    <go href="#results"/>
  </onevent>
    <p>
      Pick the ringer volume:
      <select  title="Category" name="volume">
        <option value="4"Excruciating</option>
        <option value="3">Loud</option>
        <option value="2">Soft</option>
        <option value="1">Silent</option>
      </select>
    </p>
  </card>

  <card id="results">
    <onevent  type="onenterbackward">
      <prev/>
    </onevent>
    <do  type="options" label="Clear">
      <refresh>
        <setvar  name="volume" value="1"/>
```

```
        </refresh>
      </do>
      <p>
        You selected ringer volume level $(volume).
        Your phone will be set appropriately.
      </p>
    </card>
</wml>
```

This deck produces the output shown in Figure 8-6. It demonstrates the use of the <select> tag and the <option> tag to select a ringer volume (note that we *don't* actually set a ringer volume, because there's no way to do that in WML without device control in WMLScript!). For each <option> tag, an onpick attribute will cause the browser to navigate to the card named results.

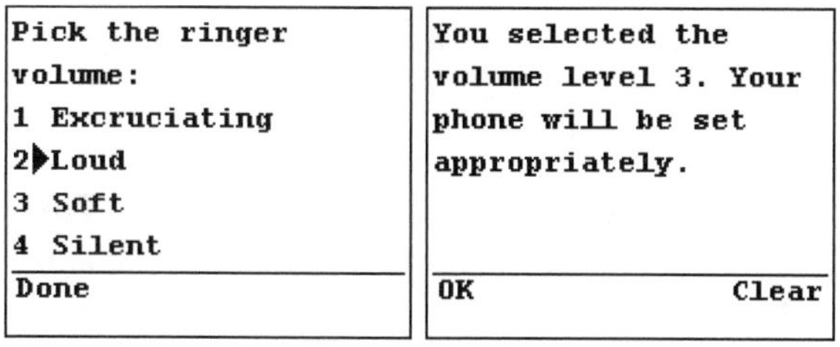

```
Pick the ringer      You selected the
volume:              volume level 3. Your
1 Excruciating       phone will be set
2▶Loud               appropriately.
3 Soft
4 Silent
Done                 OK           Clear
```

Figure 8-6. WML <select> *tag and variables*

We encountered the ontimer event in the last section. This timer event is the only event generated by an explicit tag within a WML deck. The <timer> tag declares a timer that starts when the page is viewed and counts down the specified number of tenths of a second. When the timer expires, the task associated with it is executed. The following example simply flips between two cards every two seconds using two timers:

```
<?xml version="1.0"?>
<!DOCTYPE wml PUBLIC "-//WAPFORUM//DTD WML 1.1//EN"
                            "http://www.wapforum.org/DTD/wml_1.1.xml">

<wml>
  <card  id="c1" title="Card 1" ontimer="#c2">
    <timer value="20" name="c1"/>
    <p>
      Card 1.
    </p>
```

```
    </card>

    <card id="c2" title="Card 2" ontimer="#c1">
      <timer value="20" name="c2"/>
      <p>
        Card 2.
      </p>
    </card>
  </wml>
```

This example also demonstrates the use of <card> attributes to associate navigation tasks with events. In addition to being the names of events, the words onenterforward, onenterbackward, and ontimer describe attributes of a card, as discussed earlier in this chapter. Rather than using the <onevent>/<go> tag pair to declare a task and specify a destination URL in response to an event, you can bind a task to one of these attributes.

In addition, tasks can be assigned to specific user interface elements using the <do> tag. The browser provides a set of generic interface elements, any one of which can be bound to a task using the <do> tag. When the user activates an element bound with the <do> tag, the browser executes the corresponding task. The WAP standard explicitly restricts the *names* of the user interface objects that can be customized so that developers have a set list of items that they can use. However, the WAP standard leaves open the *implementation* of those interface elements, so that hardware manufacturers can make these elements areas of a touch screen, soft keys, voice commands, or other items.

At present, the following interface elements have been defined by the WAP Forum:

- The accept element specifies the mechanism (soft key, hardware key, voice command, or so on) by which the user accepts screen data and sends it to the browser.

- The options element specifies the mechanism for providing the user with a set of options from which to make a choice.

- The delete element specifies the mechanism by which the user deletes an input.

- The help element specifies the mechanism by which the user requests help about the current deck or the phone.

- The prev element navigates to a previous card by invoking the browser's PREV mechanism.

- The reset element invokes the browser's reset mechanism, bringing the device to a "fresh from factory" state (like resetting a computer).

- The unknown element indicates an unknown interface element that has not been specified. It is equivalent to specifying a <do> task with no type attribute, or a type attribute consisting of an empty string. (This is a vendor-specific element.)

- Types of the form vnd.co-type indicate manufacturer-specific mechanisms, where co indicates the manufacturer and type indicates the type of interface element.

- Types of the form X-* and x-*, where * is any string, are reserved for future use.

(Note that unless you're developing for a specific platform with documentation indicating it's supported, you should *not* use the unknown, vnd.co-type, or X-* elements.) It's important to remember when creating content that these interface items can be *anything*—hardware keys, soft keys with a corresponding LCD labels, handwriting gestures, voice commands, or operations yet to be devised. The other attributes of the <do> tag have equally generic connotations. A <do> name attribute could represent an LCD label, a status line, a voice prompt, or virtually anything else that a device vendor could imagine.

In practice, however, most <do> types map tasks to the accept and options interface elements. We've already seen several examples of this. Consider this example:

```
...
<do  type="options" label="Clear">
  <refresh>
    <setvar  name="volume" value="1"/>
  </refresh>
</do>
...
```

This WML simply binds the operation of resetting a form's variable to a known default when the user invokes the options interface element. This element is labeled "Clear", giving the user some indication of its purpose. Within the Phone.com browser, for example, this element is the right-side soft key.

So far, we've looked only at examples of binding tasks to events over the scope of a single card. You can bind an event to a task across an entire deck by using the <template> tag. This tag resembles a simplified <card> tag with attributes, as shown here:

```
<template onenterforward="url" onenterbackward="url"
  ontimer="url">
  <!-- Arbitrary <do> or <onevent> tags -->
</template>
```

For example, a deck providing a form for a server could use the following template to override the browser's history behavior and provide a single-key mechanism for restarting a form entry from scratch:

```
...
<template onenterbackward="#home">
  <do  type="options" label="Change">
    <go href="#home"/>
  </do>
</template>
...
```

If the user pressed the PREV key while viewing any card in the deck with this `<template>`, the card identified by home would be loaded. In addition, if the user pressed the OPTIONS key, the browser would display the home card, just as if PREV were pressed, enabling the user to make changes to the data they'd input.

Tasks

In the process of exploring events and navigation, we've already encountered most of the tasks WML provides. As we've seen, the `<go>` tag specifies a navigation task:

```
<go href="url" sendreferer="boolean" method="method"
 accept-charset="charset">
  <!--Optional post field declarations here -->
</go>
```

The `<go>` tag defines both conventional navigation and form submission navigation. In the latter case, the `<go>` tag specifes the fields to be posted to the server handling the form. If desired, you can have the submitting deck's URL returned to the server using the sendreferer attribute; if you assign it a value of true, a back-end server will determine the source of the deck originating the request. Finally, for international applications, you can specify a list of supported character sets with the accept-charset attribute, indicating to the WAP gateway that content should be converted into one of the character sets named in the browser-supplied list.

The `<prev>` tag is a simplified form of navigation based on the browser's history stack. It is most often encountered as the empty tag `<prev/>`, but can also be used as a pair. Here's a bit of WML that clears a variable (presumably for a form input) with `<setvar>` before returning to a previous card:

```
...
  <prev><setvar name="last-name" value=""/></prev>
...
```

If more than one variable is to be set, use a separate <setvar> tag for each one.

Often, it becomes necessary to reset variables to a new value and redraw a page because of some user action—for example, to clear a form's input lines when the user wants to start again. The <refresh> tag implements this functionality. The volume control deck presented previously used <refresh> and <setvar> to set the volume variable to a known value:

```
...
  <do  type="options" label="Clear">
    <refresh>
      <setvar  name="volume" value="1"/>
    </refresh>
  </do>
...
```

As with the <prev> tag, more than one variable can be set within a <refresh> tag using multiple <setvar> tags.

Finally, WML provides an interesting task tag called <noop/>, which does exactly what its name implies: no operation, that is, nothing. This tag is far from useless; it can be used anywhere a task is required by syntax, but no task should occur. This is often the case when an interface element is assigned to a task across a whole deck using a template, but you want to override the task assignment on a particular card. You might want to do this, for example, on a site where you enable the user to return to the beginning of a deck to enter information, except on the page where the results are displayed (presumably because once the results are displayed, the user will want to progress to another deck). In the example that follows, the default task assigned to the options interface element is to go to the card named "home", and this element is labeled "Clear". For this one card, the task that occurs when it is invoked has been replaced with no action, and the label is suppressed.

```
<wml>
  <template onenterbackward="#home">
    <do  type="options" label="Change">
                 <go href="#home"/>
    </do>
  </template>
...
<card ...>
      <do type="options" label="">
        <noop/>
      </do>
    ...
</card>
```

Forms

While the ability of a WML deck to operate interactively with the user of the device is important, it is only half the story of WML's dynamic capabilities. WML decks also support traditional form-based operations, which enable users to submit all kinds of input data via the wireless network to Web servers and receive responses. This technology can be used for a multitude of purposes, as it is with traditional Web applications.

WML provides two kinds of input primitives: option lists and text fields. Both of these primitives use variables to store the resulting data. Once a form has been completed, a <go> task submits the variable name/value pairs to the server for processing.

Table 8-5 provides a reference to the form-related tags I detail in this section. You'll use these tags to define form elements, as well as to bind interface elements to tasks.

TAG	ATTRIBUTES	PURPOSE
<do>	type	Binds a task to an input element.
<go>	href, method, sendreferer, accept-charset	Defines a navigation task to the URL specified by href via the indicated HTTP method. Can be used to specify character transcoding for forms submissions via the accept-charset attribute.
<input>	varied	Defines an input field given attributes defining its name, input mask, length, size, and default value.
<noop/>		A task tag that does nothing.
<option>	title, value, onpick	Creates an option with the indicated title and value. The onpick attribute specifies a URL where navigation proceeds if the item is selected.
<onevent>	type	Binds an event of the specified type to a task.
<prev/>		A task tag that pops the current card from the browser's history and returns to the previously viewed card.
<postfield>	name, value	Declares the indicated value attribute to be assigned to the server-side variable.
<select>	name	Establishes a list of option tags storing the user selection in the variable specified by name.

Table 8-5   *(continued)*

Table 8-5 (continued)

TAG	ATTRIBUTES	PURPOSE
`<refresh>`		A task tag used to set the values of the indicated variables and refresh the current card.
`<template>`	`onenterforward,` `onenterbackward,` `ontimer`	Defines the default event/task bindings for a deck. Contains zero or more `<do>` or `<onevent>` tags.
	`type, label, name,` `optional`	Binds a task to an interface item identified by the type attribute given in the `name` attribute.

Table 8-5: WML Tags Used in Forms

The `<select>` and `<option>` tags allow content developers to present users with a list of items and to set a variable to a value corresponding to the option selected. For example, the following WML fragment will result in setting the variable `volume` to a value from 1 to 4 depending on which item the user chooses:

```
...
  <select  title="Category" name="volume">
    <option value="4">Excruciating</option>
    <option value="3"">Loud</option>
    <option value="2">Soft</option>
    <option value="1">Silent</option>
  </select>
...
```

Input to text fields is handled in a similar way: user input is accepted and used to assign a value to a specified variable. WML provides only a single way to request text input from the user, the `<input>` tag. This simplicity is deceiving, because its attributes provide a lot of freedom to content developers.

At a minimum, the `<input>` tag must specify the name of the variable to be set, using the `name` attribute. For example, the following line in WML results in assigning user input from that card to the variable `ticker` for access by other parts of the WML deck:

```
<input name="ticker">
```

Most requests for user input are more sophisticated, however. For one thing, good interface design for handheld clients involves restricting the user's freedom to make mistakes. The `<input>` tag provides a variety of ways to limit possible input, making errors on the client side less likely.

The most powerful of these error-avoidance techniques is the `format` attribute, which enables you to specify a string called a *type mask* that indicates, character for character, what kinds of input are valid. Each character of the type mask indicates the appropriate type for one or more characters of the input. A type mask contains integers, which indicate the number of characters that should match a specific type (an asterisk indicates that any number of characters of this type is allowed) and type declarations. Possible type declarations are shown in Table 8-6.

CHARACTER	ALLOWED INPUT TYPE
A	Any symbolic or uppercase alphabetic character.
A	Any symbolic or lowercase alphabetic character.
M	Any symbolic, numeric, or uppercase alphabetic character (changeable to lowercase)—for multiple-character input, defaults to uppercase first character.
m	Any symbolic, numeric, or lowercase alphabetic character (changeable to uppercase)—for multiple-character input, defaults to lowercase first character.
N	Any numeric character.
X	Any symbolic, numeric, or uppercase alphabetic character (not changeable to lowercase).
x	Any symbolic, numeric, or lowercase alphabetic character (not changeable to uppercase).

Table 8-6: Type Declarations for a WML Input Type Mask

Thus, the type mask `10N` indicates a ten-digit number (such as a North American phone number) with no hyphens or spaces, while the type mask `2A` restricts input to two alphanumeric characters, such as a United States state code. WML browsers may use the format attribute both to verify the data input by the user and to tune input methods to meet the indicated type. (For example, a speech recognizer may be primed to only recognize numbers when a numeric format field is encountered.)

In addition to masking the kinds of data that can be entered, you can use the `<input>` tag to specify the default and maximum size for a field. The `size` attribute gives an exact number of characters for an input field, while the `maxlength` attribute gives a maximum number. Note that these two attributes complement the format attribute. The `format` attribute defines the *actual* size and format of valid input, while the `size` and `maxlength` attributes restrict the behavior of the input element.

The value and default attributes both allow you to set a field's initial value—the two tags are functionally equivalent.

Some examples of valid input lines, which can be used as building blocks for more complicated forms, are listed here:

- A stock ticker can be accepted using `<input name="ticker" format="A4A">`.

- A domestic phone number can be accepted using `<input name="phone" format="10N">`.

- A person's name can be entered using `<input name="name" format="*M">`.

Once a form is rendered and the data collected, the browser returns the resulting variable values to the server for processing via a `<go>` tag and one or more `<postfield>` tags. The `<go>` tag, as we have seen elsewhere, specifies the destination URL where the form is to be submitted.

The `<go>` tag must also specify the method to be used to deliver form data to the server. As discussed in Chapter 4, a Web server may use one of two kinds of HTTP methods for accepting data: GET or POST. While the mechanics of these methods aren't relevant to many content authors, the net effect is. Submitting data using the GET method to a server expecting a POST method, or vice versa, can result in unexpected behavior. A server expecting the GET method expects its form data within the URL being submitted by the client, while a server using the POST method will look for form data within the object-body of the data sent to the server as part of the form request.

WML supports both kinds of form requests. You can specify which kind is to be used by setting the method attribute of the `<go>` tag to the value GET or POST, appropriately.

Within the `<go>` tag, one or more `<postfield/>` tags define the actual content to be sent in the form request. The `<postfield/>` tag is an empty tag bearing name and value attributes. The name attribute indicates the name of the server-side variable to be posted, and the value attribute gives the value of that variable. It's important to recognize that the variables being named here are *not* WML variables, but the variables defined by the *server* for processing form data. You'll use the `<postfield/>` tag to establish the correspondence between a Web server's form variables and your WML variables.

For example, a server seeking a zip code to serve location-based information might offer the following `<go>` task within a deck:

```
...
<do type="accept" label="Weather">
  <go method="post" href="?">
    <postfield name="zip" value="$zip"/>
```

```
      </go>
    </do>
    ...
```

Here, the remote server is expecting a single form variable called zip. The
tag tells the browser that the server's zip variable should be set to the
value of the WML browser's zip variable (which is evaluated using the expression
$zip) when the <go> task is executed.

An admittedly artificial example that uses WML variables, form tags, and the <go>
and tags to obtain a bank balance is shown in the following example:

```
<?xml version="1.0"?>
<!DOCTYPE wml PUBLIC "-//WAPFORUM//DTD WML 1.1//EN"
                              "http://www.wapforum.org/DTD/wml_1.1.xml">


<wml>
  <template onenterbackward="#home">
    <do  type="options" label="Change">
      <go href="#home"/>
    </do>
  </template>

  <card id="home" title="Account Type">
    <p>
      Please indicate the type of account:
      <select  title="Account" name="account">
        <option value="Savings" onpick="#number">Savings</option>
        <option value="Checking" onpick="#number">Checking</option>
      </select>
    </p>
  </card>

  <card id="number" title="Account Info" ordered="true">
    <p>
      Account Number:
      <input name="number" title="Number" type="text" format="4NA4N" emp-
tyok="false"/><br/>
      Last Four Digits of SSN:
      <input name="social" title="SSN" type="password" format="4N"
      emptyok="false"/><br/>
    </p>
    <p>
      Obtaining account balance... use Fetch to continue.
      <do type="accept" label="Fetch">
```

```
      <go href="results?" method="post">
      <postfield name="acct-type" value="$account"/>
      <postfield name="acct-num" value="$number"/>
      <postfield name="acct-ssn" value="$social"/>
    </go>
   </do>
  </p>
 </card>
</wml>
```

This example is somewhat contrived; its user interface (see Figure 8-7) fails almost anyone's test for usability and intuitiveness. But while ugly, it does give you a good idea of how to use the various user input tags together with a form to post several form variables to the server.

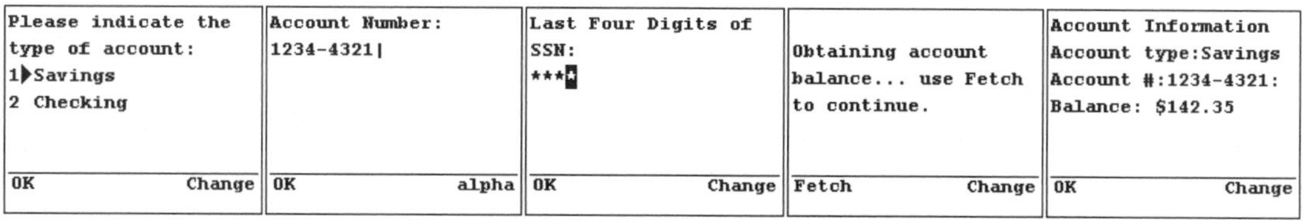

Please indicate the type of account: 1▶Savings 2 Checking	Account Number: 1234-4321|	Last Four Digits of SSN: ***■	Obtaining account balance... use Fetch to continue.	Account Information Account type:Savings Account #:1234-4321: Balance: $142.35
OK Change	OK alpha	OK Change	Fetch Change	OK Change

Figure 8-7. An example of a WML application with a poor user interface

A more complex but more user-friendly approach—assuming the device has a unique identifier—would be to use this identifier (or some other simple authenticating mechanism) with the back-end server to determine which accounts can be queried wirelessly. Then, rather than having to deal with the mechanics of entering an account number and password, the user would have to make only a single selection to indicate the desired account.

The back-end to support such a system is beyond the scope of this section (and largely dependent on both the wireless provider and bank being supported!). But the improved user interface is shown in Figure 8-8.

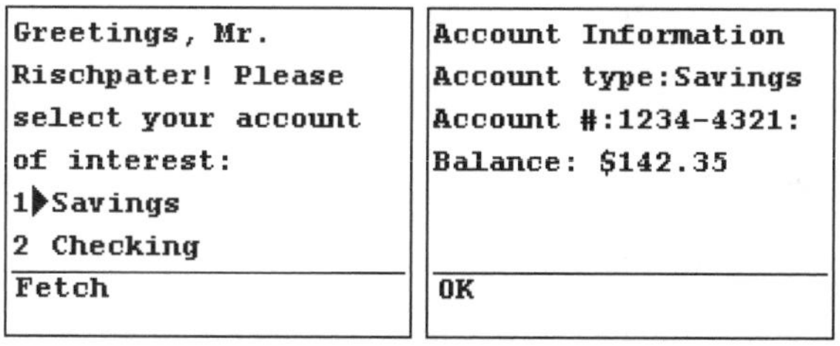

Greetings, Mr. Rischpater! Please select your account of interest: 1▶Savings 2 Checking	Account Information Account type:Savings Account #:1234-4321: Balance: $142.35
Fetch	OK

Figure 8-8. An example of a WML application with a better user interface

The WML deck returned by such a server might look like this:

```
<?xml version="1.0"?>
<!DOCTYPE wml PUBLIC "-//WAPFORUM//DTD WML 1.1//EN"
                              "http://www.wapforum.org/DTD/wml_1.1.xml">
<wml>
  <card id="home" title="Account Type">
    <p>
      Greetings,
      Mr. Doe!
      Please select your account of interest:
      <select  title="Account" name="account">
        <option value="1234-4321">Savings</option>
        <option value="4321-4321">Checking</option>
      </select>
    </p>

    <do type="accept" label="Fetch">
      <go href="results?" method="get">
        <postfield name="acct" value="$type"/>
      </go>
    </do>
  </card>
</wml>
```

Format

Despite the meager displays available on most handheld devices, WML provides a limited set of text formatting tags. These tags are largely contextual, and the WAP specifications do not specify how browsers should interpret them. Thus, their behavior can differ from device to device, and some browsers may choose to ignore some tags altogether.

As you can see in Table 8-7, most of the tags you'll use for formatting strongly resemble the HTML tags discussed in Chapter 5.

TAG	ATTRIBUTES	PURPOSE
<b>		Indicates that text should be bold face.
<big>		Indicates that text should be in a larger font than the default.

Table 8-7   *(continued)*

Table 8-7 (continued)

TAG	ATTRIBUTES	PURPOSE
` `		Indicates a line break.
`<em>`		Indicates that text should be displayed with emphasis.
`<i>`		Indicates that text should be in italics.
`<img>`	`alt, src, localsrc, height, width, align`	Indicates an image; attributes specify its text description, source (from server), source (local), height and width (in pixels), and alignment on screen (left, right, or center).
`<p>`	`align, wrap`	Defines a paragraph of text with the indicated alignment (left, right, or center) and word wrapping (wrap or nowrap).
`<small>`		Indicates that text should be in a smaller font than the default.
`<strong>`		Indicates that text should be displayed with strong emphasis.
`<table>`	`columns, align, title`	Indicates a table; attributes specify number of columns (mandatory), alignment on screen (left, right, center), and title.
`<td>`		Demarcates a cell within a table row.
`<tr>`		Demarcates a row within a table.
`<u>`		Indicates that text should be underlined.

Table 8-7: WML Tags Used for Formatting

Typography

The WML standard provides tags that alter the size of text. The `<big>` and `<small>` tags create regions of text in a larger or smaller font than the default. You can use these to create simple headings, like this:

```
...
<p><big>WAZU Fundamentals</big></p>
<p>
Last trade: 64 3/8<br/>
Today's Range 63 7/8-67 3/16<br/>
Volume: 516,200<br/>
</p>
...
```

The <strong> and <em> contextual tags provide a means to emphasize important information. Most browsers try to render these the same way HTML does: <strong> indicating boldface text, while <em> indicates italicized text. Unfortunately, many WML client devices do not have fonts with italicized text, and many don't have the ability to draw text in bold, either. (Figure 8-9 shows a client that uses bold face text for both <strong> and <em> because it cannot support an italic font.) The syntax of <strong> and <em> is the same as for the <big> and <small> tags:

```
<wml>
  <card>
    <p><big>WAZU Fundamentals</big></p>
    <p>
      Last trade: <strong>64 3/8</strong><br/>
      Today's Range <strong>63-67</strong><br/>
      <em>Volume: 516,200</em><br/>
    </p>
  </card>
</wml>
```

```
WAZU Fundamentals
Last trade: 64 3/8
Today's Range 63-67
Volume: 516,200


OK
```

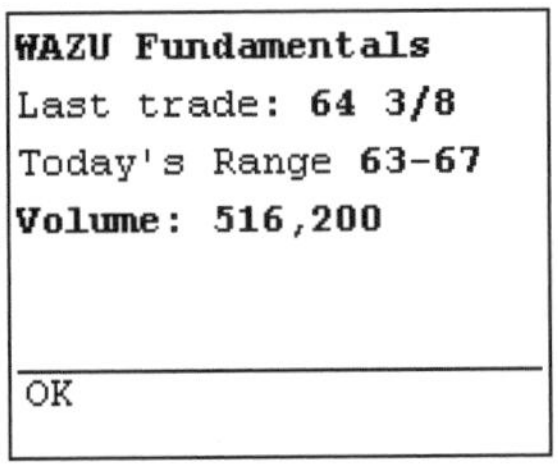

Figure 8-9. WML format tags in action

WML also has tags to produce particular effects in text, such as bold face, italics, and underlining:

- The <b> tag is used to mark text that should be **bold face**.

- The <i> tag is used to mark text that should be rendered as *italic*.

- The <u> tag is used to mark text that should be underlined.

Not all browsers support these tags, so it's important to be sure that the information called out will still appear in a clear and readable manner on devices on which these tags are not available.

Text Regions and Text Alignment

All text on a card is contained within one or more paragraphs, which are defined by the paragraph tag `<p>`. This tag has two optional attributes, `mode` and `align`. Use the `nowrap` attribute to specify whether text should be word wrapped or not. By default, paragraphs are word wrapped; if the `nowrap` attribute is equal to `nowrap`, lines will instead be truncated to fit the display. With the `align` attribute, you can specify a paragraph's alignment; possible values are `left`, `right`, and `center`.

The empty tag `<br/>` is used to break lines within a paragraph, and can be used to format simple tables, verse, or other information where the specifics of line breaks are important.

Tables

In many cases, a table can represent information more succinctly than a paragraph can. WML has a `<table>` tag, but its use, unlike the corresponding tag in HTML, is relatively restricted. A WML table must indicate the number of columns it will present using its `columns` attribute. All elements must occupy only a single cell—they cannot span multiple rows or columns. You cannot nest tables in WML.

Some formatting options exist, however. A table may be aligned to the left or right, or centered on its card using the `align` attribute, just as a paragraph can. It also may have a specific title identified by the `title` attribute, which will be presented with the table.

Within a table, rows are enclosed between opening and closing `<tr>` tags. A table may have any number of rows (within reason); this number is not specified ahead of time. A row may be empty, or it may contain table elements, which are demarcated by the `<td>` tag—one pair for each cell. There should be as many `<td>` tags per row of the table as there are columns for the table; an empty `<td>` tag indicates an empty cell in the table.

Following is an example of a WML table that generates the card shown in Figure 8-10:

```
<?xml version="1.0"?>
<!DOCTYPE wml PUBLIC "-//WAPFORUM//DTD WML 1.1//EN" "http://www.wapforum.org/DTD/
wml_1.1.xml">

<wml>
  <card title="Table">
    <p align="center">
      Weather Forecast for<br/>
      Boulder Creek, CA
      <table columns="2">
        <tr>
```

```
          <td>High</td>
          <td>68</td>
        </tr>
        <tr>
          <td>Low</td>
          <td>54</td>
        </tr>
      </table>
    </p>
  </card>
</wml>
```

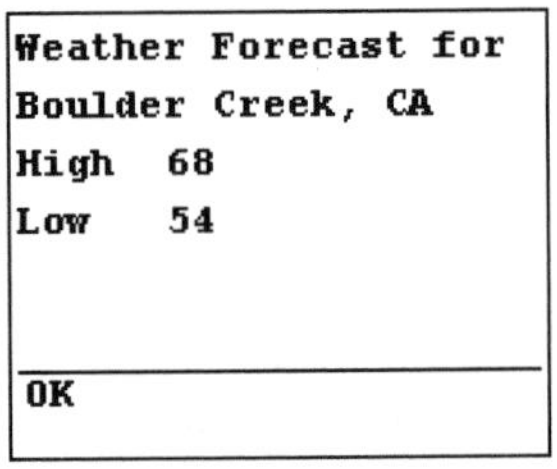

```
Weather Forecast for
Boulder Creek, CA
High   68
Low    54
_____
OK
```

Figure 8-10. A simple table in WML

Images

Although most WAP terminals are ill suited to display images, the WAP standard provides support for the presentation of image data. The <img> tag instructs the device to draw the specified image. At present, the WML specification calls for supporting the Wireless Bitmap Standard, which accommodates other graphics formats including the Graphical Interchange Format (GIF) and the standard bit-mapped graphics format (BMP). Due to the widespread adoption of Microsoft Windows by content developers, most of the WML simulators currently accept one-bit-per-pixel Windows BMP files, rather than the GIF or Joint Photographers Expert Group (JPEG) file formats most Web content developers use. The kinds of images supported by any specific device depend on the WAP gateway provided by the wireless service provider.

The alt attribute of the <img> tag specifies a text label for the image. This string should be a brief description of the image, such as "sun", "smile", "up arrow", or similar legend, which will be displayed instead of the image on any device that cannot show the image itself. In addition, all image tags must include a src attribute that specifies the source of the image the browser should display, for example:

```
<img alt="smile" src="smile.bmp"/>
```

You have the option of supplying the height and width, in pixels, that the image should occupy, although browsers are not required to honor these attributes. You can specify the vertical alignment using the `align` attribute, indicating one of `top`, `middle`, or `bottom`.

In addition to supporting Wireless Bitmap images, WML provides the ability to reference images stored on the device. These images may be in Read-Only Memory (ROM), on flash memory that can be reprogrammed by the user, or elsewhere where access isn't an issue for the browser. They can be accessed using the image tag's source attribute, `localsrc`. One image can have both a source and a local source attribute. The browser will look for the image resource named in the `localsrc` attribute first; if it doesn't find it, it will downloaded the image from the URL indicated in the `src` attribute. For example, an application targeted for Phone.com's WML browser could use the following image tag to select the cloud image within the browser:

```
<img alt=cloud" src="cloud.bmp" localsrc="cloud"/>
```

Browsers without this local resource would download and present the `cloud.bmp` image available from the server that provided the deck containing this tag.

The previous section's weather example was missing an important characteristic of weather reports—the current conditions. The following deck adds an icon to our weather report:

```
<?xml version="1.0"?>
<!DOCTYPE wml PUBLIC "-//WAPFORUM//DTD WML 1.1//EN" "http://www.wapforum.org/DTD/
wml_1.1.xml">

<wml>
  <card  title="Table">
    <p mode="nowrap" align="center">
      Weather Forecast for<br/>
      Boulder Creek, CA
    </p>
    <p align="center">
      <img alt="Sun" src="sun.bmp"/>
    </p>
    <p>
      <table columns="2">
        <tr>
          <td>High</td>
          <td>68</td>
        </tr>
        <tr>
          <td>Low</td>
```

```
            <td>54</td>
          </tr>
        </table>
      </p>
    </card>
</wml>
```

Figure 8-11 shows how this deck would appear on the screen of a browser with image support.

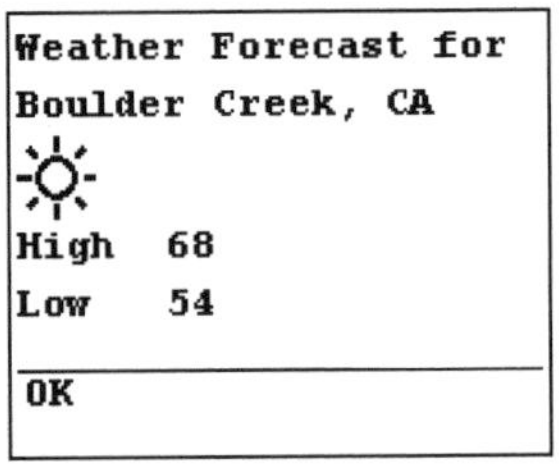

Figure 8-11. WML card including a simple image

Summary

WAP is fast becoming the international standard for wireless content distribution to smart phones. Leveraging the best of Web technologies and carefully optimized for today's wireless networks, the WAP protocols have been adopted by major manufacturers including Nokia, Ericsson, and Motorola.

The WML is an XML-based markup language for content creation for WAP applications. Web servers make WML content available via WAP gateways, enabling existing content developers to provide WML content to wireless devices. WML bears a strong resemblance to HTML, but includes facilities for dynamic content through the introduction of events, tasks, and variables.

WML treats content as decks composed of cards. Within cards, tags are provided to format and style text, as well as to accept user input. Users may interact with cards input elements or activate interface elements that generate events. A deck can use an event to trigger a task, such as loading another deck or displaying another card.

Users can make selections from lists and enter free-form text for submission to origin servers. These are performed using variables, which permit WML more flexibility than HTML can provide. Variables can be used anywhere text can, including within a card to be displayed to the user.

Another feature not found in HTML is the ability to map interface elements to tasks, such as using a key to display a specific card or setting a timer to display a new card after some period of time has elapsed. WML defines events for common browser actions and allows the developer to link these events to tasks such as navigation or redrawing the screen.

Of course, WML provides traditional markup tags as well, so that content can be presented as paragraphs or tables. While not all WML browsers support all markup tags, there are tags to present text as bold or italic, as well as to draw simple monochrome images.

Dynamic Content with WMLScript

IN THE LAST CHAPTER, I introduced the WAP standard, and specifically, the Wireless Markup Language (WML) for preparing screen phone content.

WML is only half of the story, however. One of the most exciting aspects of WML development is WMLScript, a JavaScript-like scripting environment for calculations, data validation, and interactive development. In this chapter, I talk about why WMLScript was created and show you how you can use it to develop dynamic content for wireless handsets.

Purposes of WMLScript

WMLScript was conceived by the Wireless Application Protocol (WAP) Forum in parallel with WML. Together, these two technologies are intended to provide wireless access terminals with the ability to serve dynamic content to subscribers with a minimum of wireless interaction.

While typical Web applications rely heavily on the use of back-end server applications (or memory-hungry programming platforms such as Java and JavaScript), such an approach simply isn't appropriate for wireless clients. A dependence on back-end sever solutions increases a subscriber's operating costs, while traditional scripting languages increase hardware costs because they need additional memory and horsepower from the central processing unit (CPU) to operate. Despite its lightweight nature, WMLScript is powerful enough to perform such tasks as data validation, native access to platform-specific peripherals, and dynamic content generation.

Data Validation

Form posting to back-end servers is a significant source of traffic on the Web. Unfortunately, validating form submissions in the HyperText Markup Language (HTML) is difficult without JavaScript support, making form validation with wireless HTML Web browsers a hit-or-miss affair. With WMLScript, decks can perform robust input validation prior to submitting queries to origin servers. Arithmetic and string libraries available to scripts can be used to examine inputs, make corrections, and either submit a corrected request or prompt the user for corrected input.

Native Peripheral Access

The modular nature of WMLScript enables platform vendors to make additional libraries available on individual devices. These libraries can contain device-specific code that links WML interfaces to device functions such as a calendar or address book, or interfaces to hardware elements such as a location service, infrared port, or a wireless Bluetooth module. These libraries are limited only by cost constraints and the imagination of the device manufacturers.

Dynamic Content Generation

WMLScript can access variables in the WML browser, enabling scripts to create content on the fly. This content can be simple, such as a predefined rearrangement of strings, or it can be a set of complex strings or arithmetic operations. Because these variables can be used in WMLScript with complex calculations, comparisons, and loops, costly network interactions can be avoided during the presentation of dynamic content.

Features of WMLScript

Making the decision to use WMLScript is easy. If you're using WML and want dynamic content, you should use it. Every WML browser provides a WMLScript environment, and at least for the present, no other scripting solution is available for WML browsers. There are more features to WMLScript that you should be aware of, however, to help you understand what is and isn't possible in your content.

Similarity to ECMAScript

WMLScript, in many ways, was designed to resemble ECMAScript (developed by the European Computer Manufacturers Association), which was the formalization of JavaScript. While it is not compatible with ECMAScript or JavaScript, the syntax of WMLScript is similar enough that most content developers can author WMLScript after only a little study. Moreover, automated tools can be easily retrofitted to enable developers to author and validate WMLScript quickly, bringing development tools to market quickly.

Procedural Logic

WML provides simple dynamic behavior using events, tasks, and variable substitution, but its actions are limited by its lack of procedural logic. WMLScript provides comparison and looping operators, giving content developers the ability to implement scripts that make decisions based on variables and act accordingly.

Compact Binary Representation

WMLScript is compiled by gateway servers into a compact binary representation, which is both appropriate for wireless networks and efficient for wireless terminals to interpret. In the process of sending a WMLScript to a wireless terminal, the gateway both validates and encodes the script. The process of validation makes interpreting and executing the script on the wireless terminal simpler, because common errors can be detected before they are sent to the wireless client.

A Simple WMLScript Application

While many smart phones or screen phones shipping today have integrated date books, calendars, and calculators, they are missing one feature: a Reverse Polish Notation (RPN) calculator. The following WMLScript example (see Figure 9-1) rectifies this by showing how to create a simple two-register RPN (postfix) calculator. (See the sidebar, "What Is Reverse Polish Notation?," for a brief explanation of RPN.)

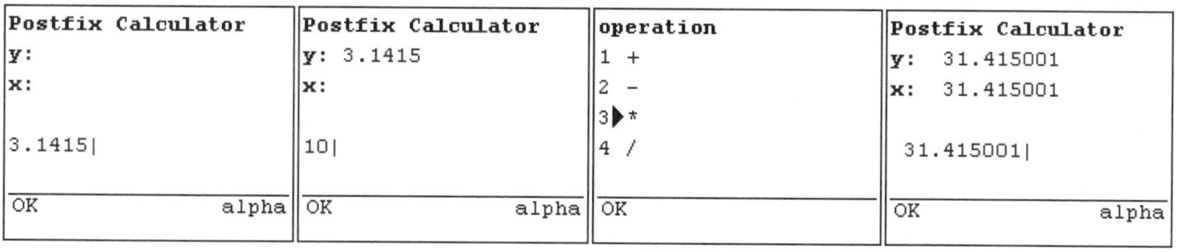

Postfix Calculator	Postfix Calculator	operation	Postfix Calculator
y:	y: 3.1415	1 +	y: 31.415001
x:	x:	2 –	x: 31.415001
		3▶ *	
3.1415|	10|	4 /	31.415001|
OK alpha	OK alpha	OK	OK alpha

Figure 9-1. A simple WMLScript application

To begin, our deck invokes a WMLScript by indicating the name of a function and the URL of a script as the href attribute of a <go> navigation task. When a WMLScript is invoked, control is passed from the WAP browser to the WMLScript interpreter; the browser is effectively frozen during the script's execution. When the function invoked by the script's <go> task is completed, control returns to the browser.

The WMLScript behind this example is stored in a hypothetical file called `s1.wmls` and looks strikingly like JavaScript or C:

```
extern function calculate() {
  var x, y, op, result;

  // Retrieve arguments from WAP Context
  x = WMLBrowser.getVar( "x" );
  y = WMLBrowser.getVar( "y" );
  op = WMLBrowser.getVar( "op" );

  // Validate arguments
  if ( x == invalid || y == invalid ||
       !Lang.isFloat(x) || !Lang.isFloat(y) ) {
    Dialogs.alert("x and y must be numbers!");
    /* Set better defaults for invalid registers */
    if ( x == invalid || !Lang.isFloat(x) ) {
      WMLBrowser.setVar( "x", "0" );
    }
    if ( y == invalid || !Lang.isFloat(y) ) {
      WMLBrowser.setVar( "y", "0" );
    }
    Lang.exit("");
  }

  /* Convert registers to floating point numbers for our use */
  x = Lang.parseFloat( x );
  y = Lang.parseFloat( y );

  /* Determine type of operation */
  if ( op == "+" ) {
    result = y + x;
  }
  else if ( op == "-" ) {
    result = y - x;
  }
  else if ( op == "*" ) {
    result = y * x;
  }
  else if ( op == "/" ) {
    result = y / x;
  }
  else {
    Dialogs.alert("An invalid operation was chosen!");
```

```
    WMLBrowser.setVar( "op", "+" );
    Lang.exit("");
  }


  /* Clean up result for display */
  result = String.format( "%10f", result );


  /* Set registers to resulting value for next calculation */
  WMLBrowser.setVar( "x", result );
  WMLBrowser.setVar( "y", result );


  /* Return the browser to the first page */
  WMLBrowser.refresh();
}
```

Our example consists of a single function, `calculate`, which takes no arguments. This function defines four variables, `x`, `y`, `op`, and `result`, to be used within the function itself.

Unlike JavaScript or Java, WMLScript does not have access to the screen. Instead, WMLScript accesses the WML browser using a library called WMLBrowser. (A *library* is simply a set of functions grouped together by functionality provided by the platform for all scripts to use). The WMLBrowser library's `getVar` and `setVar` functions enable WMLScript to get and set variables within the browser. In our example, WMLScript obtains the values of `x`, `y`, and `op` from the WML browser this way.

Once the values of `x`, `y`, and `op` are determined, the script validates `x` and `y` before proceeding. It examines the `x` and `y` variables to be sure they're valid floating-point numbers (if not, the script sets them to appropriate defaults and exits with an error message). Then the script converts the `x` and `y` variables (originally strings) to floating-point numbers and examines the `op` variable to determine which arithmetic operation to perform. (If an invalid operation is entered, unlikely though that might be, an error message is displayed, and control returns to the browser.) Once the script has performed the desired operation, it formats the results in a meaningful way for the user using the String library's `format` function. Then, the script stores the results in the WML Browser's `x` and `y` variables for the next calculation. When this function exits, it returns control to the WAP browser, which displays the original card with the results of the calculation.

The WML deck that uses this script is quite a bit shorter than the script itself, as seen here:

```
<?xml version="1.0"?>
<!DOCTYPE wml PUBLIC "-//WAPFORUM//DTD WML 1.1//EN" "http://www.wapforum.org/DTD/
wml_1.1.xml">


<wml>
```

```
<card id="calc" title="Postfix Calculator" ordered="true">
  <p>
    <big>Postfix Calculator</big>
  </p>
  <p>
    <!-- Input elements for the x and y registers -->
    <!-- The first value is really in the y register, 'cuz x -->
    <!-- gets pushed into y by the implicit enter operation -->
    <b>y: </b>$y<br/>
    <b>x: </b>$x<br/>
    <input name="y" type="text" value="0" emptyok="false"/><br/>
  </p>
  <p><big>Postfix Calculator</big></p>
  <p>
    <b>y: </b>$y<br/>
    <b>x: </b>$x<br/>
    <input name="x" type="text" value="0" emptyok="false"/><br/>
  </p>

  <p>
    <!-- Picker to select operation to be performed -->
    <b>operation</b>
    <select  title="Operation" name="op">
      <option value="+" onpick="s1.wmls#calculate()">+</option>
      <option value="-" onpick="s1.wmls#calculate()">-</option>
      <option value="*" onpick="s1.wmls#calculate()">*</option>
      <option value="/" onpick="s1.wmls#calculate()">/</option>
    </select>
  </p>

</card>
</wml>
```

This deck is straightforward except for its invocation of the WMLScript calculate.

A WML deck invokes a WMLScript using a Uniform Resource Locator (URL) that indicates both the source of the script and the WMLScript's entry point: a function name. (JavaScript developers take note: WMLScripts exist as entities apart from the content that uses them. You cannot put a WMLScript inside a WML document.) This deck uses the onpick attribute of the option selection tag to invoke the s1.wmls file's calculate function; as long as our script is stored in the location indicated by the URL given to the onpick event, control will be passed to this function.

What Is Reverse Polish Notation?

In the 1920s, the Polish mathematician Jan Lukasiewicz developed a formal logic system that allowed mathematical expressions to be specified without parentheses by placing operations symbols before (prefix notation) or after (postfix notation) the operands, rather than between them. For example the traditional expression:

$(3 + 2) * 5$

could be expressed in postfix notation as:

$3 \, 2 + 5 *$

Prefix notation came to be known as *Polish Notation* in honor of Lukasiewicz; postfix notation was later known as *Reverse Polish Notation* (RPN).

In the early years of portable calculators, computer scientists realized that RPN was efficient for computer math, as numbers could be pushed on a stack, and operations could pop arguments from the stack and push the results back on the stack.

At the time that Hewlett-Packard introduced the HP-35 calculator, competing calculators could provide only partial interpretation of algebraic order. Hewlett-Packard wisely recognized that RPN would allow them to design a calculator meeting their cost constraints that could evaluate any expression. For most users, learning this new notation was a small price to pay for the increased ability to manipulate expressions on an electronic calculator.

Today, RPN persists due to the wide user base of Hewlett-Packard computing products developed in the seventies, eighties, and nineties. Those familiar with RPN tend to see it as the "natural" way to use a calculator, and can show that it is both faster and less error-prone to use than the traditional alternatives.

A Closer Look at WMLScript

The WMLScript language is easily readable to those familiar with C, C++, Java, or JavaScript. This was done by design; programmers will have little problems picking up WMLScript.

The Basics

Like many languages, WMLScript programs are built from fundamental building blocks called *functions*. Each function consists of a group of statements, each of which instructs the computer to perform a specific operation, such as allocating a variable, performing an arithmetic operation, making a comparison, or invoking another function. Some functions may accept arguments, which are used internally by the function in its computation. Functions always return a value. Those functions that do not return an explicit value will return an empty string.

The WMLScript language is both compiled and interpreted. A WMLScript is compiled by a gateway server and provided in a compact medium to a WAP client, which in turn uses a simple virtual machine (called the *WMLScript interpreter*) to interpret the individual byte codes generated by the gateway's compiler. Each set of byte codes corresponds to an operation, variable, or similar primitive.

Structure

Within a function, *statements* determine the operation of the WMLScript interpreter. Statements are constructed from alphanumeric symbols, and are delimited by a semicolon. They can be used to declare variables, perform assignments to variables, perform operations or logical decisions, or invoke other functions. A statement may contain white spaces, such as carriage returns, spaces, or tabs, which are ignored by the scripting language. Statements are case-sensitive; not only reserved words, but variable names, function names, and other expressions must maintain a consistent case. Thus, While is *not* the while procedural statement in WMLScript, and the variable identifiers phoneNumber and phonenumber are different variables. Statements may be grouped in blocks, using the { and } characters; these blocks are used by the language to build conditional statements, functions, and loops.

Like most languages, WMLScript supports comments, which can be used to leave information for later readers about the organization, use, or other features of a function or statement. WMLScript comments follow the same style as in C++: they are delimited by /* and */ and single-line comments begin with // and end with a new line. Comments cannot be nested, so be careful if you're using them to temporarily remove a bunch of code during testing.

Some sample WMLScript statements are shown here:

```
var x;
```

declares the variable x.

```
x = x + 1;
```

adds 1 to the value of x.

```
x = Lang.random( 5 );
```

uses the lang library's random function to obtain a number between 0 and 5.

Statements may contain *literals*—primitive values such as a number or string. WMLScript defines four primitive data types for its literals: *integers, floats, strings,* and *booleans.*

Integers and floats will be familiar from high-school mathematics. An *integer* is a whole number that is less than, equal to, or greater than 0, such as –242, 13, 0, or 6. It cannot contain a fraction. The valid range for integers in WMLScript is from –2147483648 to 2147483647.

A *float* is a floating-point number—that is, any fractional number less than, equal to, or greater than 0, such as –15.0, 3.14, or 64432.155. A float in WMLScript cannot exceed $-3.40282347 \times 10^{38}$ or $3.40282347 \times 10^{38}$; the smallest float that can be expressed is $1.17549435 \times 10^{-38}$. Floats are traditionally written with a trailing decimal point if they are whole numbers, while integers are *never* written with decimal points (since they never have a fractional component).

A *string* is an aggregation of characters contained within single (') or double (") quotes, such as `"hello world"`, `"your name goes here"`, or `"this page intentionally left blank"`. Some characters are not readily expressible in strings, such as the quote marks themselves. Table 9-1 lists the special character sequences that can be used to generate characters that are otherwise reserved.

SEQUENCE	CHARACTER	SYMBOL
\'	Apostrophe (single quote)	'
\"	Double quote	"
\\	Back slash	\
\/	Forward slash	/
\t	Horizontal tab	
\b	Backspace	
\f	Form feed	
\n	New line	
\r	Carriage return	
\x*hh*	The Latin-1 ISO8859-1 character with the encoding specified by two hexadecimal digits *hh*	
ooo	The Latin-1 ISO8859-1 character with the encoding specified by the three octal digits *ooo*	
\u*hhhh*	The Unicode character with the encoding specified by the four hexadecimal digits *hhhh*.	

Table 9-1: WMLScript Character Sequences and Characters

In addition to these literal types, WMLScript defines the special value `invalid`, which indicates, not surprisingly, an invalid result or condition. Functions may return `invalid` when no reasonable result can be achieved, and comparisons can be made against the primitive value `invalid`.

The *boolean* type represents a condition that is either true or false. Two values are possible for booleans—the reserved word `true`, or the reserved word `false`. Booleans are encountered when performing logical tests, such as determining whether two statements are equal.

WMLScript uses *identifiers* to name and refer to three different elements within statements: *variables*, *functions*, and *pragmas*. An identifier is an alphanumeric string; it cannot begin with a digit, but may begin with or contain an underscore. Some valid identifiers are:

- `secret`

- `timeOfDay`

- `thx1138`

Certain identifiers are reserved by WMLScript for use as statements within WMLScript. For example, the following identifiers are reserved and may not be used by scripts as identifiers:

- `if`

- `for`

- `return`

- `while`

- `invalid`

- `true`

- `false`

- `var`

The same identifier can be used as a unique variable name and function name; thus, the variable `foo` and the function `foo()` can both exist in a script, and would refer to two different things.

Variables and Types

Variables act as placeholders for intermediate results during a script, or as the conveyors of values between one function and another. Variables must be defined before they are used. Inside a function, variables are defined using the var statement; arguments to functions are declared when a function is declared. When a variable is declared, its value is set to an empty string.

Values are assigned to variables using the = operator. This operator may be used in the context of a var declaration (setting a new variable to an initial value), or as a statement.

In the RPN calculator example shown earlier in this chapter, I defined the variables x, y, op, and result. I could have just as easily given them initial values at the same time:

```
// Retrieve arguments from WAP Context
var x = WMLBrowser.getVar( "x" );
var y = WMLBrowser.getVar( "y" );
var op = WMLBrowser.getVar( "op" );
```

A variable's scope is limited to the function in which it is declared. That is, it exists only from the time it is created until its containing function exits. Variables cannot be referenced outside the function in which they are created.

Variables may be any of the literal types defined in the previous section. WMLScript is *weakly typed*, meaning that a variable has no notion of its type internally, and may contain any of the literal data types. WMLScript attempts to convert between data types as needed, relieving the headaches caused by strongly typed languages. For example, if the following snippet were to be evaluated:

```
var x = 2.0;
var y = "3";
var res;
var display;

res = x * y;
display = "res=" + res;
```

res would be the float 6.0, and display would be the string "res=6.0". While this automatic type conversion is handy, it can catch you by mistake. Some operators act on multiple types, and using them with statements of mixed types can catch you by surprise. For example, if the variable str is a string, and x is an integer, will the result of the statement result = str + x be a string or an integer? The answer is a string, but only by careful reading of the WAP specifications or writing a test script are you likely to know this in advance.

Operators

At the heart of WMLScript are operators for arithmetic operations, string comparisons, binary manipulations, and comparisons.

In the previous example, we used the arithmetic operators +, -, /, and * to perform basic arithmetic. WMLScript also defines the `div` operator, which performs integer division, and the % operator, which performs the modulus (remainder) operation. The order of operations is algebraic, with multiplication and division taking precedence over addition and subtraction. The + operator is also used for string concatenation (combining two strings). Whether a + sign results in a string concatenation or a numeric depends on the exact expression and the combination of variables, as I will show you shortly.

Bitwise operators are also available for integers, but they will probably only be used if you write scripts to interact with hardware. The operators << and >> perform bitwise left and right shifts respectively, preserving sign, while >>> performs a bitwise shift right, filling new bits with 0. The &, |, and ^ operators perform bitwise AND, OR, and XOR operations, and can be used to apply masks for manipulating flags or other purposes.

Logical operators are available for working with the boolean values `true` and `false`. As in the C programming language and elsewhere, && corresponds to a boolean AND, while || corresponds to a boolean OR. The logical AND operator evaluates the first operand and if the result is `invalid` or `false`, does not evaluate the second operand because `false && true` and `false && false` are both `false`. Similarly, the logical OR statement will not evaluate the second operand if the first operand is `true`, because `true || false` and `true || true` are both `true`. This "short circuit" operation results in the more efficient execution of scripts.

If any value in an operation is `invalid`, the result is `invalid`.

The ! operator is also provided to perform the logical NOT operation, turning `false` to `true` and vice-versa. Logical operations can be used in both assignments and procedural statements. For example, the following two blocks are equivalent:

```
{
  var whatIf = iAmHappy && youAreHappy;
  if ( whatIf ) {
    WMLBrowser.go("#smile"):
  }
}
```

and

```
if ( iAmHappy && youAreHappy ) {
  WMLBrowser.go("#smile");
}
```

Boolean operations are often combined with comparison operations. Comparison operations test for equality or inequality, and return a boolean value. Six comparison operators exist: <, <=, ==, >=, >, and !=. The < and > operations should be familiar from grammar school math; they test whether the first operand is less than or greater than the second operand, respectively. The <= and >= perform similarly, except that the comparison returns true if the first operand is less than *or* equal to (greater than or equal to, in the case of >=). Finally, the == operand tests for equality, and the != tests for inequality.

Comparisons perform automatic type conversion. They follow the following rules when encountering different types:

- If the operand types are boolean, true is greater than false.

- If the operand types are integers, comparison is based on the given integer values.

- If the operand types are floats, comparison is based on the given floating-point values.

- If the operand types are strings, comparison is based on the order of character codes of the given string values. (Character codes are defined by the character set supported by the WMLScript interpreter.)

- If at least one of the operands is invalid, then the result of the comparison is invalid.

This behavior is intuitive, especially if you've worked with other programming languages.

Two special operators that can be helpful in debugging also exist. The typeof operator returns an integer indicating the type of its operand, while the isvalid operator returns false only if its operand is invalid. The typeof operator performs no type conversion on its operand and returns one of the integers shown in Table 9-2. The isvalid operation can be used for error checking, as shown here:

```
if ( isvalid ( x / y ) ) {
   res = x / y;
} else {
  // oh-oh. Let's handle that divide-by-zero error NOW.
  …
}
```

Of course, the above snippet could also be written

```
res = x / y
If ( res  == invalid ) {
  // oh-oh. Let's handle the divide-by-zero error here

  …

}
```

OPERAND TYPE	VALUE RETURNED
integer	0
float	1
string	2
boolean	3
invalid	4

Table 9-2: WMLScript typeof *Operator Results*

WMLScript also provides several operators that combine an assignment with an operation, such as ++, --, +=, -=, *=, and /=. The ++ and -- operators are familiar to those with previous programming experience as the increment and decrement operators. These can be used either before or after a variable to indicate that the operation should occur before or after the variable is evaluated, respectively. Operators followed immediately by = will apply the operation to the statement's right side and assign it to the left side. Thus, the following pairs of statements are equivalent:

```
x *= 10;                x = x * 10;
answer += " pounds.";   answer = answer + " pounds.";
```

Procedural Statements

For a programming language to be truly useful, it must have procedural statements that can alter the flow of a program. WMLScript follows the example of other structured languages and provides procedural statements for flow control and looping.

The if-else series of statements defines a conditional flow control based on a comparison. The expression of the if statement is evaluated, and if true, the first block is evaluated; otherwise the optional else block is evaluated. For example, the following expression sets x equal to 3.0:

```
var val = 0;
var x;
if ( val == 0 ) {
```

```
  x = 3.0;
} else {
  x = -3.0
}
```

The ?: operation is similar to the if-else conditional, but combines an assignment with the conditional behavior. If the expression before the ? is true, the value of the expression between the ? and : is returned; otherwise, the value of the expression after the : is returned. The previous example could have been written as follows:

```
var val = 0;
var x;
x = val == 0 ? 3.0 : -3.0;
```

The ?: operation is often used as shorthand in arithmetic operations. Use it sparingly; in complex expressions, it can become devilishly hard to read.

You can cause the execution of a statement to loop a number of times using the while or for keywords. The while statement evaluates an exit expression, and repeats a block of statements for as long as that expression is true:

```
while ( expression ) { statements; }
```

For example, the algebraic factorial function, *n*! (the product of all positive integers less than *n*), can be expressed using the while statement:

```
function factorial( n )
{
  var result = 1;
  if ( !Lang.isInt(n) ) {
    return invalid;
  }
  while ( n > 1 )
  {
    result = result * n;
    n = n - 1;
  }
  return result;
}
```

The for statement provides a similar looping construct, along with convenient placeholders for initialization and exit expressions and for managing a variable that tracks the status of the loop. The aforementioned factorial function could be written this way using the `for` statement:

```
function factorial( n )
{
  var result;
  if ( !Lang.isInt(n) ) {
    return invalid;
  }
  for ( result = 1; n > 1; n-- )
  {
    result = result * n;
  }
  return result;
}
```

The `for` statement takes three expressions before the block of statements: an expression indicating initial conditions, the exit expression, and an update expression to be executed every time the statement block is executed. Programmatically, the `for` statement

```
for( a; b; c) { s }
```

is functionally equivalent to the `while` statement:

```
a;
while( b ) {
  s;
  c;
}
```

Whether to use a `while` or a `for` statement is largely a matter of convention or taste. Traditionally, `for` statements are used for straightforward iteration over one variable, and `while` statements are reserved for more complex looping constructs.

You can use the `break` and `continue` constructs to exit a loop prematurely in the event that all the conditions for continuing a loop are met. The `break` statement is used to terminate the current `while` or `for` loop and continue the program execution from the statement following the terminated loop:

```
var n = 0;
while( true )
{
  n++;
  if ( n == 100) {
```

```
    break;
  }
}
// when we get here, n is 100 and the loop was run 101 times.
```

It is an error to use a break statement outside a while or a for statement.

The continue statement is used to terminate the execution of a block of statements in a while or for loop and continue the loop with the next iteration. This is often used when one or two passes of the loop encounters an exceptional circumstance, but the loop should continue anyway (hence the statement's name). The continue statement keeps control within the scope of the loop. In a while loop, control returns to the exit condition, while in a for loop, control returns to the update expression.

Functions

Functions enable script authors to organize statements into meaningful segments intended to perform a specific task; they are the fundamental organizational unit for applications. At least one function must be designated as the entry point for applications executing WMLScript from the WAP browser.

Functions are declared using the function statement. This statement specifies the name of the function and any arguments it uses to accept input from its callers:

```
function func( arg1, arg2, arg3 )
{
...
}
```

The arguments are used as variables within the function; they have scope only over the function where they are used. A function returns a value explicitly with the return statement, or implicitly when the function exits. If no value is explicitly returned, an empty string is returned to the caller. The arguments and return value enable the caller of a function to provide initial information and obtain the results of a function.

To call a function, a statement indicates the function name with its arguments in parentheses:

```
var n = factorial( 6 );
```

Function calls can be used as operands in a more complex expression:

```
howmany = factorial( 6 ) / factorial ( 3 );
```

Functions called by other scripts or from the WAP browser must be declared as external functions, indicating that they should be made available to entities outside the current script. The calculate script used in the RPN calculator example illustrated earlier in this chapter was such a function:

```
extern function calculate() {…}
```

Libraries

Libraries are collections of related functions. The WAP standard provides several standard libraries that group together functions for the management of the WML browser, language functions, string manipulation functions, and functions for URL manipulation. Platform vendors can enhance a specific device by adding specific libraries for integration between WMLScript and the device. For example, a library could be added that provided voice access or access to the device's address book.

Library functions are invoked by specifying the library's name, a . (dot) and then the desired function. This was shown in examples of how WMLScript accesses variables from the WAP browser:

```
var x = WMLBrowser.getVar( "x" );
```

The WAP requires that the manufacturers of any devices that support WML-Script supply the following standard libraries (which are detailed in Tables 9-3 through 9-8):

- The **Lang library** provides functions comprising features of WMLScript, such as type conversion operators, a random number generator, and functions specifying the minimum and maximum values for integers. (See Table 9-3.)

- The **Float library** provides functions for simple floating-point operations, including exponents and square roots. At the present time, trigonometric functions are *not* available from this library. (See Table 9-4.)

- The **String library** provides a set of primitive string functions, including a string formatter and accessory functions for manipulating individual functions within a string. (See Table 9-5.)

- The **URL library** provides routines for extracting the various parts of a URL, such as the host, protocol, and port. The URL library's functions are significantly easier to use for URL manipulations than crafting functions from the String library to do the same thing. (See Table 9-6.)

- The **Dialogs library** provides a simple modal dialog interface to present messages to the user. (See Table 9-7.)

- The **WMLBrowser library** provides routines for accessing and controlling the WML browser on the device. (See Table 9-8 and the following section.)

FUNCTION	ARGUMENTS	RETURNS	PURPOSE
abort	string		Causes the interpretation of the current script to abort and returns control to the caller of the WMLScript interpreter with the given string used to describe the error.
abs	number	number or invalid	Computes the absolute value of the given number.
characterSet		string	Returns an integer that denotes the character set supported by the WMLScript interpreter on the host platform.
exit	value		Causes the interpretation of the WMLScript to terminate and returns control to the caller of the WMLScript interpreter with the given value.
float		boolean	Returns true if the platform supports floating-point arithmetic, false otherwise.
isInt	string	boolean	Returns true if the string can be interpreted as an integer, false otherwise.
isFloat	string	boolean	Returns true if the string can be interpreted as a float, false otherwise.
min	n1, n2	number or invalid	Computes the minimum of two given numbers.
minInt		number	Returns the minimum integer value supported.
max	n1, n2	number or invalid	Computes the maximum of two given numbers.
maxInit			Returns the minimum integer value supported.

Table 9-3   *(continued)*

Table 9-3 (continued)

FUNCTION	ARGUMENTS	RETURNS	PURPOSE
parseInt	string	integer or invalid	Returns an integer interpretation of the string, or invalid if the string cannot be interpreted as an integer.
parseFloat		string	Float or invalid.
random	integer	integer	Returns a random integer between 0 and the value passed, or invalid if value is less than zero or not a number.
seed	value	string	Initializes the random number sequence and returns an empty string.

Table 9-3: Functions in the WMLScript Lang Library

FUNCTION	ARGUMENTS	RETURNS	PURPOSE
ceil	value	number	Returns the smallest integer value that is not less than the given value.
int	value	number	Returns the integer part of the given value.
floor	value	number	Returns the greatest integer value that is not greater than the given value.
maxFloat		number	Returns the maximum valid floating-point number.
minFloat		number	Returns the minimum valid floating-point number.
pow	value1, value2	number	Returns an implementation-dependent approximation of the result of computing value1^{value2}
round	value	number	Returns the integer that is closest to the value of the given number.
sqrt	value	number	Returns an implementation-dependent approximation of the square root of the given value.

Table 9-4: Functions in the WMLScript Float Library

FUNCTION	ARGUMENTS	RETURNS	PURPOSE
charAt	string, index	string	Returns a new string, one character long, containing the character in the string at the position specified by the index.
compare	string1, string2	integer	Specifies the lexicographic relation of string1 to string2 based on the character codes of the native character set.
elements	string, sep	integer or invalid	Returns the number of elements in the given string separated by the given string sep.
elementAt	string, index, sep	string	Returns the index'th element of the given string as separated by the given strings sep.
find	string, sub	value	Returns the index of sub in the given string, or invalid.
format	format, value	string	Uses the printf-style format string to format the given value and convert it to a string.
insertAt	string, element, index, separator	string	Returns a new string where new has been inserted at the index'th element of string separated by the given string sep.
isEmpty	string	boolean	Returns true if the string length is 0, false otherwise.
length	string	number	Returns the number of characters in string.
replace	string, old, new	string	Returns a new string resulting from the replacement of all occurrences in the given string of the string old by the string new.
removeAt	string, index, sep	string	Returns a new string when both the element and the corresponding given string, sep, that has the given index number, have been removed.
replaceAt	string, element, index, separator	string	Returns a new string where the index'th element of the given string separated by sep has been replaced by new.

Table 9-5   *(continued)*

Table 9-5 (continued)

FUNCTION	ARGUMENTS	RETURNS	PURPOSE
squeeze	string	string	Returns a string where all consecutive white spaces are reduced to single spaces.
subString	string, start, length	string	Returns a new string that is the substring of the given string, beginning at start and extending the number of characters indicated by length.
toString	value	string	Returns a string representation of value.
trim	string	string	Retuns a string where all leading and trailing white space has been removed.

Table 9-5: Functions in the WMLScript String Library

FUNCTION	ARGUMENTS	RETURNS	PURPOSE
escapeString	string	string	Computes a new version of string where all characters are represented in a format suitable for use in URLs.
getBase		string	Returns an absolute URL of the current WMLScript compilation unit.
getFragment		url	Returns the fragment used in the given url.
getHost	url	string	Returns the host specified in the given url.
getPath	url	string	Returns the path specified in the given url.
getParameters	url	string	Returns parameters in the last path segment of url.
getPort	url	string	Returns the port number specified in url.
getQuery	url	string	Returns the query part specified in url.

Table 9-6   *(continued)*

Table 9-6 (continued)

FUNCTION	ARGUMENTS	RETURNS	PURPOSE
getReferer		string	Returns the smallest URL relative to the base URL of the current compilation unit to the resource that called the current compilation unit.
getScheme	url	string	Returns the protocol (scheme) in url.
isValid	url	boolean	Returns true if url is a syntactically valid URL.
loadString	string, type	string, integer, or invalid	Returns the content denoted by the given absolute URL and the given content type.
resolve	base, embedded		Returns an absolute URL created from the given base and embedded URL strings.
unescapeString	string	string	Computes a new version of the given string in which URL-suitable characters have been converted to their human-readable counterparts.

Note: Relative URLs are not resolved in these functions.

Table 9-6: Functions in the WMLScript URL Library

FUNCTION	ARGUMENTS	RETURNS	PURPOSE
alert	message		Displays message and waits for user confirmation.
confirm	message, ok, cancel	Boolean	Displays message and two reply alternatives. Waits for the user to select a reply and returns true for OK, false for the second reply alternative.
prompt	message, default	string	Displays message and prompts for user input. Returns the user input.

Table 9-7: Functions in the WMLScript Dialogs Library

FUNCTION	ARGUMENTS	RETURNS	PURPOSE
getVar	var	string	Returns the value of the variable var in the browser.
go	string	string or invalid	Specifies the URL to be loaded by the browser when the script terminates.
getCurrentCard		string	Returns the smallest relative URL (relative to the script) that specifies the card being displayed by the browser.
newContext			Clears all variables and the history of the browser.
prev	string	string or invalid	Signals the browser to go to the previous card.
refresh			Signals the browser to update its UI based on the current context.
setVar	var, value	string	Sets the value of the variable var in the browser.

Table 9-8: Functions in the WMLScript WMLBrowser Library

WML Browser Integration

The WMLBrowser library provides integration between the WMLScript interpreter and the WML browser present on a device. While its semantics and use are the same as any other library, it is perhaps the most important of any of the WML-Script libraries—you'll need it every time you write a script that interacts with a WML deck. Without it, scripts would be unaware of a WML deck's variables, and could not interact with decks in any way.

The WMLBrowser library provides several functions for controlling the WML browser application. The most frequently used are getVar and setVar. The WMLBrowser.getVar function takes a string identifying the variable of interest, and returns a string containing the value of the WML variable. If the variable is not defined in the browser, the function returns invalid. The WMLBrowser.setVar function is used to set a variable within the WML context. It accepts two strings—the name of the variable and its value. It will return true if the value is correctly set, or false otherwise. To succeed, the WML variable's name must be a valid WML variable name consisting only of an underscore and alphanumeric characters.

The WMLBrowser.go function interprets its argument as the URL for a card to be shown when control returns to the WML browser. Multiple calls to this function will result in only the last specified URL being visited; no URL will be resolved if the argument's value is an empty string. The WML browser will not request the content specified by the URL until the WMLScript function exits and control returns to the WML browser.

The WMLBrowser.prev function is similar to the go function, except that it uses the last visited URL from the history stack as the URL of the card to be shown. If both WMLBrowser.go and WMLBrowser.prev are called, the last-called function has precedence. Thus, in a the following script

```
…
WMLBrowser.go("http://wap.apress.com/");
…
WMLBrowser.prev()
…
```

the WML browser performs the prev operation when control returns to the browser. In the following sequence no navigation takes place, as the last specified action was a WMLBrowser.go without a URL.

```
…
WMLBrowser.go("http://wap.apress.com/");
…
WMLBrowser.prev()
…
WMLBrowser.go("");
…
```

The WMLBrowser.refresh method is similar, but does not actually request a new URL. Rather, this function causes the browser to re-evaluate the currently displayed card based on its current content and to redisplay the card. Unlike the WMLBrowser.go and WMLBrowser.prev functions, this function is performed immediately or, if the browser cannot redraw the current card while a script is being executed, invalid is returned and the refresh will be performed when the script exits. (Some devices may not have enough memory to support both the browser and the script environments at the same time. On these devices, the script interpreter simply sets a flag telling the browser to redraw the current card when the script execution completes.)

The WMLBrowser.newContext function clears the browser's history stack and all variables. This is done without affecting any pending navigation requests, enabling the script to return the browser to a known state before returning control to the browser.

The `WMLBrowser.getCurrentCard` function returns a string indicating the source of the currently displayed card relative to the current script's base URL; it returns `invalid` if no current card is displayed. An absolute URL will be returned if no relative URL can be calculated between a script's base URL and the current card's URL. Note that the URL returned will be the *current* card, not the URL of any pending navigation actions undertaken by the `WMLBrowser.go` or `WMLBrowser.prev` functions.

Debugging Hints

Debugging a WMLScript is a little like finding your way to the bathroom at night with the lights off. Those with careful housekeeping habits and a methodical nature are likely to make the trip unscathed; those in a hurry or sloppy will eventually make it, but will bear the occasional bruise or frustrating experience.

WMLScript development debugging is best performed using a simulator, such as the Software Developer Kits (SDKs) available from Phone.com or Nokia. These simulators can greatly speed the rapid development cycle, enabling you to author and test content either on a single machine or over a high-speed local network prior to testing your application on real devices.

However, if you're expecting a robust software development toolkit with a debugger, source code editor, and breakpoints, you'll be in for a bit of a surprise. While these SDKs accurately simulate the behavior of a WAP browser and provide valuable error messages indicating problems with your content, you can perform very little true debugging with them. These environments provide a rapid cut-and-try development cycle, rather than the precision of a true software development environment.

Those who have previous experience with leading-edge (bleeding-edge?) platforms will recognize the time-tried techniques available to WMLScript authors. Foremost is the careful organization that should take place behind developing WMLScript content. Scripts should be developed iteratively and tested at regular intervals. Test functions can report values through variables displayed on test cards, and these cards kept in decks can be made available to developers and quality assurance staff. This scaffolding should be retained through the development process of an application, as it provides a valuable means for discerning the behavior of a complex script.

As functionality is added to your scripts and decks, your application should be *checkpointed* somewhere. Those with software development experience will immediately recognize the need for change control, which enables a developer to "roll back" to a previous release in the event that something goes wrong. To do this, you may not need complex tools such as Microsoft's Visual Source Safe. A careful methodology of regularly backing up working scripts and decks to a fresh directory (or to backup media such as CD-ROM, floppy, or tape) provides a convenient place to go when you find that something no longer works.

Inevitably, you'll reach a point where something doesn't work as you expect. More often than not, this will occur while you're developing a script. However, don't be surprised when you discover that a function you thought you'd finished appears to be working differently than you recall, usually because of the way it interacts with other functions within a script. The `Dialogs.alert` function can be used to display a given string while a script is running. You can pass a variable's value as a string to this function, which will let you see its value while a WMLScript function is running. For example, the code snippet below, uses a WML deck's `debug` variable to determine whether a given debug dialog should be shown displaying the value of the variable x after computations:

```
...
var debug=WMLBrowser.getVar("debug") == "true";
var x;
/* some manipulations are performed on x */
if (debug)
{
  Dialogs.alert( "x=" + x );
}
...
```

Using a WML variable to trigger debugging enables your debugging code to remain in scripts through through production. This allows you to perform rapid testing by simply altering a WML deck. Of course, the drawback is that the test code remains in your final product. This makes your scripts slightly larger and can reveal parts of the internals of your scripts to prying eyes.

Alternatively, you can use a deck's variables for debugging. You can assign intermediate results (or information regarding a function's status, for that matter) to variables as your function executes. When it returns, you can use the go function to show a card with these variables evaluated to see what your function was doing. When you've debugged your function, you can remove this card with the debugging information and the code you used to assign values to the deck's variables.

One powerful tool available to conventional developers is the `assert` macro, which verifies an assumption and presents an error if the condition is not met. WMLScript unfortunately lacks the equivalent of the `assert` macro, but there's no reason why a similar effect cannot be simulated, especially at a script's entry function between the WML browser and the WMLScript interpreter.

In the following example, I simulate an assertion using the initial verification of input values (the first conditional of the script), and call `abort` if the function is invoked with invalid arguments.

```
extern function doSomething()
{
  var debug=WMLBrowser.getVar("debug") == "true";
  var value=WMLBrowser.getVar("value");
  /* verify input value */
  if ( debug && Lang.parseFloat( value ) == invalid )
  {
    Lang.abort("doSomething Error: value should be a float!");
  }
  …
  WMLBrowser.setVar("value");
  WMLBrowser.refresh();
}
```

This defensive coding should be performed *anywhere* a mistaken value might cause your application to fail, and should perform appropriately during those failures by gracefully recovering or reporting a meaningful error to the user. Consumers are strikingly tolerant of computer failures and notoriously intolerant of failures in an embedded device such as a screen phone.

Finally, some simulator environments provide a console where messages can be logged. This simulator is accessed through a library such as Phone.com's Console library, which enables strings to be printed to the console using the console.printLn function. You can use functions like this in conjunction with a debug variable to log important events in your script, such as in the following situations:

- When a variable is initialized or assigned value.

- When your script code acts on the value of a variable.

- When your script interacts with the WMLBrowser library to load a new card or refresh the existing card.

Depending on the development environment you choose, a function such as console.printLn can be more useful than scattering various Dialogs.alert invocations throughout your script.

Summary

The WAP Forum has defined WMLScript, a dynamic scripting environment for WAP browsers. WMLScript bears a strong resemblance to JavaScript, making it easy for developers to learn. While WMLScript cannot directly interact with WML content in the browser display, its ability to interact with the WML browser's variables provides an interface between WML content and procedural scripts. These scripts can invoke libraries, perform arithmetic and string manipulations, and return content to the WML browser in the form of variables to be displayed by the WML browser itself.

WMLScript supports variables independent of the WML browser and can exchange variables with the browser. The variables within a WMLScript can be used to store strings, numbers, boolean values, and floats on platforms that support floating-point math.

WMLScript provides basic operations, including arithmetic and boolean comparisons. These can be used to manipulate variables or construct conditional expressions such as loops. Loops resembling traditional programming languages such as C and C++ can be written using both the `for` and `while` statements. These in turn can be used to create functions, at least one of which is called from a WML deck.

Debugging WMLScript is challenging, because many of the conventional debugging tools available on other platforms simply don't exist for WMLScript. However, by carefully organizing your code, you can write scripts that are easily tested. When you need to test a script, you can always save values in debugging variables that your WML deck can display, or use the `Dialogs.alert` function to display intermediate values within a long function.

CHAPTER 10

The Handheld Device Markup Language

ALTHOUGH THE WIRELESS APPLICATION PROTOCOL (WAP) is poised to be the world standard for screen phone content, the Handheld Device Markup Language (HDML) remains important in North America, and I recommend you gain at least some familiarity with it.

In this chapter, I introduce HDML and show you how you an use it to mark up your content for today's North American screen phones and wireless terminals.

Introducing HDML

HDML was the first device-specific markup language available for screen phones. Created by Phone.com, it has been widely licensed to handset manufacturers. Major wireless providers, including AT&T, offer mobile data services that carry HDML, and most North American wireless data handsets in use offer HDML browsers.

Unlike the other wireless technologies discussed in this book, the HDML standard is controlled by a single company. While the standard itself is open—anyone can develop HDML content—the direction HDML will take in the future is controlled by Phone.com.

HDML depends on the UP.Link Server, which provides server-side assistance for HDML browsers. The UP.Link Server gates HDML content from the Web to wireless terminals (see Figure 10-1 for a deployment view). As with other server-assisted wireless browsers, the UP.Link Server bridges the gap between media-rich Web content and constrained access devices. It provides network-specific services to the wireless network, converting the wireless protocols to Web protocols and making requests of the servers where the content originates—on the Web or on private enterprise networks.

Fortunately, it's not necessary to know much about the UP.Link Server to develop HDML content. Network providers make a UP.Link Server available and support it for their subscribers.

Developers who set out to learn HDML will find experience with other markup languages helpful. Understanding the Wireless Markup Language (WML) makes HDML easier to learn at a conceptual level, but knowing the HyperText Markup Language (HTML) well makes HDML easier to write. HTML and HDML tags are written using the same syntax, and some tags are the same in both markup

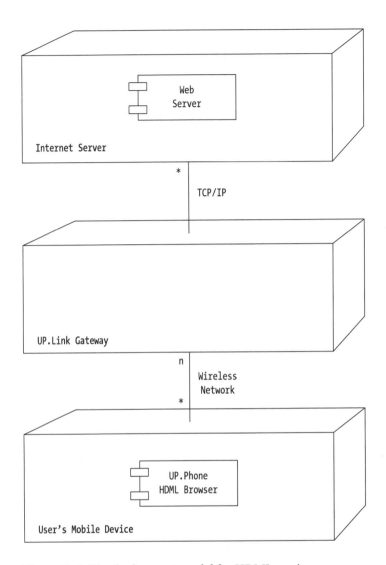

Figure 10-1. The deployment model for HDML services

languages. However, the organization of an HDML document is most similar to that of a WML document, and several important features of HDML, including tasks and variables, resemble those of WML.

This dichotomy originates in the history of HDML. Originally developed by Phone.com's previous incarnation, Unwired Planet, HDML was intended as an alternative to HTML for mobile devices on which HTML processing would be too expensive to be practical. As the adoption of HDML accelerated, a number of handset manufacturers interested in establishing an open standard for wireless data worked with Phone.com to create the WAP Forum, the industry association that developed and maintains the WAP standards. Thus, HDML has a syntax based on HTML, but includes concepts refined in WAP.

HDML or WML?

The existence of two complementary standards in a growing market leads inevitably to the question, "Which one should I use?" While there may be special cases, the basic answer in this case is simple: support both.

The differences between WML and HDML are rooted in syntax, not semantics. For conventional providers developing simple wireless applications that consist largely of information, the majority of the work involved in supporting both formats lies in selecting the appropriate tags for the language at hand. The limitations of WML and HDML devices are similar, and the two languages require similar formats and interface conventions. Content obtained from databases or other sources can be combined with HDML templates as easily as with WML templates, so there is rarely a good reason not to offer dual-platform information services.

The question becomes more difficult, however, when true interactive wireless applications are considered. These applications do more than display content; they interact with the user through client-side scripts that make them appear to be applications in their own right. Both HDML and WML provide the ability to create wireless applications, but their syntax and capabilities rule out the creation of cross-platform applications. While the organization and design of an application may be applicable to both platforms, the markup tags themselves are different, so you'll need to approach the content differently.

If you are looking to create a dynamic Web-based application, such as an e-mail client or a personal information manager, it's important to consider your target market as well as the technical differences between HDML and WML.

Wireless Web applications are much more sensitive to the market than are traditional applications. Their adoption is influenced by relationships with service providers, pricing, advertising, and a host of factors that have a much smaller impact on traditional software projects.

For example, in many cases, vertical markets already have established relationships with both kinds of vendor. For a wireless application to be useful, it must run on the end customer's selected hardware over the provider's network. You won't win points from a customer for delivering a custom WML application for an HDML-compliant Alcatel wireless handset.

Similarly, if you are targeting an application for a particular service's customers, bear in mind the limitations of that service. For example, as of this writing, AT&T's Pocket.Net service is based on HDML, so applications targeted at Pocket.Net subscribers should certainly be HDML-based.

The features of WML and HDML are technically similar. Both languages support variables, and both provide limited on-device processing of data. The WAP stack provides WMLScript, a scripting language that provides conditional evaluation and other primitives that you could only get from a server in an HDML application. However, HDML has a few other features, such as activities and iconic soft-key labels (more information about both of these appears later in this chapter), that are presently unavailable to WAP developers.

Your First HDML Page

The basic syntax of HDML is simple enough that it will become clear from a single example:

```
<HDML VERSION=3.0>
    <- A simple HDML deck ->
    <DISPLAY NAME=hello>
    Hello world!
    <BR>
    How are you today?
    </DISPLAY>
</HDML>
```

Note the strong similarities to HTML: tags are contained within angle brackets < and >. Tags that accept arguments, such as the <DISPLAY> tag, are written using the same angle brackets as empty tags such as <BR>. Comments can be included between the delimiters <-- and -->.

As in HTML, many tags have attributes that further describe their behavior. Attributes are specified as named values within the angle brackets but after the tag's name, such as the NAME attribute of the <DISPLAY> tag in the example. (These attributes are described in detail in the section "HDML for Web Developers," later in this chapter.)

You must begin any HDML document with the <HDML> tag. Because only one document can occupy a file, every HDML file begins and ends with this tag.

HDML documents are organized as decks of cards, just as WAP documents are. Thus, the <HDML> tag delimits a deck. The cards within the deck can specify actions, declare or evaluate variables, and display content. (A card may do any or all of these things, although as you'll see, one card can't prompt for both text and selection input.) Any of the tags <DISPLAY>, <NODISPLAY>, <CHOICE>, and <INPUT> can be used to specify a card.

Another feature of HDML that differs from HTML is the use of variables, which play an important role in handling user input in HDML. Variables are set using the <ACTION> tag and can be evaluated using the $ operator. The best way to explain their use is through an example:

```
<HDML VERSION="3.0">
  <NODISPLAY>
    <ACTION TYPE=ACCEPT TASK=GO DEST=#main
     VARS=pi=3.14159&e=2.71828>
  </NODISPLAY>
  <DISPLAY NAME=main>
  PI is $pi.<BR>
  e is $e.
  </DISPLAY>
<HDML>
```

This card uses a hidden card—declared with the `<NODISPLAY>` tag—to set the variables `pi` and `e`. The second card of the deck evaluates these variables in the content that is displayed using the $ operator.

Browsers, Tools, and SDKs

A variety of handset manufacturers, including Motorola, Alcatel, Mitsubishi, and Samsung, now offer phones with HDML browsers. At present, there are no HDML browsers for laptops, Personal Digital Assistants (PDAs), or hybrid phone/computing devices.

Despite its well-established presence in the industry, no proven tools exist that enable What-You-See-Is-What-You-Get (WYSIWYG) editing or validation of HDML; for the present, therefore, HDML content must be developed with a text editor. Any simple text editor will do, including WordPad, BBEdit, or emacs. Simply compose using standard ASCII characters, and save the result as a flat text file with the extension ".hdml". (When saving HDML files from a Windows text editor, don't forget to specify the filename in quotes, or your file will end up with the suffix ".hdml.txt").

To preview HDML content, you need a copy of the Phone.com UP.SDK software developer kit (SDK). It includes a simulator capable of displaying HDML files from remote servers or a local file system. Figure 10-2 shows a screen shot of the simulator in action.

HDML for Web Developers

If you have used another markup language, you'll find it easy to learn HDML. HDML draws on common concepts found in both WML and HTML, although it has some new ideas of its own as well.

Figure 10-2. The Phone.com UP.SDK HDML browser

Cards and Decks

As with WML (see Chapter 8), the fundamental organizational metaphor behind HDML is the deck of cards. *Cards* act as containers for all other objects; they describe actions, display content, and control interactions with users. There are also hidden cards, which encapsulate tasks or manage variables. Cards are grouped together into *decks*. Usually, the cards in a deck relate to a common purpose or subject, such as a stock quote, news headline, or weather report.

Attributes of Cards and Decks

Many HDML tags accept *attributes,* which control how the client renders the tag. The attributes of cards and decks are paramount, because they determine such aspects of your content as who can access your decks and how navigation between decks and cards behaves. In the following sections, I discuss the attributes of the cards and decks shown in Table 10-1.

TAG	ATTRIBUTE	PURPOSE
<DISPLAY>*	BOOKMARK	Specifies the URL that should be saved when a card is bookmarked.
	NAME	Specifies a card's name for navigation purposes.
	MARKABLE	Indicates whether a card can be bookmarked (true) or not (false).
	TITLE	Specifies a card's bookmark title.
<HDML>	VERSION	Specifies the version number of HDML being used.
	ACCESSPATH	Specifies the path of decks that are allowed to link to an access-controlled deck.
	ACCESSDOMAIN	Specifies the domain of decks that are allowed to link to an access-controlled deck.
	PUBLIC	Indicates whether access to a deck is restricted (false) or not (true).
	TTL	Specifies a deck's time to live, in seconds.

*The attributes of the <DISPLAY> tag are also attributes of the <CHOICE> and <INPUT> tags.

Table 10-1: HDML Deck and Card Attributes

Deck Attributes

The attributes of a deck determine the behavior of all the cards in it: whether they can be bookmarked or not (by default), how long they can remain in the browser's cache, and whether the deck can be linked to other decks.

The <HDML> tag, mentioned previously, introduces each deck. Its MARKABLE attribute specifies whether or not cards within the deck can, by default, be bookmarked. This flag may be overridden for specific cards.

The PUBLIC attribute indicates whether or not a deck's contents can be accessed from other decks. By default, any deck can be linked to any other deck. You can restrict access to a deck by specifying a value of FALSE for the <HDML> PUBLIC attribute.

When PUBLIC is set to FALSE for a given deck, other decks can only be linked to it if they have the base URL specified in its ACCESSDOMAIN and ACCESSPATH attributes. The default for the PUBLIC attribute is FALSE. Note that a deck with restricted access cannot have bookmarkable contents, because the bookmark would access the deck from outside the specified access domain and path.

The final attribute of the <HDML> tag is the optional TTL attribute, which specifies the amount of time a deck may remain cached. (TTL stands for "time to live" and is given in seconds.) Rapidly changing data, such as stock quotes, call for a low TTL, while content that does not change can last longer. The following example indicates that the deck should be cached for twelve hours:

```
<HDML VERSION=3.0 TTL=43200>
<DISPLAY NAME="result" MARKABLE=FALSE>
  <-- Card's content -->
</DISPLAY>
</HDML>
```

By default, content will remain cached for no more than thirty days. Even without a TTL attribute, however, data may end up cached for considerably less time, as browsers may delete content viewed infrequently.

Card Attributes

A card's behavior is primarily determined by four attributes of the <DISPLAY> tag and the <NODISPLAY>, <CHOICE>, and <INPUT> card tags. The NAME attribute specifies a name for the card. Card names are used for navigation and are similar to anchor references within HTML text. Every card can be uniquely identified by the URL of the deck, followed by the hash symbol (#), followed by the name of the card; for example, http://hdml.apress.com/example.hdml#hello. As you'll see in the next section, it's also possible to navigate among cards using relative URLs.

HDML provides control over the behavior of bookmarks via the TITLE, MARKABLE, and BOOKMARK attributes of the <DISPLAY> tag. A card can only be bookmarked if its MARKABLE attribute is TRUE. (This attribute for a specific card overrides the default, which is determined by the MARKABLE attribute for the deck.) The TITLE attribute of a bookmarkable tag specifies a short title for the card within a bookmark; the URL that a bookmark opens can be set using the BOOKMARK attribute. By default, the current card's URL becomes the bookmark's URL.

Consider a deck that is used to obtain telephone numbers from a directory. You wouldn't want users bookmarking the results card or cards where form input has already taken place. In fact, you probably would want only the first card of the deck, where users enter the desired name, to be bookmarkable. Your HDML in this situation might look like this:

```
<HDML VERSION=3.0>
<DISPLAY NAME="home" TITLE="Moonlight Directory" MARKABLE=TRUE>
  <ACTION TYPE=ACCEPT TASK=GO DEST="#InputChoice">
    <-- Card content -->
</DISPLAY>
<CHOICE MARKABLE=FALSE NAME="InputChoice"
  KEY=choice IKEY=num IDEFAULT=1>
  <-- Card content -->
</CHOICE>
<ENTRY MARKABLE=FALSE NAME="Last" KEY=last FORMAT="*a">
  <-- Card content -->
</ENTRY>
<ENTRY MARKABLE=FALSE NAME="First" KEY=first FORMAT="*a">
  <-- Card content -->
</ENTRY>
</HDML>
```

Types of Cards

There are three broad categories of HDML cards:

- Cards that display information, such as formatted text or images.

- Hidden cards that specify variables or encapsulate actions available to the user, such as navigation to other pages.

- Cards that accept input from the user.

In most decks, these categories are mixed and matched to create interactive applications. For example, an application that searches an intranet phone directory will use cards with actions and input to enable users to look up coworkers' names and dial their phone numbers. The first card would be an input card in which the user enters a name. This card would link to a hidden card, which would wrap up the form content and post it to the server for a response. The response deck would contain a content card that displayed the phone number and included an associated action that dialed the number on the card.

Displaying Information

HDML decks can display plain text (there is no support for styles or different fonts) and images. Although it is minimalist in comparison to either WML or HTML, the text formats it supports are reasonable considering the capabilities of the target devices, and quite sufficient to display a wide variety of information. Table 10-2 summarizes the tags available for formatting display information on HDML cards.

TAG	ATTRIBUTE	PURPOSE
 		Inserts a line break.
<CENTER>		Centers the current line.
<IMG>	ALT	Specifies a text label for an image.
	ICON	Specifies the source of an image found in ROM (by name).
	SRC	Specifies the source of an image obtained from the wireless Web (by URL).
<LINE>		Indicates that the current line should not be broken.
<RIGHT>		Right-justifies the text that follows.
<TAB>		Indicates that display should continue at the next tab stop.
<WRAP>		Indicates that a wrapping line follows (paragraph form).

Note: These are all empty tags.

Table 10-2: HDML Tags for Card Formatting

Text Presentation

HDML supports two kinds of text display: wrapped text (paragraphs) and a marquee scrolling text line. Wrapped text is appropriate for most purposes, including text paragraphs, news stories, and data display. The <WRAP> tag works much like the <P> tag in HTML; it goes at the beginning of each chunk of word-wrapped text.

Occasionally, you may need to display a line of text as an unbroken string, such as an e-mail header. The <LINE> tag comes in handy for this purpose:

```
<HDML VERSION=3.0>
<DISPLAY>
  <LINE>
```

```
The quick brown fox jumped over the lazy dog adroitly.
<WRAP>
After her followed four slow green turtles.
</DISPLAY>
</HDML>
```

The contents of the `<LINE>` tag will scroll repeatedly across the display so that the entire contents can be read on a single line. (Unfortunately, it's hard to demonstrate this feature in a paper book!) This tag should be used sparingly, as a host of moving lines is difficult to read.

Text Formatting

Formatting is controlled by a set of tags that specify a particlar operation for the browser, such as "center the following text," "place a tab," or "end a line." Unlike in HTML, these tags do not enclose the text being formatted; rather, they are placed at the begining of the run of text to which they apply, and the indicated formatting continues until the next formatting tag.

Line breaks can be inserted at any point using the empty `<BR>` tag. You can also use it in conjunction with the empty `<TAB>` tag, which inserts a tab, to create simple tables. By default, all text is left-justified. Text regions can be centered or right-justified using the empty `<CENTER>` and `<RIGHT>` tags.

The following example shows some of these text formatting features in action. The initial text is centered with the `<CENTER>` tag, while the weather report appears in a two-column table.

```
<DISPLAY NAME="result" MARKABLE=FALSE>
  <WRAP>
  <CENTER>
      Weather Forecast for<BR>
      Boulder Creek CA
  <LINE>
  <TAB>High<TAB>70<BR>
  <TAB>Low<TAB>54<BR>
</DISPLAY>
```

Displaying Images

Some HDML browsers can display simple monochrome images. The empty `<IMG>` tag specifies an image's text label and origin. Images may be obtained from origin servers on the Web or selected from a set of images the manufacturer stores in the smart phone's Read-Only Memory. This set of images is defined by Phone.com and consists of almost 200 clip-art images appropriate for wireless applications. These images are designed to look good on the phone, and may have been redrawn to best suit a particular phone's display. Whenever possible, select your images from the phone collection, as they will both load faster and look better on the display. The particular images available may depend on the version of the HDML browser you are using; check the Phone.com SDK documentation when selecting an image for your content.

You specify the image you want using the ICON attribute if the image is in the phone, or the SRC attribute if it originates from a Web site. Using the <IMG> ALT attribute, you should also provide an alternate text title for each image. This attribute is optional in HDML, but in practice it is essential, as some users may not be able to view the images you include with your content.

Adding an icon from the handset's ROM to our weather page gives us an HDML file like the one that follows. (The resulting display is shown in Figure 10-3.) The cloud image is only one of several weather-related images available from the HDML browser.

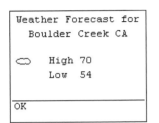

Figure 10-3. A card with formatted text and an image

```
<DISPLAY NAME="result" MARKABLE=FALSE>
  <WRAP>
  <CENTER>
      Weather Forecast for<BR>
      Boulder Creek CA
  <LINE>
  <IMG ICON="cloud" ALT="clouds"><TAB>High<TAB>70<BR>
  <TAB>Low<TAB>54<BR>
</DISPLAY>
```

If you need an icon that isn't available in the cache, you can always specify a URL source for the image, as shown here:

```
<DISPLAY NAME="result" MARKABLE=FALSE>
  <WRAP>
  <CENTER>
      Weather Forecast for<BR>
      Mercury Desert
  <LINE>
  <IMG SRC="scorched.gif" ALT="scorched"><TAB>High<TAB>680<BR>
  <TAB>Low<TAB>-354<BR>
</DISPLAY>
```

Displaying Special Characters

As in other markup languages, there are certain characters in HDML that can't be used except as parts of the markup tags themselves. HDML reserves the characters <, >, ".&, and $ to define tags and manipulate variables. If you need one of these characters elsewhere, you can insert it using one of the special empty tags shown in Table 10-3.

CHARACTER	TAG	PURPOSE
<	&lt;	"less than" symbol
>	&gt;	"greater than" symbol
"	&quot;	double quotation mark
&	&amp;	ampersand
$	$dol	dollar sign

Note: These are all empty tags.

Table 10-3: Reserved HDML Characters and Their Corresponding Entities

Tasks and Actions

<ACTION> tags assign tasks to be performed with specific interface elements. These tasks include navigating to other decks, returning to a previous deck, and making a phone call. Some interface elements on the handset itself—such as the soft keys, help key, and delete key—can be assigned to tasks. In general, these are keys, but they may be other controls, such as a rocker switch or touchscreen. (In the following discussion, I refer to them simply as *keys* for brevity.) The actual kind and nature of these buttons may change from manufacturer to manufacturer.

Actions may be defined either at the card level or at the deck level; if you assign an action at the deck level, it becomes the default for the entire deck, but may be overridden by a particular card. A card may contain one or more actions.

Assigning a Task

The empty tag <ACTION> binds an interface element to a task. Interface elements recognized by the browser include the following:

- The **ACCEPT** button.

- The **HELP** action (which may or may not be a physical control on the phone).

- The **PREV** action (which also may or may not be a physical control on the phone), used by the browser to navigate to the previously viewed deck.

- The soft keys, **SOFT1** and **SOFT2**. These keys have no physical labels; a label must be specified to appear on the liquid crystal display (LCD) in conjunction with the key.

- The **SEND** button, which initiates a voice call.

- The **DELETE** button.

Some of these elements are associated with expected phone-related behaviors, of course. For example, users expect the handset to dial a number when they press the SEND button, so assigning this button to the task of clearing a form, for example, is a bad idea. In general, it is best to assign most tasks to the ACCEPT buttons or to one of the soft keys.

The <ACTION> tag accepts the attributes listed in Table 10-4. I discuss each of these in the following sections.

ATTRIBUTE	PURPOSE
<ACTION>	Links a task to an interface element.
DEST	Specifies the destination of a GO or GOSUB task.
ICON	Specifies the name of an image from ROM to be used as a soft key label.
LABEL	Specifies a label for a soft key.
METHOD	Specifies whether a navigation should result in a GET or POST request.
NUMBER	Specifies a phone number to be called.
REL	When set to the value NEXT, "prefetches" the indicated DEST resource.
SRC	Specifies the source URL for an image to be used as a soft key label.
TASK	Specifies the task to be performed.
TYPE	Specifies the type of interface element to be assigned a task.

Table 10-4: Attributes of the <ACTION> *Tag*

Kinds of Tasks

The browser recognizes the following tasks:

- The GO task requests a URL.

- The PREV task displays the previous card to the user. (If no previous card is available, the PREV task is equivalent to the CANCEL task.)

- The CALL task initiates a voice call to a specified number.

- The NOOP task does nothing. It is used to hide the default behavior of an interface item or to override a deck's <ACTION> tag for a specific interface item.

- The GOSUB task pushes the current activity on the activity stack and requests a URL.

- The CANCEL task cancels the current activity and returns the user to the caller's CANCEL card.

- The RETURN task returns to the previous activity as specified by the caller's NEXT card.

The relationship between an interface element and a task is established through the <ACTION> TYPE and TASK attributes. TYPE specifies the interface element, and TASK, not surprisingly, indicates the task to be performed:

```
<ACTION TYPE=ACCEPT TASK=GO DEST="http://hdml.apress.com/index.hdml">
```

In this example, the ACCEPT key has been programmed to display the first card of the deck at http://hdml.apress.com/index.hdml when pressed.

Other attributes of the <ACTION> tag are used to control its behavior. The SRC and ICON attributes can specify an image resource for use as the label of a soft key's label, using the same syntax as the <IMG> tag.

The REL attribute can specify "prefetch" of the indicated resource. The following code

```
<ACTION TYPE=ACCEPT TASK=GO DEST="football.hdml" REL=NEXT>
```

causes the browser to fetch the deck at football.hdml while the current card is being viewed. Prefetching content increases the perceived speed at which a site functions—but it can increase airtime costs unnecessarily, as the user might not view some prefetched content (but must pay for the time spent fetching it anyway). The REL attribute should therefore be used carefully.

Data from HDML forms is sent to servers using the GO task in conjunction with the METHOD and POSTDATA attributes. By default, all navigation operations use the HTTP GET request method. Specifying METHOD=POST instructs the UP.Link server to send form data by using the POST method instead. (You can also specify the GET method using METHOD=GET to be explicit.) When the POST method is in use, the POSTDATA attribute should be set to an ampersand-delimited list of variables and names for the server to process, for example:

```
<ACTION TYPE=ACCEPT TASK=GO
  DEST="http://masamune.lothlorien.com/wx"
  METHOD=POST
  POSTDATA="choice=$menu&city=$city&state=$state&zip=$zip">
```

This sends the form variables named choice, city, state, and zip to the server, along with the values of the HDML variables menu, city, state, and zip.

Other attributes of the <ACTION> tag are used to specify arguments for a particular task. In the last example, we saw the DEST attribute used to specify the location of the deck to be displayed. The DEST attribute may be any partial or whole URL, evaluated relative to the current deck's URL. Similarly, the NUMBER attribute specifies a telephone number for the CALL task.

Although there is no way to set HDML variables directly, it is possible to set them from within an action using the VARS attribute. VARS accepts a list of HDML variables and their values in the same format as the one used by the POSTDATA attribute. For example, the following <ACTION> tag sets the default value for the variable ZIP:

```
<ACTION TYPE=ACCEPT TASK=GO DEST=#result VARS=ZIP=95006>
```

Actions Spanning an Entire Deck

An <ACTION> tag contained within a card assigns a task only for that card. There may be times when you'll want to replace the default action of an interface element over the scope of an entire deck. To do this, simply place the <ACTION> tag after the <HDML> tag that begins the deck. For example, you might want to offer help information that will be the same across most or all of the cards in a deck, as shown in this example:

```
<HDML VERSION=3.0>
  <ACTION TYPE=HELP TASK=GO DEST="#help">
... <-- Other deck cards here -->
<DISPLAY NAME="help">
  <ACTION TYPE=ACCEPT TASK=PREV>
  <WRAP>
You can obtain the weather by entering
```

```
   either the ZIP code of the place of
   interest, or the city and state.
</DISPLAY>
</HDML>
```

From any card in this deck, activating the browser's HELP interface will display the card named "help".

You may think that you can do the same to set variables across a deck using this method, but you can't. The VARS attribute of the <ACTION> tag isn't a task in its own right, but must be contained within another task. You can't simply embed it in any <ACTION> tag in the deck either, because the variables will only be set when the task is performed. Fortunately, there's another way to have the same effect.

The <NODISPLAY> tag creates a hidden card that immediately invokes its ACCEPT or PREV interface element. It can be used at the beginning of a deck to initialize variables that will be needed later or to encapsulate several references to a single URL. (I'll use this trick later in this chapter in "Integrating Actions and Input." The basic idea is to collect a form's various post operations on a single card, rather than scattering them across multiple cards. This makes making changes to a deck easier.)

The deck shown here demonstrates the difference between a hidden card that sets a variable and a deck <ACTION> tag:

```
<HDML VERSION=3.0>
<ACTION TYPE=HELP TASK=GO DEST="#help" VARS=setwhere=help>

<NODISPLAY>
  <ACTION TYPE=ACCEPT TASK=GO DEST="#home" VARS="setwhere=hidden">
</NODISPLAY>

<DISPLAY NAME="home" TITLE="Weather" MARKABLE=TRUE>
  <WRAP>
  The variable setwhere was set to $setwhere.
</DISPLAY>
<DISPLAY NAME="help">
 <ACTION TYPE=ACCEPT TASK=PREV>
 <WRAP>
  This card simply sets the setwhere variable.
</DISPLAY>
</HDML>
```

When this deck is first viewed, the display will read "The variable setwhere was set to hidden." But after a user has invoked HELP and then returned from the help card, the message will change to "The variable setwhere was set to help."

The <NODISPLAY> card, like any other card, occupies memory in the cache and takes the browser some time to process. Because of this, you should avoid creating paths that require the browser to navigate through more than one hidden card at a time, so that the user does not encounter a lengthy period of apparent inactivity.

Using Activities to Organize Decks

HDML activities are a mechanism for organizing decks. They provide some of the same structure as functions do in a structured programming language. An activity can be invoked, and when it is complete, control is returned to the calling activity. Control can be returned with either a RETURN task—if the activity has been successfully completed—or a CANCEL task—if the user has cancelled the activity in progress.

Within a single activity, all variables have global scope across cards and decks. Variables of the same name can be used in different activities without conflict.

Consider the process of reading e-mail. The user begins by perusing subjects, an activity unto itself. He or she then selects a particular message, initiating a new activity—reading the message. At this point, there are now two activities—one in progress (reading a message) and the other suspended (perusal of subjects). The user could then add a third activity to the queue by replying to the message. This activity can either be terminated by the user (if the reply is cancelled) or ended normally (if the message is sent). In the case of a termination, the user would return to the message display; if the message was sent, the user would return to the list of messages. Each of these activities can use a different set of variables and one or more decks. The activity stack allows a deck within a specific activity to refer to a calling activity (the preceding one) without previous knowledge of its behavior or location.

The GOSUB task invokes a new activity. Like the GO task, it navigates to the deck specified in its DEST URL. Unlike GO, however, it declares a new activity context, creating a new name space for all variables. Previous variables' names and values are saved, but cannot be accessed by the current deck or others until a RETURN or CANCEL task occurs. These variables may be shared with the previous activity using the RECEIVE and RETVALS attributes of tasks.

When a GOSUB task is specified, attributes of the <ACTION> tag are used to specify how variables are communicated between the two activities, and where the browser should go when the activity is completed.

The NEXT and CANCEL attributes of GOSUB specify which card should be shown *after* the DEST card that was specified for the GOSUB activity has been invoked and the activity completed. The RECEIVE attribute may be used to assign the variables used by the browser to contain results once the activity is completed.

Whether the activity ends with a CANCEL or a RETURN task, the variables specified by the RECEIVE attribute of the GOSUB task can be set using the value of RETVALS. Both RECEIVE and RETVALS should contain semicolon-delimited lists of variables to be set, but RECEIVE indicates the receiving variables used by the browser, while RETVALS contains the values to return. The first value in RETVALS is stored in the first variable in RECEIVE, the second value in the second variable, and so on.

Here's an example of two activities, which we'll call "home" and "sub":

```
<HDML VERSION=3.0>
<DISPLAY NAME="home">
  <ACTION TYPE=ACCEPT TASK=GOSUB DEST="#sub"
   NEXT="#next" CANCEL="#cancel"
   RECEIVE="result;">
  This is a specific activity, like reading message headers.
</DISPLAY>
<DISPLAY NAME="next">
  <ACTION TYPE=ACCEPT TASK=GO DEST="#home">
  You've returned from the nested activity.<BR>
  result is $result.
</DISPLAY>
<DISPLAY NAME="cancel">
  <ACTION TYPE=ACCEPT TASK=GO DEST="#home">
  You've cancelled the nested activity.<BR>
  result is $result.
</DISPLAY>
<DISPLAY NAME="sub">
  <ACTION TYPE=ACCEPT LABEL="Return" TASK=RETURN
   RETVALS="result;">
  <ACTION TYPE=SOFT2 LABEL="Cancel" TASK=CANCEL
   RETVALS="cancel;">
  This is a nested activity.
</DISPLAY>
</HDML>
```

Figure 10-4 shows the how the display screen will appear as the browser enters each of the activities and the variables are reset accordingly. When the deck is first opened, "home" is the current activity. Pressing the ACCEPT key creates a new activity, the "sub" activity, which will set a return value to either "result" or "cancel". The user can exit this activity by pressing either ACCEPT (which is assigned to the RETURN task) or SOFT2 (which has been assigned to the CANCEL task).

It's important to remember that HDML provides support for activities through the *activity stack*, a hidden part of the implementation of the HDML browser, and the navigation task GOSUB. This stack contains each activity's variable context, RECEIVE and RETVALS, NEXT URL, and related information. There are no explicit tags that specify an activity, its bounds, or how or where it can be invoked. Thus, using activities in HDML can be a little like object-oriented programming in a procedural language—you can do it, but the language doesn't offer any concrete mechanisms to help.

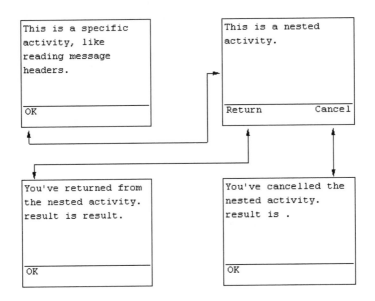

Figure 10-4. Flow between activities

The GOSUB task always starts a new activity, using system resources in the process. Be sure therefore that when a GOSUB task invokes a deck, it returns control to the calling deck with either a RETURN or a CANCEL task to end the new activity. Similarly, avoid using RETURN or CANCEL tasks in decks that can be accessed by GO, as they will end the current activity and may cause the browser to return to an activity other than one you intend.

Managing User Input

HDML supports both single-choice menu selections and text or numeric input. The <CHOICE> and <ENTRY> tags are used to present input choices and accept input from the user. Table 10-5 summarizes the tags and attributes for input handling in HDML.

TAG	ATTRIBUTE	PURPOSE
<CHOICE>		Indicates a choice card.
	DEFAULT	Specifies a default choice if the variable named by KEY is empty.
	KEY	Specifies the variable to contain the choice's VALUE attribute.

Table 10-6   *(continued)*

Table 10-6 (continued)

TAG	ATTRIBUTE	PURPOSE
	IDEFAULT	Specifies a default choice by index if the variable named by IKEY is empty.
	IKEY	Specifies what variable is to contain the choice's index.
	METHOD	Indicates whether the list should be numbered or unnumbered.
<CE>*		Indicates that this item is a choice that may be selected.
	VALUE	Specifies the value of a choice item.
<ENTRY>		Indicates an input card.
	DEFAULT	Specifies a default value if the variable named by KEY is empty.
	EMPTYOK	When true, allows empty inputs.
	FORMAT	Specifies what kind of input is valid for the entry card.
	KEY	Specifies a variable to contain the input.
	NOECHO	When TRUE, blocks input echo to the user.

*This is an empty tag.

Table 10-5: HDML Tags for Card Input

The Choice Card

The <CHOICE> tag creates a card that presents a menu of choices like the one shown in Figure 10-5. The user navigates among the choices on the screen using the arrows or number keypad, then presses the OK button when the browser highlights the desired choice. Cards featuring this tag are called *choice cards*.

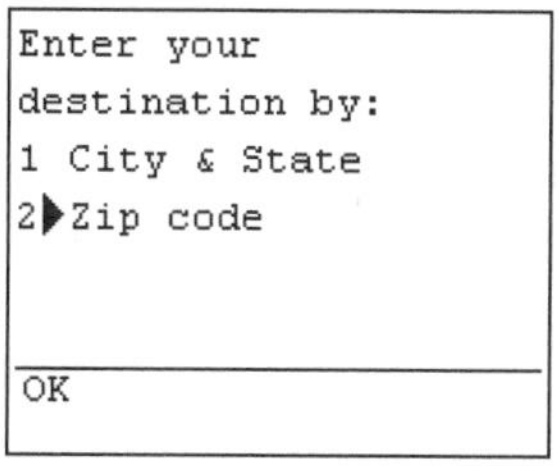

The attributes of the <DISPLAY> tag also apply to the <CHOICE> tag; it also has several more attributes that control the behavior of the choice interface.

Every <CHOICE> tag must contain at least one <CE> tag (which is short, one would guess, for "Choice Entry," but that's an unsolved mystery) to indicate choices to the user. The browser records the user's selection in a variable indicated by the <CHOICE> tag.

```
Enter your
destination by:
1 City & State
2▶Zip code


OK
```

Figure 10-5. A choice card

```
<CHOICE NAME="InputChoice"
  KEY=choice IKEY=num DEFAULT=zip>
  Enter your destination by:
  <ACTION TYPE=ACCEPT TASK=GO DEST="#$num">
  <CE VALUE=city>City & State
  <CE VALUE=zip>Zip code
</CHOICE>
```

In the previous example, the `<CE>` tag declares two choices. The first is the value `city`, and the second is the value `zip`. The user's selection is returned in the HDML variable named `choice`, and the index of the choice (counting from 1) is returned in the HDML variable `num`. Thus, if the user selects the first choice, the value of the HDML variable `choice` will be `city`, and the value of `num` will be 1. If he or she picks the second choice, `choice` will be `zip`, and `num` will be 2.

The attribute `KEY` of the `<CHOICE>` tag names a variable that will contain the user's selection. The browser will assign this variable whatever value is contained in the `VALUE` attribute of the selected choice. Similarly, the `IKEY` attribute names the variable to contain the index of the selected item. The `KEY` attribute is best used to assign human-readable values to variables for display in the current deck (such as phone numbers, prices, and so on), while the `IKEY` attribute can be used to return menu choices to a server for processing.

You can present two kinds of choices in HDML: *numbered* and *unnumbered*. By default, choices are numbered, and the user can make a selection with the appropriate number key. If, however, you set the `METHOD` `<CHOICE>` attribute to `ALPHA`, the choices will be displayed without numbers, and the only way to select an item will be to scroll to it and press the ACCEPT key. While there might be stylistic reasons for wanting a menu of choices without numbers, don't be too quick to create one. Unnumbered choices require the user to scroll through a list and enter multiple keystrokes, and therefore, are much slower to use than numbered ones.

The `DEFAULT` and `IDEFAULT` attributes of the `<CHOICE>` tag are used to select an initial choice. If the variable assigned to the `KEY` or `IKEY` attributes does not have a value when the choice card is invoked, the `DEFAULT` and `IDEFAULT` attributes are used to specify the default item by name or index, respectively. For example, the last HDML deck used the `DEFAULT=zip` attribute to indicate that, by default, the second item should be chosen. This could also have been done by providing the attribute `IDEFAULT=2`.

Choices are specified using the `<CE>` tag. At least one `<CE>` tag must be included in each `<CHOICE>` tag, although normally, of course, there are at least two. The `<CE>` tag specifies how the choice will be labeled, along with its value and, optionally, a task to be performed if it is selected.

Most `<CE>` tags are simple, like the one shown in the previous city and zip code example. The `VALUE` attribute of a `<CE>` tag specifies the value assigned to the `KEY` variable when the user makes his or her choice. This value can be used elsewhere in the stack or posted to the server where the content originates to be processed.

More complex `<CE>` tags contain a TASK attribute. The `<CE>` tag accepts all of the same attributes used to modify task behavior that the `<ACTION>` tag does (see Table 10-4). For example, the following tag causes the browser to navigate to another deck named on the current origin server if the user selects the option labeled "football".

```
...
<CE VALUE="football" TASK=GO DEST="football.hdml">
  <IMG ICON="football"><TAB>Football News
...
```

The Entry Card

The `<ENTRY>` tag creates a card, commonly called an *entry card*, with a single entry field. (See Figure 10-6 for an example.)

Input from the user may be numeric, alphanumeric, or constrained to certain kinds of characters. Like the `<CHOICE>` tag, the `<ENTRY>` tag declares a separate card and accepts all of the same attributes as does the `<DISPLAY>` tag. This card prompts the user for a zip code, then the browser downloads a new deck created by the Common Gateway Interface (CGI) weather on the remote server using that zip code:

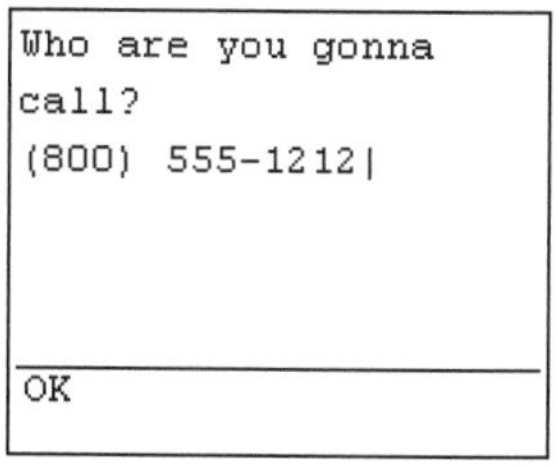

Figure 10-6. An entry card

```
<ENTRY MARKABLE=FALSE NAME="2" KEY=zip FORMAT=NNNNN>
  <ACTION TYPE=ACCEPT TASK=GO DEST="weather?zip=$zip" >
  Zip:
</ENTRY>
```

The `<ENTRY>` tag uses the KEY and DEFAULT attributes in the same manner as the `<CHOICE>` tag does. The KEY attribute specifies the destination variable for the input, and the DEFAULT attribute specifies a default value in the event that the destination variable was empty when the entry card was drawn.

By default, an entry card's action will be performed and the browser will navigate to a subsequent card only if the user provides some input. The EMPTYOK attribute—if set to TRUE—indicates that an action or that navigation may occur even if no input is made.

The NOECHO attribute prevents entered data from being displayed. The use of this tag is a matter of some debate, however. Many developers feel it is appropriate to disable the display for sensitive input such as passwords or access codes, while others argue that a phone display's small size makes it easy for the user to enter sensitive information discreetly, and that not echoing input detracts from a deck's user interface. For those who choose to use it, the tag is there. By default, the

NOECHO attribute is FALSE, but when it is set to TRUE, entered characters will not be displayed to the user.

The FORMAT attribute is the most important <ENTRY> attribute you'll use. It defines a *format specifier*, which is used by the HDML browser to validate input before allowing the user to continue. This specifier can include the following:

- **Integers**, which specify the number of characters that should match a specific type.

- **Asterisks**, which indicate that an arbitrary number of characters should be allowed.

- **One-letter type declarations**, which indicate the permitted type(s) of input. (Possible type declarations are listed in Table 10-6.)

CHARACTER	ALLOWED INPUT TYPE
A	Any symbolic or uppercase alphabetic character.
a	Any symbolic or lowercase alphabetic character.
M	Any symbolic, numeric, or uppercase alphabetic character (changeable to lowercase); for multiple character input, defaults to uppercase first character.
m	Any symbolic, numeric, or lowercase alphabetic character (changeable to uppercase); for multiple character input, defaults to lowercase first character.
N	Any numeric character.
X	Any symbolic, numeric, or uppercase alphabetic character (not changeable to lowercase).
x	Any symbolic, numeric, or lowercase alphabetic character (not changeable to uppercase).

Table 10-6: Character Type Declarations for HDML Input Specifiers

For example, if the required input was a United States zip code, you could use the input specifier 5N to restrict allowable input to exactly five numeric characters; for a United States two-letter state code, you might make the input specifier 2A.

Characters in the format specifier can also be used to format user input. For example, the input specifier \(NNN\)\ NNN\-NNNN accepts a phone number with area code. The display will begin with a leading parenthesis. Once three digits have been entered, a closing parenthesis will be displayed. After another three digits have been entered, a hyphen is displayed, and then the user can enter the final four digits.

The following card shows this input specifier in action, prompting the user for a phone number:

```
<HDML VERSION=3.0>
<ENTRY KEY=phone FORMAT="\(NNN\)\ NNN\-NNNN">
  <ACTION TYPE=ACCEPT TASK=CALL NUMBER=$phone>
  <ACTION TYPE=SEND TASK=CALL NUMBER=$phone>
  Who are you gonna call?
  </ENTRY>
</HDML>
```

Integrating Actions and Input

Actions are closely related to input. If HDML did not make it possible send user input to servers for processing, then the input would have little purpose. Similarly, actions are mostly useful for purposes of navigation based on user choices.

HDML variables provide the link between actions and input. Through variables, user input is collected and verified; these variables are then sent to the origin server through an action for processing. Origin servers can format their responses as simple HDML decks, or use variables again to enable dynamic content display. HDML's use of variables is, in a way, similar to that of programming languages, a fact best demonstrated by means of an example.

The weather example we've been using throughout this chapter uses variables to return the user's location to the origin server. Both the city/state and zip code inputs use the hidden result card to request the weather results. This hidden card merges the variables entered by the user and posts them to the origin server for processing:

```
<HDML VERSION=3.0>
<DISPLAY NAME="home" TITLE="Weather" MARKABLE=TRUE>
  <-- Home card display here -->
</DISPLAY>

<CHOICE MARKABLE=FALSE NAME="InputChoice"
  KEY=zip IKEY=num IDEFAULT=2>
  <ACTION TYPE=ACCEPT TASK=GO DEST="#$num">
  Enter your destination by:
  <CE VALUE=city>City & State
  <CE VALUE=zip>Zip code
</CHOICE>

<ENTRY MARKABLE=FALSE NAME="1" KEY=city FORMAT=*M>
```

```
     <ACTION TYPE=ACCEPT TASK=GO DEST=#state>
     City:
  </ENTRY>

  <ENTRY MARKABLE=FALSE NAME="state" KEY=state FORMAT=AA>
    <ACTION TYPE=ACCEPT TASK=GO DEST=#result>
    State:
  </ENTRY>

  <ENTRY MARKABLE=FALSE NAME="2" KEY=zip FORMAT=NNNNN>
    <ACTION TYPE=ACCEPT TASK=GO DEST=#result>
    Zip:
  </ENTRY>

  <NODISPLAY NAME="result">
   <ACTION TYPE=ACCEPT TASK=GO
     DEST="http://masamune.lothlorien.com/wx"
     METHOD=POST
     POSTDATA="choice=$numcity&city=$city&state=$state&zip=$zip">
  </NODISPLAY>
  </HDML>
```

The action for the hidden card result collects the user's input into a single POST request sent by the client to the server. The server-side script then uses the values provided to create and return a deck with the desired information.

In the last example, the deck's interaction with the server occupies its own card. The deck could also have been written with the server interacting over several cards, as shown here:

```
...
<ENTRY MARKABLE=FALSE NAME="state" KEY=state FORMAT=AA>
<ACTION TYPE=ACCEPT TASK=GO
  DEST="http://masamune.lothlorien.com/wx"
  METHOD=POST
  POSTDATA="choice=$numcity=$city&state=$state&zip=$zip">
  State:
</ENTRY>
<ENTRY MARKABLE=FALSE NAME="2" KEY=zip FORMAT=NNNNN>
<ACTION TYPE=ACCEPT TASK=GO
  DEST="http://masamune.lothlorien.com/wx"
  METHOD=POST
  POSTDATA="choice=$numcity&city=$city&state=$state&zip=$zip">
  Zip:
</ENTRY>
</HDML>
```

Although this version may make more sense at first and may be slightly faster to create, the approach chosen in the original answer is easier to debug and maintain. If you keep input cards separate from navigation, you'll only need to change a single card if your origin server URL changes.

HDML User Interface Design

One benefit of HDML is how relatively consistent the hardware on which it is available is. Most phones running the Phone.com browser share a large number of characteristics, which together represent the notion of an *abstract phone*, including the following:

- A fixed-width display, at least four lines deep and twelve characters across.

- Support for vertical scrolling.

- Support for the ASCII character set (upper and lower case).

- Numeric and alphanumeric character entry and editing.

- Choice selection via an arrow keypad, numeric keys, or rocker switch.

- Keys providing functionality for ACCEPT and PREV (to move forward and back between screens).

- One or two programmable soft keys with LCD labels.

At a minimum, all wireless handsets that can browse HDML share these characteristics. A few may have more features, but if you are developing content for these devices you shouldn't assume a higher level of functionality than what is outlined here.

In addition to the constraints of the handheld client, it is possible for the the UP.Link server to impose additional restrictions of its own. Most important, current versions of the UP.Link server cannot transmit HDML decks larger then 1492 bytes to browsers. Decks larger than this generate an error in the UP.Link server. Because the server's manipulation of a deck can change its size slightly, decks slightly smaller than this limit may also fail. To be safe, keep all uncompiled decks under 1200 bytes. If this challenge sounds onerous, bear in mind that 1KB of data on a screen phone is quite a bit—it can represent up to 21 screens of content!

Cards and Decks

Wireless handsets are not suitable tools for browsing long passages of text. Long runs of text are best avoided in HDML documents (as in WML ones), but if they are necessary, they should be broken up and presented on several different cards. When the phone shows multiple cards, the default key should move the user to the next card, and navigating backward should be accomplished by the handset's PREV key, not by a soft key.

As with HDML (and even wireless HTML), less is truly more. A deck should be simple, and content kept brief. Screens are small and bandwidth precious, making brevity necessary.

Tasks and Actions

The functions to be performed by soft keys can be assigned within HDML content. By convention, the first soft key should always be assigned to the default option, and the second one reserved for less common operations, such as "Menu," "Delete," or "Home." (One secondary action that doesn't belong on a soft key is a link to user help—use HDML's HELP action.) When many functions are available at one time, the best solution is to label the second soft key simply "Menu" and offer the user a list of choices.

Several hardware keys on a Phone.com phone can be overridden. When reassigning built-in keys, think carefully about what the user will expect when using your deck. Assign keys in such a way as to extend and integrate the phone's functionality with your deck, rather than to try to tailor its behavior to meet your expectations. Rewiring the phone's SEND button, which ordinarily is used to make a call, to display a help message will not only be difficult to use but outright annoying to most users. A better use of the SEND key, for example, is to map it to dial the number currently displayed on a card.

Data Input

HDML content must enable data entry, of course, but keep in mind that data entry on wireless phone handsets is always time-consuming and frustrating. Limit users' need to enter data by providing lists of choices wherever possible. Even these lists should be brief, preferably including less than seven items each.

Where text entry is unavoidable, use format specifiers to restrict input to the kinds of characters required. For example, since it makes no sense for a user to enter alphabetic characters in a United States zip code, it's best to use format specifiers to restrict entry to digits to prevent users from entering a letter in error.

Displaying Images

As not all HDML devices can display images, images should be used sparingly if at all in HDML-based content. Include an image in a deck only if it carries a meaning that text cannot. Every image should be accompanied (using the ALT attribute) by a text legend detailing its meaning. These guidelines are identical to the guidelines for using images with WML and HTML content.

Bookmarks

Bookmarks provide a quick way for users to return to a commonly viewed card—just as desktop-browser bookmarks enable easy access to commonly viewed pages.

Dynamic content, however, should not be bookmarked. If a page was generated using input from a form, for example, you would want to make only the beginning of the form interface markable, not the end results—this way, you can ensure users don't make the mistake of bookmarking the wrong page.

Summary

HDML provides a rich alternative to HTML for content intended for screen phones and similar devices. While HDML's organization and syntax predate WML, the two languages share many concepts, making it easy for developers to create both HDML and WML content for the same service.

Like WML, HDML presents content in decks of cards. Cards may contain text, choice lists, or user input. Navigation between cards and other activities take the form of tasks, which can be bound to interface controls such as the phone's ACCEPT key or a soft key. Using tasks and variables that contain strings, decks can collect and validate input from a user, perform simple operations, and display the results.

While HDML is a flexible language capable of some complexity, it is best to keep decks simple. Users usually interact with wireless devices, with one hand, in a variety of settings that divide their attention among several tasks. With this in mind, you should organize decks according to the activities they help a user to perform, and provide only the information that is necessary during a particular transaction.

Don't use images unless you're sure they're worth the size they occupy. When choosing an image, be sure to see if the phone provides one you can use through the <IMG> LOCAL attribute. And, of course, avoid user input unless it's absolutely necessary.

Custom Applications: When a Browser Won't Work

IN PREVIOUS CHAPTERS, we looked at three standard client technologies for delivering wireless Web content to users: standard Web pages, synchronized browsers, and screen phone markup languages. In almost every case, data can successfully be presented using one or more of these technologies.

However, there are times where a custom wireless application is necessary. In this chapter, I discuss the (very few) situations where choosing to develop such an application makes sense, and address some of the pitfalls you may encounter on the road to bringing a custom wireless application to market.

Deciding to Roll Your Own

Making the decision to write a custom wireless application is a difficult process. The development process is expensive—more expensive than most developers realize. Understanding why a custom application might be necessary, what the limitations are, and where the costs are likely to appear will help you make the choice between using an existing wireless Web application technology, such as the HyperText Markup Language (HTML) or the Wireless Application Protocol (WAP), and writing a custom application.

Motivations

The market for wireless applications can be broken down in to two categories. The largest to date has been the *vertical* market, which consists of custom applications narrowly targeted (hence "vertical") to meet the needs of customers in a specific enterprise, such as health care workers or salespeople who are always on the road. The second market, with significantly smaller numbers of applications, is the traditional consumer marketplace. Arguably, the wireless browsers themselves fall into this second category, although usually they are considered platform applications.

In many cases, vendors leap to create new wireless applications before they realize that using an existing wireless Web browsers is sufficient for their purposes.

There are many reasons why a custom application may *appear* to be more desirable than an existing wireless client technology, but most of those reasons boil down to a variation on one of these three:

1. A desire to create an interface with specific legacy systems.

2. A desire to meet specific user interface ("look-and-feel") needs, including corporate branding requirements.

3. The technological limitations of existing wireless client technology.

In fact, only the last of these reasons is usually compelling enough to justify the cost of a custom wireless application.

Legacy System Interfaces

Most vertical applications require an interface to an existing data processing system of some sort. Many of these systems predate the availability of wireless access by years, and consequently, may not make convenient hooks for even conventional Web access, let alone for wireless Web access. Most are based on large back-end databases running on Structured Query Language (SQL) database servers from major manufacturers; many have custom front ends to facilitate data access and manipulation.

In general, the amount of effort required to create a custom interface between these applications and a mobile client of any kind is comparable to the level of effort required to upgrade to a system with Web access and craft a wireless-accessible Web site. This process often involves upgrading a back-end database and wiring it to a Web application server—a common occurrence in today's intranets. There are costs associated with both options, of course, but upgrading generally provides users and operators with greater long-term flexibility than bolting a custom application onto an existing system.

Of course, upgrading a vertical system and making it compatible with a standard wireless Web interface takes a significant amount of work; you must create a custom interface between the wireless Web and the legacy system, then customize the content for wireless devices. But the techniques required are essentially only those needed for constructing a traditional Web application, with some extra attention paid to the content (whether it is HTML targeted for wireless devices, or WML or HDML for wireless terminals).

Custom Interfaces and Branding

While significant enhancements have been made recently to the user interfaces of many wireless devices, there's always room for improvement. Efforts to bring about such improvements can lead enterprise customers or platform vendors to develop applications that are specific to a particular wireless device (such as two-way messaging with a wireless mail client) or even, in some cases, to revolutionary products (such as the Palm VII wireless handheld computing device from Palm, Inc.).

Another common goal, related to this drive to improve the interface, is to closely associate a brand identity with an application. This is especially common for applications whose end customers—platform manufacturers or the recipients of a custom application—seek to differentiate between their applications and existing similar ones.

However, neither of these motivations—a desire to improve the interface or branding—is, in most cases, justification for the development of a custom application in a situation where an existing wireless Web application can do the job. The reason for this, though it may seem sadly mundane to those excited by technology, is a compelling one: cost. As I illustrate in the next section, the cost of developing a wireless application can be prohibitive.

Most of the usual goals of a custom interface—input validation, presentation of information, and innovative input schemes (such as handwriting recognition or alternate keyboards)—can be met by an existing browser running on a target platform that accepts third-party additions.

In the case of branding, the client usually seeks a series of subtle user interface changes, such as redefined controls, logos in prominent locations, and the adoption of specific guidelines for the look and feel (fonts, layout, and so on) of the various screens of the application. For example, one commonly requested feature is the ability to validate specific kinds of content, such as phone numbers, within an input line.

In almost every case, these requirements can be better met by tailoring content. Establishing a brand using content provides greater extensibility for the future and, most important, is almost infinitely easier than developing custom software to meet the same goals.

Overcoming Technological Limitations

The human drive to overcome limitations is what led to the development of wireless networking in the first place. If you're involved in developing a wireless application with the goal of overcoming one of the many existing technical limitations to wireless use (such as limited bandwidth or network roaming challenges), you are likely to be well aware of the difficulties of creating a new wireless application. The information in the remainder of this chapter may help you make some decisions as you go about developing your application, but you should probably start by considering the following basic question: *Does your business model support the work you're setting out to do?*

For some developers, business models may not be an issue. Those working in academic settings or corporate research centers are able to work at the cutting edge of technology with, at most, only a cursory examination of the business prospects for a project. (This freedom to explore without an immediately apparent business return has been an essential element in the development of some innovations that have eventually led to soaring profitability, such as the cellular radio networks now spanning the globe.) However, if your organization is a more traditional business, you should be sure that you can afford the relatively lengthy adoption period (probably measured in years, not months) necessary to build a large enough market share to recover initial investments. Many successful products in this sector have been created on the assumption that the first, or first few, iterations will be loss leaders, enabling developers to fine-tune the offering and incorporate customer feedback.

Pitfalls

The previous comments may well have dissuaded you from the desire to write a custom wireless application—and, in general, so they should. If you are still convinced that your purposes call for a custom treatment, however, read on. Your path may be fraught with danger.

Wireless application development is significantly more expensive than traditional development, for several reasons:

- The tools and technologies behind wireless devices are less robust.

- Wireless applications are more complex.

- The development cycle of a wireless application is longer.

- There is a relatively low level of integration between wireless hardware and the handheld computers necessary to support wireless software development.

Let's take a closer look at each of these challenges.

Fledgling Tools and Technologies

Mobile device technology is young and evolving quickly. Working in a new field can make a developer's life exciting—and risky.

At almost any juncture during the creation of an application, an unforeseen problem with the environment's tools or with the platform's documentation, can cause costly delays. Any experienced developer knows that this risk is part of the cost of doing business. But you should recognize that the level of risk is unusually high when you are developing for wireless devices because the tools used for wireless platforms are younger. They have significantly fewer years of constant improvement and debugging behind them than do more robust tool chains such as GNU or Microsoft Developer Studio.

While many platform vendors have worked to mitigate these risks by basing their tool chains on existing environments (for example, Microsoft Windows CE relies heavily on Microsoft Developer Studio, and the development of the PalmOS was hosted on Metrowerks' Codewarrior), the wireless developer is still in many ways on untested ground. In general, when developing a new application for the wireless Web, you can expect delays of between 10% and 20% of total development time to be spent dealing with tools, documentation, and bona fide platform bugs.

The causes of these delays are usually impossible to pinpoint except in hindsight. However, training time on new platforms, mistakes on the part of application developers, and problems with the new tools and platforms themselves are all factors you should anticipate.

Complexity

Every wireless application actually consists of at least *two* applications: the client application running on the wireless terminal and the back-end software providing content to the mobile device. Moreover, the software must communicate seamlessly across a wireless network, managing intermittent connections, signal strength fluctuations, and other issues with robust error handling in all conditions. All parts of the system must interact flawlessly in this demanding environment.

Consequently, wireless applications are more complex than traditional applications. Developers familiar with mobile development tend to underestimate this complexity, just as desktop developers with server-side experience tend to underestimate the risks of wireless terminal development.

Development Cycle Length

The development cycle for a wireless application is considerably longer than for a traditional application. The two factors discussed previously are part of the reason why. Another important factor is the need for *pilot releases*.

Developing a new application requires one or more pilot releases to test the application in the field. Pilot releases go one step further than beta tests, exposing potential markets to an entire end-to-end service while enabling product and service developers to fine-tune the offering. Usually, this fine-tuning involves changes to both the client software and the back-end network.

Pilot releases generally take calendar months of effort—besides the time required for the pilot program itself, significant lead time is needed to prepare for it. A successful pilot program may take over a staff-year to implement. Because most wireless applications require several such pilot releases before they are ready for a full-scale launch, this part of the development process adds significantly to both the cost and the schedule.

Lack of Integration

In the present marketplace, few wireless terminals are tightly coupled with devices that can support third-party applications. For a given handheld computing platform with software development kits (SDKs) available to the public, perhaps two or three devices exist that offer integrated wireless support. In turn, these devices will only link to one or two major networks, each of which is available only in a single country.

This situation is a key failing of the current state of wireless networking, and it creates several barriers to the adoption of new wireless applications. Most significant is the burden of system integration—selecting a provider and a wireless terminal and connecting the wireless terminal to the handheld device—which often remains the responsibility of the end consumer. While most wireless data users still belong to the "early adopter" segment of the consumer population, very few are able and willing to shop for and select wireless access terminals, services, and applications to support wireless access. Therefore, integrated, "all-in-one" products such as the pdQ smartphone from Qualcomm, Wireless Markup Language (WML) screen phones, and other products have typically outsold standalone "integration nightmares" many fold.

Furthermore, the relatively small number of integrated products available makes it unlikely that custom applications for them will achieve widespread adoption across a global market. Applications targeted for Palm Inc.'s Palm VII, for example, will only work in markets where the Palm VII has wireless coverage. At present, that means the United States, period. Writing software to a specific set of platforms narrows your market, limiting your application to the level of adoption of the platforms on which it runs.

Of course, like all problems, this one can be made into an opportunity by entrepreneurial businesses. For example, Omnisky offers integrated wireless expansion for the PalmOS platform via Cellular Digital Packet Data (CDPD), providing subscribers with a clip-on modem and wireless service that make any Palm Computing Platform device a wireless one.

Picking a Platform

If, even after this discussion of the pitfalls of custom wireless development, you still feel you simply can't make do with the existing wireless Web technology, your next step is to focus on selecting the best platform for hosting your wireless solution. This decision has three parts. You will need to choose all of the following:

1. A mobile computing platform.

2. A wireless network.

3. A server platform.

Handheld Platforms

Your selection of target platforms is at present limited to the "Big Three" of the handheld industry: the PalmOS, Windows CE, and Symbian. Each of these platforms has been adopted by a number of hardware manufacturers, and each has its own products.

Some developers let a preference for a particular hardware vendor drive their selection of a platform for their applications; others pick a platform first and then select from the available hardware that runs it. While the baseline features of a given platform are generally similar, hardware manufacturers have worked to differentiate themselves by adding features specific to their market expertise. Thus, for example, a company called Symbol has licensed the PalmOS and manufactures devices based on the Palm with extra hardware, such as a bar code scanner for vertical applications. Qualcomm offers the pdQ smartphone, which is based on PalmOS and offers Code Division Multiple Access (CDMA) wireless connectivity.

The handheld computing marketplace is changing so rapidly that it is difficult to predict what hardware options will be available at the time you're reading this book. However, the platforms themselves change relatively slowly (in general, most vendors release a new version of a platform every year to eighteen months, while hardware manufacturers release as many as three or four devices annually, each with different features). I have therefore opted to provide an overview of the strengths and weaknesses of the major platforms, leaving you to research the details of what hardware is currently available.

The Palm Computing Platform

The Palm Computing Platform, from Palm, Inc., is the most widely adopted hand-held operating system developed to date. The "Zen of Palm," as the company calls its approach, involves providing simple applications that connect customers with their data. These devices are targeted at a variety of users, although the most important segment that has adopted them so far are white-collar workers spending time away from their office.

Palm is arguably the simplest of the three platforms you're likely to encounter, as its creators emphasized low cost and ease of use. The environment provides networking support for IrDA (Infrared Data Association) and TCP/IP (Transmission Control Protocol/Internet Protocol) over the serial port to a modem or wireless modem, giving developers access to these facilities through a BSD socket-like interface. Application developers usually use C or C++, although several other third-party environments are available. These support high-level forms-based languages. In addition, other languages such as Forth and simple Java runtimes are also available and are used by a few developers. The leading environment is the one supported by Palm, Inc., based on Metrowerks' tool chain and available from either company. A version of the GNU tool chain has also been available to Palm developers for some time.

To date, hardware licensees include Qualcomm, which has a wireless device compatible with CDMA networks; Symbol, which offers local-area wireless devices for enterprise developers; and Handspring, which provides a low-cost–base handheld device that can be expanded with plug-in modules meeting their Springboard interface requirements. (Although they are not available as of this writing, it is speculated that there will soon be Springboard modules supporting access to both the CDMA and CDPD networks from manufacturers.) In addition, Palm, Inc.'s own Palm VII provides a wide-area wireless interface in the form of a Bell South Mobile Data modem, although this modem is not directly accessible to developers. Rather, the Palm VII offers services to obtain data via HyperText Transfer Protocol (HTTP) through its Web Clipping technology, using proxy servers stationed at Palm headquarters.

The chief advantage to the Palm platform is its wide adoption. With over five million users and a host of licensees to date, the PalmOS is virtually guaranteed continued success. Its relatively simple environment may feel primitive to many desktop developers, but it has the advantage of being simple to learn.

Microsoft Windows CE

Microsoft Windows CE has been Microsoft's most successful foray into the hand-held computing market to date. Available in a wide variety of form factors, from subnotebook to palm-sized, products running Windows CE offer an alternative to laptops, subnotebook computers, and palm computers as well.

The Windows CE programming environment will be familiar to any seasoned Windows developer, as its APIs (application program interface) are derived from existing Win32 APIs. However, the operating system was built from the ground up for portable and mobile devices, and has been streamlined to run on relatively low-cost hardware appropriate for the handheld consumer and vertical markets.

Microsoft's history in this market has been spotty at best. It came late to the table, after several successful products were already available from Symbian (formerly Psion) and Palm. Devices running Windows CE have been slow to sell, and many hardware manufacturers have already retreated from the Windows CE market after offering initial devices. However, Microsoft's ongoing commitment to the marketplace and its unparalleled ability to invest leave little doubt that it will remain an important player in handheld computing for years to come.

Not surprisingly, the tool chain for Windows CE is based on Microsoft's Developer Studio, whch enables developers to develop in C and C++ for the platform. For a time, Microsoft pledged support for Java, although as of this writing the battles between Sun Microsystems and Microsoft make a Java tool chain from Microsoft unlikely in the near future.

Very little Windows CE hardware exists with tight wireless integration. Novatel Wireless offers a representative product: a clamshell handheld device that runs Windows CE and sports an integrated CDPD wireless modem. Thus, end users of wireless applications on this platform must face the integration headaches discussed previously (choosing a wireless terminal, a service, and a hardware device appropriate for their needs and getting them to work together seamlessly). Fortunately, this problem is mitigated somewhat by the availability of wireless PC Cards that can be used with Windows CE devices to access the CDPD, ARDIS, Bell South, and other wireless networks.

Symbian

Symbian is a new name for an old player in the mobile hardware space: Psion. As a world leader in handheld computing, Psion has successfully worked to reposition itself as a leader in the wireless data marketplace. It provides platform technology to hardware manufacturers. Its chief offering, the EPOC platform, is being embraced by a number of manufacturers worldwide.

While its adoption has been relatively slow in the United States, the EPOC platform has been used worldwide for some time. It is also the only mobile device with a practical Java implementation; developers can choose either native EPOC development or the use of networked applications in Java.

Wireless hardware using the EPOC platform has been slow to appear. At present, most users of EPOC devices have third-party wireless handsets or expansion cards, rather than integrated devices. Doubtless this will change over the next year as current licensees develop and market their products.

Traditional Desktop Systems

Throughout this book, I've disregarded the traditional desktop platforms (Windows, Linux, other UNIX variants, and the MacOS) that are available on portable devices such as laptops or slates. In general, these devices offer near-desktop computing capacity, and the only limitation they are subject to regarding the wireless Web is the bandwidth of the wireless connection itself.

These devices represent the smallest consumer base for wireless access as well; few consumers are likely to tote a laptop computer for purposes that could be achieved with a handheld device. However, these devices do provide opportunities for wireless application development, and they enjoy a strong following in some vertical markets, such as outside plant management, where the need for custom wireless applications exists alongside the ability to pay. These applications can provide, for example, specialized access to corporate databases, records, and computational resources.

Before initiating the development of an application for one of these platforms, however, consider carefully whether the Web might in fact be the best option. The sophistication of desktop Web browsers can be leveraged with local content and scripts to provide a robust environment for users; only in rare circumstances is there actually a need for a custom application.

Only a very specialized application—say, one that needed to interface with portable test equipment or facility monitoring equipment—would require support a Web browser could not provide. In such a case, it might be appropriate to write a custom application or a custom component to be used by a laptop browser using ActiveX or another technology.

Wireless Networks

Selecting the right wireless network over which to deploy your new application is as important as choosing the development platform. In Chapter 2, I reviewed the major nationwide domestic networks, many of which are undergoing regular expansions and upgrades. In addition, the third generation of wireless networks will soon be up and running, increasing the alternatives available.

The choice of wireless network is driven by several factors:

- Coverage.

- Integration with the mobile client.

- Integration with the content servers hosting target content.

Availability of developer support and business relationships remains a primary concern for all wireless applications. If mobile users can't access application services at critical times, the business purposes that drove the development of the application in the first place are defeated.

Domestic developers must grapple with this problem every time they set out. Presently, no single network provides truly ubiquitous coverage across the United States. Two-way networks such as Bell South Wireless Data cover over 90% of the U.S. business population, but if your user base is in the remaining 10%, even that is simply not good enough. Many other networks, such as CDPD (available from various providers, including AT&T) have dramatically smaller coverage areas—often less than 50% of the business population. These numbers are for coverage of major commercial and industrial districts, where custom wireless applications are most likely to be used. Residential areas, especially rural ones, are likely to have even poorer coverage.

Determining whether a specific network provides the coverage needed for a particular application involves a combination of research and site surveys.

Some simple research is the starting point. Basic information regarding a network's coverage is often available directly from the service provider, usually as maps of areas covered or as lists of major metropolitan areas (MMAs) included in coverage. This information provides a general idea as to whether the network even approaches the coverage desired. Coverage maps from Metricom's Ricochet network, for example, show that it provides ideal coverage across much of the south San Francisco Bay Area, but a wireless application deployed in Boise, Idaho would likely require the deployment of a new Ricochet network to support its users.

Typically, the information obtained from service providers is insufficient to determine whether coverage will in fact be adequate for an actual application. *Site surveys* are often necessary to evaluate coverage more thoroughly. A site survey involves visiting a sample of the physical locations where the application will be used and testing the actual coverage available at those locations. In some cases an informal survey is all that's needed—the results might consist of a list of locations (usually stated as both street addresses and GPS locations to make analysis of them easier), along with signal strength measurements and logs of successful and failed communications from each one. A more serious survey would include multiple measurements at each location taken over a period of days or weeks to account for factors such as network load, radio propagation, and nearby sources of interference.

If you are developing an application that requires high reliability, it is essential to perform at least a few site surveys. In general, the information provided by service providers relates to average network availability as determined by a theoretical model of coverage. But the great number of variables that can affect wireless networks, such as signal reflections, noise sources, varying usage, the impact of weather on signal propagation, and a host of other factors, make it necessary to take direct measurements if you want to be sure of service availability in a specific area.

In addition to coverage, you must also consider such factors as throughput and latency when evaluating a network's ability to successfully deliver your wireless application. If a network doesn't reliably provide timely access to information, the application will be unable to meet users' needs.

Client Integration

Integration with the mobile client hardware is another important factor to consider when selecting a wireless network. Your coverage analysis may determine, for example, that a CDPD network such as that offered by AT&T is more than adequate for your application, but what if the hardware the client plans to use does not provide an easy physical or software interface to the network in question? Fortunately, this is problem can be solved for most combinations of network and hardware.

A wide variety of PC Cards provide access to almost every wide-area wireless network to mobile devices with PC Card slots. Even some devices that lack native PC Card interfaces can be refitted with aftermarket adapters offering PC Card support. Over time, this family of wireless PC Cards will likely be augmented by CompactFlash cards, increasing the available options for wireless connectivity. Finally, some wireless adapters are available as a snap-on cradle, which provides wireless connectivity at the expense of slight additional bulk.

These two-component solutions (mobile client with an add-on card or snap-on module) are not without their drawbacks, however. Requiring consumers to obtain an additional component shrinks the size of the consumer market considerably. In the vertical market, inventory problems can arise as devices become separated from their wireless modules and units are damaged or lost. Furthermore, nearly all mobile devices are taxed by the power a wireless access unit draws, and some subsystems require additional sources of power.

Hence, many users prefer the idea of working with an integrated device. Typically this means a super phone, although some Personal Digital Assistants (PDAs) with integral wireless access, such as the Palm VII, exist. They generally offer a wireless access unit within the case of the mobile computer and are explicitly designed with integration in mind. However, not all networks are supported by these integrated devices, so the needs of network coverage and cost must be balanced against the desire to use an integrated mobile device. Clearly, selecting a handheld device and a choosing a network are issues closely related.

The development of a large-scale application often involves trials with several combinations of hardware and network during the early concept phases, as platform capabilities, network coverage, and available network adapter configurations are evaluated.

Server Integration

A major advantage of wireless Web applications is the simplicity of integrating wireless access terminals and content servers. This simple, generally seamless integration has been achieved as a result of substantial time and effort investments on the part of network operators who realized that establishing good gateways between their wireless networks and the Internet was a key to wide adoption.

Unfortunately, anyone developing a wireless application from scratch must do much of this integration work anew. Most wireless networks use proprietary protocols tuned to the characteristics of the network itself, and creating an interface between the wireless network and these protocols is a significant task. These interfaces are often multi-hop, connecting server farms at the network operator with the wireless application host servers across a leased line—or, more recently, across Internet connections tunneling the wireless protocol.

Some companies have produced wireless middleware solutions, which mitigate much of the difficulty involved in integrating content servers with wireless network providers. This is a new and growing market involving mostly small companies, so developers should exercise some care when selecting a middleware solution. When evaluating a solution, remember to consider the following factors:

- **Longevity of the vendor**: How long will the provider be able to support a deployed solution?

- **Previous solutions to similar problems**: As with any technology acquisition, the experience of previous users can answer a lot of questions about the stability of a solution.

- **Supported platforms and networks**: Many middleware solutions are designed to provide network-independent services on both server and mobile client, but a particular option may not support the network you need.

Developer Support

The level of network support available to third party developers can be a crucial factor in the success or failure of creating a new wireless application. Few developers consider this aspect of selecting a network. This is unfortunate, because a provider's developer support staff can provide unique insight into how an application might best run on their network, and promote a deeper understanding of the wireless network's operation, capabilities, and limitations.

In addition to special pricing on products and services, most developer support organizations provide not only documentation, but also e-mail or phone support; some actually offer face-to-face consultations and reviews. Of course, these benefits are not without cost, but the up-front costs of a good developer support organization can more than offset the long-term costs of going it alone.

Server Platform

The selection of the server platform is often the most straightforward decision to make when preparing to develop a wireless application. Not only are there fewer options, all of them well understood, but in general, the wireless application will be interfacing with existing sources of data, which dictate the server platform and leave little choice for the developer in many cases.

Therefore, the questions to consider when choosing a server platform are relatively straightforward. They include the following:

- Will the platform have the longevity necessary to support the commercial rollout of your product or service? (That is, will your platform vendor be around to support you when you have both users and questions?)

- Will the platform offer the scalability necessary to support anticipated production loads of your application? (It may be easy to manage server load when your developer staff is using the single server you've installed, but what about when you have 30,000 users?)

- Will the platform support the tools you plan to use during development of the application?

In general, any of the leading platforms—Linux, other UNIX variants such as BSD or Sun Microsystems' Solaris, or Microsoft Windows 2000—are good choices for most purposes.

Summary

There's little need to develop a custom wireless application for all but a few cases. The power of wireless Web browsers, coupled with the ease of developing content for the wireless Web, make custom applications a costly choice by comparison.

It's unlikely you'll want—or need—to tackle this challenge. While the opportunity may sound exciting, few development organizations have the necessary expertise and resources. The early-stage nature of the available tools, the complexity of the task, and the need for multiple communicating applications (every client application after all, has a server someplace it will need to talk to!) all conspire to make custom wireless application development an expensive proposition.

For the few with the need and the means to create a wireless application from scratch, choosing platforms, networks, and hardware is a complex process, as each selection affects the others. Be sure to look over *all* of the available options; you will likely spend time exploring several before you find a match of hardware platform, form factor, and wireless network that meets your needs.

CHAPTER 12

Other Technologies

THE DEVELOPMENT OF TECHNOLOGIES for the wireless Web is far from over. Although the "Big Three" of wireless content— HyperText Markup Language (HTML), Wireless Markup Language (WML), and Handheld Device Markup Language (HDML)— have led the way, the opportunities available in this growing market have spawned a number of other technologies. Most of these are applicable to only a few developers, or are closely based on existing standards and require little learning by developers.

In this chapter, I discuss three of the most exciting technologies on the horizon for wireless Web developers: the eXtensible Markup Language (XML), server-assisted browsers, and Palm, Inc.'s Web Clipping Architecture.

XML

XML is positioned to become the lingua franca of the wireless Web. Its strength is in its generality: virtually any kind of structured data can be described in XML, and once described, the data can be presented in other formats. Moreover, XML is already being used for a host of server-server communication applications, which make it possible for different data servers to easily exchange information. The trend toward a common format for representing data will doubtless present new opportunities for both Web and wireless Web clients.

An XML Primer

Like documents created using other markup languages, XML documents consist of data and markup tags. XML, however, leaves the document author free to establish whatever markup tags may be appropriate for a given data set. For example, a site that manages intranet inventory content might use tags such as `<ITEMID>`, `<ITEMSERIALNO>`, `<LOCATION>`, `<OWNER>`, and `<BIN>` to represent the fields in a legacy inventory database; a site providing weather information might use markup tags such as `<WHERE>`, `<REPORTED>`, `<FORECAST>`, and `<TEMPERATURE>`. Also, unlike the previous markup languages we've seen, XML is likely to be "viewed" not by a subscriber client application, but by an application responsible for processing data in XML and returning another data set—perhaps in a different markup format, such as HTML.

An XML document consists of one or more *entities*, each containing either text or binary data (but not both). An entity's primary purpose is to hold content, such as XML or text. An entity consists of a *prologue*, which describes the entity, and the entity data itself, which consists of exactly one root entity tag and its contents.

XML entities and the documents that contain them need not be files on a host on the network. They may be, but they also may be built from aggregates of files, or constructed on the fly from databases—or virtually any other source.

A prologue must be the first entry in an XML file; it declares attributes about the XML entities that follow. The general form of the prologue is shown here:

```
<?xml attribute="value" attribute2="value"... ?>
```

Between the `<?xml` and `?>` are name/value attribute pairs describing the XML document that follows. In general, the prologue will contain at least one attribute specifying the version of XML version used within the entity. It looks like this:

```
<?xml version="1.0"?>
```

XML documents consist of character data interspersed with tags that define the document's structure. As in HTML, tags are enclosed within `<` and `>` characters; the text between these characters is the *tag name*. Tag names should begin with a letter or underscore and may contain letters, numbers, and underscores, but not spaces. Unlike HTML, XML is case-sensitive; thus, `<FORECAST>` and `<Forecast>` are interpreted as two different tags.

Following are some valid XML tags:

- `<WEATHER>`

- `<REPORTED>`

- `<FORECAST>`

- `<TEMPERATURE>`

- `<species>`

- `<Day_of_Week>`

XML tags, like those in most other markup languages, come in pairs. A closing tag has the same name as the corresponding opening tag, but with the addition of a slash at the beginning of the name, like this: `</WEATHER>`.

Some tags are *empty*—that is, they do not contain data and do not require closing tags. These are written with a slash at the *end* of the tag name. Thus, the XML equivalent of a break tag in XHTML (XML-compliant HTML) is written `<BR/>`. (In practice, HTML viewers may not handle this tag properly, and those trying to write XML-compliant HTML generally use `<BR>…</BR>`.) The correct use of the `<`, `>`, and / characters enable an XML parser to accurately construct a grammar when inspecting a new page of XML.

Tags may contain *attributes*, which describe their characteristics. Attributes are simply written as name-value pairs separated by an = after the tag name before the closing > character. For example:

```
<TEMPERATURE UNITS="Centigrade">24</TEMPERATURE>
```

describes a comfortably warm day on the California coast in centigrade. Attributes in general are used to qualify a tag and don't contain information specific about the thing being described. Thus, one could also describe the temperature as:

```
<TEMPERATURE UNITS="Centigrade" VALUE="24"/>
```

or

```
<TEMPERATURE><VALUE>24</VALUE><UNITS>Centigrade</UNITS></TEMPERATURE>
```

Note that these different examples, while they represent the same information to a human, are not equivalent in XML. The first states that VALUE and UNITS modify the absolute notion of a temperature entity (which has no content), while the other says that VALUE 24 and UNITS CENTIGRADE are *part* of a temperature entity.

Generally, if a piece of information is part of the data represented, it is best given as a tag; if it is meta data *about* the representation, it is best given using an attribute. Thus, because units and value are part of a temperature, the second is preferable for most applications.

Comments are an important part of XML, especially when a document is fleshed out during the initial development steps. A comment is flanked by the characters <!-- and -->, and can contain any text except the two-hyphen string --. XML readers ignore material contained in comments. Comments can be used to leave notes to future (human) readers of a file, to keep part of a document hidden from the XML parser during testing, or for similar purposes.

As you might guess, the < and > have special meaning within XML. In fact, there are five characters that have special meaning: <, >, ", ', and &. If you want to use any of these characters within an XML document for any other purpose, you must refer to them using their defined entity references (<, >, ", ', and < respectively). HTML defines a number of additional entities, such as &copy for the © (copyright) symbol, but these five characters are the only pre-defined entities established in XML. Thus, if explaining an XML tag *in* XML, I might write the following:

```
...use the &lt;TEMPERATURE&gt; tag to specify a temperature...
```

Of course, if you're actually writing about XML, remembering to include these entities on a regular basis can become awkward. To make things easier, you can define a special section of data in an XML document, called a CDATA section. Within a CDATA section, all text is treated as text except its closing delimiter, which is a double square bracket and a closing angle bracket:]]>. In fact, if comments appear in CDATA [code font] sections, they are not treated as comments; the XML parser essentially ignores all tags within a CDATA. Thus, the last example could also be written:

```
<![CDATA[
...use the <TEMPERATURE> tag to specify a temperature...
]]>
```

Putting this together, let's create a sample XML file that describes the current and future weather for my hometown:

```
<?xml version="1.0"?>
<WEATHER>
<LOCATION>
  <WHERE>Boulder Creek, CA</WHERE>
  <ZIP>95006</ZIP>
</LOCATION>
<REPORTED>
  <WHEN><DATE>11/25/2000</DATE><TIME>14:00</TIME></WHEN>
  <TEMPERATURE units="Centigrade">17</TEMPERATURE>
  <BAROMETER units="kPa">102.5</BAROMETER>
  <HUMIDITY>.65</HUMIDITY>
  <WIND><SPEED units="m/s">1.8</SPEED><DIRECTION>270</DIRECTION></WIND>
  <PRECIPITATION units="mm">0</PRECIPITATION>
</REPORTED>
<FORECAST>
  <WHEN><DATE>11/26/2000</DATE></WHEN>
  <PRECIPITATION_PROBABILITY>.10</PRECIPITATION_PROBABILITY>
  <NWS>A small system will move in during the afternoon,
     leading to partly cloudy skies with a chance of showers
     in the evening. Temperatures in the high teens.
  </NWS>
</FORECAST>
</WEATHER>
```

As the example shows, XML is readable, but verbose. Its strength is its flexibility: because tags can be dynamically defined for each document in a self-describing, hierarchical manner, virtually any kind of content can be marked up in XML for further processing.

Sharing Your Creation: The Document Type Definition

Of course, an XML document is of little use unless you and the others for whom it is intended agree on the tags to be used. The real strength of XML lies in the ability to not just create tags, but to describe tag sets in an interoperable way so that groups can agree to use a specific tag set to denote certain kinds of data. Already, this feature has led to the creation of specialized XML applications in several industries, including the following:

- **VoxML**: Used by telephone service providers to mark up text for speech synthesis in PBX systems, voicemail systems, and the like.

- **MathML**: Represents simple or complex mathematical formulae.

- **Open Financial Exchange (OFX)**: Used to represent financial transactions for personal finance packages.

In fact, parts of XML widely used, such as the eXtensible Style Language (XSL), are really expressed in XML itself. (Note that VoxML, MathML, OFX, and XML are not applications in the software sense, but rather, *applications of XML itself*. The distinction is confusing, but it needs to be made because the literature refers to them as XML applications.)

The Document Type Definition (DTD) is XML's means of capturing the syntax for a particular set of tags. While not required, a DTD enables others to understand what tags have been defined and how they're to be used. A group that has agreed on a standard set of tags and usage can specify them in a DTD, which will then enable XML parsers to validate documents based on its specifications. When describing a DTD, developers often refer to the XML application's tags as *entities*, so as not to confuse them with the tags used in the DTD.

Each tag in a valid XML document must be declared with an element declaration in an XML application's DTD using the `<!ELEMENT>` tag. This tag specifies the name, attributes, and usage of an element. The `<!ELEMENT>` tags are contained in a `<!DOCTYPE>` tag, which specifies the root element for a document.

In our previous example, weather information was recorded by a WEATHER element, which contained the various elements for a current measurement and forecast. Our DTD for that document would begin as follows:

```
<!DOCTYPE WEATHER[
 ... <!-- Elements for the weather information go here -->
]>
```

All of the individual elements would be included within the `<!DOCTYPE>` tag and defined with the `<!ELEMENT>` tag. An `<!ELEMENT>` tag specifies an element's valid contents, including enclosed entities, element ordering, and attributes. The syntax of the `<!ELEMENT>` tag is:

```
<!ELEMENT element-name usage>
```

where `usage` describes what may exist within the element. Thus, our element tag for the `WEATHER` element might be:

```
<!ELEMENT WEATHER (LOCATION (REPORTED | FORECAST)*)>
```

This simply says that the `WEATHER` element can contain a single `LOCATION` element and zero or more `REPORTED` elements and `FORECAST` elements, in either order.

In the usage portion of the `<!ELEMENT>` tag, you can specify that a tag can be used for any valid content by using the special keyword `ALL`. Or you can restrict usage to only character data with no markup tags (called "parsed character data") by using the special tag `#PCDATA`. In addition, a tag can specify the valid list of entities contained within itself using a simple list-based syntax. This syntax follows these rules:

- The list appears within parentheses.

- Order of the list is defined left-to-right, as the list is read.

- Elements that are required are delimited by commas.

- Elements that are exclusive are delimited by a `|`.

- An element can be followed by a * to indicate that it may appear zero or more times.

- An element can be followed by a + to indicate that t must appear one or more times.

- An element can be followed by a ? to indicate that it must appear exactly zero or one times.

- Lists can be treated as elements by containing them in parentheses (in other words, lists can be nested).

For example:

- `<!ELEMENT P ALL>` states that the P element can contain any kind of element or character text.

- `<!ELEMENT dl ( dt, dd )*>` states that a `dl` element consists of alternating matched `dt` and `dd` entities (this is simply the definition list style in HTML).

- `<!ELEMENT WHEN ( DATE, TIME? )>` states that `WHEN` must contain a `DATE` element, and no more than one time element after the `DATE` element.

- `<!ELEMENT CHAPTER (TITLE, ( PARAGRAPH, FIGURE )* )>` defines a `CHAPTER` element that must begin with exactly one `TITLE`, and then contains an alternation of `PARAGRAPH` and `FIGURE` entities.

An element can be declared empty by using the special tag `EMPTY`.

Each element must specify its attribute list (if any) using the `<!ATTLIST>` tag. This tag's structure is similar to that of the `<!ELEMENT>` tag:

```
<!ATTLIST element-name attribute-name type default-value>
```

Possible types are listed in Table 12-1. In general, the most commonly encountered type is `CDATA`, which indicates that an attribute's value may be a string. Less commonly, the `ID`, `IDREF`, or `IDREFS` attributes are used to establish relationships among specific XML entities. Another common type is the enumeration—a list of possible values from which only one may be chosen:

```
<!ATTLIST ITEM show ( true | false ) "true">
```

You have the option of specifying a default value at the end of an `<!ATTLIST>` tag like I did here; by default, `show` will be `true`. The default may be a specific value like this one, or you can use one of the following special values: `#REQUIRED`, which indicates that an entry is required but that no default is specified (the DTD may not have a good idea what the default is, but a value will be necessary); `#IMPLIED`, which shows that an entry is desirable but no default can be specified (there's no good default, but the user should supply one, and if not, the XML application is free to do what it wishes with the omission), or `#FIXED`, which indicates that the default value cannot be overridden.

ATTRIBUTE TYPE	MEANING
CDATA	Character data without markup tags.
ENTITY	The name of an entity declared in the DTD.
ENTITIES	A list of the names of entities declared in the DTD.
ID	A unique identifier (not shared by any other ID attribute in the document).

Table 12-1   *(continued)*

Table 12-1 (continued)

ATTRIBUTE TYPE	MEANING
IDREF	A reference to an ID attribute elsewhere in the document.
IDREFS	A list of references to ID attributes elsewhere in the document.
NMTOKEN	A name in XML syntax for use with other XML tags.
NMTOKENS	A list of names in XML syntax.
NOTATION	The name of a notation specified in the DTD.
enumeration	A list of possible values from which exactly one may be chosen.

Table 12-1: Attribute Types for XML Elements

Although XML comes with some predefined entities, there are many reasons why a content developer might want to add additional ones. The most common reason is a desire to create shortcuts for entering common long strings of text, such as a name, header, footer, or other information. The <!ENTITY> tag in a DTD specifies additional entities:

```
<!ENTITY entity-name value>
```

You can extend this definition capability to external entities—those defined outside the DTD—using the SYSTEM keyword. If, for example, a Web site contains a footer that is required on all pages, it could be kept in an independent location, and authors could reference it on each page using the entity &footer, which they have defined in a DTD as follows:

```
<!ENTITY footer SYSTEM "http://www.site.org/entities/footer.xml">.
```

Because the DTD serves to document a markup language for both computers and humans, comments are an essential part of it. DTD comments are written using the same syntax as elsewhere in XML, using the <!-- and --> delimiters.

A DTD can precede the XML in a document, or it can be a separate document to which the XML document refers. This second possibility is a powerful advantage of XML, as it means that one DTD can serve to describe a multitude of documents across many Web sites. In the case of a DTD specific to a single document, the <!DOCTYPE> tag simply precedes the XML content like this:

```
<!DOCTYPE WEATHER[
  ...
]>\
<DOCUMENT>
  ...
</DOCUMENT>
```

For a shared DTD, however, this `<!DOCTYPE>` tag is omitted. Instead, the DTD is denoted using a special `<!DOCTYPE>` tag with a `SYSTEM` keyword specifying the URL where the DTD can be found:

```
<!DOCTYPE WEATHER SYSTEM"http://www.lothlorien.com/weather.dtd">
```

This indicates that the DTD can be found at the URL `http://www.lothlorien.com/dtd`.
   Our weather DTD looks like this:

```
<!DOCTYPE WEATHER[

<!ELEMENT DATE #PCDATA>
<!ELEMENT TIME #PCDATA>

<!ELEMENT WHERE #PCDATA>
<!ELEMENT ZIP #PCDATA>
<!ELEMENT WEATHER (LOCATION (REPORTED | FORECAST)*)>
<!ELEMENT LOCATION ( WHERE? | ZIP? )>
<!ELEMENT DIRECTION #PCDATA>
<!ELEMENT DIRECTION #PCDATA>

<!ELEMENT WHEN ( DATE, TIME? )>

   <!ELEMENT REPORTED ( WHEN, TEMPERATURE, BAROMETER?, HUMIDITY,
                        WIND?, PRECIPITATION? ) >
   <!ELEMENT TEMPERATURE #PCDATA>
   <!ATTLIST TEMPERATURE units ( "centigrade" | "farenheight") #REQUIRED>
   <!ELEMENT BAROMETER #PCDATA>
   <!ATTLIST BAROMETER units ( "kPa" | "inHg" | "mmHg" | "mBar" ) #REQUIRED>
   <!ELEMENT HUMIDITY #PCDATA>
   <!ELEMENT WIND (SPEED, DIRECTION)>
   <!ELEMENT SPEED #PCDATA>
   <!ATTLIST SPEED units ("m/s" | "mi/h" | "km/h" | "knots" ) >
   <!ELEMENT DIRECTION #PCDATA>
   <!ELEMENT PRECIPITATION #PCDATA>
   <!ATTLIST units ("mm"| "cm"| "m"| "in"| "ft" )>

<!ELEMENT FORECAST ( WHEN, PRECIPITATION_PROBABILITY?, NWS?) >
   <!ELEMENT PRECIPITATION_PROBABILITY #PCDATA>
   <!ELEMENT NWS #PCDATA>
]>
```

The Future of Wireless XML

If all of this looks removed from the wireless Web, rest assured that it will be of more use than you may think. While an XML browser that can handle arbitrary XML, DTDs, and the style languages (such as Cascading Style Sheets or the XSL) is unlikely to be available for many years, applications of XML on the server side of the Wireless Web abound already.

One challenge every wireless developer will need to overcome is the support for multiple kinds of devices. In general, developers have three choices:

- Establish minimal-functionality content targeted to the lowest common denominator—functionality for all wireless devices.

- Establish one or more sets of content targeted to different classes of wireless devices.

- Establish a common set of data and build scripts that can format the data for different devices on the fly.

Clearly, when large volumes of data or frequently changing content are involved, the third alternative is the most appropriate. As discussed in Chapter 7, while virtually any data representation and scripting can be used to exchange content, XML provides a level of structure and interoperability that is invaluable for large sites. More important, companies such as IBM and Microsoft are developing tools to aid in the development of server-side applications capable of merging XML content with style sheets appropriate for different browsers.

XML has been widely embraced by the open source movement as well. Not only are many of the primitive tools needed for building XML solutions available as open-source projects, but XML is also being used as the framework for Cocoon, a next-generation framework for Web distribution that will enable data representation in XML and the server-side application of style sheets, so that the served data can be represented in HTML, WML, or other formats.

Server-Assisted Technologies

One set of enabling technologies for the wireless Web is the use of back-end proxy servers to preprocess data for mobile clients. These servers act as intelligent gateways, reducing Web content to a form suitable for the wireless network and the client device. The idea is not new: AllPen Software demonstrated a prototype system for the Apple Newton in 1995 and shipped a commercial system enabling handheld devices to view reduced HTML content in 1996.

Figure 12-1 shows the deployment view for a typical server-assisted browser. This diagram should look familiar if you've investigated WAP or HDML (see Chapters 8 and 10), as those two architectures are arguably specialized examples of server-assisted browsing. Most server-assisted browsers can display at least some subset of HTML—the server reduces the content to this supported HTML, formatting images appropriately, and possibly gating the wireless device's protocol on and off the Web.

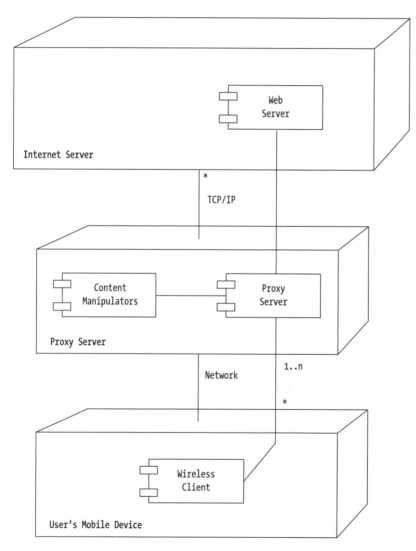

Figure 12-1. The deployment view for a server-assisted browser

Server-assisted technologies enable content providers to create a single version of content, usually one intended for minimal-functionality desktop browsers, and still allow clients with bandwidth or other constraints, who otherwise would be unable to read this content, to have access to it. As content passes through the proxy server, the HTML connected with it is modified or removed as necessary to make the Web content appropriate for the wireless terminal. Advanced examples of this technology can often perform other operations, such as manipulating images, managing HTML content by applying scripts using the Document Object Model (DOM), or actually converting content from one format to another.

Server-assisted technologies historically have been used by network operators to improve network throughput and make pages available to subscribers who would otherwise be unable to view (or perhaps could only view poorly) the content. Increasingly, they will be of use to content providers, who will be expected to serve a wide variety of devices with differing capabilities. With these technologies, developers can create their content only once, but ensure it is available on a diverse number of devices without using scripting or conditional access (see Chapter 7), or XML.

ProxiWare

Proxinet.com, a subsidiary of Puma Technology, Inc., offers ProxiWare, a content conversion server for mobile devices. This server, used by the company's ProxiWeb mobile Web browser for the Palm Computing Platform, is also available for enterprise developers.

The ProxiWare server solution streamlines HTML by removing unsupported content and reduces images as necessary to fit client devices.

Unless you're an enterprise developer planning on the wide deployment of devices running the ProxiWeb browser and a ProxiWare server, you're not likely to directly use the services of the ProxiWare proxy server. However, the ProxiWeb Web browser with Puma's ProxiWare public proxy server remains the mobile browser of choice for many consumers using Palm Computing Platform.

IBM's Transcoding Proxy

IBM's Transcoding Proxy, developed by the alphaWorks team at IBM, provides server-side bandwidth reduction and translates data from one representation to another. The technology is based on a transcoding proxy module inserted into the popular Apache server. This proxy is capable of discarding HTML comments or JavaScript tags to reduce bandwidth, reducing tables to lists, or creating a text summary of a page by discarding formatting tags. In addition, it can reduce color images to monochrome or grayscale ones, or reduce the amount of color to "thin" an image.

This transcoding proxy technology can also be seen in conjunction with wiredAnywhere, a WML browser in the proof-of-concept stage for the Palm Computing Platform.

Spyglass Prism

Probably the most widely used server-assist technology is Spyglass Prism, which is available from Spyglass, Inc. for Windows or Solaris servers. Currently used by companies such as Fujitsu, Seiko Epson, and Riverbed Technologies, Inc., Spyglass Prism provides Internet access to various kinds of mobile clients.

Spyglass Prism enables operators to apply two key techniques to reduce the demands their content places on wireless access terminals. The most widely understood technique is *content reduction*, in which the proxy manipulates content from origin servers. Spyglass Prism can remove a set of HTML tags, render tables as fixed regions, and reduce image size by both scaling and removing colors.

One particularly powerful converter module performs *text extraction*, whereby portions of a document are extracted from the source page and used to synthesize a new document. A client's request for a desktop URL causes the converter module to perform a set of text extraction operations and create a new page to be returned to the client. Typically, this page is formatted more simply than the original, making it more appropriate for the wireless access terminal.

Each manipulation is performed by a different component of Spyglass Prism. This feature enables developers to create specific components for new applications— for example, a component could be crafted to convert Graphical Interchange Format (GIF) and Joint Photographers Expert Group (JPEG) images to Portable Network Graphic (PNG) for a new minimal-functionality mobile browser that supported only PNG images. These components can placed in any sequence required by a particular translation, which makes this technology highly efficient and flexible. For example, an image conversion module need only worry about the mechanics of actual conversion; if an operation is needed that also rescales the image, an image-scaling module can be placed ahead of the converter.

Administrators can configure a database of devices and users, and instruct the proxy server to select an appropriate translation sequence based on the device type in question and a user identifier. An interface can also be constructed for network operators that allows different subscribers to select different levels of filtering based on such variables as the device being used, the cost of the wireless connection, and network bandwidth. Hooks are available for both authentication-based user logins and logins based on device authentication through a network ID or other mechanism. In addition, devices can identify themselves via a set of standard HyperText Transfer Protocol (HTTP) headers, which can be used to trigger specific translations based on configurations in the user and device databases.

One innovative development Spyglass has created based on this is an HTML-WAP gateway, which enables WAP browsers that allow for only limited formatting within a screen-based user metaphor to access HTML content. The possibilities for developers to combine Spyglass Prism with their content in similar ways are limitless.

Palm, Inc.'s Web Clipping Application Architecture

Palm, Inc.'s Web Clipping Application Architecture is a unique use of server-assisted browsing that allows certain devices, such as the Palm Computing Platform Palm VII with wireless interfaces, to access specific Web sites via a wireless network. Responsibility for the content is shared between the server where it originates and a Web Clipping Application (WCA), which resides in the handheld device. (Historically, the WCA was known as the "Palm Query Application," and it is so referred to in a great deal of the existing literature.)

The WCA residing on the wireless terminal isn't truly an application; rather, it's a highly compressed aggregation of Web content formatted for the Palm device created by a desktop tool. This content may contain both images and HTML content, including links to external Web resources. The Palm WCA viewer displays this package of data with an interface reminiscent of a simple Web browser. (Figure 12-2 shows a sample of what this browser looks like.)

Figure 12-2. The Palm Computing Web Clipping Application viewer

When a link in a WCA is selected, the viewer application finds the content—either in the WCA itself, or by making a wireless connection to a proxy server—which in turn requests the information from the origin server and returns it to the proxy. The proxy then performs a strenuous series of operations on the content, reducing it through tag stripping, representing tags as individual (and shorter) words called *tokens*, and compressing the content before returning it to the device.

WCAs can be used for a variety of purposes: transferring formatted static content to a device, aggregating a number of Web links, or tailoring content for a Palm device.

The first of these—viewing static information on a handheld device—is surprisingly underused. The desktop program responsible for creating a WCA can be used to aggregate content from a number of devices, and the fact that content can be provided in HTML means that sophisticated typesetting, including the embedding of links and graphics, is possible. WCAs are thus a good distribution mechanism for such static content as e-texts, corporate handbooks, and other information.

Then there is content that is relatively dynamic, but formatted simply, such as remote log files or other measurements. A WCA can be constructed as an index page that leads to an array of content, with links to the content that changes. Care should be taken when establishing this kind of application, however, to be sure that the amount of data to be obtained over the wireless network is not excessively large.

The most typical WCA is a *query application*, in which a targeted page in the WCA connects with a server CGI over the wireless network to obtain a specific piece of information—in other words, to answer a query (hence the name). Most users will gain the greatest value from these query applications, as their mode of operation most closely mirrors the needs of most mobile users, who want specific pieces of information quickly, such as "What was that stock's price?" and "What does 'lachrymose' mean?," and so on.

Query applications can manage both clear text (not requiring encryption) and secure data. Because the Palm Web Clipping Architecture provides over-the-air encryption of HTTP requests, it can be useful in e-commerce solutions.

WCA Fundamentals

Developing content for a WCA is similar to developing content for other mobile devices. However, the Palm platform has some specific limitations that must be met:

- Images must fit the Palm's screen (153×144 pixels).

- Images are best viewed if produced in one or two bits per pixel. (Animated GIF images are not supported.)

- Here are some significant HTML restrictions: Rendering of nested tables, frames, or scripting is not supported, nor are small font sizes, superscripts, subscripts, or vertical alignment using the `VALIGN` attribute, the `LINK` tag, or the `ISINDEX` tag.

Generally, content that resides on the device is written first, followed by any content and/or scripts to be provided by a Web server. Content for the device could be a single HTML page or multiple pages with links between them. These links should be relative—in other words, in the form `file.html`.

WCA-Specific HTML Entities

There are several entities unique to WCAs that developers can use to do the following:

- Create new interface components not found in traditional HTML.

- Uniquely identify the device in form submissions.

- Control the operation of the viewing application.

These entities are presented in Table 12-2, and discussed in more detail in the section to follow.

ENTITY	PURPOSE
`<META NAME="PalmComputingPlatform" VALUE="true">`	Indicates that content is tailored for WCAs.
`<META NAME="HistoryListText" VALUE="value">`	Creates a custom label for a page's history entry in the history picker.
`<META NAME="PalmLauncherRevision" VALUE="value">`	Establishes the WCA's version number for PalmOS dialog boxes.
`<META NAME="LocalIcon" value="file.gif">`	Adds a local image to a WCA for over-the-air references.
`%DEVICEID`	Is replaced with a device's unique ID, if available.
`%ZIPCODE`	Is replaced by the zip code of the base station serving the wireless terminal.
`DATEPICKER`	Presents a form element for validated date entry that adheres to Palm's user interface constraints.
`TIMEPICKER`	Presents a form element for validated time entry that adheres to Palm's user interface constraints.
`<SMALLSCREENIGNORE>`	Causes the Palm proxy server to disregard the content between the start and end of the entity.

Table 12-2: WCA-Specific HTML Entities

As a developer, the most important of these for you to remember is the meta entity with the name/value pair of `PalmComputingPlatform` and `TRUE` (that is, `<META NAME="PalmComputingPlatform" VALUE="true">`), which indicates to the Palm proxy server that a given page was authored with the Palm in mind. If you do not include this tag, the proxy server may manipulate the content in ways you don't expect in the hopes of making the page smaller or more presentable on the device.

Another meta entity you're likely to use is `HistoryListText`. Each WCA keeps a separate history, which users can access by tapping the pick list in the upper right corner of the screen. By default, a page in the history list is named after the title of the page, but you have the option of creating a separate name for each item in the list. If you include the special keywords `&DATE` and `&TIME` in a name, they will be replaced by the viewer with the page's download date and time.

Two other special meta entities are used only in the main page of a WCA. `<META NAME=PalmLauncherRevision" value="value">` enables a WCA to specify a recognized PalmOS version number. `<META NAME="LocalIcon" value="file.gif">` enables a WCA to include an image that is not referenced anywhere else in the WCA, but referenced from content downloaded by the device over the wireless network. Developers can use this feature to cache commonly used graphics such as company logos and weather symbols within the WCA, minimizing the application's reliance on over-the-air image transmission. WCAs can access such local content using URLs in the form `file://myapp.pqa/file.gif`, where `myapp.pqa` is the file name of the WCA and `file.gif` is the name of the image to be accessed specified in a `LocalIcon` `<META>` entity.

Two form entities are used to help developers identify a query's origin: the `%DEVICEID` and `%ZIPCODE` tags. The `%DEVICEID` tag provides a unique identifier for the wireless access terminal. While the identifier is guaranteed to be unique, you should not attempt to use its contents to determine additional information about the device. This identifier will be prefixed by a −1, 0, or 1, which indicate, respectively, that the device was positively identified, that no positive identification could take place, or that the validation failed and no identification was available.

The `%ZIPCODE` entity provides the U.S. zip code of the base station from which the wireless terminal is accessing the network, allowing query applications to provide a medium-scale notion of their location. Typically, this entity is used in a form, where the service provider's proxy servers replace `%ZIPCODE` with the base station's zip code.

The following example obtains both a device's unique identifier and location of use. The identifier is returned in the form variable `userid`, while the zip code of the base station is returned in the form variable `location`:

```
<FORM>
  <INPUT TYPE="hidden" NAME="location" VALUE=%ZIPCODE>
  <INPUT TYPE="hidden" NAME="userid" VALUE=%DEVICEID>
</FORM>
```

Note that these entities are not foolproof. A user with a Web browser can spoof the %DEVICEID value, and the %ZIPCODE entities may not give a very precise location because in some areas one base station serves several ZIP code regions.

WCAs also support specialized entities for entering dates and times in forms. The DATEPICKER and TIMEPICKER entities produce custom input items that match the Palm Computing Platform's user interface guidelines for the entry of date and time in forms (several examples are shown in Figure 12-3). They are used as follows:

```
<P><FORM ACTION="…" METHOD=POST>
   Birthdate: <INPUT TYPE=datepicker NAME=date><BR>
   Birth time: <INPUT TYPE=timepicker NAME=time><BR>
   <INPUT TYPE=submit NAME=Go VALUE="Submit">
</FORM></P>
```

Figure 12-3. The DatePicker and TimePicker form items in action

Any developer constructing one set of content for both a WCA and a conventional Web browser will want to take advantage of the <SMALLSCREENIGNORE> entity. As its name implies, this entity causes the Palm servers to ignore any content referred to within this tag.

The PQA Builder

Once you have created new Web pages suitable for your WCA, they are bundled using a tool from Palm, Inc., called the PQA Builder, which is short for "Palm Query Application," the name used by Palm staff before the term "Web Clipping Application" originated. This tool creates a Palm database from the HTML files and some additional information, which can then be installed on Palm devices and viewed using the Web Clipping browser application.

The application is simple to use. Figure 12-4 shows how it appears on a Macintosh (the interface is similar under Microsoft Windows). The main list shows the HTML files to be included in the WCA and their sizes. Below this list, you see the uncompressed and compressed sizes of your WCA (the compressed size tells you how big it will be on the device). You can pick small and large icons that appear in the Palm's application launcher using the buttons on the right. To use the tool, follow these steps:

1. Open the main page of your WCA—the one viewers should see when launching the WCA itself—with the menu File ➤ Open command.

2. If you want to choose different custom icons to appear in the Palm's Launcher application (where installed WCAs are displayed), click the Small Icon or Large Icon button and select the appropriate images.

3. Enter a name for the resulting WCA (usually a file in the form name.pqa) and press the Build button.

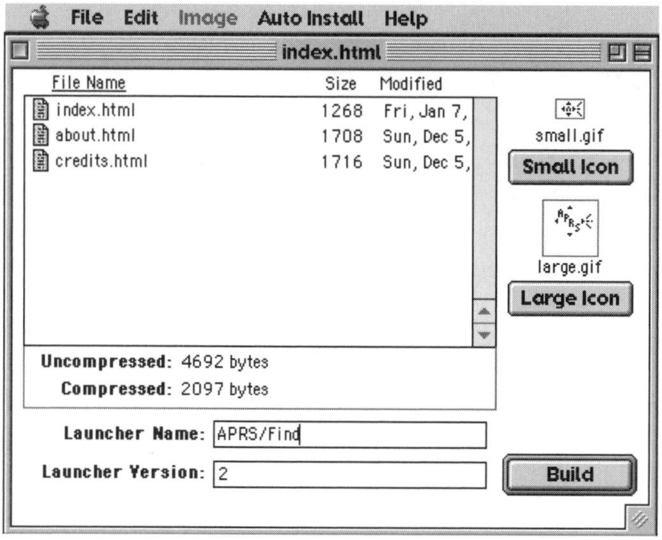

Figure 12-4. The Palm Computing PQA Builder application

Once the WCA has been produced, it can be tested on either a Web-Clipping–enabled device or on the Palm emulator. Freely available from Palm Computing Web's site, this emulator needs a Read-Only Memory (ROM) image from a device such as the Palm VII in order to operate. For testing purposes, you can either obtain the ROM from Palm, Inc., or use the emulator's own tools to extract it from a Palm VII.

A Sample WCA

For a sample WCA, I've returned to the example used in Chapters 6 and 7 involving applications that indicate the position of APRS stations. Figure 12-5 shows a WCA that accomplishes the same task—presenting a map showing a station's location when the user enters its call letters. The WCA has three parts:

- An index page that prompts the user for the station's call letters and the desired map scale.

- A page about APRS and amateur radio.

- A page acknowledging contributors to the data stream behind the WCA.

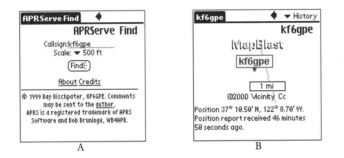

Figure 12-5. The APRS/Find WCA (A) input form and (B) results

The files for each of these three pages and the logo for the application are stored in a single folder, and PQA Builder is run with the index.html page as its opening argument. The resulting WCA can be installed on the Palm VII in the same manner as any application, and made available through channels familiar to Palm users.

The HTML for the index page is shown here:

```
<HTML>
<HEAD>
  <TITLE>APRServe Find</TITLE>
  <META NAME="PalmComputingPlatform" CONTENT="true">
  </HEAD>
<BODY>
<H1 ALIGN=right>APRServe Find</H1>
<P>
<FORM ACTION="http://www.lothlorien.com/cgi-bin/palmvii-aprs-map.pl"
    METHOD=POST>
  <CENTER>
    Callsign:
```

```
    <INPUT TYPE=text NAME=CALLSIGN VALUE="" SIZE=10 MAXLENGTH=10>
    <BR>
    Scale:
    <SELECT NAME=SCALE>
      <OPTION VALUE=50000>500 ft
      <OPTION VALUE=100000>1 mi
      <OPTION VALUE=200000>2 mi
      <OPTION VALUE=800000>5 mi
      <OPTION VALUE=3200000>50 mi
    </SELECT>
    <INPUT TYPE=hidden NAME=version VALUE=2>
    <INPUT TYPE=submit NAME=Submit VALUE="Find">
  </CENTER>
</FORM>
</P>
<CENTER>
<A HREF="about.html">About</A><BR>
<A HREF="credits.html">Credits</A>
</CENTER>
<HR>
<FONT SIZE="-1">&copy; 1999 Ray Rischpater, KF6GPE. Comments may be
sent to the </FONT><A HREF="mailto:dove@lothlorien.com"><FONT SIZE="-1">author</
FONT></A><FONT SIZE="-1">.<BR>
APRS is a registered trademark of APRS Software and Bob Bruninga,
WB4APR.</FONT></CENTER>
</BODY>
</HTML>
```

Note that this HTML file strongly resembles one that might be used by a Web page on a server to provide the same information, or the AvantGo channel described in Chapter 6. This similarity demonstrates the utility of a WCA for a content developer: it requires little work beyond creating a standard HTML file. Note that the form contains a hidden version number that can be used by the server to differentiate among different versions of the WCA in the event that you decide to extend its functionality on the back end.

The server behind the content returns a page with content such as the following:

```
<!DOCTYPE HTML PUBLIC "-//IETF//DTD HTML//EN">
<HTML><HEAD><TITLE>kf6gpe</TITLE>
<META NAME="palmcomputingplatform" CONTENT="true">
</HEAD><BODY>
<H1 ALIGN=right>kf6gpe</H1>
<CENTER>
<P>
```

```
<IMG SRC="http://www.vicinity.com/gif?&CT=37.17500:122.14500:200000:&IC=37.17500:-
122.14500:100:kf6gpe&FAM=mapblast&W=80&H=80">
</P>
</CENTER>
Position 37&#176 10.50' N, 122&#176 8.70' W.<BR>
Position report received 9 minutes 41 seconds ago.
</BODY></HTML>
```

Chapter 7 shows the server scripts behind this response. Note that the page includes the `PalmComputingPlatform` `<META>` entity, indicating to the Palm proxy servers that this page was formatted for Palm handheld computing devices. The image itself is intentionally small, both to respect the 153-pixel screen-size limit and to keep the average size of the data for the page with the image to a kilobyte or so.

Summary

While the triad of HTML/WML/HDML undoubtedly dominates wireless content today, the environment behind the wireless Web remains a dynamic one; a host of other technologies exist as alternatives or adjuncts to the Big Three.

Some of these technologies, such as XML, fuel the server-to-server communication that supports mobile client access. Used in conjunction with industry-specific conventions, XML enables content from varied servers and enterprises to be manipulated, formatted, and presented as a cohesive whole.

Server-assisted browsers use proxy servers that intervene between the origin and destination of content, modifying it to meet the constraints of a particular device. Images can be reduced, text formatted, and other operations performed to enhance the experience of users accessing existing Web content using a handheld device.

A specific implementation of the server-assist concept, Palm Inc.'s Web Clipping Application Architecture is an example of the many kinds of niche wireless access technologies that will doubtless appear in the coming years as standards settle and vendors produce hardware meeting these standards. A WCA can access Web content wirelessly via proxy servers that reduce wireless content to a tightly compressed, proprietary format targeted for the Palm Computing Platform, where it is displayed as simple images and HTML text.

Resources for Wireless Web Developers

THE FACE OF THE WIRELESS WEB is changing so quickly that a book such as this cannot keep up. By the time this reaches your hands, it's likely that a host of new tools for wireless Web development will be available. Moreover, there are more tools currently available than can be used in a work such as this.

Throughout the production of this book, I've used tools with which I've had prior experience; these are tools I recommend to friends or colleagues setting out to create their own wireless Web sites.

In this appendix, I provide pointers to sources for and information on these and other emerging tools for wireless Web development. This is not an exhaustive list, but I set out to include the most widely used and deployed tools and solutions for wireless developers. For each resource, I've included the vendor, a URL where more information can be found, the platforms it supports, and a brief description.

HTML

Wireless HTML Viewing

AvantGo

> **Vendor**: AvantGo, Inc.
> **URL**: http://www.avantgo.com
> **Platforms**: Microsoft Windows CE, Palm Computing Platform
> The market's leading synchronizing HTML browser.

Foliage iBrowser

> **Vendor**: Foliage Software Systems, Inc.
> **URL**: http://www.foliage.com/
> **Platforms**: Microsoft Windows CE
> Browser for palm-sized PCs running Microsoft Windows CE.

iSiloWeb

Vendor: DC & Co.
URL: http://www.isilo.com/
Platform: Palm Computing Platform
A general-purpose synchronizing browser and electronic book viewer.

Microsoft Mobile Channels

Vendor: Microsoft, Inc.
URL: http://www.microsoft.com/windowsce/
Platform: Microsoft Windows CE
A synchronizing browser based on Microsoft Mobile Channels for Microsoft Windows CE. Available in ROM on palm-sized PCs and as an aftermarket package for handheld PCs running Microsoft Windows CE.

Palm Computing Platform Palm Clipping Viewer

Vendor: Palm, Inc.
URL: http://www.palm.com/
Platform: Selected palm computing Platform devices, including the Palm VII and the Palm V with the Omnisky service.

pdqSuite

Vendor: Qualcomm, Inc.
URL: http://register.qualcomm.com/pdQsuite/
Platform: Palm Computing Platform
Text-only HTML browser and email client for Palm Computing Platform devices, including Qualcomm's pdQ phones running the Palm Operating System.

Pocket Internet Explorer

Vendor: Microsoft, Inc.
URL: http://www.microsoft.com/windowsce/
Platform: Microsoft Windows CE
The native browser shipped with many Windows CE devices.

ProxiWeb

Vendor: Puma Technology, Inc.
URL: http://www.proxiware.com/
Platform: Palm Computing Platform
Server-assisted browser with free proxy support for Palm Computing Platform handhelds; supports both text and image display.

HTML Validation

HTML::Clean

Vendor: Paul Lindner
URL: http://people.itu.int/~lindner/
Platform: Anywhere Perl is available, including Microsoft Windows, Macintosh, UNIX, and Linux
Not a true validation tool, but useful for reducing the size of HTML files.

Tidy

Vendor: W3C, written by David Raggett
URL: http://www.w3.org/People/Raggett/tidy/
Platform: Microsoft Windows, Unix, Macintosh ports available. Source code available.
One of the oldest—and best—HTML validation tools available.

cjpeg

Vendor: Independent JPEG Group
URL: ftp.uu.net://graphics/jpeg/jpegsrc.v6b.tar.gz
Platfom: Microsoft Windows, Unix, Linux, Macintosh ports available. Source code available.
Not a validation tool but a free tool enabling developers to experiment with different compression levels within JPEG images.

WML

The primary source of information from the companies that bring you WAP is, of course, the WAP Forum, `http://www.wapforum.org`.

Nokia WAP Server

Vendor: Nokia Corporation
URL: `http://www.nokia.com/wap/index.html`
Platform: Varied
Provides developers and networks with a reference server for gating WAP content from HTTP to WAP browsers using UDP over TCP/IP.

Nokia WAP Tookit

Vendor: Nokia Corporation
URL: `http://www.nokia.com/wap/index.html`
Platform: Microsoft Windows
Toolkit for developing and debugging WAP content on a PC-based simulator.

Up.SDK

Vendor: Phone.com
URL: `http://www.phone.com/`
Platform: Windows, Solaris
Provides a simulator for viewing and debugging HDML and WML applications, along with copious documentation and examples.

HDML

Up.SDK

Vendor: Phone.com
URL: `http://www.phone.com/`
Platform: Microsoft Windows, Solaris
Provides a simulator for viewing and debugging HDML and WML applications, along with copious documentation and examples.

Active Server Tools

Apache

Vendor: The Apache Software Foundation
URL: http://www.apache.org/
Platform: Varied (UNIX, Linux)
The Web's most widely-used Web server.

PHP: Hypertext Processor

Vendor: Open-source project led by Rasmus Lerdorf
URL: http://www.php.net/
Platform: UNIX (including Linux), Microsoft Windows
Robust, widely deployed, active scripting toolkit.

Software Development

EPOC Platform

Vendor: Symbian, Inc.
URL: http://www.epocworld.com/
Platform: Microsoft Windows
Software development kits for developing EPOC applications for devices
running the EPOC platform, including Psion devices.

Palm Computing Platform

Metrowerks CodeWarrior for Palm

Vendor: Metrowerks, Inc.
URL: http://www.metrowerks.com/
Platform: Macintosh, Microsoft Windows, some flavors of UNIX
Provides C and C++ development environment for the Palm Computing Platform.

PRC-Tools

Vendor: Open source project maintained by Palm, Inc.
URL: http://www.palm.com/devzone/tools/gcc/
Platform: UNIX (including Linux), Microsoft Windows
An open-source tool chain for the Palm Computing Platform.

Microsoft Windows CE

Microsoft Windows CE SDKs

Vendor: Microsoft, Inc.
URL: http://msdn.microsoft.com/cetools/platform/support.asp
Platform: Microsoft Windows targeting Microsoft Windows CE
Software development kits based on Visual Studio for Windows CE handhelds, including handheld PCs and palm-sized PCs.

The Unified Modeling Language for Web Developers

Why UML?

THE UNIFIED MODELING LANGUAGE (UML) is being widely adopted as the visual language of choice for modeling relationships within computer systems. An outgrowth of the modeling languages used to represent object-oriented systems, UML is fast becoming the standard way to represent almost any aspect of a computing system.

Traditionally, developers have used modeling languages to explain the inner workings of software, or the arrangement of a cluster of computers performing some task. In many cases, their use has been highly technical, and the resulting diagrams have not necessarily been of interest to Webmasters, content developers, or those in related occupations. UML, however, is different, for two key reasons.

First, its capabilities are far more general than many visual modeling languages. UML is capable of capturing many views of a system, and in fact can be applied to non-computer problems as well, such as business and process flow.

Second, its wide adoption makes it a useful common language that stakeholders, including business users, user advocates, software architects, software developers, system operators, and content developers can all share.

If you currently represent your network as little cylinders connected by arrows and lines pointing to a big cloud simply labeled "The Web," UML is for you.

UML for Wireless Web Developers

Developers use UML to capture and visualize the structure and operation of a system from different *views*. In general, views are related but different ways of thinking about how a system functions. For example, one view may represent a network topology for a service, while another may show the flow of a request through the service. While these views are related, the emphasis of each is different.

There are three broad categories of view: structural, dynamic, and meta.

- *Structural views* represent a system's static structure, such as the organization of software components.

- *Dynamic views* represent a view of the system over time, such as how the parts of a system interact to exchange information.

- *Meta views* are used to represent relationships within the model itself, such as the organizational hierarchy of a system's models.

UML uses various shapes to denote different aspects of a system, and lines between the shapes to denote relationships. One shape used by all diagrams is the comment shape, which is reminiscent of a sticky note. Solid lines indicate a named relationship, while dashed lines indicate a dependency of some kind. For example, for a component diagram—which shows how different components relate to one another—I would choose a solid line to indicate the flow of data and a dotted line to connect a comment with a component. When direction is important, arrows indicate the flow of information or dependency. Finally, numbers placed at the intersection of a line and a shape indicate the number of entities on each side of a connection. (An asterisk means that it could be any number.)

Consider Figure B-1, a deployment diagram representing a server-assisted browser service. Each host on the network is represented by the three-dimensional block and relies on another host for its correct operation (this relationship is shown by the solid lines). The asterisk (*) tells us that there are an arbitrary number of clients; these interact with between one and four proxy servers, which are represented by the three-dimensional block beside which is written 1…4; these proxy servers rely in turn on servers on the Internet for their content.

Views for Representing Web Relationships

One of UML's great strengths is its flexibility. UML brings many different visual modeling views together under one common standard. Let's take a closer look at the four views that you'll find the most helpful in wireless Web development.

Deployment Views

A *deployment view* is responsible for conveying the configuration of a running system. Deployment views indicate the hosts on which programs run, the networks between hosts, and other details. These views are valuable tools for network designers and administrators, for whom an understanding of the physical layout of a network and its relationship is of paramount importance.

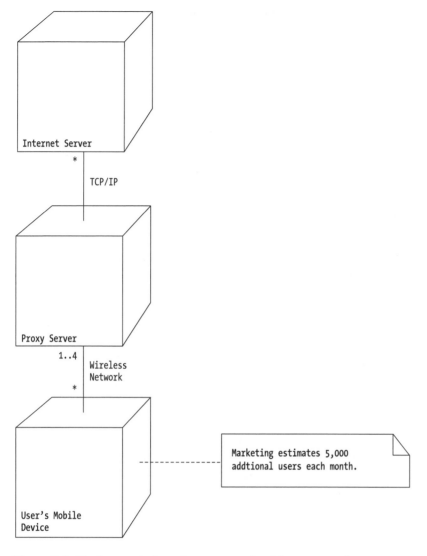

Figure B-1. Deployment view of a server-assisted browser service

One essential component of a deployment view is a *node*, a computational resource being used by the system. This is most often a server or workstation, although in some systems it may represent other computational resources, such as a single processor or a cluster of redundant servers containing the same set of components. A node is drawn as a three-dimensional block.

Nodes are drawn containing components, which are represented using the same symbols as in component diagrams. The relationships among components within a node are shown using solid and dashed lines, again as in component diagrams.

Solid lines are also used to represent the communications infrastructure that links two or more nodes.

Figure B-1, shown previously, illustrates a deployment view of a server-assisted browser service.

Use Case Views

A *use case view* captures how a system interacts as seen from the perspective of an outside user performing a task. Use cases represent users as *actors*, idealized users of a system who exchange a series of messages between the system and its components. Use cases can be fundamental in establishing how a subscriber will use a wireless system; they help everyone understand how a system can be organized to meet users' needs.

In developing use cases, it's important to focus on establishing the actors that will characterize the various users of the system. Each actor represents a class of users with similar goals and backgrounds. In essence, actors are profiles of user populations, such as subscribers, administrators, network support staff, and so on.

An actor invokes a *use case*, a functional unit of the system performing a particular task. A use case defines *what* happens in a system, without needing to worry about *how* it will happen. It captures all aspects of the operation, including appropriate behavior, abnormal operation, and input, as well as the appropriate response. Use cases are used by system designers to determine the kinds of classes, objects, and relationships needed to construct a system.

Figure B-2 shows the use case view for a simple wireless Web service that provides location information for its users. Stick figures denote actors, while use cases themselves are represented as labeled ovals. Interaction between actors and use cases is shown by solid lines.

Component Views

A *component view* captures how the various components of a system interact. A *component* is a physical entity, such as an aggregation of software that performs a well-defined task. Examples include a logging subsystem, a Web server, an eXtensible Markup Language (XML) parser, or a client application. Components can contain simpler components responsible for particular smaller tasks, such as caching or logging. Components are drawn as rectangles with two smaller rectangles overlapping the left side; the component's name is within the larger rectangle. Components may be drawn with the objects they contain shown inside them, although often the objects are omitted for clarity.

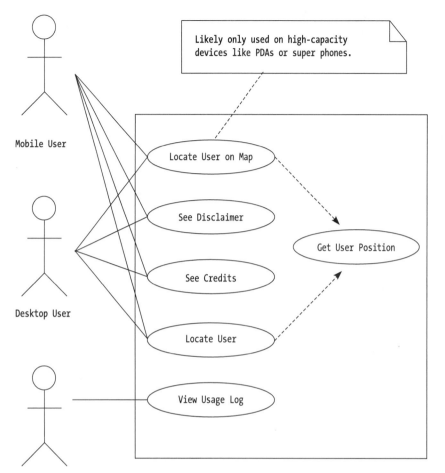

Likely only used on high-capacity devices like PDAs or super phones.

Mobile User

Locate User on Map

See Disclaimer

Get User Position

See Credits

Desktop User

Locate User

View Usage Log

Site Administrator

Figure B-2. A simple use case view

Each component has an *interface,* which must be well defined. Components share the details of their interfaces, while keeping other aspects of their implementation hidden from view by other components. All interactions with a component occur through an interface, which can be viewed as a kind of contract: promising what a component will do and deliver, but leaving just *how* it does this as a private matter for the component and its developers. Interfaces are drawn as circles; a solid line joins an interface with its provider. A dotted line indicates that one component relies upon another's interface. When the direction of dependence is not relevant, components are drawn connected with straight lines, indicating simply that they communicate through some interface and have a dependence on each other. When direction of dependence is relevant, the arrow points to the provider of the interface.

Figure B-3 shows a hypothetical component view diagram for the wireless Web service detailed in the previous section.

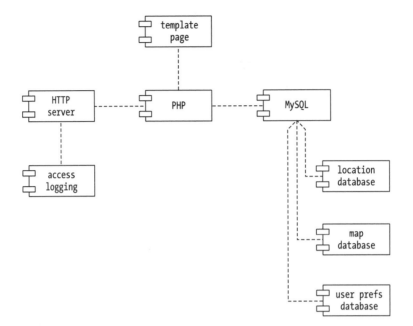

Figure B-3. A component view

Sequence Views

A *sequence view* shows the interaction of components via messages sent between them through interfaces, arranged by time. Typically, software developers use sequence views to visualize the interaction of objects, although there's no reason why they can't also capture interactions among components, network nodes, or other parts of a model. Another common use more relevant to Web developers is showing the sequence of the events that make up a use case.

Participants—components or hosts in a network—in a sequence view are drawn as rectangles along the top of the diagram; time is shown along the vertical axis. Interactions (typically messages) between items are drawn as arrows.

Figure B-4 shows the sequence view representing the locate user case shown in Figure B-2.

Other Views

When working with software developers, you'll often encounter other views. These views may not be directly pertinent to your own work if you're only responsible for

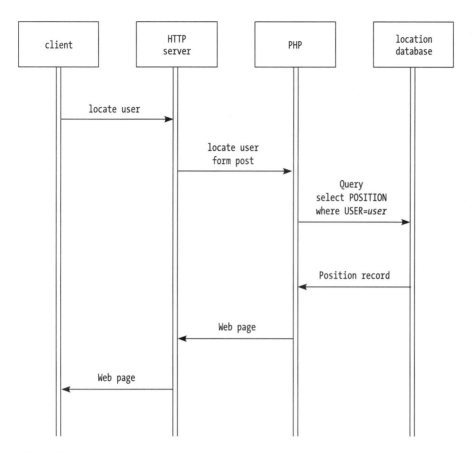

Figure B-4. A sequence view

creating and managing content, but understanding what these other views are will help you communicate with software developers.

Model Management Views

Developers use the *model management* view to subdivide a set of views into manageable units. It breaks a model into packages, and describes the relationships among these packages within the model.

A *package* is simply a logical segment of a model. Just as a file folder contains files (or as the icon of a file folder on your computer contains icons of files), a package contains other models. You shouldn't confuse a package with other things in views; unlike most views, the model management view represents the model itself; the notion of a package doesn't have a real-world equivalent within the system, but rather represents a group of other model components.

Every part of a model must belong to exactly one package. Modelers organize packages by conceptual categories, such as the responsibilities of subsystems. Each package contains top-level views in a particular category, such as use cases, component views, deployment views, or other entities. A view can belong to only one package, although a view in one package can point to a view in another package for more information. In large environments with document control, packages may be used as the organizational focus of change control—for example, you could divide models into packages stored in separate directories on a server and keep each directory under change control.

Packages are drawn with an icon resembling a desktop folder icon—a larger rectangle with a smaller rectangle on top. Figure B-5 shows a model management view of the relationship among the user management, user location, and client services packages of the location wireless Web site we illustrated in Figure B-2.

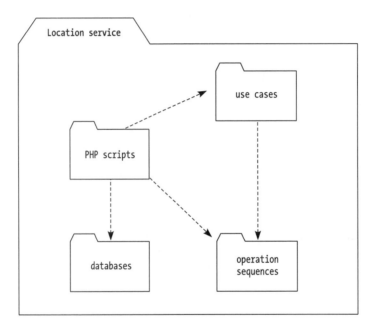

Figure B-5. A model management view

Class Views

UML's roots are in a number of object-oriented modeling systems developed over the last decade. Hence, UML provides a robust set of symbols for representing the relationships between classes and objects. If you're a software developer with a background in object-oriented design, especially if you're familiar with another modeling system, you'll find UML's static class views largely intuitive. (If not, consult one of the many good books on the topic available today.)

Within a *class view*, classes—abstractions of related object representations—are represented as rectangles, divided in to sections containing the class name, attributes, and operations within the class. Class relationships are shown by solid lines; constraints by dotted lines. Named relationships between classes, called *roles*, are used to demonstrate important interactions between classes.

Collaboration Views

Developers attempting to visualize the relationships among classes over time use *collaboration views* to represent the flow of messages through *objects*. (An object is a specific thing that behaves according to the description of a class.) Objects are denoted as named rectangles, with links between objects shown as solid lines. Messages passed between objects are numbered and written with arrows next to the links between objects. Collaboration views are similar to sequence diagrams, but generally of more use to software developers than Web designers, as they can be used to represent the inner workings of a program performing a specific operation.

State Machine and Activity Views

The *state machine view* also represents dynamic behavior. It shows a system's operation as a series of states, with lines indicating the transitions between them. A *state* can be thought of as some period in the life of an object, and a *transition* can be thought of as the process of moving to a new state after certain conditions have been met. State machine views are often used to show the inner workings of protocol stacks.

An *activity view* is a variation of the state machine view that focuses on portraying the computational activities required to meet the conditions necessary for transitions and describing the behavior of a transition.

Further Reading

UML is a valuable tool for anyone working in the field of information technology and processing. Although to date it has been largely used only by software developers, an understanding of UML can be of great use to Web designers and content developers working in a technical environment.

For those interested in learning more about UML, one of the best works to date is *UML Distilled, Second Edition: A Brief Guide to the Standard Object Modeling Language* by Martin Fowler, Kendall Scott, and Grady Booch. For the stalwart, *The Unified Modeling Language Reference Manual*, by James Rumbaugh, Ivar Jacobson, and Grady Booch is the definitive reference to UML.

Index

Symbols

/* and */ for PHP comments, 150
<!ATTLIST> tag (XML), 301
<!DOCTYPE> tag (XML), 299, 302–303
<!ELEMENT> tag (XML), 299–301
<!ENTITY> tag (XML), 302
(hash) for server directives, 137
$ (dollar sign) for PHP variable names, 151
%DEVICEID entity, for WCA, 310, 311–312
%ZIPCODE entity, for WCA, 310, 311–312
// for PHP comments, 150
< and > (angle brackets)
   for HTML tags, 51
   for WML tags, 183
   for XML tags, 122
<?php and ?> tags, 149
?: operator (WMLScript), 233

A

<A HREF> tag (HTML), 88–89
<A NAME> tag (HTML), 88
<a> tag (WML), 191, 194
abort function (WMLScript), 237
abs function (WMLScript), 237
abstract phone, 275
<ABSTRACT> tag (XML), 121
accept element (WAP), 200
<access> tag (WML), 188
ACCESSDOMAIN attribute for <HDML> tag,
   255, 256
ACCESSPATH attribute for <HDML> tag,
   255, 256
ACTION attribute for <FORM> tag, 96, 97–98
<ACTION> tag (HDML), 252–253, 261–265
   attributes, 262
actions in HDML, integration with input,
   273–275
Active Server Pages (ASP), 130
active server technology. *See* server-side
   scripting
activities, in HDML, 266–268
activity stack in HDML, 267
activity view (UML), 331
actors in use case view, 326
Adobe, PDF (Portable Document Format), 19
Advanced Mobile Phone System (AMPS), 2
alert function (WMLScript), 241, 245
ALIGN attribute
   for <INPUT> tag, 96
   for table elements, 94, 95

alignment of text
   in HDML, 259
   in HTML, 87–88
AllPen Software, 304
ALT attribute for <IMG> tag, 88, 89
   in HDML, 260
alt attribute of WML <img> tag, 214
amateur radio stations, 113–114
American Factfinder, 167
American Standard Code for Information
   Interchange (ASCII), 51
AMPS (Advanced Mobile Phone System), 2
analog cellular service, 2
<anchor> tag (WML), 191, 193–194
angle brackets (< and >)
   for HTML tags, 51
   for WML tags, 183
   for XML tags, 122
Apache Web HTTP server
   client-specific content with server
      directives, 143–148
   server directives, 137–143
      config, 137, 138
      echo, 137, 142, 146
      elif, 137, 142–143, 147
      else, 137, 142–143
      endif, 138, 142–143
      exec, 138, 141
      flastmod, 138, 139–140
      fsize, 138, 138–139
      if, 138, 142–143, 146
      include, 138, 140–141, 146, 147
      printenv, 138
      set, 138
      variables, 141–142
apparent throughput, 34–35
APRS (Automatic Position Reporting
   System), 113
aprs-image.php3 script, 157
ARDIS, 2, 5
array function (PHP), 152
arrays for PHP variables, 152
ASCII (American Standard Code for
   Information Interchange), 51
ASP (Active Server Pages), 130
assert macro, 245–246
associative arrays, 152
AT&T
   cellular telephone network proposal, 2
   Pocket.Net service, 251

A HISTORY OF
WESTERN SOCIETY

A HISTORY OF WESTERN SOCIETY

THIRD EDITION

VOLUME B:
FROM THE RENAISSANCE
TO 1815

JOHN P. McKAY
BENNETT D. HILL
JOHN BUCKLER

University of Illinois at Urbana-Champaign

HOUGHTON MIFFLIN COMPANY   BOSTON
Dallas   Geneva, Illinois
Lawrenceville, New Jersey   Palo Alto

Text Credits   Discussion of European family life during the Reformation based on S. Ozment, *When Fathers Ruled: Family Life in Reformation Europe* (Cambridge, Mass.: Harvard University Press), pp. 9–14, 50–99. Published by permission of Harvard University Press. Excerpts from "Richard II," Act II, Scene 1, and "Hamlet," Act III, Scene 1, from *The Complete Plays and Poems of William Shakespeare,* eds., William A. Neilson and C. J. Hill. Copyright © 1942 by Houghton Mifflin Company. Copyright renewed, 1969. Reprinted by permission of the publisher.

Chapter opener credits appear on page 697.

Cover: *Portrait of a Member of the Wedigh Family,* by Hans Holbein the Younger, German, 1497/1498. The Metropolitan Museum of Art, Bequest of Edward S. Harkness, 1940.

Printed in the U.S.A.

Library of Congress Catalog Card Number: 86–81470

ISBN: 0–395–42412–7

CDEFGHIJ–RM–8987

About the Authors

John P. McKay   Born in St. Louis, Missouri, John P. McKay received his B.A. from Wesleyan University (1961), his M.A. from the Fletcher School of Law and Diplomacy (1962), and his Ph.D. from the University of California, Berkeley (1968). He began teaching history at the University of Illinois in 1966 and became a professor there in 1976. John won the Herbert Baxter Adams Prize for his book *Pioneers for Profit: Foreign Entrepreneurship and Russian Industrialization, 1885–1913* (1970). He has also written *Tramways and Trolleys: The Rise of Urban Mass Transport in Europe* (1976) and has translated Jules Michelet's *The People* (1973). His research has been supported by fellowships from the Ford Foundation, the Guggenheim Foundation, the National Endowment for the Humanities, and IREX. His articles and reviews have appeared in numerous journals, including *The American Historical Review, Business History Review, The Journal of Economic History,* and *Slavic Review.* He edits *Industrial Development and the Social Fabric: An International Series of Historical Monographs.*

Bennett D. Hill   A native of Philadelphia, Bennett D. Hill earned an A.B. at Princeton (1956) and advanced degrees from Harvard (A.M., 1958) and Princeton (Ph.D., 1963). He taught history at the University of Illinois at Urbana, where he was department chairman from 1978 to 1981. He has published *English Cistercian Monasteries and Their Patrons in the Twelfth Century* (1968) and *Church and State in the Middle Ages* (1970); and articles in *Analecta Cisterciensia, The New Catholic Encyclopaedia, The American Benedictine Review,* and *The Dictionary of the Middle Ages.* His reviews have appeared in *The American Historical Review, Speculum, The Historian, The Catholic Historical Review,* and *Library Journal.* He has been a fellow of the American Council of Learned Societies and has served on committees for the National Endowment for the Humanities. Now a Benedictine monk at St. Anselm's Abbey, Washington, D.C., he is also a Lecturer at the University of Maryland at College Park.

John Buckler   Born in Louisville, Ky., John Buckler received his B.A. from the University of Louisville in 1967. Harvard University awarded him the Ph.D. in 1973. From 1984 to 1986 he was the Alexander von Humboldt Fellow at Institut für Alte Geschichte, University of Munich. He is currently an associate professor at the University of Illinois, and is serving on the Subcommittee on Cartography of the American Philological Association. In 1980 Harvard University Press published his *The Theban Hegemony, 371–362 B.C.* His articles have appeared in journals both here and abroad, like the *American Journal of Ancient History, Classical Philology, Rheinisches Museum für Philologie, Classical Quarterly, Wiener Studien,* and *Symbolae Osloenses.*

CONTENTS

MAPS

PREFACE

A HISTORY OF WESTERN SOCIETY grew out of the authors' desire to infuse new life into the study of Western civilization. We knew full well that historians were using imaginative questions and innovative research to open up vast new areas of historical interest and knowledge. We also recognized that these advances had dramatically affected the subject of European economic, intellectual, and, especially, social history, while new research and fresh interpretations were also revitalizing the study of the traditional mainstream of political, diplomatic, and religious development. Despite history's vitality as a discipline, however, it seemed to us that both the broad public and the intelligentsia were generally losing interest in the past. The mathematical economist of our acquaintance who smugly quipped "What's new in history?"—confident that the answer was nothing and that historians were as dead as the events they examine—was not alone.

It was our conviction, based on considerable experience introducing large numbers of students to the broad sweep of Western civilization, that a book reflecting current trends could excite readers and inspire a renewed interest in history and our Western heritage. Our strategy was twofold. First, we made social history the core element of our work. Not only did we incorporate recent research by social historians, but also we sought to re-create the life of ordinary people in appealing human terms. At the same time we were determined to give great economic, political, intellectual, and cultural developments the attention they unquestionably deserve. We wanted to give individual readers and instructors a balanced, integrated perspective, so that they could pursue on their own or in the classroom those themes and questions that they found particularly exciting and significant. In an effort to realize fully the potential of our fresh yet balanced approach, we made many changes, large and small, in the second edition.

In preparing the third edition we have worked hard to keep our book up-to-date and to make it still more effective. First, every chapter has been carefully revised to incorporate recent scholarship. Many of our revisions relate to the ongoing explosion in social history, and once again important findings on such subjects as class relations, population, women, and the family have been integrated into the text. New scholarship also led to substantial revisions on many other questions, such as the Neolithic agricultural revolution, political and economic growth in ancient Greece, the rise and spread of Christianity, the Germanic nobility, medieval feudalism, the origins of the Renaissance, Louis XIV and the French nobility, eighteenth-century absolutism, the French Revolution and Napoleon, nationalism, life in the postwar era, and events of the recent past. We believe that the incorporation of newer interpretations of the main political developments in the medieval, early modern, and French revolutionary periods is a particularly noteworthy change in this edition. Better integration of political and social development contributes to this improvement.

Second, we have carefully examined each chapter for organization and clarity. Chapters 7, 8, 9, 11, 14, and 15 have been thoroughly reorganized, while Chapters 17, 18, 21, and 23 have been reordered to a lesser extent. The result of these changes is a more logical presentation of material and a clearer chronological sequence. Similarly, the reorganization of Chapters 30 and 31 and the addition of Chapter 32 have permitted a more complete discussion of changes since World War Two and an innovative interpretation of this complicated era. We have also taken special care to explain terms and concepts as soon as they are introduced.

Third, we have added or expanded material on previously neglected topics to help keep our work fresh and appealing. Coverage of religious developments, with special emphasis on their popular and social aspects, now extends from ancient to modern times and includes several new sections. The reader will also find new material on many other topics, notably the Minoans, Greek and Roman wars, medieval Germany, the Hanseatic League, the African slave trade, Hume and d'Holbach, the pre-revolutionary French elite, Mill, and events since the late 1960s.

Finally, the illustrative component of our work has been completely revised. There are many new illustrations, including a tripling of the color plates that let both great art and earlier times come alive. Twenty new maps containing social as well as political material have also been added, while maps from the second edition have been re-edited and placed in a more effective format. As in earlier editions, all il-

lustrations have been carefully selected to complement the text, and all carry captions that enhance their value. Artwork remains an integral part of our book, for the past can speak in pictures as well as words.

Distinctive features from earlier editions remain in the third. To help guide the reader toward historical understanding we have posed specific historical questions at the beginning of each chapter. These questions are then answered in the course of the chapter, each of which concludes with a concise summary of the chapter's findings. The timelines added in the second edition have proved useful, and still more are found in this edition.

We have also tried to suggest how historians actually work and think. We have quoted extensively from a wide variety of primary sources and have demonstrated in our use of these quotations how historians sift and weigh evidence. We want the reader to realize that history is neither a list of cut-and-dried facts nor a senseless jumble of conflicting opinions. It is our further hope that the primary quotations, so carefully fitted into their historical context, will give the reader a sense that even in the earliest and most remote periods of human experience history has been shaped by individual men and women, some of them great aristocrats, others ordinary folk.

Each chapter concludes with carefully selected suggestions for further reading. These suggestions are briefly described in order to help readers know where to turn to continue thinking and learning about the Western world. The chapter bibliographies have been revised and expanded in order to keep them current with the vast and complex new work being done in many fields.

Western civilization courses differ widely in chronological structure from one campus to another. To accommodate the various divisions of historical time into intervals that fit a two-quarter, three-quarter, or two-semester period, *A History of Western Society* is being published in three versions, each set embracing the complete work:

One-volume hardcover edition, A HISTORY OF WESTERN SOCIETY; two-volume paperback, A HISTORY OF WESTERN SOCIETY *Volume I: From Antiquity to the Enlightenment* (Chapters 1–17), *Volume II: From Absolutism to the Present* (Chapters 16–32); three-volume paperback, A HISTORY OF WESTERN SOCIETY *Volume A: From Antiquity to the Reforma-*

tion (Chapters 1–13), *Volume B: From the Renaissance to 1815* (Chapters 12–21), *Volume C: From the Revolutionary Era to the Present* (Chapters 21–32).

Note that overlapping chapters in both the two- and the three-volume sets permit still wider flexibility in matching the appropriate volume with the opening and closing dates of a course term. Furthermore, for courses beginning with the Renaissance rather than antiquity or the medieval period, the reader can begin study with Volume B.

Learning and teaching ancillaries, including a *Study Guide, Computerized Study Guide, Instructor's Manual, Test Items, Computerized Test Items,* and *Map Transparencies,* also contribute to the usefulness of the text. The excellent *Study Guide* has been revised by Professor James Schmiechen of Central Michigan University. Professor Schmiechen has been a tower of strength ever since he critiqued our initial prospectus, and he has continued to give us many valuable suggestions and his warmly appreciated support. His *Study Guide* contains chapter summaries, chapter outlines, review questions, extensive multiple-choice exercises, self-check lists of important concepts and events, and a variety of study aids and suggestions. One innovation in the *Study Guide* that has proved useful to the student is the step-by-step Reading with Understanding exercises, which take the reader by ostensive example through reading and studying activities like underlining, summarizing, identifying main points, classifying information according to sequence, and making historical comparisons. To enable both students and instructors to use the *Study Guide* with the greatest possible flexibility, the guide is available in two volumes, with considerable overlapping of chapters. Instructors and students who use only Volumes A and B of the text have all the pertinent study materials in a single volume, *Study Guide, Volume 1* (Chapters 1–21); likewise, those who use only Volumes B and C of the text also have all the necessary materials in one volume, *Study Guide, Volume 2* (Chapters 12–32). The multiple-choice sections of the *Study Guide* are also available in a computerized version that provides the student with tutorial instruction.

The *Instructor's Manual,* prepared by Professor Philip Adler of East Carolina University, contains learning objectives, chapter synopses, suggestions for lectures and discussion, paper and class activity topics, and lists of audio-visual resources. The accompanying *Test Items,* also by Professor Adler, offers more than 1100 multiple-choice and essay questions and approximately 500 identification terms. The test items are available to adopters on computer tape and disk. In addition, a set of forty color map transparencies is available on adoption.

It is a pleasure to thank the many instructors who have read and critiqued the manuscript through its development: James W. Alexander, University of Georgia; Susan D. Amussen, Connecticut College; Jack M. Balcer, Ohio State University; Ronald M. Berger, State University College at Oneonta, New York; Charles R. Berry, Wright State University; Shirley J. Black, Texas A & M University; John W. Bohnstedt, California State University at Fresno; Paul Bookbinder, University of Massachusetts— Boston, Harbor Campus; Jerry H. Brookshire, Middle Tennessee State University; Thomas S. Burns, Emory University; Robert Clouse, Indiana State University; Norman H. Cooke, Rhode Island College; Charles E. Daniel, University of Rhode Island; Gary S. Cross, Pennsylvania State University; Lawrence G. Duggan, University of Delaware; J. Rufus Fears, Indiana University; John B. Freed, Illinois State University; James Friguglietti, Eastern Montana College; Charles L. Geddes, University of Denver; James Gump, University of San Diego; Charles D. Hamilton, San Diego State University; Barbara Hanawalt, Indiana University; Thomas J. Heston, West Chester State College; Edward J. Kealey, College of the Holy Cross; Isabel F. Knight, Pennsylvania State University; Charles A. Le Guin, Portland State University; Richard Lyman, Simmons College; Rhoda McFadden, Montgomery County Community College; Christian D. Nokkentved, University of Illinois at Chicago; John E. Roberts, Jr., Lincoln Land Community College; William J. Roosen, Northern Arizona University; Lawrence Silverman, University of Colorado; Armstrong Starkey, Adelphi University; Robert E. Stebbins, Eastern Kentucky University; Bailey S. Stone, University of Houston; C. Mary Taney, Glassboro State College; Allen M. Ward, University of Connecticut; and Donald Wilcox, University of New Hampshire.

Many of our colleagues at the University of Illinois kindly provided information and stimulation for our book, often without even knowing it. N. Frederick

Nash, Rare Book Librarian, gave freely of his time and made many helpful suggestions for illustrations. The World Heritage Museum at the University continued to allow us complete access to its sizable holdings. James Dengate kindly supplied information on objects from the museum's collection. Caroline Buckler took many excellent photographs of the museum's objects and generously helped us at crucial moments in production. Such wide-ranging expertise was a great asset for which we are very appreciative. Bennett Hill wishes to express his sincere appreciation to Ramón de la Fuente of Washington, D.C., for his support, encouragement, and research assistance in the preparation of this third edition. John Buckler extends his thanks to Elke Bernlocher.

Each of us has benefited from the generous criticism of his co-authors, although each of us assumes responsibility for what he has written. John Buckler has written the first six chapters; Bennett Hill has continued the narrative through Chapter 16; and John McKay has written Chapters 17 through 32. Finally, we continue to welcome from our readers comments and suggestions for improvements, for they have helped us greatly in this ongoing endeavor.

JOHN P. MCKAY
BENNETT D. HILL
JOHN BUCKLER

12

THE CRISIS OF THE LATER MIDDLE AGES

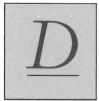

*D*URING the later Middle Ages, the closing book of the New Testament, the Book of Revelation, inspired thousands of sermons and hundreds of religious tracts. The Book of Revelation deals with visions of the end of the world, with disease, war, famine, and death. It is no wonder this part of the Bible was so popular. Between 1300 and 1450, Europeans experienced a frightful series of shocks: economic dislocation, plague, war, social upheaval, and increased crime and violence. Death and preoccupation with death make the fourteenth century one of the gloomiest periods in Western civilization.

The miseries and disasters of the later Middle Ages bring to mind a number of questions. What economic difficulties did Europe experience? What were the social and psychological effects of repeated attacks of plague and disease? Some scholars maintain that war is often the catalyst for political, economic, and social change. Does this theory have validity for the fourteenth century? What political and social developments do new national literatures express? What provoked the division of the church in the fourteenth century? What other ecclesiastical difficulties was the schism a sign of, and what impact did it have on the faith of the common people? How can we characterize the dominant features in the lives of ordinary people? This chapter will focus on these questions.

PRELUDE TO DISASTER

Economic difficulties originating in the later thirteenth century were fully manifest by the start of the fourteenth. In the first decade, the countries of northern Europe experienced a considerable price inflation. The costs of grain, livestock, and dairy products rose sharply. Bad weather made a serious situation worse. An unusual number of storms brought torrential rains, ruining the wheat, oats, and hay crops on which people and animals depended almost everywhere. Since long-distance transportation of food was expensive and difficult, most urban areas depended for bread and meat on areas no more than a day's journey away. Poor harvests—and one in four was likely to be poor—led to scarcity and starvation.

Almost all of northern Europe suffered a terrible famine in the years 1315 to 1317.

Hardly had western Europe begun to recover from this disaster when another struck. An epidemic of typhoid fever carried away thousands. In 1316, 10 percent of the population of the city of Ypres may have died between May and October alone. Then in 1318 disease hit cattle and sheep, drastically reducing the herds and flocks. Another bad harvest in 1321 brought famine, starvation, and death.

The province of Languedoc in France presents a classic example of agrarian crisis. For over 150 years Languedoc had enjoyed continual land reclamation, steady agricultural expansion, and enormous population growth. Then the fourteenth century opened with four years of bad harvests. Torrential rains in 1310 ruined the harvest and brought on terrible famine. Harvests failed again in 1322 and 1329. In 1332 desperate peasants survived the winter on raw herbs. In the half-century from 1302 to 1348, poor harvests occurred twenty times. The undernourished population was ripe for the Grim Reaper, who appeared in 1348 in the form of the Black Death.

These catastrophes had grave social consequences. Population had steadily increased in the twelfth and thirteenth centuries, and large amounts of land had been put under cultivation. The amount of food yielded, however, did not match the level of population growth. Bad weather had disastrous results. Poor harvests meant that marriages had to be postponed. Later marriages and the deaths caused by famine and disease meant a reduction in population. Meanwhile, the international character of trade and commerce meant that a disaster in one country had serious implications elsewhere. For example, the infection that attacked English sheep in 1318 caused a sharp decline in wool exports in the following years. Without wool, Flemish weavers could not work, and thousands were laid off. Without woolen cloth, the businesses of Flemish, French, and English merchants suffered. Unemployment encouraged many men to turn to crime.

To none of these problems did governments have any solutions. In fact, they even lacked policies. After the death of Edward I in 1307, England was governed by the incompetent and weak Edward II (1307–1327), whose reign was dominated by a series of baronial conflicts. In France the three sons of Philip the Fair, who followed their father to the French throne

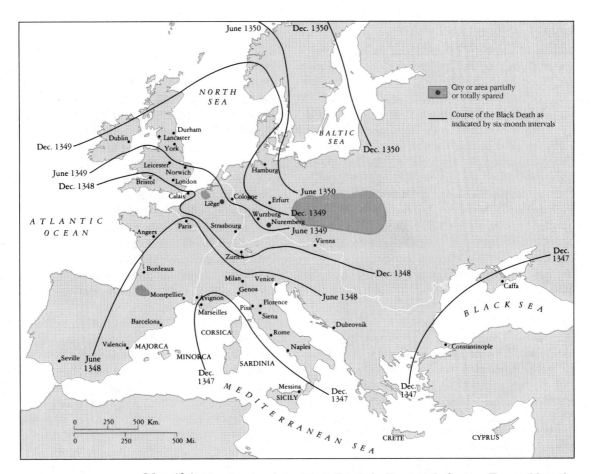

MAP 12.1 The Course of the Black Death in Fourteenth-Century Europe Note the routes that the bubonic plague took across Europe. How do you account for the fact that several regions were spared the "dreadful death"?

between 1314 and 1328, took no interest in the increasing economic difficulties. In Germany power drifted into the hands of local rulers. The only actions the governments took tended to be in response to the demands of the upper classes. Economic and social problems were aggravated by the appearance in western Europe of a frightful disease.

THE BLACK DEATH

In 1291 Genoese sailors had opened the Straits of Gibraltar to Italian shipping by defeating the Moroccans. Then, shortly after 1300, important advances were made in the design of Italian merchant ships. A square rig was added to the mainmast, and ships began to carry three masts instead of just one. Additional sails better utilized wind power to propel the ship. The improved design permitted year-round shipping for the first time, and Venetian and Genoese merchant ships could sail the dangerous Atlantic coast even in the winter months. With ships continually at sea, the rats that bore the disease spread rapidly beyond the Mediterranean to Atlantic and North Sea ports.

Around 1331 the bubonic plague broke out in China. In the course of the next fifteen years, merchants, traders, and soldiers carried the disease across the Asian caravan routes until in 1346 it reached the

Crimea in southern Russia. From there the plague had easy access to the Mediterranean lands and western Europe.

In October 1347, Genoese ships brought the plague to Messina, from which it spread to Sicily. Venice and Genoa were hit in January 1348, and from the port of Pisa the disease spread south to Rome and east to Florence and all Tuscany. By late spring, southern Germany was attacked. Frightened French authorities chased a galley bearing the disease from the port of Marseilles, but not before plague had infected the city, from which it spread to Languedoc and Spain. In June 1348, two ships entered the Bristol Channel and introduced it into England. All Europe felt the scourge of this horrible disease (see Map 12.1).

PATHOLOGY

Modern understanding of the bubonic plague rests on the research of two bacteriologists, one French and one Japanese, who in 1894 independently identified the bacillus that causes the plague, *Pasteurella pestis* (so labeled after the French scientist's teacher, Louis Pasteur). The bacillus liked to live in the bloodstream of an animal or, ideally, in the stomach of a flea. The flea in turn resided in the hair of a rodent, sometimes a squirrel but preferably the hardy, nimble, and vagabond black rat. Why the host black rat moved so much, scientists still do not know, but it often traveled by ship. There the black rat could feast for months on a cargo of grain or live snugly among bales of cloth. Fleas bearing the bacillus also had no trouble nesting in saddlebags.[1] Comfortable, well fed, and often having greatly multiplied, the black rats ended their ocean voyage and descended on the great cities of Europe.

Although by the fourteenth century urban authorities from London to Paris to Rome had begun to try to achieve a primitive level of sanitation, urban conditions remained ideal for the spread of disease. Narrow streets filled with mud, refuse, and human excrement were as much cesspools as thoroughfares. Dead animals and sore-covered beggars greeted the traveler. Houses whose upper stories projected over the lower ones eliminated light and air. And extreme overcrowding was commonplace. When all members of an aristocratic family lived and slept in one room, it should not be surprising that six or eight persons in

a middle-class or poor household slept in one bed—if they had one. Closeness, after all, provided warmth. Houses were beginning to be constructed of brick, but many remained of wood, clay, and mud. A determined rat had little trouble entering such a house.

Standards of personal hygiene remained frightfully low. Since water was considered dangerous, partly for good reasons, people rarely bathed. Skin infections, consequently, were common. Lack of personal cleanliness, combined with any number of temporary ailments such as diarrhea and the common cold, naturally weakened the body's resistance to serious disease. Fleas and body lice were universal afflictions: everyone from peasants to archbishops had them. One more bite did not cause much alarm. But if that nibble came from a bacillus-bearing flea, an entire household or area was doomed.

The symptoms of the bubonic plague started with a growth the size of a nut or an apple in the armpit, in the groin, or on the neck. This was the boil, or *buba,* that gave the disease its name and caused agonizing pain. If the buba was lanced and the pus thoroughly drained, the victim had a chance of recovery. The secondary stage was the appearance of black spots or blotches caused by bleeding under the skin. (This syndrome did not give the disease its common name; contemporaries did not call the plague the Black Death. Sometime in the fifteenth century, the Latin phrase *atra mors,* meaning "dreadful death" was translated "black death," and the phrase stuck.) Finally the victim began to cough violently and spit blood. This stage, indicating the presence of thousands of bacilli in the bloodstream, signaled the end, and death followed in two or three days. Rather than evoking compassion for the victim, a French scientist has written, everything about the bubonic plague provoked horror and disgust: "All the matter which exuded from their bodies let off an unbearable stench; sweat, excrement, spittle, breath, so fetid as to be overpowering; urine turbid, thick, black or red."[2]

Medieval people had no rational explanation for the disease nor any effective medical treatment for it. Fourteenth-century medical literature indicates that physicians could sometimes ease the pain, but they had no cure. Most people—lay, scholarly, and medical—believed that the Black Death was caused by some "vicious property in the air" that carried the disease from place to place. When ignorance was joined to fear and ancient bigotry, savage cruelty

The Plague-Stricken Even as the dead were wrapped in shrouds and collected in carts for mass burial, the disease struck others. The man collapsing has the symptomatic buba on his neck. As Saint Sebastian pleads for mercy (above), a winged devil, bearer of the plague, attacks an angel. *(Walters Art Gallery, Baltimore)*

sometimes resulted. Many people believed that the Jews had poisoned the wells of Christian communities and thereby infected the drinking water. This charge led to the murder of thousands of Jews across Europe. According to one chronicler, sixteen thousand were killed at the imperial city of Strasbourg alone in 1349.

The Italian writer Giovanni Boccaccio (1313–1375), describing the course of the disease in Florence in the preface to his book of tales, *The Decameron,* pinpointed the cause of the spread:

Moreover, the virulence of the pest was the greater by reason that intercourse was apt to convey it from the sick to the whole, just as fire devours things dry or greasy when they are brought close to it. Nay, the evil went yet further, for not merely by speech or association with the sick was the malady communicated to the healthy with consequent peril of common death, but any that touched the clothes of the sick or aught else that had been touched or used by them, seemed thereby to contract the disease.[3]

The highly infectious nature of the plague, especially in areas of high population density, was recognized by a few sophisticated Arabs. When the disease struck the town of Salé in Morocco, Ibu Abu Madyan shut in his household with sufficient food and water and allowed no one to enter or leave until the plague had passed. Madyan was entirely successful. The rat that carried the disease-bearing flea avoided travel outside the cities. Thus the countryside was relatively safe. City dwellers who could afford to move fled to the country districts.

The mortality rate cannot be specified, because population figures for the period before the arrival of the plague do not exist for most countries and cities. The largest amount of material survives for England, but it is difficult to use and, after enormous scholarly controversy, only educated guesses can be made. Of a total population of perhaps 4.2 million, probably 1.4 million died of the Black Death in its several visits.[4] Densely populated Italian cities endured incredible losses. Florence lost between half and two-thirds of its 1347 population of 85,000 when the plague visited in 1348. The disease recurred intermittently in the 1360s and 1370s and reappeared many times down to 1700. There have been twentieth-century outbreaks in such places as Hong Kong, Bombay, and Uganda.

SOCIAL AND PSYCHOLOGICAL CONSEQUENCES

Predictably, the poor died more rapidly than the rich, because the rich enjoyed better health to begin with; but the powerful were not unaffected. In England, two archbishops of Canterbury fell victim to the plague in 1349, King Edward III's daughter Joan died, and many leading members of the London guilds followed her to the grave.

It is noteworthy that, in an age of mounting criticism of clerical wealth, the behavior of the clergy during the plague was often exemplary. Priests, monks, and nuns cared for the sick and buried the dead. In places like Venice, from where even physicians fled, priests remained to give what ministrations they could. Consequently, their mortality rate was phenomenally high. The German clergy, especially, suffered a severe decline in personnel in the years after 1350. With the ablest killed off, the wealth of the German church fell into the hands of the incompetent and weak. The situation was ripe for reform.

The plague accelerated the economic decline begun in the early part of the fourteenth century. In many parts of Europe, there had not been enough work for people to do. The Black Death was a grim remedy to this problem. Population decline, however, led to an increased demand for labor and to considerable mobility among the peasant and working classes. Wages rose sharply. The shortage of labor and steady requests for higher wages put landlords on the defensive. They retaliated with such measures as the English Statute of Laborers (1351), which attempted to freeze salaries and wages at pre-1347 levels. The statute could not be enforced and therefore was largely unsuccessful.

Even more frightening than the social effects were the psychological consequences. The knowledge that the disease meant almost certain death provoked the most profound pessimism. Imagine an entire society in the grip of the belief that it was at the mercy of a frightful affliction about which nothing could be done, a disgusting disease from which family and friends would flee, leaving one to die alone and in agony. It is not surprising that some sought release in orgies and gross sensuality while others turned to the severest forms of asceticism and frenzied religious fervor. Some extremists joined groups of *flagellants,* who collectively whipped and scourged themselves as

Danse Macabre Naked, rotting corpses dance with the living of different social classes —who are frozen with shock. The purpose of the painting is to remind the viewer of the uncertainty of the hour of death's appearance and of everyone's equality before it. *Macabre* probably means corpse. *(Giraudon/Art Resource)*

penance for their and society's sins, in the belief that the Black Death was God's punishment for humanity's wickedness.

The literature and art of the fourteenth century reveal a terribly morbid concern with death. One highly popular artistic motif, the Dance of Death, depicted a dancing skeleton leading away a living person. No wonder survivors experienced a sort of shell shock and a terrible crisis of faith. Lack of confidence in the leaders of society, lack of hope for the future, defeatism, and malaise wreaked enormous anguish and contributed to the decline of the Middle Ages. A long international war added further misery to the frightful disasters of the plague.

THE HUNDRED YEARS' WAR
(CA 1337–1453)

In January 1327, Queen Isabella of England, her lover Mortimer, and a group of barons, having deposed and murdered Isabella's incompetent husband, King Edward II, proclaimed his fifteen-year-old son king as Edward III. Isabella and Mortimer, however, held real power until 1330, when Edward seized the reins of government. In 1328 Charles IV of France, the last surviving son of the French king Philip the Fair, died childless. With him ended the Capetian dynasty. An assembly of French barons, mean-

THE FRENCH AND ENGLISH SUCCESSIONS

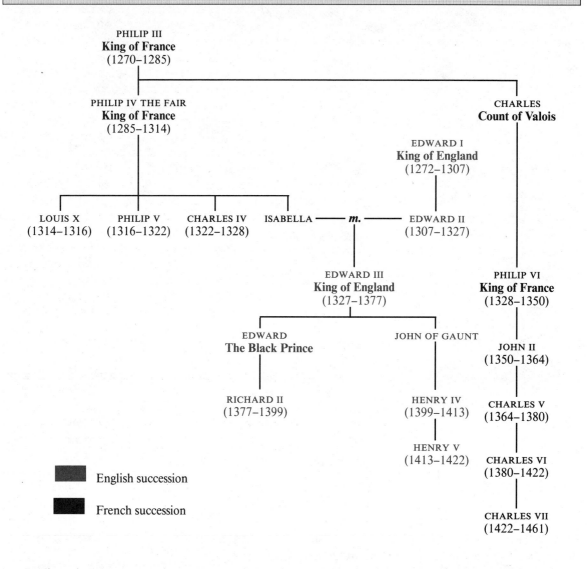

PHILIP III
King of France
(1270–1285)

PHILIP IV THE FAIR
King of France
(1285–1314)

CHARLES
Count of Valois

EDWARD I
King of England
(1272–1307)

LOUIS X
(1314–1316)

PHILIP V
(1316–1322)

CHARLES IV
(1322–1328)

ISABELLA —— *m.* —— **EDWARD II**
(1307–1327)

EDWARD III
King of England
(1327–1377)

PHILIP VI
King of France
(1328–1350)

EDWARD
The Black Prince

JOHN OF GAUNT

JOHN II
(1350–1364)

RICHARD II
(1377–1399)

HENRY IV
(1399–1413)

CHARLES V
(1364–1380)

HENRY V
(1413–1422)

CHARLES VI
(1380–1422)

CHARLES VII
(1422–1461)

◼ English succession

◼ French succession

In discussing the causes of the Hundred Years' War, modern scholars emphasize economic factors or the French-English dispute over the province of Gascony. Fourteenth-century Englishmen, however, believed they were fighting because King Edward III was denied his legal right to the French crown. He was the eldest surviving male descendant of Philip the Fair.

ing to exclude Isabella—who was Charles's sister and the daughter of Philip the Fair—and her son Edward III from the French throne, proclaimed that "no woman nor her son could succeed to the [French] monarchy." The barons passed the crown to Philip VI of Valois (1328–1350), a nephew of Philip the Fair. In these actions lie the origins of another phase of the centuries-old struggle between the English and French monarchies, one that was fought intermittently from 1337 to 1453.

CAUSES

The Hundred Years' War had both distant and immediate causes. In 1259 France and England signed the Treaty of Paris, in which the English king agreed to become—for himself and his successors—vassal of the French crown for the duchy of Aquitaine. The English claimed Aquitaine as an ancient inheritance. French policy, however, was strongly expansionist, and the French kings resolved to absorb the duchy into the kingdom of France. In 1329 Edward III paid homage to Philip VI for Aquitaine. In 1337 Philip, determined to exercise full jurisdiction there, confiscated the duchy. This action was the immediate cause of the war. Edward III maintained that the only way he could exercise his rightful sovereignty over Aquitaine was by assuming the title of king of France.[5] As the eldest surviving male descendant of Philip the Fair, he believed he could rightfully make this claim. Moreover, the dynastic argument had feudal implications: in order to increase their independent power, French vassals of Philip VI used the excuse that they had to transfer their loyalty to a more legitimate overlord, Edward III. Consequently, one reason the war lasted so long was that it became a French civil war, with French barons supporting English monarchs in order to thwart the centralizing goals of the French crown.

Economic factors involving the wool trade and the control of the Flemish towns had served as justifications for war between France and England for centuries. The causes of the conflicts known as the Hundred Years' War were thus dynastic, feudal, political, and economic. Recent historians have stressed economic factors. The wool trade between England and Flanders served as the cornerstone of both countries' economies; they were closely interdependent. Flanders was a fief of the French crown, and the Flemish aristocracy was highly sympathetic to the monarchy in Paris. But the wealth of Flemish merchants and cloth manufacturers depended on English wool, and Flemish burghers strongly supported the claims of Edward III. The disruption of commerce with England threatened their prosperity.

It is impossible to measure the precise influence of the Flemings on the cause and course of the war. Certainly Edward could not ignore their influence, because it represented money he needed to carry on the war. Although the war's impact on commerce fluctuated, over the long run it badly hurt the wool trade and the cloth industry.

Why did the struggle last so long? One historian has written in jest that, if Edward III had been locked away in a castle with a pile of toy knights and archers to play with, he would have done far less damage.[6] The same might be said of Philip VI. Both rulers glorified war and saw it as the perfect arena for the realization of their chivalric ideals. Neither king possessed any sort of policy for dealing with his kingdom's social, economic, or political ills.

THE POPULAR RESPONSE

The governments of both England and France manipulated public opinion to support the war. Whatever significance modern students ascribe to the economic factor, public opinion in fourteenth-century England held that the war was waged for one reason: to secure for King Edward the French crown he had been denied.[7] Edward III issued letters to the sheriffs describing in graphic terms the evil deeds of the French and listing royal needs. Royal letters instructed the clergy to deliver sermons filled with patriotic sentiment. Frequent assemblies of Parliament —which in the fourteenth century were meetings of representatives of the nobility, clergy, counties, and towns, as well as royal officials summoned by the king to provide information or revenue or to do justice—spread royal propaganda for the war. The royal courts sensationalized the wickedness of the other side and stressed the great fortunes to be made from the war. Philip VI sent agents to warn communities about the dangers of invasion and to stress the French crown's revenue needs to meet the attack.

The royal campaign to rally public opinion was highly successful, at least in the early stage of the war. Edward III gained widespread support in the 1340s and 1350s. The English developed a deep hatred of the French and feared that King Philip intended "to have seized and slaughtered the entire realm of England." As England was successful in the field, pride in the country's military proficiency increased.

Most important of all, the war was popular because it presented unusual opportunities for wealth and advancement. Poor and unemployed knights were promised regular wages. Criminals who enlisted were granted pardons. The great nobles expected to be rewarded with estates. Royal exhortations to the troops before battles repeatedly stressed that, if victorious, the men might keep whatever they seized. The French chronicler Jean Froissart wrote that, at the time of Edward III's expedition of 1359, men of all ranks flocked to the king's banner. Some came to acquire honor, but many came in order "to loot and pillage the fair and plenteous land of France."[8]

The Indian Summer of Medieval Chivalry

The period of the Hundred Years' War witnessed the final flowering of the aristocratic code of medieval chivalry. Indeed, the enthusiastic participation of the nobility in both France and England was in response primarily to the opportunity the war provided to display chivalric behavior. What better place to display chivalric qualities than on the field of battle?

War was considered an ennobling experience; there was something elevating, manly, fine, and beautiful about it. When Shakespeare in the sixteenth century wrote of "the pomp and circumstance of glorious war," he was echoing the fourteenth- and fifteenth-century chroniclers who had glorified the trappings of war. Describing the French army before the battle of Poitiers (1356), a contemporary said:

Then you might see banners and pennons unfurled to the wind, whereon fine gold and azure shone, purple, gules and ermine. Trumpets, horns and clarions—you might hear sounding through the camp; the Dauphin's [title borne by the eldest son of the king of France] great battle made the earth ring.[9]

At Poitiers it was marvelous and terrifying to hear the thundering of the horses' hooves, the cries of the wounded, the sound of the trumpets and clarions, and the shouting of war cries. The tumult was heard at a distance of more than three leagues. And it was a great grief to see and behold the flower of all the nobility and chivalry of the world go thus to destruction, death, and martyrdom.

This romantic and "marvelous" view of war holds little appeal for modern men and women, who are more conscious of the slaughter, brutality, dirt, and blood that war inevitably involves. Also, modern thinkers are usually conscious of the broad mass of people, while the chivalric code applied only to the aristocratic military elite. Chivalry had no reference to those outside the knightly class.

The knight was supposed to show courtesy, graciousness, and generosity to his social equals, but certainly not to his social inferiors. When English knights fought French ones, they were social equals fighting according to a mutually accepted code of behavior. The infantry troops were looked on as inferior beings. When a peasant force at Longueil destroyed a contingent of English knights, their comrades mourned them because "it was too much that so many good fighters had been killed by mere peasants."[10]

The Course of the War to 1419

Armies in the field were commanded by rulers themselves; by princes of the blood such as Edward III's son Edward, the Black Prince—so called because of the color of his armor—or by great aristocrats. Knights formed the cavalry; the peasantry served as infantrymen, pikemen, and archers. Edward III set up recruiting boards in the counties to enlist the strongest peasants. Perhaps 10 percent of the adult population of England was involved in the actual fighting or in supplying and supporting the troops. The French contingents were even larger. By medieval standards, the force was astronomically large, especially considering the difficulty of transporting men, weapons, and horses across the English Channel. The costs of these armies stretched French and English resources to the breaking point.

MAP 12.2 English Holdings in France During the Hundred Years' War The year 1429 marked the greatest extent of English holdings in France. Why was it unlikely that England could have held these territories permanently?

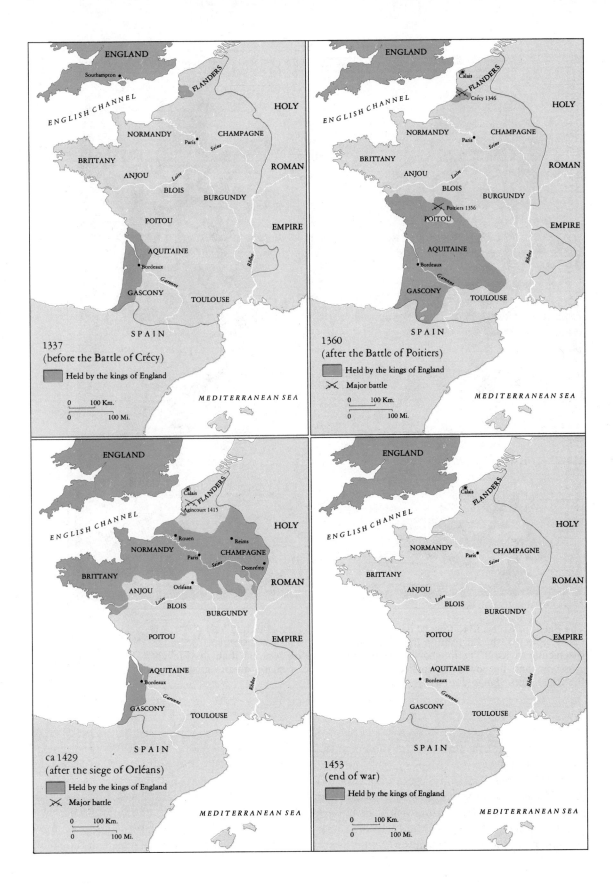

ENGLAND

Southampton

ENGLISH CHANNEL

FLANDERS

HOLY

NORMANDY   CHAMPAGNE

BRITTANY   Paris   Seine   ROMAN

ANJOU   Loire

BLOIS

BURGUNDY

POITOU   EMPIRE

AQUITAINE

Bordeaux   Rhône

Garonne

GASCONY   TOULOUSE

SPAIN

1337
(before the Battle of Crécy)

Held by the kings of England

0   100 Km.
0   100 Mi.

MEDITERRANEAN SEA

ENGLAND

Calais
ENGLISH CHANNEL   FLANDERS
Crécy 1346   HOLY

NORMANDY   CHAMPAGNE

BRITTANY   Paris   Seine   ROMAN

ANJOU   Loire

BLOIS

BURGUNDY

Poitiers 1356

POITOU   EMPIRE

AQUITAINE

Bordeaux   Rhône

Garonne

GASCONY   TOULOUSE

SPAIN

1360
(after the Battle of Poitiers)

Held by the kings of England

✕ Major battle

0   100 Km.
0   100 Mi.

MEDITERRANEAN SEA

ENGLAND

Calais
ENGLISH CHANNEL   FLANDERS
Agincourt 1415

HOLY

Rouen   Reims
NORMANDY   CHAMPAGNE
Paris   Seine   Domrémy

BRITTANY   ROMAN

ANJOU   Orléans
Loire
BLOIS

BURGUNDY

POITOU   EMPIRE

AQUITAINE

Bordeaux   Rhône

Garonne

GASCONY   TOULOUSE

SPAIN

ca 1429
(after the siege of Orléans)

Held by the kings of England

✕ Major battle

0   100 Km.
0   100 Mi.

MEDITERRANEAN SEA

ENGLAND

Calais
ENGLISH CHANNEL   FLANDERS

HOLY

NORMANDY   CHAMPAGNE

BRITTANY   Paris   Seine   ROMAN

ANJOU   Loire

BLOIS

BURGUNDY

POITOU   EMPIRE

AQUITAINE

Bordeaux   Rhône

Garonne

GASCONY   TOULOUSE

SPAIN

1453
(end of war)

Held by the kings of England

0   100 Km.
0   100 Mi.

MEDITERRANEAN SEA

The Battle of Crécy, 1346 Pitched battles were unusual in the Hundred Years' War. At Crécy, however, the English (on the right with lions on their royal standard) scored a spectacular victory. The longbow proved a more effective weapon than the French crossbow, and the low-born English archers withstood a charge of the aristocratic French knights. *(Photo: Larousse)*

The war was fought almost entirely in France and the Low Countries (see Map 12.2). It consisted mainly of a series of random sieges and cavalry raids. In 1335 the French began supporting Scottish incursions into northern England, ravaging the countryside in Aquitaine, and sacking and burning English coastal towns, such as Southampton. Naturally such

tactics lent weight to Edward III's propaganda campaign. In fact, royal propaganda on both sides fostered a kind of early nationalism.

During the war's early stages, England was highly successful. At Crécy in northern France in 1346, English longbowmen scored a great victory over French knights and crossbowmen. Although the fire of the longbow was not very accurate, it allowed for rapid reloading, and English archers could send off three arrows to the French crossbowmen's one. The result was a blinding shower of arrows that unhorsed the French knights and caused mass confusion. The firing of cannon—probably the first use of artillery in the West—created further panic. Thereupon the English horsemen charged and butchered the French.

This was not war according to the chivalric rules that Edward III would have preferred. The English victory at Crécy rested on the skill and swiftness of the yeomen archers, who had nothing at all to do with the chivalric ideals for which the war was being fought. Ten years later, Edward the Black Prince, using the same tactics as at Crécy, smashed the French at Poitiers, captured the French king, and held him for ransom. Again, at Agincourt near Arras in 1415, the chivalric English soldier-king Henry V (1413–1422) gained the field over vastly superior numbers. Henry followed up his triumph at Agincourt with the reconquest of Normandy. By 1419 the English had advanced to the walls of Paris (see Map 12.2).

But the French cause was not lost. Though England had scored the initial victories, France won the war.

JOAN OF ARC AND FRANCE'S VICTORY

The ultimate French success rests heavily on the actions of an obscure French peasant girl, Joan of Arc, whose vision and work revived French fortunes and led to victory. A great deal of pious and popular legend surrounds Joan the Maid, because of her peculiar appearance on the scene, her astonishing success, her martyrdom, and her canonization by the Catholic church. The historical fact is that she saved the French monarchy, which was the embodiment of France.

Born in 1412 to well-to-do peasants in the village of Domrémy in Champagne, Joan of Arc grew up in a religious household. During adolescence she began to hear voices, which she later said belonged to Saint Michael, Saint Catherine, and Saint Margaret. In 1428 these voices spoke to her with great urgency, telling her that the dauphin (the uncrowned King Charles VII) had to be crowned and the English expelled from France. Joan went to the French court, persuaded the king to reject the rumor that he was il-

Fifteenth-Century Armor This kind of expensive plate armor was worn by the aristocratic nobility in the fifteenth and sixteenth centuries. The use of gunpowder gradually made armor outmoded. *(Courtesy, World Heritage Museum. Photo: Caroline Buckler)*

legitimate, and secured his support for her relief of the besieged city of Orléans.

The astonishing thing is not that Joan the Maid overcame serious obstacles to see the dauphin, not even that Charles and his advisers listened to her. What is amazing is the swiftness with which they were convinced. French fortunes had been so low for so long that the court believed only a miracle could save the country. Because Joan cut her hair short and dressed like a man, she scandalized the court. But hoping she would provide the necessary miracle, Charles allowed her to accompany the army that was preparing to raise the English siege of Orléans.

In the meantime Joan, herself illiterate, dictated the following letter calling on the English to withdraw:

Jhesus Maria

King of England, and you Duke of Bedford, calling yourself regent of France, you William Pole, Count of Suffolk John Talbot, and you Thomas Lord Scales, calling yourselves Lieutenants of the said Duke of Bedford, do right in the King of Heaven's sight. Surrender to The Maid *sent hither by God the King of Heaven, the keys of all the good towns you have taken and laid waste in France. She comes in God's name to establish the Blood Royal, ready to make peace if you agree to abandon France and repay what you have taken. And you, archers, comrades in arms, gentles and others, who are before the town of Orléans, retire in God's name to your own country. If you do not, expect to hear tidings from* The Maid *who will shortly come upon you to your very great hurt.*[11]

Joan arrived before Orléans on April 28, 1429. Seventeen years old, she knew little of warfare and believed that if she could keep the French troops from swearing and frequenting whorehouses, victory would be theirs. On May 8, the English, weakened by disease and lack of supplies, withdrew from Orléans. Ten days later, Charles VII was crowned king at Rheims. These two events marked the turning point in the war.

In 1430 England's allies, the Burgundians, captured Joan and sold her to the English. When the English handed her over to the ecclesiastical authorities for trial, the French court did not intervene. While the English wanted Joan eliminated for obvious political reasons, sorcery (witchcraft) was the ostensible charge at her trial. Witch persecution was in-

creasing in the fifteenth century, and Joan's wearing of men's clothes appeared not only aberrant but indicative of contact with the devil.

Joan of Arc's political impact on the course of the Hundred Years' War and on the development of the kingdom of France has led scholars to examine her character and behavior very closely. Besides being an excellent athlete and a superb rider, she usually dressed like a rich and elegant young nobleman. Some students maintain that Joan's manner of dress suggests uncertainty about her own sexual identity. She did not menstruate—very rare in a healthy girl of eighteen—though she was female in every external respect: many men, including several dukes, admired her beautiful breasts. Perhaps, as Joan herself said, wearing men's clothes meant nothing at all. On the other hand, as some writers believe, she may have wanted to assume a completely new identity. Joan always insisted that God had specially chosen her for her mission. The richness and masculinity of her clothes, therefore, emphasized her uniqueness and made her highly conspicuous.[12] In 1431 the court condemned her as a heretic—her claim of direct inspiration from God, thereby denying the authority of church officials, constituted heresy—and burned her at the stake in the marketplace at Rouen. A new trial in 1456 rehabilitated her name. In 1920 she was canonized and declared a holy maiden, and today she is revered as the second patron saint of France. The nineteenth-century French historian Jules Michelet extolled Joan of Arc as a symbol of the vitality and strength of the French peasant classes.

The relief of Orléans stimulated French pride and rallied French resources. As the war dragged on, loss of life mounted, and money appeared to be flowing into a bottomless pit, demands for an end increased in England. The clergy and intellectuals pressed for peace. Parliamentary opposition to additional war grants stiffened. Slowly the French reconquered Normandy and, finally, ejected the English from Aquitaine. At the war's end in 1453, only the town of Calais remained in English hands.

COSTS AND CONSEQUENCES

For both France and England, the war proved a disaster. In France, the English had slaughtered thousands of soldiers and civilians. In the years after the sweep of the Black Death, this additional killing meant a

Joan of Arc Later considered the symbol of the French state in its struggle against the English, Joan of Arc here carries a sword in one hand and a banner with the royal symbol of fleur-de-lis in the other. Her face, which scholars believe to be a good resemblance, shows inner strength and calm determination. *(Archives Nationales, Paris)*

grave loss of population. The English had laid waste to hundreds of thousands of acres of rich farmland, leaving the rural economy of many parts of France a shambles. The war had disrupted trade and the great fairs, resulting in the drastic reduction of French participation in international commerce. Defeat in battle and heavy taxation contributed to widespread dissatisfaction and aggravated peasant grievances.

In England, only the southern coastal ports experienced much destruction; yet England fared little better than France. The costs of war were tremendous: England spent over £5 million in the war effort, a huge sum in the fourteenth and fifteenth centuries. The worst loss was in manpower. From 10 to 15 percent of the adult male population between the ages of fifteen and forty-five fought in the army or navy. In the decades after the plague, when the country was already suffering a severe manpower shortage, war losses made a bad situation frightful. Peasants serving in France as archers and pikemen were desperately needed to till the fields. The knights who ordinarily handled the work of local government as sheriffs, coroners, jurymen, and justices of the peace were abroad, and their absence contributed to the breakdown of order at the local level. The English government attempted to finance the war effort by raising taxes on the wool crop. Because of steadily increasing costs, the Flemish and Italian buyers could not afford English wool. Consequently, raw wool exports slumped drastically between 1350 and 1450.

Many men of all social classes had volunteered for service in France in the hope of acquiring booty and becoming rich. The chronicler Walsingham, describing the period of Crécy, tells of the tremendous prosperity and abundance resulting from the spoils of war: "For the woman was of no account who did not possess something from the spoils of . . . cities overseas in clothing, furs, quilts, and utensils . . . tablecloths and jewels, bowls of murra [semiprecious stone] and silver, linen and linen cloths."[13] Walsingham is referring to 1348, in the first generation of war. As time went on, most fortunes seem to have been squandered as fast as they were made.

If English troops returned with cash, they did not invest it in land. In the fifteenth century, returning soldiers were commonly described as beggars and vagabonds, roaming about making mischief. Even the large sums of money received from the ransom of the great—such as the £250,000 paid to Edward III

for the freedom of King John of France—and the money paid as indemnities by captured towns and castles did not begin to equal the more than £5 million spent. England suffered a serious net loss.[14]

The long war also had a profound impact on the political and cultural lives of the two countries. Most notably, it stimulated the development of the English Parliament. Between 1250 and 1450, representative assemblies from several classes of society flourished in many European countries. In the English parliaments, French Estates, German diets, and Spanish Cortes, deliberative practices developed that laid the foundations for the representative institutions of modern liberal-democratic nations. While representative assemblies declined in most countries after the fifteenth century, the English Parliament endured. Edward III's constant need for money to pay for the war compelled him to summon not only the great barons and bishops, but knights of the shires and burgesses from the towns as well. Between the outbreak of the war in 1337 and the king's death in 1377, parliamentary assemblies met twenty-seven times. Parliament met in thirty-seven of the fifty years of Edward's reign.[15]

The frequency of the meetings is significant. Representative assemblies were becoming a habit, a tradition. Knights and burgesses—or the "Commons," as they came to be called—recognized their mutual interests and began to meet apart from the great lords. The Commons gradually realized that they held the country's purse strings, and a parliamentary statute of 1341 required that all nonfeudal levies have parliamentary approval. When Edward III signed the law, he acknowledged that the king of England could not tax without Parliament's consent. Increasingly, during the course of the war, money grants were tied to royal redress of grievances: if the government was to raise money, it had to correct the wrongs its subjects protested.

As the Commons met in a separate chamber—the House of Commons—it also developed its own organization. The Speaker came to preside over debates in the House of Commons and to represent the Commons before the House of Lords and the king. Clerks kept a record of what transpired during discussions in the Commons.

In England, theoretical consent to taxation and legislation was given in one assembly for the entire country. France had no such single assembly; instead, there were many regional or provincial assemblies. Why did a national representative assembly fail to develop in France? The initiative for convening assemblies rested with the king, who needed revenue almost as much as the English ruler. But the French monarchy found the idea of representative assemblies thoroughly distasteful. Large gatherings of the nobility potentially or actually threatened his power. The advice of a counselor to King Charles VI (1380–1422), "above all things be sure that no great assemblies of nobles or of *communes* take place in your kingdom,"[16] was accepted. Charles VII (1422–1461) even threatened to punish those proposing a national assembly.

The English Parliament was above all else a court of law, a place where justice was done and grievances remedied. No French assembly (except that of Brittany) had such competence. The national assembly in England met frequently. In France, general assemblies were so rare that they never got the opportunity to develop precise procedures or to exercise judicial functions.

No one in France wanted a national assembly. Linguistic, geographic, economic, legal, and political differences were very strong. People tended to think of themselves as Breton, Norman, Burgundian, or whatever, rather than French. Through much of the fourteenth and early fifteenth centuries, weak monarchs lacked the power to call a national assembly. Provincial assemblies, highly jealous of their independence, did not want a national assembly. The costs of sending delegates to it would be high, and the result was likely to be increased taxation. Finally, the Hundred Years' War itself hindered the growth of a representative body of government. Possible violence on dangerous roads discouraged people from travel.

In both countries, however, the war did promote the growth of *nationalism*—the feeling of unity and identity that binds together a people who speak the same language, have a common ancestry and customs, and live in the same area. In the fourteenth century, nationalism largely took the form of hostility toward foreigners. Both Philip VI and Edward III drummed up support for the war by portraying the enemy as an alien, evil people. Edward III sought to justify his personal dynastic quarrel by linking it with England's national interests. As the Parliament Roll of 1348 states:

The Knights of the shires and the others of the Commons were told that they should withdraw together and take good counsel as to how, for withstanding the malice of the said enemy and for the salvation of our said lord the King and his Kingdom of England . . . the King could be aided.[17]

After victories, each country experienced a surge of pride in its military strength. Just as English patriotism ran strong after Crécy and Poitiers, so French national confidence rose after Orléans. French national feeling demanded the expulsion of the enemy not merely from Normandy and Aquitaine but from French soil. Perhaps no one expressed this national consciousness better than Joan of Arc, when she exulted that the enemy had been "driven out of *France.*"

VERNACULAR LITERATURE

Few developments expressed the emergence of national consciousness more vividly than the emergence of national literatures. Across Europe people spoke the language and dialect of their particular locality and class. In England, for example, the common people spoke regional English dialects, while the upper classes conversed in French. Official documents and works of literature were written in Latin or French. Beginning in the fourteenth century, however, national languages—the vernacular—came into widespread use not only in verbal communication but in literature as well. Three masterpieces of European culture, Dante's *Divine Comedy* (1321), Chaucer's *Canterbury Tales* (1387–1400), and Villon's *Grand Testament* (1461), brilliantly manifest this new national pride.

Dante Alighieri (1265–1321) descended from an aristocratic family in Florence, where he held several positions in the city government. Dante called his work a "comedy" because he wrote it in Italian and in a different style from the "tragic" Latin; a later generation added the adjective "divine," referring both to its sacred subject and to Dante's artistry. The *Divine Comedy* is an allegorical trilogy of one hundred cantos (verses) whose three equal parts (1 + 33 + 33 + 33) each describe one of the realms of the next world, Hell, Purgatory, and Paradise. Dante re-

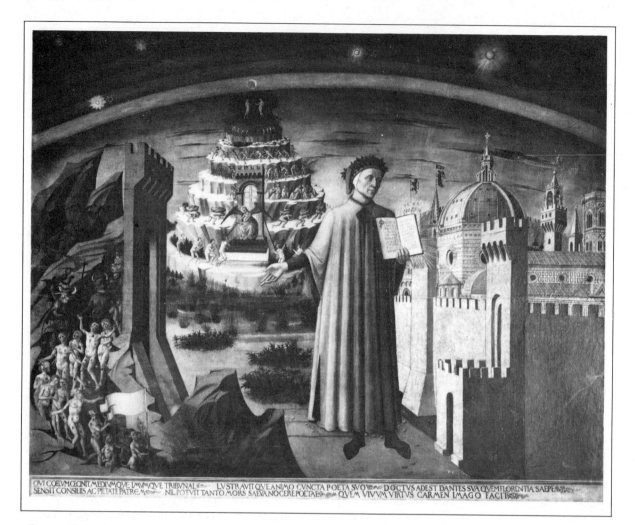

Dante Alighieri In this fifteenth-century fresco the poet, crowned with the wreath of poet laureate, holds the book containing the opening lines of his immortal *Commedia*. On the left is Hell and the mountain of purgatory; on the right, the city of Florence. *(Alinari/Art Resource)*

counts his imaginary journey through these regions toward God. The Roman poet Virgil, representing reason, leads Dante through Hell where he observes the torments of the damned and denounces the disorders of his own time, especially ecclesiastical ambition and corruption. Passing up into Purgatory, Virgil shows the poet how souls are purified of their disordered inclinations. In Paradise, home of the angels and saints, Saint Bernard—representing mystic contemplation—leads Dante to the Virgin Mary. Through her intercession he at last attains a vision of God.

The *Divine Comedy* portrays contemporary and historical figures, comments on secular and ecclesiastical affairs, and draws on scholastic philosophy. Within the framework of a symbolic pilgrimage to the City of God, the *Divine Comedy* embodies the psychological tensions of the age. A profoundly Christian poem, it also contains bitter criticism of some church authorities. In its symmetrical structure and use of figures from the ancient world, such as Virgil, the poem perpetuates the classical tradition, but as the first major work of literature in the Italian vernacular, it is distinctly modern.

Geoffrey Chaucer (1340–1400), the son of a London wine merchant, was an official in the administrations of the English kings Edward III and Richard II and wrote poetry as an avocation. Chaucer's *Canterbury Tales* is a collection of stories in lengthy, rhymed narrative. On a pilgrimage to the shrine of Saint Thomas Becket at Canterbury (see page 321), thirty people of various social backgrounds each tell a tale. The Prologue sets the scene and describes the pilgrims, whose characters are further revealed in the story each one tells. For example, the gentle Christian Knight relates a chivalric romance; the gross Miller tells a vulgar story about a deceived husband; the earthy Wife of Bath, who has buried five husbands, sketches a fable about the selection of a spouse; and the elegant Prioress, who violates her vows by wearing jewelry, delivers a homily on the Virgin. In depicting the interests and behavior of all types of people, Chaucer presents a rich panorama of English social life in the fourteenth century. Like the *Divine Comedy, Canterbury Tales* reflects the cultural tensions of the times. Ostensibly Christian, many of the pilgrims are also materialistic, sensual, and worldly, suggesting the ambivalence of the broader society's concern for the next world and frank enjoyment of this one.

Our knowledge of François Villon (1431–1463), probably the greatest poet of late medieval France, derives from Paris police records and his own poetry. Born to desperately poor parents in the year of Joan of Arc's execution, Villon was sent by his guardian to the University of Paris, where he earned the Master of Arts degree. A rowdy and free-spirited student, he disliked the stuffiness of academic life. In 1455 Villon killed a man in a street brawl; banished from Paris, he joined one of the bands of wandering thieves that harassed the countryside after the Hundred Years' War. For his fellow bandits he composed ballads in thieves' jargon.

Villon's *Lais* (1456), a pun on the word *legs* ("legacy"), is a series of farcical bequests to friends and enemies. "Ballade des Pendus" ("Ballad of the Hanged") was written while contemplating that fate in prison. (His execution was commuted.) Villon's greatest and most self-revealing work, the *Grand Testament,* contains another string of bequests, including a legacy to a prostitute, and describes his unshakeable faith in the beauty of life on earth. The *Grand Testament* possesses elements of social rebellion, bawdy humor, and rare emotional depth. While

the themes of Dante's and Chaucer's poetry are distinctly medieval, Villon's celebration of the human condition brands him as definitely modern. While he used medieval forms of versification, Villon's language was the despised vernacular of the poor and the criminal.

THE DECLINE OF THE CHURCH'S PRESTIGE

In times of crisis or disaster, people of all faiths have sought the consolation of religion. In the fourteenth century, however, the official Christian church offered very little solace. In fact, the leaders of the church added to the sorrow and misery of the times.

THE BABYLONIAN CAPTIVITY

From 1309 to 1376, the popes lived in the city of Avignon in southeastern France. In order to control the church and its policies, Philip the Fair of France pressured Pope Clement V to settle in Avignon (page 350). Clement, critically ill with cancer, lacked the will to resist Philip. This period in church history is often called the Babylonian Captivity (referring to the seventy years the ancient Hebrews were held captive in Mesopotamian Babylon).

The Babylonian Captivity badly damaged papal prestige. The Avignon papacy reformed its financial administration and centralized its government. But the seven popes at Avignon concentrated on bureaucratic matters to the exclusion of spiritual objectives. Though some of the popes led austere lives there, the general atmosphere was one of luxury and extravagance. The leadership of the church was cut off from its historic roots and the source of its ancient authority, the city of Rome. In the absence of the papacy, the Papal States in Italy lacked stability and good government. The economy of Rome had long been based on the presence of the papal court and the rich tourist trade the papacy attracted. The Babylonian Captivity left Rome poverty-stricken. As long as the French crown dominated papal policy, papal influence in England (with whom France was intermittently at war) and in Germany declined.

Many devout Christians urged the popes to return to Rome. The Dominican mystic Catherine of Siena, for example, made a special trip to Avignon to plead with the pope to return. In 1377 Pope Gregory XI brought the papal court back to Rome. Unfortunately, he died shortly after the return. At Gregory's death, Roman citizens demanded an Italian pope who would remain in Rome. Determined to influence the *papal conclave* (the assembly of cardinals who choose the new pope) to elect an Italian, a Roman mob surrounded Saint Peter's Basilica, blocked the roads leading out of the city, and seized all boats on the Tiber River. Between the time of Gregory's death and the opening of the conclave, great pressure was put on the cardinals to elect an Italian. At the time, none of them protested this pressure.

Sixteen cardinals—eleven Frenchmen, four Italians, and one Spaniard—entered the conclave on April 7, 1378. After two ballots they unanimously chose a distinguished administrator, the archbishop of Bari, Bartolomeo Prignano, who took the name Urban VI. Each of the cardinals swore that Urban had been elected, "sincerely, freely, genuinely, and canonically."

Urban VI (1378–1389) had excellent intentions for church reform. He wanted to abolish simony, *pluralism* (holding several church offices at the same time), absenteeism, clerical extravagance, and ostentation. These were the very abuses being increasingly criticized by Christian peoples across Europe. Unfortunately, Pope Urban went about the work of reform in a tactless, arrogant, and bullheaded manner. The day after his coronation he delivered a blistering attack on cardinals who lived in Rome while drawing their income from benefices elsewhere. His criticism was well founded but ill timed and provoked opposition among the hierarchy before Urban had consolidated his authority.

In the weeks that followed, Urban stepped up attacks on clerical luxury, denouncing individual cardinals by name. He threatened to strike the cardinal archbishop of Amiens. Urban even threatened to excommunicate certain cardinals, and when he was advised that such excommunications would not be lawful unless the guilty had been warned three times, he shouted, "I can do anything, if it be my will and judgment."[18] Urban's quick temper and irrational behavior have led scholars to question his sanity. Whether he was medically insane or just drunk with power is a moot point. In any case, Urban's actions brought on disaster.

In groups of two and three, the cardinals slipped away from Rome and met at Anagni. They declared Urban's election invalid because it had come about under threats from the Roman mob, and they asserted that Urban himself was excommunicated. The cardinals then proceeded to the city of Fondi between Rome and Naples and elected Cardinal Robert of Geneva, the cousin of King Charles V of France, as pope. Cardinal Robert took the name Clement VII. There were thus two popes—Urban at Rome and the antipope Clement VII (1378–1394), who set himself up at Avignon in opposition to the legally elected Urban. So began the Great Schism, which divided Western Christendom until 1417.

The Great Schism

The powers of Europe aligned themselves with Urban or Clement along strictly political lines. France naturally recognized the French antipope, Clement. England, France's historic enemy, recognized Pope Urban. Scotland, whose attacks on England were subsidized by France, followed the French and supported Clement. Aragon, Castile, and Portugal hesitated before deciding for Clement at Avignon. The emperor, who bore ancient hostility to France, recognized Urban VI. At first the Italian city-states recognized Urban; when he alienated them, they opted for Clement.

John of Spoleto, a professor at the law school at Bologna, eloquently summed up intellectual opinion of the schism:

The longer this schism lasts, the more it appears to be costing, and the more harm it does; scandal, massacres, ruination, agitations, troubles and disturbances . . . this dissention is the root of everything: divers tumults, quarrels between kings, seditions, extortions, assassinations, acts of violence, wars, rising tyranny, decreasing freedom, the impunity of villains, grudges, error, disgrace, the madness of steel and of fire given license.[19]

The scandal "rent the seamless garment of Christ," as the church was called, and provoked horror and vigorous cries for reform. The common people, wracked by inflation, wars, and plague, were thoroughly confused about which pope was legitimate. The schism weakened the religious faith of many Christians and gave rise to instability and religious excesses. It brought the church leadership into serious disrepute.

At a time when ordinary Christians needed the consolation of religion and confidence in religious leaders, church officials were fighting among themselves for power.

THE CONCILIAR MOVEMENT

Calls for church reform were not new. A half century before the Great Schism, in 1324, Marsiglio of Padua, then rector of the University of Paris, had published *Defensor Pacis (The Defender of the Peace).* Dealing as it did with the authority of state and church, *Defensor Pacis* proved to be one of the most controversial works written in the Middle Ages.

Marsiglio argued that the state was the great unifying power in society and that the church was subordinate to the state. He put forth the revolutionary ideas that the church had no inherent jurisdiction and should own no property. Authority in the Christian church, according to Marsiglio, should rest in a general council, made up of laymen as well as priests and superior to the pope. These ideas directly contradicted the medieval notion of a society governed by the church and the state, with the church supreme.

Defensor Pacis was condemned by the pope, and Marsiglio was excommunicated. But the idea that a general council representing all of the church had a higher authority than the pope was repeated by John Gerson (1363–1429), a later chancellor of the University of Paris and influential theologian.

Even more earthshaking than the theories of Marsiglio of Padua were the ideas of the English scholar and theologian John Wyclif (1329–1384). Wyclif wrote that papal claims of temporal power had no foundation in the Scriptures, and that the Scriptures alone should be the standard of Christian belief and practice. He urged the abolition of such practices as the veneration of saints, pilgrimages, pluralism, and absenteeism. Every sincere Christian, according to Wyclif, should read the Bible for himself. Wyclif's views had broad social and economic significance. He urged that the church be stripped of its property. His idea that every Christian free of mortal sin possessed lordship was seized on by peasants in England during a revolt in 1381 and used to justify their goals.

In advancing these views, Wyclif struck at the roots of medieval church structure and religious practices. Consequently, he has been hailed as the precursor of the Reformation of the sixteenth century. Although Wyclif's ideas were vigorously condemned by ecclesiastical authorities, they were widely disseminated by humble clerics and enjoyed great popularity in the early fifteenth century. Wyclif's followers were called "Lollards." The term, which means "mumblers of prayers and psalms," refers to what they criticized. After Anne, sister of Wenceslaus, king of Germany and Bohemia, married Richard II of England, members of Queen Anne's household carried Lollard principles back to Bohemia, where they were spread by John Hus, rector of the University of Prague.

While John Wyclif's ideas were being spread, two German scholars at the University of Paris, Henry of Langenstein and Conrad of Gelnhausen, produced treatises urging the summoning of a general council. Conrad wrote that the church, as the congregation of all the faithful, was superior to the pope. Although canon law held that only a pope might call a council, a higher law existed: the common good. The common good required the convocation of a council.

In response to continued Europewide calls for a council, the two colleges of cardinals—one at Rome, the other at Avignon—summoned a council at Pisa in 1409. A distinguished gathering of prelates and theologians deposed both popes and selected another. Neither the Avignon pope nor the Roman pope would resign, however, and the appalling result was a threefold schism.

Finally, due to the pressure of the German emperor Sigismund, a great council met at the imperial city of Constance (1414–1418). It had three objectives: to end the schism, to reform the church "in head and members" (from top to bottom), and to wipe out heresy. The council condemned the Lollard ideas of John Hus, and he was burned at the stake. The council eventually deposed both the Roman pope and the successor of the pope chosen at Pisa, and it isolated the Avignonese antipope. A conclave elected a new leader, the Roman cardinal Colonna, who took the name Martin V (1417–1431).

Martin proceeded to dissolve the council. Nothing was done about reform. The schism was over, and though councils subsequently met at Basel and at Ferrara-Florence, in 1450 the papacy held a jubilee, celebrating its triumph over the conciliar movement. In the later fifteenth century, the papacy concentrated on Italian problems to the exclusion of universal Christian interests. But the schism and the conciliar movement had exposed the crying need for ecclesiastical reform, thus laying the foundations for the great reform efforts of the sixteenth century.

The Burning of John Hus The Council of Constance executed Hus as a demonstration of conciliar authority within the church. Persons burned at the stake usually died of smoke inhalation. *(Yale University Library)*

THE LIFE OF THE PEOPLE

In the fourteenth century, economic and political difficulties, disease, and war profoundly affected the lives of European peoples. Decades of slaughter and destruction, punctuated by the decimating visits of the Black Death, made a grave economic situation virtually disastrous. In many parts of France and the Low Countries, fields lay in ruin or untilled for lack of manpower. In England, as taxes increased, criticism of government policy and mismanagement multiplied. Crime, always a factor in social history, aggravated economic troubles, and throughout Europe the frustrations of the common people erupted into widespread revolts. For most people, marriage and the local parish church continued to be the center of their lives.

MARRIAGE

Marriage and the family provided such peace and satisfaction as most people attained. In fact, life for those who were not clerics or nuns meant marriage. Apart from sexual and emotional urgency, the community expected people to marry. For a girl, childhood was a preparation for marriage. In addition to the thousands of chores involved in running a household, girls learned obedience, or at least subordination. Adulthood meant living as a wife or widow.

However, sweeping statements about marriage in the Middle Ages have limited validity. Most peasants were illiterate and left slight record of their feelings toward their spouses or about marriage as an institution. The gentry, however, often could write, and the letters exchanged between Margaret and John Paston, upper-middle-class people who lived in Norfolk, England, in the fifteenth century, provide important evidence of the experience of one couple.

John and Margaret Paston were married about 1439, after an arrangement concluded entirely by their parents. John spent most of his time in London fighting through the law courts to increase his family properties and business interests; Margaret remained in Norfolk to supervise the family lands. Her enormous responsibilities involved managing the Paston estates, hiring workers, collecting rents, ordering supplies for the large household, hearing complaints and settling disputes among tenants, and marketing her crops. In these duties she proved herself a remarkably shrewd businessperson. Moreover, when an army of over a thousand men led by the aristocratic thug Lord Moleyns attacked her house, she successfully withstood the siege. When the Black Death entered her area, Margaret moved her family to safety.

Margaret Paston did all this on top of raising eight children (there were probably other children who did not survive childhood). Her husband died before she was forty-three, and she later conducted the negotiations for the children's marriages. Her children's futures, like her estate management, were planned with an eye toward economic and social advancement. When one daughter secretly married the estate bailiff, an alliance considered beneath her, the girl was cut off from the family as if she were dead.[20]

The many letters surviving between Margaret and John reveal slight tenderness toward their children. They seem to have reserved their love for each other, and during many of his frequent absences they wrote to express mutual affection and devotion. How typical the Paston relationship was modern historians cannot say, but the marriage of John and Margaret, although completely arranged by their parents, was based on respect, responsibility, and love.[21]

At what age did people usually marry? The largest amount of evidence on age at first marriage survives from Italy, and a comparable pattern probably existed in northern Europe. For girls, population surveys at Prato in 1372 place the age at 16.3 years in 1372 and 21.1 in 1470. Chaucer's wife of Bath says that she married first in her twelfth year. Among the German nobility recent research has indicated that in the Hohenzollern family in the later Middle Ages "five brides were between 12 and 13; five about 14, and five about 15."

Men were older. An Italian chronicler writing about 1354 says that men did not marry before the age of 30. At Prato in 1371, the average age of men at first marriage was 24 years, very young for Italian men, but this data may represent an attempt to regain population losses due to the recent attack of the plague. In England, Chaucer's wife of Bath describes her first three husbands as "goode men, and rich, and old." Among 17 males in the noble Hohenzollern family, eleven were over 20 years when married, five between 18 and 19, one 16. The general pattern in late medieval Europe was marriage between men in their middle or late twenties and women under twenty.[22]

In the later Middle Ages, as earlier—indeed, until the late nineteenth century—economic factors, rather than romantic love or physical attraction, determined whom and when a person married. The young agricultural laborer on the manor had to wait until he had sufficient land. Thus most men had to wait until their fathers died or yielded the holding. The age of marriage was late, which in turn affected the number of children a couple had. The journeyman craftsman in the urban guild faced the same material difficulties. Prudent young men selected (or their parents selected for them) girls who would bring the most land or money to the union. Once a couple married, the union ended only with the death of one partner.

Deep emotional bonds knit members of medieval families. Parents delighted in their children, and the church encouraged a cult of paternal care. The church stressed its right to govern and sanctify marriage, and emphasized monogamy. Tighter moral and emotional unity within marriages resulted.[23]

Divorce—complete dissolution of the contract between a woman and man lawfully married—did not exist in the Middle Ages. The church held that a marriage validly entered into could not be dissolved. A valid marriage consisted of the mutual oral consent or promise of two parties. Church theologians of the day urged that marriage be publicized by *banns,* or announcements made in the parish church, and that the couple's union be celebrated and witnessed in a church ceremony and blessed by a priest.

Domestic Brawl In all ages the hen-pecked husband has been a popular subject for jests. This elaborate woodcarving from a fifteenth-century English choir stall shows the husband holding distaff and ball of thread, symbolic of wife's work as "spinster," while his wife thrashes him. *(Royal Commission on the Historical Monuments of England)*

A great number of couples did not observe the church's regulations. Some treated marriage as a private act—they made the promise and spoke the words of marriage to each other without witnesses and then proceeded to enjoy the sexual pleasures of marriage. This practice led to a great number of disputes, because one or the other of the two parties could later deny having made a marriage agreement. The records of the ecclesiastical courts reveal many cases arising from privately made contracts. Here is a typical case heard by the ecclesiastical court at York in 1372:

[The witness says that] one year ago on the feast day of the apostles Philip and James just past, he was present in the house of William Burton, tanner of York. . . . when and where John Beke, saddler . . . called the said Marjory to him and said to her, "Sit with me." Acquiescing in this, she sat down. John said to her, "Marjory, do you wish to be my wife?" And she replied, "I will if you wish." And taking at once the said Marjory's right hand, John said, "Marjory, here I take you as my wife, for better or worse, to have and to hold until the end of my life; and of this I give you my faith." The said Marjory replied to him, "Here I take you John as my husband, to have and

to hold until the end of my life, and of this I give you my faith." And then the said John kissed the said Marjory."[24]

This was a private arrangement, made in secret and without the presence of clergy. Evidence survives of marriages contracted in a garden, in a blacksmith's shop, at a tavern, and, predictably, in a bed. Church courts heard a great number of similar cases. The records of those courts that relate to marriage reveal that, rather than suits for divorce, the great majority of petitions asked the court to enforce the marriage contract that one of the parties believed she or he had validly made. Annulments were granted in extraordinary circumstances, such as male impotence, on the grounds that a lawful marriage had never existed.

LIFE IN THE PARISH

In the later Middle Ages, the land and the parish remained the focus of life for the European peasantry. Work on the land continued to be performed collectively. All men, for example, cooperated in the annual tasks of planting and harvesting. The close association of the cycle of agriculture and the liturgy of the Christian calendar endured. The parish priest blessed the fields before the annual planting, offering prayers on behalf of the people for a good crop. If the harvest was a rich one, the priest led the processions and celebrations of thanksgiving.

How did the common people feel about their work? Since the vast majority were illiterate and inarticulate, it is difficult to say. It is known that the peasants hated the ancient services and obligations on the lords' lands and tried to get them commuted for money rents. When lords attempted to reimpose service duties, the peasants revolted.

In the thirteenth century, the craft guilds provided the small minority of men living in towns and cities with the psychological satisfaction of involvement in the manufacture of a superior product. The guild member also had economic security. The craft guilds set high standards for their merchandise. The guilds looked after the sick, the poor, the widowed, and the orphaned. Masters and journeymen worked side by side.

In the fourteenth century, those ideal conditions began to change. The fundamental objective of the craft guild was to maintain a monopoly on its product, and to do so recruitment and promotion were carefully restricted. Some guilds required a high entrance fee for apprentices; others admitted only the sons or relatives of members. Apprenticeship increasingly lasted a long time, seven years. Even after a young man had satisfied all the tests for full membership in the guild and had attained the rank of master, other hurdles had to be passed, such as finding the funds to open his own business or special connections just to get in a guild. Restrictions limited the number of apprentices and journeymen to the anticipated openings for masters. The larger a particular business was, the greater was the likelihood that the master did not know his employees. The separation of master and journeyman and the decreasing number of openings for master craftsmen created serious frustrations. Strikes and riots occurred in the Flemish towns, in France, and in England.

The recreation of all classes reflected the fact that late medieval society was organized for war and that violence was common. The aristocracy engaged in tournaments or jousts; archery and wrestling had great popularity among ordinary people. Everyone enjoyed the cruel sports of bullbaiting and bearbaiting. The hangings and mutilations of criminals were exciting and well-attended events, with all the festivity of a university town before a Saturday football game. Chroniclers exulted in describing executions, murders, and massacres. Here a monk gleefully describes the gory execution of William Wallace in 1305:

Wilielmus Waleis, a robber given to sacrilege, arson and homicide . . . was condemned to most cruel but justly deserved death. He was drawn through the streets of London at the tails of horses, until he reached a gallows of unusual height, there he was suspended by a halter; but taken down while yet alive, he was mutilated, his bowels torn out and burned in a fire, his head then cut off, his body divided into four, and his quarters transmitted to four principal parts of Scotland.[25]

Violence was as English as roast beef and plum pudding, as French as bread, cheese, and *potage.*

Alcohol, primarily beer or ale, provided solace to the poor, and the frequency of drunkenness reflects their terrible frustrations.

During the fourteenth and fifteenth centuries, the laity began to exercise increasing control over parish affairs. Churchmen were criticized. The constant quarrels of the mendicant orders (the Franciscans and Dominicans), the mercenary and grasping attitude of the parish clergy, the scandal of the Great Schism and a divided Christendom—all these did much to weaken the spiritual mystique of the clergy in the popular mind. The laity steadily took responsibility for the management of parish lands. Lay people organized associations to vote on and purchase furnishings for the church. And ordinary lay people secured jurisdiction over the structure of the church building, its vestments, books, and furnishings. These new responsibilities of the laity reflect the increased dignity of parishioners in the late Middle Ages.[26]

FUR-COLLAR CRIME

The Hundred Years' War had provided employment and opportunity for thousands of idle and fortune-seeking knights. But during periods of truce and after the war finally ended, many nobles once again had little to do. Inflation also hurt them. Although many were living on fixed incomes, their chivalric code demanded lavish generosity and an aristocratic lifestyle. Many nobles turned to crime as a way of raising money. The fourteenth and fifteenth centuries witnessed a great deal of "fur-collar crime," so called for the miniver fur the nobility alone were allowed to wear on their collars. England provides a good case study of upper-class crime.

Fur-collar crime rarely involved such felonies as homicide, robbery, rape, and arson. Instead, nobles used their superior social status to rob and extort from the weak and then to corrupt the judicial process. Groups of noble brigands roamed the English countryside stealing from both rich and poor. Sir John de Colseby and Sir William Bussy led a gang of thirty-eight knights who stole goods worth £3,000 in various robberies. Operating exactly like modern urban racketeers, knightly gangs demanded that peasants pay "protection money" or else have their hovels burned and their fields destroyed. Members of the household of a certain Lord Robert of Payn beat up a victim and then demanded money for protection from future attack.

Attacks on the rich often took the form of kidnapping and extortion. Individuals were grabbed in their homes, and wealthy travelers were seized on the highways and held for ransom. In northern England a gang of gentry led by Sir Gilbert de Middleton abducted Sir Henry Beaumont; his brother, the bishop-elect of Durham; and two Roman cardinals in England on a peacemaking visit. Only after a ransom was paid were the victims released.[27]

Fur-collar criminals were terrorists, but like some twentieth-century white-collar criminals who commit nonviolent crimes, medieval aristocratic criminals got away with their outrages. When accused of wrongdoing, fur-collar criminals intimidated witnesses. They threatened jurors. They used "pull" or cash to bribe judges. As a fourteenth-century English judge wrote to a young nobleman, "For the love of your father I have hindered charges being brought against you and have prevented execution of indictment actually made."[28]

The ballads of Robin Hood, a collection of folk legends from the late medieval England, describe the adventures of the outlaw hero and his band of followers, who lived in Sherwood Forest and attacked and punished those who violated the social system and the law. Most of the villains in these simple tales are fur-collar criminals—grasping landlords, wicked sheriffs such as the famous sheriff of Nottingham, and mercenary churchmen. Robin and his merry men performed a sort of retributive justice. Robin Hood was a popular figure, because he symbolized the deep resentment of aristocratic corruption and abuse; he represented the struggle against tyranny and oppression.

Criminal activity by nobles continued decade after decade because governments were too weak to stop it. Then, too, much of the crime was directed against a lord's own serfs, and the line between a noble's legal jurisdiction over his peasants and criminal behavior was a fine one indeed. Persecution by lords, on top of war, disease, and natural disaster, eventually drove long-suffering and oppressed peasants all across Europe to revolt.

PEASANT REVOLTS

Peasant revolts occurred often in the Middle Ages. Early in the thirteenth century, the French preacher Jacques de Vitry asked rhetorically, "How many serfs have killed their lords or burnt their castles?"[29] Social and economic conditions in the fourteenth and fifteenth centuries caused a great increase in peasant uprisings (see Map 12.3).

The Jacquerie Because social revolt on the part of war-weary, frustrated poor seemed to threaten the natural order of Christian society during the fourteenth and fifteenth centuries, the upper classes everywhere exacted terrible vengeance on peasants and artisans. In this scene some *jacques* are cut down, some beheaded, and others drowned. *(Bibliothèque Nationale, Paris)*

In 1358, when French taxation for the Hundred Years' War fell heavily on the poor, the frustrations of the French peasantry exploded in a massive uprising called the *Jacquerie,* after a supposedly happy agricultural laborer, Jacques Bonhomme (Good Fellow). Peasants in Picardy and Champagne went on the rampage. Crowds swept through the countryside slashing the throats of nobles, burning their castles, raping their wives and daughters, killing or maiming their horses and cattle. Peasants blamed the nobility for oppressive taxes, for the criminal brigandage of the countryside, for defeat in war, and for the general misery. Artisans, small merchants, and parish priests joined the peasants. Urban and rural groups committed terrible destruction, and for several weeks the nobles were on the defensive . Then the upper class united to repress the revolt with merciless ferocity. Thousands of the "Jacques," innocent as well as guilty, were cut down.

This forcible suppression of social rebellion, without some effort to alleviate its underlying causes, could only serve as a stopgap measure and drive protest underground. Between 1363 and 1484, serious peasant revolts swept the Auvergne; in 1380, uprisings occurred in the Midi; and in 1420, they erupted in the Lyonnais region of France.

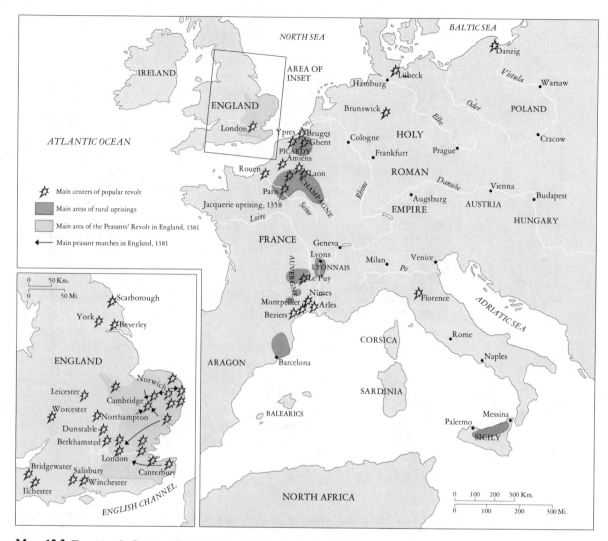

MAP 12.3 Fourteenth-Century Peasant Revolts In the later Middle Ages and early modern times, peasant and urban uprisings were endemic, as common as factory strikes in the industrial world. The threat of insurrection served to check unlimited exploitation.

The Peasants' Revolt in England in 1381, involving perhaps a hundred thousand people, was probably the largest single uprising of the entire Middle Ages (see Map 12.3). The causes of the rebellion were complex and varied from place to place. In general, though, the thirteenth century had witnessed the steady commutation of labor services for cash rents, and the Black Death had drastically cut the labor supply. As a result, peasants demanded higher wages and fewer manorial obligations. Thirty years earlier the parliamentary Statute of Laborers of 1351 (see page 360) had declared:

Whereas to curb the malice of servants who after the pestilence were idle and unwilling to serve without securing excessive wages, it was recently ordained . . . that such servants, both men and women, shall be bound to serve in return for salaries and wages that were customary . . . five or six years earlier. [30]

This statute was an attempt by landlords to freeze wages and social mobility.

The statute could not be enforced. As a matter of fact, the condition of the English peasantry steadily improved in the course of the fourteenth century. Some scholars believe that the peasantry in most places was better off in the period 1350 to 1450 than it had been for centuries before or was to be for four centuries after.

Why then was the outburst in 1381 so serious? It was provoked by a crisis of rising expectations. The relative prosperity of the laboring classes led to demands that the upper classes were unwilling to grant. Unable to climb higher, the peasants' frustration found release in revolt. Economic grievances combined with other factors. Decades of aristocratic violence, much of it perpetrated against the weak peasantry, had bred hostility and bitterness. In France frustration over the lack of permanent victory increased. In England the social and religious agitation of the popular preacher John Ball fanned the embers of discontent. Such sayings as Ball's famous couplet

When Adam delved and Eve span
Who was then the gentleman?

reflect real revolutionary sentiment. But the lords of England believed that God had permanently fixed the hierarchical order of society and that nothing man could do would change that order. Moreover, the south of England, where the revolt broke out, had been subjected to frequent and destructive French raids. The English government did little to protect the south, and villages grew increasingly scared and insecure. Fear erupted into violence.

The straw that broke the camel's back in England was the re-imposition of a head tax on all adult males. Although it met widespread opposition in 1380, the royal council ordered the sheriffs to collect it again in 1381 on penalty of a huge fine. Beginning with assaults on the tax collectors, the uprising in England followed much the same course as had the Jacquerie in France. Castles and manors were sacked; manorial records were destroyed. Many nobles, including the archbishop of Canterbury, who had ordered the collection of the tax, were murdered.

Although the center of the revolt lay in the highly populated and economically advanced south and east, sections of the north and the Midlands also witnessed rebellions. Violence took different forms in different places. The townspeople of Cambridge expressed their hostility toward the university by sacking one of the colleges and building a bonfire of academic property. In towns containing skilled Flemish craftsmen, fear of competition led to their attack and murder. Urban discontent merged with rural violence. Apprentices and journeymen, frustrated because the highest positions in the guilds were closed to them, rioted.

The boy-king Richard II (1377–1399) met the leaders of the revolt, agreed to charters ensuring peasants' freedom, tricked them with false promises, and then proceeded to crush the uprising with terrible ferocity. Although the nobility tried to restore ancient duties of serfdom, virtually a century of freedom had elapsed, and the commutation of manorial services continued. Rural serfdom had disappeared in England by 1550.

Conditions in England and France were not unique. In Florence in 1378, the *ciompi,* the poor propertyless workers, revolted. Serious social trouble occurred in Lübeck, Brunswick, and other German cities. In Spain in 1391, aristocratic attempts to impose new forms of serfdom, combined with demands for tax relief, led to massive working-class and peasant uprisings in Seville and Barcelona. These took the form of vicious attacks on Jewish communities. Rebellions and uprisings everywhere reveal deep peasant and working-class frustration and the general socioeconomic crisis of the time.

———————

Late medieval preachers likened the crises of their times to the Four Horsemen of the Apocalypse in the Book of Revelation, who brought famine, war, disease, and death. The crises of the fourteenth and fifteenth centuries were acids that burned deeply into the fabric of traditional medieval European society. Bad weather brought poor harvests, which contributed to the international economic depression. Disease, over which people also had little control, fostered widespread depression and dissatisfaction. Population losses caused by the Black Death and the Hundred Years' War encouraged the working classes to try to profit from the labor shortage by selling their services higher: they wanted to move up the economic ladder. The ideas of thinkers like John Wyclif, John Hus, and John Ball fanned the flames of social discontent. When peasant frustrations exploded in uprisings, the frightened nobility and upper-middle class joined to crush the revolts and condemn

Albrecht Dürer: The Four Horsemen of the Apocalypse From right to left, representatives of war, strife, famine, and death gallop across Christian society leaving thousands dead or in misery. The horrors of the age made this subject extremely popular in art, literature, and sermons. *(Courtesy, Museum of Fine Arts, Boston)*

heretical preachers as agitators of social rebellion. But the war had heightened social consciousness among the poor.

The Hundred Years' War served as a catalyst for the development of representative government in England. The royal policy of financing the war through parliament-approved taxation gave the middle classes an increased sense of their economic power. They would pay taxes in return for some influence in shaping royal policies.

In France, on the other hand, the war stiffened opposition to national assemblies. The disasters that wracked France decade after decade led the French people to believe that the best solutions to complicated problems lay not in an assembly but in the hands of a strong monarch. France became the model for continental countries in the evolution toward royal absolutism.

The war also stimulated technological experimentation, especially with artillery. After about 1350, the cannon, although highly inaccurate, was commonly used all over Europe.

Religion remained the cement that held society together. European culture was a Christian culture. But the Great Schism weakened the prestige of the church and people's faith in papal authority. The conciliar movement, by denying the church's universal sovereignty, strengthened the claims of secular government to jurisdiction over all their peoples. The later Middle Ages witnessed a steady shift of basic loyalty from the church to the emerging national states.

NOTES

1. W. H. McNeill, *Plagues and Peoples,* Doubleday, New York, 1976, pp. 151–168.
2. Quoted by P. Ziegler, *The Black Death,* Pelican Books, Harmondsworth, England, 1969, p. 20.
3. J. M. Rigg, trans., *The Decameron of Giovanni Boccaccio,* J. M. Dent & Sons, London, 1903, p. 6.
4. Ziegler, pp. 232–239.
5. See G. P. Cuttino, "Historical Revision: The Causes of the Hundred Years' War," *Speculum* 31:3 (July 1956):463–472.
6. N. F. Cantor, *The English: A History of Politics and Society to 1760,* Simon & Schuster, New York, 1967, p. 260.
7. J. Barnie, *War in Medieval English Society: Social Values and the Hundred Years' War,* Cornell University Press, Ithaca, N.Y., 1974, p. 6.
8. Quoted by Barnie, p. 34.
9. Ibid., p. 73.
10. Ibid, pp. 72–73.
11. W. P. Barrett, trans., *The Trial of Jeanne d'Arc,* George Routledge, London, 1931, pp. 165–166.
12. Quoted by Edward A. Lucie-Smith, *Joan of Arc,* W. W. Norton, New York, 1977, pp. 32–35.
13. Quoted by Barnie, pp. 36–37.
14. M. M. Postan, "The Costs of the Hundred Years' War," *Past and Present* 27 (April 1964): 34–53.
15. See G. O. Sayles, *The King's Parliament of England,* W. W. Norton & Co., New York, 1974, Appendix, pp. 137–141.
16. Quoted by P. S. Lewis, "The Failure of the Medieval French Estates," *Past and Present* 23 (November 1962):6.
17. C. Stephenson and G. F. Marcham, eds., *Sources of English Constitutional History,* rev. ed., Harper & Row, New York, 1972, p. 217.
18. Quoted by J. H. Smith, *The Great Schism 1378: The Disintegration of the Papacy,* Weybright & Talley, New York, 1970, p. 141.
19. Ibid., p. 15.
20. A. S. Haskell, "The Paston Women on Marriage in Fifteenth Century England," *Viator* 4 (1973):459–469.
21. Ibid., p. 471.
22. See David Herlihy, *Medieval Households,* Harvard University Press, Cambridge, Mass., 1985, pp. 103–111.
23. Ibid., pp. 118–130.
24. Quoted by R. H. Helmholz, *Marriage Litigation in Medieval England,* Cambridge University Press, Cambridge, Eng., 1974, pp. 28–29.
25. A. F. Scott, ed., *Everyone a Witness: The Plantagenet Age,* Thomas Y. Crowell, New York, 1976, p. 263.
26. See E. Mason, "The Role of the English Parishioner, 1000–1500," *Journal of Ecclesiastical History* 27:1 (January 1976):17–29.
27. B. A. Hanawalt, "Fur Collar Crime: The Pattern of Crime Among the Fourteenth-Century English Nobility," *Journal of Social History* 8 (Spring 1975):1–14.
28. Ibid., p. 7.
29. Quoted by M. Bloch, *French Rural History,* trans. Janet Sondeimer, University of California Press, Berkeley, 1966, p. 169.
30. Stephenson and Marcham, p. 225.

SUGGESTED READING

Students who wish further elaboration of the topics covered in this chapter should consult the following studies, on which the chapter leans extensively. For the Black Death, see R. S. Gottfried, *The Black Death* (1983), a fresh and challenging work, and P. Ziegler, *The Black Death* (1969), a fascinating and highly readable study. For the social implications of disease, see W. H. McNeill, *Plagues and Peoples* (1976); F. F. Cartwright, *Disease and History* (1972); and H. E. Sigerist, *Civilization and Disease* (1970).

The standard study of the long military conflicts of the fourteenth and fifteenth centuries remains that of E. Perroy, *The Hundred Years' War* (1959). J. Henneman, *Royal Taxation in Fourteenth Century France: The Development of War Financing, 1322–1356* (1971), is an important technical work by a distinguished historian. J. Barnie's *War in Medieval English Society: Social Values and the Hundred Years' War* (1974), treats the attitude of patriots, intellectuals, and the general public. D. Seward, *The Hundred Years' War: The English in France, 1337–1453* (1981), tells an exciting story, and J. Keegan, *The Face of Battle* (1977), Chapter 2, "Agincourt," describes what war meant to the ordinary soldier. B. Tuchman, *A Distant Mirror: The Calamitous 14th Century* (1980), gives a vivid picture of many facets of fourteenth-century life, while concentrating on the war. The best treatment of the financial costs of the war is probably M. M. Postan, "The Costs of the Hundred Years' War," *Past and Present* 27 (April 1964):34–53. E. Searle and R. Burghart, "The Defense of England and the Peasants' Revolt," *Viator* 3 (1972), is a fascinating study of the peasants' changing social attitudes. For strategy, tactics, armaments, and costumes of war, see H. W. Koch, *Medieval Warfare* (1978), a beautifully illustrated book, while R. Barber, *The Knight and Chivalry* (1982), and M. Keen, *Chivalry* (1984), give fresh interpretations of the cultural importance of chivalry.

For political and social conditions in the fourteenth and fifteenth centuries, the following studies are all useful: P. S. Lewis, *Later Medieval France: The Polity* (1968) and "The Failure of the French Medieval Estates," *Past and Present* 23 (November 1962); L. Romier, *A History of France* (1962); G. O. Sayles, *The King's Parliament of England* (1974); A. R. Meyers, *Parliaments and Estates in Europe to 1789* (1975); *The English Parliament in the Middle Ages,* eds. R. G. Davies and J. H. Denton (1981); M. Bloch, *French Rural History* (1966); I. Kershaw, "The Great Famine and Agrarian Crisis in England, 1315–1322," *Past and Present* 59 (May 1973); B. A. Hanawalt, "Fur Collar Crime: The Pattern of Crime Among the Fourteenth-Century English Nobility," *Journal of Social History* 8 (Spring 1975):1–17, a fascinating discussion; K. Thomas, "Work and Leisure in Pre-Industrial Society," *Past and Present* 29 (December 1964); R. Hilton, *Bond Men Made Free: Medieval Peasant Movements and the English Rising of 1381* (1973), a comparative study; M. Keen, *The Outlaws of Medieval Legend* (1961) and "Robin Hood—Peasant or Gentleman?" *Past and Present* 19 (April 1961):7–18; P. Wolff, "The 1391 Pogrom in Spain, Social Crisis or Not?" *Past and Present* 50 (February 1971):4–18; and R. H. Helmholz, *Marriage Litigation in Medieval England* (1974). Students are especially encouraged to consult the brilliant achievement of E. L. Ladurie, *The Peasants of Languedoc* (trans. John Day, 1976). R. H. Hilton, ed., *Peasants, Knights, and Heretics: Studies in Medieval English Social History* (1976), contains a number of valuable articles primarily on the social implications of agricultural change. J. C. Holt, *Robin Hood* (1982), is a soundly researched and highly readable study of the famous outlaw. For the Pastons, see R. Barber, ed., *The Pastons: Letters of a Family in the Wars of the Roses* (1984).

The poetry of Dante, Chaucer, and Villon may be read in the following editions: D. Sayers, trans., *Dante: The Divine Comedy,* 3 vols. (1963); N. Coghill, trans., *Chaucer's Canterbury Tales* (1977); P. Dale, trans., *The Poems of Villon* (1973). The social setting of *Canterbury Tales* is brilliantly evoked in D. W. Robertson, Jr., *Chaucer's London* (1968).

For the religious history of the period, F. Oakley, *The Western Church in the Later Middle Ages* (1979), is an excellent introduction. S. Ozment, *The Age of Reform, 1250–1550* (1980), discusses the schism and the conciliar movement in the intellectual context of the ecclesiopolitical tradition of the Middle Ages. Students seeking a highly detailed and comprehensive work should consult H. Beck et al., *From the*

High Middle Ages to the Eve of the Reformation, trans. A. Biggs, vol. IV in the History of the Church series edited by H. Jedin and J. Dolan (1980). J. Bossy, "The Mass as a Social Institution, 1200–1700," *Past and Present* 100 (August 1983):29–61, provides a technical study of the central public ritual of the Latin Church and its importance to Christian practice, while E. Mason, "The Role of the English Parishioner, 1000–1500," *Journal of Ecclesiastical History* 27 (January 1976):17–29, describes the influence of lay people on church organization and practice. The older study of J. H. Smith, *The Great Schism 1378: The Disintegration of the Medieval Papacy* (1970), is still valuable.

13

**EUROPEAN SOCIETY IN
THE AGE OF THE
RENAISSANCE**

HILE THE FOUR HORSEMEN of the Apocalypse carried war, plague, famine, and death across the Continent, a new culture was emerging in southern Europe. The fourteenth century witnessed the beginnings of remarkable changes in many aspects of Italian society. In the fifteenth century, these phenomena spread beyond Italy and gradually influenced society in northern Europe. These cultural changes have been collectively labeled the Renaissance. What does the term *Renaissance* mean? How did the Renaissance manifest itself in politics, government, and social organization? What were the intellectual and artistic hallmarks of the Renaissance? Did the Renaissance involve shifts in religious attitudes? What developments occurred in the evolution of the nation-state? This chapter will concentrate on these questions.

THE EVOLUTION OF THE ITALIAN RENAISSANCE

The Italian Renaissance evolved in two broad and slightly overlapping movements. The first stage, extending roughly from 1050 to 1300, witnessed phenomenal economic development, the growing political power of self-governing cities, and remarkable population expansion. The second phase, lasting from the late thirteenth to the late sixteenth century, was characterized by an incredible efflorescence of artistic energies.[1] Scholars commonly use the term *renaissance* to describe the cultural achievements of the fourteenth through sixteenth centuries; those achievements rest on the political and economic developments of earlier centuries.

In the great commercial revival of the eleventh century, northern Italian cities led the way. By the middle of the twelfth century, Venice, supported by a huge merchant marine, had grown enormously rich through overseas trade. It profited tremendously from the diversion of the Fourth Crusade to Constantinople (page 275). Genoa and Milan also enjoyed the benefits of a large volume of trade with the Middle East and northern Europe. These cities fully exploited their geographical positions as natural crossroads for mercantile exchange between the East and

West. In the early fourteenth century, furthermore, Genoa and Venice made important strides in shipbuilding, allowing their ships for the first time to sail all year long. Most goods were purchased directly from the producers and sold a good distance away. For example, Italian merchants bought fine English wool directly from the Cistercian abbeys of Yorkshire in northern England. The wool was transported to the bazaars of North Africa either overland or by ship through the Straits of Gibraltar. The risks in such an operation were great, but the profits were enormous. These profits were continually reinvested to earn more.

Scholars tend to agree that the first artistic and literary manifestations of the Italian Renaissance appeared in Florence. Florence possessed enormous wealth despite geographical constraints: it was an inland city without easy access to water transportation. But toward the end of the thirteenth century, Florentine merchants and bankers acquired control of papal banking. From their position as tax collectors for the papacy, Florentine mercantile families began to dominate European banking on both sides of the Alps. These families had offices in Paris, London, Bruges, Barcelona, Marseilles, Tunis and the North African ports, and, of course, Naples and Rome. The profits from loans, investments, and money exchanges that poured back to Florence were pumped into urban industries. Such profits contributed to the city's economic vitality.

The Florentine wool industry, however, was the major factor in the city's financial expansion and population increase. Florence purchased the best-quality wool from England and Spain, developed remarkable techniques for its manufacture, and employed thousands of workers to turn it into cloth. Florentine weavers produced immense quantities of superb woolen cloth, which brought the highest prices in the fairs, markets, and bazaars of Europe, Asia, and Africa.

By the first quarter of the fourteenth century, the economic foundations of Florence were so strong that even severe crises could not destroy the city. In 1344 King Edward III of England repudiated his huge debts to Florentine bankers and forced some of them into bankruptcy. Florence suffered frightfully from the Black Death, losing perhaps half of its population. Serious labor unrest, such as the Ciompi revolts of 1378 (see Chapter 12), shook the political establishment. Still, the basic Florentine economic structure remained stable. Driving enterprise, technical know-how, and competitive spirit saw Florence through the difficult economic period of the late fourteenth century.

COMMUNES AND REPUBLICS

The northern Italian cities were *communes,* sworn associations of free men seeking complete political and economic independence from local nobles. The merchant guilds that formed the communes built and maintained the city walls, regulated trade, raised taxes, and kept civil order. In the course of the twelfth century, communes at Milan, Florence, Genoa, Siena, and Pisa fought for and won their independence from surrounding feudal nobles. The nobles, attracted by the opportunities of long-distance and maritime trade, the rising value of urban real estate, the new public offices available in the expanding communes, and the chances for advantageous marriages into rich commercial families, frequently settled within the cities. Marriage vows often sealed business contracts between the rural nobility and the mercantile aristocracy. This merger of the northern Italian feudal nobility and the commercial aristocracy constituted the formation of a new social class, an urban nobility. Within the nobility, groups tied by blood, economic interests, and social connections formed tightly knit alliances to defend and expand their rights.

This new class made citizenship in the communes dependent on a property qualification, years of residence within the city, and social connections. Only a tiny percentage of the male population possessed these qualifications and thus could hold office in the commune's political councils. The *popolo,* or middle class, bitterly resented their exclusion from power. The popolo wanted places in the communal government and equality of taxation. Throughout most of the thirteenth century, in city after city, the popolo used armed forces and violence to take over the city governments. Republican governments were established in Bologna, Siena, Parma, Florence, Genoa, and other cities. The victory of the popolo, however, proved temporary. Because they practiced the same sort of political exclusivity as had the noble communes—denying influence to the classes below them, whether the poor, the unskilled, or new immigrants—the popolo never won their support. Moreover, the popolo could not establish civil order within

Business Activities in a Florentine Bank The Florentines early developed new banking devices. One man (left) presents a letter of credit or a bill of exchange, forerunners of the modern check, which allowed credit in distant places. A foreign merchant (right) exchanges one kind of currency for another. The bank profited from the fees it charged for these services. *(Prints Division; New York Public Library; Astor, Lenox and Tilden Foundation)*

their cities. Consequently, these movements for republican government failed. By 1300 *signori* (despots, or one-man ruler) or *oligarchies* (the rule of merchant aristocracies) had triumphed everywhere.[2]

For the next two centuries, the Italian city-states were ruled by signori or by constitutional oligarchies. Despots predominated in cities with strong agricultural bases, such as Verona, Mantua, and Ferrara; oligarchies governed in cities with strong commercial or industrial bases, such as Venice, Florence, Genoa, and Bologna. In the signories, despots pretended to observe the law while actually manipulating it to conceal their basic illegality. Oligarchic regimes possessed constitutions, but through a variety of schemes, a small, restricted class of wealthy merchants exercised the judicial, executive, and legislative functions of government. Thus in 1422 Venice had a population of 84,000, but 200 men held all power; Florence had about 40,000 people, but 600 men ruled. Oligarchic regimes maintained only a facade of republican government, in which political power theoretically resides in the people and is exercised by their chosen representatives. The Renaissance nostalgia for the Roman form of government, combined with calculating shrewdness, prompted the leaders of Venice, Milan, and Florence to use the old forms.

In the fifteenth century, political power and elite culture centered at the princely courts of despots and oligarchs. "A court was the space and personnel around a prince as he made laws, received ambassadors, made appointments, took his meals, and proceeded through the streets."[3] At his court a prince flaunted his patronage of learning and the arts by munificent gifts to writers, philosophers, and artists. The princely court afforded the despot or oligarch the opportunity to display his wealth. Ceremonies connected with family births, baptisms, marriages, funerals, or triumphant entrances into the city served as occasions for magnificent pageantry and elaborate ritual—all designed to assert the ruler's wealth and power.

THE BALANCE OF POWER AMONG THE ITALIAN CITY-STATES

Renaissance Italians had a passionate attachment to their individual city-states: political loyalty and feeling centered on the local city. This intensity of local feeling perpetuated the dozens of small states and hindered the development of one unified state. Italy, consequently, was completely disunited.

In the fifteenth century, five powers dominated the Italian peninsula—Venice, Milan, Florence, the Papal States, and the kingdom of Naples (see Map 13.1). The rulers of the city-states—whether despots in Milan, patrician elitists in Florence, or oligarchs in Venice—governed as monarchs. They crushed urban revolts, levied taxes, killed their enemies, and used massive building programs to employ, and the arts to overawe, the masses.

Venice, with enormous trade and vast colonial empire, ranked as an international power. Though Venice had a sophisticated constitution and was a republic in name, an oligarchy of merchant-aristocrats actually ran the city. Milan was also called a republic, but despots of the Sforza family ruled harshly and dominated the smaller cities of the north. Likewise in Florence the form of government was republican, with authority vested in several councils of state. In reality, between 1434 and 1494, power in Florence was held by the great Medici banking family. Though not public officers, Cosimo (1434–1464) and Lorenzo (1469–1492) ruled from behind the scenes.

Central Italy consisted mainly of the Papal States, which during the Babylonian Captivity had come under the sway of important Roman families. Pope Alexander VI (1492–1503), aided militarily and politically by his son Cesare Borgia, reasserted papal authority in the papal lands. Cesare Borgia became the hero of Machiavelli's *The Prince* because he began the work of uniting the peninsula by ruthlessly conquering and exacting total obedience from the principalities making up the Papal States.

South of the Papal States was the kingdom of Naples, consisting of virtually all of southern Italy and, at times, Sicily. The kingdom of Naples had long been disputed by the Aragonese and by the French. In 1435 it passed to Aragon.

The major Italian city-states controlled the smaller ones, such as Siena, Mantua, Ferrara, and Modena, and competed furiously among themselves for territory. The large cities used diplomacy, spies, paid informers, and any other means to get information that could be used to advance their ambitions. While the states of northern Europe were moving toward centralization and consolidation, the world of Italian politics resembled a jungle where the powerful dominated the weak.

In one significant respect, however, the Italian city-states anticipated future relations among competing European states after 1500. Whenever one Italian state appeared to gain a predominant position within the peninsula, other states combined to establish a balance of power against the major threat. In 1450, for example, Venice went to war against Milan in protest against Francesco Sforza's acquisition of the title of duke of Milan. Cosimo de' Medici of Florence, a long-time supporter of a Florentine-Venetian alliance, switched his position and aided Milan. Florence and Naples combined with Milan against powerful Venice and the papacy. In the peace treaty signed at Lodi in 1454, Venice received territories in return for recognizing Sforza's right to the duchy. This pattern of shifting alliances continued until 1494. In the formation of these alliances, Renaissance Italians invented the machinery of modern diplomacy; permanent embassies with resident ambassadors in capitals where political relations and commercial ties needed continual monitoring. The resident ambassador is one of the great achievements of the Italian Renaissance.

MAP 13.1 The Italian City-States, ca 1494 In the fifteenth century the Italian city-states represented great wealth and cultural sophistication. The political divisions of the peninsula invited foreign intervention.

At the end of the fifteenth century, Venice, Florence, Milan, and the papacy possessed great wealth and represented high cultural achievement. However, their imperialistic ambitions at each other's expense and their inability to form a common alliance against potential foreign enemies, made Italy an inviting target for invasion. When Florence and Naples

entered into an agreement to acquire Milanese territories, Milan called on France for support.

At Florence the French invasion had been predicted by the Dominican friar Girolamo Savonarola (1452–1498). In a number of fiery sermons between 1491 and 1494, Savonarola attacked what he considered the paganism and moral vice of the city, the

Palazzo Vecchio, Florence Built during the late thirteenth and early fourteenth centuries as a fortress of defense against both popular uprising and foreign attack, the building housed the *podesta,* the city's highest magistrate, and all the offices of the government. *(Alinari/Art Resource)*

career also illustrates the internal instability of Italian cities such as Florence, an instability that invited foreign invasion.

The invasion of Italy in 1494 by the French king Charles VIII (1483–1498) inaugurated a new period in Italian and European power politics. Italy became the focus of international ambitions and the battleground of foreign armies. Charles swept down the peninsula with little opposition, and Florence, Rome, and Naples soon bowed before him. When Piero de' Medici, Lorenzo's son, went to the French camp seeking peace, the Florentines exiled the Medicis and restored republican government.

Charles's success simply whetted French appetites. In 1508 his son Louis XII formed the League of Cambrai with the pope and the German emperor Maximilian for the purpose of stripping rich Venice of its mainland possessions. Pope Leo X soon found the French a dangerous friend, and in a new alliance called on the Spanish and Germans to expel the French from Italy. This anti-French combination was temporarily successful. In 1519 Charles V succeeded his grandfather Maximilian as Holy Roman emperor. When the French returned to Italy in 1522, there began the series of conflicts called the Habsburg-Valois wars (named for the German and French dynasties), whose battlefield was Italy.

In the sixteenth century, the political and social life of Italy was upset by the relentless competition for dominance between France and the empire. The Italian cities suffered severely from the continual warfare, especially in the frightful sack of Rome in 1527 by imperial forces under Charles V. Thus the failure of the city-states to form some federal system, or to consolidate, or at least to establish a common foreign policy, led to the continuation of the centuries-old subjection of the peninsula by outside invaders. Italy was not to achieve unification until 1870.

INTELLECTUAL HALLMARKS OF THE RENAISSANCE

The Renaissance was characterized by self-conscious awareness among fourteenth- and fifteenth-century Italians that they were living in a new era. The realization that something new and unique was happening first came to men of letters in the fourteenth century, especially to the poet and humanist Francesco

undemocratic government of Lorenzo de' Medici, and the corruption of Pope Alexander VI. For a time Savonarola enjoyed wide popular support among the ordinary people; he became the religious leader of Florence and as such contributed to the fall of the Medici. Eventually, however, people wearied of his moral denunciations, and he was excommunicated by the pope and executed. Savonarola stands as proof that the common people did not share the worldly outlook of the commercial and intellectual elite. His

Petrarch (1304–1374). Petrarch thought that he was living at the start of a new age, a period of light following a long night of Gothic gloom. He believed that the first two centuries of the Roman Empire represented the peak in the development of human civilization. The Germanic invasions had caused a sharp cultural break with the glories of Rome and inaugurated what Petrarch called the "Dark Ages." Medieval people had believed that they were continuing the glories that had been ancient Rome and had recognized no cultural division between the world of the emperors and their own times. But for Petrarch and many of his contemporaries, the thousand-year period between the fourth and the fourteenth centuries constituted a barbarian, or Gothic, or "middle" age. The sculptors, painters, and writers of the Renaissance spoke contemptuously of their medieval predecessors and identified themselves with the thinkers and artists of Greco-Roman civilization. Petrarch believed he was witnessing a new golden age of intellectual achievement—a rebirth or, to use the French word that came into English, a *renaissance*. The division of historical time into periods is often arbitrary and done for the convenience of historians. In terms of the way most people lived and thought, no sharp division exists between the Middle Ages and the Renaissance. Some important poets, writers, and artists, however, believed they were living in a new golden age.

The Renaissance also manifested itself in a new attitude toward men, women, and the world—an attitude that may be described as individualism. A humanism characterized by a deep interest in the Latin classics and the deliberate attempt to revive antique lifestyles emerged, as did a bold new secular spirit.

INDIVIDUALISM

Though the Middle Ages had seen the appearance of remarkable individuals, recognition of such persons was limited. The examples of Saint Augustine in the fifth century and Peter Abelard and Guibert of Nogent in the twelfth—men who perceived of themselves as unique and produced autobiographical statements—stand out for that very reason: Christian humility discouraged self-absorption. In the fourteenth and fifteenth centuries, moreover, such characteristically medieval and corporate attachments as the guild and the parish continued to provide strong support for the individual and to exercise great social influence. Yet, in the Renaissance intellectuals developed a new sense of historical distance from earlier periods. A large literature specifically concerned with the nature of individuality emerged. This literature represented the flowering of a distinctly Renaissance individualism.

The Renaissance witnessed the emergence of many distinctive personalities who gloried in their uniqueness. Italians of unusual abilities were self-consciously aware of their singularity and unafraid to be unlike their neighbors; they had enormous confidence in their ability to achieve great things. Leon Battista Alberti (1404–1474), a writer, architect, and mathematician remarked, "Men can do all things if they will."[4] The Florentine goldsmith and sculptor Benvenuto Cellini (1500–1574) prefaced his *Autobiography* with a sonnet that declares:

My cruel fate hath warr'd with me in vain:
Life, glory, worth, and all unmeasur'd skill,
Beauty and grace, themselves in me fulfill
That many I surpass, and to the best attain.[5]

Cellini, certain of his genius, wrote so that the whole world might appreciate it.

Individualism stressed personality, genius, uniqueness, and the fullest development of capabilities and talents. Artist, athlete, painter, scholar, sculptor, whatever—a person's potential should be stretched until fully realized. Thirst for fame, a driving ambition, a burning desire for success drove such people to the complete achievement of their potential. The quest for glory was a central component of Renaissance individualism.

THE REVIVAL OF ANTIQUITY

In the cities of Italy, especially Rome, civic leaders and the wealthy populace showed phenomenal archaeological zeal for the recovery of manuscripts, statues, and monuments. Pope Nicholas V (1447–1455), a distinguished scholar, planned the Vatican Library for the nine thousand manuscripts he had collected. Pope Sixtus IV (1471–1484) built that library, which remains one of the richest repositories of ancient and medieval documents.

Patrician Italians consciously copied the lifestyle of the ancients and even searched out pedigrees dating back to ancient Rome. Aeneas Silvius Piccolomini, a native of Siena who became Pope Pius II (1458–

Michelangelo: Medici Tomb, Florence Between two busy periods in Rome, Michelangelo visited Florence (1516–1534) to work on the tombs of Guiliano and Lorenzo de' Medici. Here Lorenzo, dressed in Roman armor, presides over the sensuous figures of Dawn and Twilight. Of this monument, the contemporary Vasari asked, "Where in the world's history has any statue shown such art?" *(Giraudon/Art Resource)*

1464), once pretentiously declared, "Rome is as much my home as Siena, for my House, the Piccolomini, came in early times from the capital to Siena, as is proved by the constant use of the names Aeneas and Silvius in my family."[6]

The revival of antiquity also took the form of profound interest in and study of the Latin classics. This feature of the Renaissance became known as the "new learning," or simply "humanism," the term of the Florentine rhetorician and historian Leonardo Bruni (1370–1444). The words *humanism* and *humanist* derived ultimately from the Latin *humanitas,* which Cicero used to mean the literary culture needed by anyone who would be considered educated and civilized. Humanists studied the Latin classics to learn what they reveal about human nature. Humanism emphasized human beings, their

achievements, interests, and capabilities. Although churchmen supported the new learning, by the later fifteenth century Italian humanism was increasingly a lay phenomenon.

Appreciation for the literary culture of the Romans had never died in the West. Bede, Alcuin, and Einhard in the eighth century and Ailred of Rievaulx, Bernard of Clairvaux, and John of Salisbury in the twelfth century had all studied and imitated the writings of the ancients. Medieval writers, however, had studied the ancients in order to come to know God. Medieval thinkers held that human beings are the noblest of God's creatures and that, though they have fallen, they are still capable of regeneration and thus deserving of respect. Medieval scholars interpreted the classics in a Christian sense and invested the ancients' poems and histories with Christian meaning.

Renaissance humanists approached the classics differently. Where medieval writers accepted pagan and classical authors uncritically, Renaissance humanists were skeptical of their authority, conscious of the historical distance separating themselves from the ancients, and fully aware that classical writers often disagreed among themselves. Like their medieval predecessors, Renaissance humanists were deeply Christian. They studied the classics to understand human nature, and while they fully grasped the moral thought of pagan antiquity, Renaissance humanists viewed man from a strongly Christian perspective: he was made in the image and likeness of God. For example, in a remarkable essay, *On the Dignity of Man,* the Florentine writer Pico della Mirandola stressed that man possesses great dignity, because he was made as Adam in the image of God before the Fall and as Christ after the Resurrection. Man's place in the universe is somewhere between the beasts and the angels, but because of the divine image planted in him, there are no limits to what he can accomplish. Humanists rejected classical ideas that were opposed to Christianity. Or they sought through reinterpretation an underlying harmony between the pagan and secular and the Christian faith. The fundamental difference between Renaissance humanists and medieval ones is that the former were more self-conscious about what they were doing.[7]

The fourteenth- and fifteenth-century humanists loved the language of the classics and considered it superior to the corrupt Latin of the medieval schoolmen. Renaissance writers were very excited by the purity of ancient Latin. They eventually became concerned more about form than about content, more about the way an idea was expressed than about the significance and validity of the idea. Literary humanists of the fourteenth century wrote each other highly stylized letters imitating ancient authors, and they held witty philosophical dialogues in conscious imitation of the Platonic Academy of the fourth century B.C. Whenever they could, Renaissance humanists heaped scorn on the "barbaric" Latin style of the medievalists. The leading humanists of the early Renaissance were rhetoricians, seeking effective and eloquent communication, both oral and written.

SECULAR SPIRIT

Secularism involves a basic concern with the material world instead of eternal and spiritual interests. A secular way of thinking tends to find the ultimate explanation of everything and the final end of human beings within the limits of what the senses can discover. Medieval businesspeople ruthlessly pursued profits while medieval monks fought fiercely over property. Renaissance people often held strong and deep spiritual interests. Yet in a religious society, such as the medieval, the dominant ideals focused on the other-worldly, on life after death. In a secular society, attention is concentrated on the here and now, often on the acquisition of material things. The fourteenth and fifteenth centuries witnessed the slow but steady growth of secularism in Italy.

The economic changes and rising prosperity of the Italian cities in the thirteenth century worked a fundamental change in social and intellectual attitudes and values. In the Middle Ages, the feudal nobility and the higher clergy had determined the dominant patterns of culture. The medieval aristocracy expressed disdain for money making. Christian ideas and values infused literature, art, politics, and all other aspects of culture. In the Renaissance, by contrast, the business concerns of the urban bourgeoisie required constant and rational attention.

Worries about shifting rates of interest, shipping routes, personnel costs, and employee relations did not leave much time for thoughts about penance and purgatory. The busy bankers and merchants of the Italian cities calculated ways of making and increasing their money. Money allowed greater material pleasures, a more comfortable life, the leisure time to appreciate and patronize the arts. Money could buy many sensual gratifications, and the rich, social-climbing patricians of Venice, Florence, Genoa, and Rome came to see life more as an opportunity to be enjoyed than as a painful pilgrimage to the City of God.

In *On Pleasure,* the humanist Lorenzo Valla (1406–1457) defended the pleasures of the senses as the highest good. Scholars praise Valla as a father of modern historical criticism. His study *On the False Donation of Constantine* (1444) demonstrated by careful textual examination that an anonymous eighth-century document supposedly giving the papacy jurisdiction over vast territories in western Europe was a forgery. Medieval people had accepted the Donation of Constantine as a reality, and the proof that it was an invention weakened the foundations of papal claims to temporal authority. Lorenzo Valla's work exemplifies the application of critical scholarship to old and almost-sacred writings, as well as the new secular spirit of the Renaissance. The tales in the *Decameron* by the Florentine Boccaccio, which describe ambitious merchants, lecherous friars, and cuckolded husbands, portray a frankly acquisitive, sensual, and worldly society. The "contempt of the world" theme, so pervasive in medieval literature, had disappeared. Renaissance writers justified the accumulation and enjoyment of wealth with references to ancient authors.

Nor did church leaders do much to combat the new secular spirit. In the fifteenth and early sixteenth centuries, the papal court and the households of the cardinals were just as worldly as those of great urban patricians. Of course, most of the popes and higher church officials had come from the bourgeois aristocracy. The Medici pope Leo X (1513–1521), for example, supported artists and men of letters because patronage was an activity he had learned in the household of his father, Lorenzo the Magnificent. Renaissance popes beautified the city of Rome and patronized the arts. They expended enormous enthusiasm and huge sums of money on the re-embellishment of the city. A new papal chancellery, begun in 1483 and finished in 1511, stands as one of the architectural masterpieces of the High Renaissance. Pope Julius II (1503–1513) tore down the old Saint Peter's Basilica and began work on the present structure in 1506. Michelangelo's dome for Saint Peter's is still considered his greatest work. Papal interests, far removed from spiritual concerns, fostered rather than discouraged the new worldly attitude.

The broad mass of the people and the intellectuals and leaders of society remained faithful to the Christian church. Few people questioned the basic tenets of the Christian religion. Italian humanists and their aristocratic patrons were antiascetic, antischolastic, and ambivalent, but they were not agnostics or skeptics. The thousands of pious paintings, sculptures, processions, and pilgrimages of the Renaissance period prove that strong religious feeling persisted.

ART AND THE ARTIST

No feature of the Renaissance evokes greater admiration than its artistic masterpieces. The 1400s (*quattrocento*) and 1500s (*cinquecento*) bore witness to a dazzling creativity in painting, architecture, and sculpture. In all the arts, the city of Florence led the way. According to the Renaissance art historian Giorgio Vasari (1511–1574), the painter Perugino once asked why it was in Florence and not elsewhere that men achieved perfection in the arts. The first answer he received was, "There were so many good critics there, for the air of the city makes men quick and perceptive and impatient of mediocrity."[8] But Florence was not the only artistic center. In the period art historians describe as the "High Renaissance" (1500–1527), Rome took the lead. The main characteristics of High Renaissance art—classical balance, harmony, and restraint—are revealed in the masterpieces of Leonardo da Vinci (1452–1519), Raphael (1483–1520), and Michelangelo (1475–1564), all of whom worked in Rome at this time.

Some historians and art critics have maintained that the Renaissance "rediscovered" the world of nature and of human beings. This is nonsense, as a quick glance at a Gothic cathedral reveals. The enormous detail applied to the depiction of animals' bodies, the careful carving of leaves, flowers, and all kinds of vegetation, the fine sensitivity shown in human faces—these clearly show medieval and ancient people's appreciation for nature in all its manifestations. Saint Francis of Assisi encouraged throughout his life an awareness of nature. No historical period has a monopoly on the appreciation of nature or beauty.

ART AND POWER

Significant changes in the realm of art did occur in the fourteenth century, however. In early Renaissance Italy, art manifested corporate power. Powerful urban groups such as guilds or religious confraternities commissioned works of art. The Florentine cloth merchants, for example, delegated Brunelleschi to build the magnificent dome on the Cathedral of Florence and selected Ghiberti to design the bronze doors of the Baptistry. These works represented the merchants' dominant influence in the community. Corporate patronage is also reflected in the Florentine government's decision to hire Michelangelo to sculpt David, the great Hebrew hero and king. The subject matter of art through the early fifteenth century, as in the Middle Ages, remained overwhelmingly religious. Religious themes appeared in all media—wood carvings, painted frescoes, stone sculptures, paintings. As in the Middle Ages, art served an educational

Botticelli: Adoration of the Magi According to the Florentine artist, biographer, and Medici courtier Giorgio Vasari (1511–1574), this painting contains the most faithful likenesses portrayed of Cosimo (kneeling before the Christ-Child) and Lorenzo (far right). Though the subject is Christian, the painting has a secular spirit, introduces individual portraits (Botticelli himself is in the far left-hand corner), and serves to glorify the Medici family. *(Alinari/Scala/Art Resource)*

purpose. A religious picture or statue was intended to spread a particular doctrine, act as a profession of faith, or recall sinners to a moral way of living.

Increasingly in the later fifteenth century, individuals and oligarchs, rather than corporate groups, sponsored works of art. Patrician merchants and bankers, popes and princes supported the arts as a means of glorifying themselves and the families. Vast sums were spent on family chapels, frescoes, religious panels, and tombs. Writing about 1470, the Florentine oligarch Lorenzo de' Medici declared that over the past thirty-five years his family had spent the astronomical sum of 663,755 gold florins for artistic

and architectural commissions. Yet, "I think it casts a brilliant light on our estate [public reputation] and it seems to me that the monies were well spent and I am very pleased with this." Powerful men wanted to glorify themselves, their families, and their offices. A magnificent style of living, enriched by works of art, served to prove the greatness and the power of the despot or oligarch.[9]

As the fifteenth century advanced, the subject matter of art became steadily more secular. The study of classical texts brought deeper understanding of ancient ideas. Classical themes and motifs, such as the lives and loves of pagan gods and goddesses, figured increasingly in painting and sculpture. Religious topics, such as the Annunciation of the Virgin and the Nativity, remained popular among both patrons and artists, but frequently the patron had himself and his family portrayed. In Botticelli's *Adoration of the Magi,* for example, Cosimo de' Medici appears as one of the Magi kneeling before the Christ child. People were conscious of their physical uniqueness and wanted their individuality immortalized. Paintings cost money and thus were also means of displaying wealth. Although many Renaissance paintings have classical or Christian themes, the appearance of the patron reflects the new spirit of individualism and secularism.

The style of Renaissance art was decidedly different from that of the Middle Ages. The individual portrait emerged as a distinct artistic genre. In the fifteenth century, members of the newly rich middle class often had themselves painted in a scene of romantic chivalry or in courtly society. Rather than reflecting a spiritual ideal, as medieval painting and sculpture tended to do, Renaissance portraits mirrored reality. The Florentine painter Giotto (1276–1337) led the way in the depiction of realism; his treatment of the human body and face replaced the formal stiffness and artificiality that had for so long characterized the representation of the human body. The sculptor Donatello (1386–1466) probably exerted the greatest influence of any Florentine artist before Michelangelo. His many statues express an appreciation of the incredible variety of human nature. While medieval artists had depicted the nude human body only in a spiritualized and moralizing context, Donatello revived the classical figure with its balance and self-awareness. The short-lived Florentine Masaccio (1401–1428), sometimes called the father of modern painting, inspired a new style characterized by great realism, narrative power, and remarkably effective use of light and dark. As important as realism was the new "international style," so called because of the wandering careers of influential artists, the close communications and rivalry of princely courts, and the increased trade in works of art. Rich color, decorative detail, curvilinear rhythms, and swaying forms characterized the international style. As the term "international" implies, this style was European, not merely Italian.

Narrative artists depicted the body in a more scientific and natural manner. The female figure is voluptuous and sensual. The male body, as in Michelan-

gelo's *David* and *The Last Judgment,* is strong and heroic. Renaissance glorification of the human body reveals the secular spirit of the age. Filippo Brunelleschi (1377–1446), together with Piero della Francesca (1420–1492), seem to have pioneered *perspective* in painting, the linear representation of distance and space on a flat surface. *The Last Supper* of Leonardo da Vinci, with its stress on the tension between Christ and the disciples, is an incredibly subtle psychological interpretation.

THE STATUS OF THE ARTIST

In the Renaissance the social status of the artist improved. The lower-middle-class medieval master mason had been viewed in the same light as a mechanic. The artist in the Renaissance was considered a free intellectual worker. An artist did not produce unsolicited pictures or statues for the general public; that could mean loss of status. He usually worked on commission from a powerful prince. The artist's reputation depended on the support of powerful patrons, and through them some artists and architects achieved not only economic security but very great wealth. All aspiring artists received a practical (not theoretical) education in a recognized master's workshop. For example, Michelangelo (1475–1564) was apprenticed at age thirteen to the artist Ghirlandaio (1449–1494), although he later denied the fact to make it appear he never had any formal training. The more famous the artist, the more he attracted assistants or apprentices. Lorenzo Ghiberti (1378–1455) had twenty assistants during the period he was working on the bronze doors of the Baptistry in Florence, his most famous achievement.

Ghiberti's salary of two hundred florins a year compared very favorably with that of the head of the city government, who earned five hundred florins. Moreover, at a time when a man could live in a princely fashion on three hundred ducats a year, Leonardo da Vinci was making two thousand annually. Michelangelo was paid three thousand ducats for painting the ceiling of the Sistine Chapel. When he agreed to work on Saint Peter's Basilica, he refused a salary; he was already a wealthy man.[10]

Renaissance society respected and rewarded the distinguished artist. In 1537 the prolific letter writer, humanist, and satirizer of princes Pietro Aretino (1492–1556) wrote to Michelangelo while he was painting the Sistine Chapel:

Hans Memling: Tommaso and Maria Portinari A Florentine citizen, Tommaso Portinari earned (and later lost) a fortune as representative of the Medici banking interests in Bruges, Flanders. Husband and wife are dressed in rich but durable black broadcloth; Maria's necklace displays their wealth. Both faces reflect the calm restrained Flemish piety of Membling's work. *(The Metropolitan Museum of Art: Bequest of Benjamin Altman, 1913)*

To the Divine Michelangelo:

Sir, just as it is disgraceful and sinful to be unmindful of God so it is reprehensible and dishonourable for any man of discerning judgement not to honour you as a brilliant and venerable artist whom the very stars use as a target at which to shoot the rival arrows of their favour. You are so accomplished, therefore, that hidden in your hands lives the idea of a new king of creation, whereby the most challenging and subtle problem of all in the art of painting, namely that of outlines, has been so mastered by you that in the contours of the human body you express and contain the purpose of art. . . . And it is surely my duty to honour you with this salutation, since the world has many kings but only one Michelangelo.[11]

When the Holy Roman emperor Charles V (1519–1556) visited the workshop of the great Titian (1477–1576) and stooped to pick up the artist's dropped paintbrush, the emperor was demonstrating that the patron himself was honored in the act of honoring the artist. The social status of the artist of genius was immortally secured.

Renaissance artists were not only aware of their creative power; they boasted about it. Describing his victory over five others, including Brunelleschi, in the competition to design the bronze doors of Florence's Baptistry, Ghiberti exulted, "The palm of victory was conceded to me by all the experts and by all my fellow-competitors. By universal consent and without a single exception the glory was conceded to me."[12] Some medieval painters and sculptors had signed their works; Renaissance artists almost universally did so, and many of them incorporated self-portraits, usually as bystanders, in their paintings.

The Renaissance, in fact, witnessed the birth of the concept of the artist as genius. In the Middle Ages, people believed that only God created, albeit through

School of Luca della Robbia: Virgin and Child In the late fifteenth century, della Robbia's invention of the process of making polychrome-glazed terracottas led contemporaries to consider him a great artistic innovator. The warm humanity of the roundel (circular panel) is characteristic of della Robbia's art. *(Marion Gray. By permission of St. Anselm's Abbey, Washington, D.C.)*

individuals; the medieval conception recognized no particular value in artistic originality. Renaissance artists and humanists came to think that a work of art was the deliberate creation of a unique personality, of an individual who transcended traditions, rules, and theories. A genius had a peculiar gift, which ordinary laws should not inhibit. Cosimo de' Medici described a painter, because of his genius, as "divine," implying that the artist shared in the powers of God. The word *divine* was widely applied to Michelangelo. The Renaissance thus bequeathed the idea of genius to the modern world.

But the student must guard against interpreting Italian Renaissance culture in twentieth-century democratic terms. The culture of the Renaissance was that of a small mercantile elite, a business patriciate with aristocratic pretensions. Renaissance culture did not directly affect the broad middle classes, let alone the vast urban proletariat. The typical small tradesman or craftsman could not read the sophisticated Latin essays of the humanists, even if he had the time to do so. He could not afford to buy the art works of the great masters. A small, highly educated minority of literary humanists and artists created the culture of and for an exclusive elite. They cared little for ordinary people. Castiglione, Pico, and Vergerio, for example, thoroughly despised the masses. Renaissance humanists were a smaller and narrower group than the medieval clergy had ever been. High churchmen had commissioned the construction of the Gothic cathedrals, but once finished, the buildings were for all to enjoy. The modern visitor can still see the deep ruts in the stone floors of Chartres and Canterbury where the poor pilgrims slept at night. Nothing comparable was built in the Renaissance. Insecure, social-climbing merchant princes were hardly egalitarian.[13] The Renaissance continued the gulf between the learned minority and the uneducated multitude that has survived for many centuries.

SOCIAL CHANGE

The Renaissance changed many aspects of Italian, and subsequently European, society. The new developments brought about real breaks with the medieval past. What impact did the Renaissance have on educational theory and practice, on political thought? How did printing, the era's most stunning technolog-

ical discovery, affect fifteenth- and sixteenth-century society? How did Renaissance culture affect the experience of women? What roles did blacks play in Renaissance society?

EDUCATION AND POLITICAL THOUGHT

One of the central preoccupations of the humanists was education and moral behavior. Humanists poured out treatises, often in the form of letters, on the structure and goals of education and the training of rulers. In one of the earliest systematic programs for the young, Peter Paul Vergerio (1370–1444) wrote Ubertinus, the ruler of Carrara:

For the education of children is a matter of more than private interest; it concerns the State, which indeed regards the right training of the young as, in certain aspects, within its proper sphere. . . . In order to maintain a high standard of purity all enticements of dancing, or suggestive spectacles, should be kept at a distance: and the society of women as a rule carefully avoided. A bad companion may wreck the character. Idleness, of mind and body, is a common source of temptation to indulgence, and unsociable, solitary temper must be disciplined, and on no account encouraged. Tutors and comrades alike should be chosen from amongst those likely to bring out the best qualities, to attract by good example, and to repress the first signs of evil. . . . Above all, respect for Divine ordinances is of the deepest importance; it should be inculcated from the earliest years. Reverence towards elders and parents is an obligation closely akin. In this, antiquity offers us a beautiful illustration. For the youth of Rome used to escort the Senators, the Fathers of the City, to the Senate House: and awaiting them at the entrance, accompany them at the close of their deliberations on their return to their homes. In this the Romans saw an admirable training in endurance and in patience. This same quality of reverence will imply courtesy towards guests, suitable greeting to elders, to friends and to inferiors. . . .

We call those studies liberal *which are worthy of a free man; those studies by which we attain and practise virtue and wisdom; that education which calls forth, trains and develops those highest gifts of body and of mind which ennoble men, and which are rightly judged to rank next in dignity to virtue only.*[14]

Part of Vergerio's treatise specifies subjects for the instruction of young men in public life: history teaches virtue by examples from the past; ethics focuses on virtue itself; and rhetoric or public speaking trains for eloquence.

No book on education had broader influence than Baldassare Castiglione's *The Courtier* (1528). This treatise sought to train, discipline, and fashion the young man into the courtly ideal, the gentleman. According to Castiglione, the educated man of the upper class should have a broad background in many academic subjects, and his spiritual and physical, as well as intellectual, capabilities should be trained. The courtier should have easy familiarity with dance, music, and the arts. Castiglione envisioned a man who could compose a sonnet, wrestle, sing a song and accompany himself on an instrument, ride expertly, solve difficult mathematical problems, and above all speak and write eloquently. With these accomplishments, he would be the perfect Renaissance man. Whereas the medieval chivalric ideal stressed the military virtues of bravery and loyalty, the Renaissance man had to develop his artistic and intellectual potential as well as his fighting skills.

In contrast to the pattern of medieval education, the Renaissance courtier had the aristocrat's hostility toward specialization and professionalism. Medieval higher education, as offered by the universities, had aimed at providing a practical grounding in preparation for a career. After exposure to the rudiments of grammar, rhetoric, and logic, which the medieval student learned mainly through memorization, he was trained for a profession—usually law—in the government of the state or the church. Education was very functional and, by later standards, middle class.

In manner and behavior, the Renaissance courtier had traits his medieval predecessor probably had not had time to acquire. The gentleman was supposed to be relaxed, controlled, always composed and cool, elegant but not ostentatious, doing everything with a casual and seemingly effortless grace. In the sixteenth and seventeenth centuries, *The Courtier* was widely read. It influenced the social mores and patterns of conduct of elite groups in Renaissance and early modern Europe. The courtier became the model of the European gentleman.

No Renaissance book on any topic, however, has been more widely read and studied in all the centuries since its publication than the short political treatise *The Prince*, by Nicolò Machiavelli (1469–1527). Some political scientists maintain that Machiavelli was describing the actual competitive framework of the Italian states with which he was familiar. Other thinkers praise *The Prince* because it revolutionized political theory and destroyed medieval views of the nature of the state. Still other scholars consider this work a classic because it deals with eternal problems of government and society.

Born to a modestly wealthy Tuscan family, Machiavelli received a good education in the Latin classics. He entered the civil service of the Florentine government and served on thirty diplomatic missions. When the exiled Medicis returned to power in the city in 1512, they expelled Machiavelli from his position as officer of the city government. In exile he wrote *The Prince*.

The subject of *The Prince* is political power: how the ruler should gain, maintain, and increase it. In this, Machiavelli implicitly addresses the question of the citizen's relationship to the state. As a good humanist, he explores the problems of human nature and concludes that human beings are selfish and out to advance their own interests. This pessimistic view of humanity leads him to maintain that the prince may have to manipulate the people in any way he finds necessary:

The manner in which men live is so different from the way in which they ought to live, that he who leaves the common course for that which he ought to follow will find that it leads him to ruin rather than to safety. For a man who, in all respects, will carry out only his professions of good, will be apt to be ruined amongst so many who are evil. A prince therefore who desires to maintain himself must learn to be not always good, but to be so or not as necessity may require.[15]

The prince should combine the cunning of a fox with the ferocity of a lion to achieve his goals. Asking rhetorically whether it is better for a ruler to be loved or feared, Machiavelli wrote:

A prince, therefore, should not mind the ill repute of cruelty, when he can thereby keep his subjects united and loyal; for a few displays of severity will really be more merciful than to allow, by an excess of clemency, disorders to occur, which are apt to result in rapine and murder; for these injure a whole community, whilst the executions ordered by the prince fall only upon a few individuals. And, above all others, the new prince will find it almost impossible to avoid the reputation of cruelty, because new states are generally exposed to many dangers. . . .

. . . This, then, gives rise to the question "whether it be better to be loved than feared, or to be feared than loved." It will naturally be answered that it would be desirable to be both the one and the other; but as it is difficult to be both at the same time, it is much more safe to be feared than to be loved, when you have to choose between the two. For it may be said of men in general that they are ungrateful and fickle, dissemblers, avoiders of danger, and greedy of gain. So long as you shower benefits upon them, they are all yours. [16]

Medieval political theory derived ultimately from Saint Augustine's view that the state arose as a consequence of Adam's fall and people's propensity to sin. The test of good government was whether it provided justice, law, and order. Political theorists and theologians from Alcuin to Marsiglio of Padua had stressed the way government *ought* to be; they set high moral and Christian standards for the ruler's conduct.

Machiavelli maintained that the ruler should be concerned *not* with the way things ought to be but with the way things actually are. The sole test of a "good" government was whether it was effective, whether the ruler increased his power. Machiavelli did not advocate amoral behavior, but he believed that political action cannot be restricted by moral considerations. While amoral action might be the most effective approach in a given situation, he did not argue for generally amoral behavior over the moral. In the *Discources of the Ten Books of Titus Livy,* Machiavelli even showed his strong commitment to republican government. Nevertheless, on the basis of a crude interpretation of *The Prince,* the word *machiavellian* entered the language as a synonym for devious, corrupt, and crafty politics in which the end justifies the means. The ultimate significance of Machiavelli rests on two ideas: first, that one permanent social order reflecting God's will cannot be established and, second, that politics has its own laws and ought to be a science. [17]

THE PRINTED WORD

Sometime in the thirteenth century, paper money and playing cards from China reached the West. They were *block-printed*—that is, Chinese characters or pictures were carved into a wooden block, inked, and the words or illustrations put on paper. Since each word, phrase, or picture was on a separate block, this method of reproduction was extraordinarily expensive and time-consuming.

Around 1455, probably through the combined efforts of three men—Johann Gutenberg, Johann Fust, and Peter Schöffer, all experimenting at Mainz —movable type came into being. The mirror image of each letter (rather than entire words or phrases) was carved on relief on a small block. Individual letters, easily movable, were put together to form words; words separated by blank spaces formed lines of type; and lines of type were brought together to make up a page. Once the printer had placed wooden pegs around the type for a border and locked the whole in a frame, the page was ready for printing. Since letters could be arranged into any format, an infinite variety of texts could be printed by reusing and rearranging pieces of type.

By the middle of the fifteenth century, paper was no problem. The technologically advanced but extremely isolated Chinese knew how to manufacture paper as early as the first century A.D. This knowledge reached the West in the twelfth century, when the Arabs introduced the process into Spain. Europeans quickly learned that old rags could be shredded, mixed with water, placed in a mold, squeezed, and dried to make a durable paper, far less expensive than the vellum (calfskin) and parchment (sheepskin) on which medieval scribes had relied for centuries.

The effects of the invention of movable-type printing were not felt overnight. Nevertheless, within a half-century of the publication of Gutenberg's Bible of 1456, movable type brought about radical changes. The costs of reproducing books were drastically reduced. It took less time and money to print a book by machine than to make copies by hand. The press also reduced the chances of error. If the type had been accurately set, all the copies would be correct no matter how many were reproduced. The greater the number of pages a scribe copied, the greater the chances for human error.

Between the sixteenth and eighteenth centuries, printing brought about profound changes in European society and culture. Printing transformed both the private and the public lives of Europeans. Governments that "had employed the cumbersome methods of manuscripts to communicate with their subjects switched quickly to print to announce declarations of war, publish battle accounts, promulgate treaties or argue disputed points in pamphlet form. Theirs was an effort 'to win the psychological war.' " Printing made propaganda possible, emphasizing

The Print Shop Sixteenth-century printing involved a division of labor. Two persons (left) at separate benches set the pieces of type. Another (center, rear) inks the chase (or locked plate containing the set type). Another (right) operates the press, which prints the sheets. The boy removes the printed pages and sets them to dry. Meanwhile, a man carries in fresh paper on his head. *(BBC Hulton Picture Library/The Bettmann Archive)*

differences between various groups, such as crown and nobility, church and state. These differences laid the basis for the formation of distinct political parties. Printed materials reached an invisible public, allowing silent individuals to join causes and groups of individuals widely separated by geography to form a common identity; this new group consciousness could compete with older, localized loyalties. Book shops, coffee shops, and public reading rooms gradually appeared and, together with print shops, provided sanctuaries and meeting places for intellectuals and wandering scholars. Historians have yet to assess the degree to which such places contributed to the rise of intellectuals as a distinct social class.

Printing also stimulated the literacy of lay people and eventually came to have a deep effect on their private lives. Although most of the earliest books and pamphlets dealt with religious subjects, students, housewives, businessmen, and upper- and middle-class people sought books on all subjects. Printers re-

sponded with moralizing, medical, practical, and travel manuals. Pornography as well as piety assumed new forms. Broadsides and flysheets allowed great public festivals, religious ceremonies, and political events to be experienced vicariously by the stay-at-home. Since books and printed materials were read aloud to the illiterate, print bridged the gap between written and oral cultures.[18]

WOMEN IN RENAISSANCE SOCIETY

The status of upper-class women declined during the Renaissance. If women in the High Middle Ages are compared with those of fifteenth- and sixteenth-century Italy with respect to the kind of work they performed, their access to property and political power, and the role they played in shaping the outlook of their society, it is clear that ladies in the Renaissance ruling classes generally had less power than comparable ladies of the feudal age.

In the cities of Renaissance Italy, girls received a similar education to boys. Young ladies learned their letters and studied the classics. Many read Greek as well as Latin, knew the poetry of Ovid and Virgil, and could speak one or two "modern" languages, such as French or Spanish. In this respect, Renaissance humanism represented a real educational advance for women. Girls also received some training in painting, music, and dance. What were they to do with this training? They were to be gracious, affable, charming —in short, decorative. Renaissance women were better educated than their medieval counterparts. But whereas education trained a young man to rule and to participate in the public affairs of the city, it prepared a woman for the social functions of the home. An educated lady was supposed to know how to attract artists and literati to her husband's court and grace her husband's household.

Whatever the practical reality, a striking difference also exists between the medieval literature of courtly love, the etiquette books and romances, and the widely studied Renaissance manual on courtesy and good behavior, Castiglione's *The Courtier*. In the medieval books, manners shaped the man to please the lady; in *The Courtier* the lady was to make herself pleasing to the man. With respect to love and sex, the Renaissance witnessed a downward shift in women's status. In contrast to the medieval tradition of relative sexual equality, Renaissance humanists laid the foundations for the bourgeois double standard. Men, and men alone, operated in the public sphere; women belonged in the home. Castiglione, the foremost spokesman of Renaissance love and manners, completely separated love from sexuality. For women, sex was restricted entirely to marriage. Ladies were bound to chastity, to the roles of wife and mother in a politically arranged marriage. Men, however, could pursue sensual indulgence outside marriage.[19]

Official attitudes toward rape provide another index of the status of women in the Renaissance. A careful study of the legal evidence from Venice in the years 1338–1358 is informative. The Venetian shipping and merchant elite held economic and political power and made the laws. Those laws reveal that rape was not considered a particularly serious crime against either the victim or society. Noble youths committed a higher percentage of rapes than their small numbers in Venetian society would imply, despite government-regulated prostitution. The rape of

a young girl of marriageable age or a child under twelve was considered a graver crime than the rape of a married woman. Still, the punishment for rape of a noble, marriageable girl was only a fine or about six months' imprisonment. In an age when theft and robbery were punished by mutilation, and forgery and sodomy by burning, this penalty was very mild indeed. When a youth of the upper class was convicted of the rape of a nonnoble girl, his punishment was even lighter.

By contrast, the sexual assault on a noblewoman by a man of working-class origin, which was extraordinarily rare, resulted in severe penalization because the crime had social and political overtones.

In the eleventh century, William the Conqueror had decreed that rapists be castrated, implicitly according women protection and a modicum of respect. But in the early Renaissance, Venetian laws and their enforcement show that the governing oligarchy believed that rape damaged, but only slightly, men's property—women.[20]

Evidence from Florence in the fifteenth century also sheds light on infanticide, which historians are only now beginning to study in the Middle Ages and the Renaissance. Early medieval penitentials and church councils had legislated against abortion and infanticide, though it is known that Pope Innocent III (1198–1216) was moved to establish an orphanage "because so many women were throwing their children into the Tiber."[21] In the fourteenth and early fifteenth centuries, a considerable number of children died in Florence under suspicious circumstances. Some were simply abandoned outdoors. Some were said to have been crushed to death while sleeping in the same bed with their parents. Some died from "crib death" or suffocation. These deaths occurred too frequently to have all been accidental. And far more girls than boys died thus, reflecting societal discrimination against girl children as inferior and less useful than boys. The dire poverty of parents led them to do away with unwanted children.

The gravity of the problem of infanticide, which violated both the canon law of the church and the civil law of the state, forced the Florentine government to build the Foundling Hospital. Supporters of the institution maintained that, without public responsibility, "many children would soon be found dead in the rivers, sewers, and ditches, unbaptized."[22] The city fathers commissioned Filippo Brunelleschi, who had recently completed the dome over the Cathedral of

Titian: The Rape of Europa According to Greek myth, the Phoenician princess Europa was carried off to Crete by the god Zeus disguised as a white bull. The story was highly popular in the Renaissance with its interests in the classics. In this masterpiece, the erotic and voluptuous female figure reveals the new interest in the human form and the secular element in Renaissance art. *(Isabella Stewart Gardner Museum, Boston)*

Florence, to design the building. (Interestingly enough, the Foundling Hospital—completed in 1445—was the very first building to use the revitalized Roman classic design that characterizes Renaissance architecture.) The unusually large size of the hospital suggests that great numbers of children were abandoned.

BLACKS IN RENAISSANCE SOCIETY

Ever since the time of the Roman republic, a few black people had lived in western Europe. They had come, along with white slaves, as the spoils of war.

Even after the collapse of the Roman Empire, Muslim and Christian merchants continued to import them. The evidence of medieval art attests to the presence of Africans in the West and Europeans' awareness of them. In the twelfth and thirteenth centuries, a large cult surrounded Saint Maurice, martyred in the fourth century for refusing to renounce his Christian faith, who was portrayed as a black knight. Saint Maurice received the special veneration of the nobility. The numbers of blacks, though, had always been small.

Beginning in the fifteenth century, however, hordes of black slaves entered Europe. Portuguese explorers imported perhaps a thousand a year and sold them at the markets of Seville, Barcelona, Marseilles, and Genoa. The Venetians specialized in the import of white slaves, but blacks were so greatly in demand at the Renaissance courts of northern Italy that the Venetians defied papal threats of excommunication to secure them. What roles did blacks play in Renaissance society? What image did Europeans have of Africans?

The medieval interest in curiosities, the exotic, and the marvelous continued into the Renaissance. Because of their rarity, black servants were highly prized and much sought after. In the late fifteenth century, Isabella, the wife of Gian Galazzo Sforza, took pride in the fact that she had ten blacks, seven of them females; a black lady's-maid was both a curiosity and a symbol of wealth. In 1491 Isabella of Este, duchess of Mantua, instructed her agent to secure a black girl between four and eight years old, "shapely and as black as possible." The duchess saw the child as a source of entertainment: "we shall make her very happy and shall have great fun with her." She hoped that the little girl would become "the best buffoon in the world."[23] The cruel ancient tradition of a noble household retaining a professional "fool" for the family's amusement persisted through the Renaissance—and even down to the twentieth century.

Adult black slaves filled a variety of positions. Many served as maids, valets, and domestic servants. Italian aristocrats such as the Marchesa Elena Grimaldi had their portraits painted with their black page boys to indicate their wealth. The Venetians employed blacks—slave and free—as gondoliers and stevedores on the docks. Tradition, stretching back at least as far as the thirteenth century, connected blacks with music and dance. In Renaissance Spain and Italy, blacks performed as dancers, as actors and actresses in courtly dramas, and as musicians, sometimes composing full orchestras.[24]

Before the sixteenth-century "discoveries" of the non-European world, Europeans had little concrete knowledge of Africans and African culture. Europeans knew little about them beyond biblical accounts. The European attitude toward Africans was ambivalent. On the one hand, Europeans perceived Africa as a remote place, the home of strange people isolated by heresy and Islam from superior European civilization. Africans' contact even as slaves with Christian Europeans could only "improve" the blacks. Most Europeans' knowledge of the black as a racial type was based entirely on theological speculation. Theologians taught that God is light. Blackness, the opposite of light, therefore represented the hostile forces of the underworld: evil, sin, and the devil. Thus the devil was commonly represented as a black man in medieval and early Renaissance art. Blackness, however, also possessed certain positive qualities. It symbolized the emptiness of worldly goods, the humility of the monastic way of life. Black clothes permitted a conservative and discreet display of wealth. Black vestments and funeral trappings indicated grief, and Christ had said that those who mourn are blessed. Until the exploration and observation of the sixteenth, seventeenth, and nineteenth centuries allowed, ever so slowly, for the development of more scientific knowledge, the Western conception of Africa and black people remained bound up with religious notions.[25] In Renaissance society, blacks, like women, were signs of wealth; both were used for display.

THE RENAISSANCE IN THE NORTH

In the last quarter of the fifteenth century, Italian Renaissance thought and ideals penetrated northern Europe. Students from the Low Countries, France, Germany, and England flocked to Italy, imbibed the "new learning," and carried it back to their countries. Northern humanists interpreted Italian ideas about and attitudes toward classical antiquity, individualism, and humanism in terms of their own traditions. The cultural traditions of northern Europe tended to remain more distinctly Christian, or at least pietistic, than those of Italy. Italian humanists certainly were strongly Christian, as the example of Pico della Mirandola shows. But in Italy secular and pagan themes and Greco-Roman motifs received more humanistic attention. North of the Alps, the Renaissance had a distinctly religious character, and humanists stressed biblical and early Christian themes. What fundamentally distinguished Italian humanists from northern ones is that the latter had a program for broad social reform based on Christian ideals.

Christian humanists were interested in the development of an ethical way of life. To achieve it, they

Baldung: Adoration of the Magi Early sixteenth-century German artists produced thousands of adoration scenes depicting a black man as one of the three kings: these paintings were based on direct observation, reflecting the increased presence of blacks in Europe. The elaborate costumes, jewelry, and landscape expressed royal dignity, Christian devotion, and oriental luxury. *(Gemälde galerie. Staatliche Museen, Bildarchiv Preussischer Kulturbesitz, Berlin [West])*

believed that the best elements of classical and Christian cultures should be combined. For example, the classical ideals of calmness, stoical patience, and broad-mindedness should be joined in human conduct with the Christian virtues of love, faith, and hope. Northern humanists also stressed the use of reason, rather than acceptance of dogma, as the foundation for an ethical way of life. Like the Italians, they were impatient with scholastic philosophy. Christian humanists had a profound faith in the power of human intellect to bring about moral and institutional reform. They believed that, although human nature had been corrupted by sin, it was fundamentally good and capable of improvement through education, which would lead to piety and an ethical way of life.

This optimistic viewpoint found expression in scores of lectures, treatises, and collections of precepts. Treatises such as Erasmus's *The Education of a Christian Prince* express the naive notion that peace, harmony among nations, and a truly ethical society will result from a new system of education. This hope has been advanced repeatedly in Western history—by the ancient Greeks, by the sixteenth-century Christian humanists, by the eighteenth-century philosophers of the Enlightenment, and by nineteenth-century advocates of progress. The proposition remains highly debatable, but each time the theory has reappeared, education has been further extended.

The work of the French priest Jacques Lefèvre d'Etaples (ca 1455–1536) is one of the early attempts to apply humanistic learning to religious problems. A brilliant thinker and able scholar, he believed that more accurate texts of the Bible would lead people to live better lives. According to Lefèvre, a solid education in the Scriptures would increase piety and raise the level of behavior in Christian society. Lefèvre produced an edition of the Psalms and a commentary on Saint Paul's Epistles. In 1516, when Martin Luther lectured to his students at Wittenberg on Paul's Letter to the Romans, he relied on Lefèvre's texts.

Lefèvre's English contemporary John Colet (1466–1519) also published lectures on Saint Paul's Epistles, approaching them in the new critical spirit. Unlike medieval theologians, who studied the Bible for allegorical meanings, Colet, a priest, interpreted the Pauline letters historically—that is, in the social and political context of the times when they were written. Both Colet and Lefèvre were later suspected of heresy, as humanistic scholarship got entangled with the issues of the Reformation.

Colet's friend and countryman Thomas More (1478–1535) towers above other figures in sixteenth-century English social and intellectual history. More's political stance later, at the time of the Reformation (page 450), a position that in part flowed from his humanist beliefs, got him into serious trouble with King Henry VIII and has tended to obscure his contribution to Christian humanism.

The early career of Thomas More presents a number of paradoxes that reveal the marvelous complexity of the man. Trained as a lawyer, More lived as a student in the London Charterhouse, a Carthusian monastery. He subsequently married and practiced law, but became deeply interested in the classics, and his household served as a model of warm Christian family life and a mecca for foreign and English humanists. In the career pattern of such Italian humanists as Petrarch, he entered government service under Henry VIII and was sent as ambassador to Flanders. There More found the time to write *Utopia* (1516), which presented a revolutionary view of society.

Utopia, which literally means "nowhere," describes an ideal socialistic community on an island somewhere off the mainland of the New World. All its children receive a good education, primarily in the Greco-Roman classics, and learning does not cease with maturity, for the goal of all education is to develop rational faculties. Adults divide their days equally between manual labor or business pursuits and various intellectual activities.

Because the profits from business and property are held strictly in common, there is absolute social equality. The Utopians use gold and silver to make chamber pots or to prevent wars by buying off their enemies. By this casual use of precious metals, More meant to suggest that the basic problems in society were caused by greed. Utopian law exalts mercy above justice. Citizens of Utopia lead an ideal, nearly perfect existence because they live by reason; their institutions are perfect. More punned on the word *Utopia*—which he termed "a good place. A good place which is no place."

More's ideas were profoundly original in the sixteenth century. Contrary to the long-prevailing view that vice and violence exist because women and men are basically corrupt, More maintained that acquisitiveness and private property promoted all sorts of vices and civil disorders. Since society protected

The Later Middle Ages, Renaissance, and Protestant and Catholic Reformations, 1300–1600

As is evident in this chronology, early manifestations of the Renaissance and Protestant Reformation coincided in time with major events of the Later Middle Ages.

1300–1321	Dante, *The Divine Comedy*
1304–1374	Petrarch
1309–1372	Babylonian Captivity of the papacy
1337–1453	Hundred Years' War
1347–1351	The Black Death
ca 1350	Boccaccio, *The Decameron*
1356	Golden Bull: transforms the Holy Roman Empire into an aristocratic federation
1358	The Jacquerie
ca 1376	John Wyclif publishes *Civil Dominion* attacking the church's temporal power and asserting the supremacy of Scripture
1377–1417	The Great Schism
1378	Laborers' revolt in Florence
1381	Peasants' Revolt in England
1385–1400	Chaucer, *Canterbury Tales*
1414–1418	Council of Constance: ends the schism, postpones reform, executes John Hus
1431	Joan of Arc is burned at the stake
1434	Medici domination of Florence begins
1438	Pragmatic Sanction of Bourges: declares autonomy of the French church from papal jurisdiction
1453	Capture of Constantinople by the Ottoman Turks, ending the Byzantine Empire
1453–1471	Wars of the Roses in England
1456	Gutenberg Bible
1492	Columbus reaches the Americas
	Unification of Spain under Ferdinand and Isabela; expulsion of Jews from Spain
1494	France invades Italy, inaugurating sixty years of war on Italian soil
	Florence expels the Medici and restores republican government
1509	Erasmus, *The Praise of Folly*
1512	Restoration of the Medici in Florence

private property, *society's* flawed institutions were responsible for corruption and war. Today people take this view so much for granted that it is difficult to appreciate how radical it was in the sixteenth century. According to More, the key to improvement and reform of the individual was reform of the social institutions that mold the individual.

Better known by his contemporaries than Thomas More was the Dutch humanist Desiderius Erasmus of Rotterdam (1466?–1536). Orphaned as a small boy, Erasmus was forced to enter a monastary. Although he intensely disliked the monastic life, he developed there an excellent knowledge of the Latin language and a deep appreciation for the Latin classics. During a visit to England in 1499, Erasmus met John Colet, who decisively influenced his life's work: the application of the best humanistic learning to the study and explanation of the Bible. As a mature scholar with an international reputation stretching from Crakow to London, Erasmus could boast with truth, "I brought it about that humanism, which among the Italians ... savored of nothing but pure paganism, began nobly to celebrate Christ."[26]

Erasmus's long list of publications includes *The Adages* (1500), a list of Greek and Latin precepts on ethical behavior: *The Education of a Christian Prince* (1504), which combines idealistic and practical suggestions for the formation of a ruler's character through the careful study of Plutarch, Aristotle, Cicero, and Plato; *The Praise of Folly* (1509), a satire of worldly wisdom and a plea for the simple and spontaneous Christian faith of children; and, most important of all, a critical edition of the Greek New Testament (1516). In the preface to the New Testament, Erasmus explained the purpose of his great work:

Only bring a pious and open heart, imbued above all things with a pure and simple faith. . . . For I utterly dissent from those who are unwilling that the sacred Scriptures should be read by the unlearned translated into their vulgar tongue, as though Christ had taught such subtleties that they can scarcely be understood even by a few theologians. . . . Christ wished his mysteries to be published as openly as possible. I wish that even the weakest woman should read the Gospel—should read the epistles of Paul. And I wish these were translated into all languages, so that they might be read and understood, not only by Scots and Irishmen, but also by Turks and Saracens. To make them understood is surely the first step. It may be that they might be ridiculed by many, but some would take them to heart. I long that the husbandman should sing portions of them to himself as he follows the plough, that the weaver should hum them to the tune of his shuttle, that the traveller should beguile with their stories the tedium of his journey. . . .

Why do we prefer to study the wisdom of Christ in men's writings rather than in the writing of Christ himself?[27]

Two fundamental themes run through all of Erasmus's scholarly work. First, education was the means to reform, the key to moral and intellectual improvement. The core of education ought to be study of the Bible and the classics. Second, the essence of Erasmus's thought is, in his own phrase, "the philosophy of Christ." By this Erasmus meant that Christianity is an inner attitude of the heart or spirit. Christianity is not formalism, special ceremonies, or law; Christianity is Christ—his life and what he said and did, not what theologians have written about him. The Sermon on the Mount, for Erasmus, expressed the heart of the Christian message.

While the writings of Colet, Erasmus, and More have strong Christian themes and have drawn the attention primarily of scholars, the stories of the French humanist François Rabelais (1490?–1553) possess a distinctly secular flavor and have attracted broad readership among the literate public. Rabelais' *Gargantua* and *Pantagruel* (serialized between 1532 and 1552) belong among the great comic masterpieces of world literature. These stories' gross and robust humor introduced the adjective *Rabelaisian* into the language.

Gargantua and *Pantagruel* can be read on several levels: as comic romances about the adventures of the giant Gargantua and his son, Pantagruel; as a spoof on contemporary French society; as a program for educational reform; or as illustrations of Rabelais' prodigious learning. The reader enters a world of Renaissance vitality, ribald joviality, and intellectual curiosity. On his travels Gargantua meets various absurd characters, and within their hilarious exchanges there occur serious discussions on religion, politics, philosophy, and education. Rabelais had received an excellent humanistic education in a monastery, and Gargantua discusses the disorders of contemporary religious and secular life. Like More and Erasmus, Rabelais did not denounce institutions directly. Like Erasmus, Rabelais satirized hypocritical monks, pedantic academics, and pompous lawyers. But where Erasmus employed intellectual cleverness and sophisticated wit, Rabelais applied wild and gross humor. Like Thomas More, Rabelais believed that institutions molded individuals and that education was the key to a moral and healthy life. While the middle-class inhabitants of More's *Utopia* lived lives of restrained moderation, the aristocratic residents of Rabelais' Thélème lived for the full gratification of their physical instincts and rational curiosity.

Thélème, the abbey Gargantua establishes, parodies traditional religion and other social institutions. Thélème, whose motto is "Do as Thou Wilt," admits women *and* men; allows all to eat, drink, sleep, and work when they choose; provides excellent facilities for swimming, tennis, and football; and encourages sexual experimentation and marriage. Rabelais believed profoundly in the basic goodness of human beings and the rightness of instinct.

The most roguishly entertaining Renaissance writer, Rabelais was convinced that "laughter is the essence of manhood." A convinced believer in the Roman Catholic faith, he included in Gargantua's education an appreciation for simple and reasonable prayer. Rabelais combined the Renaissance zest for life and enjoyment of pleasure with a classical insistence on the cultivation of the body and the mind.

The distinctly religious orientation of the literary works of the Renaissance in the north also characterized northern art and architecture. Some Flemish painters, notably Jan van Eyck (1366–1441), were the equals of Italian painters. One of the earliest artists successfully to use oil-based paints, van Eyck, in paintings such as *Ghent Altarpiece* and the portrait of *Giovanni Arnolfini and His Bride,* shows the Flemish love for detail; the effect is great realism. Van Eyck's paintings also demonstrate remarkable attention to human personality, as do those of Hans Memling (d. 1494) in his studies of *Tommaso Portinari and His Wife.* Typical of northern piety, the Portinari are depicted in an attitude of prayer (see p. 401).

Another Flemish painter, Jerome Bosch (ca 1450–1516), frequently used religious themes, but in combination with grotesque fantasies, colorful imagery, and peasant folk legends. Many of Bosch's paintings reflect the confusion and anguish often associated with the end of the Middle Ages. In *Death and the Miser,* Bosch's dramatic treatment of the Dance of Death theme, the miser's gold, increased by usury, is ultimately controlled by diabolical rats and toads, while his guardian angel urges him to choose the crucifix.

A quasi-spiritual aura likewise infuses architectural monuments in the north. The city halls of wealthy Flemish towns like Bruges, Brussels, Louvain, and Ghent strike the viewer more as shrines to house the bones of saints than as settings for the mundane decisions of politicians and businessmen. Northern architecture was little influenced by the classical revival so obvious in Renaissance Rome and Florence.

POLITICS AND THE STATE IN THE RENAISSANCE (CA 1450–1521)

The High Middle Ages had witnessed the origins of many of the basic institutions of the modern state. Sheriffs, inquests, juries, circuit judges, professional bureaucracies, and representative assemblies all trace their origins to the twelfth and thirteenth centuries (pages 310–324). The linchpin for the development of states, however, was strong monarchy, and during the period of the Hundred Years' War, no ruler in western Europe was able to provide effective leadership. The resurgent power of feudal nobilities weakened the centralizing work begun earlier.

Beginning in the fifteenth century, rulers utilized the aggressive methods implied by Renaissance political ideas to rebuild their governments. First in Italy, then in France, England, and Spain, rulers began the work of reducing violence, curbing unruly nobles and troublesome elements and establishing domestic order. Within the Holy Roman Empire of Germany, the lack of centralization helps to account for the later German distrust of the Roman papacy. Divided into scores of independent principalities Germany could not deal with the Roman church as an equal.

Jerome Bosch: Death and the Miser Netherlandish painters frequently used symbolism, and Bosch (ca 1450–1516) is considered the master artist of symbolism and fantasy. Here rats, which because of their destructiveness symbolize evil, control the miser's gold. Bosch's imagery appealed strongly to twentieth-century surrealist painters. *(National Gallery of Art, Washington, D.C., Samuel H. Kress Collection)*

The dictators and oligarchs of the Italian city-states, however, together with Louis XI of France, Henry VII of England, and Ferdinand of Aragon, were tough, cynical, calculating rulers. In their ruthless push for power and strong governments, they subordinated morality to hard results. They preferred to be secure, if feared, rather than loved. They could not have read Machiavelli's *The Prince,* but they acted as if they understood its ideas.

Some historians have called Louis XI (1461–1483), Henry VII (1485–1509), and Ferdinand and Isabella in Spain (1474–1516) "new monarchs." The term is only partly appropriate. These monarchs were new in that they invested kingship with a strong sense of royal authority and national purpose. They stressed that monarchy was the one institution that linked all classes and peoples within definite territorial boundaries. Rulers emphasized the royal majesty and royal sovereignty and insisted that all must respect and be loyal to them. They ruthlessly suppressed opposition and rebellion, especially from the nobility. They loved the business of kingship and worked hard at it.

In other respects, however, the methods of these rulers, which varied from country to country, were not so new. They reasserted long-standing ideas and practices of strong monarchs in the Middle Ages. The Holy Roman emperor Frederick Barbarossa, the English Edward I, and the French King Philip the Fair had all applied ideas drawn from Roman law in the High Middle Ages. Renaissance princes also did so. They seized on the maxim of the Justinian Code, "What pleases the prince has the force of law," to advance their authority. Some medieval rulers such as Henry I of England, had depended heavily on middle-class officials. Renaissance rulers, too, tended to rely on middle-class civil servants. With tax revenues, medieval rulers had built armies to crush feudal anarchy. Renaissance townspeople with commercial and business interests naturally wanted a reduction of violence and usually were willing to be taxed in order to achieve it.

Scholars have often described the fifteenth-century "new monarchs" as crafty, devious, and thoroughly Machiavellian in their methods. Yet contemporaries of the Capetian Philip the Fair considered him every bit as devious and crafty as his Valois successors, Louis XI and Francis I, were considered in the fifteenth and sixteenth centuries. Machiavellian politics were not new in the age of the Renaissance. What

was new was a marked acceleration of politics, whose sole rationalization was the acquisition and expansion of power. Renaissance rulers spent precious little time seeking a religious justification for their actions. With these qualifications of the term "new monarchs" in mind, let us consider the development of national monarchies in France, England, and Spain in the period 1450 to 1521.

FRANCE

The Hundred Years' War left France badly divided, drastically depopulated, commercially ruined, and agriculturally weak. Nonetheless, the ruler whom Joan of Arc had seen crowned at Rheims, Charles VII (1422–1461), revived the monarchy and France. He seemed an unlikely person to do so. Frail, ugly, feeble, hypochondriacal, mistrustful, called the "son of a madman and a loose woman," Charles VII began France's long recovery.

Charles reconciled the Burgundians and Armagnacs, who had been waging civil war for thirty years. By 1453 French armies had expelled the English from French soil except in Calais. Charles reorganized the royal council, giving increased influence to the middle-class men, and strengthened royal finances through such taxes as the *gabelle* (on salt) and the taille land tax. These taxes remained the crown's chief sources of state income until the Revolution of 1789.

Charles also reformed the justice system and remodeled the army. By establishing regular companies of cavalry and archers—recruited, paid, and inspected by the state—Charles created the first permanent royal army. (In the victory over the English in 1453, however, French artillery played the decisive role.) In 1438 Charles published the Pragmatic Sanction of Bourges, asserting the superiority of a general council over the papacy, giving the French crown major control over the appointment of bishops, and depriving the pope of French ecclesiastical revenues. The Pragmatic Sanction established the Gallican (or French) liberties, because it affirmed the special rights of the French crown over the French church. Greater control over the church, the army, and justice helped to consolidate the authority of the French crown.

Charles's son Louis XI, called the "Spider King" by his subjects because of his treacherous and cruel character, was very much a Renaissance prince. Fac-

French Tradesmen A bootmaker, a cloth merchant (with bolts of material on shelves), and a dealer in gold plate and silver share a stall. Through sales taxes, the French crown received a portion of the profits. *(Bibliothèque Municipale, Rouen/Giraudon/Art Resource)*

ing the perpetual French problems of unification of the realm and reduction of feudal disorder, he saw money as the answer. Louis promoted new industries, such as silk weaving at Lyons and Tours. He welcomed tradesmen and foreign craftsmen, and he entered into commercial treaties with England, Portugal, and the towns of the Hanseatic League (see Chapter 11). The revenues raised through these economic activities and severe taxation were used to improve the army. With the army Louis stopped aristocratic brigandage and slowly cut into urban independence.

Luck favored his goal of expanding royal authority and unifying the kingdom. On the timely death of Charles the Bold, duke of Burgundy, in 1477 Louis invaded Burgundy and gained some territories. Three years later, the extinction of the house of Anjou brought Louis the counties of Anjou, Bar, Maine, and Provence.

Some scholars have credited Louis XI with laying the foundations for later French royal absolutism. Louis summoned only one meeting of the Estates General, and the delegates requested that they not be summoned in the future. Thereafter the king would decide. Building on the system begun by his father, Louis XI worked tirelessly to remodel the government following the disorders of the fourteenth and fifteenth centuries. In his reliance on finances supplied by the middle classes to fight the feudal nobility, Louis was typical of the new monarchs.

Two further developments strengthened the French monarchy. The marriage of Louis XII and Anne of Brittany added the large western duchy of Brittany to the state. Then the French king Francis I and Pope Leo X reached a mutually satisfactory agreement in 1516. The new treaty, the Concordat of Bologna, rescinded the Pragmatic Sanction's assertion of the superiority of a general council over the papacy and approved the pope's right to receive the first year's income of new bishops and abbots. In return, Leo X recognized the French ruler's right to select French bishops and abbots. French kings thereafter effectively controlled the appointment and thus the policies of church officials within the kingdom.

ENGLAND

English society suffered severely from the disorders of the fifteenth century. The aristocracy dominated the government of Henry IV (1399–1413) and indulged in mischievous violence at the local level. Population, decimated by the Black Death, continued to decline. While Henry V (1413–1422) gained chivalric prestige for his military exploits in France, he was totally dependent on the feudal magnates who controlled the royal council and Parliament. Henry V's death, leaving a nine-month-old son, the future Henry VI (1422–1461), gave the barons a perfect opportunity to entrench their power. Between 1455 and 1471, adherents of the ducal houses of York and Lancaster waged civil war, commonly called the Wars of the Roses because the symbol of the Yorkists was a white rose and that of the Lancastrians a red one. Although only a small minority of the nobility participated, the chronic disorder hurt trade, agriculture, and domestic industry. Under the pious but mentally disturbed Henry VI, the authority of the monarchy sank lower than it had been in centuries.

Edward IV (1461–1483) began establishing domestic tranquility. He succeeded in defeating the Lancastrian forces and after 1471 began to reconstruct the monarchy and consolidate royal power. Edward, his brother Richard III (1483–1485), and

Henry VII of the Welsh house of Tudor worked to restore royal prestige, to crush the power of the nobility, and to establish order and law at the local level. All three rulers used methods that Machiavelli himself would have praised—ruthlessness, efficiency, and secrecy.

The Hundred Years' War had cost the nation dearly, and the money to finance it had been raised by Parliament. Dominated by various baronial factions, Parliament had been the arena where the nobility exerted its power. As long as the monarchy was dependent on the lords and the commons for revenue, the king had to call Parliament. Thus Edward IV revived the medieval ideal that he would "live of his own," meaning on his own financial resources. He reluctantly established a policy the monarchy was to follow with rare exceptions down to 1603. Edward, and subsequently the Tudors, excepting Henry VIII, conducted foreign policy on the basis of diplomacy, avoiding expensive wars. Thus the English monarchy did not depend on Parliament for money, and the crown undercut that source of aristocratic influence.

Henry VII did, however, summon several meetings of Parliament in the early years of his reign. He used these assemblies primarily to confirm laws. Parliament remained the highest court in the land, and a statute registered (approved) there by the lords, bishops, and Commons gave the appearance of broad national support plus thorough judicial authority.

The center of royal authority was the royal council, which governed at the national level. There, too, Henry VII revealed his distrust of the nobility: though they were not completely excluded, very few great lords were among the king's closest advisers. Regular representatives on the council numbered between twelve and fifteen men, and while many gained high ecclesiastical rank (the means, as it happened, by which the crown paid them), their origins were the lesser landowning class and their education was in law. They were, in a sense, middle class.

The royal council handled any business the king put before it—executive, legislative, judicial. For example, the council conducted negotiations with foreign governments and secured international recognition of the Tudor dynasty through the marriage in 1501 of Henry VII's eldest son Arthur to Catherine of Aragon, the daughter of Ferdinand and Isabella of Spain. The council prepared laws for parliamentary ratification. The council dealt with real or potential aristocratic threats through a judicial offshoot, the court of Star Chamber, so called because of the stars painted on the ceiling of the room.

The court of Star Chamber applied principles of Roman law, and its methods were sometimes terrifying: the accused was not entitled to see evidence against him; sessions were secret; torture could be applied to extract confessions; and juries were not called. These procedures ran directly counter to English common-law precedents, but they effectively reduced aristocratic troublemaking.

Unlike the continental countries of Spain and France, England had no standing army or professional civil service bureaucracy. The Tudors relied on the support of unpaid local officials, the justices of the peace. These influential landowners in the shires handled all the work of local government. They apprehended and punished criminals, enforced parliamentary statutes, supervised conditions of service, fixed wages and prices, maintained proper standards of weights and measures, and even checked up on moral behavior. Justices of the peace were appointed and supervised by the council. From the royal point of view, the justices were an inexpensive method of government.

The Tudors won the support of the influential upper-middle class because the crown linked government policy with their interests. A commercial or agricultural upper class fears and dislikes few things more than disorder and violence. If the Wars of the Roses served any useful purpose, it was killing off dangerous nobles and thus making the Tudors' work easier. The Tudors promoted peace and social order, and the gentry did not object to arbitrary methods, like the institution of the court of Star Chamber, because the government had halted the long period of anarchy.

Grave, secretive, cautious, and always thrifty, Henry VII rebuilt the monarchy. He encouraged the cloth industry and built up the English merchant marine. Both English exports of wool and the royal export tax on that wool steadily increased. Henry crushed an invasion from Ireland and secured peace with Scotland through the marriage of his daughter Margaret to the Scottish king. When Henry VII died in 1509, he left a country at peace both domestically and internationally, a substantially augmented treasury, and the dignity and role of the royal majesty much enhanced.

SPAIN

Political development in Spain followed a pattern different from that of France and England. The central theme in the history of medieval Spain—or, more accurately, of the separate kingdoms Spain comprised—was disunity and plurality. The various peoples who lived in the Iberian Peninsula lacked a common cultural tradition. Different languages, laws, and religious communities made for a rich diversity. Complementing the legacy of Hispanic, Roman, and Visigothic peoples, Muslims and Jews had significantly affected the course of Spanish society.

The centuries-long *reconquista*—the attempts of the northern Christian kingdoms to control the entire peninsula—had both military and religious objectives: expulsion or conversion of the Arabs and Jews and political control of the south. By the middle of the fifteenth century, the kingdoms of Castile and Aragon dominated the weaker Navarre, Granada, and Portugal, and with the exception of Granada, the Iberian Peninsula had been won for Christianity. The wedding in 1469 of the dynamic and aggressive Isabella, heiress of Castile, and the crafty and persistent Ferdinand, heir of Aragon, was the final major step in the unification and christianization of Spain. This marriage, however, constituted a dynastic union of two royal houses, not the political union of two peoples. Although Ferdinand and Isabella pursued a common foreign policy, Spain under their rule remained a loose confederation of separate states. Each kingdom continued to maintain its own cortes (parliament), laws, courts, bureaucracies, and systems of coinage and taxation.

Isabella and Ferdinand determined to strengthen royal authority. In order to curb rebellious and warring aristocracy, they revived an old medieval institution. Popular groups in the towns called *hermandades,* or "brotherhoods," were given the authority to act both as local police forces and as judicial tribunals. Local communities were made responsible for raising troops and apprehending and punishing criminals. The hermandades repressed violence with such savage punishments that by 1498 they could be disbanded.

The decisive step Ferdinand and Isabella took to curb aristocratic power was the restructuring of the royal council. Aristocrats and great territorial magnates were rigorously excluded; thus the influence of the nobility on state policy was greatly reduced. Ferdinand and Isabella intended the council to be the cornerstone of their governmental system, with full executive, judicial, and legislative power under the monarchy. The council was also to be responsible for the supervision of local authorities. The king and queen, therefore, appointed to the council only people of middle-class background. The council and various government boards recruited men trained in Roman law, a system that exalted the power of the crown as the embodiment of the state.

In the extension of royal authority and the consolidation of the territories of Spain, the church was the linchpin. The church possessed vast power and wealth, and churchmen enjoyed exemption from taxation. Most of the higher clergy were descended from great aristocratic families, controlled armies and strategic fortresses, and fully shared the military ethos of their families.

The major issue confronting Isabella and Ferdinand was the appointment of bishops. If the Spanish crown could select the higher clergy, then the monarchy could influence ecclesiastical policy, wealth, and military resources. Through a diplomatic alliance with the papacy, especially with the Spanish pope Alexander VI, the Spanish monarchs secured the right to appoint bishops in Spain and in the Hispanic territories in America. This power enabled the "Catholic Kings of Spain," a title granted Ferdinand and Isabella by the papacy, to establish, in effect, a national church.[28]

The Spanish rulers used their power to reform the church, and they used some of its wealth for national purposes. For example, they appointed a learned and zealous churchman, Cardinal Francisco Jiménez (1436–1517), to reform the monastic and secular clergy. Jiménez proved effective in this task and established the University of Alcalá in 1499 for the education of the clergy, although instruction did not actually begin until 1508. A highly astute statesman, Jiménez twice served as regent of Castile.

Revenues from ecclesiastical estates provided the means to raise an army to continue the reconquista. The victorious entry of Ferdinand and Isabella into Granada on January 6, 1492, signaled the culmination of eight centuries of Spanish struggle against the Arabs in southern Spain and the conclusion of the reconquista (see Map 13.2). Granada in the south was incorporated into the Spanish kingdom, and in 1512 Ferdinand conquered Navarre in the north.

Although the Arabs had been defeated, there still remained a sizable and, in the view of the Catholic sovereigns, potentially dangerous minority, the Jews. Since ancient times, governments had never tolerated religious pluralism; religious faiths that differed from the official state religion were considered politically dangerous. Medieval writers quoted the fourth-century Byzantine theologian Saint John Chrysostom, who had asked rhetorically, "Why are the Jews degenerate? Because of their odious assassination of Christ." John Chrysostom and his admirers in the Middle Ages chose to ignore two facts: that it was the Romans who had killed Christ (because they considered him a *political* troublemaker) and that Christ had forgiven his executioners from the cross. France and England had expelled their Jewish populations in the Middle Ages, but in Spain Jews had been tolerated. In fact, Jews had played a decisive role in the economic and intellectual life of the several Spanish kingdoms.

Anti-Semitic riots and pogroms in the late fourteenth century had led many Jews to convert; they were called *conversos*. By the middle of the fifteenth century, many conversos held high positions in Spanish society as financiers, physicians, merchants, tax collectors, and even officials of the church hierarchy. Numbering perhaps 200,000 in a total population of about 7.5 million, Jews exercised an influence quite disproportionate to their numbers. Aristocratic grandees who borrowed heavily from Jews resented their financial dependence, and churchmen questioned the sincerity of Jewish conversions. At first, Isabella and Ferdinand continued the policy of royal toleration—Ferdinand himself had inherited Jewish blood from his mother. But many conversos apparently reverted to the faith of their ancestors, prompting Ferdinand and Isabella to secure Rome's permission to revive the Inquisition, a medieval judicial procedure for the punishment of heretics.

Although the Inquisition was a religious institution established to ensure the Catholic faith, it was controlled by the crown and served primarily as a politically unifying force in Spain. Because the Spanish Inquisition commonly applied torture to extract confessions, first from lapsed conversos, then from Muslims, and later from Protestants, it gained a notorious reputation. Thus, the word *inquisition,*

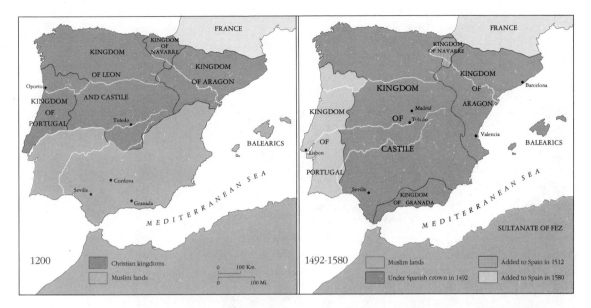

MAP 13.2 **The Christianization and Unification of Spain** The political unification of Spain was inextricably tied up with conversion or expulsion of the Muslims and the Jews. Why?

meaning "any judicial inquiry conducted with ruthless severity," came into the English language. The methods of the Spanish Inquisition were cruel, though not as cruel as the investigative methods of some twentieth-century governments. In 1478 the deeply pious Ferdinand and Isabella introduced the Inquisition into their kingdoms to handle the problem of backsliding conversos. They solved the problem in a dire and drastic manner. Shortly after the reduction of the Moorish stronghold at Granada in 1492, Isabella and Ferdinand issued an edict expelling all practicing Jews from Spain. Of the community of perhaps 200,000 Jews, 150,000 fled. (Efforts were made, through last-minute conversions, to retain good Jewish physicians.) Absolute religious orthodoxy served as the foundation of the Spanish national state.

The diplomacy of the Catholic rulers of Spain achieved a success they never anticipated. Partly out of hatred for the French and partly to gain international recognition for their new dynasty, Ferdinand and Isabella in 1496 married their second daughter, Joanna, heiress to Castile, to the archduke Philip, heir through his mother to the Burgundian Netherlands and through his father to the Holy Roman Empire. Philip and Joanna's son, Charles V (1519–1556), thus succeeded to a vast patrimony on two

continents. When Charles's son Philip II united Portugal to the Spanish crown in 1580, the Iberian Peninsula was at last politically united.

The Italian Renaissance, spanning the period from the eleventh through sixteenth centuries, developed in two broad stages. In the first stage, from about 1050 to 1300, a new economy emerged, based on Venetian and Genoese shipping and long-distance trade and on Florentine banking and cloth manufactures. These commercial activities, combined with the struggle of urban communes for political independence from surrounding feudal lords, led to the appearance of a new wealthy aristocratic class. The second stage, extending roughly from 1300 to 1600, witnessed a remarkable intellectual efflorescence. Based on a strong interest in the ancient world, the Renaissance had a classicizing influence on many facets of culture: law, literature, government, education, religion, and art. In the city-states of fifteenth- and sixteenth-century Italy, oligarchic or despotic powers governed; Renaissance culture was manipulated to enhance the power of those rulers.

Jews at Prayer in Spanish Synagogue The presence of the frequently educated and sometimes wealthy Jewish and Muslim populations in medieval Spain led to some degree of religious toleration and promoted a highly sophisticated culture. The architecture of the synagogue shows obvious Middle Eastern influences. *(The British Library)*

Expanding outside Italy, the intellectual features of this movement affected the culture of all Europe. The intellectual characteristics of the Renaissance were a secular attitude toward life, a belief in individual potential, and a serious interest in the Latin classics. The printing press revolutionized communication. Meanwhile, the status of women in society declined, and black people entered Europe in sizable numbers for the first time since the collapse of the Roman empire. In northern Europe, city merchants and rural gentry allied with rising monarchies. With taxes provided by businesspeople, kings provided a greater degree of domestic peace and order, conditions essential for trade. In Spain, France, and England, rulers also emphasized royal dignity and authority, and they utilized Machiavellian ideas to ensure the preservation and continuation of their governments. Feudal monarchies gradually evolved in the direction of nation-states.

NOTES

1. See Lauro Martines, *Power and Imagination. City-States in Renaissance Italy*, Vintage Books, New York, 1980, esp. pp. 332–333.
2. Ibid., pp. 22–61.
3. Ibid., pp. 221–237, esp. p. 221.

4. Quoted by J. Burckhardt, *The Civilization of the Renaissance in Italy,* Phaidon Books, London, 1951, p. 89.

5. *Memoirs of Benvenuto Cellini; A Florentine Artist; Written by Himself,* Everyman's Library, J. M. Dent & Sons, London, 1927, p. 2.

6. Quoted by Burckhardt, p. 111.

7. See Charles Trinkaus, *In Our Image and Likeness: Humanity and Divinity in Italian Humanist Thought,* 2 vols., Constable, London, 1970, vol. 2, pp. 505–529.

8. B. Burroughs, ed., *Vasari's Lives of the Artists,* Simon & Schuster, New York, 1946, pp. 164–165.

9. See Martines, chap. 13, esp. pp. 241, 243.

10. See "The Social Status of the Artists," in A. Hauser, *The Social History of Art,* Vintage Books, New York, 1959, vol. 2, chap. 3, esp. pp. 60, 68.

11. G. Bull, trans., *Aretino: Selected Letters,* Penguin Books, Baltimore, 1976, p. 109.

12. Quoted by Peter and Linda Murray, *A Dictionary of Art and Artists,* Penguin Books, Baltimore, 1963, p. 125.

13. Hauser, pp. 48–49.

14. Quoted by W. H. Woodward, *Vittorino da Feltre and Other Humanist Educators,* Cambridge University Press, Cambridge, Eng., 1897, pp. 96–97.

15. C. E. Detmold, trans., *The Historical, Political and Diplomatic Writings of Niccolo Machiavelli,* J. R. Osgood & Co., Boston, 1882, pp. 51–52.

16. Ibid., pp. 54–55.

17. See Felix Gilbert, *Machiavelli and Guicciardini: Politics and History in Sixteenth Century Florence,* W. W. Norton & Co., New York, 1985, pp. 197–200.

18. Quoted in Elizabeth L. Eisenstein, *The Printing Press as an Agent of Change: Communications and Cultural Transformations in Early Modern Europe,* Cambridge University Press, New York, 1979, vol. I, pp. 126–159, esp. p. 135.

19. This account rests on J. Kelly-Gadol, "Did Women Have a Renaissance?" in R. Bridenthal and C. Koontz, eds., *Becoming Visible: Women in European History,* Houghton Mifflin, Boston, 1977, pp. 137–161, esp. p. 161.

20. G. Ruggerio, "Sexual Criminality in Early Renaissance Venice, 1338–1358," *Journal of Social History* 8 (Spring 1975): 18–31.

21. Quoted by R. C. Trexler, "Infanticide in Florence: New Sources and First Results," *History of Childhood Quarterly* 1:1 (Summer 1973): 99.

22. Ibid., p. 100.

23. See Jean Devisse and Michel Mollat, *The Image of the Black in Western Art,* trans. William Granger Ryan, William Morrow and Company, New York, 1979, vol. II, part 2, pp. 187–188.

24. Ibid., pp. 190–194.

25. Ibid., pp. 255–258.

26. Quoted by E. H. Harbison, *The Christian Scholar and His Calling in the Age of the Reformation,* Charles Scribner's Sons, New York, 1956, p. 109.

27. Quoted by F. Seebohm, *The Oxford Reformers,* Everyman's Library, J. M. Dent & Sons, London, 1867, p. 256.

28. See J. H. Elliott, *Imperial Spain, 1469–1716,* Mentor Books, New York, 1963, esp. pp. 75, 97–108.

SUGGESTED READING

There are scores of exciting studies available on virtually all aspects of the Renaissance. In addition to the titles given in the Notes, the curious student interested in a broad synthesis should see J. H. Plumb, *The Italian Renaissance* (1965), a superbly written book based on deep knowledge and understanding; this book is probably the best starting point. J. R. Hale, *Renaissance Europe: The Individual and Society, 1480–1520* (1978), is an excellent treatment of individualism by a distinguished authority. J. R. Hale, ed., *A Concise Encyclopaedia of the Italian Renaissance* (1981), is a useful reference tool. F. H. New, *The Renaissance and Reformation: A Short History* (1977), gives a concise, balanced, and up-to-date account. M. P. Gilmore, *The World of Humanism* (1962), is an older but sound study that recent scholarship has not superseded on many subjects. Students interested in the problems the Renaissance has raised for historians should see K. H. Dannenfeldt, ed., *The Renaissance: Medieval or Modern* (1959), an anthology with a variety of interpretations, and W. K. Ferguson, *The Renaissance in Historical Thought* (1948), a valuable but difficult book. For the city where much of it originated, G. A. Brucker, *Renaissance Florence* (1969), gives a good description of Florentine economic, political, social, and cultural history. Learned, provocative, beautifully written, and the work on which this chapter leans heavily, L. Martines, *Power and Imagination: City-States in Renaissance Italy* (1980), is probably the best broad appreciation of the period produced in several decades.

J. R. Hale, *Machiavelli and Renaissance Italy* (1966), is a sound short biography, while G. Bull, trans., *Machiavelli: The Prince* (1975), provides a readable and easily accessible edition of the political thinker's major work. F. Gilbert, *Machiavelli and Guicciardini* (1984), places the two thinkers in their intellectual and social context. C. Singleton, trans., *The Courtier* (1959), presents an excellent picture of Renaissance court life.

The best introduction to the Renaissance in northern Europe and a book that has greatly influenced twentieth-century scholarship is J. Huizinga, *The Waning of the Middle Ages: A Study of the Forms of Life, Thought, and Art in France and the Netherlands in the Dawn of the Renaissance* (1954). This book challenges the whole idea of Renaissance. L. Febvre, *Life in Renaissance France* (trans. and ed., M. Rothstein, 1977), is a brilliant evocation of French Renaissance civilization by an international authority. The leading northern humanist is sensitively treated in M. M. Phillips, *Erasmus and the Northern Renaissance* (1956), and J. Huizinga, *Erasmus of Rotterdam* (1952). R. Marius, *Thomas More: A Biography* (1984), is an original study of the great English humanist and statesman, but the student may also want to consult E. E. Reynolds, *Thomas More* (1962), and R. W. Chambers, *Thomas More* (1935). J. Leclercq, trans., *The Complete Works of Rabelais* (1963), is easily available.

The following titles should prove useful for various aspects of Renaissance social history: E. L. Eisenstein, *The Printing Press as an Agent of Change: Communications and Cultural Transformations in Early Modern Europe*, 2 vols. (1979), a fundamental work; G. Ruggerio, *Violence in Early Renaissance Venice* (1980), a pioneering study of crime and punishment in a stable society; D. Weinstein and R. M. Bell, *Saints and Society: The Two Worlds of Christendom, 1000–1700* (1982), an essential book for an understanding of the perception of holiness and of the social origins of saints in early modern Europe; J. C. Brown, *Immodest Acts: The Life of A Lesbian Nun in Renaissance Italy* (1985), which is helpful for an understanding of the role and status of women; and I. Maclean, *The Renaissance Notion of Women* (1980).

Renaissance art has understandably inspired vast researches. In addition to Vasari's volume of biographical sketches on the great masters referred to in the Notes, A. Martindale, *The Rise of the Artist in the Middle Ages and Early Renaissance* (1972), is a splendidly illustrated introduction. B. Berenson, *Italian Painters of the Renaissance* (1957), the work of an American expatriate

who was an internationally famous art historian, has become a classic. W. Sypher, *Four Stages of Renaissance Style* (1956), relates drama and poetry to the visual arts of painting and sculpture. One of the finest appreciations of Renaissance art, written by one of the greatest art historians of this century, is E. Panofsky, *Meaning in the Visual Arts* (1955). Both Italian and northern painting are treated in the brilliant study of M. Meiss, *The Painter's Choice: Problems in the Interpretation of Renaissance Art* (1976), a collection of essays dealing with Renaissance style, form, and meaning. The splendidly illustrated work of M. McCarthy, *The Stones of Florence* (1959), celebrates the energy and creativity of the greatest Renaissance city. L. Steinberg, *The Sexuality of Christ in Renaissance Art and in Modern Oblivion* (1983), is a brilliant work that relates Christ's sexuality to incarnational theology. Students interested in the city of Rome and its architectural history should consult the elegantly illustrated and entertaining study of C. Hibbert, *Rome: The Biography of a City* (1985). Da Vinci's scientific and naturalist ideas and drawings are available in I. A. Richter, ed., *The Notebooks of Leonardo da Vinci* (1985). The magisterial achievement of J. Pope-Hennessy, *Cellini* (1985), is a superb evocation of that artist's life and work.

The student who wishes to study blacks in medieval and early modern European society should see the rich and original achievement of J. Devisse and M. Mollat, *The Image of the Black in Western Art*, vol. II: part I, *From the Demonic Threat to the Incarnation of Sainthood*, and part 2, *Africans in the Christian Ordinance of the World: Fourteenth to Sixteenth Century* (trans. W. G. Ryan, 1979.)

The following works are not only useful for the political and economic history of the age of the Renaissance but also contain valuable bibliographical information: A. J. Slavin, ed., *The "New Monarchies" and Representative Assemblies* (1965), a collection of interpretations, and R. Lockyer, *Henry VII* (1972), a biography with documents illustrative of the king's reign. For Spain, see M. Defourneaux, *Daily Life in Spain in the Golden Age* (1970); B. Bennasar, *The Spanish Character: Attitudes and Mentalities from the Sixteenth to the Nineteenth Century* (trans. B. Keen, 1979); J. H. Elliott, *Imperial Spain: 1469–1716* (1966), and H. Kamen, *The Spanish Inquisition* (1965). For the Florentine business classes, see I. Origo, *The Merchant of Prato* (1957), and G. Brucker, *Two Memoirs of Renaissance Florence: The Diaries of Buonaccorso Pitti and Gregorio Dati* (trans. J. Martines, 1967).

14

**REFORM AND
RENEWAL IN THE
CHRISTIAN CHURCH**

*T*HE IDEA OF REFORM is as old as Christianity itself. In his letter to the Christians of Rome, Saint Paul exhorted: "Do not model yourselves on the behavior of the world around you, but let your behavior change, reformed by your new mind. That is the only way to discover the will of God and know what is good, what it is that God wants, what is the perfect thing to do."[1] In the early fifth century, Saint Augustine of Hippo, describing the final stage of world history, wrote, "In the sixth age of the world our reformation becomes manifest, in newness of mind, according to the image of Him who created us." In the middle of the twelfth century, Saint Bernard of Clairvaux complained about the church of his day: "There is as much difference between us and the men of the primitive Church as there is between muck and gold."

The need for reform of the individual Christian and of the institutional church is central to the Christian faith. The Christian humanists of the late fifteenth and early sixteenth centuries—More, Erasmus, Colet, and Lefèvre d'Etaples—urged reform of the church on the pattern of the early church, primarily through educational and social change. Men and women of every period believed the early Christian church represented a golden age, and critics in every period called for reform.

Sixteenth-century cries for reformation, therefore, were hardly new. In fact, many scholars today interpret the sixteenth-century Reformation against the background of reforming trends begun in the fifteenth century. What late medieval religious developments paved the way for the adoption and spread of Protestant thought? What role did political and social factors play in the several reformations? What were the consequences of religious division? Why did the theological ideas of Martin Luther trigger political, social, and economic reactions? What response did the Catholic Church make to the movements for reform? This chapter will explore these questions.

THE CONDITION OF THE CHURCH (CA 1400–1517)

The papal conflict with the German emperor Frederick II in the thirteenth century, followed by the Babylonian Captivity and then the Great Schism, badly damaged the prestige of church leaders. In the fourteenth and fifteenth centuries, conciliarists reflected educated public opinion when they called for the reform of the church "in head and members." The secular humanists of Italy and the Christian humanists of the north denounced corruption in the church. As Machiavelli put it, "We Italians are irreligious and corrupt above others, because the Church and her representatives set us the worst example."[2] In *The Praise of Folly,* Erasmus condemned the absurd superstitions of the parish clergy and the excessive rituals of the monks. The records of episcopal visitations of parishes, civil court records, and even such literary masterpieces as Chaucer's *Canterbury Tales* and Boccaccio's *Decameron* tend to confirm the sarcasms of the humanists.

Concrete evidence of disorder is spotty. Since a great deal of corruption may have gone unreported, the moral situation may have been worse than the evidence suggests. On the other hand, bishops' registers and public court records mention the exceptional, not the typical. The thousands of priests who quietly and conscientiously went about their duties received no mention in the documents.

The religious life of most people in early sixteenth-century Europe took place at the village or local level. Any assessment of the moral condition of the parish clergy must take into account one fundamental fact: parish priests were peasants, and they were poor. All too frequently, the spiritual quality of their lives was not much better than that of the people to whom they ministered. The clergy identified religion with life; that is, they injected religious symbols and practices into everyday living. Some historians, therefore, have accused the clergy of vulgarizing religion. But if the level of belief and practice was vulgarized, still the lives of rural, isolated, and semipagan people were spiritualized.

SIGNS OF DISORDER

In the early sixteenth century, critics of the church concentrated their attacks on three disorders: clerical immorality, clerical ignorance, and clerical pluralism, with the related problem of absenteeism. There was little pressure for doctrinal change; the emphasis was on moral and administrative reform.

Since the fourth century, church law had required that candidates for the priesthood accept absolute celibacy. It had always been difficult to enforce.

Many priests, especially those ministering to country people, had concubines, and reports of neglect of the rule of celibacy were common. Immorality, of course, included more than sexual transgressions. Clerical drunkenness, gambling, and indulgence in fancy dress were frequent charges. There is no way of knowing how many priests were guilty of such behavior. But because such conduct was so much at odds with the church's rules and moral standards, it scandalized the educated faithful.

The bishops enforced regulations regarding the education of priests casually. As a result, standards for ordination were shockingly low. Many priests could barely read and write, and critics laughed at the illiterate priest mumbling the Latin words to the mass, which he could not understand. Predictably, this was the disorder that the Christian humanists, with their concern for learning, particularly condemned.

Absenteeism and pluralism constituted the third major abuse. Many clerics, especially higher ecclesiastics, held several *benefices* (or offices) simultaneously but seldom visited their benefices, let alone performed the spiritual responsibilities those offices entailed. Instead, they collected revenues from all of them and paid a poor priest a fraction of the income to fulfill the spiritual duties of a particular local church.

Many Italian officials in the papal curia held benefices in England, Spain, and Germany. Revenues from those countries paid the Italian priests' salaries, provoking not only charges of absenteeism but nationalistic resentment. King Henry VIII's chancellor Thomas Wolsey was archbishop of York for fifteen years before he set foot in his diocese. The French king Louis XII's famous diplomat Antoine du Prat is perhaps the most notorious example of absenteeism: as archbishop of Sens, the first time he entered his cathedral was in his own funeral procession. Critics condemned pluralism, absenteeism, and the way money seemed to change hands when a bishop entered into his office.

Although royal governments strengthened their positions and consolidated their territories in the fifteenth and sixteenth centuries, rulers lacked sufficient revenues to pay and reward able civil servants. The Christian church, with its dioceses and abbeys, possessed a large proportion of the wealth of the countries of Europe. What better way to reward government officials than with high church offices? After all, the practice was sanctioned by centuries of tradition. Thus in Spain, France, England, and the Holy Roman Empire—in fact, all over Europe—because church officials served their monarchs, those officials were allowed to govern the church.

The broad mass of the people, in financially supporting the church, supported everything that churchmen did. Bishops and abbots did a lot of work for secular governments. Churchmen served as royal councilors, diplomats, treasury officials, chancellors, viceroys, and judges. These positions had nothing whatsoever to do with spiritual matters. Bishops worked for their respective states as well as for the church, and they were paid by the church for their services to the state. It is astonishing that so many conscientiously tried to carry out their religious duties on top of their public burdens.

The prodigious wealth of the church inevitably stimulated criticism. For centuries devout laypeople had bequeathed land, money, rights, and privileges to religious institutions. By the sixteenth century, these gifts and shrewd investments had resulted in vast treasure. Some was spent in the service of civil governments. Much of it was used to help the poor. But some also provided a luxurious lifestyle for the clergy.

In most countries except England, members of the nobility occupied the highest church positions. The sixteenth century was definitely not a democratic age. The spectacle of proud, aristocratic prelates living in magnificent splendor contrasted very unfavorably with the simple fishermen who were Christ's disciples. Nor did the popes of the period 1450 to 1550 set much of an example. They lived like secular Renaissance princes. Pius II (1458–1464), although deeply learned and a tireless worker, enjoyed a reputation as a clever writer of love stories and Latin poetry. Sixtus IV (1471–1484) beautified the city of Rome, built the famous Sistine Chapel, and generously supported several artists. Innocent VIII (1484–1492) made the papal court a model of luxury and scandal. All three popes used papal power and wealth in order to advance the material interests of their own families.

The court of the Spanish pope Rodrigo Borgia, Alexander VI (1492–1503), who publicly acknowledged his mistress and children, reached new heights of impropriety. Because of the prevalence of intrigue, sexual promiscuity, and supposed poisonings, the name Borgia became a synonym for moral corruption. Julius II (1503–1513), the nephew of Sixtus IV, donned military armor and personally led papal

The Church Contrasted Satirical woodcuts as well as the printed word attacked conditions in the church. Here the mercenary spirit of the sixteenth-century papacy is contrasted with the attitude of Christ toward money changers: Christ drove them from the temple, but the pope kept careful records of revenues owed to the church. *(Pierpont Morgan Library)*

troops against the French invaders of Italy in 1506. After him, Giovanni de' Medici, the son of Lorenzo the Magnificent, carried on as Pope Leo X (1513–1521) the Medicean tradition of being a great patron of the arts.

Through the centuries, papal prestige and influence had rested heavily on the moral quality of the popes' lives—that is, on their strong fidelity to Christian teaching as revealed in the Gospel. The lives of the Renaissance popes revealed little of this Gospel message.

SIGNS OF VITALITY

Calls for reform testify to the spiritual vitality of the church as well as to its problems. Before a patient can be cured of sickness, he or she must acknowledge that a problem exists. In the late fifteenth and early sixteenth centuries, both individuals and groups within the church were working actively for reform. In Spain, Cardinal Jiménez visited religious houses, en-

couraged the monks and friars to keep their rules and constitutions, and set high standards for the training of the diocesan clergy.

Lefèvre d'Etaples in France and John Colet in England called for a return to the austere Christianity of the early church. Both men stressed the importance of sound preaching of the Scriptures.

In Holland, beginning in the late fourteenth century, a group of pious laypeople called the "Brethren of the Common Life" lived in stark simplicity while daily carrying out the Gospel teaching of feeding the hungry, clothing the naked, and visiting the sick. The Brethren also taught in local schools with the goal of preparing devout candidates for the priesthood and the monastic life. Through prayer, meditation, and the careful study of Scripture, the Brethren sought to make religion a personal, inner experience. The spirituality of the Brethren of the Common Life found its finest expression in the classic *The Imitation of Christ* by Thomas à Kempis. As its title implies, *The Imitation* urges ordinary lay Christians to take Christ as their model and to seek perfection in a simple way of life. Like later Protestants, the Brethren stressed the centrality of Scripture in the spiritual life.[3] In the mid-fifteenth century, the movement had houses in the Netherlands, central Germany, and the Rhineland; it was a true religious revival.

So, too, were the activities of the Oratories of Divine Love in Italy. The oratories were groups of priests living in communities who worked to revive the church through prayer and preaching. They did not withdraw from the world as medieval monks had done but devoted themselves to pastoral and charitable activities such as founding hospitals and orphanages. Oratorians served God in an active ministry.

If external religious observances are a measure of depth of heartfelt conviction, Europeans in the early sixteenth century remained deeply pious and loyal to the Roman Catholic church. Villagers participated in processions honoring the local saints. Middle-class people made pilgrimages to the great national shrines, as the enormous wealth of Saint Thomas Becket's tomb at Canterbury in England and the shrine of Saint James de Compostella in Spain testify. The upper classes continued to remember the church in their wills. In England, for example, between 1480 and 1490 almost £30,000, a prodigious sum in those days, was bequeathed to religious foundations. People of all social classes devoted an enormous amount of their time and income to religious causes and foundations. Sixteenth-century society remained deeply religious; all across Europe people sincerely yearned for salvation.

The papacy also expressed concern for reform. Pope Julius II summoned an ecumenical (universal) council, which met in the church of Saint John Lateran in Rome from 1512 to 1517. Since most of the bishops were Italian and did not represent a broad cross-section of international opinion, the term *ecumenical* is not appropriate. Nevertheless, the bishops and theologians present strove earnestly to reform the church. They criticized the ignorance of priests, lamenting that only 2 percent of the clergy could understand the Latin of the liturgical books. The Lateran Council also condemned superstitions believed by many of the laity. The council recommended higher standards for education of the clergy and instruction of the common people. The bishops placed the responsibility for eliminating bureaucratic corruption squarely on the papacy and suggested significant doctrinal reforms. But many obstacles stood in the way of ecclesiastical change. Nor did the actions of an obscure German friar immediately force the issue.

MARTIN LUTHER AND THE BIRTH OF PROTESTANTISM

As the result of a personal religious struggle, a German Augustinian friar, Martin Luther (1483–1546), launched the Protestant Reformation of the sixteenth century. Luther was not a typical person of his time; miners' sons who become professors of theology are never typical. But Luther is representative of his time in the sense that he articulated the widespread desire for reform of the Christian church and a deep yearning for salvation. In the sense that concern for salvation was an important motivating force for Luther and other reformers, the sixteenth-century Reformation was in part a continuation of the medieval religious search.

LUTHER'S EARLY YEARS

Martin Luther was born at Eisleben in Saxony, the second son of a hardworking and ambitious copper miner. At considerable sacrifice, his father sent him to school and then to the University of Erfurt, where Martin earned a master's degree with distinction at the young age of twenty-one. Hans Luther intended his son to proceed to the study of law and a legal career, which had since Roman times been the steppingstone to public office and material success. Badly frightened during a thunderstorm, however, Martin Luther vowed to become a friar. Without consulting his father, he entered the monastery of the Augustinian friars at Erfurt in 1505. Luther was ordained a priest in 1507 and after additional study earned the doctorate of theology. From 1512 until his death in 1546, he served as professor of Scripture at the new University of Wittenberg.

Martin Luther was exceedingly scrupulous in his monastic observances and devoted to prayer, penances, and fasting; nevertheless, the young friar's conscience troubled him constantly. The doubts and conflicts felt by any sensitive young person who has just taken a grave step were especially intense in

Young Luther Lucas Cranach, court painter to Elector Frederick of Saxony and a friend of Luther, captured the piety, the strength, and the intense struggle of the young friar. *(Photo: Caroline Buckler)*

young Luther. He had terrible anxieties about sin and worried continually about his salvation. Luther intensified his monastic observances but still found no peace of mind.

A recent psychological interpretation of Luther's early life suggests that he underwent a severe inner crisis in the years 1505 to 1515. Luther had disobeyed his father, thus violating one of the Ten Commandments, and serious conflict persisted between them. The religious life seemed to provide no answers to his mental and spiritual difficulties. Three fits that he suffered in the monastic choir during those years may have been outward signs of his struggle.[4] Luther was grappling, as had thousands of medieval people before him, with the problem of salvation and thus the meaning of life. He was also searching for his life's work.

Luther's wise and kindly professor, Staupitz, directed him to the study of Saint Paul's letters. Gradually, Luther arrived at a new understanding of the Pauline letters and of all Christian doctrine. He came to believe that salvation comes not through external observances and penances but through a simple faith in Christ. Faith is the means by which God sends humanity his grace, and faith is a free gift that cannot be earned. Thus Martin Luther discovered himself, God's work for him, and the centrality of faith in the Christian life.

THE NINETY-FIVE THESES

An incident illustrative of the condition of the church in the early sixteenth century propelled Martin Luther onto the stage of history and brought about the Reformation in Germany. The University of Wittenberg lay within the ecclesiastical jurisdiction of the archdiocese of Magdeburg. The twenty-seven-year-old archbishop of Magdeburg, Albert, was also administrator of the see of Halberstadt and had been appointed archbishop of Mainz. To hold all three offices simultaneously—blatant pluralism—required papal dispensation. At that moment, Pope Leo X was anxious to continue the construction of Saint Peter's Basilica but was hard pressed for funds. Archbishop Albert borrowed money from the Fuggers, a wealthy banking family of Augsburg, to pay for the papal dispensation allowing him to hold the several episcopal benefices. Only a few powerful financiers and churchmen knew the details of the arrangement, but Leo X authorized Archbishop Albert to sell indulgences in Germany to repay the Fuggers.

Wittenberg was in the political jurisdiction of Frederick of Saxony, one of the seven electors of the Holy Roman Empire. When Frederick forbade the sale of indulgences within his duchy, people of Wittenberg, including some of Professor Luther's students, streamed across the border from Saxony into Jütenborg in Thuringia to buy indulgences (see Chapter 10).

What was an *indulgence*? According to Catholic theology, individuals who sin alienate themselves from God and his love. In order to be reconciled to God, the sinner must confess his or her sins to a priest and do the penance assigned. For example, the man who steals must first return the stolen goods and then perform the penance given by the priest, usually certain prayers or good works. This is known as the tem-

Grünewald: Crucifixion (ca 1510) The bloodless hands, tortured face, and lacerated body contain an unprecedented depiction of the horrors of physical suffering and reflect the deep emotional piety of northern Europe. Court painter to Albert of Brandenburg, Grünewald later was strongly attracted to Luther's ideas. *(National Gallery of Art, Washington, D.C., Samuel H. Kress Collection)*

poral (or earthly) penance, since no one knows what penance God will ultimately require.

The doctrine of indulgence rested on three principles. First, God is merciful, but he is also just. Second, Christ and the saints, through their infinite virtue, established a "treasury of merits," on which the church, through its special relationship with Christ and the saints, can draw. Third, the church has the authority to grant sinners the spiritual benefits of those merits. Originally an indulgence was a remission of the temporal (priest-imposed) penalties for sin. Beginning in the twelfth century, the papacy and bishops had given Crusaders such indulgences. By the later Middle Ages people widely believed that an indulgence secured total remission of penalties for sin—on earth or in purgatory—an assured swift entry into heaven.

Archbishop Albert hired the Dominican friar John Tetzel to sell the indulgences. Tetzel mounted a blitz advertising campaign. One of his slogans—"As soon as coin in coffer rings, the soul from purgatory springs"—brought phenomenal success. Men and women could buy indulgences not only for themselves but for deceased parents, relatives, or friends. Tetzel even drew up a chart with specific prices for the forgiveness of particular sins. One should be cautious in condemning the gullibility of sixteenth-century Germans, however. Who would not want a spiritual insurance policy?

Luther was severely troubled that ignorant people believed that they had no further need for repentance once they had purchased an indulgence. Thus, according to historical tradition, in the academic fashion of the times, on the eve of All Saints' Day (October 31), 1517, he attached to the door of the church at Wittenberg Castle a list of ninety-five theses (or propositions) on indulgences. By this act Luther intended only to start a theological discussion of the subject and to defend the theses publicly.

Luther firmly rejected the notion that salvation could be achieved by good works, such as indulgences. Some of his theses challenged the pope's power to grant indulgences, and others criticized papal wealth: "Why does not the Pope, whose riches are at this day more ample than those of the wealthiest of the wealthy, build the one Basilica of St. Peter's with his own money, rather than with that of poor believers . . . ?"[5] Luther at first insisted that the pope had not known about the traffic in indulgences, for if he had known, he would have put a stop to it.

The theses were soon translated into German, printed, and read throughout the empire. Immediately, broad theological issues were raised. When questioned, Luther insisted that Scripture persuaded him of the invalidity of indulgences. He rested his fundamental argument on the principle that there was no biblical basis for indulgences. But, replied Luther's opponents, to deny the legality of indulgences was to deny the authority of the pope who had authorized them. The issue was drawn: where did authority lie in the Christian church?

Through 1518 and 1519 Luther studied the history of the papacy. Gradually he gained the conviction, as had Marsiglio and Hus before him (page 375), that ultimate authority in the church belonged, not to the papacy, but to a general council. Then, in 1519, in a large public disputation with the Catholic debater John Eck at Leipzig, Luther denied both the authority of the pope and the infallibility of a general council. The Council of Constance, he said, had erred when it condemned John Hus.

The papacy responded with a letter condemning some of Luther's propositions, ordering that his books be burned, and giving him two months to recant or be excommunicated. Luther retaliated by publicly burning the letter. By January 3, 1521, when the excommunication was supposed to become final, the controversy involved more than theological issues. The papal legate wrote, "All Germany is in revolution. Nine-tenths shout 'Luther' as their warcry; and the other tenth cares nothing about Luther, and cries 'Death to the court of Rome.' "[6]

In this highly charged atmosphere the twenty-one-year-old emperor Charles V held his first diet (assembly of the Estates of the empire) at Worms and summoned Luther to appear before it. When ordered to recant, Luther replied in language that rang all over Europe:

Unless I am convinced by the evidence of Scripture or by plain reason—for I do not accept the authority of the Pope or the councils alone, since it is established that they have often erred and contradicted themselves—I am bound by the Scriptures I have cited and my conscience is captive to the Word of God. I cannot and will not recant anything, for it is neither safe nor right to go against conscience. God help me. Amen.[7]

Luther was declared an outlaw of the empire: he was denied legal protection.

Jerome Bosch: Christ Before Pilate Pilate (right) grasps the pitcher of water as he prepares to wash his hands. The peasant faces around Christ are vicious, grotesque, even bestial, perhaps signifying humanity's stupidity and blindness. *(Princeton University)*

PROTESTANT THOUGHT

Between 1520 and 1530, Luther worked out the basic theological tenets that became the articles of faith for his new church and subsequently for all Protestant groups. The word *Protestant* derives from the protest drawn up by a small group of reforming German princes at the Diet of Speyer in 1529. The princes "protested" the decisions of the Catholic majority. At first *Protestant* meant "Lutheran," but with the appearance of many protesting sects, it became a general term applied to all non-Catholic Christians. Lutheran Protestant thought was officially formulated in the Confession of Augsburg in 1530.

Ernst Troeltsch, a German student of the sociology of religion, has defined Protestantism as a "modification of Catholicism, in which the Catholic formulation of questions was retained, while a different answer was given to them." Luther provided new answers to four old, basic theological issues.

First, how is a person to be saved? Traditional Catholic teaching held that salvation was achieved by both faith *and* good works. Luther held that salvation comes by *faith alone.* Women and men are saved, said Luther, by the arbitrary decision of God, irrespective of good works or the sacraments. God, not people, initiates salvation.

Second, where does religious authority reside?

Christian doctrine had long maintained that authority rests both in the Bible and in the traditional teaching of the church. Luther maintained that authority rests in the Word of God as revealed in the Bible alone and as interpreted by an individual's conscience. He urged that each person read and reflect on the Scriptures.

Third, what is the church? Luther re-emphasized the Catholic teaching that the church consists of the entire community of Christian believers. Medieval churchmen, however, had tended to identify the church with the clergy.

Finally, what is the highest form of Christian life? The medieval church had stressed the superiority of the monastic and religious life over the secular. Luther argued that all vocations have equal merit, whether ecclesiastical or secular, and that every person should serve God in his or her individual calling.[8] Protestantism, in sum, represented a reformulation of the Christian heritage.

THE SOCIAL IMPACT OF LUTHER'S BELIEFS

As early as 1521, Luther had a vast following. Every encounter with ecclesiastical or political authorities attracted attention to him. Pulpits and printing presses spread his message all over Germany. By the time of his death, people of all social classes had become Lutheran. What was the immense appeal of Luther's religious ideas?

Recent historical research on the German towns has shown that two significant late medieval developments prepared the way for Luther's ideas. First, since the fifteenth century, city governments had expressed resentment at clerical privileges and immunities. Priests, monks, and nuns paid no taxes and were exempt from civic responsibilities, such as defending the city. Yet religious orders frequently held large amounts of urban property. At Zurich in 1467, for example, religious orders held one-third of the city's taxable property. City governments were determined to integrate the clergy into civic life by reducing their privileges and giving them public responsibilities. Accordingly, the Zurich magistracy subjected the religious to taxes; inspected wills so that legacies to the church and legacies left by churchmen could be controlled; and placed priests and monks under the jurisdiction of the civil courts.

Preacherships also spread Luther's ideas. Critics of the late medieval church, especially informed and intelligent townspeople, condemned the irregularity and poor quality of sermons. As a result, prosperous burghers in many towns established preacherships. Preachers were men of superior education who were required to deliver about a hundred sermons a year, each lasting about forty-five minutes. Endowed preacherships had important consequences after 1517. Luther's ideas attracted many preachers, and in such towns as Stuttgart, Reutlingen, Eisenach, and Jena, preachers became Protestant leaders. Preacherships also encouraged the Protestant form of worship, in which the sermon, not the Eucharist, was the central part of the service.[9]

In the countryside the attraction of the German peasants to Lutheran beliefs was almost predictable. Luther himself came from a peasant background, and he admired their ceaseless toil. Peasants respected Luther's defiance of church authority. Moreover, they thrilled to the words Luther used in his treatise *On Christian Liberty* (1520): "A Christian man is the most free lord of all and subject to none." Taken by themselves, these words easily contributed to social unrest. In the early sixteenth century, the economic condition of the peasantry varied from place to place but was generally worse than it had been in the fifteenth century and deteriorating. Crop failures in 1523 and 1524 aggravated an explosive situation. In 1525 representatives of the Swabian peasants met at the city of Memmingen and drew up the Twelve Articles, which expressed their grievances. The Twelve Articles condemn lay and ecclesiastical lords and summarize the agrarian crisis of the early sixteenth century. The articles complain that nobles had seized village common lands, which traditionally had been used by all; that they had imposed new rents on manorial properties and new services on the peasants working those properties; and that they had forced the poor to pay unjust death duties in the form of the peasants' best horses or cows. Wealthy, socially mobile peasants especially resented these burdens, which they emphasized as new.[10] The peasants believed their demands conformed to Scripture and cited Luther as a theologian who could prove that they did.

Luther wanted to prevent rebellion. Initially he sided with the peasants, and in a tract *On Admonition to Peace,* he blasted the lords:

We have no one on earth to thank for this mischievous rebellion, except you lords and princes, especially you blind bishops and mad priests and monks. . . . In your government you do nothing but flay and rob your subjects in order that you may lead a life of splendor and pride, until the poor common folk can bear it no longer.[11]

But, Luther warned the peasants, nothing justified the use of armed force: "The fact that rulers are unjust and wicked does not excuse tumult and rebellion; to punish wickedness does not belong to everybody, but to the worldly rulers who bear the sword." As for biblical support for the peasants' demands, he maintained that Scripture had nothing to do with earthly justice or material gain.[12]

Massive revolts first broke out near the Swiss frontier and then swept through Swabia, Thuringia, the Rhineland, and Saxony. The crowds' slogans came directly from Luther's writings. "God's righteousness" and the "Word of God" were invoked in the effort to secure social and economic justice. The peasants who expected Luther's support were soon disillusioned. He had written of the "freedom" of the Christian, but he had meant the freedom to obey the Word of God, for in sin men and women lose their freedom and break their relationship with God. Freedom for Luther meant independence from the authority of the Roman church; it did *not* mean opposition to legally established secular powers. Firmly convinced that rebellion hastened the end of civilized society, he tossed off a tract *Against the Murderous, Thieving Hordes of the Peasants*:

Let everyone who can smite, slay, and stab [the peasants], secretly and openly, remembering that nothing can be more poisonous, hurtful or devilish than a rebel. It is just as when one must kill a mad dog; if you do not strike him, he will strike you and the whole land with you.[13]

The Peasants' Revolt The peasants were attracted to Luther's faith because it seemed to give religious support to their economic grievances. Carrying the banner of the Peasants' League and armed with pitchforks and axes, a group of peasants surround a knight. *(Photo: Caroline Buckler)*

The nobility ferociously crushed the revolt. Historians estimate that over 75,000 peasants were killed in 1525.

Luther took literally these words of Saint Paul's letter to the Romans: "Let every soul be subject to the higher powers. For there is no power but of God: the powers that be are established by God. Whosoever resists the power, resists the ordinance of God: and they that resist shall receive to themselves damnation."[14] As it developed, Lutheran theology exalted the state, subordinated the church to the state, and everywhere championed "the powers that be." The consequences for German society were profound and have redounded into the twentieth century. The revolt of 1525 strengthened the authority of lay rulers. Peasant economic conditions, however, moderately improved. For example, in many parts of Germany enclosed fields, meadows, and forests were returned to common use.

Scholars in many disciplines have attributed Luther's fame and success to the invention of the printing press, which rapidly reproduced and made known his ideas. Equally important was Luther's incredible skill with language. Some thinkers have lavished praise on the Wittenberg reformer; others have bitterly condemned him. But, in the words of psychologist Erik Erikson:

The one matter on which professor and priest, psychiatrist and sociologist, agree is Luther's immense gift for language: his receptivity for the written word; his memory for the significant phrase; and his range of verbal expression (lyrical, biblical, satirical, and vulgar) which in English is paralleled only by Shakespeare.[15]

Language proved to be the weapon with which this peasant's son changed the world.

Educated people and humanists, like the peasants, were much attracted by Luther's words. He advocated a simpler, personal religion based on faith, a return to the spirit of the early church, the centrality of the Scriptures in the liturgy and in Christian life, abolition of elaborate ceremonial—precisely the reforms the northern humanists had been calling for. Ulrich Zwingli (1483–1531), for example, a humanist of Zurich, was strongly influenced by Luther's bold stand; they stimulated Zwingli's reforms in that Swiss city. The nobleman Ulrich von Hutten (1488–1523), who had published several humanistic tracts, in 1519 dedicated his life to the advancement of Luther's reformation. And the Frenchman John Calvin (1509–1564), often called the organizer of Protestantism, owed a great deal to Luther's thought.

The publication of Luther's German translation of the New Testament in 1523 democratized religion. His insistence that everyone should read and reflect on the Scriptures attracted the literate and thoughtful middle classes partly because Luther appealed to their intelligence. Moreover, the business classes, preoccupied with making money, envied the church's wealth, disapproved of the luxurious lifestyle of some churchmen, and resented tithes and ecclesiastical taxation. Luther's doctrines of salvation by faith and the priesthood of all believers not only raised the religious status of the commercial classes but protected their pocketbooks as well.

For his time Luther held enlightened views on matters of sexuality and marriage. He wrote to a young man, "Dear lad, be not ashamed that you desire a girl, nor you my maid, the boy. Just let it lead you into matrimony and not into promiscuity, and it is no more cause for shame than eating and drinking."[16] Luther was confident that God took delight in the sexual act and denied that original sin affected the goodness of creation. He believed, however, that marriage was a woman's career. A student recorded Luther as saying, early in his public ministry, "Let them bear children until they are dead of it; that is what they are for." A happy marriage to the ex-nun Katharine von Bora mellowed him, and another student later quoted him as saying, "Next to God's Word there is no more precious treasure than holy matrimony. God's highest gift on earth is a pious, cheerful, God-fearing, home-keeping wife, with whom you may live peacefully, to whom you may entrust your goods, and body and life."[17] Though Luther deeply loved his "dear Katie," he believed that women's concerns revolved exclusively around the children, the kitchen, and the church. A happy woman was a patient wife, an efficient manager, and a good mother. Kate was an excellent financial manager (which Luther—much inclined to give money and goods away—was not). Himself a stern if often indulgent father, Luther held that the father should rule his household while the wife controlled its economy. With many relatives and constant visitors, Luther's was a large and happy household, a model for Protestants if an abomination for Catholics.

GERMANY AND THE PROTESTANT REFORMATION

The history of the Holy Roman Empire in the later Middle Ages is a story of dissension, disintegration, and debility. Unlike Spain, France, and England, the empire lacked a strong central power. The Golden Bull of 1356 legalized what had long existed—government by an aristocratic federation. Each of seven electors—the archbishops of Mainz, Trier, and Cologne, the margrave of Brandenburg, the duke of Saxony, the count palatine of the Rhine, and the king of Bohemia—gained virtual sovereignty in his own territory. The agreement ended disputed elections in the empire; it also reduced the central authority of the emperor. Germany was characterized by weak borders, localism, and chronic disorder. The nobility strengthened their territories, while imperial power declined.

Against this background of decentralization and strong local power, Martin Luther had launched a movement to reform the church. Two years after Luther posted the Ninety-Five Theses, the electors chose as emperor a nineteen-year-old Habsburg prince who ruled as Charles V. Luther's interests and motives were primarily religious, but many people responded to his teachings for political, social, or economic reasons. How did the goals and interests of the emperor influence the course of the Reformation in Germany? What impact did the upheaval in the Christian church have on the political condition in Germany?

THE RISE OF THE HABSBURG DYNASTY

The marriage in 1477 of Maximilian I of the house of Habsburg and Mary of Burgundy was a decisive event in early modern European history. Through this union with the rich and powerful duchy of Burgundy, the Austrian house of Habsburg, already the strongest ruling family in the empire, became an international power.

In the fifteenth and sixteenth centuries, as in the Middle Ages, relations among states continued to be greatly affected by the connections of royal families. Marriage often determined the diplomatic status of states. The Habsburg-Burgundian marriage angered the French, who considered Burgundy part of French territory. Louis XI of France repeatedly ravaged parts of the Burgundian Netherlands until he was able to force Maximilian to accept French terms: the Treaty of Arras (1482) emphatically declared Burgundy a part of the kingdom of France. The Habsburgs, however, never really renounced their claim to Burgundy, and intermittent warfare over it continued between France and Maximilian. Within the empire, German principalities that resented Austria's pre-eminence began to see that they shared interests with France. The marriage of Maximilian and Mary inaugurated centuries of conflict between the Austrian house of Habsburg and the kings of France. And Germany was to be the chief arena of the struggle.

"Other nations wage war; you, Austria, marry." Historians dispute the origins of the adage, but no one questions its accuracy. The heir of Mary and Maximilian, Philip of Burgundy, married Joanna of Castile, daughter of Ferdinand and Isabella of Spain. Philip and Joanna's son Charles V (1500–1558) fell heir to a vast conglomeration of territories. Through a series of accidents and unexpected deaths, Charles inherited Spain from his mother, together with her possessions in the New World and the Spanish dominions in Italy, Sicily, Sardinia, and Naples. From his father he inherited the Habsburg lands in Austria, southern Germany, the Low Countries, and Franche-Comté in east central France.

Charles's inheritance was an incredibly diverse collection of states and peoples, each governed in a different manner and held together only by the person of the emperor (see Map 14.1). Charles's Italian adviser, the grand chancellor Gattinara, told the young ruler: "God has set you on the path toward world monarchy." Charles not only believed this; he was convinced that it was his duty to maintain the political and religious unity of Western Christendom. In this respect Charles V was the last medieval emperor.

Charles needed and in 1519 secured the imperial title. Forward-thinking Germans proposed governmental reforms. They urged placing the administration in the hands of an imperial council whose president, the emperor's appointee, would have ultimate executive power. Reforms of the imperial finances, the army, and the judiciary were also recommended. Such ideas did not interest the young emperor at all. When he finally arrived in Germany from Spain and opened his first diet at Worms in January 1521, he naively announced that "the empire from of old has

Titian: The Emperor Charles V (1548) Court painter to Charles V, Titian portrayed him shortly after the emperor's defeat of the league of German Protestant princes at the battle of Mühlberg near Leipzig. In this idealization, one of the earliest equestrian portraits, Charles appears as heroic victor, chivalric knight, and defender of the church. *(The Mansell Collection)*

THE POLITICAL IMPACT OF LUTHER'S BELIEFS

In the sixteenth century, the practice of religion remained a public matter. Everyone participated in the religious life of the community, just as almost everyone shared in the local agricultural work. Whatever spiritual convictions individuals held in the privacy of their consciences, the emperor, king, prince, magistrate, or other civil authority determined the official form of religious practice within his jurisdiction. Almost everyone believed that the presence of a faith different from that of the majority represented a political threat to the security of the state. Only a tiny minority, and certainly none of the princes, believed in religious liberty.

Against this background, the religious storm launched by Martin Luther swept across Germany. Several elements in his religious reformation stirred patriotic feelings. Anti-Roman sentiment ran high. Humanists lent eloquent intellectual support. And Luther's translation of the New Testament into German evoked national pride.

For decades devout laymen and churchmen had called on the German princes to reform the church. In 1520 Luther took up the cry in his *Appeal to the Christian Nobility of the German Nation.* Unless the princes destroyed papal power in Germany, Luther argued, reform was impossible. He urged the princes to confiscate ecclesiastical wealth and to abolish indulgences, dispensations, pardons, and clerical celibacy. He told them that it was their public duty to bring about the moral reform of the church. Luther based his argument in part on the papacy's financial exploitation of Germany:

Now that Italy is sucked dry, they come into Germany, and begin, oh so gently. But let us beware, or Germany will soon become like Italy. . . . They skim the cream off the bishoprics, monasteries, and benefices, and because they do not yet venture to turn them all to shameful use, as they have done in Italy, they only practice for the present the sacred trickery of coupling together ten or twenty prelacies and taking a yearly portion from each of them so as to make a tidy sum after all. The priory of Würzburg yields a thousand gulden; that of Bamberg, something; Mainz, Trier, and the others, something more; and so . . . that a cardinal might live at Rome like a rich king.

had not many masters, but one, and it is our intention to be that one." Charles went on to say that he was to be treated as of greater account than his predecessors because he was more powerful than they had been. In view of the long history of aristocratic power, Charles's notions were pure fantasy.

Charles continued the Burgundian policy of his grandfather Maximilian. That is, German revenues and German troops were subordinated to the needs of other parts of the empire, first Burgundy and then Spain. Habsburg international interests came before the need for reform in Germany.

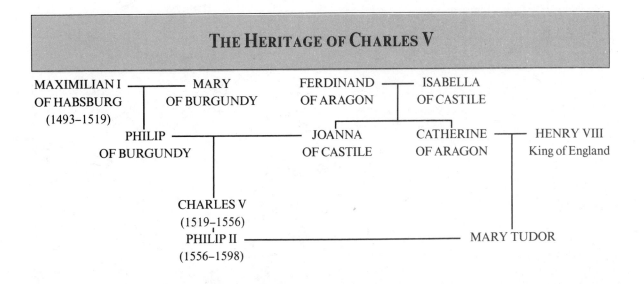

THE HERITAGE OF CHARLES V

MAXIMILIAN I OF HABSBURG (1493–1519) —— MARY OF BURGUNDY

FERDINAND OF ARAGON —— ISABELLA OF CASTILE

PHILIP OF BURGUNDY —— JOANNA OF CASTILE

CATHERINE OF ARAGON —— HENRY VIII King of England

CHARLES V (1519–1556)
PHILIP II (1556–1598) —————————— MARY TUDOR

From his Habsburg grandparents, Charles V inherited claims to the imperial title, Austria, and Burgundy; through his mother, Charles acquired Spain, the Spanish territories in Italy, and the vast uncharted Spanish possessions in the New World.

How comes it that we Germans must put up with such robbery and such extortion of our property at the hands of the pope? If the Kingdom of France has prevented it, why do we Germans let them make such fools and apes of us? It would all be more bearable if in this way they only stole our property; but they lay waste the churches and rob Christ's sheep of their pious shepherds, and destroy the worship and the Word of God. As it is they do nothing for the good of Christendom; they only wrangle about the incomes of bishoprics and prelacies, and that any robber could do.[18]

These words fell on welcome ears and itchy fingers. Luther's appeal to German patriotism gained him strong support, and national feeling influenced many princes otherwise confused by or indifferent to the complexities of the religious issues.

The church in Germany possessed great wealth. And, unlike other countries, Germany had no strong central government to check the flow of gold to Rome. Rejection of Roman Catholicism and adoption of Protestantism would mean the legal confiscation of lush farmlands, rich monasteries, and wealthy shrines. Some German princes, such as the prince-

archbishop of Cologne, Hermann von Wied, were sincerely attracted to Lutheranism, but many civil authorities realized that they had a great deal to gain by embracing the new faith. A steady stream of duchies, margraviates, free cities, and bishoprics secularized church property, accepted Lutheran theological doctrines, and adopted simpler services conducted in German. The decision reached at Worms in 1521 to condemn Luther and his teaching was not enforced because the German princes did not want to enforce it.

Charles V was a vigorous defender of Catholicism, and contemporary social and political theory denied the possibility of two religions coexisting peacefully in one territory. Thus many princes used the religious issue to extend their financial and political independence. When doctrinal differences became linked to political ambitions and financial receipts, the results proved unfortunate for the improvement of German government. The Protestant movement ultimately proved a political disaster for Germany.

Charles V must share blame with the German princes for the disintegration of imperial authority in the empire. He neither understood nor took an inter-

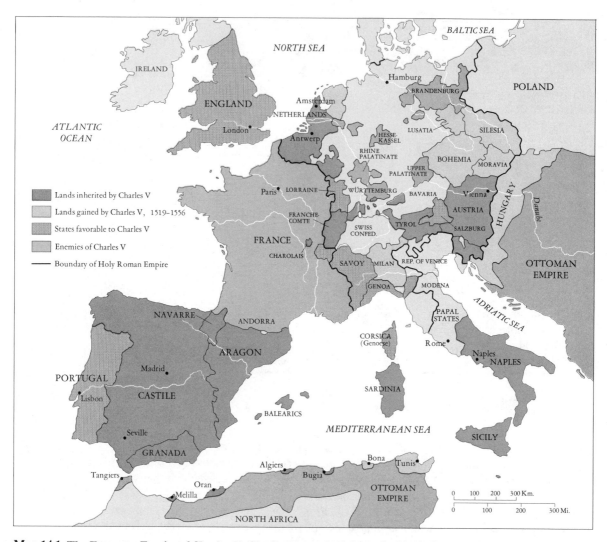

MAP 14.1 The European Empire of Charles V Charles V exercised theoretical jurisdiction over more territory than anyone since Charlemagne. This map does not show his Latin American and Asian possessions.

est in the constitutional problems of Germany, and he lacked the material resources to oppose Protestantism effectively there. Throughout his reign he was preoccupied with his Flemish, Spanish, Italian, and American territories. Moreover, the Turkish threat prevented him from acting effectively against the Protestants; Charles's brother Ferdinand needed Protestant support against the Turks who besieged Vienna in 1529.

Five times between 1521 and 1555, Charles V went to war with the Valois kings of France. The issue each time was the Habsburg lands acquired by the marriage of Maximilian and Mary of Burgundy. Much of the fighting occurred in Germany. The cornerstone of French foreign policy in the sixteenth and seventeenth centuries was the desire to keep the German states divided. Thus Europe witnessed the paradox of the Catholic king of France supporting the Lutheran princes in their challenge to his fellow Catholic, Charles V. French policy was successful. The long dynastic struggle commonly called the Habsburg-Valois Wars advanced the cause of Protestantism and promoted the political fragmentation of the German empire.

Charles's efforts to crush the Lutheran states were unsuccessful. Finally, in 1555, he agreed to the Peace of Augsburg, which, in accepting the status quo, officially recognized Lutheranism. Each prince was permitted to determine the religion of his territory. Most of northern and central Germany became Lutheran, while the south remained Roman Catholic. There was no freedom of religion, however. Princes or town councils established state churches to which all subjects of the area had to belong. Dissidents, whether Lutheran or Catholic, had to convert or leave. The political difficulties Germany inherited from the Middle Ages had been compounded by the religious crisis of the sixteenth century.

THE GROWTH OF THE PROTESTANT REFORMATION

The printing press publicized Luther's defiance of the Roman church and spread his theological ideas all over Europe. People discovered in Luther's ideas the economic theories they wanted to find. Christian humanists believed initially that Luther supported their own educational and intellectual goals. Princes steadily read in Luther's theories an expansion of state power and authority. What began as one man's religious search in a small corner of Germany soon became associated with many groups' interests and aspirations.

By 1555 much of northern Europe had broken with the Roman Catholic church. All of Scandinavia, England, Scotland, and such self-governing cities as Geneva and Zurich in Switzerland and Strasbourg in Germany had rejected the religious authority of Rome and adopted new faiths. In that a common religious faith had been the one element uniting all of Europe for almost a thousand years, the fragmentation of belief led to profound changes in European life and society. The most significant new form of Protestantism was Calvinism, of which the Peace of Augsburg had made no mention at all.

John Calvin The lean, ascetic face with the strong jaw reflects the iron will and determination of the organizer of Protestantism. The fur collar represents his training in law. *(Photo: Caroline Buckler)*

CALVINISM

In 1509, while Luther was studying for the doctorate at Wittenberg, John Calvin (1509–1564) was born in Noyon in northwestern France. Luther inadvertently launched the Protestant Reformation. Calvin, however, had the greater impact on future generations. His theological writings profoundly influenced the social thought and attitudes of Europeans and English-speaking peoples all over the world, especially in Canada and the United States. Although he had originally intended to have an ecclesiastical career, Calvin studied law, which had a decisive impact on his mind and later thought. In 1533 he experienced a religious crisis, as a result of which he converted to Protestantism.

Convinced that God selects certain people to do his work, Calvin believed that God had specifically called him to reform the church. Accordingly, he accepted an invitation to assist in the reformation of the city of Geneva. There, beginning in 1541, Calvin worked assiduously to establish a Christian society ruled by God through civil magistrates and reformed ministers. Geneva, "a city that was a Church," became the model of a Christian community for sixteenth-century Protestant reformers.

To understand Calvin's Geneva, it is necessary to understand Calvin's ideas. These he embodied in *The Institutes of the Christian Religion,* first published in 1536 and definitively issued in 1559. The cornerstone of Calvin's theology was his belief in the absolute sovereignty and omnipotence of God and the total weakness of humanity. Before the infinite power of God, he asserted, men and women are as insignificant as grains of sand:

Our souls are but faint flickerings over against the infinite brilliance which is God. We are created, he is without beginning. We are subject to ignorance and shame. God in his infinite majesty is the summation of all virtues. Whenever we think of him we should be ravished with adoration and astonishment . . . The chief end of man is to enjoy the fellowship of God and the chief duty of man is to glorify God.[19]

Calvin did not ascribe free will to human beings, because that would detract from the sovereignty of God. Men and women cannot actively work to achieve salvation; rather, God in his infinite wisdom decided at the beginning of time who would be saved and who damned. This viewpoint constitutes the theological principle called *predestination:*

Predestination we call the eternal decree of God, by which he has determined in himself, what he would have become of every individual of mankind. For they are not all created with a similar destiny; but eternal life is foreordained for some, and eternal damnation for others. . . .

In conformity, therefore, to the clear doctrine of the Scripture, we assert, that by an eternal and immutable counsel, God has once for all determined, both whom he would admit to salvation, and whom he would condemn to destruction. We affirm that this counsel, as far as concerns the elect, is founded on his gratuitous mercy, totally irrespective of human merit; but that to those whom he devotes to condemnation, the gate of life is closed by a just and irreprehensible, but incomprehensible, judgment.

How exceedingly presumptuous it is only to inquire into the causes of the Divine will; which is in fact, and is justly entitled to be, the cause of everything that exists. . . . For the will of God is the highest justice; so that what he wills must be considered just, for this very reason, because he wills it.[20]

Though the doctrine of predestination dates back to Saint Augustine and Saint Paul, many people have found it a pessimistic view of the nature of God, who they feel, revealed himself in the Old and New Testaments as merciful as well as just. Calvin maintained that while individuals cannot know whether they will be saved—and the probability is that they will be damned—good works are a "sign" of election. In any case, people should concentrate on worshiping God and doing his work and not waste time worrying about salvation. Though the doctrine of predestination may strike *us* as pessimistic, it inspired an enormous amount of energy in the sixteenth and later centuries.

Calvin aroused Genevans to a high standard of morality. He had two remarkable assets: complete mastery of Scripture and exceptional fluency in French. Through his sermons and a program of religious education, God's laws and man's were enforced in Geneva. Calvin's powerful sermons delivered the Word of God and thereby monopolized the strongest contemporary means of communication, preaching. Through his *Genevan Catechism,* published in 1541, children and adults memorized set questions and answers and acquired a summary of their faith and a guide for daily living. Calvin's sermons and his *Catechism* gave a whole generation of Genevans thorough instruction in the reformed religion.[21]

In the reformation of the city, the Genevan Consistory also exercised a powerful role. This body consisted of twelve laymen, plus the Company of Pastors of which Calvin was the permanent Moderator (presider). The duties of the Consistory were "to keep watch over every man's life [and] to admonish amiably those whom they see leading a disorderly life." While Calvin emphasized that the Consistory's activities should be thorough and "its eyes may be everywhere," corrections were only "medicine to turn sinners to the Lord."[22] Thus austere living, public fasting, and evening curfew became the order of the day. Fashionable clothes, dancing, card playing, and heavy drinking were absolutely prohibited. The Consistory investigated the private morals of citizens but were unwilling to punish the town prostitutes as severely as Calvin would have preferred. Calvin exercised some political influence through the Consistory, but the civil magistrates in Geneva maintained firm control.

Calvin reserved his harshest condemnation for religious dissenters. He declared:

If anybody slanders a mortal man he is punished and shall we permit a blasphemer of the living God to go unscathed? If a prince is injured, death appears to be insufficient for vengeance. And now when God, the sovereign Emperor, is reviled by a word, is nothing to be done? God's glory and our salvation are so conjoined that a traitor to God is also an enemy to the human race and worse than a murderer because he brings souls to perdition. Some object that since the offense consists only in words, there is no need for severity. But we muzzle dogs, and shall we leave men free to open their mouths as they please? Those who object are dogs and swine. They murmur that they will go to America where nobody will bother them.

God makes plain that the false prophet is to be stoned without mercy. We are to crush beneath our heel all affections of nature when His honor is concerned. The father should not spare his child, nor brother his brother, nor husband his own wife or the friend who is dearer to him than life. No human relationship is more than animal unless it be grounded in God.[23]

Calvin translated his words into action. In the 1550s the Spanish humanist Michael Servetus had gained international notoriety for his publications denying the Christian dogma of the Trinity, which holds that God is three divine persons, Father, Son, and Holy Spirit. Servetus had been arrested by the Spanish Inquisition but escaped to Geneva, where he hoped for support. He was promptly re-arrested. At his trial he not only held to his belief that there is no scriptural basis for the Trinity but rejected child baptism and insisted that a person under twenty cannot commit a mortal sin. The city fathers considered this last idea dangerous to public morality, "especially in these days when the young are so corrupted." Though Servetus begged that he be punished by ban-

ishment, Calvin and the town council maintained that the denial of child baptism and the Trinity amounted to a threat to all society. Whispering "Jesus, Son of the eternal God, have pity on me," Servetus was burned at the stake.

To many sixteenth-century Europeans, Calvin's Geneva seemed "the most perfect school of Christ since the days of the Apostles." Religious refugees from France, England, Spain, Scotland, and Italy poured into the city. Subsequently, the reformed church of Calvin served as the model for the Presbyterian church in Scotland, the Huguenot church in France, and the Puritan churches in England and New England.

Calvinism became the compelling force in international Protestantism. The Calvinist ethic of the "calling" dignified all work with a religious aspect. Hard work, well done, was pleasing to God. This doctrine encouraged an aggressive, vigorous activism. In the *Institutes* Calvin provided a systematic theology for Protestantism. The reformed church of Calvin had a strong and well-organized machinery of government. These factors, together with the social and economic applications of Calvin's theology, made Calvinism the most dynamic force in sixteenth- and seventeenth-century Protestantism.

THE ANABAPTISTS

The name *Anabaptist* derives from a Greek word meaning "to baptize again." The Anabaptists, sometimes described as the "left wing of the Reformation," believed that only adults could make a free choice about religious faith, baptism, and entry into the Christian community. Thus they considered the practice of baptizing infants and children preposterous and claimed there was no scriptural basis for it. They wanted to rebaptize believers who had been baptized as children. Anabaptists took the Gospel and, at first, Luther's teachings absolutely literally and favored a return to the kind of church that had existed among the earliest Christians—a voluntary association of believers who had experienced an inner light.

Anabaptists maintained that only a few people would receive the inner light. This position meant that the Christian community and the Christian state were not identical. In other words, Anabaptists believed in the separation of church and state and in re-

Calvinist Church, Nuremburg Stark simplicity and the absence of all ornamentation characterized early Protestant churches. With the men seated on the left and right, women in the center, all eyes and ears focus on the Bible on the lectern, the preacher in the pulpit, and the bare communion table. *(Germanisches National Museum, Nürnberg)*

ligious tolerance. They almost never tried to force their values on others. In an age that believed in the necessity of state-established churches, Anabaptist views on religious liberty were thought to undermine that concept.

Each Anabaptist community or church was entirely independent; it selected its own ministers and ran its own affairs. In 1534 the community at Münster in Germany, for example, established a legal code that decreed the death penalty for insubordinate wives. Moreover, the Münster community also practiced polygamy and forced all women under a certain age to marry or face expulsion or execution.

Anabaptists introduced polygamy but admitted women to the priesthood. They shared goods as the early Christians had done, refused all public offices, and would not serve in the armed forces. In fact, they laid great stress on pacifism. A favorite Anabaptist scriptural quotation was "By their fruits you shall know them," meaning that if Christianity was a religion of peace, the Christian should not fight. Good deeds were the sign of Christian faith, and to be a Christian meant to imitate the meekness and mercy of Christ. With such beliefs Anabaptists were inevitably a minority. Anabaptism later attracted the poor, the unemployed, the uneducated. Geographically,

Anabaptists drew their members from depressed urban areas—from among the followers of Zwingli in Zurich and from Basel, Augsburg, and Nuremberg.

Ideas such as absolute pacifism and the distinction between the Christian community and the state brought down on these unfortunate people fanatical hatred and bitter persecution. Zwingli, Luther, Calvin, and Catholics all saw—quite correctly—the separation of church and state as leading ultimately to the complete secularization of society. The powerful rulers of Swiss and German society immediately saw the connection between religious heresy and economic dislocation. Civil authorities feared that the combination of religious differences and economic grievances would lead to civil disturbances. In Saxony, in Strasbourg, and in the Swiss cities, Anabaptists were either banished or cruelly executed by burning, beating, or drowning. Their community spirit and the edifying example of their lives, however, contributed to the survival of Anabaptist ideas.

Later, the Quakers with their gentle pacifism; the Baptists with their emphasis on an inner spiritual light; the Congregationalists with their democratic church organization; and, in 1789, the authors of the United States Constitution with their concern for the separation of church and state—all these trace their origins in part to the Anabaptists of the sixteenth century.

THE ENGLISH REFORMATION

As on the Continent, the Reformation in England had social and economic causes as well as religious ones. As elsewhere, too, Christian humanists had for decades been calling for the purification of the church. When the political matter of the divorce of King Henry VIII (1509–1547) became enmeshed with other issues, a complete break with Rome resulted.

Demands for ecclesiastical reform dated back at least to the fourteenth century. The Lollards (see page 375) had been driven underground in the fifteenth century but survived in parts of London, East Anglia, west Kent, and southern England. Working-class people, especially cloth workers, were attracted to their ideas. The Lollards stressed the individual's reading and interpretation of the Bible, which they considered the only standard of Christian faith and holiness. Consequently, they put no stock in the

value of the sacraments and were vigorously anticlerical. Lollards opposed ecclesiastical wealth, the veneration of the saints, prayers for the dead, and all war. Although they had no notion of justification by faith, like Luther they insisted on the individual soul's direct responsibility to God.

The work of the English humanist William Tyndale (ca 1494–1536) stimulated cries for reform. Tyndale visited Luther at Wittenberg in 1524, and a year later at Antwerp he began printing an English translation of the New Testament. From Antwerp, merchants carried the New Testament into England, where it was distributed by Lollards. Fortified with copies of Tyndale's English Bible and some of Luther's ideas, the Lollards represented the ideal of "a personal, scriptural, non-sacramental, and lay-dominated religion."[24] In this manner, doctrines that would later be called Protestant flourished underground in England before any official or state-approved changes.

In the early sixteenth century, the ignorance of much of the parish clergy, and the sexual misbehavior of some, compared unfavorably with the education and piety of lay people. In 1510 Dr. William Melton, an official of York Cathedral, exhorted the newly ordained priests of the diocese:

. . . from this darkness of ignorance . . . arises that great and deplorable evil throughout the whole Church of God, that everywhere throughout town and countryside there exists a crop of oafish and boorish priests, some of whom are engaged in ignoble and servile tasks, while others abandon themselves to tavern haunting, swilling and drunkenness. Some cannot get along without their wenches; others pursue their amusement in dice and gambling and other such trifling all day long. . . . This is inevitable, for since they are completely ignorant of good literature, how can they obtain improvement or enjoyment in reading and study. . . . We must avoid and keep far from ourselves that grasping, deadly plague of avarice for which practically every priest is accused and held in disrepute before the people, when it is said that we are greedy for rich promotions, or harsh and grasping in retaining and amassing money.[25]

Even more than the ignorance of the lower clergy, the wealth of the English church fostered resentment and anticlericalism. The church controlled perhaps 20 percent of the land and also received an annual tithe of the produce of lay people's estates. Since the church had jurisdiction over wills, the clergy also received mortuary fees, revenues paid by the deceased's relatives. Mortuary fees led to frequent lawsuits, since the common lawyers nursed a deep jealousy of the ecclesiastical courts.

The career of Thomas Wolsey (1474?–1530) provides an extreme example of pluralism in the English church in the early sixteenth century. The son of a butcher, Wolsey became a priest and in 1507 secured an appointment as chaplain to Henry VII. In 1509 Henry VIII made Wolsey a privy councillor, where his remarkable ability and energy won him rapid advancement. In 1515 he became a cardinal and lord chancellor, and in 1518 papal legate. As chancellor, Wolsey dominated domestic and foreign policy, prosecuted the rich in the royal courts, and attacked the nobility in Parliament. As papal legate he ruled the English church, with final authority in all matters relating to marriage, wills, the clergy, and ecclesiastical appointments. Wolsey had more power than any previous royal minister, and he used that power to amass a large number of church offices, including the archbishopric of York, the rich bishoprics of Winchester and Lincoln, and the abbacy of Saint Albans. He displayed the vast wealth these positions brought him with ostentation and arrogance, which in turn fanned the embers of anticlericalism. The divorce of Henry VIII ignited all these glowing coals.

Having fallen in love with Anne Boleyn, sister of his cast-off mistress Mary Boleyn, Henry wanted to divorce his wife, Catherine of Aragon. Legal, diplomatic, and theological problems stood in his way, however. Catherine had first been married to Henry's brother Arthur. Contemporaries doubted that Arthur's union with Catherine had been consummated during the short time Arthur lived, and theologians therefore believed that no true marriage existed between them. When Henry married Catherine in 1509, he boasted that she was a virgin. According to custom, and in order to eliminate all doubts and legal technicalities about Catherine's marriage to Arthur, Henry secured a dispensation from Pope Julius II. For eighteen years Catherine and Henry lived together in what contemporaries thought a happy marriage. Catherine produced six children, but only the princess Mary survived childhood.

Precisely when Henry lost interest in his wife as a woman is unknown, but around 1527 he began to quote from a passage in the Old Testament Book of Leviticus: "You must not uncover the nakedness of your brother's wife; for it is your brother's nakedness. . . . The man who takes to wife the wife of his brother: that is impurity; he has uncovered his brother's nakedness, and they shall be childless."[26] Henry insisted that God was denying him a male heir to punish him for marrying his brother's widow. Henry claimed that he wanted to spare England the dangers of a disputed succession. The anarchy and disorders of the Wars of the Roses would surely be repeated if a woman, the princess Mary, inherited the throne. Although Henry contended that the succession was the paramount issue in his mind, his behavior suggests otherwise.

Henry went about the business of ensuring a peaceful succession in an extraordinary manner. He petitioned Pope Clement VII for an annulment of his marriage to Catherine. Henry wanted the pope to declare that a legal marriage with Catherine had never existed, in which case Princess Mary was illegitimate and thus ineligible to succeed to the throne. The pope was an indecisive man, whose attention at the time was focused on the Lutheran revolt in Germany and the Habsburg-Valois struggle for control of Italy. But there is a stronger reason why Clement could not grant Henry's petition. Henry argued that Pope Julius's dispensation had contradicted the law of God—that a man may not marry his brother's widow. The English king's request reached Rome at the very time that Luther was widely publishing tracts condemning the papacy as the core of wickedness subverting the law of God. Had Clement granted Henry's annulment and thereby admitted that his recent predecessor, Julius II, had erred, Clement would have given support to the Lutheran assertion that popes substitute their own evil judgments for the law of God. This Clement could not do, so he delayed acting on Henry's request.[27] The capture and sack of Rome in 1527 by the emperor Charles V, Queen Catherine's nephew, thoroughly tied the pope's hands. Charles could hardly allow the pope to grant the annulment, thereby acknowledging that his aunt, the queen of England, was a loose woman who had lived in sin with Henry VIII.

Accordingly, Henry determined to get his divorce in England. The convenient death of the archbishop of Canterbury allowed Henry to appoint a new archbishop, Thomas Cranmer (1489–1556). Cranmer heard the case in his archiepiscopal court, granted the annulment, and thereby paved the way for Henry's

Henry VIII's "Victory" This cartoon shows Henry VIII, assisted by Cromwell and Cranmer, triumphing over Pope Clement VII. Though completely removed from the historical facts, such illustrations were effectively used to promote antipapal feeling in late sixteenth-century England. *(Photo: Caroline Buckler)*

marriage to Anne Boleyn. English public opinion was against this marriage and strongly favored Queen Catherine as a woman much wronged. By rejecting Catherine, Henry ran serious political risks, and all for a woman whom contemporaries found neither very intelligent nor very attractive. The only distinguishing feature they noticed was a sixth finger on her right hand. The marriage between Henry and Anne was publicly announced on May 28, 1533. In September the princess Elizabeth was born.

Since Rome had refused to support Henry's matrimonial plans, he decided to remove the English church from papal jurisdiction. Henry used Parliament to legalize the Reformation in England. The Act in Restraint of Appeals (1533) declared that:

Where, by divers sundry old authentic histories and chronicles, it is manifestly declared and expressed that this realm of England is an empire, . . . governed by one supreme head and king having the dignity and royal estate of the imperial crown of the same (he being also institute and furnished by the goodness and sufferance of Almighty God with plenary, whole, and entire power, . . . to render and yield justice and final determination to all manner of folk residents or subjects within this his realm.[28]

The act went on to forbid all judicial appeals to the papacy, thus establishing the crown as the highest legal authority in the land. In effect, the Act in Restraint of Appeals placed sovereign power in the king.

Holbein: Sir Thomas More This powerful portrait (1527), revealing More's strong character and humane sensitivity, shows Holbein's complete mastery of detail —down to the stubble on More's chin. The chain was an emblem of More's service to Henry VIII. (© *The Frick Collection, New York*)

The Act for the Submission of the Clergy (1534) required churchmen to submit to the king and forbade the publication of all ecclesiastical laws without royal permission. The Supremacy Act of 1534 declared the king the supreme head of the Church of England. Both the Act in Restraint of Appeals and the Supremacy Act led to heated debate in the House of Commons. An authority on the Reformation Parliament has written that probably only a small number of those who voted the Restraint of Appeals actually knew they were voting for a permanent break with

Rome.[29] Some opposed the king. John Fisher, the bishop of Rochester, a distinguished scholar and humanist who had preached the oration at the funeral of Henry VII, lashed the clergy with scorn for their cowardice. Another humanist, Thomas More, resigned the chancellorship to protest the passage of the Act for the Submission of the Clergy and would not take an oath recognizing Anne's daughter as heir. Fisher, More, and other dissenters were beheaded.

When Anne Boleyn failed in her second attempt to produce a male child, Henry VIII charged her with adulterous incest and in 1536 had her beheaded. Parliament promptly proclaimed the princess Elizabeth illegitimate and, with the royal succession thoroughly confused, left the throne to whomever Henry chose. His third wife, Jane Seymour, gave Henry the desired son, Edward, and then died in childbirth. Henry went on to three more wives. Before he passed to his reward in 1547, he got Parliament to reverse the decision of 1536, relegitimating Mary and Elizabeth and fixing the succession first in his son and then in his daughters.

Between 1535 and 1539, under the influence of his chief minister, Thomas Cromwell, Henry decided to dissolve the English monasteries because, he charged, they were economically mismanaged and morally corrupt. Actually, he wanted their wealth. Justices of the peace and other local officials who visited religious houses throughout the land found the contrary. Ignoring their reports, the king ended nine hundred years of English monastic life, dispersed the monks and nuns, and confiscated their lands. Hundreds of properties were later sold to the middle and upper classes and the proceeds spent on war. The dissolution of the monasteries did not achieve a more equitable distribution of land and wealth or advance the cause of social justice. Rather, the "bare ruined choirs where late the sweet birds sang"—as Shakespeare described the desolate religious houses—testified to the loss of a valuable aesthetic and cultural force in English life. The redistribution of land, however, greatly strengthened the upper classes and tied them to the Tudor dynasty.

Did the religious changes have broad popular support? Englishmen had long criticized ecclesiastical abuses. Sentiment for reform was strong, though only a minority held distinctly Protestant doctrinal views. Recent scholarly research has emphasized that the English Reformation came from above. The surviving evidence does not allow us to gauge the degree of opposition to (or support for) Henry's break with Rome. Certainly, many laypeople wrote to the king, begging him to spare the monasteries. Most laypeople "acquiesced in the Reformation because they hardly knew what was going on, were understandably reluctant to jeopardise life or limb, a career or the family's good name."[30] But all did not quietly acquiesce. In 1536 popular opposition in the North to the religious changes led to the Pilgrimage of Grace, a massive multi-class rebellion that proved the largest in English history. In 1546 serious rebellions in East Anglia and in the West, despite possessing economic and Protestant components, reflected considerable public opposition to the state-ordered religious changes.[31]

Henry's motives combined personal, political, social, and economic elements. Theologically he retained such traditional Catholic practices and doctrines as auricular confession, clerical celibacy, and *transubstantiation* (the doctrine of the real presence of Christ in the bread and wine of the Eucharist). Meanwhile, Protestant literature circulated, and Henry approved the selection of men with known Protestant sympathies as tutors for his son. Until late in the century, the religious situation remained fluid.

The nationalization of the church and the dissolution of the monasteries led to important changes in governmental administration. Vast tracts of land came temporarily under the crown's jurisdiction, and new bureaucratic machinery had to be developed to manage those properties. New departments had to be coordinated with old ones. Medieval government had been household government: all branches of the state were associated with the person and personality of the monarch. In finances, for example, no distinction was made between the king's personal income and state revenues. Each branch of government was supported with funds from a specific source; if the source had a bad year, that agency suffered, while other branches of government were well in the black. Massive confusion and overlapping of responsibilities existed.

Thomas Cromwell reformed and centralized the king's household, the council, the secretariats, and the Exchequer. New departments of state were set up. Surplus funds from all departments went into a liquid fund to be applied to areas where there were deficits. This balancing resulted in greater efficiency and

economy. In Henry VIII's reign can be seen the growth of the modern centralized bureaucratic state.

For several decades after Henry's death in 1547, the English church shifted left and right. In the short reign of Henry's sickly son Edward VI (1547–1553), the strongly Protestant ideas of Archbishop Thomas Cranmer exerted a significant influence on the religious life of the country. Cranmer simplified the liturgy, invited Protestant theologians to England, and prepared the first *Book of Common Prayer* (1549). In stately and dignified English, the *Book of Common Prayer* included, together with the Psalter, the order for all services of the Church of England.

The equally brief reign of Mary Tudor (1553–1558) witnessed a sharp move back to Catholicism. The devoutly Catholic daughter of Catherine of Aragon, Mary rescinded the Reformation legislation of her father's reign and fully restored Roman Catholicism. Mary's marriage to her cousin Philip of Spain, son of the emperor Charles V, proved highly unpopular in England, and her persecution and execution of several hundred devout Protestants further alienated her subjects. During her reign, many Protestants fled to the Continent. Mary's death raised to the throne her sister Elizabeth (1558–1603) and inaugurated the beginnings of religious stability.

At the beginning of her reign, Elizabeth's position was insecure. Although the populace cheered her accession, many questioned her legitimacy. On the one hand, Catholics wanted a Roman Catholic ruler. On the other hand, a vocal number of returned English exiles wanted all Catholic elements in the Church of England destroyed. The latter, because they wanted to "purify" the church, were called "Puritans."

Elizabeth had been raised a Protestant, but if she had genuine religious convictions she kept them to herself. Probably one of the shrewdest politicians in English history, Elizabeth chose a middle course between Catholic and Puritan extremes. She insisted on dignity in church services and political order in the land. She did not care what people believed as long as they kept quiet about it. Avoiding precise doctrinal definitions, Elizabeth had herself styled "Supreme Governor of the Church of England, Etc.," and left it to her subjects to decide what the "Etc." meant.

The parliamentary legislation of the early years of Elizabeth's reign—laws sometimes labeled the "Elizabethan Settlement"—required outward conformity to the Church of England and uniformity in all ceremonies. Everyone had to attend Church of England services; those who refused were fined. In 1563 a convocation of bishops approved the Thirty-Nine Articles, a summary in thirty-nine short statements of the basic tenets of the Church of England. During Elizabeth's reign, the Anglican church (for the Latin *Ecclesia Anglicana*), as the Church of England was called, moved in a moderately Protestant direction. Services were conducted in English, monasteries were not re-established, and the clergy were allowed to marry. But the bishops remained as church officials, and apart from language, the services were quite traditional.

THE ESTABLISHMENT OF THE CHURCH OF SCOTLAND

Reform of the church in Scotland did not follow the English model. In the early sixteenth century, the church in Scotland presented an extreme case of clerical abuse and corruption, and Lutheranism initially attracted sympathetic support. In Scotland as elsewhere, political authority was the decisive influence in reform. The monarchy was very weak, and factions of virtually independent nobles competed for power. King James V and his daughter Mary, Queen of Scots (1560–1567), staunch Catholics and close allies of Catholic France, opposed reform. The Scottish nobles supported it. One man, John Knox (1505?–1572), dominated the movement for reform in Scotland.

In 1559 Knox, a dour, narrow-minded, and fearless man with a reputation as a passionate preacher, set to work reforming the church. He had studied and worked with Calvin in Geneva and was determined to structure the Scottish church after the model of Calvin's Geneva. In 1560 Knox persuaded the Scottish parliament, which was dominated by reform-minded barons, to enact legislation ending papal authority. The Mass was abolished and attendance at it forbidden under penalty of death. Knox then established the Presbyterian Church of Scotland, so named because *presbyters,* or ministers—not bishops—governed it. The Church of Scotland was strictly Calvinist in doctrine, adopted a simple and dignified service of worship, and laid great emphasis on preaching. Knox's *Book of Common Order* (1564) became the liturgical directory for the church. The

Presbyterian Church of Scotland was a national, or state, church, and many of its members maintained close relations with English Puritans.

PROTESTANTISM IN IRELAND

To the ancient Irish hatred of English political and commercial exploitation, the Reformation added the bitter antagonism of religion. Henry VIII wanted to "reduce that realm to the knowledge of God and obedience to us." English rulers in the sixteenth century regarded the Irish as barbarians, and a policy of complete extermination was rejected only because "to enterprise [attempt] the whole extirpation and total destruction of all the Irishmen in the land would be a marvelous sumptious charge and great difficulty."[32] In other words, it would have cost too much.

In 1536, on orders from London, the Irish parliament, which represented only the English landlords and the people of the Pale (the area around Dublin), approved the English laws severing the church from Rome and making the English king sovereign over ecclesiastical organization and practice. The Church of Ireland was established on the English pattern, and the (English) ruling class adopted the new reformed faith. Most of the Irish, probably for political reasons, defiantly remained Roman Catholic. Monasteries were secularized. Catholic property was confiscated and sold and the profits shipped to England. With the Roman church driven underground, the Catholic clergy acted as national as well as religious leaders.

LUTHERANISM IN SWEDEN, NORWAY, AND DENMARK

In Sweden, Norway, and Denmark, the monarchy took the initiative in the religious Reformation. The resulting institutions were Lutheran state churches. Since the late fourteenth century, the Danish kings had ruled Sweden and Norway as well as Denmark. In 1520 the Swedish nobleman Gustavus Vasa led a successful revolt against Denmark, and Sweden became independent. As king, Gustavus Vasa seized church lands and required the bishops' loyalty to the Swedish crown. The Wittenberg-educated Swedish reformer Olaus Petri (1493–1552) translated the New Testament into Swedish and, with the full support of

Gustavus Vasa, organized the church along strict Lutheran lines. This consolidation of the Swedish monarchy in the sixteenth century was to have a profound effect on Germany in the seventeenth century.

In Denmark, King Christian III (1534–1559) secularized church property and set up a Lutheran church. Norway, which was governed by Denmark until 1814, adopted Lutheranism as its state religion under Danish influence.

THE CATHOLIC AND THE COUNTER-REFORMATIONS

Between 1517 and 1547, the reformed versions of Christianity known as Protestantism made remarkable advances. All of England, Scotland, Scandinavia, much of Germany, and sizable parts of France and Switzerland adopted the creeds of Luther, Calvin, and other reformers. Still, the Roman Catholic church made a significant comeback. After about 1540, no new large areas of Europe, except for the Netherlands, accepted Protestant beliefs (see Map 14.2).

Historians distinguish between two types of reform within the Catholic church in the sixteenth and seventeenth centuries. The Catholic Reformation began before 1517 and sought renewal basically through the stimulation of a new spiritual fervor. The Counter-Reformation started in the 1530s as a reaction to the rise and spread of Protestantism. The Counter-Reformation involved Catholic efforts to convince or coerce dissidents or heretics to return to the church lest they corrupt the entire community of Catholic believers. Fear of the "infection" of all Christian society by the religious dissident was a standard sixteenth-century attitude. If the heretic could not be persuaded to reconvert, counter-reformers believed it necessary to call on temporal authorities to defend Christian society by expelling or eliminating the dissident. The Catholic Reformation and the Counter-Reformation were not mutually exclusive; in fact, after about 1540 they progressed simultaneously.

What factors influenced the attitudes and policies of the papacy? Why did church leaders wait so long before leading with the issues of schism and reform? How did the Catholic church succeed in reforming itself and in stemming the tide of Protestantism?

THE SLOWNESS OF INSTITUTIONAL REFORM

The Renaissance princes who sat on the throne of Saint Peter were not blind to the evils that existed. Modest reform efforts had begun with the Lateran Council called in 1512 by Pope Julius II. The Dutch pope Adrian VI (1522–1523) had instructed his legate in Germany to

say that we frankly confess that God permits this [Lutheran] persecution of his church on account of the sins of men, especially those of the priests and prelates. . . . We know that in this Holy See now for some years there have been many abominations, abuses in spiritual things, excesses in things commanded, in short that all has become perverted. . . . We have all turned aside in our ways, nor was there, for a long time, any who did right—no, not one.[33]

Why did the popes, spiritual leaders of the Western church, move so slowly? The answers lie in the personalities of the popes themselves, their preoccupation with political affairs in Italy, and the awesome difficulty of reforming so complicated a bureaucracy as the Roman curia.

Pope Leo X (1513–1521), who reputedly opened his pontificate with the words "Now that God has given us the papacy, let us enjoy it," typified the attitude of the Renaissance papacy. Leo concerned himself with artistic beauty and sensual pleasures. He first dismissed the Lutheran revolution as "a monkish quarrel," and by the time he finally acted with a letter condemning Luther, much of northern Germany had already rallied around the sincere Augustinian.

Adrian VI tried desperately to reform the church and to check the spread of Protestantism. His reign lasted only thirteen months, however, and the austerity of his life and his Dutch nationality provoked the hostility of pleasure-loving Italian curial bureaucrats.

Clement VII, a true Medicean, was far more interested in elegant tapestries and Michelangelo's painting of the Last Judgment than in theological disputes in barbaric Germany. Indecisive and vacillating, Pope Clement must bear much of the responsibility for the great spread of Protestantism. While Emperor Charles V and the French king Francis I competed for the domination of divided Italy, the papacy wor-

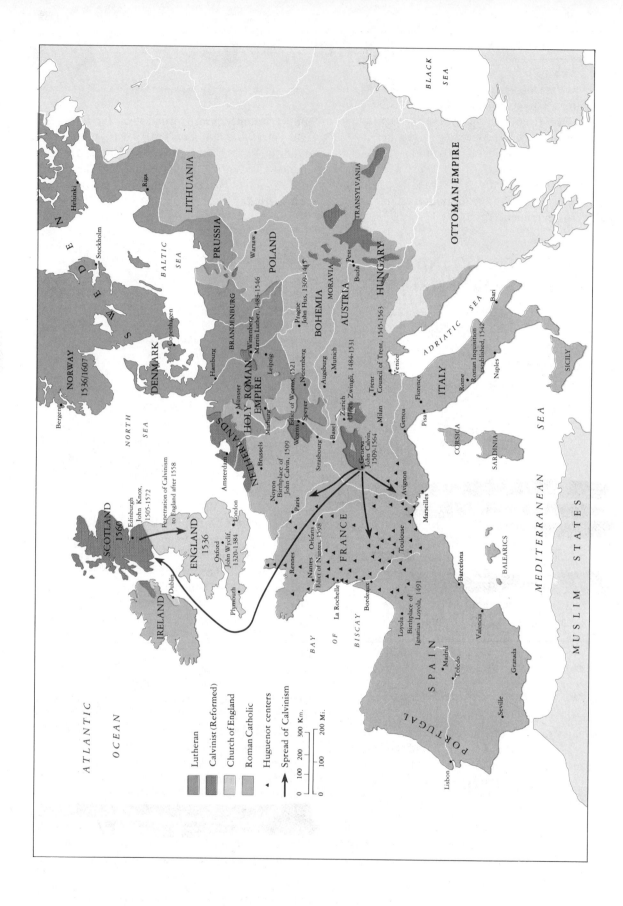

ATLANTIC
OCEAN

Lutheran
Calvinist (Reformed)
Church of England
Roman Catholic
◄ Huguenot centers
Spread of Calvinism

0 100 200 300 Km.
0   100   200 Mi.

NORWAY
1536/1607

Bergen

N
O
R
T
H

S
E
A

SWEDEN

Stockholm

Helsinki

Riga

BALTIC
SEA

LITHUANIA

PRUSSIA

BRANDENBURG

Warsaw

POLAND

DENMARK
Copenhagen

Hamburg

Leipzig

Wittenberg
Martin Luther, 1483–1546

Prague
John Hus, 1369–1415

BOHEMIA

MORAVIA

Pest
Buda

HUNGARY

TRANSYLVANIA

OTTOMAN EMPIRE

BLACK
SEA

HOLY ROMAN
EMPIRE

Münster

Marburg

Worms
Edict of Worms, 1521

Nuremberg

Augsburg

Munich

AUSTRIA

Trent
Council of Trent, 1545–1563

Venice

ADRIATIC
SEA

Bari

NETHERLANDS

Brussels

Amsterdam

Speyer

Basel

Zurich
Ulrich Zwingli, 1484–1531

Milan

Genoa

Florence

Pisa

ITALY

Rome
Roman Inquisition
established, 1542

Naples

SICILY

SCOTLAND
1560

Edinburgh
John Knox,
1505–1572

Penetration of Calvinism
to England after 1558

ENGLAND
1536

Oxford
John Wyclif,
1320–1384

London

IRELAND

Dublin

Plymouth

Noyon
Birthplace of
John Calvin, 1509

Geneva
John Calvin,
1509–1564

Avignon

Marseilles

CORSICA

SARDINIA

BALEARICS

MEDITERRANEAN
SEA

Paris

Strasbourg

Rennes

Nantes
Edict of Nantes, 1598

Orléans

FRANCE

Toulouse

BAY

OF

BISCAY

La Rochelle

Bordeaux

Loyola
Birthplace of
Ignatius Loyola, 1491

Barcelona

Valencia

SPAIN

Madrid

Toledo

Granada

Seville

PORTUGAL

Lisbon

MUSLIM STATES

MAP 14.2 The Protestant and the Catholic reformations The reformations shattered the religious unity of Western Christendom. What common cultural traits predominated in regions where a particular branch of the Christian faith was maintained or took root?

ried about the security of the Papal States. Clement tried to follow a middle course, backing first the emperor and then the French ruler. At the battle of Pavia in 1525, Francis I suffered a severe defeat and was captured. In a reshuffling of diplomatic alliances, the pope switched from Charles and the Spaniards to Francis I. The emperor was victorious once again, however, and in 1527 his Spanish and German mercenaries sacked and looted Rome and captured the pope. Obviously, papal concern about Italian affairs and the Papal States diverted attention from reform.

The idea of reform was closely linked to the idea of a general council representing the entire church. Early in the sixteenth century, Ferdinand of Spain appointed a committee of Spanish bishops to draft materials for conciliar reform. In France, the University of Paris also pressed for a council. (French monarchs subsequently used this academic demand to support their military intervention in Italy.) The emperor Charles V, increasingly disturbed by the Lutheran threat, called for "a free Christian council in German lands." German Catholic bishops drew up lists of "oppressive disorders" that needed reform. A strong contingent of countries beyond the Alps—Spain, Germany, and France—wanted to reform the vast bureaucracy of Latin officials, reducing offices, men, and revenues.

Popes from Julius II to Clement VII, remembering fifteenth-century conciliar attempts to limit papal authority, resisted calls for a council. The papal bureaucrats who were the popes' intimates warned the popes against a council, fearing loss of power, revenue, and prestige. Five centuries before, Saint Bernard of Clairvaux had anticipated the situation: "The most grievous danger of any Pope lies in the fact that, encompassed as he is by flatterers, he never hears the truth about his own person and ends by not wishing to hear it."[34]

THE COUNCIL OF TRENT

In the papal conclave that followed the death of Clement VII, Cardinal Alexander Farnese promised two German cardinals that if he were elected pope he would summon a council. He won the election and ruled as Pope Paul III (1534–1549). This Roman aristocrat, humanist, and astrologer, who immediately made his teenage grandsons cardinals, seemed an unlikely person to undertake serious reform. Yet Paul III appointed as cardinals several learned churchmen, such as Caraffa (later Pope Paul IV); established the Inquisition in the Papal States; and—true to his word—called a council, which finally met at Trent, an imperial city close to Italy.

The Council of Trent met intermittently from 1545 to 1563. It was called not only to reform the church but to secure reconciliation with the Protestants. Lutherans and Calvinists were invited to participate, but their insistence that the Scriptures be the sole basis for discussion made reconciliation impossible. Other problems bedeviled all the sessions of the council. International politics repeatedly cast a shadow over the theological debates. Charles V opposed discussions on any matter that might further alienate his Lutheran subjects, fearing the loss of additional imperial territory to Lutheran princes. Meanwhile, the French kings worked against the reconciliation of Roman Catholicism and Lutheranism. As long as religious issues divided the German states, the empire would be weakened, and a weak and divided empire meant a stronger France.

Trent had been selected as the site for the council in the hope it would attract Protestants. In fact, Italian bishops predominated, and no Protestants attended. The city's climate, small size, and poor accommodations; the advanced age of many bishops; the difficulties of travel in the sixteenth century; and the refusal of Charles V and Henry II of France to allow their national bishops to attend certain sessions—these factors drastically reduced attendance. Portugal, Poland, Hungary, and Ireland sent representatives, but very few German bishops attended.

Another problem was the persistence of the conciliar theory of church government. Some bishops wanted a concrete statement asserting the supremacy of a church council over the papacy. The adoption of the conciliar principle could have led to a divided church. The bishops had a provincial and national

The Council of Trent This seventeenth-century engraving depicts one of the early and sparsely attended sessions of the Council of Trent. The tridentine sessions of 1562–1563 drew many more bishops and laymen, but there were never many representatives from northern Europe. *(Photo: Caroline Buckler)*

outlook; only the papacy possessed an international religious perspective. The centralizing tenet was established that all acts of the council required papal approval.

In spite of the obstacles, the achievements of the Council of Trent are impressive. It dealt with both doctrinal and disciplinary matters. The council gave equal validity to the Scriptures and to tradition as sources of religious truth and authority. It reaffirmed the seven sacraments and the traditional Catholic teaching on transubstantiation. Thus, Lutheran and Calvinist positions were rejected.

The council tackled the problems arising from ancient abuses by strengthening ecclesiastical discipline. Tridentine (from *Tridentum,* the Latin word for Trent) decrees required bishops to reside in their own dioceses, suppressed pluralism and simony, and forbade the sale of indulgences. Clerics who kept concubines were to be warned to give them up and, if they refused, stripped of all ecclesiastical income. The jurisdiction of bishops over all the clergy of their dioceses was made almost absolute, and bishops were ordered to visit every religious house within the diocese at least once every two years. In a highly original canon, the council required every diocese to establish a seminary for the education and training of the clergy; the council even prescribed the curriculum and insisted that preference for admission be given to sons of the poor. Finally, great emphasis was laid on preaching and instructing the laity, especially the uneducated.

The Council of Trent did not meet everyone's expectations. Reconciliation with Protestantism was not achieved, nor was reform brought about immediately. Nevertheless, the Tridentine decrees laid a solid basis for the spiritual renewal of the church and for the enforcement of correction. For four centuries, the doctrinal and disciplinary legislation of Trent served as the basis for Roman Catholic faith, organization, and practice.

New Religious Orders

The establishment of new religious orders within the church reveals a central feature of the Catholic Reformation. These new orders developed in response to one crying need: to raise the moral and intellectual level of the clergy and people. Education was a major goal of them all.

The Ursuline order of nuns founded by Angela Merici (1474–1540) attained enormous prestige for the education of women. The daughter of a country gentleman, Angela Merici worked for many years among the poor, sick, and uneducated around her native Brescia in northern Italy. In 1535 she established the Ursuline order to combat heresy through Christian education. The first religious order concentrating exclusively on teaching young girls, the Ursulines sought to re-Christianize society by training future wives and mothers. Approved as a religious community by Paul III in 1544, the Ursulines rapidly grew and spread to France and the New World. Their schools in North America, stretching from Quebec to New Orleans, provided superior education for young women and inculcated the spiritual ideals of the Catholic Reformation.

The Society of Jesus, founded by Ignatius Loyola (1491–1556), a former Spanish soldier, played a powerful international role in resisting the spread of Protestantism, converting Asians and Latin American Indians to Catholicism, and spreading Christian education all over Europe. While recuperating from a severe battle wound in his legs, Loyola studied a life of Christ and other religious books and decided to give up his military career and become a soldier of Christ. During a year spent in seclusion, prayer, and personal mortification, he gained the religious insights that went into his great classic, *Spiritual Exercises.*

This work, intended for study during a four-week period of retreat, directed the individual imagination and will to the reform of life and a new spiritual piety.

Loyola was apparently a man of considerable personal magnetism. After study at the universities in Salamanca and Paris, he gathered a group of six companions and in 1540 secured papal approval of the new Society of Jesus, whose members were called "Jesuits." Their goals were the reform of the church primarily through education, preaching the Gospel to pagan peoples, and fighting Protestantism. Within a short time, the Jesuits had attracted many recruits.

The Society of Jesus was a highly centralized, tightly knit organization. Candidates underwent a two-year novitiate, in contrast to the usual one-year probation. Although new members took the traditional vows of poverty, chastity, and obedience, the emphasis was on obedience. Carefully selected members made a fourth vow of obedience to the pope and the governing members of the society. As faith was the cornerstone of Luther's life, so obedience became the bedrock of the Jesuit tradition.

The Jesuits had a modern, quasi-military quality; they achieved phenomenal success for the papacy and the reformed church. Jesuit schools adopted modern teaching methods, and while they first concentrated on the children of the poor, they were soon educating the sons of the nobility. As confessors and spiritual directors to kings, Jesuits exerted great political influence. Operating on the principle that the end sometimes justifies the means, they were not above spying. Indifferent to physical comfort and personal safety, they carried Christianity to India and Japan before 1550, to Brazil and the Congo in the seventeenth century. Within Europe, the Jesuits brought southern Germany and much of eastern Europe back to Catholicism.

The Sacred Congregation of the Holy Office

In 1542 Pope Paul III established the Sacred Congregation of the Holy Office with jurisdiction over the Roman Inquisition, a powerful instrument of the Counter-Reformation. The Inquisition was a committee of six cardinals with judicial authority over all Catholics and the power to arrest, imprison, and execute. Under the direction of the fanatical Cardinal Caraffa, it vigorously attacked heresy.

Pope Paul III's Confirmation of the Jesuit Constitutions On the right Ignatius Loyola receives the constitutions of the Society of Jesus by direct illumination from God. At left the pope approves them. When the constitutions were read to him, Paul III supposedly murmured, "There is the finger of God." *(Historical Picture Service, Chicago)*

The Roman Inquisition operated under the principles of Roman law. It accepted hearsay evidence, was not obliged to inform the accused of charges against them, and sometimes applied torture. Echoing one of Calvin's remarks about heresy, Cardinal Caraffa wrote, "No man is to lower himself by showing toleration towards any sort of heretic, least of all a Calvinist."[35] The Holy Office published the *Index of Prohibited Books,* a catalog of forbidden reading that included the publications of many printers.

Within the Papal States, the Inquisition effectively destroyed heresy (and many heretics). Outside the papal territories, however, its influence was slight. Governments had their own judicial systems for the suppression of treasonable activities, as religious heresy was then considered. The republic of Venice is a good case in point.

In the sixteenth century, Venice was one of the great publishing centers of Europe. The Inquisition and the Index could have badly damaged the Venetian book trade. Authorities there cooperated with the Holy Office only when heresy became a great threat to the security of the republic. The Index had no influence on scholarly research in nonreligious areas, such as law, classical literature, and mathematics. As a result of the Inquisition, Venetians and Italians were *not* cut off from the main currents of European learning.[36]

The age of the Reformation presents very real paradoxes. The break with Rome and the rise of Lutheran, Anglican, Calvinist, and other faiths destroyed the unity of Europe as an organic Christian society. Saint Paul's exhortation, "There should be no schism in the body [of the church]. . . . You are all one in Christ,"[37] was gradually ignored. On the other hand, religious belief remained tremendously strong.

In fact, the strength of religious convictions caused political fragmentation. In the later sixteenth century and through most of the seventeenth, religion and religious issues continued to play a major role in the lives of individuals and in the policies and actions of governments. Religion, whether Protestant or Catholic, decisively influenced the growth of national states. While most reformers rejected religious toleration, they helped pave the way for it.

For almost a thousand years, the church had taught Europeans "to believe in order that you may know." In the seventh through ninth centuries, European peoples had been led in massive numbers to the waters of Christian baptism. The Christian faith and Christian practices, however, meant little to the pagan barbarians of the early Middle Ages. Many centuries passed before the church had a significantly christianizing impact on those peoples. Therein lies another paradox. At the moment when literature, sermons, and especially art were expressing the widespread desire for individual and emotional experience within a common spiritual framework, the schism brought confusion, divisiveness, and destruction. The Reformation was, ironically, a tribute to the successful educational work of the medieval church.

Finally, scholars have maintained that the sixteenth century witnessed the beginnings of the modern world. They are both right and wrong. The sixteenth-century revolt from the church paved the way for the eighteenth-century revolt from the Christian God, one of the strongest supports of life in Western culture. In this respect, the Reformation marked the beginning of the modern world, with its secularism and rootlessness. At the same time, it can equally be argued that the sixteenth century represented the culmination of the Middle Ages. Martin Luther's anxieties about salvation show him to be very much a medieval man. His concerns had deeply troubled serious individuals since the time of Saint Augustine. Modern people tend to be less troubled by this issue.

NOTES

1. Romans 12:2–3.
2. Quoted by J. Burckhardt, *The Civilization of the Renaissance in Italy,* Phaidon Books, London, 1951, p. 262.
3. See R. R. Post, *The Modern Devotion: Confrontation with Reformation and Humanism,* E. J. Brill, Leiden, 1968, esp. pp. 237–238, 255, 323–348.
4. E. Erikson, *Young Man Luther: A Study in Psychoanalysis and History,* W. W. Norton, New York, 1962.
5. T. C. Mendenhall et al., eds., *Ideas and Institutions in European History: 800–1715,* Henry Holt, New York, 1948, p. 220.
6. Quoted by O. Chadwick, *The Reformation,* Penguin Books, Baltimore, 1976, p. 55.
7. Quoted by E. H. Harbison, *The Age of Reformation,* Cornell University Press, Ithaca, N.Y., 1963, p. 52.
8. Based heavily on Harbison, pp. 52–55.
9. See Steven E. Ozment, *The Reformation in the Cities: The Appeal of Protestantism to Sixteenth-Century Germany and Switzerland,* Yale University Press, New Haven, Conn., 1975, pp. 32–45.
10. See Steven E. Ozment, *The Age of Reform, 1250–1550: An Intellectual and Religious History of Late Medieval and Reformation Europe,* Yale University Press, New Haven, Conn., 1980, pp. 273–279.
11. Cited in ibid., p. 280.
12. Ibid., p. 281.
13. Ibid., p. 284.
14. Romans 13: 1–2.
15. Erikson, p. 47.
16. H. G. Haile, *Luther: An Experiment in Biography,* Doubleday, Garden City, N.Y., 1980, p. 272.
17. Quoted by J. Atkinson, *Martin Luther and the Birth of Protestantism,* Penguin Books, Baltimore, 1968, pp. 247–248.
18. *Martin Luther: Three Treatises,* Muhlenberg Press, Philadelphia, 1947, pp. 28–31.
19. Quoted by R. Bainton, *The Travail of Religious Liberty,* Harper & Brothers, New York, 1958, p. 65.
20. J. Allen, trans., *John Calvin: The Institutes of the Christian Religion,* Westminster Press, Philadelphia, 1930, book 3, chap. 21, paras. 5, 7.
21. E. William Monter, *Calvin's Geneva,* John Wiley & Sons, New York, 1967, pp. 98–108.
22. Ibid., p. 137.
23. Quoted by Bainton, pp. 69–70.
24. A. G. Dickens, *The English Reformation,* Schocken Books, New York, 1964, p. 36.
25. A. G. Dickens and Dorothy Carr, eds., *The Reformation in England to the Accession of Elizabeth I,* Edward Arnold, London, 1969, pp. 15–16.
26. Leviticus 18:16, 20, 21.

27. See Richard Marius, *Thomas More: A Biography,* Knopf, New York, 1984, pp. 215–216.

28. C. Stephenson and G. F. Marcham, *Sources of English Constitutional History,* Harper & Row, New York, 1937, p. 304.

29. See Stanford E. Lehmberg, *The Reformation Parliament 1529–1536,* Cambridge University Press, Cambridge, Eng., 1970, pp. 174–176 and 204–205.

30. See J. J. Scarisbrick, *The Reformation and the English People,* Basil Blackwell, Oxford, 1984, pp. 81–84, esp. p. 81.

31. Ibid.

32. Quoted by P. Smith, *The Age of the Reformation,* rev. ed., Henry Holt, New York, 1951, p. 346.

33. Ibid., p. 84.

34. Quoted by H. Jedin, *A History of the Council of Trent,* Nelson & Sons, London, 1957, I. 126.

35. Quoted by Chadwick, p. 270.

36. See P. Grendler, *The Roman Inquisition and the Venetian Press, 1540–1605,* Princeton University Press, Princeton, N.J., 1977.

37. I Corinthians I:25, 27.

SUGGESTED READING

There are many easily accessible and lucidly written general studies of the reformations of the sixteenth century. O. Chadwick, *The Reformation* (1976); H. Hillerbrand, *Men and Ideas in the Sixteenth Century* (1969); and E. H. Harbison, *The Age of Reformation* (1963), are all good general introductions. P. Smith, *The Age of the Reformation,* rev. ed. (1951), is an older, broad, and often anecdotal treatment. L. W. Spitz, *The Protestant Reformation, 1517–1559* (1985), provides a sound and comprehensive survey, which incorporates the latest scholarly research. For the current trend in scholarship, interpreting the Reformation against the background of fifteenth-century reforming developments, see J. H. Overfield, *Humanism and Scholasticism in Late Medieval Germany* (1984), which portrays the intellectual life of the German universities, the milieu from which the Protestant Reformation emerged; H. A. Oberman, *The Harvest of Medieval Theology: Gabriel Biel and Late Medieval Nominalism* (1963); R. R. Post, *The Modern Devotion* (1968); G. Strauss, ed., *Manifestations of Discontent in Germany on the Eve of the Reformation*

(1971), a useful and exciting collection of documents; and S. Ozment, *The Age of Reform, 1250–1550: An Intellectual and Religious History of Late Medieval and Reformation Europe* (1980), which combines intellectual and social history.

For the central figure of the early reformation, Martin Luther, students should see, in addition to the titles in the Notes, J. Atkinson, *Martin Luther and the Birth of Protestantism* (1968); H. Boehmer, *Martin Luther: Road to Reformation* (1960), a well-balanced work treating Luther's formative years; and the perceptive study of H. G. Haile, *Luther: An Experiment in Biography* (1980), which focuses on the character of the mature and aging reformer. Students may expect thorough and sound treatments of Luther's theology in the following distinguished works: H. Bornkamm, *Luther in Mid-Career, 1521–1530* (1983); A. E. McGrath, *Luther's Theology of the Cross: Martin Luther on Justification* (1985); M. Brecht, *Martin Luther: His Road to Reformation* (trans. J. L. Schaaf, 1985), which includes an exploration of Luther's background and youth; and J. Pelikan, *Reformation of Church and Dogma, 1300–1700* (1986).

The best introduction to Calvin as a man and theologian is probably the balanced account of F. Wendel, *Calvin: The Origins and Development of His Thought* (trans. P. Mairet, 1963). J. T. McNeill, *History and Character of Calvinism* (1954), presents useful information, while W. E. Monter, *Calvin's Geneva* (1967), is an excellent account of Calvin's reforms on the social life of that Swiss city. R. T. Kendall, *Calvinism and English Calvinism to 1649* (1981), presents English conditions, while R. M. Mitchell, *Calvin and the Puritan's View of the Protestant Ethic* (1979), interprets the socioeconomic implications of Calvin's thought. Students interested in the left wing of the Reformation should see the profound though difficult work of G. H. Williams, *The Radical Reformers* (1962).

For various aspects of the social history of the period, see L. P. Buck and J. W. Zophy, eds., *The Social History of the Reformation* (1972), and K. von Greyerz, ed., *Religion and Society in Early Modern Europe, 1500–1800* (1984), both of which contain interesting essays on religion, society, and popular culture; S. Ozment, *The Reformation in the Cities: The Appeal of Protestantism to Sixteenth-Century Germany and Switzerland* (1975), an especially important book; and G. Strauss, *Luther's House of Learning: The Indoctrination of the Young in the German Reformation* (1978), which describes how plain people were imbued with Reformation ideals and patterns of behavior. R. L. De Molen, *Leaders of the Reformation* (1984), contains provocative portraits of several figures including Zwingli, Loyola, Cromwell, and Calvin. For women's studies in the German context, see M. Wiesner, *Women in the Sixteenth Century: A Bibliography* (1983), a useful reference tool; while S. M. Wyntjes, "Women in the Reformation Era," in *Becoming Visible: Women in European History,* ed. R. Blumenthal and C. Koonz (1977), is an interesting general survey. The best recent treatment of marriage and the family is S. Ozment, *When Fathers Ruled: Family Life in Reformation Europe* (1983). Ozment's edition of *Reformation Europe: A Guide to Research* (1982), contains not only helpful references but valuable articles on such topics as "The German Peasants," "The Anabaptists," and "The Confessional Age: The Late Reformation in Germany." For Servetus, see R. H. Bainton, *Hunted Heretic: The Life and Death of Michael Servetus* (1953), which remains valuable.

For England, in addition to the fundamental works by Dickens cited in the Notes, see for pre-Reformation popular religion K. Thomas, *Religion and the Decline of Magic* (1971), J. J. Scarisbrick, *The Reformation and the English People* (1985), and S. T. Bindoff, *Tudor England* (1959), a good short synthesis. The marital trials of Henry VIII are treated in both the sympathetic study of G. Mattingly, *Catherine of Aragon* (1949), and H. A. Kelly, *The Matrimonial Trials of Henry VIII* (1975). A persuasive treatment of Henry VIII's possible syphilis and its effects on his children is given in F. S. Cartwright,

Disease and History (1972). The legal implications of Henry VIII's divorces have been thoroughly analyzed by J. J. Scarisbrick, *Henry VIII* (1968), an almost definitive biography. On the dissolution of the English monasteries, see D. Knowles, *The Religious Orders in England,* vol. 3 (1959), one of the finest examples of historical prose in English written in the twentieth century. Knowles's *Bare Ruined Choirs* (1976) is an attractively illustrated abridgment of *Religious Orders.* G. R. Elton, *The Tudor Revolution in Government* (1959), discusses the modernization of English government under Thomas Cromwell, while the same author's *Reform and Reformation: England, 1509–1558* (1977), combines political and social history in a broad study. Many aspects of English social history are discussed in J. Youings, *Sixteenth Century England* (1984), a beautifully written work, which is highly recommended. R. Marius, *Thomas More: A Biography* (1984), provides a thorough and perceptive study of the great humanist, lord chancellor, and saint.

P. Janelle, *The Catholic Reformation* (1951), is a comprehensive treatment of the Catholic Reformation from a Catholic point of view, and A. G. Dickens, *The Counter Reformation* (1969), gives the Protestant standpoint in a beautifully illustrated book. The definitive study of the Council of Trent was written by H. Jedin, *A History of the Council of Trent,* 3 vols. (1957–1961). For the Jesuits, see M. Foss, *The Founding of the Jesuits, 1540* (1969), and W. B. Bangert, *A History of the Society of Jesus* (1972).

15

E.NOSE

THE AGE OF EUROPEAN EXPANSION AND RELIGIOUS WARS

*B*ETWEEN 1560 AND 1648 two developments dramatically altered the world in which Europeans lived: overseas expansion and the reformations of the Christian churches. Overseas expansion broadened Europeans' geographical horizons and brought them into confrontation with ancient civilizations in Africa, Asia, and the Americas. These confrontations led first to conquest, then to exploitation, and finally to profound social changes in both Europe and the conquered territories. Likewise, the Renaissance and the reformations drastically changed intellectual, political, religious, and social life in Europe. War and religious issues dominated the politics of European states. Though religion was commonly used to rationalize international conflict, wars were fought for power and territorial expansion.

Meanwhile, Europeans carried their political, religious, and social attitudes to the territories they subdued. Why, in the sixteenth and seventeenth centuries, did a relatively small number of people living on the edge of the Eurasian land mass gain control of the major sea lanes of the world and establish political and economic hegemony on distant continents? What effect did overseas expansion have on Europe and on conquered societies? What were the causes and consequences of the religious wars in France, the Netherlands, and Germany? How did the religious crises of this period affect the status of women? How and why did African slave labor become the dominant form of labor organization in the New World? What religious and intellectual developments led to the growth of skepticism? What literary masterpieces of the English-speaking world did this period produce? This chapter will address these questions.

DISCOVERY, RECONNAISSANCE, AND EXPANSION

Historians have variously called the period from 1450 to 1650 the "Age of Discovery," "Age of Reconnaissance," and "Age of Expansion." All three labels are appropriate. The "Age of Discovery" refers to the era's phenomenal advances in geographical

World Map of Vesconte Maggioli, 1511 Renaissance geographers still accepted the Greco-Egyptian Ptolemy's theory (second century A.D.) that the earth was one continuous land mass. Cartographers could not subscribe to the idea of a new, separate continent. Thus, this map, inaccurate when it was drawn, shows America as an extension of Asia. *(John Carter Brown Library, Brown University, Providence)*

knowledge and technology, often achieved through trial and error. In 1350 it took as long to sail from the eastern end of the Mediterranean to the western end as it had taken a thousand years earlier. Even in the fifteenth century, Europeans knew little more about the earth's surface than the Romans had. By 1650, however, Europeans had made an extensive reconnaissance—or preliminary exploration—and had sketched fairly accurately the physical outline of the whole earth. Much of the geographical information they had gathered was tentative and not fully understood—hence the appropriateness of the term the "Age of Reconnaissance."

The designation of the era as the "Age of Expansion" refers to the migration of Europeans to other parts of the world. This colonization resulted in political control of much of South and North America; coastal regions of Africa, India, China, and Japan; and many Pacific islands. Political hegemony was accompanied by economic exploitation, religious domination, and the introduction of European patterns of social and intellectual life. The sixteenth-century expansion of European society launched a new age in world history.

OVERSEAS EXPLORATION AND CONQUEST

The outward expansion of Europe began with the Viking voyages across the Atlantic in the ninth and tenth centuries. Under Eric the Red and Leif Ericson, the Vikings discovered Greenland and the eastern coast of North America. The Crusades of the eleventh through thirteenth centuries were another phase in Europe's attempt to explore, christianize, and exploit peoples on the periphery of the Continent. But these early thrusts outward resulted in no permanent settlements. The Vikings made only quick raids in search of booty. Lacking stable political institutions

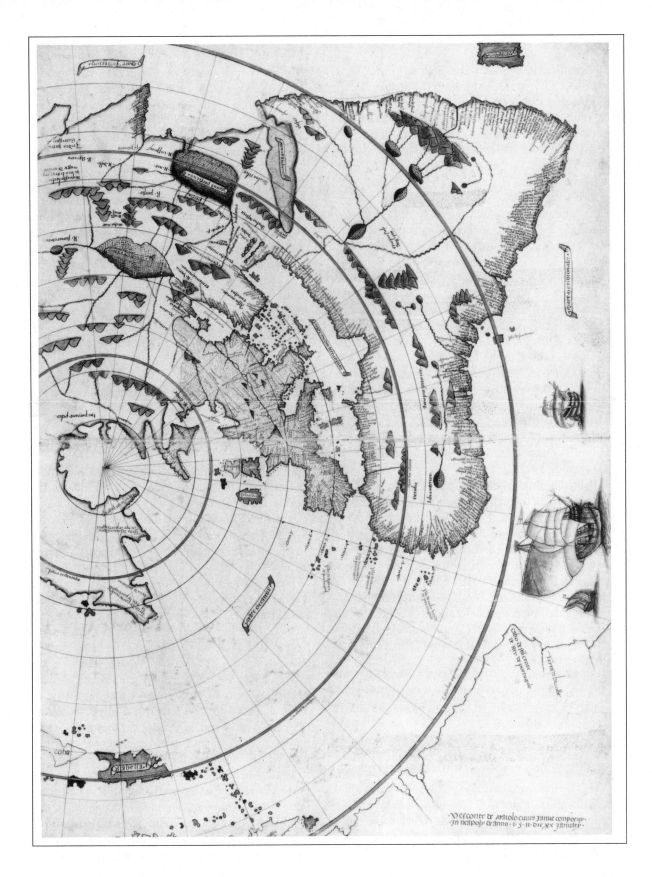

465

in Scandinavia, they had no workable forms of government to impose on distant continents. In the twelfth and thirteenth centuries, the lack of a strong territorial base, weak support from the West, and sheer misrule combined to make the medieval Crusader kingdoms short-lived. Even in the mid-fifteenth century, Europe seemed ill prepared for international ventures. By 1450 a grave new threat had appeared in the East—the Ottoman Turks.

Combining excellent military strategy with efficient administration of their conquered territories, the Turks had subdued most of Asia Minor and begun to settle on the western side of the Bosporus. The Ottoman Turks under Sultan Mohammed II (1451–1481) captured Constantinople in 1453, pressed southwest into the Balkans, and by the early sixteenth century controlled the eastern Mediterranean. The Turkish menace badly frightened Europeans. In France in the fifteenth and sixteenth centuries, twice as many books were printed about the Turkish threat as about the American discoveries. The Turks imposed a military blockade on eastern Europe, thus forcing Europeans' attention westward. Yet the fifteenth and sixteenth centuries witnessed a fantastic continuation, on a global scale, of European expansion.

Political centralization in Spain, France, and England helps to explain those countries' outward push. In the fifteenth century, Isabella and Ferdinand had consolidated their several kingdoms to achieve a more united Spain. The Catholic rulers reduced the powers of the nobility, revamped the Spanish bureaucracy, and humbled dissident elements, notably the Muslims and the Jews. The Spanish monarchy was stronger than ever before and in a position to support foreign ventures; it could bear the costs and dangers of exploration. But Portugal, situated on the extreme southwestern edge of the European continent, got the start on the rest of Europe. Still insignificant as a European land power despite its recently secured frontiers, Portugal sought greatness in the unknown world overseas.

Portugal's taking of Ceuta, an Arab city in northern Morocco, in 1415 marked the beginning of European exploration and control of overseas territory. The objectives of Portuguese policy included the historic Iberian Crusade to christianize Muslims and the search for gold, for an overseas route to the spice markets of India, and for the mythical Christian ruler of Ethiopia, Prester John.

In the early phases of Portuguese exploration, Prince Henry (1394–1460), called "the Navigator" because of the annual expeditions he sent down the western coast of Africa, played the leading role. In the fifteenth century, most of the gold that reached Europe came from the Sudan in West Africa and from Ashanti blacks living near the area of present-day Ghana. Muslim caravans brought the gold from the African cities of Niani and Timbuktu and carried it north across the Sahara to Mediterranean ports. Then the Portuguese muscled in on this commerce in gold. Prince Henry's carefully planned expeditions succeeded in reaching Guinea, and under King John II (1481–1495), the Portuguese established trading posts and forts on the Guinea coast and penetrated into the continent all the way to Timbuktu (see Map 15.1). Portuguese ships transported gold to Lisbon, and by 1500 Portugal controlled the flow of gold to Europe. The golden century of Portuguese prosperity had begun.

Still the Portuguese pushed farther south down the west coast of Africa. In 1487 Bartholomew Diaz rounded the Cape of Good Hope at the southern tip, but storms and a threatened mutiny forced him to turn back. On a second expedition (1497–1499), the Portuguese mariner Vasco da Gama reached India and returned to Lisbon loaded with samples of Indian wares. King Manuel (1495–1521) promptly dispatched thirteen ships under the command of Pedro Alvares Cabral, assisted by Diaz, to set up trading posts in India. On April 22, 1500, the coast of Brazil in South America was sighted and claimed for the crown of Portugal. Cabral then proceeded south and east around the Cape of Good Hope and reached India. Half the fleet was lost on the return voyage, but the six spice-laden vessels that dropped anchor in Lisbon harbor in July 1501 more than paid for the entire expedition. Thereafter, convoys were sent out every March. Lisbon became the entrance port for Asian goods into Europe—but not without a fight.

For centuries the Muslims had controlled the rich spice trade of the Indian Ocean, and they did not surrender it willingly. Portuguese commercial activities were accompanied by the destruction or seizure of strategic Muslim coastal forts, which later served Portugal as both trading posts and military bases. Alfonso de Albuquerque, whom the Portuguese crown appointed as governor of India (1509–1515), decided that these bases and not inland territories should control the Indian Ocean. Accordingly, his cannon

Columbus Lands on San Salvador The printed page and illustrations, such as this German woodcut, spread reports of Columbus's voyage all over Europe. According to Columbus, a group of naked Indians greeted the Spaniards' arrival. Pictures of the Indians as "primitive" and "uncivilized" instilled prejudices which centuries have not erased. *(New York Public Library)*

blasted open the ports of Calicut, Ormuz, Goa, and Malacca, the vital centers of Arab domination of south Asian trade. This bombardment laid the foundation for Portuguese imperialism in the sixteenth and seventeenth centuries: a strange way to bring Christianity to "those who were in darkness." As one scholar wrote about the opening of China to the West, "while Buddha came to China on white elephants, Christ was borne on cannon balls."[1]

In March 1493, between the first and second voyages of Vasco da Gama, Spanish ships entered Lisbon harbor bearing a triumphant Italian explorer in the service of the Spanish monarchy. Christopher Columbus (1451–1506), a Genoese mariner, had secured Spanish support for an expedition to the East. He sailed from Palos, Spain, to the Canary Islands and crossed the Atlantic to the Bahamas, landing in October 1492 on an island that he named "San Salvador" and believed to be the coast of India.

Columbus explained his motives in *Book of the First Navigation and Discovery of the Indies:*

And Your Highnesses, as Catholic Christians and Princes devoted to the Holy Christian Faith and the propagators thereof, and enemies of the sect of Mahomet and of all idolatries and heresies, resolved to send me Christopher Columbus to the said regions of India, to see the said princes and peoples and lands and [to observe] the disposition of them and of all, and the manner in which may be undertaken their conversion to our Holy Faith, and ordained that I should not go by land (the usual way) to the Orient, but by the route of the Occident, by which no one to this day knows for sure that anyone has gone.[2]

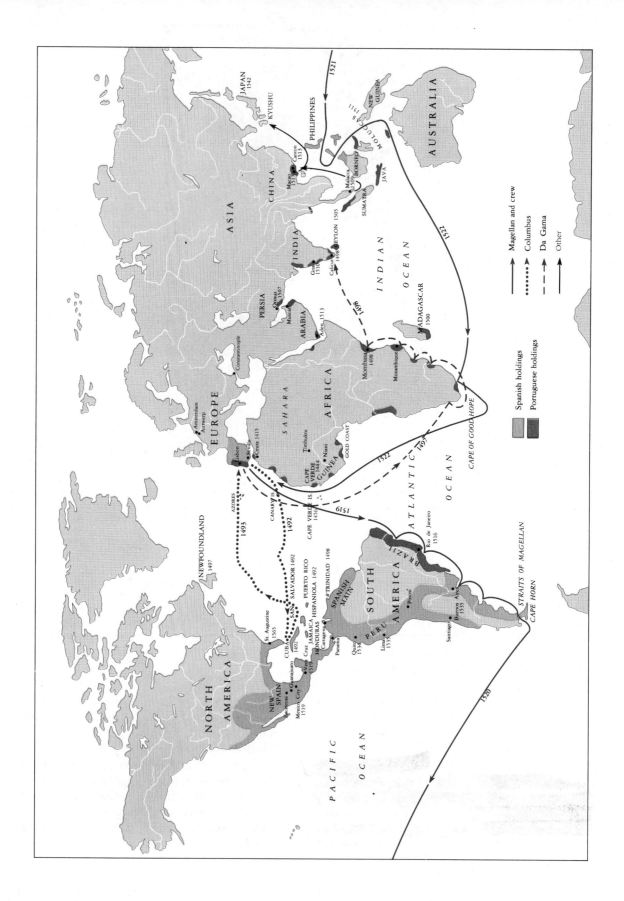

NORTH AMERICA

NEW SPAIN

Zacatecas • Guanajuato
Mexico City 1519
Vera Cruz 1519

St. Augustine 1565

CUBA 1492
JAMAICA
HONDURAS
Cartagena
Panama

SPANISH MAIN

SOUTH AMERICA

PERU

Potosí

Quito 1534
Lima 1535

Santiago

Buenos Aires 1535

BRAZIL

Rio de Janeiro 1516

STRAITS OF MAGELLAN

CAPE HORN

PACIFIC OCEAN

NEWFOUNDLAND 1497

PUERTO RICO 1492
HISPANIOLA 1492
TRINIDAD 1498

SAN SALVADOR 1492

AZORES 1493
CANARY IS. 1492
CAPE VERDE IS. 1456

1519

1520

ATLANTIC OCEAN

1522

1497

CAPE OF GOOD HOPE

EUROPE

Amsterdam
Antwerp

Lisbon
Seville
Ceuta 1415

AFRICA

SAHARA

Timbuktu
Niani
CAPE VERDE 1444
GUINEA
GOLD COAST

Constantinople

PERSIA
Ormuz 1507
ARABIA
Muscat
Aden 1513

MADAGASCAR 1500

Mombasa 1498
Mozambique

1498

ASIA

CHINA

INDIA

Goa 1510
Calicut 1498
CEYLON 1505

INDIAN OCEAN

1521

1511

JAPAN 1542
KYUSHU

PHILIPPINES

Canton 1513
Macao 1511
Malacca 1509
BORNEO
SUMATRA
JAVA
MOLUCCAS

NEW GUINEA

AUSTRALIA

Magellan and crew
Columbus
Da Gama
Other

Spanish holdings
Portuguese holdings

MAP 15.1 Overseas Exploration and Conquest. Fifteenth and Sixteenth Centuries The voyages of discovery marked another phase in the centuries-old migrations of European peoples. Consider the major contemporary significance of each of the three voyages depicted on the map.

Like most people of his day, Christopher Columbus was deeply religious. The crew of his flagship, *Santa Maria,* recited vespers every night and sang a hymn to the Virgin before going to bed. Nevertheless, the Spanish fleet, sailing westward to find the East, sought wealth as well as souls to convert.

Between 1492 and 1502, Columbus made four voyages to America, discovering all the major islands of the Caribbean—Haiti (which he called "Dominica" and the Spanish named "Hispaniola"), San Salvador, Puerto Rico, Jamaica, Cuba, Trinidad—and Honduras in Central America. Columbus believed until he died that the islands he found were off the coast of India. In fact, he had opened up for the rulers of Spain a whole new world. The Caribbean islands —the West Indies—represented to Spanish missionary zeal millions of Indian natives for conversion to Christianity. Hispaniola, Cuba, and Puerto Rico also offered gold.

Forced labor, disease, and starvation in the Spaniards' gold mines rapidly killed off the Indians of Hispaniola. When Columbus arrived in 1493, the population had been approximately 100,000; in 1570, 300 people survived. Indian slaves from the Bahamas and black Africans from Guinea were then imported to do the mining.

The search for precious metals determined the direction of Spanish exploration and expansion into South America. When it became apparent that placer mining (in which ore is separated from soil by panning) in the Caribbean islands was slow and the rewards slim, new routes to the East and new sources of gold and silver were sought.

In 1519 the Spanish ruler Charles V commissioned Ferdinand Magellan (1480–1521) to find a direct route to the Moluccan Islands off the southeast coast of Asia. Magellan sailed southwest across the Atlantic to Brazil and proceeded south around Cape Horn into the Pacific Ocean (see Map 15.1). He crossed the Pacific, sailing west, to the Malay Archipelago, which he called the "Western Isles." (These islands were conquered in the 1560s and named the "Philippines" for Philip II of Spain.)

Though Magellan was killed, the expedition continued, returning to Spain in 1522 from the east by way of the Indian Ocean, the Cape of Good Hope, and the Atlantic. Terrible storms, mutiny, starvation, and disease haunted this voyage. Nevertheless, it verified Columbus's theory that the earth was round and brought information about the vastness of the Pacific. Magellan also proved that the earth was much larger than Columbus and others had believed.

In the West Indies, the slow recovery of gold, the shortage of a healthy labor force, and sheer restlessness speeded up Spain's search for wealth. In 1519, the year Magellan departed on his worldwide expedition, a brash and determined Spanish adventurer, Hernando Cortez (1485–1547), crossed from Hispaniola to mainland Mexico with six hundred men, seventeen horses, and ten cannon. Within three years, Cortez had conquered the fabulously rich Aztec empire, taken captive the Aztec emperor Montezuma, and founded Mexico City as the capital of New Spain. The subjugation of northern Mexico took longer, but between 1531 and 1550 the Spanish gained control of Zacatecas and Guanajuato, where rich silver veins were soon tapped.

Another Spanish conquistador, Francisco Pizarro (1470–1541), repeated Cortez's feat in Peru. Between 1531 and 1536, with even fewer resources, Pizarro crushed the Inca empire in northern South America and established the Spanish viceroyalty of Peru with its center at Lima. In 1545 Pizarro opened at Potosí in the Peruvian highlands what became the richest silver mines in the New World.

Between 1525 and 1575, the riches of the Americas poured into the Spanish port of Seville and the Portuguese capital of Lisbon. For all their new wealth, however, Lisbon and Seville did not become important trading centers. It was the Flemish city of Antwerp, although controlled by the Spanish Habsburgs, that developed into the great entrepôt for overseas bullion and Portuguese spices and served as the commercial and financial capital of the entire European world.

Since the time of the great medieval fairs, cities of the Low Countries (so called because much of the land lies below sea level) had been important sites for the exchange of products from the Baltic and Italy. Antwerp, ideally situated on the Scheldt River at the intersection of many trading routes, steadily expanded as the chief intermediary for international commerce and finance. English woolens; Baltic wheat, fur, and timber; Portuguese spices; German iron and copper; Spanish fruit, French wines and dyestuffs; Italian silks, marbles, and mirrors; together with vast amounts of cash—all were exchanged at Antwerp. The city's harbor could dock 2,500 vessels at once, and 5,000 merchants from many nations gathered daily in the *bourse* (or exchange). Spanish silver was drained to the Netherlands to pay for food and luxury goods. Even so, the desire for complete economic independence from Spain was to play a major role in the Netherlands' revolt in the late sixteenth century.

By the end of the sixteenth century, Amsterdam had overtaken Antwerp as the financial capital of Europe (see page 532). The Dutch had also embarked on foreign exploration and conquest. The Dutch East India Company, founded in 1602, became the major organ of Dutch imperialism and within a few decades expelled the Portuguese from Ceylon and other East Indian islands. By 1650 the Dutch West India Company had successfully intruded on the Spanish possessions in America and gained control of much of the African and American trade.

English and French explorations lacked the immediate, sensational results of the Spanish and Portuguese. In 1497 John Cabot, a Genoese merchant living in London, sailed for Brazil but discovered Newfoundland. The next year he returned and explored the New England coast and perhaps as far south as Delaware. Since these expeditions found no spices or gold, King Henry VII lost interest in exploration. Between 1534 and 1541, the Frenchman Jacques Cartier made several voyages and explored the Saint Lawrence region of Canada, but the first permanent French settlement, at Quebec, was not founded until 1608.

THE EXPLORERS' MOTIVES

The expansion of Europe was not motivated by demographic pressures. The Black Death had caused serious population losses from which Europe had not recovered in 1500. Few Europeans emigrated to North or South America in the sixteenth century.

Half of those who did sail to begin a new life in America died en route; half of those who reached the New World eventually returned to their homeland. Why, then, did explorers brave the Atlantic and Pacific oceans, risking their lives to discover new continents and spread European culture?

The reasons are varied and complex. People of the sixteenth century were still basically medieval in the sense that their attitudes and values were shaped by religion and expressed in religious terms. In the late fifteenth century, crusading fervor remained a basic part of the Portuguese and Spanish national ideal. The desire to christianize Muslims and pagan peoples played a central role in European expansion. Queen Isabella of Spain, for example, showed a fanatical zeal for converting the Muslims to Christianity, but she concentrated her efforts on the Arabs in Granada. After the abortive crusading attempts of the thirteenth century, Isabella and other rulers realized full well that they lacked the material resources to mount the full-scale assault on Islam necessary for victory. Crusading impulses thus shifted from the Muslims to the pagan peoples of Africa and the Americas.

Moreover, after the reconquista, enterprising young men of the Spanish upper classes found economic and political opportunities severely limited. As a recent study of the Castilian city of Ciudad Real shows, the ancient aristocracy controlled the best agricultural land and monopolized urban administrative posts. Great merchants and a few nobles (surprisingly, since Spanish law forbade noble participation in commercial ventures) dominated the textile and leather glove manufacturing industries. Consequently, many ambitious men emigrated to the Americas to seek their fortunes.[3]

Government sponsorship and encouragement of exploration also help to account for the results of the various voyages. Mariners and explorers could not afford, as private individuals, the massive sums needed to explore mysterious oceans and to control remote continents. The strong financial support of Prince Henry the Navigator led to Portugal's phenomenal success in the spice trade. Even the grudging and modest assistance of Isabella and Ferdinand eventually brought untold riches—and complicated problems—to Spain. The Dutch in the seventeenth century, through such government-sponsored trading companies as the Dutch East India Company, reaped enormous wealth, and although the Nether-

Market at Cartagena Founded in 1533 as a port on the Caribbean Sea, Cartagena (modern Colombia) became the storage depot for precious metals waiting shipment to Spain. In this fanciful woodcut, male Indians wearing tunics composed of overlapping feathers and nude females sell golden necklaces, fish, fruit, and grain. *(New York Public Library)*

lands was a small country in size, it dominated the European economy in 1650. In England, by contrast, Henry VII's lack of interest in exploration delayed English expansion for a century.

Scholars have frequently described the European discoveries as a manifestation of Renaissance curiosity about the physical universe, the desire to know more about the geography and peoples of the world. There is truth to this explanation. Cosmography, natural history, and geography aroused enormous interest among educated people in the fifteenth and sixteenth centuries. Just as science fiction and speculation about life on other planets excite readers today, quasi-scientific literature about Africa, Asia, and the Americas captured the imaginations of literate Europeans. Oviedo's *General History of the Indies* (1547), a detailed eyewitness account of plants, animals, and peoples, was widely read.

Spices were another important incentive to voyages of discovery. Introduced into western Europe by the Crusaders in the twelfth century, nutmeg, mace, ginger, cinnamon, and pepper added flavor and variety to the monotonous diet of Europeans. Spices were also used in the preparation of medicinal drugs and incense for religious ceremonies. In the late thirteenth century, the Venetian Marco Polo

(1254?–1324?), the greatest of medieval travelers, had visited the court of the Chinese emperor. The widely publicized account of his travels in the *Book of Various Experiences* stimulated a rich trade in spices between Asia and Italy. The Venetians came to hold a monopoly of the spice trade in western Europe.

Spices were grown in India and China, shipped across the Indian Ocean to ports on the Persian Gulf, and then transported by Arabs across the Arabian Desert to Mediterranean ports. But the rise of the Ming dynasty in China in the late fourteenth century resulted in the expulsion of foreigners. And the steady penetration of the Ottoman Turks into the eastern Mediterranean and of hostile Muslims across North Africa forced Europeans to seek a new route to the Asian spice markets.

The basic reason for European exploration and expansion, however, was the quest for material profit. Mariners and explorers frankly admitted this. As Bartholomew Diaz put it, his motives were "to serve God and His Majesty, to give light to those who were in darkness and to grow rich as all men desire to do." When Vasco da Gama reached the port of Calicut, India, in 1498, a native asked what the Portuguese wanted. Da Gama replied, "Christians and spices."[4] The bluntest of the Spanish conquistadors, Hernando Cortez, announced as he prepared to conquer Mexico, "I have come to win gold, not to plow the fields like a peasant."[5]

Spanish and Portuguese explorers carried the fervent Catholicism and missionary zeal of the Iberian Peninsula to the New World, and once in America they urged home governments to send clerics. At bottom, however, wealth was the driving motivation. A sixteenth-century diplomat, Ogier Gheselin de Busbecq, summed up this paradoxical attitude well: in expeditions to the Indies and the Antipodes, he said, "religion supplies the pretext and gold the motive."[6] The mariners, explorers, and conquistadors were religious and "medieval" in justifying their actions, while remaining materialistic and "modern" in their behavior.

TECHNOLOGICAL STIMULI TO EXPLORATION

Technological developments were the key to Europe's remarkable outreach. By 1350 *cannon*—iron or bronze guns that fired iron or stone balls—had been fully developed in western Europe. These pieces of artillery emitted frightening noises and great flashes of fire and could batter down fortresses and even city walls. Sultan Mohammed II's siege of Constantinople in 1453 provides a classic illustration of the effectiveness of cannon fire.

Constantinople had very strong walled fortifications. The sultan secured the services of a Western technician who built fifty-six small cannon and a gigantic gun that could hurl stone balls weighing about eight hundred pounds. The gun could be moved only by several hundred oxen and loaded and fired only by about a hundred men working together. Reloading took two hours. This awkward but powerful weapon breached the walls of Constantinople before it cracked on the second day of the bombardment. Lesser cannon finished the job.

Early cannon posed serious technical difficulties. Iron cannon were cheaper than bronze to construct, but they were difficult to cast effectively and were liable to crack and injure the artillerymen. Bronze guns, made of copper and tin, were less subject than iron to corrosion, but they were very expensive. All cannon were extraordinarily difficult to move, required considerable time for reloading, and were highly inaccurate. They thus proved inefficient for land warfare. However, they could be used at sea.

The mounting of cannon on ships and improved techniques of shipbuilding gave impetus to European expansion. Since ancient times, most seagoing vessels had been narrow, open boats called *galleys,* propelled by manpower. Slaves or convicts who had been sentenced to the galleys manned the oars of the ships that sailed the Mediterranean, and both cargo and warships carried soldiers for defense. Though well suited to the placid and thoroughly explored waters of the Mediterranean, galleys could not withstand the rough winds and uncharted shoals of the Atlantic. The need for sturdier craft, as well as population losses caused by the Black Death, forced the development of a new style of ship that would not require soldiers for defense.

In the course of the fifteenth century, the Portuguese developed the *caravel,* a small, light, three-masted sailing ship. Though somewhat slower than the galley, the caravel held more cargo and was highly maneuverable. When fitted with cannon, it could dominate larger vessels, such as the round ships commonly used as merchantmen. The substitution of windpower for manpower, and artillery fire for soldiers, signaled a great technological advance and gave Europeans navigational and fighting ascendancy over the rest of the world.[7]

Other fifteenth-century developments in navigation helped make possible the conquest of the Atlantic. The magnetic compass enabled sailors to determine their direction and position at sea. The *astrolabe,* an instrument developed by Muslim navigators in the twelfth century and used to determine the altitude of the sun and other celestial bodies, permitted mariners to plot their *latitude,* or position north or south of the equator. Steadily improved maps and sea charts provided information about distance, sea depths, and general geography.

THE ECONOMIC EFFECTS OF SPAIN'S DISCOVERIES IN THE NEW WORLD

The sixteenth century has often been called the "Golden Century" of Spain. The influence of Spanish armies, Spanish Catholicism, and Spanish wealth was felt all over Europe. This greatness rested largely on the influx of precious metals from the New World.

The mines at Zacatecas and Guanajuato in Mexico and Potosí in Peru poured out huge quantities of precious metals. To protect this treasure from French and English pirates, armed convoys transported it each year to Spain. Between 1503 and 1650, 16 million kilograms of silver and 185,000 kilograms of gold entered the port of Seville. Scholars have long debated the impact of all this bullion on the economies of Spain and Europe as a whole. Spanish predominance, however, proved temporary.

In the sixteenth century, Spain experienced a steady population increase, creating a sharp rise in the demand for food and goods. Spanish colonies in the Americas also represented a demand for products. Since Spain had expelled some of the best farmers and businessmen, the Muslims and the conversos, in the fifteenth century, the Spanish economy was already suffering and could not meet the new demands. Prices rose. Because the costs of manufacturing cloth and other goods increased, Spanish products could not compete in the international market with cheaper products made elsewhere. The textile industry was badly hurt. Prices spiraled upward, faster than the government could levy taxes to dampen the economy. (Higher taxes would have cut the public's buying power; with fewer goods sold, prices would have come down.)

Did the flood of American silver bullion *cause* the inflation? Prices rose most steeply before 1565, but bullion imports reached their peak between 1580 and 1620. Thus there is no direct correlation between silver imports and the inflation rate. Did the substantial population growth accelerate the inflation rate? Perhaps, since when the population pressure declined after 1600, prices gradually stabilized. One fact is certain: the price revolution severely strained governmental budgets. Several times between 1557 and 1647, Philip II and his successors were forced to repudiate the state debt, which in turn undermined confidence in the government. By the seventeenth century, the economy was a shambles.

As Philip II paid his armies and foreign debts with silver bullion, the Spanish inflation was transmitted to the rest of Europe. Between 1560 and 1600, much of Europe experienced large price increases. Prices doubled and in some cases quadrupled. Spain suffered most severely, but all European countries were affected. People who lived on fixed incomes, such as the continental nobles, were badly hurt because their money bought less. Those who owed fixed sums of money, such as the middle class, prospered: in a time of rising prices, debts had less value each year. Food costs rose most sharply, and the poor fared worst of all.

COLONIAL ADMINISTRATION

Columbus, Cortez, and Pizarro claimed the lands they had "discovered" for the crown of Spain. How were they to be governed? According to the Spanish theory of absolutism, the crown was entitled to exercise full authority over all imperial lands. In the sixteenth century the crown divided its New World territories into four viceroyalties or administrative divisions: New Spain, which consisted of Mexico, Central America, and present-day California, Arizona, New Mexico, and Texas, with the capital at Mexico City; Peru, originally all the lands in continental South America, later reduced to the territory of modern Peru, Chile, Bolivia, and Ecuador, with the viceregal seat at Lima; New Granada, including present-day Venezuela, Colombia, Panama, and after 1739 Ecuador, with Bogotá as its administrative center; and La Plata, consisting of Argentina, Uruguay, and Paraguay, with Buenos Aires as the capital. Within each territory, the viceroy or imperial governor exercised broad military and civil authority as the direct representative of the sovereign in Madrid. The viceroy presided over the *audiencia,* a board of twelve to fifteen judges, which served as his advisory council and the highest judicial body. The enlightened Spanish king Charles III (1716–1788) introduced the system of *intendants.* These royal officials possessed broad military, administrative, and financial authority within their intendancy and were responsible, not to the viceroy, but to the monarchy in Madrid.

From the early sixteenth century to the beginning of the nineteenth, the Spanish monarchy acted on the mercantilist principle that the colonies existed for the financial benefit of the mother country. The mining of gold and silver was always the most important industry in the colonies. The crown claimed the *quinto,* one-fifth of all precious metals mined in South America. Gold and silver yielded the Spanish monarchy 25 percent of its total income. In return, it shipped manufactured goods to America and discouraged the development of native industries.

The Portuguese governed their colony of Brazil in a similar manner. After the union of the crowns of Portugal and Spain in 1580, Spanish administrative forms were introduced. Local officials called *corregidores* held judicial and military powers. Mercantilist policies placed severe restrictions on Brazilian industries that might compete with those of Portugal. In the seventeenth century the use of black slave labor made possible the cultivation of coffee and cotton, and in the eighteenth century Brazil led the world in the production of sugar. The unique feature of colonial Brazil's culture and society was its thoroughgoing intermixture of Indians, whites, and blacks.

POLITICS, RELIGION, AND WAR

In 1559 France and Spain signed the Treaty of Cateau-Cambrésis, which ended the long conflict known as the Habsburg-Valois Wars. This event marks a decisive watershed in early modern European history. Spain was the victor. France, exhausted by the struggle, had to acknowledge Spanish dominance in Italy, where much of the war had been fought. Spanish governors ruled in Sicily, Naples, and Milan, and Spanish influence was strong in the Papal States and Tuscany.

Emperor Charles V had divided his attention between the Holy Roman Empire and Spain. Under his son Philip II (1556–1598), however, the center of the Habsburg empire and the political center of gravity for all of Europe shifted westward to Spain. Before 1559, Spain and France had fought bitterly for control of Italy; after 1559, the two Catholic powers aimed their guns at Protestantism. The Treaty of Cateau-Cambrésis ended an era of strictly dynastic wars and initiated a period of conflicts in which politics and religion played the dominant roles.

Because a variety of issues were stewing, it is not easy to generalize about the wars of the late sixteenth century. Some were continuations of struggles between the centralizing goals of monarchies and the feudal reactions of nobilities. Some were crusading battles between Catholics and Protestants. Some were struggles for national independence or for international expansion.

These wars differed considerably from earlier wars. Sixteenth- and seventeenth-century armies were bigger than medieval ones; some forces numbered as many as fifty thousand men. Because large armies were expensive, governments had to reorganize their administrations to finance them. The use of gunpowder altered both the nature of war and popular attitudes toward it. Guns and cannon killed and wounded from a distance, indiscriminately. Writers scorned gunpowder as a coward's weapon that allowed a common soldier to kill a gentleman. The Italian poet Ariosto lamented:

Through thee is martial glory lost, through
Thee the trade of arms becomes a worthless art:
And at such ebb are worth and chivalry that
The base often plays the better part.[8]

Gunpowder weakened the notion, common during the Hundred Years' War, that warfare was an ennobling experience. Governments had to utilize propaganda, pulpits, and the printing press to arouse public opinion to support war.[9]

Late sixteenth-century conflicts fundamentally tested the medieval ideal of a unified Christian society governed by one political ruler, the emperor, to whom all rulers were theoretically subordinate, and one church, to which all people belonged. The Protestant Reformation had killed this ideal, but few people recognized it as dead. Catholics continued to believe that Calvinists and Lutherans could be reconverted; Protestants persisted in thinking that the Roman church should be destroyed. Most people believed that a state could survive only if its members shared the same faith. Catholics and Protestants alike feared people of the other faith living in their midst. The settlement finally achieved in 1648, known as the Peace of Westphalia, signaled the end of the medieval ideal.

THE ORIGINS OF DIFFICULTIES IN FRANCE (1515–1559)

In the first half of the sixteenth century, France continued the recovery begun under Louis XI (page 417). The population losses caused by the plague and the disorders accompanying the Hundred Years' War had created such a labor shortage that serfdom virtually disappeared. Cash rents replaced feudal rents and servile obligations. This development clearly benefited the peasantry. Meanwhile, the declining buying power of money hurt the nobility. The steadily increasing French population brought new lands under cultivation, but the division of property among sons meant that most peasant holdings were very small. Domestic and foreign trade picked up; mercantile centers such as Rouen and Lyons expanded; and in 1517 a new port city was founded at Le Havre.

The charming and cultivated Francis I (1515–1547) and his athletic, emotional son Henry II (1547–1559) governed through a small, efficient council. Great nobles held titular authority in the provinces as governors, but Paris-appointed baillis and seneschals continued to exercise actual fiscal and judicial responsibility (page 315). In 1539 Francis issued an ordinance that placed the whole of France under the jurisdiction of the royal law courts and made French the language of those courts. This act had a powerful centralizing impact. The taille, a tax on land, provided what strength the monarchy had and supported a strong standing army. Unfortunately, the tax base was too narrow for France's extravagant promotion of the arts and ambitious foreign policy.

Deliberately imitating the Italian Renaissance princes, the Valois monarchs lavished money on a magnificent court, a vast building program, and Italian artists. Francis I commissioned the Paris architect Pierre Lescot to rebuild the palace of the Louvre. Francis secured the services of Michelangelo's star pupil, Il Rosso, who decorated the wing of the Fontainebleau chateau, subsequently called the Gallery Francis I, with rich scenes of classical and mythological literature. After acquiring Leonardo da Vinci's Mona Lisa, Francis brought Leonardo himself to France, where he soon died. Henry II built a castle at Dreux for his mistress, Diana de Poitiers, and a palace in Paris, the Tuileries, for his wife, Catherine de' Medici. Art historians credit Francis I and Henry II with importing Italian Renaissance art and architecture to France. Whatever praise these monarchs deserve for their cultural achievement, they spent far more than they could afford.

The Habsburg-Valois Wars, waged intermittently through the first half of the sixteenth century, also cost more than the government could afford. Financing the war posed problems. In addition to the time-honored practices of increasing taxes and heavy borrowing, Francis I tried two new devices to raise revenue: the sale of public offices and a treaty with the papacy. The former proved to be only a temporary source of money. The offices sold tended to become hereditary within a family, and once a man bought an office he and his heirs were tax-exempt. The sale of public offices thus created a tax-exempt class called the "nobility of the robe," which held positions beyond the jurisdiction of the crown.

The treaty with the papacy was the Concordat of Bologna (page 418), in which Francis agreed to recognize the supremacy of the papacy over a universal council. In return, the French crown gained the right to appoint all French bishops and abbots. This understanding gave the monarchy a rich supplement of money and offices and a power over the church that lasted until the Revolution of 1789. The Concordat of Bologna helps to explain why France did not later become Protestant: in effect, it established Catholicism as the state religion. Because they possessed control over appointments and had a vested financial interest in Catholicism, French rulers had no need to revolt from Rome.

However, the Concordat of Bologna perpetuated disorders within the French church. Ecclesiastical offices were used primarily to pay and reward civil servants. Churchmen in France, as elsewhere, were promoted to the hierarchy not for any special spiritual qualifications but because of their services to the

Rosso and Primaticcio: The Gallery of Frances I Flat paintings alternating with rich sculpture provide a rhythm that directs the eye down the long gallery at Fontainebleau, constructed between 1530 and 1540. Francis I sought to re-create in France the elegant Renaissance lifestyle found in Italy. *(Giraudon/Art Resource)*

state. Such bishops were unlikely to work to elevate the intellectual and moral standards of the parish clergy. Few of the many priests in France devoted scrupulous attention to the needs of their parishioners. The teachings of Luther and Calvin, as the presses disseminated them, found a receptive audience.

Luther's tracts first appeared in France in 1518, and his ideas attracted some attention. After the publication of Calvin's *Institutes* in 1536, sizable numbers of French people were attracted to the "reformed religion," as Calvinism was called. Because Calvin wrote in French rather than Latin, his ideas gained wide circulation. Initially, Calvinism drew converts from among reform-minded members of the Catholic clergy, the industrious middle classes, and artisan groups. Most Calvinists lived in major cities, such as Paris, Lyons, Meaux, and Grenoble.

In spite of condemnation by the universities, government bans, and massive burnings at the stake, the numbers of Protestants grew steadily. When Henry II died in 1559, there were 40 well-organized and 2,150 mission churches in France. Perhaps one-tenth of the population had become Calvinist.

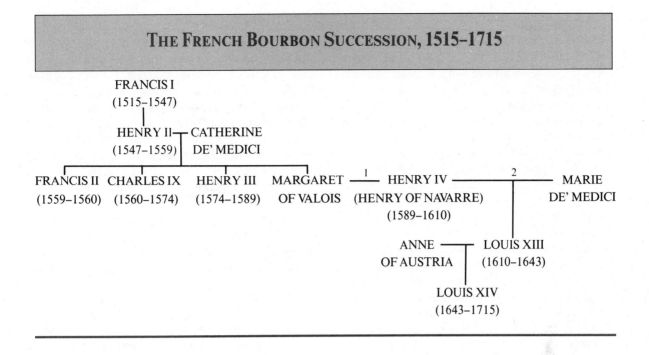

FRANCIS I
(1515–1547)

HENRY II — CATHERINE
(1547–1559)   DE' MEDICI

FRANCIS II   CHARLES IX   HENRY III   MARGARET —¹— HENRY IV —²— MARIE
(1559–1560) (1560–1574) (1574–1589) OF VALOIS   (HENRY OF NAVARRE)   DE' MEDICI
                                                    (1589–1610)

ANNE —— LOUIS XIII
OF AUSTRIA   (1610–1643)

LOUIS XIV
(1643–1715)

RELIGIOUS RIOTS AND CIVIL WAR IN FRANCE (1559–1589)

For thirty years, from 1559 to 1589, violence and civil war divided and shattered France. The feebleness of the monarchy was the seed from which the weeds of civil violence germinated The three weak sons of Henry II who occupied the throne could not provide the necessary leadership. Francis II (1559–1560) died after seventeen months. Charles IX (1560–1574) succeeded at the age of ten and was thoroughly dominated by his opportunistic mother, Catherine de' Medici, who would support any party or position to maintain her influence. The intelligent and cultivated Henry III (1574–1589) divided his attention between debaucheries with his male lovers and frantic acts of repentance.

The French nobility took advantage of this monarchial weakness. In the second half of the sixteenth century, between two-fifths and one-half of the nobility at one time or another became Calvinist. Just as German princes in the Holy Roman Empire had adopted Lutheranism as a means of opposition to the emperor Charles V, so French nobles frequently adopted the "reformed religion" as a religious cloak for their independence. No one believed that peoples of different faiths could coexist peacefully within the same territory. The Reformation thus led to a resurgence of feudal disorder. Armed clashes between Catholic royalist lords and Calvinist antimonarchial lords occurred in many parts of France.

Among the upper classes the Catholic-Calvinist conflict was the surface issue, but the fundamental object of the struggle was power. Working-class crowds composed of skilled craftsmen and the poor wreaked terrible violence on people and property. Both Calvinists and Catholics believed that the others' books, services, and ministers polluted the community. Preachers incited violence, and ceremonies like baptisms, marriages, and funerals triggered it. Protestant pastors encouraged their followers to destroy statues and liturgical objects in Catholic churches. Catholic priests urged their flocks to shed the blood of the Calvinist heretics.

In 1561 in the Paris church of Saint-Médard, a Protestant crowd cornered a baker guarding a box containing the consecrated Eucharistic bread. Taunting "Does your God of paste protect you now from the pains of death?"[10] the mob proceeded to kill the poor man. Calvinists believed that the Catholic emphasis on symbols in their ritual desecrated what was truly sacred and promoted the worship of images. In scores of attacks on Catholic churches, religious statues were knocked down, stained-glass windows smashed, and sacred vestments, vessels, and Eucharistic elements defiled. In 1561 a Catholic crowd charged a group of just-released Protestant prisoners, killed them, and burned their bodies in the street. Hundreds of Huguenots, as French Calvinists were called, were tortured, had their tongues or throats slit, were maimed or murdered.

In the fourteenth and fifteenth centuries, crowd action—attacks on great nobles and rich prelates—had expressed economic grievances. Religious rioters of the sixteenth century believed that they could assume the power of public magistrates and rid the community of corruption. Municipal officials criticized the crowds' actions, but the participation of pastors and priests in these riots lent them some legitimacy.[11]

A savage Catholic attack on Calvinists in Paris on August 24, 1572 (Saint Bartholomew's Day), followed the usual pattern. The occasion was a religious ceremony, the marriage of the king's sister Margaret of Valois to the Protestant Henry of Navarre, which was intended to help reconcile Catholics and Huguenots. Among the many Calvinists present for the wedding festivities was the admiral of Coligny, head of one of the great noble families of France and leader of the Huguenot party. Coligny had recently replaced Catherine in influence over the young king Charles IX. When, the night before the wedding, the leader of the Catholic aristocracy, Henry of Guise, had Coligny murdered, rioting and slaughter followed. The Huguenot gentry in Paris were massacred, and religious violence spread to the provinces. Between August 25 and October 3, perhaps twelve thousand Huguenots perished at Meaux, Lyons, Orléans, and Paris. The contradictory orders of the unstable Charles IX worsened the situation.

The Saint Bartholomew's Day massacre launched the War of the Three Henrys, a civil conflict among factions led by the Catholic Henry of Guise, the Protestant Henry of Navarre, and King Henry III, who succeeded the tubercular Charles IX. Though he remained Catholic, King Henry realized that the Catholic Guise group represented his greatest danger. The Guises wanted, through an alliance of Catholic nobles called the "Holy League," not only to destroy Calvinism but to replace Henry III with a member of the Guise family. France suffered fifteen more years of religious rioting and domestic anarchy. Agriculture in many areas was destroyed; commercial life declined severely; starvation and death haunted the land.

What ultimately saved France was a small group of Catholic moderates called *politiques* who believed that only the restoration of strong monarchy could reverse the trend toward collapse. No religious creed was worth the incessant disorder and destruction. Therefore the politiques supported religious toleration. The death of Catherine de' Medici, followed by the assassinations of Henry of Guise and King Henry III, paved the way for the accession of Henry of Navarre, a politique who became Henry IV (1589–1610).

This glamorous prince, "who knew how to fight, to make love, and to drink," as a contemporary remarked, wanted above all a strong and united France. He knew, too, that the majority of the French were Roman Catholics. Declaring "Paris is worth a Mass," Henry knelt before the archbishop of Bourges and was received into the Roman Catholic church. Henry's willingness to sacrifice religious principles to political necessity saved France. The Edict of Nantes, which Henry published in 1598, granted to Huguenots liberty of conscience and liberty of public worship in two hundred fortified towns, such as La Rochelle. The reign of Henry IV and the Edict of Nantes prepared the way for French absolutism in the seventeenth century by helping to restore internal peace in France.

THE NETHERLANDS UNDER CHARLES V

In the last quarter of the sixteenth century, the political stability of England, the international prestige of Spain, and the moral influence of the Roman papacy all became mixed up with the religious crisis in the Low Countries. The Netherlands was the pivot around which European money, diplomacy, and war revolved. What began as a movement for the reformation of the church developed into a struggle for Dutch independence.

The emperor Charles V (1519–1556) had inherited the seventeen provinces that compose present-day Belgium and Holland (pages 439–440). Ideally situated for commerce between the Rhine and Scheldt rivers, the great towns of Bruges, Ghent, Brussels, Arras, and Amsterdam made their living by trade and industry. The French-speaking southern towns produced fine linens and woolens, while the wealth of the Dutch-speaking northern cities rested on fishing, shipping, and international banking. The city of Antwerp was the largest port and the greatest money market in Europe. In the cities of the Low Countries, trade and commerce had produced a vibrant cosmopolitan atmosphere, which was well personified by the urbane Erasmus of Rotterdam.

Each of the seventeen provinces of the Netherlands possessed historical liberties: each was self-governing and enjoyed the right to make its own laws and collect its own taxes. Only the recognition of a common ruler in the person of the emperor Charles V united the provinces. Delegates from each province met together in the Estates General, but important decisions had to be referred back to each province for approval. In the middle of the sixteenth century, the provinces of the Netherlands had a limited sense of federation.

In the Low Countries, as elsewhere, corruption in the Roman church and the critical spirit of the Renaissance provoked pressure for reform. Lutheran tracts and Dutch translations of the Bible flooded the seventeen provinces in the 1520s and 1530s, attracting many people to Protestantism. Charles V's government responded with condemnation and mild repression. This policy was not particularly effective, however, because ideas circulated freely in the cosmopolitan atmosphere of the commercial centers. But Charles's Flemish loyalty checked the spread of Lutheranism. Charles had been born in Ghent and raised in the Netherlands; he was Flemish in language and culture. He identified with the Flemish and they with him.

In 1556, however, Charles V abdicated, dividing his territories between his brother Ferdinand, who received Austria and the Holy Roman Empire, and his son Philip, who inherited Spain, the Low Countries, Milan and the kingdom of Sicily, and the Spanish possessions in America. Charles delivered his abdication speech before the Estates General at Brussels. The emperor was then fifty-five years old, white-haired, and so crippled in the legs that he had to lean for support on the young Prince William of Orange. According to one contemporary account of the emperor's appearance:

His under lip, a Burgundian inheritance, as faithfully transmitted as the duchy and county, was heavy and hanging, the lower jaw protruding so far beyond the upper that it was impossible for him to bring together the few fragments of teeth which still remained, or to speak a whole sentence in an intelligible voice.[12]

Charles spoke in Flemish. His small, shy, and sepulchral son Philip responded in Spanish; he could speak neither French nor Flemish. The Netherlanders had always felt Charles one of themselves. They were never to forget that Philip was a Spaniard.

THE REVOLT OF THE NETHERLANDS (1566–1587)

By the 1560s, there was a strong, militant minority of Calvinists in most of the cities of the Netherlands. The seventeen provinces possessed a large middle-class population, and the "reformed religion," as a contemporary remarked, had a powerful appeal "to those who had grown rich by trade and were therefore ready for revolution."[13] Calvinism appealed to the middle classes because of its intellectual seriousness, moral gravity, and emphasis on any form of labor well done. It took deep root among the merchants and financiers in Amsterdam and the northern provinces. Working-class people were also converted, partly because their employers would hire only fellow Calvinists. Well organized and with the backing of rich merchants, Calvinists quickly gained a wide following. Lutherans taught respect for the powers that be. The "reformed religion," however, tended to encourage opposition to "illegal" civil authorities.

In 1559 Philip II appointed his half-sister Margaret as regent of the Netherlands (1559–1567). A proud, energetic, and strong-willed woman, who once had Ignatius Loyola as her confessor, Margaret pushed Philip's orders to wipe out Protestantism. She introduced the Inquisition. Her more immediate problem, however, was revenue to finance the government of the provinces. Charles V had steadily increased taxes in the Low Countries. When Margaret appealed to the Estates General, they claimed that the Low Countries were more heavily taxed than Spain. Nevertheless, Margaret raised taxes. In so doing, she quickly succeeded in uniting the opposition to the government's fiscal policy with the opposition to official repression of Calvinism.

In August of 1566, a year of very high grain prices, fanatical Calvinists, primarily of the poorest classes, embarked on a rampage of frightful destruction. As in France, Calvinist destruction in the Low Countries was incited by popular preaching, and attacks were aimed at religious images as symbols of false doctrines, not at people. The Cathedral of Notre Dame at

To Purify the Church The destruction of pictures and statues representing biblical events, Christian doctrine, or sacred figures was a central feature of the Protestant Reformation. Here Dutch Protestant soldiers destroy what they consider idols in the belief that they are purifying the church. *(Fotomas Index)*

Antwerp was the first target. Begun in 1124 and finished only in 1518, this church stood as a monument to the commercial prosperity of Flanders, the piety of the business classes, and the artistic genius of centuries. On six successive summer evenings, crowds swept through the nave. While the town harlots held tapers to the greatest concentration of art works in northern Europe, people armed with axes and sledge-hammers smashed altars, statues, paintings, books, tombs, ecclesiastical vestments, missals, manuscripts, ornaments, stained-glass windows, and sculptures. Before the havoc was over, thirty more churches had been sacked and irreplaceable libraries burned. From Antwerp the destruction spread to

Brussels and Ghent and north to the provinces of Holland and Zeeland.

From Madrid, Philip II sent twenty thousand Spanish troops under the duke of Alva to pacify the Low Countries. Alva interpreted "pacification" to mean the ruthless extermination of religious and political dissidents. On top of the Inquisition he opened his own tribunal, soon called the Council of Blood. On March 3, 1568, fifteen hundred men were executed. Even Margaret was sickened and resigned her regency. Alva resolved the financial crisis by levying a 10 percent sales tax on every transaction, which in a commercial society caused widespread hardship and confusion.

For ten years, between 1568 and 1578, civil war raged in the Netherlands between Catholics and Protestants and between the seventeen provinces and Spain. A series of Spanish generals could not halt the fighting. In 1576 the seventeen provinces united under the leadership of Prince William of Orange, called "the Silent" because of his remarkable discretion. In 1578 Philip II sent his nephew Alexander Farnese, duke of Parma, to crush the revolt once and for all. A general with a superb sense of timing, an excellent knowledge of the geography of the Low Countries, and a perfect plan, Farnese arrived with an army of German mercenaries. Avoiding pitched battles, he fought by patient sieges. One by one the cities of the south fell—Maastricht, Tournai, Bruges, Ghent, and finally the financial capital of northern Europe, Antwerp. Calvinism was forbidden in these territories, and Protestants were compelled to convert or leave. The collapse of Antwerp marked the farthest extent of Spanish jurisdiction and ultimately the religious division of the Netherlands.

The ten southern provinces, the Spanish Netherlands (the future Belgium), remained under the control of the Spanish Habsburgs. The seven northern provinces, led by Holland, formed the Union of Utrecht and in 1581 declared their independence from Spain. Thus was born the United Provinces of the Netherlands (see Map 15.2).

Geography and sociopolitical structure differentiated the two countries. The northern provinces were ribboned with sluices and canals and therefore were highly defensible. Several times the Dutch had broken the dikes and flooded the countryside to halt the advancing Farnese. In the Southern provinces the Ardennes Mountains interrupt the otherwise flat terrain. In the north the commercial aristocracy possessed the predominant power; in the south the landed nobility had the greater influence. The north was Protestant; the south remained Catholic.

Philip II and Alexander Farnese did not accept this geographical division, and the struggle continued after 1581. The United Provinces repeatedly begged the Protestant Queen Elizabeth of England for assistance.

The crown on the head of Elizabeth I (pages 451–452) did not rest easily. She had steered a moderately Protestant course between the Puritans, who sought the total elimination of Roman Catholic elements in the English church, and the Roman Catholics, who wanted full restoration of the old religion. Elizabeth

MAP 15.2 The Netherlands, 1578–1609 Though small in geographical size, the Netherlands held a strategic position in the religious struggles of the sixteenth century. Why?

survived a massive uprising by the Catholic north in 1569 to 1570. She survived two serious plots against her life. In the 1570s the presence in England of Mary, Queen of Scots, a Roman Catholic and the legal heir to the English throne, produced a very embarrassing situation. Mary was the rallying point of all opposition to Elizabeth, yet the English sovereign hesitated to set the terrible example of regicide by ordering Mary executed.

Elizabeth faced a grave dilemma. If she responded favorably to Dutch pleas for military support against the Spanish, she would antagonize Philip II. The Spanish king had the steady flow of silver from the Americas at his disposal, and Elizabeth, lacking such treasure, wanted to avoid war. But if she did not help the Protestant Netherlands and they were crushed by Farnese, the likelihood was that the Spanish would invade England.

Three developments forced Elizabeth's hand. First, the wars in the Low Countries—the chief market for English woolens—badly hurt the English economy. When wool was not exported, the crown lost valuable customs revenues. Second, the murder of William the Silent in July 1584 eliminated not only a great Protestant leader but the chief military check on the Farnese advance. Third, the collapse of Antwerp appeared to signal a Catholic sweep through the Netherlands. The next step, the English feared, would be a Spanish invasion of their island. For these reasons, Elizabeth pumped £250,000 and two thousand troops into the Protestant cause in the Low Countries between 1585 and 1587. Increasingly fearful of the plots of Mary, Queen of Scots, Elizabeth finally signed her death warrant. Mary was beheaded on February 18, 1587. Sometime between March 24 and 30, the news of Mary's death reached Philip II.

Philip II and the Spanish Armada

Philip pondered the Dutch and English developments at the Escorial northwest of Madrid. Begun in 1563 and completed under the king's personal supervision in 1584, the Monastery of Saint Lawrence of the Escorial served as a monastery for Jeromite monks, a tomb for the king's Habsburg ancestors, and a royal palace for Philip and his family. The vast buildings resemble a gridiron, the instrument on which Saint Lawrence (d. 258) had supposedly been roasted alive. The royal apartments were in the center of the Italian Renaissance building complex. King Philip's tiny bedchamber possessed a concealed sliding window that opened directly onto the high altar of the monastery church so he could watch the services and pray along with the monks. In this somber atmosphere, surrounded by a community of monks and close to the bones of his ancestors, the Catholic ruler of Spain and much of the globe passed his days.

Philip of Spain considered himself the international defender of Catholicism and the heir to the medieval imperial power. Hoping to keep England within the Catholic church when his wife Mary Tudor died, Philip had asked Elizabeth to marry him; she had emphatically refused. Several popes had urged him to move against England. When Pope Sixtus V (1585–1590) heard of the death of the queen of Scots, he promised to pay Philip one million gold ducats the moment Spanish troops landed in Eng-

land. Alexander Farnese had repeatedly warned that to subdue the Dutch, he would have to conquer England and cut off the source of Dutch support. Philip worried that the vast amounts of South American silver he was pouring into the conquest of the Netherlands seemed to be going down a bottomless pit. Two plans for an expedition were considered. Philip's naval adviser recommended that a fleet of 150 ships sail from Lisbon, attack the English navy in the Channel, and invade England. Another proposal had been to assemble a collection of barges and troops in Flanders to stage a cross-Channel assault. With the expected support of English Catholics, Spain would achieve a great victory. Parma opposed this plan as militarily unsound.

Philip compromised. He prepared a vast armada to sail from Lisbon to Flanders, fight off Elizabeth's navy *if* it attacked, rendezvous with Farnese, and escort his barges across the English Channel. The expedition's purpose was to transport the Flemish army.

On May 9, 1588, *la felicissima armada*—"the most fortunate fleet," as it was ironically called in official documents—sailed from Lisbon harbor on the last medieval crusade. The Spanish fleet of 130 vessels carried 123,790 cannon balls and perhaps 30,000 men, every one of whom had confessed his sins and received the Eucharist. An English fleet of about 150 ships met the Spanish in the Channel. It was composed of smaller, faster, more maneuverable ships, many of which had greater firing power. A combination of storms and squalls, spoiled food and rank water, inadequate Spanish ammunition, and to a lesser extent, English fire ships that caused the Spanish to panic and scatter, gave England the victory. Many Spanish ships went to the bottom of the ocean; perhaps 65 managed to crawl home by way of the North Sea.

The battle in the Channel has frequently been described as one of the decisive battles in the history of the world. In fact, it had mixed consequences. Spain soon rebuilt its navy, and after 1588 the quality of the Spanish fleet improved. The destruction of the Armada did not halt the flow of silver from the New World. More silver reached Spain between 1588 and 1603 than in any other fifteen-year period. The war between England and Spain dragged on for years.

The defeat of the Spanish Armada was decisive, however, in the sense that it prevented Philip II from reimposing unity on western Europe by force. He did

Defeat of the Spanish Armada The crescent-shaped Spanish formation was designed to force the English to fight at close quarters—by ramming and boarding. When the English sent burning ships against the Spaniards, the crescent broke up, the English pounced on individual ships, and an Atlantic gale swept the Spaniards into the North Sea, finishing the work of destruction. *(National Maritime Museum, London)*

not conquer England, and Elizabeth continued her financial and military support of the Dutch. In the Netherlands, however, neither side gained significant territory. The borders of 1581 tended to become permanent. In 1609 Philip III of Spain (1598–1621) agreed to a truce, in effect recognizing the independence of the United Provinces. In seventeenth-century Spain memory of the defeat of the Armada contributed to a spirit of defeatism.

THE THIRTY YEARS' WAR
(1618–1648)

While Philip II dreamed of building a second armada and Henry IV began the reconstruction of France, the political-religious situation in central Europe deteriorated. An uneasy truce had prevailed in the Holy Roman Empire since the Peace of Augsburg of 1555

(page 443). The Augsburg settlement, in recognizing the independent power of the German princes, had destroyed the authority of the central government. The Habsburg ruler in Vienna enjoyed the title of emperor but had no power.

According to the Augsburg settlement, the faith of the prince determined the religion of his subjects. Later in the century, though, Catholics grew alarmed because Lutherans, in violation of the Peace of Augsburg, were steadily acquiring north German bishoprics. The spread of Calvinism further confused the issue. The Augsburg settlement had pertained only to Lutheranism and Catholicism, but Calvinists ignored it and converted several princes. Lutherans feared that the Augsburg principles would be totally undermined by Catholic and Calvinist gains. Also, the militantly active Jesuits had reconverted several Lutheran princes to Catholicism. In an increasingly

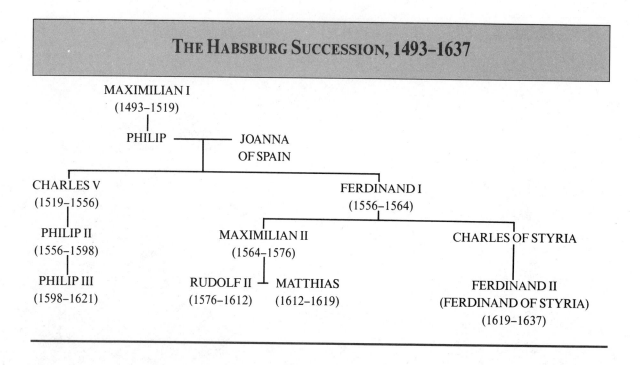

THE HABSBURG SUCCESSION, 1493–1637

MAXIMILIAN I
(1493–1519)

PHILIP —— JOANNA
OF SPAIN

CHARLES V
(1519–1556)

FERDINAND I
(1556–1564)

PHILIP II
(1556–1598)

MAXIMILIAN II
(1564–1576)

CHARLES OF STYRIA

PHILIP III
(1598–1621)

RUDOLF II
(1576–1612)

MATTHIAS
(1612–1619)

FERDINAND II
(FERDINAND OF STYRIA)
(1619–1637)

tense situation, Lutheran princes formed the Protestant Union (1608) and Catholics retaliated with the Catholic League (1609). Each alliance was determined that the other should make no religious (that is, territorial) advance. The empire was composed of two armed camps.

Dynastic interests were also involved in the German situation. When Charles V abdicated in 1556, he had divided his possessions between his son Philip II and his brother Ferdinand I. This partition began the Austrian and Spanish branches of the Habsburg family. Ferdinand inherited the imperial title and the Habsburg lands in central Europe, including Austria. Ferdinand's grandson Matthias had no direct heirs and promoted the candidacy of his fiercely Catholic cousin, Ferdinand of Styria. The Spanish Habsburgs strongly supported the goals of their Austrian relatives: the unity of the empire and the preservation of Catholicism within it.

In 1617 Ferdinand of Styria secured election as king of Bohemia, a title that gave him jurisdiction over Silesia and Moravia as well as Bohemia. The Bohemians were Czech and German in nationality, and Lutheran, Calvinist, Catholic, and Hussite in religion; all these faiths enjoyed a fair degree of religious freedom. When Ferdinand proceeded to close some Protestant churches, the heavily Protestant Estates of

Bohemia protested. On May 23, 1618, Protestants hurled two of Ferdinand's officials from a castle window in Prague. They fell seventy feet but survived: Catholics claimed that angels had caught them; Protestants said the officials fell on a heap of soft horse manure. Called the "defenestration of Prague," this event marked the beginning of the Thirty Years' War (1618–1648).

Historians traditionally divide the war into four phases. The first, or Bohemian, phase (1618–1625) was characterized by civil war in Bohemia between the Catholic League, led by Ferdinand, and the Protestant Union, headed by Prince Frederick of the Palatinate. The Bohemians fought for religious liberty and independence from Habsburg rule. In 1618 the Bohemian Estates deposed Ferdinand and gave the crown of Bohemia to Frederick, thus uniting the interests of German Protestants with those of the international enemies of the Habsburgs. Frederick wore his crown only a few months. In 1620 he was totally defeated by Catholic forces at the battle of the White Mountain. Ferdinand, who had recently been elected Holy Roman emperor as Ferdinand II, followed up his victories by wiping out Protestantism in Bohemia through forcible conversions and the activities of militant Jesuit missionaries. Within ten years, Bohemia was completely Catholic.

Gustavus Adolphus at Breitenfeld (1631) This dramatic engraving shows the Swedish king (second from right on the left) directing the battle, probably the most decisive of the war. He followed up his victory with an invasion of Catholic regions of south-central Germany. *(BBC Hulton/The Bettmann Archive)*

The second, or Danish, phase of the war (1625–1629)—so called because of the participation of King Christian IV of Denmark (1588–1648), the ineffective leader of the Protestant cause—witnessed additional Catholic victories. The Catholic imperial army led by Albert of Wallenstein scored smashing victories. It swept through Silesia, north through Schleswig and Jutland to the Baltic, and east into Pomerania. Wallenstein had made himself indispensable to the emperor Ferdinand, but he was an unscrupulous opportunist who used his vast riches to build an army loyal only to himself. The general seemed interested more in carving out an empire for himself than in aiding the Catholic cause. He quarreled with the League, and soon the Catholic forces were divided. Religion was eclipsed as a basic issue of the war.

The year 1629 marked the peak of Habsburg power. The Jesuits persuaded the emperor to issue the Edict of Restitution, whereby all Catholic properties lost to Protestantism since 1552 were to be restored and only Catholics and Lutherans (*not* Calvinists, Hussites, or other sects) were to be allowed to practice their faiths. Ferdinand appeared to be embarked on a policy to unify the empire. When Wallenstein began ruthless enforcement of the edict, Protestants throughout Europe feared collapse of the balance of power in north central Europe.

The third, or Swedish, phase of the war (1630–1635) began with the arrival in Germany of the Swedish king Gustavus Adolphus (1594–1632). The ablest administrator of his day and a devout Lutheran, Gustavus Adolphus intervened to support

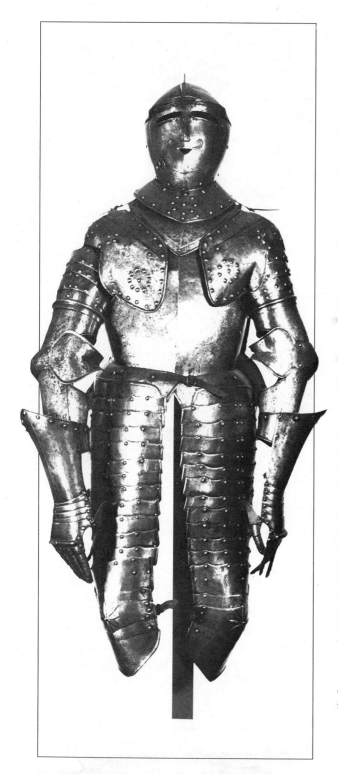

Seventeenth-Century Battle Armor Armor remained a symbol of the noble's high social status and military profession, though armor gave much less protection after the invention of gun powder. The maker had a sense of humor. *(Photo: Caroline Buckler)*

the oppressed Protestants within the empire and to assist his relatives, the exiled dukes of Mecklenburg. Cardinal Richelieu, the chief minister of King Louis XIII of France (1610–1643), subsidized the Swedes, hoping to weaken Habsburg power in Europe. In 1631, with a small but well-disciplined army equipped with superior muskets and warm uniforms, Gustavus Adolphus won a brilliant victory at Breitenfeld. Again in 1632 he was victorious at Lützen, though he was fatally wounded in the battle.

The participation of the Swedes in the Thirty Years' War proved decisive for the future of Protestantism and later German history. When Gustavus Adolphus landed on German soil, he had already brought Denmark, Poland, Finland, and the smaller Baltic states under Swedish influence. The Swedish victories ended the Habsburg ambition of uniting all the German states under imperial authority.

The death of Gustavus Adolphus, followed by the defeat of the Swedes at the Battle of Nördlingen in 1634, prompted the French to enter the war on the side of the Protestants. Thus began the French, or international, phase of the Thirty Years' War (1635–1648). For almost a century, French foreign policy had been based on opposition to the Habsburgs, because a weak empire divided into scores of independent principalities enhanced France's international stature. In 1622, when the Dutch had resumed the war against Spain, the French had supported Holland. Now, in 1635, Cardinal Richelieu declared war on Spain and again sent financial and military assistance to the Swedes and the German Protestant princes. The war dragged on. French, Dutch, and Swedes, supported by Scots, Finns, and German mercenaries, burned, looted, and destroyed German agriculture and commerce. The Thirty Years' War lasted so long because neither side had the resources to win a quick, decisive victory. Finally, in October 1648, peace was achieved.

The treaties signed at Münster and Osnabrück, commonly called the "Peace of Westphalia," mark a turning point in European political, religious, and

MAP 15.3 Europe in 1648 Which country emerged from the Thirty Years' War as the strongest European power? What dynastic house was that country's major rival in the early modern period?

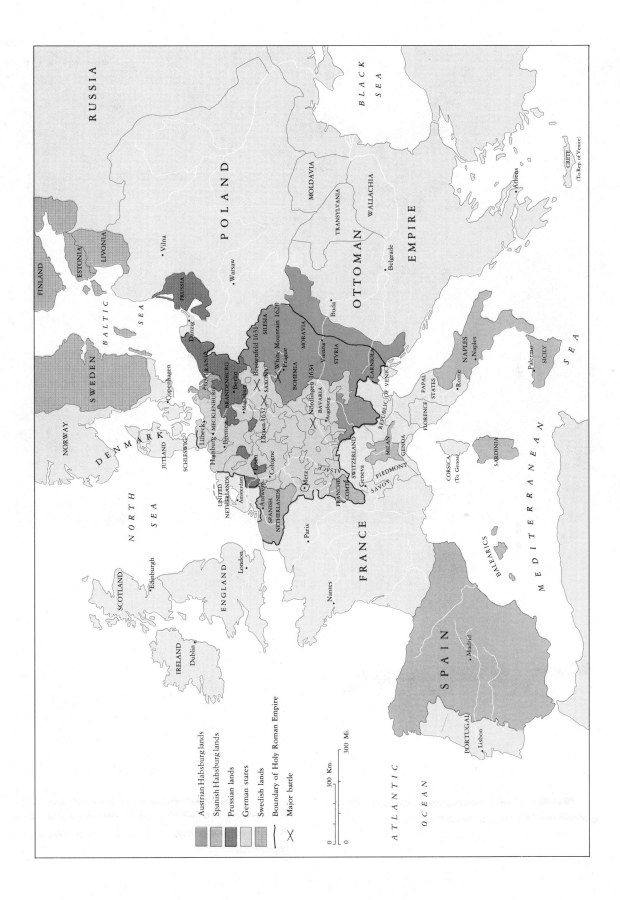

Soldiers Pillage a Farmhouse Billeting troops on civilian populations caused untold hardships. In this late seventeenth-century Dutch illustration, brawling soldiers take over a peasant's home, eat his food, steal his possessions, and insult his family. Peasant retaliation sometimes proved swift and bloody. *(Rijksmuseum, Amsterdam)*

social history. The treaties recognized the sovereign, independent authority of the German princes. Each ruler could govern his particular territory and make war and peace as well. With power in the hands of more than three hundred princes, with no central government, courts, or means of controlling unruly rulers, the Holy Roman Empire as a real state was effectively destroyed (see Map 15.3).

The independence of the United Provinces of the Netherlands was acknowledged. The international stature of France and Sweden was also greatly improved. The political divisions within the empire, the

weak German frontiers, and the acquisition of the province of Alsace increased France's size and prestige. The treaties allowed France to intervene at will in German affairs. Sweden received a large cash indemnity and jurisdiction over German territories along the Baltic Sea. The powerful Swedish presence in northeastern Germany subsequently posed a major threat to the future kingdom of Brandenburg-Prussia. The treaties also denied the papacy the right to participate in German religious affairs—a restriction symbolizing the reduced role of the church in European politics.

In religion, the Westphalian treaties stipulated that the Augsburg agreement of 1555 should stand permanently. The sole modification was that Calvinism, along with Catholicism and Lutheranism, would become a legally permissible creed. In practice, the north German states remained Protestant, the south German states Catholic. The war settled little. Both sides had wanted peace, and with remarkable illogic they fought for thirty years to get it.

GERMANY AFTER THE THIRTY YEARS' WAR

The Thirty Years' War was a disaster for the German economy and society, probably the most destructive event in German history before the twentieth century. Population losses were frightful. Perhaps one-third of the urban residents and two-fifths of the inhabitants of rural areas died. Entire areas of Germany were depopulated, partly by military actions, partly by disease—typhus, dysentery, bubonic plague, and syphilis accompanied the movements of armies—and partly by the thousands of refugees who fled to safer areas.

In the late sixteenth and early seventeenth centuries, all Europe experienced an economic crisis primarily caused by the influx of silver from South America. Because the Thirty Years' War was fought on German soil, these economic difficulties were badly aggravated in the empire. Scholars still cannot estimate the value of losses in agricultural land and livestock, in trade and commerce. The trade of southern cities like Augsburg, already hard hit by the shift in transportation routes from the Mediterranean to the Atlantic, was virtually destroyed by the fighting in the south. Meanwhile, towns like Lübeck, Hamburg, and Bremen in the north and Essen in the Ruhr area actually prospered because of the many refugees they attracted. The destruction of land and foodstuffs, compounded by the flood of Spanish silver, brought on a severe price rise. During and after the war, inflation was worse in Germany than anywhere else in Europe.

Agricultural areas suffered catastrophically. The population decline caused a rise in the value of the labor, and owners of great estates had to pay more for agricultural workers. Farmers who needed only small amounts of capital to restore their lands started over again. Many small farmers, however, lacked the revenue to rework their holdings and had to become day laborers. Nobles and landlords bought up many small holdings and acquired great estates. In some parts of Germany, especially east of the Elbe in areas like Mecklenburg and Pomerania, peasants' loss of land led to the rise of a new serfdom.[14] Thus the Thirty Years' War contributed to the legal and economic decline of the largest segment of German society.

CHANGING ATTITUDES

The age of religious wars revealed extreme and violent contrasts. It was a deeply religious period in which men fought passionately for their beliefs; 70 percent of the books printed dealt with religious subjects. Yet the times saw the beginnings of religious skepticism. Europeans explored new continents, partly with the missionary aim of christianizing the peoples they encountered. Yet the Spanish, Portuguese, Dutch, and English proceeded to enslave the Indians and blacks they encountered. While Europeans indulged in gross sensuality, the social status of women declined. The exploration of new continents reflects deep curiosity and broad intelligence, yet Europeans irrationally believed in witches and burned thousands at the stake. Sexism, racism, and skepticism had all originated in ancient times. But late in the sixteenth century they began to take on their familiar modern forms.

THE STATUS OF WOMEN

Do new ideas about women appear in this period? Theological and popular literature on marriage in Reformation Europe helps to answer this question. These manuals emphasize the qualities expected of each partner. A husband was obliged to provide for the material welfare of his wife and children. He should protect his family while remaining steady and self-controlled. Especially was a husband and father to rule his household, firmly but justly. But he was not to behave like a tyrant, a guideline counselors repeated frequently. A wife should be mature, a good household manager, and subservient and faithful to her spouse. The husband also owed fidelity, and both Protestant and Catholic moralists rejected the double standard of sexual morality as a threat to family unity. Counselors believed that marriage should

490

Veronese: Mars and Venus United by Love Taking a theme from classical mythology, the Venetian painter Veronese celebrates in clothing, architecture, and landscape the luxurious wealth of the aristocracy (painted ca 1580). The lush and curvaceous Venus and the muscular and powerfully built Mars suggest the anticipated pleasures of sexual activity and the frank sensuality of the age. *(Metropolitan Museum of Art, New York, John Stewart Kennedy Fund, 1910)*

be based on mutual respect and trust. While they discouraged impersonal unions arranged by parents, they did not think romantic attachments—based on physical attraction and emotional love—a sound basis for an enduring relationship.

Moralists held that the household was a woman's first priority. She might assist in her own or her husband's business and do charitable work. Involvement in social or public activities, however, was inappropriate because it distracted the wife from her primary responsibility, her household. If women suffered under their husbands' yoke, writers explained that submission was their punishment inherited from Eve, penance for man's fall, like the pain of childbearing. Moreover, they said, a woman's lot was no worse than a man's: he must earn the family's bread by the sweat of his brow.[15]

Catholics viewed marriage as a sacramental union, which, validly entered into, could not be dissolved. Protestants stressed the contractual form of marriage, whereby each partner promised the other support, companionship, and the sharing of mutual goods. Protestants recognized a mutual right to divorce and remarriage for various reasons, including adultery and irreparable breakdown.[16] Society in the early modern period was patriarchal. While women neither lost their identity nor lacked meaningful work, the all-pervasive assumption was that men ruled. Leading students of the Lutherans, Catholics, French Calvinists, and English Puritans tend to concur that there was no amelioration in women's definitely subordinate status.

There are some remarkable success stories, however. Elizabeth Hardwick, the orphaned daughter of an obscure English country squire, made four careful marriages, each of which brought her more property

and carried her higher up the social ladder. She managed her estates, amounting to more than a hundred thousand acres, with a degree of business sense rare in any age. The two great mansions she built, Chatsworth and Hardwick, stand today as monuments to her acumen. As countess of Shrewsbury, "Bess of Hardwick" so thoroughly enjoyed the trust of Queen Elizabeth that Elizabeth appointed her jailer of Mary, Queen of Scots. Having established several aristocratic dynasties, the countess of Shrewsbury died in 1608, past her eightieth year, one of the richest people in England.[17]

Artists' drawings of plump, voluptuous women and massive, muscular men reveal the contemporary standards of physical beauty. It was a sensual age that gloried in the delights of the flesh. Some people, such as the humanist-poet Aretino, found sexual satisfaction with both sexes. Reformers and public officials simultaneously condemned and condoned sexual "sins." The oldest profession had many practitioners, and when in 1566 Pope Pius IV expelled all the prostitutes from Rome, so many people left and the city suffered such a loss of revenue that in less than a month the pope was forced to rescind the order. Scholars debated Saint Augustine's notion that whores serve a useful social function by preventing worse sins. Prostitution was common because desperate poverty forced women and young men into it. The general public took it for granted. Consequently, civil authorities in both Catholic and Protestant countries licensed houses of public prostitution. These establishments were intended for the convenience of single men, and some Protestant cities, such as Geneva and Zurich, installed officials in the brothels with the express purpose of preventing married men from patronizing them. Moralists naturally railed against prostitution. For example, Melchior Ambach, the Lutheran editor of many tracts against adultery and whoring, wrote in 1543 that if "houses of women" for single and married men were allowed, why not provide a "house of boys" for womenfolk who lack a husband to service them? "Would whoring be any worse for the poor, needy female sex?" Ambach, of course, was not being serious: by treating infidelity from the perspective of female rather than male customers, he was still insisting that prostitution destroyed the family and society.[18]

What became of the thousands of women who left their convents and nunneries during the Reforma-

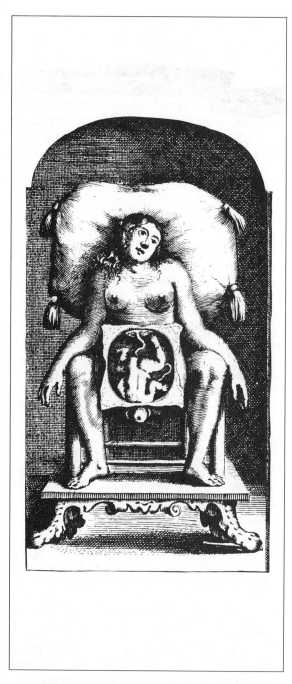

Woman in Labor The production of male heirs was women's major social responsibility. Long into modern times a sitting or squatting position for the delivery of babies was common, because it allowed the mother to push. The calm and wistful look on the mother's face suggests a remarkably easy delivery; it is the artist's misconception of the process. *(Photo: Caroline Buckler)*

tion? The question concerns primarily women of the upper classes, who formed the dominant social group in the religious houses of late medieval Europe. (Single women of the middle and working classes in the sixteenth and seventeenth centuries worked in many occupations and professions—as butchers, shopkeepers, nurses, goldsmiths, and midwives and in the weaving and printing industries. Those who were married normally assisted in their husbands' businesses.) Luther and the Protestant reformers believed that the monastic cloister symbolized antifeminism, that young girls were forced by their parents into convents and once there were bullied by men into staying. Therefore reformers favored the suppression of women's religious houses and encouraged ex-nuns to marry. Marriage, the reformers maintained, not only gave women emotional and sexual satisfaction, it freed them from clerical domination, cultural deprivation, and sexual repression.[19] It would appear, consequently, that women passed from clerical domination to subservience to husbands.

If some nuns in the Middle Ages lacked a genuine religious vocation and if some religious houses witnessed financial mismanagement and moral laxness, convents nevertheless provided women of the upper classes with scope for their literary, artistic, medical, or administrative talents if they could not or would not marry. Marriage became virtually the only occupation for Protestant women. This helps explain why Anglicans, Calvinists, and Lutherans established communities of religious women, such as the Lutheran one at Kaiserwerth in the Rhineland, in the eighteenth and nineteenth centuries.[20]

THE GREAT EUROPEAN WITCH HUNT

The great European witch scare reveals something about contemporary attitudes toward women. The period of the religious wars witnessed a startling increase in the phenomenon of witch-hunting, whose prior history was long but sporadic. "A witch," according to Chief Justice Coke of England, "was a person who hath conference with the Devil to consult with him or to do some act." This definition by the highest legal authority in England demonstrates that educated people, as well as the ignorant, believed in witches. Belief in *witches*—individuals who could mysteriously injure other people, for instance by causing them to become blind or impotent, and who could harm animals, for example, by preventing cows from giving milk—dates back to the dawn of time. For centuries, tales had circulated about old women who made nocturnal travels on greased broomsticks to *sabbats,* or assemblies of witches, where they participated in sexual orgies and feasted on the flesh of infants. In the popular imagination witches had definite characteristics. The vast majority were married women or widows between fifty and seventy years old, crippled or bent with age, with pockmarked skin. They often practiced midwifery or folk medicine, and most had sharp tongues and were quick to scold.

In the sixteenth century, religious reformers' extreme notions of the devil's powers and the insecurity created by the religious wars contributed to the growth of belief in witches. The idea developed that witches made pacts with the devil in return for the power to work mischief on their enemies. Since pacts with the devil meant the renunciation of God, witchcraft was considered heresy, and all religions persecuted it.

Fear of witches took a terrible toll of innocent lives in parts of Europe. In southwestern Germany, 3,229 witches were executed between 1561 and 1670, most by burning. The communities of the Swiss Confederation tried 8,888 persons between 1470 and 1700 and executed 5,417 of them as witches. In all the centuries before 1500, witches in England had been suspected of causing perhaps "three deaths, a broken leg, several destructive storms and some bewitched genitals." Yet between 1559 and 1736, witches were thought to have caused thousands of deaths, and in that period almost 1,000 witches were executed in England.[21]

Historians and anthropologists have offered a variety of explanations for the great European witch hunt. Some scholars maintain that charges of witchcraft were a means of accounting for inexplicable misfortunes. Just as the English in the fifteenth century had blamed their military failures in France on Joan of Arc's sorcery, so in the seventeenth century the English Royal College of Physicians attributed undiagnosable illnesses to witchcraft. Some scholars hold that in small communities, which typically insisted on strict social conformity, charges of witchcraft were a means of attacking and eliminating the nonconformist; witches, in other words, served the collective need for scapegoats. The evidence of witches' trials, some writers suggest, shows that women were not accused because they harmed or threatened their neighbors; rather, their communities believed such women worshiped the devil, engaged in wild sexual activities with him, and ate infants. Other scholars argue the exact opposite: that people were tried and executed as witches because their neighbors feared their evil powers. Finally, there is the theory that the unbridled sexuality attributed to witches was a psychological projection on the part of their accusers, resulting from Christianity's repression of sexuality. The reasons for the persecution of women as witches probably varied from place to place.

Though several hypotheses exist, scholars still cannot fully understand the phenomenon. Nevertheless, given the broad strand of misogyny (hatred of women) in Western religion, the ancient belief in the susceptibility of women (so-called weaker vessels) to the devil's allurements, and the pervasive seventeenth-century belief about women's multiple and demanding orgasms and thus their sexual insatiability, it is not difficult to understand why women were accused of all sorts of mischief and witchcraft. Charges of witchcraft provided a legal basis for the execution of tens of thousands of women. As the most important capital crime for women in early modern times, witchcraft has considerable significance for the history and status of women.[22]

EUROPEAN SLAVERY AND THE ORIGINS OF AMERICAN RACISM

Almost all peoples in the world have engaged in the enslavement of other human beings at some time in their histories. Since ancient times, victors in battle have enslaved conquered peoples. In the later Middle Ages slavery was deeply entrenched in southern Italy, Sicily, Crete, and Mediterranean Spain. The bubonic plague, famines, and other epidemics created a severe shortage of agricultural and domestic workers in parts of northern Europe, encouraging Italian merchants to buy slaves from the Balkans, Thrace, southern Russia, and central Anatolia for sale in the

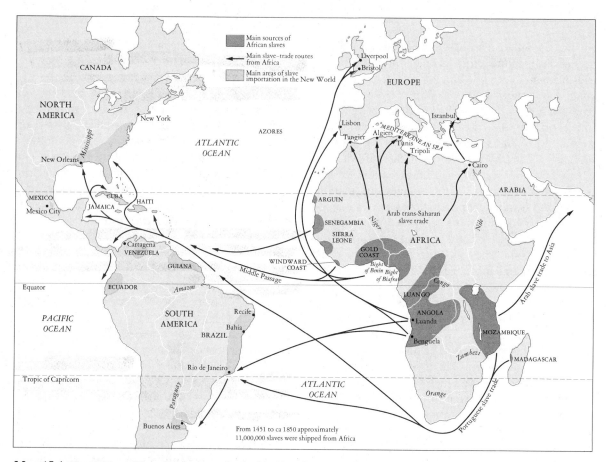

MAP 15.4 The African Slave Trade Decades before the discovery of America, Greek, Russian, Bulgarian, Armenian, and then black slaves worked the plantation economies of southern Italy, Sicily, Portugal, and Mediterranean Spain—thereby serving as models for the American form of slavery.

West. In 1364 the Florentine government allowed the unlimited importation of slaves, so long as they were not Roman Catholics. Between 1414 and 1423, at least ten thousand slaves were sold in Venice alone. The slave trade represents one aspect of Italian business enterprise during the Renaissance: where profits were lucrative, papal threats of excommunication completely failed to stop Genoese slave traders. The Genoese set up colonial stations in the Crimea and along the Black Sea, and according to an international authority on slavery, these outposts were "virtual laboratories" for the development of slave plantation agriculture in the New World.[23] This form of slavery had nothing to do with race; almost all slaves were white. How, then, did black African slavery enter the European picture and take root in the New World?

In 1453 the Ottoman capture of Constantinople halted the flow of white slaves from the Black Sea region and the Balkans. Mediterranean Europe, cut off from her traditional source of slaves, had no alternative source for slave labor but sub-Saharan Africa. The centuries-old trans-Saharan trade was greatly stimulated by finding a ready market in the vineyards and sugar plantations of Sicily and Majorca. By the later fifteenth century, the Mediterranean had developed an "American" form of slavery before the discovery of America.

Meanwhile, the Genoese and other Italians had colonized the Canary Islands in the western Atlantic.

African Slave and Indian Woman A black slave approaches an Indian prostitute. Unable to explain what he wants, he points with his finger; she eagerly grasps for the coin. The Spanish caption above moralizes on the black man using stolen money—yet the Spaniards ruthlessly expropriated all South American mineral wealth. *(New York Public Library)*

Prince Henry the Navigator's sailors (pages 466–467) discovered the Madeira Islands and made settlements there. In this stage of European expansion, "the history of slavery became inextricably tied up with the history of sugar." Though an expensive luxury that only the affluent could afford, population increases and monetary expansion in the fifteenth century led to an increasing demand for sugar. Resourceful Italians provided the capital, cane, and technology for sugar cultivation on plantations in southern Portugal, Madeira, and the Canary Islands. In Portugal alone between 1490 and 1520, 302,000 black slaves entered the port of Lisbon (see Map 15.4). From Lisbon, where African slaves performed most of the manual labor and constituted 10 percent of the city's population, slaves were transported to the sugar plantations of the Madeira, Azores, Cape Verdes, and then Brazil. Sugar and the small Atlantic islands gave New World slavery its distinctive shape. Columbus himself, who spent a decade in Madeira, took sugar plants on his voyages to "the Indies."[24]

As already discussed, European expansion across the Atlantic led to the economic exploitation of the Americas. In the New World the major problem settlers faced was a shortage of labor. As early as 1495 the Spanish solved the problem by enslaving the native Indians. In the next two centuries, the Portuguese, Dutch, and English followed suit.

Unaccustomed to any form of manual labor, certainly not to panning gold for more than twelve hours a day in the broiling sun, the Indians died "like fish in a bucket," as one Spanish settler reported.[25] In 1515 a Spanish missionary, Bartholomé de Las Casas (1474–1566), who had seen the evils of Indian slavery, urged Emperor Charles V to end Indian slavery in his American dominions. Las Casas recommended the importation of blacks from Africa, both because church law did not strictly forbid black slavery and because blacks could better survive under South American conditions. The emperor agreed, and in 1518 the African slave trade began. Columbus's introduction of sugar plants, moreover, stimulated the need for black slaves; and the experience and model of plantation slavery in Portugal and the Atlantic islands encouraged the establishment of a similar agricultural pattern in the New World.

Several European nations participated in the African slave trade. Spain brought the first slaves to Brazil; by 1600, 44,000 were being imported annually. Between 1619 and 1623, the Dutch West India Company, with the full support of the government of the United Provinces, transported 15,430 Africans to Brazil. Only in the late seventeenth century, with the chartering of the Royal African Company, did the English get involved. Thereafter, large numbers of African blacks poured into North America (see Map 15.4). In 1790 there were 757,181 blacks in a total U.S. population of 3,929,625. When the first census was taken in Brazil in 1798, blacks numbered about 2 million in a total population of 3.25 million.

Settlers brought to the Americas the racial attitudes they had absorbed in Europe. Settlers' beliefs and attitudes toward blacks derive from two basic sources: Christian theological speculation (pages 408–409) and Muslim ideas. In the sixteenth and seventeenth centuries, the English, for example, were extremely curious about Africans' lives and customs, and slavers' accounts were extraordinarily popular. Travel literature depicted Africans as savages because of their eating habits, morals, clothing, and social customs; as barbarians because of their language and methods of war; and as heathens because they were not Christian. English people saw similarities between apes and Africans; thus, the terms *bestial* and *beastly* were frequently applied to Africans. Africans were believed to possess a potent sexuality. One seventeenth-century observer considered Africans "very

lustful and impudent, . . . (for a Negroes hiding his members, their extraordinary greatness) is a token of their lust." African women were considered sexually aggressive with a "temper hot and lascivious."[26]

"At the time when Columbus sailed to the New World, Islam was the largest world religion, and the only world religion that showed itself capable of expanding rapidly in areas as far apart and as different from each other as Senegal [in northwest Africa], Bosnia [in the Balkans], Java, and the Philippines." Medieval Arabic literature emphasized blacks' physical repulsiveness, mental inferiority, and primitivism. In contrast to civilized peoples from the Mediterranean to China, Muslim writers claimed, sub-Saharan blacks were the only peoples who had produced no sciences or stable states. The fourteenth-century Arab historian Ibn Khaldun wrote that "the only people who accept slavery are the Negroes, owing to their low degree of humanity and their proximity to the animal stage." Though black kings, Khaldun alleged, sold their subjects without even a pretext of crime or war, the victims bore no resentment because they gave no thought to the future and have "by nature few cares and worries; dancing and rhythm are for them inborn."[27] It is easy to see how such absurd images developed into the classic stereotypes used to justify black slavery in South and North America in the seventeenth, eighteenth, and nineteenth centuries. Medieval Christians and Muslims had similar notions of blacks as inferior and primitive people ideally suited to enslavement. Perhaps centuries of commercial contacts between Muslim and Mediterranean peoples had familiarized the latter with Muslim racial attitudes. In any case, these were the racial beliefs that the Portuguese, Spanish, Dutch, and English brought to the New World.

THE ORIGIN OF MODERN SKEPTICISM: MICHEL DE MONTAIGNE

The decades of religious fanaticism, bringing in their wake death, famine, and civil anarchy, caused both Catholics and Protestants to doubt that any one faith contained absolute truth. The late sixteenth and early seventeenth centuries witnessed the beginnings of modern skepticism. *Skepticism* is a school of thought

ART: A MIRROR OF SOCIETY

Art reveals the interests and values of society and frequently gives intimate and unique glimpses of how people actually lived. In portraits and statues, it preserves the memory and fame of men and women who shaped society. In paintings, drawings, and carvings, it also shows how people worked, played, relaxed, suffered, and triumphed. Art, therefore, is extremely useful to the historian, especially for periods when written records are scarce. Every work of art and every part of it has meaning and has something of its own to say.

Art also manifests the changes and continuity of European life; as values changed in Europe, so did major artistic themes. The art of the later Middle Ages, a time that saw the emergence of a rich urban middle class, increasingly displayed secular rather than spiritual interests. In the Renaissance portraiture became a common artistic form, a reflection of the new era's emphasis on individualism. Rich and prominent patrons used art to display their wealth, power, and social status, as is evident in Gozzoli's *Journey of the Magi* (following). Europeans remained deeply religious but wished to reconcile Christian revelation with classical art's emphasis on human dignity and their own concern with the material world.

These interests are evident in the *Triptych of the Annunciation* by the Flemish master Robert Campin, below. The Annunciation scene in the center panel occurs in the house of a Flemish burgher. Every detail has significance. For example, the serious, modestly dressed middle-class donors in the left panel are memorialized observing the mystery, the lilies on the table in the center panel represent the Virgin's chastity, and Joseph in front of a delicately painted view of a fifteenth-century city in the right panel carves a mousetrap, symbolizing Christ, the bait set to catch the devil. (The Metropolitan Museum of Art; The Cloisters Collection.)

In the sixteenth and seventeenth centuries, some artists produced masterpieces of peasant life in which rustic simplicity and nature itself are celebrated. Others, notably the Dutch painters of the seventeenth century, focused on middle-class people and the ordinary activities of everyday life. Though secular and even lowlife subjects had been painted before, their purpose had usually been to illustrate a moral. The subjects of seventeenth-century genre painting were considered interesting in themselves and worthy of high art. Again, art reflected shifting social attitudes and values.

Feast of the Gods *(above)* Giovanni Bellini (1430?–1516). In this pastoral scene based on a story of the Roman poet Ovid, Olympian gods picnic in a wooded grove as satyrs and nymphs serve them. The peacock in the tree symbolizes the gods' immortality. The pagan theme, the appreciation for perspective and nature, and the sensual atmosphere make this painting a fine example of Italian Renaissance classicism. *(National Gallery of Art, Washington; Widener Collection)*

Journey of the Magi *(above)* Benozzo Gozzoli (ca 1459). Few Renaissance paintings better illustrate art in the service of the princely court, in this case the Medici. Commissioned by Piero de' Medici to adorn his palace chapel, everything in this fresco—the large crowd, the feathers and diamonds adorning many of the personages, the black servant in front—serve to flaunt the power and wealth of the House of Medici. There is nothing especially religious about it; the painting could more appropriately be called "Journey of the Medici." The artist has discreetly placed himself in the crowd, the name Benozzo embroidered on his cap. *(Scala/Art Resource)*

The Bury St. Edmunds' Cross *(right)* Probably made for the English abbot Samson of Bury St. Edmunds (1181–1211) and used in ceremonial processions, this walrus ivory cross, 2 feet high and 14 inches across, contains 8 scenes, 108 figures, and 60 inscriptions from the Old and New Testaments. It is a superb example of late twelfth-century craftsmanship, piety, and, some inscriptions imply, anti-Semitic attitudes. *(The Metropolitan Museum of Art; The Cloisters Collection)*

Madonna of Chancellor Rodin *(above)* Jan van Eyck (ca 1435). The tough and shrewd chancellor who ordered this rich painting visits the Virgin and Christ-Child (though they seem to be visiting him). An angel holds the crown of heaven over the Virgin's head while Jesus, the proclaimed savior of the world, holds it in his left hand and raises his right hand in blessing. Through the colonnade, sculpted with scenes from Genesis, is the city of Bruges. Van Eyck's achievement in portraiture is extraordinary; his treatment of space and figures and his ability to capture the infinitely small and very large prompted the art historian Erwin Panofsky to write that "his eye was at one and the same time a microscope and a telescope." *(Cliché des Musées Nationaux-Paris)*

Lo Sposalizio (Wedding) *(right)* Raphael (ca 1504). Younger than Michelangelo and Leonardo da Vinci but considered their equal as a master of the High Renaissance, Raphael produced this visionary painting of the marriage of Mary and Joseph when he was just 21 years old. It reflects a super grasp of symmetry and perspective. The temple in the background suggests Raphael's genius as an architect. *(Scala/Art Resource)*

The Wedding Dance *(above)* Pieter Bruegel the Elder (ca 1565). This painting is characteristic of many scenes of country life that earned the elder Bruegel the title "Peasant Bruegel." Though we are prevented from identifying the groom by the custom at Dutch weddings to separate bride and groom as much as possible, the bride as always wears her hair loose with a wreath. Varied activities, rich colors, humor, and sadness appear in many of Bruegel's works. *(Walters Art Gallery, Baltimore)*

A Woman Peeling Apples *(above)* Pieter de Hooch (1629–1677). Stability, seriousness, and thrift are idealized in this Dutch domestic scene. The light filtering through the window represents a technical achievement and the secular appreciation for the world of nature and of man. *(Reproduced by permission of the Trustees of the Wallace Collection)*

Portrait of a Merchant *(above)* Jan Gossaert (ca 1530). The banker sits at his desk with quills and inkstand; business letters hang on racks behind him. The skeptical, suspicious expression on his face probably attests to the cautious nature of someone whose business is lending money. Gossaert did much of his work at Antwerp, then the commercial capital of Europe. *(National Gallery of Art, Washington; Ailsa Mellon Bruce Fund)*

Ginevra de Benci *(above)* Leonardo da Vinci (1452–1519). Da Vinci painted this Florentine lady early in his career, when he was about 22. Her pale face stands out against
the dark juniper tree—*ginepra* in Italian, a reference to her name—in the background.
Its expression of a deep inner life led the contemporary Vasari to say of the painting, "it
looked like Ginevra herself and not like a picture." Work on this portrait perhaps prepared da Vinci for the *Mona Lisa,* which he started soon afterward. *(National Gallery of
Art, Washington; Ailsa Mellon Bruce Fund)*

founded on doubt that total certainty or definitive knowledge is ever attainable. The skeptic is cautious and critical and suspends judgment. Perhaps the finest representative of early modern skepticism is the Frenchman Michel de Montaigne (1533–1592).

Montaigne came from a bourgeois family that had made a fortune selling salted herring and in 1477 had purchased the title and property of Montaigne in Gascony. Montaigne received a classical education before studying law and securing a judicial appointment in 1554. Though a member of the nobility, in embarking on a judicial career he identified with the new nobility of the robe. He condemned the ancient nobility of the sword for being more concerned with war and sports than with the cultivation of the mind.

At the age of thirty-eight, Montaigne resigned his judicial post, retired to his estate, and devoted the rest of his life to study, contemplation, and the effort to understand himself. Like the Greeks, he believed that the object of life was to "know thyself," for self-knowledge teaches men and women how to live in accordance with nature and God. Montaigne developed a new literary genre, the essay—from the French *essayer,* meaning "to test or try"—to express his thoughts and ideas.

Montaigne's *Essays* provide insight into the mind of a remarkably humane, tolerant, and civilized man. He was a humanist; he loved the Greek and Roman writers and was always eager to learn from them. In his essay "On Solitude," he quoted the Roman poet Horace:

Reason and sense remove anxiety,
Not villas that look out upon the sea

Some said to Socrates that a certain man had grown no better by his travels. "I should think not," he said; "he took himself along with him. . . ."
We should have wife, children, goods, and above all health, if we can; but we must not bind ourselves to them so strongly that our happiness depends on them. We must reserve a back shop all our own, entirely free, in which to establish our real liberty and our principal retreat and solitude. [28]

From the ancient authors, especially the Roman Stoic Seneca, Montaigne acquired a sense of calm, inner peace, and patience. The ancient authors also inculcated in him a tolerance and broad-mindedness.

Montaigne had grown up during the French civil wars, perhaps the worst kind of war. Religious ideology had set family against family, even brother against brother. He wrote:

In this controversy . . . France is at present agitated by civil wars, the best and soundest side is undoubtedly that which maintains both the old religion and the old government of the country. However, among the good men who follow that side . . . we see many whom passion drives outside the bounds of reason, and makes them sometimes adopt unjust, violent, and even reckless courses. [29]

Though he remained a Catholic, Montaigne possessed a detachment, an independence, an openness of mind, and a willingness to look at all sides of a question. As he wrote, "I listen with attention to the judgment of all men; but so far as I can remember, I have followed none but my own. Though I set little value upon my own opinion, I set no more on the opinions of others."

In a violent and cruel age, Montaigne was a gentle and sensitive man. In his famous essay "On Cruelty," he stated:

Among other vices I cruelly hate cruelty, both by nature and by judgment, as the extreme of all vices. . . .
I live in a time when we abound in incredible examples of this vice, through the license of our civil wars; and we see in the ancient histories nothing more extreme than what we experience of this every day. But that has not reconciled me to it at all. [30]

In the book-lined tower where Montaigne passed his days, he became a deeply learned man. Yet he was not ignorant of the world of affairs, and he criticized scholars and bookworms who ignored the life around them. Montaigne's essay "On Cannibals" reflects the impact of overseas discoveries on Europeans' consciousness. His tolerant mind rejected the notion that one culture is superior to another:

I long had a man in my house that lived ten or twelve years in the New World, discovered in these latter days, and in that part of it where Villegaignon landed [Brazil]. . . .
I find that there is nothing barbarous and savage in [that] nation, by anything that I can gather, excepting,

that every one gives the title of barbarism to everything that is not in use in his own country. As, indeed, we have no other level of truth and reason, than the example and idea of the opinions and customs of the place wherein we live. [31]

In his belief in the nobility of human beings in the state of nature, uncorrupted by organized society, and in his cosmopolitan attitude toward different civilizations, Montaigne anticipated many eighteenth-century thinkers.

The thought of Michel de Montaigne marks a sharp break with the past. Faith and religious certainty had characterized the intellectual attitudes of Western society for a millennium. Montaigne's rejection of any kind of dogmatism, his secularism, and his skepticism thus represent a basic change. In his own time and throughout the seventeenth century, few would have agreed with him. The publication of his ideas, however, anticipated a basic shift in attitudes. Montaigne inaugurated an era of doubt. "Wonder," he said, "is the foundation of all philosophy, research is the means of all learning, and ignorance is the end." [32]

ELIZABETHAN AND JACOBEAN LITERATURE

The age of the religious wars and European expansion also experienced an extraordinary degree of intellectual ferment. In addition to the development of the essay as a distinct literary genre, the late sixteenth and early seventeenth centuries fostered remarkable creativity in other branches of literature. England especially, in the latter part of Elizabeth's reign and the first years of her successor James I (1603–1625), witnessed unparalleled brilliance. The terms *Elizabethan* and *Jacobean* (referring to the reign of James) are used to designate the English music, poetry, prose, and drama of this period. The poetry of Sir Philip Sidney (1554–1586), such as *Astrophel and Stella,* strongly influenced later poetic writing. *The Faerie Queene* of Edmund Spenser (1552–1599) endures as one of the greatest moral epics in any language. The rare poetic beauty of the plays of Christopher Marlowe (1564–1593), such as *Tamburlaine* and *The Jew of Malta,* paved the way for the work of Shakespeare. Above all, the immortal dramas of Shakespeare and the stately prose of the Authorized or King James Bible mark the Elizabethan and Jacobean periods as the golden age of English literature.

William Shakespeare (1564–1616), the son of a successful glove manufacturer who rose to the highest municipal office in the Warwickshire town of Stratford-on-Avon, chose a career on the London stage. By 1592 he had gained recognition as an actor and playwright. Between 1599 and 1603, Shakespeare performed in the Lord Chamberlain's Company and became co-owner of the Globe Theatre, which after 1603 presented his plays.

Shakespeare's genius lies in the originality of his characterizations, the diversity of his plots, his understanding of human psychology, and his unexcelled gift for language. Shakespeare was a Renaissance man in his deep appreciation for classical culture, individualism, and humanism. Such plays as *Julius Caesar, Pericles,* and *Antony and Cleopatra* deal with classical subjects and figures. Several of his comedies have Italian Renaissance settings. The nine history plays, including *Richard II, Richard III,* and *Henry IV,* enjoyed the greatest popularity among Shakespeare's contemporaries. Written during the decade after the defeat of the Spanish Armada, the history plays express English national consciousness. Lines such as these from *Richard II* reflect this sense of national greatness with unparalleled eloquence:

This royal Throne of Kings, this sceptre'd Isle,
This earth of Majesty, this seat of Mars,
This other Eden, demi-paradise,
This fortress built by Nature for herself,
Against infection and the hand of war:
This happy breed of men, this little world,
This precious stone, set in the silver sea,
Which serves it in the office of a wall,
Or as a moat defensive to a house,
Against the envy of less happier Lands,
This blessed plot, this earth, this Realm, this England.

Shakespeare's later plays, above all the tragedies *Hamlet, Othello,* and *Macbeth,* explore an enormous range of human problems and are open to an almost infinite variety of interpretations. *Othello,* which the nineteenth-century historian Thomas Macaulay called "perhaps the greatest work in the world," portrays an honorable man destroyed by a flaw in his

A Royal Picnic The English court imitated the Italian Renaissance devotion to nature. (See Bellini's *Feast of the Gods* in Color Insert III.) In this pastoral atmosphere suggestive of classical mythology, Queen Elizabeth interrupts the day's hunt for a picnic in the forest. The meal seems to consist of fowl, bread, wine, and perhaps pastries. *(The Huntington Library, San Marino, California)*

Velazquez: Juan de Pareja This portrait (1650) of the Spanish painter Velazquez's one-time assistant, a black man of obvious intellectual and sensual power and himself a renowned religious painter, suggests the integration of some blacks in seventeenth-century society. The elegant lace collar attests to his middle-class status. *(Metropolitan Museum of Art, purchased with purchase funds and special contributions bequeathed or given by friends of the museum)*

own character and the satanic evil of his supposed friend Iago. *Macbeth's* central theme is exorbitant ambition. Shakespeare analyzes the psychology of sin in the figures of Macbeth and Lady Macbeth, whose mutual love under the pressure of ambition leads to their destruction. The central figure in *Hamlet,* a play suffused with individuality, wrestles with moral problems connected with revenge and with man's relationship to life and death. The soliloquy in which Hamlet debates suicide is perhaps the most widely quoted passage in English literature:

To be, or not to be: that is the question:
Whether 'tis nobler in the mind to suffer
The slings and arrows of outrageous fortune,
Or to take arms against a sea of troubles,
And by opposing end them? . . .

Hamlet's sad cry, "There is nothing either good or bad but thinking makes it so," expresses the anguish and uncertainty of modern man. *Hamlet* has always enjoyed great popularity, because in his many faceted personality people have seen an aspect of themselves.

Shakespeare's dynamic language bespeaks his extreme sensitivity to the sounds and meanings of words. Perhaps no phrase better summarizes the reason for his immortality than this line, slightly modified, from *Antony and Cleopatra*: "Age cannot wither [him], nor custom stale/ [his] infinite variety."

The other great masterpiece of the Jacobean period was the *Authorized Bible.* At a theological conference in 1604, a group of Puritans urged James I to support a new translation of the Bible. The king in turn assigned the task to a committee of scholars, who published their efforts in 1611. Based on the best scriptural research of the time and divided into chapters and verses, the Authorized Version is actually a revision of earlier Bibles more than an original work. Yet it provides a superb expression of the mature English vernacular in the early seventeenth century. Thus Psalm 37:

Fret not thy selfe because of evill doers, neither bee thou
envious against the workers of iniquitie.
For they shall soone be cut downe like the grasse; and
wither as the greene herbe.
Trust in the Lord, and do good, so shalt thou dwell in the
land, and verely thou shalt be fed.
Delight thy selfe also in the Lord; and he shall give thee
the desires of thine heart.
Commit thy way unto the Lord: trust also in him, and he
shall bring it to passe.
And he shall bring forth thy righteousness as the light,
and thy judgement as the noone day.

The Authorized Version, so called because it was produced under royal sponsorship—it had no official ecclesiastical endorsement—represented the Anglican and Puritan desire to encourage lay people to read the Scriptures. It quickly achieved great popularity and displaced all earlier versions. British settlers carried this Bible to the North American colonies, where it became known as the *King James Bible.* For centuries the *King James Bible* has had a profound influence on the language and lives of English-speaking peoples.

In the sixteenth and seventeenth centuries, Europeans for the first time gained access to large parts of the globe. European peoples had the intellectual curiosity, driving ambition, and scientific technology to attempt feats that were as difficult and expensive then as going to the moon is today. Exploration and exploitation contributed to a more sophisticated standard of living, in the form of spices and Asian luxury goods, and to a terrible international inflation resulting from the influx of South American silver and gold. Governments, the upper classes, and the peasantry were badly hurt by the inflation. Meanwhile the middle class of bankers, shippers, financiers, and manufacturers prospered for much of the seventeenth century.

European expansion and colonization took place against a background of religious conflict and rising national consciousness. The seventeenth century was by no means a secular period. Though the medieval religious framework had broken down, people still thought largely in religious terms. Europeans explained what they did politically and economically in terms of religious doctrine. Religious ideology served as a justification for a variety of goals: the French nobles' opposition to the crown, the Dutch struggle for political and economic independence from Spain. In Germany, religious pluralism and foreign ambitions added to political difficulties. After 1648 the divisions between Protestant and Catholic tended to become permanent. Religious skepticism and racial attitudes were harbingers of developments to come.

NOTES

1. Quoted by C. M. Cipolla, *Guns, Sails, and Empires: Technological Innovation and the Early Phases of European Expansion, 1400–1700,* Minerva Press, New York, 1965, pp. 115–116.
2. Quoted by S. E. Morison, *Admiral of the Ocean Sea: A Life of Christopher Columbus,* Little, Brown, Boston, 1946, p. 154.
3. See Carla Rahn Phillips, *Ciudad Real, 1500–1750: Growth, Crisis, and Readjustment in the Spanish Economy,* Harvard University Press, Cambridge, Mass., 1979, pp. 103–104, 115.
4. Quoted by Cipolla, p. 132.
5. Quoted by F. H. Littell, *The Macmillan Atlas History of Christianity,* Macmillan, New York, 1976, p. 75.
6. Quoted by Cipolla, p. 133.
7. J. H. Parry, *The Age of Reconnaissance,* Mentor Books, New York, 1963, chaps. 3 and 5.
8. Quoted by J. Hale, "War and Public Opinion in the Fifteenth and Sixteenth Centuries," *Past and Present* 22 (July 1962):29.
9. See ibid., pp. 18–32.
10. Quoted by N. Z. Davis, "The Rites of Violence: Religious Riot in Sixteenth Century France," *Past and Present* 59 (May 1973):59.
11. See ibid., pp. 51–91.
12. Quoted by J. L. Motley, *The Rise of the Dutch Republic,* David McKay, Philadelphia, 1898, 1.109.
13. Quoted by P. Smith, *The Age of the Reformation,* Henry Holt, New York, 1951, p. 248.
14. H. Kamen, "The Economic and Social Consequences of the Thirty Years' War," *Past and Present* 39 (April 1968):44–61.
15. Based heavily on Steven Ozment, *When Fathers Ruled: Family Life in Reformation Europe,* Harvard University Press, Cambridge, Mass., 1983, pp. 50–99.
16. Ibid., pp. 85–92.
17. See D. Durant, *Bess of Hardwick: Portrait of an Elizabethan Dynast,* Weidenfeld & Nicolson, London, 1977.
18. Quoted by Ozment, p. 56.
19. Ozment, pp. 9–14.
20. See Francois Biot, *The Rise of Protestant Monasticism,* Helicon Press, Baltimore, 1968, pp. 74–78.
21. Norman Cohn, *Europe's Inner Demons: An Enquiry Inspired by the Great Witch-Hunt,* Basic Books, New York, 1975, pp. 253–254; Keith Thomas, *Religion and the Decline of Magic,* Charles Scribner's Sons, New York, 1971, pp. 450–455.
22. See E. William Monter, "The Pedestal and the Stake: Courtly Love and Witchcraft," in R. Bridenthal and C. Koonz, eds., *Becoming Visible: Women in European History,* Houghton Mifflin, Boston, 1977, pp. 132–135, and A. Fraser, *The Weaker Vessel,* Random House, New York, 1985, pp. 100–103.
23. See Charles Verlinden, *The Beginnings of Modern Colonization,* trans. Y. Freccero, Cornell University Press, Ithaca, New York, 1970, pp. 5–6, 80–97.
24. This section leans heavily on David Brion Davis, *Slavery and Human Progress,* Oxford University Press, New York, 1984, pp. 54–62.
25. Quoted by D. P. Mannix, *Black Cargoes: A History of the Atlantic Slave Trade,* Viking, New York, 1968, p. 5.
26. Ibid., p. 19.
27. Quoted by Davis, *op. cit.,* pp. 43–44.
28. Quoted by D. M. Frame, trans., *The Complete Works of Montaigne,* Stanford University Press, Stanford, Calif., 1958, pp. 175–176.
29. Ibid., p. 177.
30. Ibid., p. 306.
31. Quoted by C. Cotton, trans., *The Essays of Michel de Montaigne,* A. L. Burt, New York, 1893, pp. 207, 210.
32. Ibid., p. 523.

SUGGESTED READING

Perhaps the best starting point for the study of European society in the age of exploration is J. H. Parry, *The Age of Reconnaissance* (1963), which treats the causes and consequences of the voyages of discovery. Parry's splendidly illustrated *The Discovery of South America* (1979) examines Europeans' reactions to the maritime discoveries and treats the entire concept of new *discoveries.* The urbane studies of C. M. Cipolla present fascinating material on technological and sociological developments written in a lucid style: *Guns, Sails, and Empires: Technological Innovation and the Early Phases of European Expansion, 1400–1700* (1965); *Clocks and Culture, 1300–1700* (1967); *Cristofano and the Plague: A Study in the History of Public Health in the Age of Galileo* (1973); and *Public Health and the Medical Profession in the Renaissance* (1976). S. E. Morison, *Admiral of the Ocean Sea: A Life of Christopher Columbus* (1946), is the standard biography of the great discoverer. The advanced student should consult F. Braudel, *Civilization and Capitalism, 15th–18th Century,* trans. Sian Reynolds, vol. 1: *The Structures of Everyday Life* (1981); vol. 2: *The Wheels of Commerce* (1982); and vol. 3: *The Perspective of the World* (1984). These three fat volumes combine vast erudition, a global perspective, and remarkable illustrations.

For the religious wars, in addition to the references in the Suggested Reading for Chapter 14 and the Notes to this chapter, see J. H. M. Salmon, *Society in Crisis: France in the Sixteenth Century* (1975), which traces the fate of French institutions during the civil wars. A. N. Galpern, *The Religions of the People in Sixteenth-Century Champagne* (1976), is a useful case study in religious anthropology, and W. A. Christian, Jr., *Local Religion in Sixteenth Century Spain* (1981), traces the attitudes and practices of ordinary people.

A cleverly illustrated introduction to the Low Countries is K. H. D. Kaley, *The Dutch in the Seventeenth Ceetury* (1972). The old study of J. L. Motley, *The Rise of the Dutch Republic* (1898), still provides a good comprehensive treatment and makes fascinating reading. For Spanish military operations in the Low Countries, see G. Parker, *The Army of Flanders and the Spanish Road, 1567–1659: The Logistics of Spanish Victory and Defeat in the Low Countries' Wars* (1972). The same author's *Spain and the Netherlands, 1559–1659: Ten Studies* (1979) contains useful essays, of which students may

especially want to consult "Why Did the Dutch Revolt Last So Long?" For the later phases of the Dutch-Spanish conflict, see J. I. Israel, *The Dutch Republic and the Hispanic World, 1606–1661* (1982), which treats the struggle in global perspective.

Of the many biographies of Elizabeth of England, W. T. MacCaffrey, *Queen Elizabeth and the Making of Policy, 1572–1588* (1981), examines the problems posed by the Reformation and how Elizabeth solved them. J. E. Neale, *Queen Elizabeth I* (1957), remains valuable, and L. B. Smith, *The Elizabethan Epic* (1966), is a splendid evocation of the age of Shakespeare with Elizabeth at the center. The best recent biography is C. Erickson, *The First Elizabeth* (1983), a fine, psychologically resonant portrait.

Nineteenth- and early twentieth-century historians described the defeat of the Spanish Armada as a great victory for Protestantism, democracy, and capitalism, which those scholars tended to link together. Recent historians have treated the event in terms of its contemporary significance. For a sympathetic but judicious portrait of the man who launched the Armada, see G. Parker, *Philip II* (1978). D. Howarth, *The Voyage of the Armada* (1982), discusses the expedition largely in terms of the individuals involved, while G. Mattingly, *The Armada* (1959), gives the diplomatic and political background; both Howarth and Mattingly tell very exciting tales. M. Lewis, *The Spanish Armada* (1972), also tells a good story, but strictly from the English perspective. Significant aspects of Portuguese culture are treated in A. Hower and R. Preto-Rodas, eds., *Empire in Transition. The Portuguese World in the Time of Camões* (1985).

C. V. Wedgwood, *The Thirty Years' War* (1961), must be qualified in light of recent research on the social and economic effects of the war, but it is still a good (if detailed) starting point on a difficult period. A variety of opinions on the causes and results of the war are given in T. K. Rabb's anthology, *The Thirty Years' War* (1981). The following articles, all of which appear in the scholarly journal *Past and Present,* provide some of the latest important findings: H. Kamen, "The Economic and Social Consequences of the Thirty Years' War," 39 (1968); J. Hale, "War and Public Opinion in the Fifteenth and Sixteenth Centuries," 22 (1962); J. V. Polišenský, "The Thirty Years' War and the Crises and Revolutions of Sixteenth Century Europe," 39 (1968); and for the overall significance of Sweden, M. Roberts, "Queen Christina and the General Crisis of the Seventeenth Century," 22 (1962).

As background to the intellectual changes instigated by the Reformation, D. C. Wilcox, *In Search of God and Self: Renaissance and Reformation Thought* (1975), contains a perceptive analysis, and T. Ashton, ed., *Crisis in Europe, 1560–1660* (1967), is fundamental. On witches and witchcraft see, in addition to the titles by N. Cohn and K. Thomas in the Notes, J. B. Russell, *Witchcraft in the Middle Ages* (1976) and *Lucifer: The Devil in the Middle Ages* (1984), M. Summers, *The History of Witchcraft and Demonology* (1973), and H. R. Trevor-Roper, *The European Witch-Craze of the Sixteenth and Seventeenth Centuries* (1967), an important collection of essays.

For women, marriage, and the family, see L. Stone, *The Family, Sex, and Marriage in England, 1500–1800* (1977), an important but controversial work; D. Underdown, "The Taming of the Scold," and S. Amussen, "Gender, Family, and the Social Order," in *Order and Disorder in Early Modern England* (eds. A. Fletcher and J. Stevenson, 1985); A. Macfarlane, *Marriage and Love in England. Modes of Reproduction, 1300–1848* (1986); C. R. Boxer, *Women in Iberian Expansion Overseas, 1415–1815* (1975), an invaluable study of women's role in overseas emigration; S. M. Wyntjes, "Women in the Reformation Era," in *Becoming Visible: Women in European History* (eds. R. Bridenthal and C. Koonz, 1977), a quick survey of conditions in different countries, and S. Ozment, *When Fathers Ruled: Family Life in Reformation Europe* (1983), a seminal study concentrating on Germany and Switzerland.

As background to slavery and racism in North and South America, students should see J. L. Watson, ed., *Asian and African Systems of Slavery* (1980), a valuable collection of essays, and D. B. Davis, *Slavery and Human Progress* (1984), which shows how slavery was viewed as a progressive force in the expansion of the Western world. For North American conditions, interested students should consult W. D. Jordan, *The White Man's Burden: Historical Origins of Racism in the United States* (1974), and D. P. Mannix in collaboration with M. Cowley, *Black Cargoes: A History of the Atlantic Slave Trade* (1968), a hideously fascinating account. For Caribbean and South American developments, see F. P. Bowser, *The African Slave in Colonial Peru* (1974); J. S. Handler and F. W. Lange, *Plantation Slavery in Barbados: An Archeological and Historical Investigation* (1978); and R. E. Conrad, *Children of God's Fire: A Documentary History of Black Slavery in Brazil* (1983).

The leading authority on Montaigne is D. M. Frame. See his *Montaigne's Discovery of Man* (1955) and his translation *The Complete Works of Montaigne* (1958).

16

ABSOLUTISM AND CONSTITUTIONALISM IN WESTERN EUROPE (CA 1589–1715)

*T*HE SEVENTEENTH century was a period of revolutionary transformation. Some of its most profound developments were political: it has been called the century when government became modern. The sixteenth century had witnessed the emergence of the nation-state. The long series of wars fought in the name of religion—actually contests between royal authority and aristocratic power—brought social dislocation and agricultural and commercial disaster. Increasingly, strong national monarchy seemed the only solution. Spanish and French monarchs gained control of the major competing institution in their domains, the Roman Catholic church. Rulers of England and some of the German principalities, who could not completely regulate the church, set up national churches. In the German empire, the Treaty of Westphalia placed territorial sovereignty in the princes' hands. The kings of France, England, and Spain claimed the basic loyalty of their subjects. Monarchs made laws, to which everyone within their borders was subject. These powers added up to something close to sovereignty.

A state may be termed *sovereign* when it possesses a monopoly over the instruments of justice and the use of force within clearly defined boundaries. In a sovereign state no system of courts, such as ecclesiastical tribunals, competes with state courts in the dispensation of justice; and private armies, such as those of feudal lords, present no threat to royal authority because the state's army is stronger. Royal law touches all persons within the country. Sovereignty had been evolving during the late sixteenth century. Seventeenth-century governments now needed to address the problem of *which* authority within the state would possess sovereignty—the crown or the nobility.

In the period between roughly 1589 and 1715, two basic patterns of government emerged in Europe: absolute monarchy and the constitutional state. Almost all subsequent governments have been modeled on one of these patterns. In what sense were they "modern"? How did these forms of government differ from the feudal and dynastic monarchies of earlier centuries? Which Western countries most clearly illustrate the new patterns of political organization? This chapter will be concerned with these political questions.

ABSOLUTISM

In the *absolutist* state, sovereignty is embodied in the person of the ruler. The ruler is not restrained by legal authority. Absolute kings claim to rule by divine right, meaning they are responsible to God alone. (Medieval kings governed "by the grace of God," but invariably they acknowledged that they had to respect and obey the law.) Absolute monarchs in the seventeenth and eighteenth centuries had to respect the divine law and the fundamental laws of the land. But they were not checked by national assemblies. Estates general and parliaments met at the wish and in response to kings' needs. Because these meetings provided opportunities for opposition to the crown to coalesce, absolute monarchs eventually stopped summoning them.

Absolute rulers effectively controlled all competing jurisdictions, institutions, or interest groups in their territories. They regulated religious sects. They abolished the liberties long held by certain areas, groups, or provinces. Absolute kings also secured mastery over the one class that historically had posed the greatest threat to monarchy, the nobility. Medieval governments, restrained by the church, the feudal nobility, and their own financial limitations, had been able to do none of these.

In some respects, the key to the power and success of absolute monarchs lay in how they solved their financial problems. Kings frequently found temporary financial support through bargains with the nobility: the nobility agreed to an ad hoc grant in return for freedom from future taxation. The absolutist solution was the creation of new state bureaucracies, which directed the economic life of the country in the interests of the king, either increasing taxes ever higher or devising alternative methods of raising revenue.

Bureaucracies were composed of career officials, appointed by and solely accountable to the king. The backgrounds of these civil servants varied. Absolute monarchs sometimes drew on the middle class, as in France, or utilized members of the nobility, as in Spain and eastern Europe. Where there was no middle class or an insignificant one, as in Austria, Prussia, Spain, and Russia, the government of the absolutist state consisted of an interlocking elite of monarchy, aristocracy, and bureaucracy.

Royal agents in medieval kingdoms had used their public offices and positions to benefit themselves and their families. In England, for example, crown servants from Thomas Becket to Thomas Wolsey had treated their high offices as their private property and reaped considerable profit from them. The most striking difference between seventeenth-century bureaucracies and their medieval predecessors was that seventeenth-century civil servants served the state as represented by the king. Bureaucrats recognized that the offices they held were public, or state, positions. The state paid them salaries to handle revenues that belonged to the crown, and they were not supposed to use their positions for private gain. Bureaucrats gradually came to distinguish between public duties and private property.

Absolute monarchs also maintained permanent standing armies. Medieval armies had been raised by feudal lords for particular wars or campaigns, after which the troops were disbanded. In the seventeenth century, monarchs alone recruited and maintained armies—in peacetime as well as during war. Kings deployed their troops both inside and outside the country in the interests of the monarchy. Armies became basic features of absolutist, and modern, states. Absolute rulers also invented new methods of compulsion. They concerned themselves with the private lives of potentially troublesome subjects, often through the use of secret police.

Rule of absolute monarchs was not all-embracing, because they lacked the financial and military resources and the technology to make it so. Thus the absolutist state was not the same as a totalitarian state. *Totalitarianism* is a twentieth-century phenomenon; it seeks to direct all facets of a state's culture—art, education, religion, the economy, and politics—in the interests of the state. By definition totalitarian rule is *total* regulation. By twentieth-century standards, the ambitions of an absolute monarch were quite limited: he sought the exaltation of himself as the embodiment of the state. Whether or not Louis XIV of France actually said, "L'état, c'est moi!" ("I am the state!"), the remark expresses his belief that he personified the French nation. Yet the absolutist state did foreshadow recent totalitarian regimes in two fundamental respects: in the glorification of the state over all other aspects of the state's culture, and in the use of war and an expansionist foreign policy to divert attention from domestic ills.

All of this is best illustrated by the experience of France, aptly known as the model of absolute monarchy.

THE FOUNDATIONS OF ABSOLUTISM IN FRANCE: HENRY IV AND SULLY

The ingenious Huguenot-turned-Catholic Henry IV (pages 477–478) ended the French religious wars with the Edict of Nantes. The first of the Bourbon dynasty, and probably the first French ruler since Louis IX in the thirteenth century genuinely to care about the French people, Henry IV and his great minister Sully (1560–1641) laid the foundations of later French absolutism. Henry denied influence on the royal council to the nobility, which had harassed the countryside for half a century. Maintaining that "if we are without compassion for the people, they must succumb and we all perish with them," Henry also lowered the severe taxes on the overburdened peasantry.

Sully proved himself a financial genius. He not only reduced the crushing royal debt but began to build up the treasury. He levied an annual tax, the *paulette,* on people who had purchased financial and judicial offices and had consequently been exempt from royal taxation. One of the first French officials to appreciate the significance of overseas trade, Sully subsidized the Company for Trade with the Indies. He started a countrywide highway system and even dreamed of an international organization for the maintenance of peace.

In twelve short years, Henry IV and Sully restored public order in France and laid the foundations for economic prosperity. By late sixteenth-century standards, Henry IV's government was progressive and promising. His murder in 1610 by a crazed fanatic led to a severe crisis.

THE CORNERSTONE OF FRENCH ABSOLUTISM: LOUIS XIII AND RICHELIEU

After the death of Henry IV, the queen-regent Marie de' Medici led the government for the child-king Louis XIII (1610–1643), but in fact feudal nobles and princes of the blood dominated the political scene. In 1624 Marie de' Medici secured the appointment of Armand Jean du Plessis—Cardinal Richelieu (1585–1642)—to the council of ministers. It was a remarkable appointment. The next year Richelieu became president of the council, and after 1628 he was first minister of the French crown. Richelieu used his strong influence over King Louis XIII to exalt the French monarchy as the embodiment of the French state. One of the greatest servants of the French state, Richelieu set in place the cornerstone of French absolutism, and his work served as the basis for France's cultural domination of Europe in the later seventeenth century.

Richelieu's policy was the total subordination of all groups and institutions to the French monarchy. The French nobility, with its selfish and independent interests, had long constituted the foremost threat to the centralizing goals of the crown and to a strong national state. Therefore, Richelieu tried to break the power of the nobility. He leveled castles, long the symbol of feudal independence. He crushed aristocratic conspiracies with quick executions. For example, when the duke of Montmorency, the first peer of France and the godson of Henry IV, became involved in a revolt in 1632, he was summarily put to death.

The constructive genius of Cardinal Richelieu is best reflected in the administrative system he established. He extended the use of the royal commissioners called intendants. France was divided into thirty-two *généralités* ("districts"), in each of which a royal intendant had extensive responsibility for justice, police, and finances. The intendants were authorized "to decide, order and execute all that they see good to do." Usually members of the upper middle class or minor nobility, the intendants were appointed directly by the monarch, to whom they were solely responsible. The intendants recruited men for the army, supervised the collection of taxes, presided over the administration of local law, checked up on the local nobility, and regulated economic activities —commerce, trade, the guilds, marketplaces—in their districts. They were to use their power for two related purposes: to enforce royal orders in the généralités of their jurisdiction and to weaken the power and influence of the regional nobility. The system of government by intendants derived from Philip Augustus's baillis and seneschals, and ultimately from Charlemagne's missi dominici. As the intendants' power grew during Richelieu's administration, so did the power of the centralized state.

Though Richelieu succeeded in building a rational and centralized political machine in the intendant system, he was not the effective financial administrator Sully had been. France lacked a sound system of taxation, a method of raising sufficient revenue to meet the needs of the state. Richelieu reverted to the old device of selling offices. He increased the number of sinecures, tax exemptions, and benefices that were purchasable and inheritable. In 1624 this device brought in almost 40 percent of royal revenues.

The rising cost of foreign and domestic policies led to the auctioning of *tax farms,* a system whereby a man bought the right to collect taxes. Tax farmers kept a very large part of the receipts they collected. The sale of offices and this antiquated system of tax collection were improvisations that promoted confusion and corruption. Even worse, state offices, once purchased, were passed on to heirs, which meant that a family that held a state office was eternally exempt from taxation. Richelieu's inadequate and temporary solutions created grave financial problems for the future.

The cardinal perceived that Protestantism all too often served as a cloak for the political intrigues of ambitious lords. When the Huguenots revolted in 1625, under the duke of Rohan, Richelieu personally supervised the siege of their walled city, La Rochelle, and forced it to surrender. Thereafter, fortified places of security were abolished. Huguenots were allowed to practice their faith, but they no longer possessed armed strongholds or the means to be an independent party in the state. Another aristocratic prop was knocked down.

French foreign policy under Richelieu was aimed at the destruction of the fence of Habsburg territories that surrounded France. Consequently Richelieu supported the Habsburgs' enemies. In 1631 he signed a treaty with the Lutheran king Gustavus Adolphus, promising French support against the Catholic Habsburgs in what has been called the Swedish phase of the Thirty Years' War (page 485). French influence became an important factor in the political future of the German empire. Richelieu acquired for France extensive rights in Alsace in the east and Arras in the north.

Richelieu's efforts at centralization extended even to literature. In 1635 he gave official recognition to a group of philologists who were interested in grammar and rhetoric. Thus was born the French Academy. With Richelieu's encouragement, the Academy began the preparation of a dictionary to standardize the French language; it was completed in 1694. The French Academy survives as a prestigious learned society, whose membership has been broadened to include people outside the field of literature.

Richelieu personified the increasingly secular spirit of the seventeenth century. Though a bishop of the Roman Catholic church, he gave his first loyalty to the French state. Though a Roman Catholic cardinal, he gave strong support to the Protestant Lutherans of Germany.

Richelieu had persuaded Louis XIII to appoint his protegé Jules Mazarin (1602–1661) as his successor. An Italian diplomat of great charm, Mazarin served on the Council of State under Richelieu, acquiring considerable political experience. He became a cardinal in 1641 and a French citizen in 1643. When Louis XIII followed Richelieu to the grave in 1643 and a regency headed by Queen Anne of Austria governed for the child-king Louis XIV, Mazarin became the dominant power in the government. He continued the antifeudal and centralizing policies of Richelieu, but his attempts to increase royal revenues led to the civil wars known as the "Fronde."

The word *fronde* means "slingshot" or "catapult," and a *frondeur* was originally a street urchin who threw mud at the passing carriages of the rich. The term came to be used for anyone who opposed the policies of the government. Richelieu had stirred up the bitter resentment of the aristocracy, who felt its constitutional status and ancient privileges threatened. He also bequeathed to the crown a staggering debt, and when Mazarin tried to impose financial reforms, the monarchy incurred the enmity of the middle classes. Both groups plotted against Anne and Mazarin. Most historians see the Fronde as the last serious effort by the French nobility to oppose the monarchy by force. When in 1648 Mazarin proposed new methods for raising income, bitter civil war ensued between the monarchy on the one side and the frondeurs (the nobility and the upper-middle classes) on the other. Riots and public turmoil wracked Paris and the nation. The violence continued intermittently for almost twelve years. Factional disputes among the nobles led to their ultimate defeat.

The conflicts of the Fronde had two significant results for the future: a badly disruptive effect on the French economy and a traumatic impact on the young Louis XIV. The king and his mother were frequently threatened and sometimes treated as pris-

Philippe de Champaigne: Cardinal Richelieu This portrait, with its penetrating eyes, expression of haughty and imperturable cynicism, and dramatic sweep of red robes, suggests the authority, grandeur, and power that Richelieu wished to convey as first minister of France. *(Reproduced by courtesy of the Trustees, The National Gallery, London)*

oners by aristocratic factions. On one occasion a mob broke into the royal bedchamber to make sure the king was actually there; it succeeded in giving him a bad fright. Louis never forgot such humiliations. The period of the Fronde formed the cornerstone of his political education and of his conviction that the sole alternative to anarchy was absolute monarchy.

Coysevox: Louis XIV (1687–1689) The French court envisioned a new classical age with the Sun King as emperor and his court a new Rome. This statue depicts Louis in a classical pose, clothed (except for the wig) as for a Roman military triumph. *(Caisse Nationale des Monuments Historiques et des Sites, Paris)*

THE ABSOLUTE MONARCHY OF LOUIS XIV

According to the court theologian Bossuet, the clergy at the coronation of Louis XIV in Rheims Cathedral asked God to cause the splendors of the French court to fill all who beheld it with awe. God subsequently granted that prayer. In the reign of Louis XIV (1643–1715), the longest in European history, the French monarchy reached the peak of absolutist development. In the magnificence of his court, in his absolute power, in the brilliance of the culture over which he presided and which permeated all of Europe, and in his remarkably long life, the "Sun King" dominated his age. No wonder scholars have characterized the second half of the seventeenth century as the "Grand Century," the "Age of Magnificence," and echoing the eighteenth-century philosopher Voltaire, the "Age of Louis XIV."

Who was this phenomenon, of whom it was said that when Louis sneezed, all Europe caught cold? Born in 1638, king at the age of five, he entered into personal, or independent, rule in 1661. One of the first tales recorded about him gained wide circulation during his lifetime. Taken as a small child to his father's deathbed, he identified himself as *Louis Quatorze* ("Louis the fourteenth"). Since neither Louis nor his father referred to themselves with numerals, the story is probably untrue. But it reveals the incredible sense of self that contemporaries, both French and foreign, believed that Louis possessed throughout his life.

In old age, Louis claimed that he had grown up learning very little, but recent historians think he was being modest. True, he knew little Latin and only the rudiments of arithmetic and thus by Renaissance standards was not well educated. On the other hand, he learned to speak Italian and Spanish fluently; he spoke and wrote elegant French; he knew some French history and more European geography than the ambassadors accredited to his court. He imbibed the devout Catholicism of his mother, Anne of Austria, and throughout his long life scrupulously performed his religious duties. Religion, Anne, and Mazarin all taught Louis that God had established kings as his rulers on earth. The royal coronation consecrated him to God's service, and he was certain—to use Shakespeare's phrase—that there was a divinity that doth hedge a king. Though kings were a race apart, they could not do as they pleased: they must obey God's laws and rule for the good of the people.

Louis's education was more practical than formal. Under Mazarin's instruction, he studied state papers as they arrived, and he attended council meetings and sessions at which French ambassadors were dispatched abroad and foreign ambassadors received. He learned by direct experience and gained professional training in the work of government. Above all, the misery he suffered during the Fronde gave Louis an eternal distrust of the nobility and a profound sense of his own isolation. Accordingly, silence, caution, and secrecy became political tools for the achievement of his goals. His characteristic answer to requests of all kinds became the enigmatic "Je verrai" ("I shall see").

Louis grew up with an absolute sense of his royal dignity. Tall and distinguished in appearance, he was inclined to heaviness because of the gargantuan meals in which he indulged. Seduced by one of his mother's maids when he was sixteen, the king matured into a highly sensual man easily aroused by an attractive female face and figure. It is to his credit, however, that neither his wife, Queen Maria Theresa, whom he married as the result of a diplomatic agreement with Spain, nor his mistresses ever possessed any political influence. One contemporary described him this way: "He has an elevated, distinguished, proud, intrepid, agreeable air . . . a face that is at the same time sweet and majestic. . . . His manner is cold; he speaks little except to people with whom he is familiar . . . (and then) he speaks well and effectively,

and says what is apropos. . . . he has natural goodness, is charitable, liberal, and properly acts out the role of king."[1] Louis XIV was a consummate actor, and his "terrifying majesty" awed all who saw him. He worked extremely hard and succeeded in being "every moment and every inch a king." Because he so relished the role of monarch, historians have had difficulty distinguishing the man from the monarch.

Recent scholarship indicates that Louis XIV introduced significant governmental innovations: the most significant was his acquisition of absolute control over the French nobility. Indeed it is often said that Louis achieved complete "domestication of the nobility." In doing so, Louis XIV turned the royal court into a fixed institution. In the past, the king of France and the royal court had traveled constantly, visiting the king's properties, the great noblemen, and his *bonnes villes* or "good towns." Since the time of Louis IX, or even Charlemagne, rulers had traveled to maintain order in distant parts of the realm, to impress humbler subjects with the royal dignity and magnificence, and to bind the country together through loyalty to the king. Since the early Middle Ages, the king's court had consisted of his family, trusted advisers and councilors, a few favorites, and servants. Except for the very highest officials of the state, members of the council had changed constantly.

Louis XIV installed the court at Versailles, a small town ten miles from Paris. He required all the great nobility of France, at the peril of social, political, and sometimes economic disaster, to come live at Versailles for at least part of the year. Today Versailles stands as the best surviving museum of a vanished society on earth. In the seventeenth century, it became a model of rational order, the center of France and thus the center of Western civilization, the perfect symbol of the king's absolute power.

Louis XIII began Versailles as a hunting lodge, a retreat from a queen he did not like. His son's architects, Le Nôtre and Le Vau, turned what Saint-Simon called "the most dismal and thankless of sights" into a veritable paradise. Wings were added to the original building to make the palace U-shaped. Everywhere at Versailles the viewer had a sense of grandeur, vastness, and elegance. Enormous state rooms became display galleries for inlaid tables, Italian marble statuary, Gobelin tapestries woven at the state factory in Paris, silver ewers, and beautiful (if uncomfortable) furniture. If genius means attention to detail, Louis XIV and his designers had it: the décor was perfected down to the last doorknob and keyhole. In the gigantic Hall of Mirrors, later to reflect so much of German as well as French history, hundreds of candles illuminated the domed ceiling, where allegorical paintings celebrated the king's victories.

The Ambassador's Staircase was of brilliantly colored marble, with part of the railing gold-plated. The staircase was dominated by a great bust of the king, which, when completed, so overwhelmed a courtier that he exclaimed to the sculptor, Bernini, "Don't do anything more to it, it's so good I'm afraid you might spoil it." The statue, like the staircase—and the entire palace—succeeded from the start in its purpose: it awed.

The formal, carefully ordered, and perfectly landscaped gardens at Versailles express at a glance the spirit of the age of Louis XIV. Every tree, every bush, every foot of grass, every fountain, pool, and piece of statuary within three miles is perfectly laid out. The vista is of the world made rational and absolutely controlled. Nature itself was subdued to enhance the greatness of the king.

The art and architecture of Versailles served as fundamental tools of state policy under Louis XIV. Architecture was another device the king used to overawe his subjects and foreign visitors. Versailles was seen as a reflection of French genius. Thus the Russian tsar Peter the Great imitated Versailles in the construction of his palace, Peterhof, as did the Prussian emperor Frederick the Great in his palace at Potsdam outside Berlin.

As in architecture, so too in language. Beginning in the reign of Louis XIV, French became the language of polite society and the vehicle of diplomatic exchange. French also gradually replaced Latin as the language of international scholarship and learning. The wish of other kings to ape the courtly style of Louis XIV and the imitation of French intellectuals and artists spread the language all over Europe. The royal courts of Sweden, Russia, Poland, and Germany all spoke French. In the eighteenth century, the great Russian aristocrats were more fluent in French than in Russian. In England, the first Hanoverian king, George I, spoke fluent French, halting English. France inspired a cosmopolitan European culture in the late seventeenth century, and that culture was inspired by the king. That is why the French today revere Louis XIV as one of their greatest national heroes: because of the culture that he inspired and symbolized.

Against this background of magnificent splendor, as Saint-Simon describes him, Louis XIV

reduced everyone to subjection, and brought to his court those very persons he cared least about. Whoever was old enough to serve did not dare demur. It was still another device to ruin the nobles by accustoming them to equality and forcing them to mingle with everyone indiscriminately. . . .

To keep everyone assiduous and attentive, the King personally named the guests for each festivity, each stroll through Versailles, and each trip. These were his rewards and punishments. He knew there was little else he could distribute to keep everyone in line. He substituted idle rewards for real ones and these operated through jealousy, the petty preferences he showed many times a day, and his artfulness in showing them. . . .

Upon rising, at bedtime, during meals, in his apartments, in the gardens of Versailles, everywhere the courtiers had a right to follow, he would glance right and left to see who was there; he saw and noted everyone; he missed no one, even those who were hoping they would not be seen. . . .

Louis XIV took great pains to inform himself on what was happening everywhere, in public places, private homes, and even on the international scene. . . . Spies and informers of all kinds were numberless. . . .

But the King's most vicious method of securing information was opening letters.[2]

Though this passage was written by one of Louis's severest critics, all agree that the king used court ceremonial to undermine the power of the great nobility. By excluding the highest nobles from his councils, he destroyed their ancient right to advise the king and to participate in government; they became mere instruments of royal policy. Operas, fetes, balls, gossip, and trivia occupied the nobles' time and attention. Through painstaking attention to detail and precisely calculated showmanship, Louis XIV emasculated the major threat to his absolute power. He separated power from status and grandeur: the nobility enjoyed the status and grandeur in which they lived; the king alone held the power.

Louis dominated the court, and in his scheme of things, the court was more significant than the government. In government Louis utilized several councils of state, which he personally attended, and the intendants, who acted for the councils throughout France. A stream of questions and instructions flowed between local districts and Versailles, and under Louis XIV a uniform and centralized administration was imposed on the country. In 1685 France was the strongest and most highly centralized state in Europe.

Councilors of state and intendants came from the recently ennobled or the upper-middle class. Royal service provided a means of social mobility. These professional bureaucrats served the state in the person of the king, but they did not share power with him. Louis stated that he chose bourgeois officials because he wanted "people to know by the rank of the men who served him that he had no intention of sharing power with them."[3] If great ones were the king's advisers, they would seem to share the royal authority; professional administrators from the middle class would not.

Throughout his long reign and despite increasing financial problems, he never called a meeting of the Estates General. The nobility, therefore, had no means of united expression or action. Nor did Louis have a first minister, freeing him from worry about the inordinate power of a Richelieu. Louis's use of terror—a secret police force, a system of informers, and the practice of opening private letters—foreshadowed some of the devices of the modern state. French government remained highly structured, bureaucratic, centered at Versailles, and responsible to Louis XIV.

FINANCIAL AND ECONOMIC MANAGEMENT UNDER LOUIS XIV: COLBERT

Finance was the grave weakness of Louis XIV's absolutism. An expanding professional bureaucracy, the court of Versailles, and extensive military reforms (see above) cost a great amount of money. The French method of collecting taxes consistently failed to produce enough revenue. Tax farmers pocketed the difference between what they raked in and what they handed over to the state. Consequently, the tax farmers profited, while the government got far less than the people paid. Then, by an old agreement between the crown and the nobility, the king could freely tax the common people, provided he did not

tax the nobles. The nobility thereby relinquished a role in government: since they did not pay taxes, they could not legitimately claim a say in how taxes were spent. Louis, however, lost enormous potential revenue. The middle classes, moreover, secured many tax exemptions. With the rich and prosperous classes exempt, the tax burden fell heavily on those least able to pay, the poor peasants.

The king named Jean-Baptiste Colbert (1619–1683), the son of a wealthy merchant-financier of Rheims, as controller-general of finances. Colbert came to manage the entire royal administration and proved himself a financial genius. Colbert's central principle was that the wealth and the economy of France should serve the state. He did not invent the system called "mercantilism," but he rigorously applied it to France.

Mercantilism is a collection of governmental policies for the regulation of economic activities, especially commercial activities, by and for the state. In seventeenth- and eighteenth-century economic theory, a nation's international power was thought to be based on its wealth, specifically its gold supply. To accumulate gold, a country should always sell abroad more than it bought. Colbert believed that a successful economic policy meant more than a favorable balance of trade. He insisted that the French sell abroad and buy *nothing* back. France should be self-sufficient, able to produce within its borders everything the subjects of the French king needed. Consequently, the outflow of gold would be halted, debtor states would pay in bullion, and with the wealth of the nation increased, its power and prestige would be enhanced.

Colbert attempted to accomplish self-sufficiency through state support for both old industries and newly created ones. He subsidized the established cloth industries at Abbeville, Saint-Quentin, and Carcassonne. He granted special royal privileges to the rug and tapestry industries at Paris, Gobelin, and Beauvais. New factories at Saint-Antoine in Paris manufactured mirrors to replace Venetian imports. Looms at Chantilly and Alençon competed with English lacemaking, and foundries at Saint-Etienne made steel and firearms that cut Swedish imports. To ensure a high-quality finished product, Colbert set up a system of state inspection and regulation. To ensure order within every industry, he compelled all craftsmen to organize into guilds, and within every guild he gave the masters absolute power over their workers. Colbert encouraged skilled foreign craftsmen and manufacturers to immigrate to France, and he gave them special privileges. To improve communications, he built roads and canals, the most famous linking the Mediterranean and the Bay of Biscay. To protect French goods, he abolished many domestic tariffs and enacted high foreign tariffs, which prevented foreign products from competing with French ones.

Colbert's most important work was the creation of a powerful merchant marine to transport French goods. He gave bonuses to French ship owners and builders and established a method of maritime conscription, arsenals, and academies for the training of sailors. In 1661 France possessed 18 unseaworthy vessels; by 1681 it had 276 frigates, galleys, and ships of the line. Colbert tried to organize and regulate the entire French economy for the glory of the French state as embodied in the king.

Colbert hoped to make Canada—rich in untapped minerals and some of the best agricultural land in the world—part of a vast French empire. He gathered four thousand peasants from western France and shipped them to Canada, where they peopled the province of Quebec. (In 1608, one year after the English arrived at Jamestown, Virginia, Sully had established the city of Quebec, which became the capital of French Canada.) Subsequently, the Jesuit Marquette and the merchant Joliet sailed down the Mississippi River and took possession of the land on both sides, as far south as present-day Arkansas. In 1684 the French explorer La Salle continued down the Mississippi to its mouth and claimed vast territories and the rich delta for Louis XIV. The area was called, naturally, "Louisiana."

How successful were Colbert's policies? His achievement in the development of manufacturing was prodigious. The textile industry, especially in woolens, expanded enormously, and "France. . . had become in 1683 the leading nation of the world in industrial productivity."[4] The commercial classes prospered, and between 1660 and 1700 their position steadily improved. The national economy, however, rested on agriculture. Although French peasants did not become serfs, as did the peasants of eastern Europe, they were mercilessly taxed. After 1685 other hardships afflicted them: poor harvests, continuing deflation of the currency, and fluctuation in the price of grain. Many peasants emigrated. With the decline in population and thus in the number of taxable people (the poorest), the state's resources fell. A totally inadequate tax base and heavy expenditure for war in the later years of the reign nullified Colbert's goals.

THE REVOCATION OF THE EDICT OF NANTES

We now see with the proper gratitude what we owe to God . . . for the best and largest part of our subjects of the so-called reformed religion have embraced Catholicism, and now that, to the extent that the execution of the Edict of Nantes remains useless, we have judged that we can do nothing better to wipe out the memory of the troubles, of the confusion, of the evils that the progress of this false religion has caused our kingdom . . . than to revoke entirely the said Edict.[5]

Thus in 1685, Louis XIV revoked the Edict of Nantes, by which his grandfather Henry IV had granted liberty of conscience to French Huguenots. The new law ordered the destruction of churches, the closing of schools, the Catholic baptism of Huguenots, and the exile of Huguenot pastors who refused to renounce their faith. Why? There had been so many mass conversions during previous years (many of them forced) that Madame de Maintenon, Louis's second wife, could say that "nearly all the Huguenots were converted." Some Huguenots had emigrated. Richelieu had already deprived French Calvinists of political rights. Why, then, did Louis, by revoking the Edict, persecute some of his most loyal and industrially skilled subjects, force others to flee abroad, and provoke the outrage of Protestant Europe?

Recent scholarship has convincingly shown that Louis XIV was basically tolerant. He insisted on religious unity not for religious but for political reasons. His goal was "one king, one law, one faith." He hated division within the realm and insisted that religious unity was essential to his royal dignity and to the security of the state. The seventeenth century, moreover, was not a tolerant one. While France in the early years of Louis's reign permitted religious liberty, it was not a popular policy. Revocation was solely the king's decision, and it won him enormous praise. "If the flood of congratulation means anything, it . . . was probably the one act of his reign that, at the time, was popular with the majority of his subjects."[6]

Plus on a de moyens, plus on e[...]
Ce pauure apporte tout, bled, fruit, [...]
Ce gros Milord assis, prest a tou[...]
Ne luy veut pas donner la douceu[...]

a la mouche
qui volle
il ne faut
point daisia

Il faut
paier ou
agreer.

A tous
Seigneurs
tous
honneurs.

Maigre
comme vn
leurier
datache

Plus a le Diable
plus il en veut auoir.
I lagnit [...]

Le Noble est l'araignée et
le Paisan la mouche.

The Spider and the Fly In reference to the insect symbolism (upper left), the caption on the lower left side of this illustration states, "The noble is the spider, the peasant the fly." The other caption (upper right) notes, "The more people have, the more they want. The poor man brings everything—wheat, fruit, money, vegetables. The greedy lord sitting there ready to take everything will not even give him the favor of a glance." This satirical print summarizes peasant grievances. *(New York Public Library)*

While contemporaries applauded Louis XIV, scholars since the eighteenth century damned him for the adverse impact that revocation had on the economy and foreign affairs. Tens of thousands of Huguenot craftsmen, soldiers, and businesspeople emigrated, depriving France of their skills and tax revenues and carrying their bitterness to Holland, England, and Prussia. Modern scholarship has greatly modified this picture. While Huguenot settlers in northern Europe aggravated Protestant hatred for Louis, the revocation of the Edict of Nantes had only minor and scattered effects on French economic development.[7]

FRENCH CLASSICISM

Scholars characterize the art and literature of the age of Louis XIV as "French classicism." By this they mean that the artists and writers of the late seventeenth century deliberately imitated the subject matter and style of classical antiquity; that their work resembled that of Renaissance Italy; and that French art possessed the classical qualities of discipline, balance, and restraint. Classicism was the official style of Louis's court. In painting, however, French classicism had already reached its peak before 1661, the beginning of the king's personal government.

Poussin: The Rape of the Sabine Women Considered the greatest French painter of the seventeenth century, Poussin in this dramatic work (ca 1636) shows his complete devotion to the ideals of classicism. The heroic figures are superb physical specimens, but hardly life-like. *(Metropolitan Museum of Art, New York, Harris Brisbane Dick Fund, 1946)*

Nicholas Poussin (1593–1665) is generally considered the finest example of French classicist painting. Poussin spent all but eighteen months of his creative life in Rome because he found the atmosphere in Paris uncongenial. Deeply attached to classical antiquity, he believed that the highest aim of painting was to represent noble actions in a logical and orderly, but not realistic, way. His masterpiece, "The Rape of the Sabine Women," exhibits these qualities. Its subject is an incident in Roman history; the figures of people and horses are ideal representations, and the emotions expressed are studied, not spontaneous. Even the buildings are exact architectural models of ancient Roman structures.

While Poussin selected grand and "noble" themes, Louis Le Nain (1593–1648) painted genre scenes of peasant life. At a time when artists favored Biblical and classical allegories, Le Nain's paintings are unique for their depiction of peasants. The highly realistic group assembled in "The Peasant Family" has great human dignity. The painting itself is reminiscent of portrayals of peasants by seventeenth-century Dutch painters.

Le Nain and Poussin, whose paintings still had individualistic features, did their work before 1661. After Louis's accession to power, the principles of absolutism molded the ideals of French classicism. Individualism was not allowed, and artists' efforts were directed to the glorification of the state as personified by the king. Precise rules governed all aspects of culture, with the goal of formal and restrained perfection.

Contemporaries said that Louis XIV never ceased playing the role of grand monarch on the stage of his court. If the king never fully relaxed from the pressures and intrigues of government, he did enjoy music and theater and used them as a backdrop for court ceremonial. Louis favored Jean-Baptiste Lully (1632–1687), whose orchestral works combine lively animation with the restrained austerity typical of French classicism. Lully also composed court ballets, and his operatic productions achieved a powerful influence throughout Europe. Louis supported François Couperin (1668–1733), whose harpsichord and organ works possess the regal grandeur the king loved, and Marc-Antoine Charpentier (1634–1704), whose solemn religious music entertained him at meals. Charpentier received a pension for the *Te Deums,* hymns of thanksgiving, he composed to celebrate French military victories.

Louis XIV loved the stage, and in the plays of Molière and Racine his court witnessed the finest achievements in the history of the French theater. When Jean-Baptiste Poquelin (1622–1673), the son of a prosperous tapestry maker, refused to join his father's business and entered the theater, he took the stage name "Molière." As playwright, stage manager, director, and actor, Molière produced comedies that exposed the hypocrisies and follies of society through brilliant caricature. *Tartuffe* satirized the religious hypocrite, *Le Bourgeois Gentilhomme (The Would-Be Gentleman)* attacked the social parvenu, and *Les Femmes Savantes (The Learned Women)* mocked the fashionable pseudo-intellectuals of the day. In structure Molière's plays followed classical models, but they were based on careful social observation. Molière made the bourgeoisie the butt of his ridicule; he stopped short of criticizing the nobility, thus reflecting the policy of his royal patron.

While Molière dissected social mores, his contemporary Jean Racine (1639–1699) analyzed the power of love. Racine based his tragic dramas on Greek and Roman legends, and his persistent theme is the conflict of good and evil. Several plays—*Andromache, Berenice, Iphigenie,* and *Phèdre*—bear the names of women and deal with the power of passion in women. Louis preferred *Mithridates* and *Britannicus* because of the "grandeur" of their themes. For sim-

plicity of language, symmetrical structure, and calm restraint, the plays of Racine represent the finest examples of French classicism. His tragedies and Molière's comedies are still produced today.

LOUIS XIV'S WARS

Just as the architecture and court life at Versailles served to reflect the king's glory, and as the economy of the state under Colbert was managed to advance the king's prestige, so did Louis XIV use war to exalt himself above the other rulers and nations of Europe. He visualized himself as a great military hero. "The character of a conqueror," he remarked, "is regarded as the noblest and highest of titles." Military glory was his aim. In 1666 Louis appointed François le Tellier (later marquis of Louvois) as secretary of war. Louvois created a professional army, which was modern in the sense that the French state, rather than private nobles, employed the soldiers. The king himself took personal command of the army and directly supervised all aspects and details of military affairs.

A commissariat was established to feed the troops, in place of their ancient practice of living off the countryside. An ambulance corps was designed to look after the wounded. Uniforms and weapons were standardized. Finally, a rational system of recruitment, training, discipline, and promotion was imposed. With this new military machine, for the first time in Europe's history one national state, France, was able to dominate the politics of Europe.

Louis continued on a broader scale the expansionist policy begun by Cardinal Richelieu. In 1667, using a dynastic excuse, he invaded Flanders, part of the Spanish Netherlands, and Franche-Comté in the east. In consequence he acquired twelve towns, including the important commercial centers of Lille and Tournai (see Map 16.1). Five years later, Louis personally led an army of over a hundred thousand men into Holland, and the Dutch ultimately saved themselves only by opening the dikes and flooding the countryside. This war, which lasted six years and eventually involved the Holy Roman Empire and Spain, was concluded by the Treaty of Nijmegen (1678). Louis gained additional Flemish towns and the whole of Franche-Comté.

Encouraged by his successes, by the weakness of the German Empire, and by divisions among the other European powers, Louis continued his aggression. In 1681 he seized the city of Strasbourg and three years later sent his armies into the province of Lorraine. At that moment the king seemed invincible. In fact, Louis had reached the limit of his expansion at Nijmegen. The wars of the 1680s and 1690s brought him no additional territories. In 1689 the Dutch prince William of Orange, a bitter foe of Louis XIV, became king of England. William joined the League of Augsburg—which included the Habsburg emperor, the kings of Spain and Sweden, and the electors of Bavaria, Saxony, and the Palatinate—adding British resources and men to the alliance. Neither the French nor the league won any decisive victories. The alliance served instead as preparation for the long-expected conflict known as the War of the Spanish Succession.

This struggle (1701–1713), provoked by the territorial disputes of the past century, also involved the dynastic question of the succession to the Spanish throne. It was an open secret in Europe that the king of Spain, Charles II (1665–1700), was mentally defective and sexually impotent. In 1698 the European powers, including France, agreed by treaty to partition, or divide, the vast Spanish possessions between the king of France and the Holy Roman emperor, who were Charles II's brothers-in-law. When Charles died in 1700, however, his will left the Spanish crown and the worldwide Spanish empire to Philip of Anjou, Louis XIV's grandson. While the will specifically rejected union of the French and Spanish crowns, Louis was obviously the power in France, not his seventeen-year-old grandson. Louis reneged on the treaty and accepted the will.

The Dutch and the English would not accept French acquisition of the Spanish Netherlands and of the rich trade with the Spanish colonies. The union of the Spanish and French crowns, moreover, would have totally upset the European balance of power. The Versailles declaration that "the Pyrenees no longer exist" provoked the long-anticipated crisis. In 1701 the English, Dutch, Austrians, and Prussians formed the Grand Alliance against Louis XIV. They claimed that they were fighting to prevent France from becoming too strong in Europe, but during the previous half-century, overseas maritime rivalry among France, Holland, and England had created serious international tension. The secondary motive of the allied powers was to check France's expanding commercial power in North America, Asia, and Africa. In the ensuing series of conflicts, two great soldiers dominated the alliance against France: Eu-

gene, prince of Savoy, representing the Holy Roman Empire, and the Englishman John Churchill, subsequently duke of Marlborough. Eugene and Churchill inflicted a severe defeat on Louis in 1704 at Blenheim in Bavaria. Marlborough followed with another victory at Ramillies near Namur in Brabant.

The war was finally concluded at Utrecht in 1713, where the principle of partition was applied. Louis's grandson Philip remained the first Bourbon king of Spain on the understanding that the French and Spanish crowns would never be united. France surrendered Newfoundland, Nova Scotia, and the Hudson Bay territory to England, which also acquired Gibraltar, Minorca, and control of the African slave trade from Spain. The Dutch gained little because Austria received the former Spanish Netherlands.

The Peace of Utrecht had important international consequences. It represented the balance-of-power principle in operation, setting limits on the extent to which any one power, in this case France, could expand. The treaty completed the decline of Spain as a great power. It vastly expanded the British Empire. Finally, Utrecht gave European powers experience in international cooperation, thus preparing them for the alliances against France at the end of the century.

The Peace of Utrecht marked the end of French expansionist policy. In Louis's thirty-five-year quest for military glory, his main territorial acquisition was Strasbourg. Even revisionist historians, who portray the aging monarch as responsible in negotiation and moderate in his demands, acknowledge "that the widespread misery in France during the period was in part due to royal policies, especially the incessant wars."[8] To raise revenue for the wars, forty thousand additional offices had been sold, thus increasing the number of families exempt from future taxation. Constant war had disrupted trade, which meant the state could not tax the profits of trade. Widespread starvation in the provinces provoked peasant revolts, especially in Brittany. In 1714 France hovered on the brink of financial bankruptcy. Louis had exhausted the country without much compensation. It is no wonder that when he died on September 1, 1715, Saint-Simon wrote, "Those . . . wearied by the heavy and oppressive rule of the King and his ministers, felt a delighted freedom. . . . Paris . . . found relief in the hope of liberation. . . . The provinces . . . quivered with delight . . . [and] the people, ruined, abused, despairing, now thanked God for a deliverance which answered their most ardent desires."[9]

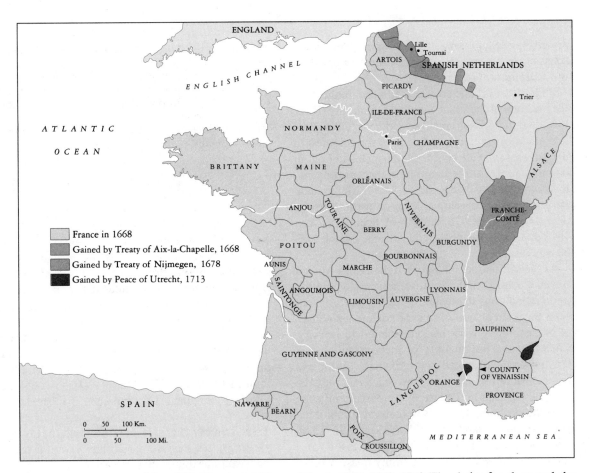

MAP 16.1 The Acquisitions of Louis XIV, 1668–1713 The desire for glory and the weakness of his German neighbors encouraged Louis' expansionist policy. But he paid a high price for his acquisitions.

THE DECLINE OF ABSOLUTIST SPAIN IN THE SEVENTEENTH CENTURY

Spanish absolutism and greatness had preceded that of the French. In the sixteenth century, Spain (or, more precisely, the kingdom of Castile) had developed the standard features of absolute monarchy: a permanent bureaucracy staffed by professionals employed in the various councils of state, a standing army, and national taxes, the *servicios,* which fell most heavily on the poor.

France depended on financial and administrative unification within its borders; Spain had developed an international absolutism on the basis of silver bullion from Peru. Spanish gold and silver, armies, and glory had dominated the Continent for most of the sixteenth century, but by the 1590s the seeds of disas-

ter were sprouting. While France in the seventeenth century represented the classic model of the modern absolute state, Spain was experiencing steady decline. Fiscal disorder, political incompetence, population decline, intellectual isolation, and psychological malaise—all combined to reduce Spain, by 1715, to the rank of a second-rate power.

The fabulous and seemingly inexhaustible flow of silver from Mexico and Peru had led Philip II (page 482) to assume the role of defender of Roman Catholicism in Europe. In order to humble the Protestant Dutch and to control the Spanish Netherlands, Philip believed that England, the Netherlands' greatest supporter, had to be crushed. He poured millions of Spanish ducats and all of Spanish hopes into the vast fleet that sailed in 1588. When the "Invincible Armada" went down in the North Sea, a century of

Spanish pride and power went with it. After 1590 a spirit of defeatism and disillusionment crippled almost all efforts at reform.

Philip II's Catholic crusade had been financed by the revenues of the Spanish-Atlantic economy. These included, in addition to silver and gold bullion, the sale of cloth, grain, oil, and wine to the colonies. In the early seventeenth century, the Dutch and English began to trade with the Spanish colonies, cutting into the revenues that had gone to Spain. Mexico and Peru themselves developed local industries, further lessening their need to buy from Spain. Between 1610 and 1650, Spanish trade with the colonies fell 60 percent.

At the same time, the native Indians and African slaves, who worked the South American silver mines under conditions that would have disgraced the ancient Egyptian pharaohs, suffered frightful epidemics of disease. Moreover, the lodes started to run dry. Consequently, the quantity of metal produced for Spain steadily declined. Nevertheless, in Madrid royal expenditures constantly exceeded income. The remedies applied in the face of a mountainous state debt and declining revenues were devaluation of the coinage and declarations of bankruptcy. In 1596, 1607, 1627, 1647, and 1680, Spanish kings found no solution to the problem of an empty treasury other than to cancel the national debt. Given the frequency of cancellation, naturally public confidence in the state deteriorated.

Spain, in contrast to the other countries of western Europe, had only a tiny middle class. Disdain for money, in a century of increasing commercialism and bourgeois attitudes, reveals a significant facet of the Spanish national character. Public opinion, taking its cue from the aristocracy, condemned money-making as vulgar and undignified. Those with influence or connections sought titles of nobility and social prestige. Thousands entered economically unproductive professions and became priests, monks, and nuns: there were said to be nine thousand monasteries in the province of Castile alone. The flood of gold and silver had produced severe inflation, pushing the costs of production in the textile industry higher and higher, to the point that Castilian cloth could not compete in colonial and international markets. Many manufacturers and businessmen found so many obstacles in the way of profitable enterprise that they simply gave up.[10]

Velazquez: The Maids of Honor The Infanta Margarita painted in 1656 with her maids and playmates has invaded the artist's studio, while her parents' image is reflected in the mirror on the back wall. Velazquez (extreme left), who powerfully influenced nineteenth-century impressionist painters, imbued all of his subjects, including the pathetic dwarf (right, in black) with a sense of dignity. *(Giraudon/Art Resource)*

Spanish aristocrats, attempting to maintain an extravagant lifestyle they could no longer afford, increased the rents on their estates. High rents and heavy taxes in turn drove the peasants from the land. Agricultural production suffered and the peasants departed for the large cities, where they swelled the ranks of unemployed beggars.

Their most Catholic majesties, the kings of Spain, had no solutions to these dire problems. If one can discern personality from pictures, the portraits of Philip III (1598–1622), Philip IV (1622–1665), and Charles II hanging in the Prado, the Spanish national museum in Madrid, reflect the increasing weakness of the dynasty. Their faces—the small, beady eyes, the long noses, the jutting Habsburg jaws, the pathetically stupid expressions—tell a story of excessive inbreeding and decaying monarchy. The Spanish kings all lacked force of character. Philip III, a pallid, melancholy, and deeply pious man "whose only virtue appeared to reside in a total absence of vice," handed the government over to the lazy duke of Lerma, who used it to advance his personal and familial wealth. Philip IV left the management of his several kingdoms to Count Olivares.

Olivares was an able administrator. He did not lack energy and ideas; he devised new sources of revenue. But he clung to the grandiose belief that the solution to Spain's difficulties rested in a return to the imperial tradition. Unfortunately, the imperial tradition demanded the revival of war with the Dutch, at the expiration of a twelve-year truce in 1622, and a long war with France over Mantua (1628–1659). Spain thus became embroiled in the Thirty Years' War. These conflicts, on top of an empty treasury, brought disaster.

In 1640 Spain faced serious revolts in Naples and Portugal, and in 1643 the French inflicted a crushing defeat on a Spanish army in Belgium. By the Treaty

of the Pyrenees of 1659, which ended the French-Spanish wars, Spain was compelled to surrender extensive territories to France. This treaty marked the end of Spain as a great power.

Seventeenth-century Spain was the victim of its past. It could not forget the grandeur of the sixteenth century and look to the future. The bureaucratic councils of state continued to function as symbols of the absolute Spanish monarchy. But because those councils were staffed by aristocrats, it was the aristocracy that held the real power. Spanish absolutism had been built largely on slave-produced gold and silver. When the supply of bullion decreased, the power and standing of the Spanish state declined.

The most cherished Spanish ideals were military glory and strong Roman Catholic faith. In the seventeenth century, Spain lacked the finances and the manpower to fight the expensive wars in which it foolishly got involved. Spain also ignored the new mercantile ideas and scientific methods, because they came from heretical nations, Holland and England. The incredible wealth of South America destroyed the tiny Spanish middle class and created contempt for business and manual labor.

The decadence of the Habsburg dynasty and the lack of effective royal councilors also contributed to Spanish failure. Spanish leaders seemed to lack the will to reform. Pessimism and fatalism permeated national life. In the reign of Philip IV, a royal council was appointed to plan the construction of a canal linking the Tagus and Manzanares rivers in Spain. After interminable debate, the committee decided that "if God had intended the rivers to be navigable, He would have made them so."

In the brilliant novel *Don Quixote,* the Spanish writer Cervantes (1547–1616) produced one of the great masterpieces of world literature. *Don Quixote* —on which the modern play *The Man of La Mancha* is based—delineates the whole fabric of sixteenth-century Spanish society. The main character, Don Quixote, lives in a world of dreams, traveling about the countryside seeking military glory. From the title of the book, the English language has borrowed the word *quixotic.* Meaning "idealistic but impractical," the term characterizes seventeenth-century Spain. As a leading scholar has written, "The Spaniard convinced himself that reality was what he felt, believed, imagined. He filled the world with heroic reverberations. Don Quixote was born and grew."[11]

CONSTITUTIONALISM

The seventeenth century, which witnessed the development of absolute monarchy, also saw the appearance of the constitutional state. While France and later Prussia, Russia, and Austria solved the question of sovereignty with the absolutist state, England and Holland evolved toward the constitutional state. What is constitutionalism? Is it identical to democracy?

Constitutionalism is the limitation of government by law. Constitutionalism also implies a balance between the authority and power of the government on the one hand and the rights and liberties of the subjects on the other. The balance is often very delicate.

A nation's constitution may be written or unwritten. It may be embodied in one basic document, occasionally revised by amendment or judicial decision, like the Constitution of the United States. Or a constitution may be partly written and partly unwritten and include parliamentary statutes, judicial decisions, and a body of traditional procedures and practices, like the English and Canadian constitutions. Whether written or unwritten, a constitution gets its binding force from the government's acknowledgment that it must respect that constitution—that is, that the state must govern according to the laws. Likewise, in a constitutional state, the people look on the law and the constitution as the protectors of their rights, liberties, and property.

Modern constitutional governments may take either a republican or a monarchial form. In a constitutional republic, the sovereign power resides in the electorate and is exercised by the electorate's representatives. In a constitutional monarchy, a king or queen serves as the head of state and possesses some residual political authority, but again the ultimate or sovereign power rests in the electorate.

A constitutional government is not, however, quite the same as a democratic government. In a complete democracy, *all* the people have the right to participate either directly, or indirectly through their elected representatives, in the government of the state. Democratic government, therefore, is intimately tied up with the *franchise* (the vote). Most men could not vote until the late nineteenth century. Even then, women—probably the majority in Western societies —lacked the franchise; they gained the right to vote only in the twentieth century. Consequently, although constitutionalism developed in the seventeenth century, full democracy was achieved only in very recent times.

THE DECLINE OF ROYAL ABSOLUTISM IN ENGLAND (1603–1649)

In the late sixteenth century the French monarchy was powerless; a century later, the king's power was absolute. In 1588 Queen Elizabeth I of England exercised very great personal power; by 1689 the English monarchy was severely circumscribed. Change in England was anything but orderly. Seventeenth-century England displayed as much political stability as some modern African states. It executed one king, experienced a bloody civil war, experimented with military dictatorship, then restored the son of the murdered king, and finally, after a bloodless revolution, established constitutional monarchy. Political stability came only in the 1690s. How do we account for the fact that, after such a violent and tumultuous century, England laid the foundations for constitutional monarchy? What combination of political, socioeconomic, and religious factors brought on a civil war in 1642 to 1649 and then the constitutional settlement of 1688 to 1689?

The extraordinary success of Elizabeth I had rested on her political shrewdness and flexibility, her careful management of finances, her wise selection of ministers, her clever manipulation of Parliament, and her sense of royal dignity and devotion to hard work. The aging queen had always refused to discuss the succession. After her Scottish cousin James Stuart succeeded her as James I (1603–1625), Elizabeth's strengths seemed even greater than they actually had been. The Stuarts lacked every quality Elizabeth had possessed.

King James was well educated and learned but lacking in common sense—he was once called "the wisest fool in Christendom." He also lacked the common touch. Urged to wave at the crowds who waited to greet their new ruler, James complained that he was tired and threatened to drop his breeches "so they can cheer at my arse." Having left barbarous and violent Scotland for rich and prosperous England, James believed he had entered the "Promised Land." As soon as he got to London, the new English king went to see the crown jewels.

Abysmally ignorant of English law and of the English Parliament, but sublimely arrogant, James was devoted to the theory of the divine right of kings. He expressed his ideas about divine right in his essay "The Trew Law of Free Monarchy." According to James I, a monarch has a divine (or God-given) right to his authority and is responsible only to God. Rebellion is the worst of political crimes. If a king orders something evil, the subject should respond with passive disobedience but should be prepared to accept any penalty for noncompliance.

James substituted political theorizing and talk for real work. He lectured the House of Commons: "There are no privileges and immunities which can stand against a divinely appointed King." This notion, implying total royal jurisdiction over the liberties, persons, and properties of English men and women, formed the basis of the Stuart concept of absolutism. Such a view ran directly counter to the long-standing English idea that a person's property could not be taken away without due process of law. James's expression of such views before the English House of Commons constituted a grave political mistake.

The House of Commons guarded the state's pocketbook, and James and later Stuart kings badly needed to open that pocketbook. Elizabeth had bequeathed to James a sizable royal debt. Through prudent management the debt could have been gradually reduced, but James I looked on all revenues as a happy windfall to be squandered on a lavish court and favorite courtiers. In reality, the extravagance displayed in James's court, as well as the public flaunting of his male lovers, weakened respect for the monarchy.

Elizabeth had also left to her Stuart successors a House of Commons that appreciated its own financial strength and intended to use that strength to acquire a greater say in the government of the state. The knights and burgesses who sat at Westminster in the late sixteenth and early seventeenth centuries wanted to discuss royal expenditures, religious reform, and foreign affairs. In short, the Commons wanted what amounted to sovereignty.

Profound social changes had occurred since the sixteenth century. The English House of Commons during the reigns of James I and his son Charles I (1625–1649) was very different from the assembly Henry VIII had terrorized into passing his Reformation legislation. A social revolution had brought about the change. The dissolution of the monasteries and the sale of monastic land had enriched many people. Agricultural techniques like the draining of wasteland and the application of fertilizers improved the land and its yield. Old manorial common land had been enclosed and turned into sheep runs; breeding was carefully supervised, and the size of the flocks increased. In these activities, as well as in renting and leasing parcels of land, precise accounts were kept.

Many men invested in commercial ventures at home, such as the expanding cloth industry, and in partnerships and joint stock companies engaged in foreign enterprises. They made prudent marriages. All these developments led to a great deal of social mobility. Both in commerce and in agriculture, the English in the late sixteenth and early seventeenth centuries were capitalists, investing their profits to make more money. Though the international inflation of the period hit everywhere, in England commercial and agricultural income rose faster than prices. Wealthy country gentry, rich city merchants, and financiers invested abroad.

The typical pattern was for the commercially successful to set themselves up as country gentry, thus creating an elite group that possessed a far greater proportion of land and of the national wealth in 1640 than had been the case in 1540. Small wonder that in 1640 someone could declare in the House of Commons, probably accurately, "We could buy the House of Lords three times over." Increased wealth had also produced a better-educated and more articulate House of Commons. Many members had acquired at least a smattering of legal knowledge, and they used that knowledge to search for medieval precedents from which to argue against the king. The class that dominated the Commons wanted political power corresponding to its economic strength.

In England, unlike France, there was no social stigma attached to paying taxes. Members of the House of Commons were willing to tax themselves provided they had some say in the expenditure of those taxes and in the formulation of state policies. The Stuart kings, however, considered such ambitions intolerable presumption and a threat to their divine-right prerogative. Consequently, at every parliament between 1603 and 1640, bitter squabbles erupted between the crown and the wealthy, articulate, and legal-minded Commons. Charles I's attempt to govern without Parliament (1629–1640), and to finance his government by arbitrary nonparliamentary levies, brought the country to a crisis.

An issue graver than royal extravagance and Parliament's desire to make law also disturbed the English and embittered relations between the king and the House of Commons. That problem was religion. In the early seventeenth century, increasing numbers of English people felt dissatisfied with the Church of England established by Henry VIII and reformed by Elizabeth. Many Puritans (see Chapter 14) believed that Reformation had not gone far enough. They wanted to "purify" the Anglican church of Roman Catholic elements—elaborate vestments and ceremonial, the position of the altar in the church, even the giving and wearing of wedding rings.

It is very difficult to establish what proportion of the English population was Puritan. According to the present scholarly consensus, the dominant religious groups in the early seventeenth century were Calvinist and Protestant; their more zealous members were Puritans. It also seems clear that many English men and women were attracted by the socioeconomic implications of John Calvin's theology. Calvinism emphasized hard work, sobriety, thrift, competition, and postponement of pleasure, and tended to link sin and poverty with weakness and moral corruption. These attitudes fit in precisely with the economic approaches and practices of many (successful) businessmen and farmers. These values have frequently been called the "Protestant," "middle class," or "capitalist ethic." While it is hazardous to identify capitalism and progress with Protestantism—there were many successful Catholic capitalists—the "Protestant virtues" represented the prevailing values of members of the House of Commons.

James I and Charles I both gave the impression of being highly sympathetic to Roman Catholicism. Charles supported the policies of William Laud, archbishop of Canterbury (1573–1645), who tried to impose elaborate ritual and rich ceremonial on all churches. Laud insisted on complete uniformity of church services and enforced that uniformity through an ecclesiastical court called the "Court of High Commission." People believed the country was being led back to Roman Catholicism. When in 1637 Laud attempted to impose a new prayer book, modeled on the Anglican Book of Common Prayer, on the Presbyterian Scots, the Scots revolted. In order to finance an army to put down the Scots, King Charles was compelled to summon Parliament in November 1640.

For eleven years Charles I had ruled without Parliament, financing his government through extraordinary stopgap levies, considered illegal by most English people. For example, the king revived a medieval law requiring coastal districts to help pay the cost of ships for defense, but levied the tax, called "ship money," on inland as well as coastal counties. When the issue was tested in the courts, the judges, having been suborned, decided in the king's favor.

Most members of Parliament believed that such taxation without consent amounted to arbitrary and absolute despotism. Consequently, they were not willing to trust the king with an army. Accordingly, this parliament, commonly called the "Long Parliament" because it sat from 1640 to 1660, proceeded to enact legislation that limited the power of the monarch and made arbitrary government impossible.

In 1641 the Commons passed the Triennial Act, which compelled the king to summon Parliament every three years. The Commons impeached Archbishop Laud and abolished the House of Lords and the Court of High Commission. It went further and threatened to abolish the institution of episcopacy. King Charles, fearful of a Scottish invasion—the original reason for summoning Parliament—accepted these measures. Understanding and peace were not achieved, however, partly because radical members of the Commons pushed increasingly revolutionary propositions, partly because Charles maneuvered to rescind those he had already approved. An uprising in Ireland precipitated civil war.

Ever since Henry II had conquered Ireland in 1171, English governors had mercilessly ruled the Irish, and English landlords had ruthlessly exploited them. The English Reformation had made a bad situation worse: because the Irish remained Catholic, religious differences became united with economic and political oppression. Without an army, Charles I could neither come to terms with the Scots nor put down the Irish rebellion, and the Long Parliament remained unwilling to place an army under a king it did not trust. Charles thus instigated military action against parliamentary forces. He recruited an army drawn from the nobility and their cavalry staff, the rural gentry, and mercenaries. The parliamentary army was composed of the militia of the city of Lon-

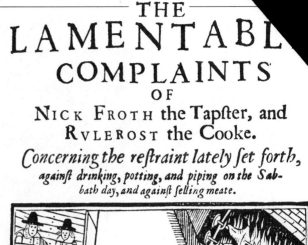

Printed in the yeare, 1641.

Puritan Ideals Opposed The Puritans preached sober living and abstention from alcoholic drink, rich food, and dancing. This pamphlet reflects the common man's hostility to such restraints. "Potting" refers to tankards of beer; "piping" means making music. *(The British Library)*

don, country squires with business connections, and men with a firm belief in the spiritual duty of serving.

The English civil war (1642–1646) tested whether sovereignty in England was to reside in the king or in Parliament. The civil war did not resolve that problem, although it ended in 1649 with the execution of King Charles on the charge of high treason—a severe blow to the theory of divine right monarchy. The period between 1649 and 1660, called the "Interregnum" because it separated two monarchial periods, witnessed England's solitary experience of military dictatorship.

The House of Commons This seal of the Commonwealth shows the small House of Commons in session with the speaker presiding: the legend "in the third year of freedom" refers to 1651, three years after the abolition of the monarchy. In 1653, however, Cromwell abolished this "Rump Parliament"—so-called because it consisted of the few surviving members elected before the Civil War—and he and the army governed the land. *(The British Library)*

PURITANICAL ABSOLUTISM IN ENGLAND: CROMWELL AND THE PROTECTORATE

The problem of sovereignty was vigorously debated in the middle years of the seventeenth century. In *Leviathan,* the English philosopher and political theorist Thomas Hobbes (1588–1679) maintained that sovereignty is ultimately derived from the people, who transfer it to the monarchy by implicit contract. The power of the ruler is absolute, but kings do not hold their power by divine right. This view pleased no one in the seventeenth century.

When Charles I was beheaded on January 30, 1649, the kingship was abolished. A *commonwealth,* or republican form of government, was proclaimed. Theoretically, legislative power rested in the surviving members of Parliament and executive power in a council of state. In fact, the army that had defeated the royal forces controlled the government, and Oliver Cromwell controlled the army. Though called the "Protectorate," the rule of Cromwell (1653–1658) constituted military dictatorship.

Oliver Cromwell (1599–1658) came from the country gentry, the class that dominated the House of Commons in the early seventeenth century. He himself had sat in the Long Parliament. Cromwell rose in the parliamentary army and achieved nationwide fame by infusing the army with his Puritan convictions and molding it into the highly effective military machine, called the "New Model Army," that defeated the royalist forces.

Parliament had written a constitution, the Instrument of Government (1653), that invested executive power in a lord protector (Cromwell) and a council of state. The instrument provided for triennial parliaments and gave Parliament the sole power to raise taxes. But after repeated disputes, Cromwell tore the document up. He continued the standing army and proclaimed quasi-martial law. He divided England into twelve military districts, each governed by a major general. The major-generals acted through the justices of the peace, though they sometimes overrode them. On the issue of religion, Cromwell favored broad toleration, and the Instrument of

Government gave all Christians, except Roman Catholics, the right to practice their faith. Toleration meant state protection of many different Protestant sects, and most English people had no enthusiasm for such a notion; the idea was far ahead of its time. Cromwell identified Irish Catholicism with sedition. In 1649 he crushed rebellion there with merciless savagery, leaving a legacy of Irish hatred for England that has not yet subsided. The state rigorously censored the press, forbade sports, and kept the theaters closed.

Cromwell's regulation of the nation's economy had features typical of seventeenth-century absolutism. The lord protector's policies were mercantilist, similar to those Colbert established in France. Cromwell enforced a navigation act requiring that English goods be transported on English ships. The navigation act was a great boost to the development of an English merchant marine and brought about a short but successful war with the commercially threatened Dutch. Cromwell also welcomed the immigration of Jews, because of their skills, and they began to return to England in larger numbers after four centuries of absence.

Absolute government collapsed when Cromwell died in 1658 because the English got fed up with military rule. They longed for a return to civilian government, restoration of the common law, and social stability. Moreover, the strain of creating a community of puritanical saints proved too psychologically exhausting. Government by military dictatorship was an unfortunate experiment that the English never forgot nor repeated. By 1660 they were ready to restore the monarchy.

THE RESTORATION OF THE ENGLISH MONARCHY

The Restoration of 1660 re-established the monarchy in the person of Charles II (1660–1685), eldest son of Charles I. At the same time both houses of Parliament were restored, together with the established Anglican church, the courts of law, and the system of local government through justices of the peace. The Restoration failed to resolve two serious problems. What was to be the attitude of the state toward Puritans, Catholics, and dissenters from the established church? And what was to be the constitutional position of the king—that is, what was to be the relationship between the king and Parliament?

About the first of these issues, Charles II, a relaxed, easygoing, and sensual man, was basically indifferent. He was not interested in doctrinal issues. Parliamentarians were, and they proceeded to enact a body of laws that sought to compel religious uniformity. Those who refused to receive the sacrament of the Church of England could not vote, hold public office, preach, teach, attend the universities, or even assemble for meetings, according to the Test Act of 1673. These restrictions could not be enforced. When the Quaker William Penn held a meeting of his Friends and was arrested, the jury refused to convict him.

In politics, Charles II was determined "not to set out in his travels again," which meant that he intended to get along with Parliament. Charles II's solution to the problem of the relationship between the king and the House of Commons had profound importance for later constitutional development. Generally good rapport existed between the king and the strongly royalist Parliament that had restored him. This rapport was due largely to the king's appointment of a council of five men who served both as his major advisers and as members of Parliament, thus acting as liason agents between the executive and the legislature. This body—known as the "Cabal" from the names of its five members (Clifford, Arlington, Buckingham, Ashley-Cooper and Lauderdale)—was an ancestor of the later cabinet system (see page 532). It gradually came to be accepted that the Cabal was answerable in Parliament for the decisions of the king. This development gave rise to the concept of ministerial responsibility: royal ministers must answer to the Commons.

Harmony between the crown and Parliament rested on the understanding that Charles would summon frequent parliaments and that Parliament would vote him sufficient revenues. However, although Parliament believed Charles had a virtual divine right to govern, it did not grant him an adequate income. Accordingly, in 1670 Charles entered into a secret agreement with Louis XIV. The French king would give Charles £200,000 annually, and in return Charles would relax the laws against Catholics, gradually re-Catholicize England, and support French policy against the Dutch.

When the details of this secret treaty leaked out, a great wave of anti-Catholic fear swept England. This fear was compounded by a crucial fact: although Charles had produced several bastards, he had no legitimate children. It therefore appeared that his brother and heir, James, duke of York, who had publicly acknowledged his Catholicism, would inaugurate a Catholic dynasty. The combination of hatred for the French absolutism embodied in Louis XIV, hostility to Roman Catholicism, and fear of a permanent Catholic dynasty produced virtual hysteria. The Commons passed an exclusion bill denying the succession to a Roman Catholic, but Charles quickly dissolved Parliament and the bill never became law.

James II (1685–1688) did succeed his brother. Almost at once the worst English anti-Catholic fears, already aroused by Louis XIV's revocation of the Edict of Nantes, were realized. In direct violation of the Test Act, James appointed Roman Catholics to positions in the army, the universities, and local government. When these actions were tested in the courts, the judges, whom James had appointed, decided for the king. The king was suspending the law at will and appeared to be reviving the absolutism of his father and grandfather. He went further. Attempting to broaden his base of support with Protestant dissenters and nonconformists, James issued a declaration of indulgence granting religious freedom to all.

Two events gave the signals for revolution. First, seven bishops of the Church of England petitioned the king that they not be forced to read the declaration of indulgence because of their belief that it was an illegal act. They were imprisoned in the Tower of London but subsequently acquitted amid great public enthusiasm. Second, in June 1688 James's second wife produced a male heir. A Catholic dynasty seemed assured. The fear of a Roman Catholic monarchy, supported by France and ruling outside the law, prompted a group of eminent persons to offer the English throne to James's Protestant daughter, Mary, and her Dutch husband, Prince William of Orange. In November 1688 James II, his queen, and their infant son fled to France and became pensioners of Louis XIV.

THE TRIUMPH OF ENGLAND'S PARLIAMENT: CONSTITUTIONAL MONARCHY AND CABINET GOVERNMENT

The English call the events of 1688 the "Glorious Revolution." The revolution was indeed glorious in the sense that it replaced one king with another with a minimum of bloodshed. It also represented the de-

struction, once and for all, of the idea of divine-right monarchy. William and Mary accepted the English throne from Parliament and in so doing explicitly recognized the supremacy of Parliament. The revolution of 1688 established the principle that sovereignty, the ultimate power in the state, rested in Parliament and that the king ruled with the consent of the governed.

The men who brought about the revolution quickly framed their intentions in the Bill of Rights, the cornerstone of the modern British constitution. The basic principles of the Bill of Rights were formulated in direct response to Stuart absolutism. Law was to be made in Parliament; once made, it could not be suspended by the crown. Parliament had to be called at least every three years. Both elections to and debate in Parliament were to be free, in the sense that the Crown was not to interfere in them; this aspect of the bill was widely disregarded in the eighteenth century. Judges would hold their offices "during good behavior," which assured the independence of the judiciary. No longer could the crown get the judicial decisions it wanted by threats of removal. There was to be no standing army in peacetime—a limitation designed to prevent the repetition of either Stuart or Cromwellian military government. The Bill of Rights granted "that the subjects which are Protestants may have arms for their defense suitable to their conditions and as allowed by law,"[12] meaning that Catholics could not possess firearms because the Protestant majority feared them. Additional legislation granted freedom of worship to Protestant dissenters and nonconformists and required that the English monarch always be Protestant.

The Glorious Revolution found its best defense in the political philosopher John Locke's "Second Treatise on Civil Government" (1690). Locke (1632–1704) maintained that men set up civil governments in order to protect their property. The purpose of government, therefore, is to protect life, liberty, and property. A government that oversteps its proper function—protecting the natural rights of life, liberty, and property—becomes a tyranny. (By "natural" rights, Locke meant rights basic to all men because all have the ability to reason.) Under a tyrannical government, the people have the natural right to rebellion. Rebellion can be avoided if the government carefully respects the rights of citizens and if the people zealously defend their liberty. Recognizing the close relationship between economic

THE RISE OF WESTERN ABSOLUTISM AND CONSTITUTIONALISM

1581	Formation of the United Provinces of the Netherlands
1588	Defeat of the Spanish Armada
1589–1610	Reign of Henry IV of France; economic reforms help to restore public order, lay foundation for absolutist rule
1598	Edict of Nantes: Henry IV ends the French Wars of Religion
1608	France establishes its first Canadian settlement, at Quebec
1609	Philip III of Spain recognizes Dutch independence
1610–1650	Spanish trade with the New World falls by 60 percent
1618–1648	Thirty Years' War
1624–1643	Richelieu dominates French government
1625	Huguenot revolt in France; siege of La Rochelle
1629–1640	Eleven Years' Tyranny: Charles I attempts to rule England without the aid of Parliament
1640–1660	Long Parliament in England
1642–1646	English Civil War
1643–1661	Mazarin dominates France's regency government
1643–1715	Reign of Louis XIV
1648–1660	The Fronde: French nobility opposes centralizing efforts of monarchy
1648	Peace of Westphalia confirms Dutch independence from Spain
1649	Execution of Charles I; beginning of the Interregnum in England
1653–1658	The Protectorate: Cromwell heads military rule of England
1659	Treaty of the Pyrenees forces Spain to cede extensive territories to France, marks end of Spain as a great power
1660	Restoration of the English monarchy: Charles II returns from exile
1661	Louis XIV enters into independent rule
ca 1663–1683	Colbert directs Louis XIV's mercantilist economic policy
1673	Test Act excludes Roman Catholics from public office in England
	France invades Holland
1678	Treaty of Nijmegen: Louis XIV acquires Franche-Comté
1680	Treaty of Dover: Charles II secretly agrees with Louis XIV to re-catholicize England
1681	France acquires Strasbourg
1685	Revocation of the Edict of Nantes
1685–1688	James II rules England, attempts to restore Roman Catholicism as state religion
1688–1689	The Glorious Revolution establishes a constitutional monarchy under William III in England; enactment of the Bill of Rights
1701–1713	War of the Spanish Succession
1713	Peace of Utrecht ends French territorial acquisitions, expands the British Empire

and political freedom, Locke linked economic liberty and private property with political freedom. Locke served as the great spokesman for the liberal English revolution of 1689 and for representative government. His idea, inherited from ancient Greece and Rome (see page 120), that there are natural or universal rights, equally valid for all peoples and societies, played a powerful role in eighteenth-century Enlightenment thought. His ideas on liberty and tyranny were especially popular in colonial America.

The events of 1688 to 1690 did not constitute a *democratic* revolution. The revolution placed sovereignty in Parliament, and Parliament represented the upper classes. The great majority of English people acquired no say in their government. The English revolution established a constitutional monarchy; it also inaugurated an age of aristocratic government, which lasted at least until 1832 and probably until 1914.

In the course of the eighteenth century, the cabinet system of government evolved. The term *cabinet* refers to the small private room in which English rulers consulted their chief ministers. In a cabinet system, the leading ministers, who must have seats in and the support of a majority of the House of Commons, formulate common policy and conduct the business of the country. During the administration of one royal minister, Sir Robert Walpole (1721–1742), the idea developed that the cabinet was responsible to the House of Commons. The Hanoverian king George I (1714–1727) normally presided at cabinet meetings throughout his reign, but his son and heir George II (1727–1760) discontinued the practice. The influence of the crown in decision making accordingly declined. Walpole enjoyed the favor of the monarchy and of the House of Commons and came to be called the king's first, or "prime," minister. In the English cabinet system, both legislative and executive power are held by the leading ministers, who form the government.

THE DUTCH REPUBLIC IN THE SEVENTEENTH CENTURY

The seventeenth century witnessed an unparalleled flowering of Dutch scientific, artistic, and literary achievement. In this period, often called the "golden age of the Netherlands," Dutch ideas and attitudes played a profound role in shaping a new and modern world-view. At the same time, the Republic of the United Provinces of the Netherlands represents another model of the development of the modern state.

In the late sixteenth century, the seven northern provinces of the Netherlands, of which Holland and Zeeland were the most prosperous, succeeded in throwing off Spanish domination. This success was based on their geographical lines of defense, the wealth of their cities, the brilliant military strategy of William the Silent, the preoccupation of Philip II of Spain with so many other concerns, and the northern provinces' vigorous Calvinism. In 1581 the seven provinces of the Union of Utrecht had formed the United Provinces (page 520). Philip II continued to try to crush the Dutch with the Armada, but in 1609 his son Philip III agreed to a truce that implicitly recognized the independence of the United Provinces. At the time neither side expected the peace to be permanent. The Peace of Westphalia in 1648, however, confirmed the Dutch republic's independence.

Within each province an oligarchy of wealthy merchants called "regents" handled domestic affairs in the local Estates. The provincial Estates held virtually all the power. A federal assembly, or States General, handled matters of foreign affairs, such as war. But the States General did not possess sovereign authority, since all issues had to be referred back to the local Estates for approval. The States General appointed a representative, the *stadholder,* in each province. As the highest executive there, the stadholder carried out ceremonial functions and was responsible for defense and good order. The sons of William the Silent, Maurice and William Louis, held the office of stadholder in all seven provinces. As members of the House of Orange, they were closely identified with the struggle against Spain and Dutch patriotism. The regents in each province jealously guarded local independence and resisted efforts at centralization. Nevertheless, Holland, which had the largest navy and the most wealth, dominated the republic and the States General. Significantly, the Estates assembled at Holland's capital, The Hague.

The government of the United Provinces fits none of the standard categories of seventeenth-century political organization. The Dutch were not monarchial, but fiercely republican. The government was controlled by wealthy merchants and financiers. Though rich, their values were not aristocratic but strongly

Vermeer: Young Woman with a Water Jug The mystery of light fascinated seventeenth-century scientists and artists, perhaps the Dutch painter Vermeer most of all. His calm quiet scenes of ordinary people performing everyday tasks in rooms suffused with light has given him the title "poet of domesticity." The map on the wall suggests curiosity about the wider world. *(Metropolitan Museum of Art, Gift of Henry G. Marquand, 1889)*

Amsterdam Harbor, 1663 Amsterdam's great wealth depended on the presence of a relatively large number of people with capital for overseas investment *and* on its fleet, which brought goods from all over the world. As shown here, because the inner docks were often jammed to capacity, many ships had to be moored outside the city. *(Scheepvaart Museum, Amsterdam)*

middle class, emphasizing thrift, hard work, and simplicity in living. The Dutch republic was not a strong federation but a confederation—that is, a weak union of strong provinces. The provinces were a temptation to powerful neighbors, yet the Dutch resisted the long Spanish effort at reconquest and withstood both French and English attacks in the second half of the century. Louis XIV's hatred of the Dutch was proverbial. They represented all that he despised —middle-class values, religious toleration, and political independence.

The political success of the Dutch rested on the phenomenal commercial prosperity of the Netherlands. The moral and ethical bases of that commercial wealth were thrift, frugality, and religious toleration. John Calvin had written, "From where do the merchant's profits come except from his own diligence and industry." This attitude undoubtedly encouraged a sturdy people who had waged a centuries-old struggle against the sea.

Alone of all European peoples in the seventeenth century, the Dutch practiced religious toleration. Peoples of all faiths were welcome within their borders. It is a striking testimony to the urbanity of Dutch society that in a century when patriotism was closely identified with religious uniformity, the Calvinist province of Holland allowed its highest official, Jan van Oldenbarneveldt, to continue to practice his Roman Catholic faith. As long as a businessman conducted his religion in private, the government did not interfere with him.

Toleration also paid off: it attracted a great deal of foreign capital and investment. Deposits at the Bank of Amsterdam were guaranteed by the city council, and in the middle years of the century the bank became Europe's best source of cheap credit and commercial intelligence and the main clearing-house for bills of exchange. Men of all races and creeds traded in Amsterdam, at whose docks on the Amstel River five thousand ships, half the merchant marine of the United Provinces, were berthed. Joost van den Vondel, the poet of Dutch imperialism, exulted:

God, God, the Lord of Amstel cried, hold every conscience free;
And Liberty ride, on Holland's tide, with billowing sails to sea,
And run our Amstel out and in; let freedom gird the bold,
And merchant in his counting house stand elbow deep in gold.[13]

The fishing industry was the cornerstone of the Dutch economy. For half the year, from June to December, fishing fleets combed the dangerous English coast and the North Sea, raking in tiny herring. Profits from herring stimulated shipbuilding, and even before 1600 the Dutch were offering the lowest shipping rates in Europe. Although Dutch cities became famous for their exports—diamonds, linen from Haarlem, pottery from Delft—Dutch wealth depended less on exports than on transport. The merchant marine was the largest in Europe.

In 1602 a group of the regents of Holland formed the Dutch East India Company, a joint stock company. Each investor received a percentage of the profits proportional to the amount of money he had put in. Within half a century, the Dutch East India Company had cut heavily into Portuguese trading in the Far East. The Dutch seized the Cape of Good Hope, Ceylon, and Malacca and established trading posts in each place. In the 1630s the Dutch East India Company was paying its investors about a 35-percent annual return on their investments. The Dutch West India Company, founded in 1621, traded extensively with Latin America and Africa.

Although the initial purpose of both companies was commercial—the import of spices and silks to Europe—the Dutch found themselves involved in the imperialistic exploitation of large parts of the Pacific and Latin America. Amsterdam, the center of a worldwide Dutch empire, became the commercial and financial capital of Europe. During the seventeenth century, the Dutch translated their commercial acumen and flexibility into political and imperialist terms with striking success. But war with France and England in the 1670s hurt the United Provinces. The long War of the Spanish Succession, in which the Dutch supported England against France, was a costly drain on Dutch manpower and financial resources. The peace signed in 1715 to end the war marked the beginning of Dutch economic decline.

According to Thomas Hobbes, the central drive in every man is "a perpetual and restless desire of Power, after Power, that ceaseth only in Death." The seventeenth century solved the problem of sovereign power in two fundamental ways: absolutism and constitutionalism. The France of Louis XIV witnessed the emergence of the fully absolutist state. The king commanded all the powers of the state: judicial, military, political, and to a great extent, ecclesiastical. France developed a centralized bureaucracy, a professional army, a state-directed economy, all of which Louis personally supervised. For the first time in history all the institutions and powers of the national state were effectively controlled by a single person. The king saw himself as the representative of God on earth, and it has been said that "to the seventeenth century imagination God was a sort of image of Louis XIV."[14]

As Louis XIV personifies absolutism, so Stuart England exemplifies the evolution of the first modern constitutional state. The conflicts between Parliament and the first two Stuart rulers, James I and Charles I, tested where sovereign power would rest in the state. The resulting civil war did not solve the problem. The Instrument of Government, the document produced in 1653 by the victorious parliamentary army, provided for a balance of governmental authority and recognition of popular rights; as such, the Instrument has been called the first modern constitution. Unfortunately, it lacked public support. James II's absolutist tendencies brought on the Revolution of 1688, and the people who made that revolution settled three basic issues. Sovereign power was divided between king and Parliament, with Parliament enjoying the greater share. Government was to be based on the rule of law. And the liberties of English people were made explicit in written form, in the Bill of Rights. The framers of the English constitution left to later generations the task of making constitutional government work.

The models of governmental power established by seventeenth-century England and France strongly influenced other states then and ever since. As the American novelist William Faulkner wrote, "The past isn't dead; it's not even past."

NOTES

1. Cited in John Wolf, *Louis XIV,* W. W. Norton & Co., New York, 1968, p. 115.
2. S. de Gramont, ed., *The Age of Magnificence: Memoirs of the Court of Louis XIV by the Duc de Saint-Simon,* Capricorn Books, New York, 1964, pp. 141–145.

3. Cited in Wolf, p. 146.

4. Cited in Andrew Trout, *Jean-Baptiste Colbert,* Twayne Publishers, Boston, 1978, p. 128.

5. Cited in Wolf, p. 394.

6. Ibid.

7. See Warren C. Scoville, *The Persecution of the Huguenots and French Economic Development: 1680–1720,* University of California Press, Berkeley, 1960.

8. See William F. Church, *Louis XIV in Historical Thought: From Voltaire to the Annales School,* W. W. Norton & Co., New York, 1976, p. 92.

9. S. de Gramont, ed., *The Age of Magnificence,* p. 183.

10. J. H. Elliott, *Imperial Spain, 1469–1716,* Mentor Books, New York, 1963, pp. 306–308.

11. See Bartolome Bennassar, *The Spanish Character: Attitudes and Mentalities from the Sixteenth to the Nineteenth Century,* trans. Benjamin Keen, University of California Press, Berkeley, 1979, p. 125.

12. C. Stephenson and G. F. Marcham, *Sources of English Constitutional History,* Harper & Row, New York, 1937, p. 601.

13. Cited in D. Maland, *Europe in the Seventeenth Century,* Macmillan, New York, 1967, pp. 198–199.

14. Cited in Carl J. Friedrich and Charles Blitzer, *The Age of Power,* Cornell University Press, Ithaca, N.Y., 1957, p. 112.

SUGGESTED READING

Students who wish to explore the problems presented in this chapter in greater depth will easily find a rich and exciting literature, with many titles available in paperback editions. The following surveys all provide good background material. G. Parker, *Europe in Crisis, 1598–1618* (1980), provides a sound introduction to the social, economic, and religious tensions of the period. R. S. Dunn, *The Age of Religious Wars, 1559–1715,* 2nd ed. (1979), examines the period from the perspective of the confessional strife between Protestants and Catholics, but there is also stimulating material on absolutism and constitutionalism. T. Aston, ed., *Crisis in Europe, 1560–1660* (1967), contains essays by leading historians. P. Anderson, *Lineages of the Absolutist State* (1974), is a Marxist interpretation of absolutism in western and eastern Europe. M. Beloff, *The Age of Absolutism* (1967), concentrates on the social forces that underlay administrative change. H. Rosenberg, "Absolute Monarchy and Its Legacy," in *Early Modern Europe, 1450–1650,* ed. N. F. Cantor and S. Werthman (1967), is a seminal study. The classic treatment of constitutionalism remains that of C. H. McIlwain, *Constitutionalism: Ancient and Modern* (1940), written by a great scholar during the rise of German fascism. S. B. Crimes, *English Constitutional History* (1967), is an excellent survey with useful chapters on the sixteenth and seventeenth centuries.

Louis XIV and his age have predictably attracted the attention of many scholars. J. Wolf, *Louis XIV* (1968), remains the best available biography. Two works of W. H. Lewis, *The Splendid Century* (1957) and *The Sunset of the Splendid Century* (1963), make delightful light reading, especially for the beginning student. The advanced student will want to consult W. F. Church, *Louis XIV in Historical Thought: From Voltaire to the Annales School* (1976), an excellent historiographical analysis. Perhaps the best work of the Annales school on the period is P. Goubert, *Louis XIV and Twenty Million Frenchmen* (1972), and his heavily detailed *The Ancien Regime: French Society, 1600–1750,* 2 vols. (1969–1973), which contains invaluable material on the lives and work of ordinary people. For the French economy and financial conditions, the old study of C. W. Cole, *Colbert and a Century of French Mercantilism,* 2 vols. (1939), is still valuable but should be supplemented by R. Bonney, *The King's Debts: Finance and Politics in France, 1589–1661* (1981); A. Trout, *Jean-Baptiste Colbert* (1978); and especially W. Scoville, *The Persecution of the Huguenots and French Economic Decline, 1680–1720* (1960), a significant book in revisionist history. For Louis XIV's foreign policy and wars, see R. Hatton, "Louis XIV: Recent Gains in Historical Knowledge," *Journal of Modern History* 45 (1973), and her edited work *Louis XIV and Europe* (1976), an important collection of essays. Hatton's *Europe in the Age of Louis XIV* (1979) is a splendidly illustrated survey of many aspects of seventeenth-century European culture. O. Ranum, *Paris in the Age of Absolutism* (1968), describes the geographical, political, economic, and architectural significance of the cultural capital of Europe, while V. L. Tapie, *The Age of Grandeur: Baroque Art and Architecture* (1960), also emphasizes the relationship between art and politics with excellent illustrations.

For Spain and Portugal, see J. H. Elliott, *Imperial Spain, 1469–1716,* rev. ed. (1977), a sensitively written and authoritative study; B. Bennassar, *The Spanish Character: Attitudes and Mentalities from the Sixteenth to the Nineteenth Century* (trans. B. Keen, 1979); and M. Defourneaux, *Daily Life in Spain in the Golden Age* (1976), which are highly useful for an understanding of ordinary people and of Spanish society; and C. R. Phillips, *Ciudad Real, 1570–1750: Growth, Crisis, and Readjustment in the Spanish Economy* (1979), a significant case study.

The following works all offer solid material on English political and social issues of the seventeenth century: M. Ashley, *England in the Seventeenth Century,* rev. ed. (1980) and *The House of Stuart: Its Rise and Fall* (1980); C. Hill, *A Century of Revolution* (1961); J. P. Kenyon, *Stuart England* (1978); and K. Wrightson, *English Society, 1580–1680* (1982). Perhaps the most comprehensive treatment of Parliament is C. Russell, *Crisis of Parliaments, 1509–1660* (1971) and *Parliaments and English Politics, 1621–1629* (1979). On the background of the English civil war, L. Stone, *The Crisis of the Aristocracy* (1965) and *The Causes of the English Revolution* (1972) are standard works, while both B. Manning, *The English People and the English Revolution* (1976), and D. Underdown, *Revel, Riot, and Rebellion* (1985), discuss the extent of popular involvement; Underdown's is the more sophisticated treatment. For English intellectual currents, see J. O. Appleby, *Economic Thought and Ideology in Seventeenth Century England* (1978); and C. Hill, *Intellectual Origins of the English Revolution* (1966) and *Society and Puritanism in Pre-Revolutionary England* (1964).

For the several shades of Protestant sentiment in the early seventeenth century, see P. Collinson, *The Religion of Protestants* (1982). C. M. Hibbard, *Charles I and the Popish Plot* (1983), treats Roman Catholic influence; like Collinson's work, it is an excellent, fundamental reference for religious issues, though the older work of W. Haller, *The Rise of Puritanism* (1957), is still valuable. For women, see R. Thompson, *Women in Stuart England and America* (1974), and A. Fraser, *The Weaker Vessel* (1985). For Cromwell and the Interregnum, C. Firth, *Oliver Cromwell and the Rule of the Puritans in England* (1956), C. Hill, *God's Englishman* (1972), and A. Fraser, *Cromwell, The Lord Protector* (1973), are all valuable. J. Morrill, *The Revolt of the*

Provinces, 2nd ed. (1980), is the best study of neutralism, while C. Hill, *The World Turned Upside Down* (1972), discusses radical thought during the period.

For the Restoration and the Glorious Revolution, see A. Fraser, *Royal Charles: Charles II and the Restoration* (1979), a highly readable biography; R. Ollard, *The Image of the King: Charles I and Charles II* (1980), which examines the nature of monarchy; J. Miller, *James II: A Study in Kingship* (1977); J. Childs, *The Army, James II, and the Glorious Revolution* (1980); J. R. Jones, *The Revolution of 1688 in England* (1972); and L. G. Schwoerer, *The Declaration of Rights, 1689* (1981), a fine assessment of that fundamental document. The ideas of John Locke are analyzed by J. P. Kenyon, *Revolution Principles: The Politics of Party, 1689–1720* (1977). R. Hutton, *The Restoration, 1658–1667* (1985), is a thorough if somewhat difficult narrative.

On Holland, K. H. D. Haley, *The Dutch Republic in the Seventeenth Century* (1972), is a splendidly illustrated appreciation of Dutch commercial and artistic achievements, while J. L. Price, *Culture and Society in the Dutch Republic During the Seventeenth Century* (1974), is a sound scholarly work. R. Boxer, *The Dutch Seaborne Empire* (1980), and the appropriate chapters of D. Maland, *Europe in the Seventeenth Century* (1967), are useful for Dutch overseas expansion and the reasons for Dutch prosperity. The following works focus on the economic and cultural life of the leading Dutch city: V. Barbour, *Capitalism in Amsterdam in the Seventeenth Century* (1950), and D. Regin, *Traders, Artists, Burghers: A Cultural History of Amsterdam in the Seventeenth Century* (1977). J. M. Montias, *Artists and Artisans in Delft: A Socio-Economic Study of the Seventeenth Century* (1982), examines another major city. The leading statesmen of the period may be studied in these biographies: H. H. Rowen, *John de Witt, Grand Pensionary of Holland, 1625–1672* (1978); S. B. Baxter, *William the III and the Defense of European Liberty, 1650–1702* (1966); and J. den Tex, *Oldenbarnevelt,* 2 vols. (1973).

Many facets of the lives of ordinary French, Spanish, English, and Dutch people are discussed by P. Burke, *Popular Culture in Early Modern Europe* (1978), an important and provocative study.

17

**ABSOLUTISM IN
EASTERN EUROPE
TO 1740**

*T*HE SEVENTEENTH century witnessed a struggle between constitutionalism and absolutism in eastern Europe. With the notable exception of the kingdom of Poland, monarchial absolutism was everywhere triumphant in eastern Europe; constitutionalism was decisively defeated. Absolute monarchies emerged in Austria, Prussia, and Russia. This was a development of great significance: these three monarchies exercised enormous influence until 1918, and they created a strong authoritarian tradition that is still dominant in eastern Europe.

Although the monarchs of eastern Europe were greatly impressed by Louis XIV and his model of royal absolutism, their states differed in several important ways from their French counterpart. Louis XIV built French absolutism on the heritage of a well-developed medieval monarchy and a strong royal bureaucracy. And when Louis XIV came to the throne, the powers of the nobility were already somewhat limited, the French middle-class was relatively strong, and the peasants were generally free from serfdom. Eastern absolutism rested on a very different social reality: a powerful nobility, a weak middle class, and an oppressed peasantry composed of serfs.

These differences in social conditions raise three major questions. First, why did the basic structure of society in eastern Europe move away from that of western Europe in the early modern period? Second, how and why, in their different social environments, did the rulers of Austria, Prussia, and Russia manage to build powerful absolute monarchies, which proved more durable than that of Louis XIV? Finally, how did the absolute monarchs' interaction with artists and architects contribute to the splendid achievements of baroque culture? These are the questions that will be explored in this chapter.

LORDS AND PEASANTS IN EASTERN EUROPE

When absolute monarchy took shape in eastern Europe in the seventeenth century, it built on social and economic foundations laid between roughly 1400 and 1650. In those years, the princes and the landed

nobility of eastern Europe rolled back the gains made by the peasantry during the High Middle Ages and reimposed a harsh serfdom on the rural masses. The nobility also reduced the importance of the towns and the middle classes. This process differed profoundly from developments in western Europe at the same time. In the west, peasants won greater freedom and the urban capitalistic middle class continued its rise. Thus, the east that emerged contrasted sharply with the west—another aspect of the shattered unity of medieval Latin Christendom.

THE MEDIEVAL BACKGROUND

Between roughly 1400 and 1650, nobles and rulers re-established serfdom in the eastern lands of Bohemia, Silesia, Hungary, eastern Germany, Poland, Lithuania, and Russia. The east—the land east of the Elbe River in Germany, which historians often call "East Elbia"—gained a certain social and economic unity in the process. But eastern peasants lost their rights and freedoms. They became bound first to the land they worked and then, by degrading obligations, to the lords they served.

This development was a tragic reversal of trends in the High Middle Ages. The period from roughly 1050 to 1300 had been a time of general economic expansion characterized by the growth of trade, towns, and population. Expansion also meant clearing the forests and colonizing the frontier beyond the Elbe River. Anxious to attract German settlers to their sparsely populated lands, the rulers and nobles of eastern Europe had offered potential newcomers attractive economic and legal incentives. Large numbers of incoming settlers obtained land on excellent terms and gained much greater personal freedom. These benefits were also gradually extended to the local Slavic populations, even those of central Russia. Thus by 1300 there had occurred a very general improvement in peasant conditions in eastern Europe. Serfdom all but disappeared. Peasants bargained freely with their landlords and moved about as they pleased. Opportunities and improvements east of the Elbe had a positive impact on western Europe, where the weight of serfdom was also reduced between 1100 and 1300.

After about 1300, however, as Europe's population and economy both declined grievously, mainly because of the Black Death, the east and the west went in different directions. In both east and west there occurred a many-sided landlord reaction, as lords sought to solve their tough economic problems by more heavily exploiting the peasantry. Yet this reaction generally failed in the west. In many western areas by 1500 almost all of the peasants were completely free, and in the rest of western Europe serf obligations had declined greatly. East of the Elbe, however, the landlords won. By 1500 eastern peasants were well on their way to becoming serfs again.

Throughout eastern Europe, as in western Europe, the drop in population and prices in the fourteenth and fifteenth centuries caused severe labor shortages and hard times for the nobles. Yet rather than offer better economic and legal terms to keep old peasants and attract new ones, eastern landlords used their political and police power to turn the tables on the peasants. They did this in two ways.

First, the lords made their kings and princes issue laws that restricted or eliminated the peasants' precious, time-honored right of free movement. Thus, a peasant could no longer leave to take advantage of better opportunities elsewhere without the lord's permission, and the lord had no reason to make such concessions. In Prussian territories by 1500, the law required that runaway peasants be hunted down and returned to their lords; a runaway servant was to be nailed to a post by one ear and given a knife to cut himself loose. Until the middle of the fifteenth century, medieval Russian peasants had been free to move wherever they wished and seek the best landlord. Thereafter this freedom was gradually curtailed, so that by 1497 a Russian peasant had the right to move only during a two-week period after the fall harvest. Eastern peasants were losing their status as free and independent men and women.

Second, lords steadily took more and more of their peasants' land and imposed heavier and heavier labor obligations. Instead of being independent farmers paying reasonable, freely negotiated rents, peasants tended to become forced laborers on the lords' estates. By the early 1500s, lords in many territories could command their peasants to work for them without pay as many as six days a week. A German writer of the mid-sixteenth century described peasants in eastern Prussia who "do not possess the heritage of their holdings and have to serve their master whenever he wants them."[1]

The gradual erosion of the peasantry's economic position was bound up with manipulation of the legal system. The local lord was also the local prosecutor, judge, and jailer. As a matter of course, he ruled in his own favor in disputes with his peasants. There were no independent royal officials to provide justice or uphold the common law.

THE CONSOLIDATION OF SERFDOM

Between 1500 and 1650, the social, legal, and economic conditions of peasants in eastern Europe continued to decline. Free peasants lost their freedom and became serfs. In Poland, for example, nobles gained complete control over their peasants in 1574, after which they could legally inflict the death penalty on their serfs whenever they wished. In Prussia a long series of oppressive measures reached their culmination in 1653. Not only were all the old privileges of the lords reaffirmed, but peasants were assumed to be in "hereditary subjugation" to their lords unless they could prove the contrary in the lords' courts, which was practically impossible. Prussian peasants were serfs tied to their lords as well as to the land.

In Russia the right of peasants to move from a given estate was "temporarily" suspended in the 1590s and permanently abolished in 1603. In 1649 a new law code completed the process. At the insistence of the lower nobility, the Russian tsar lifted the nine-year time limit on the recovery of runaways. Henceforth runaway peasants were to be returned to their lords whenever they were caught, as long as they lived. The last small hope of escaping serfdom was gone. Control of serfs was strictly the lords' own business, for the new law code set no limits on the lords' authority over their peasants. Although the political development of the various eastern states differed, the legal re-establishment of permanent hereditary serfdom was the common fate of peasants in the east by the middle of the seventeenth century.

The consolidation of serfdom between 1500 and 1650 was accompanied by the growth of estate agriculture, particularly in Poland and eastern Germany. In the sixteenth century, European economic expansion and population growth resumed after the great declines of the late Middle Ages. Prices for agricultural commodities also rose sharply as gold and silver flowed in from the New World. Thus Polish and German lords had powerful economic incentives to increase the production of their estates. And they did.

Lords seized more and more peasant land for their own estates and then demanded and received ever more unpaid serf labor on those enlarged estates.

Punishing Serfs This seventeenth-century illustration from Olearius's famous *Travels to Moscovy* suggests what eastern serfdom really meant. The scene is set in eastern Poland. There, according to Olearius, a common command of the lord was, "Beat him till the skin falls from the flesh." *(University of Illinois, Champaign)*

Even when the estates were inefficient and technically backward, as they generally were, the great Polish nobles and middle-rank German lords squeezed sizable, cheap, and thus very profitable surpluses out of their impoverished peasants. These surpluses in wheat and timber were easily sold to big foreign merchants, who exported them to the growing cities of the west. The poor east helped feed the much wealthier west.

The re-emergence of serfdom in eastern Europe in the early modern period was clearly a momentous human development, and historians have advanced a variety of explanations for it. As always, some scholars have stressed the economic interpretation. Agricultural depression and population decline in the fourteenth and fifteenth centuries led to a severe labor shortage, they have argued, and thus eastern landlords naturally tied their precious peasants to the land. With the return of prosperity and the development of export markets in the sixteenth century, the landlords finished the job, grabbing the peasants' land and making them work as unpaid serfs on the enlarged estates. This argument by itself is not very convincing, for almost identical economic developments "caused" the opposite result in the west. Indeed, some historians have maintained that labor shortage and subsequent expansion were key factors in the virtual disappearance of Western serfdom.

It seems fairly clear, therefore, that political rather than economic factors were crucial in the simultaneous rise of serfdom in the east and decline of serfdom in the west. Specifically, eastern lords enjoyed much greater political power than their western counterparts. In the late Middle Ages, when much of eastern

Europe experienced innumerable wars and general political chaos, the noble landlord class greatly increased its political power at the expense of the ruling monarchs. There were, for example, many disputed royal successions, so that weak kings were forced to grant political favors to win the support of the nobility. Thus while strong "new monarchs" were rising in Spain, France, and England and providing effective central government, kings were generally losing power in the east. Such weak kings could not resist the demands of the lords regarding their peasants.

Moreover, most eastern monarchs did not want to resist even if they could. The typical king was only first among equals in the noble class. He, too, thought mainly in private rather than public terms. He, too, wanted to squeeze as much as he could out of *his* peasants and enlarge *his* estates. The western concept and reality of sovereignty, as embodied in a king who protected the interests of all his people, was not well developed in eastern Europe before 1650.

The political power of the peasants was also weaker in eastern Europe and declined steadily after about 1400. Although there were occasional bloody peasant uprisings against the oppression of the landlords, they never succeeded. Nor did eastern peasants effectively resist day-by-day infringements on their liberties by their landlords. Part of the reason was that the lords, rather than the kings, ran the courts—one of the important concessions nobles extorted from weak monarchs. It has also been suggested that peasant solidarity was weaker in the east, possibly reflecting the lack of long-established village communities on the eastern frontier.

Finally, with the approval of weak kings, the landlords systematically undermined the medieval privileges of the towns and the power of the urban classes. Instead of selling their products to local merchants in the towns, as required in the Middle Ages, the landlords sold directly to big foreign capitalists. For example, Dutch ships sailed up the rivers of Poland and eastern Germany to the loading docks of the great estates, completely short-circuiting the local towns. Moreover, "town air" no longer "made people free," for the eastern towns lost their medieval right of refuge and were compelled to return runaways to their lords. The population of the towns and the importance of the urban middle classes declined greatly. This development both reflected and promoted the supremacy of noble landlords in most of eastern Europe in the sixteenth century.

THE RISE OF AUSTRIA AND PRUSSIA

Despite the strength of the nobility and the weakness of many monarchs before 1600, strong kings did begin to emerge in many lands in the course of the seventeenth century. War and the threat of war aided rulers greatly in their attempts to build absolute monarchies. There was an endless struggle for power, as eastern rulers not only fought each other but also battled with hordes of Asiatic invaders. In this atmosphere of continuous wartime emergency, monarchs reduced the political power of the landlord nobility. Cautiously leaving the nobles the unchallenged masters of their peasants, the absolutist monarchs of eastern Europe gradually gained and monopolized political power in three key areas. They imposed and collected permanent taxes without consent. They maintained permanent standing armies, which policed their subjects in addition to fighting abroad. And they conducted relations with other states as they pleased.

As with all general historical developments, there were important variations on the absolutist theme in eastern Europe. The royal absolutism created in Prussia was stronger and more effective than that established in Austria. This advantage gave Prussia a thin edge over Austria in the struggle for power in east-central Europe in the eighteenth century. That edge had enormous long-term political significance, for it was a rising Prussia that unified the German people in the nineteenth century and imposed on them a fateful Prussian stamp.

AUSTRIA AND THE OTTOMAN TURKS

Like all the peoples and rulers of central Europe, the Habsburgs of Austria emerged from the Thirty Years' War (pages 483–489) impoverished and exhausted. The effort to root out Protestantism in the German lands had failed utterly, and the authority of the Holy Roman Empire and its Habsburg emperors had declined almost to the vanishing point. Yet defeat in central Europe also opened new vistas. The Habsburg monarchs were forced to turn inward and eastward in the attempt to fuse their diverse holdings into a strong unified state.

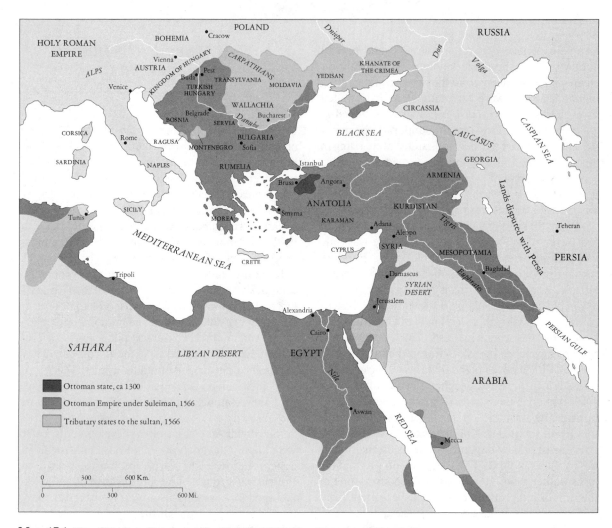

MAP 17.1 The Ottoman Empire at Its Height, 1566 The Ottomans, like their great rivals the Habsburgs, rose to rule a far-flung dynastic empire encompassing many different peoples and ethnic groups. The army and the bureaucracy served to unite the disparate territories into a single state.

An important step in this direction had actually been taken in Bohemia during the Thirty Years' War. Protestantism had been strong among the Czechs of Bohemia, and in 1618 the Czech nobles who controlled the Bohemian Estates—the semiparliamentary body of Bohemia—had risen up against their Habsburg king. Not only was this revolt crushed, but the old Czech nobility was wiped out as well. Those Czech nobles who did not die in 1620 at the Battle of the White Mountain (page 484), a momentous turning point in Czech history, had their estates confiscated. The Habsburg king, Ferdinand II (1619–

1637), then redistributed the Czech lands to a motley band of aristocratic soldiers of fortune from all over Europe.

In fact, after 1650, 80 to 90 percent of the Bohemian nobility was of recent foreign origin and owed everything to the Habsburgs. With the help of this new nobility, the Habsburgs established strong direct rule over reconquered Bohemia. The condition of the enserfed peasantry worsened: three days per week of unpaid labor—the *robot*—became the norm, and a quarter of the serfs worked for their lords every day but Sundays and religious holidays. Serfs also paid

The Ottoman Slave Tax This contemporary drawing shows Ottoman officials rounding up male Christian children in the Balkans. The children became part of a special slave corps, which served the sultan for life as soldiers and administrators. The slave tax and the slave corps were of great importance to the Ottoman Turks in the struggle with Austria. *(The British Library)*

the taxes, which further strengthened the alliance between the Habsburg monarch and the Bohemian nobility. Protestantism was also stamped out, in the course of which a growing unity of religion was brought about. The reorganization of Bohemia was a giant step toward absolutism.

After the Thirty Years' War, Ferdinand III centralized the government in the old hereditary provinces of Austria proper, the second part of the Habsburg holdings. For the first time, under Ferdinand was created a permanent standing army, which stood ready to put down any internal opposition. The Habsburg

monarchy was then ready to turn toward the vast plains of Hungary, which it claimed as the third and largest part of its dominion, in opposition to the Ottoman Turks.

The Ottomans came out of the Anatolia, in present-day Turkey, to create one of history's greatest military empires. At their peak in the middle of the sixteenth century under Suleiman the Magnificent (1520–1566), they ruled the most powerful empire in the world. Their possessions stretched from western Persia across North Africa and up into the heart of central Europe (see Map 17.1). Apostles of Islam, the

Ottoman Turks were old and determined foes of the Catholic Habsburgs. Their armies had almost captured Vienna in 1529, and for more than 150 years thereafter the Ottomans ruled all of the Balkan territories, almost all of Hungary, and part of southern Russia.

The Ottoman Empire was originally built on a fascinating and very non-European conception of state and society. There was an almost complete absence of private landed property. All the agricultural land of the empire was the personal hereditary property of the sultan, who exploited the land as he saw fit according to Ottoman political theory. There was, therefore, no security of landholding and no hereditary nobility. Everyone was dependent on the sultan and virtually his slave.

Indeed, the top ranks of the bureaucracy were staffed by the sultan's slave corps. Every year the sultan levied a "tax" of one to three thousand male children on the conquered Christian populations in the Balkans. These and other slaves were raised in Turkey as Muslims and trained to fight and to administer. The most talented slaves rose to the top of the bureaucracy; the less fortunate formed the brave and skillful core of the sultan's army, the so-called janissary corps.

As long as the Ottoman Empire expanded, the system worked well. As the sultan won more territory, he could impose his slave tax on larger populations. Moreover, he could amply reward loyal and effective servants by letting them draw a carefully defined income from conquered Christian peasants on a strictly temporary basis. For a long time, Christian peasants in eastern Europe were economically exploited less by the Muslim Turks than by Christian nobles, and they were not forced to convert to Islam. After about 1570, however, the powerful, centralized Ottoman system slowly began to disintegrate as the Turks' western advance was stopped. Temporary landholders became hard-to-control permanent oppressors. Weak sultans left the glory of the battlefield for the delights of the harem, and the army lost its dedication and failed to keep up with European military advances.

Yet in the late seventeenth century, under vigorous reforming leadership, the Ottoman Empire succeeded in marshaling its forces for one last mighty blow at Christian Europe. After wresting territory from Poland, fighting a long inconclusive war with Russia, and establishing an alliance with Louis XIV

of France, the Turks turned again on Austria. A huge Turkish army surrounded Vienna and laid siege to it in 1683. But after holding out against great odds for two months, the city was relieved by a mixed force of Habsburg, Saxon, Bavarian, and Polish troops, and the Ottomans were forced to retreat. Soon the retreat became a rout. As their Russian and Venetian allies attacked on other fronts, the Habsburgs conquered all of Hungary and Transylvania (part of present-day Rumania) by 1699 (see Map 17.2).

The Turkish wars and this great expansion strengthened the Habsburg army and promoted some sense of unity in the Habsburg lands. The Habsburgs moved to centralize their power and make it as absolute as possible. These efforts to create a fully developed, highly centralized, absolutist state were only partly successful.

The Habsburg state was composed of three separate and distinct territories—the old "hereditary provinces" of Austria, the kingdom of Bohemia, and the kingdom of Hungary. These three parts were tied together primarily by their common ruler—the Habsburg monarch. Each part had its own laws and political life, for the three noble-dominated Estates continued to exist, though with reduced powers. The Habsburgs themselves were well aware of the fragility of the union they had forged. In 1713 Charles VI (1711–1740) proclaimed the so-called Pragmatic Sanction, which stated that the Habsburg possessions were never to be divided and were always to be passed intact to a single heir, who might be female since Charles had no sons. Charles spent much of his reign trying to get this principle accepted by the various branches of the Habsburg family, by the three different Estates of the realm, and by the states of Europe. His fears turned out to be well founded.

The Hungarian nobility, despite its reduced strength, effectively thwarted the full development of Habsburg absolutism. Time and again throughout the seventeenth century, Hungarian nobles—the most numerous in Europe, making up 5 to 7 percent of the Hungarian population—rose in revolt against the attempts of Vienna to impose absolute rule. They never triumphed decisively, but neither were they ever crushed and replaced, as the Czech nobility had been in 1620.

Hungarians resisted because many of them were Protestants, especially in the area long ruled by the more tolerant Turks, and they hated the heavy-handed attempts of the conquering Habsburgs to re-

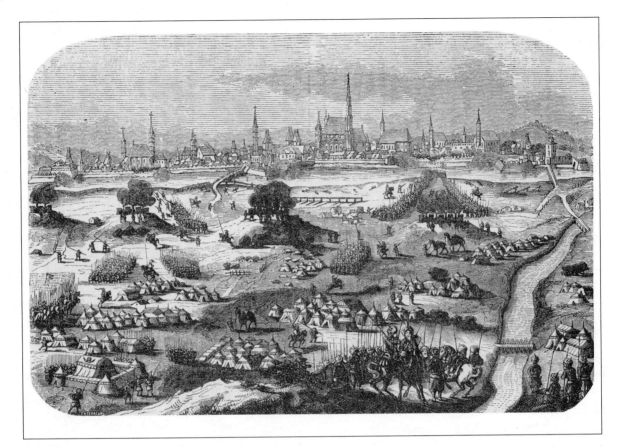

The Siege of Vienna, 1683 The Turks dreamed of establishing a western Muslim Empire in the heart of Europe. But their army of nearly 300,000 men failed to pierce the elaborate fortifications that protected the old city walls from cannon fire and underground mines. *(The Mansell Collection)*

Catholicize everyone. Moreover, the lords of Hungary often found a powerful military ally in Turkey. Finally, the Hungarian nobility, and even part of the peasantry, had become attached to a national ideal long before most of the peoples of Europe. They were determined to maintain as much independence and local control as possible. Thus when the Habsburgs were bogged down in the War of the Spanish Succession (page 520), the Hungarians rose in one last patriotic rebellion under Prince Francis Rakoczy in 1703. Rakoczy and his forces were eventually defeated, but this time the Habsburgs had to accept a definitive compromise. Charles VI restored many of the traditional privileges of the Hungarian aristocracy in return for Hungarian acceptance of hereditary Habsburg rule. Thus Hungary, unlike Austria or Bohemia, never came close to being fully integrated into a centralized, absolute Habsburg state.

PRUSSIA IN THE SEVENTEENTH CENTURY

After 1400 the status of east German peasants declined steadily; their serfdom was formally spelled out in the early seventeenth century. While the local princes lost political power and influence, a revitalized landed nobility became the undisputed ruling class. The Hohenzollern family, which ruled through its senior and junior branches as the electors of Brandenburg and the dukes of Prussia, had little real princely power. The Hohenzollern rulers were nothing more than the first among equals, the largest landowners in a landlord society.

Nothing suggested that the Hohenzollerns and their territories would ever play an important role in European or even German affairs. The elector of Brandenburg's right to help choose the Holy Roman

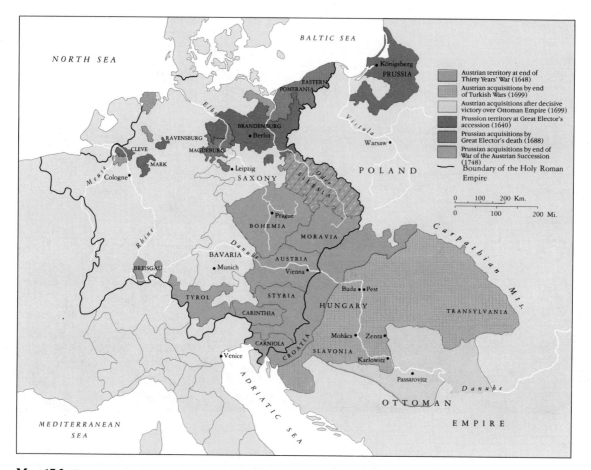

MAP 17.2 The Growth of Austria and Brandenburg-Prussia to 1748 Austria expanded to the southwest into Hungary and Transylvania at the expense of the Ottoman Empire. It was unable to hold the rich German province of Silesia, however, which was conquered by Brandenburg-Prussia.

emperor with six other electors was of little practical value, and the elector had no military strength whatsoever. The territory of his cousin, the duke of Prussia, was actually part of the kingdom of Poland. Moreover, geography conspired against the Hohenzollerns. Brandenburg, their power base, was completely cut off from the sea (see Map 17.2). A tiny part of the vast north European plain that stretches from France to Russia, Brandenburg lacked natural frontiers and lay open to attack from all directions. The land was poor, a combination of sand and swamp. Contemporaries contemptuously called Brandenburg the "sand-box of the Holy Roman Empire."[2]

Brandenburg was a helpless spectator in the Thirty Years' War, its territory alternately ravaged by Swedish and Habsburg armies. Population fell drastically,

and many villages disappeared. The power of the Hohenzollerns reached its lowest point. Yet the devastation of the country prepared the way for Hohenzollern absolutism, because foreign armies dramatically weakened the political power of the Estates—the representative assemblies of the realm. This weakening of the Estates helped the very talented young elector Frederick William (1640–1688), later known as the "Great Elector," to ride roughshod over traditional parliamentary liberties and to take a giant step toward royal absolutism. This constitutional struggle, often unjustly neglected by historians, was the most crucial in Prussian history for hundreds of years, until that of the 1860s.

When he came to power in 1640, the twenty-year-old Great Elector was determined to unify his three quite separate provinces and to add to them by diplomacy and war. These provinces were historic Brandenburg, the area around Berlin; Prussia, inherited in 1618 when the junior branch of the Hohenzollern family died out; and completely separate, scattered holdings along the Rhine in western Germany, inherited in 1614 (see Map 17.2). Each of the three provinces was inhabited by Germans; but each had its own Estates, whose power had increased until about 1600 as the power of the rulers declined. Although the Estates had not met regularly during the chaotic Thirty Years' War, they still had the power of the purse in their respective provinces. Taxes could not be levied without their consent. The Estates of Brandenburg and Prussia were dominated by the nobility and the landowning classes, known as the "Junkers." But it must be remembered that this was also true of the English Parliament before and after the civil war. Had the Estates successfully resisted the absolutist demands of the Great Elector, they, too, might have evolved toward more broadly based constitutionalism.

The struggle between the Great Elector and the provincial Estates was long, complicated, and intense. After the Thirty Years' War, the representatives of the nobility zealously reasserted the right of the Estates to vote taxes, a right the Swedish armies of occupation had simply ignored. Yet first in Brandenburg in 1653 and then in Prussia between 1661 and 1663, the Great Elector eventually had his way.

To pay for the permanent standing army he first established in 1660, Frederick William forced the Estates to accept the introduction of permanent taxation without consent. Moreover, the soldiers doubled as tax collectors and policemen, becoming the core of the rapidly expanding state bureaucracy. The power of the Estates declined rapidly thereafter, for the Great Elector had both financial independence and superior force. He turned the screws of taxation: the state's total revenue tripled during his reign. The size of the army leaped about tenfold. In 1688 a population of one million was supporting a peacetime standing army of thirty thousand. Many of the soldiers were French Huguenot immigrants, whom the Great Elector welcomed as the talented, hardworking citizens they were.

In accounting for the Great Elector's fateful triumph, two factors appear central. As in the formation of every absolutist state, war was a decisive factor. The ongoing struggle between Sweden and Poland for control of the Baltic after 1648 and the wars

of Louis XIV in western Europe created an atmosphere of permanent crisis. The wild Tartars of southern Russia swept through Prussia in the winter of 1656 to 1657, killing and carrying off as slaves more than fifty thousand people, according to an old estimate. This invasion softened up the Estates and strengthened the urgency of the elector's demands for more money for more soldiers. It was no accident that, except for commercially minded Holland, constitutionalism won out only in England, the only major country to escape devastating foreign invasions in the seventeenth century.

Second, the nobility had long dominated the government through the Estates, but only for its own narrow self-interest. When the crunch came, the Prussian nobles proved unwilling to join the representatives of the towns in a consistent common front against royal pretensions. The nobility was all too concerned with its own rights and privileges, especially its freedom from taxation and its unlimited control over the peasants. When, therefore, the Great Elector reconfirmed these privileges in 1653 and after, even while reducing the political power of the Estates, the nobility growled but did not bite. It accepted a compromise whereby the bulk of the new taxes fell on towns, and royal authority stopped at the landlords' gates. The elector could and did use naked force to break the liberties of the towns. The main leader of the urban opposition in the key city of Königsberg, for example, was simply arrested and imprisoned for life without trial.

THE CONSOLIDATION OF PRUSSIAN ABSOLUTISM

By the time of his death in 1688, the Great Elector had created a single state out of scattered principalities. But his new creation was still small and fragile. All the leading states of Europe had many more people—France with 20 million was fully twenty times as populous—and strong monarchy was still a novelty. Moreover, the Great Elector's successor, Elector Frederick III, "the Ostentatious" (1688–1713), was weak of body and mind.

Like so many of the small princes of Germany and Italy at the time, Frederick III imitated Louis XIV in every possible way. He built his own very expensive version of Versailles. He surrounded himself with cultivated artists and musicians and basked in the praise of toadies and sycophants. His only real politi-

Molding the Prussian Spirit Discipline was strict and punishment brutal in the Prussian army. This scene, from an eighteenth-century book used to teach school children, shows one soldier being flogged while another is being beaten with canes as he walks between rows of troops. The officer on horseback proudly commands. *(University of Illinois, Champaign)*

cal accomplishment was to gain the title of king from the Holy Roman emperor, a Habsburg, in return for military aid in the War of the Spanish Succession, and in 1701 he was crowned King Frederick I.

This tendency toward luxury-loving, happy, and harmless petty tyranny was completely reversed by Frederick William I, "the Soldiers' King" (1713–1740). A crude, dangerous psychoneurotic, Frederick William I was nevertheless the most talented reformer ever produced by the Hohenzollern family. It was he who truly established Prussian absolutism and

gave it its unique character. It was he who created the best army in Europe, for its size, and who infused military values into a whole society. In the words of a leading historian of Prussia:

For a whole generation, the Hohenzollern subjects were victimized by a royal bully, imbued with an obsessive bent for military organization and military scales of value. This left a deep mark upon the institutions of Prussiandom and upon the molding of the "Prussian spirit."[3]

Frederick William's passion for the army and military life was intensely emotional. He had, for example, a bizarre, almost pathological love for tall soldiers, whom he credited with superior strength and endurance. Austere and always faithful to his wife, he confided to the French ambassador: "The most beautiful girl or woman in the world would be a matter of indifference to me, but tall soldiers—they are my weakness." Like some fanatical modern-day basketball coach in search of a championship team, he sent his agents throughout both Prussia and all of Europe, tricking, buying, and kidnapping top recruits. Neighboring princes sent him their giants as gifts to win his gratitude. Prussian mothers told their sons: "Stop growing or the recruiting agents will get you."[4]

Profoundly military in temperament, Frederick William always wore an army uniform, and he lived the highly disciplined life of the professional soldier. He began his work by five or six in the morning; at ten he almost always went to the parade ground to drill or inspect his troops. A man of violent temper, Frederick William personally punished the most minor infractions on the spot: a missing button off a soldier's coat quickly provoked a savage beating with his heavy walking stick.

Frederick William's love of the army was also based on a hardheaded conception of the struggle for power and a dog-eat-dog view of international politics. Even before ascending the throne, he bitterly criticized his father's ministers: "They say that they will obtain land and power for the king with the pen; but I say it can be done only with the sword." Years later he summed up his life's philosophy in his instructions to his son: "A formidable army and a war chest large enough to make this army mobile in times of need can create great respect for you in the world, so that you can speak a word like the other powers."[5] This unshakable belief that the welfare of king and state depended on the army above all else reinforced Frederick William's personal passion for playing soldier.

The cult of military power provided the rationale for a great expansion of royal absolutism. As the king himself put it with his characteristic ruthlessness: "I must be served with life and limb, with house and wealth, with honour and conscience, everything must be committed except eternal salvation—that belongs to God, but all else is mine."[6] To make good these extraordinary demands, Frederick William cre-

A Prussian Giant Grenadier Frederick William I wanted tall handsome soldiers, and he dressed them in tight bright uniforms to distinguish them from the peasant population from which most soldiers emerged. Grenadiers wore the distinctive mitre cap instead of an ordinary hat so that they could hurl their heavy hand grenades unimpeded by a broad brim. *(The Bettmann Archive)*

ated a strong centralized bureaucracy. More commoners probably rose to top positions in the civil government than at any other time in Prussia's history. The last traces of the parliamentary Estates and local self-government vanished.

The king's grab for power brought him into considerable conflict with the noble landowners, the Junkers. In his early years, he even threatened to destroy them; yet in the end, the Prussian nobility was not destroyed but enlisted—into the army. Responding to a combination of threats and opportunities, the Junkers became the officer caste. By 1739 all but 5 of 245 officers with the rank of major or above were aristocrats, and most of them were native Prussians. A new compromise had been worked out, whereby the proud nobility imperiously commanded the peasantry in the army as well as on its estates.

Coarse and crude, penny-pinching and hardworking, Frederick William achieved results. Above all, he built a first-rate army on the basis of third-rate resources. The standing army increased from 38,000 to 83,000 during his reign. Prussia, twelfth in Europe in population, had the fourth largest army by 1740. Only the much more populous states of France, Russia, and Austria had larger forces, and even France's army was only twice as large as Prussia's. Moreover, soldier for soldier, the Prussian army became the best in Europe, astonishing foreign observers with its precision, skill, and discipline. For the next two hundred years, Prussia and then prussianized Germany would almost always win the crucial military battles.

Frederick William and his ministers also built an exceptionally honest and conscientious bureaucracy, which not only administered the country but tried with some success to develop it economically. Finally, like the miser he was, living very frugally off the income of his own landholdings, the king loved his "blue boys" so much that he hated to "spend" them. This most militaristic of kings was, paradoxically, almost always at peace.

Nevertheless, the Prussian people paid a heavy and lasting price for the obsessions of the royal drillmaster. Civil society became rigid and highly disciplined. Prussia became the "Sparta of the North"; unquestioning obedience was the highest virtue. As a Prussian minister later summed up, "To keep quiet is the first civic duty."[7] Thus the policies of Frederick William I combined with harsh peasant bondage and Junker tyranny to lay the foundations for probably the most militaristic country of modern times.

THE DEVELOPMENT OF RUSSIA

One of the favorite parlor games of nineteenth-century Russian (and non-Russian) intellectuals was debating whether Russia was a western European or a nonwestern Asiatic society. This question was particularly fascinating because it was unanswerable. To this day Russia differs fundamentally from the West in some basic ways, though Russian history has paralleled that of the West in other ways. A good case can be made for either position: thus the hypnotic attraction of Russian history.

The differences between Russia and the West were particularly striking before 1700, when Russia's overall development began to draw progressively closer to that of its western neighbors. These early differences and Russia's long isolation from Europe explain why little has so far been said here about Russia. Yet it is impossible to understand how Russia has increasingly influenced and been influenced by western European civilization since roughly the late seventeenth century without looking at the course of early Russian history. Such a brief survey will also help explain how, when absolute monarchy finally and decisively triumphed under the rough guidance of Peter the Great in the early eighteenth century, it was a quite different type of absolute monarchy from that of France or even Prussia.

THE VIKINGS AND THE KIEVAN PRINCIPALITY

In antiquity the Slavs lived as a single people in central Europe. With the start of the mass migrations of the late Roman Empire, the Slavs moved in different directions and split into three groups. Between the fifth and ninth centuries, the eastern Slavs, from whom the Ukrainians, the Russians, and the White Russians descend, moved into the vast and practically uninhabited area of present-day European Russia and the Ukraine (see Map 17.3).

This enormous area consisted of an immense virgin forest to the north, where most of the eastern Slavs settled, and an endless prairie grassland to the south. Probably organized as tribal communities, the eastern Slavs, like many North American pioneers much later, lived off the great abundance of wild game and a crude "slash and burn" agriculture. After clearing a piece of the forest to build log cabins, they burned the stumps and brush. The ashes left a rich deposit of potash and lime, and the land gave several good crops before it was exhausted. The people then moved on to another untouched area and repeated the process.

In the ninth century, the Vikings, those fearless warriors from Scandinavia, appeared in the lands of the eastern Slavs. Called "Varangians" in the old Russian chronicles, the Vikings were interested primarily in international trade, and the opportunities were good, since the Muslim conquests of the eighth century had greatly reduced Christian trade in the Mediterranean. Moving up and down the rivers, the Vikings soon linked Scandinavia and northern Europe with the Black Sea and the Byzantine Empire with its capital at Constantinople. They built a few strategic forts along the rivers, from which they raided the neighboring Slavic tribes and collected tribute. Slaves were the most important article of tribute, and *Slav* even became the word for "slave" in several European languages.

In order to increase and protect their international commerce, the Vikings declared themselves the rulers of the eastern Slavs. According to tradition, the semilegendary chieftain Ruirik founded the princely dynasty about 860. In any event, the Varangian ruler Oleg (878–912) established his residence at Kiev. He and his successors ruled over a loosely united confederation of Slavic territories—the Kievan state—until 1054. The Viking prince and his clansmen quickly became assimilated into the Slavic population, taking local wives and emerging as the noble class.

Assimilation and loss of Scandinavian ethnic identity was speeded up by the conversion of the Vikings and local Slavs to Eastern Orthodox Christianity by missionaries from the Byzantine Empire. The written language of these missionaries, Slavic—church Slavonic—was subsequently used in all religious and nonreligious documents in the Kievan principality. Thus the rapidly slavified Vikings left two important legacies for the future. They created a loose unification of Slavic territories under a single ruling prince and a single ruling dynasty. And they imposed a basic religious unity by accepting Orthodox Christianity, as opposed to Roman Catholicism, for themselves and the eastern Slavs.

Even at its height under Great Prince Iaroslav the Wise (1019–1054), the unity of the Kievan principality was extremely tenuous. Trade, rather than government, was the main concern of the rulers. Moreover, the slavified Vikings failed to find a way of peacefully transferring power from one generation to the next. In medieval western Europe this fundamental problem of government was increasingly resolved by resort to the principle of primogeniture: the king's eldest son received the crown as his rightful inheritance when his father died. Civil war was thus averted; order was preserved. In early Kiev, however, there were apparently no fixed rules and much strife accompanied each succession.

Possibly to avoid such chaos, before his death in 1054 Great Prince Iaroslav divided the Kievan principality among his five sons, who in turn divided their properties when they died. Between 1054 and 1237, Kiev disintegrated into more and more competing units, each ruled by a prince claiming to be a descendant of Ruirik. Even when only one prince was claiming to be the great prince, the whole situation was very unsettled.

The princes divided their land like private property because they thought of it as private property. A given prince owned a certain number of farms or landed estates and had them worked directly by his people, mainly slaves, called *kholops* in Russian. Outside of these estates, which constituted the princely domain, the prince exercised only very limited authority in his principality. Excluding the clergy, two kinds of people lived there: the noble *boyars* and the commoner peasants.

The boyars were the descendants of the original Viking warriors, and they also held their lands as free and clear private property. And although the boyars normally fought in princely armies, the customary law declared they could serve any prince they wished. The ordinary peasants were also truly free. The peasants could move at will wherever opportunities were greatest. In the touching phrase of the times, theirs was "a clean road, without boundaries."[8] In short, fragmented princely power, private property, and personal freedom all went together.

THE MONGOL YOKE AND THE RISE OF MOSCOW

The eastern Slavs, like the Germans and the Italians, might have emerged from the Middle Ages weak and politically divided, had it not been for a development of extraordinary importance—the Mongol conquest of the Kievan state. Wild nomadic tribes from

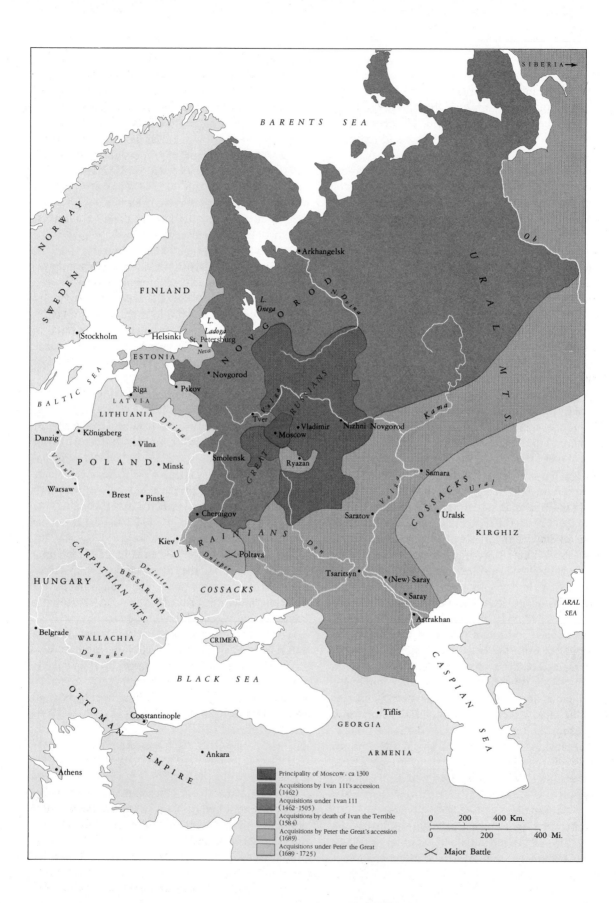

BARENTS SEA

SIBERIA →

NORWAY

SWEDEN

FINLAND

• Arkhangelsk

N O V G O R O D

U R A L   M T S.

Ob

L. Onega

Dvina

• Stockholm

• Helsinki

L. Ladoga

St. Petersburg

Neva

ESTONIA

BALTIC SEA

• Riga

• Pskov

• Novgorod

LATVIA

LITHUANIA

Dvina

GREAT RUSSIANS

Volga

• Tver

• Vladimir

Nizhni Novgorod

Kama

Danzig •

• Königsberg

• Vilna

• Moscow

POLAND

• Minsk

• Smolensk

• Ryazan

GREAT

Volga

• Samara

COSSACKS

Ural

• Warsaw

• Brest

• Pinsk

Vistula

Saratov •

Don

• Uralsk

KIRGHIZ

• Chernigov

UKRAINIANS

Dnieper

✕ Poltava

Tsaritsyn •

• (New) Saray

ARAL SEA

Kiev •

Dniester

BESSARABIA

CARPATHIAN MTS.

COSSACKS

• Saray

• Astrakhan

HUNGARY

CRIMEA

• Belgrade

WALLACHIA

Danube

BLACK SEA

CASPIAN SEA

OTTOMAN

• Constantinople

GEORGIA

• Tiflis

ARMENIA

EMPIRE

• Ankara

• Athens

Principality of Moscow. ca 1300

Acquisitions by Ivan III's accession (1462)

Acquisitions under Ivan III (1462–1505)

Acquisitions by death of Ivan the Terrible (1584)

Acquisitions by Peter the Great's accession (1689)

Acquisitions under Peter the Great (1689–1725)

0   200   400 Km.

0   200   400 Mi.

✕ Major Battle

MAP 17.3 **The Expansion of Russia to 1725** After the disintegration of the Kievan state and the Mongol conquest, the princes of Moscow and their descendants gradually extended their rule over an enormous territory.

present-day Mongolia, the Mongols were temporarily unified in the thirteenth century by Jenghiz Khan (1162–1227), one of history's greatest conquerors. In five years his armies subdued all of China. His successors then wheeled westward, smashing everything in their path and reaching the plains of Hungary victorious before they pulled back in 1242. The Mongol army—the Golden Horde—was savage in the extreme, often slaughtering the entire populations of cities before burning them to the ground. En route to Mongolia, Archbishop John of Plano Carpini, the famous papal ambassador to Mongolia, passed through Kiev in southern Russia in 1245 to 1246 and wrote an unforgettable eyewitness account:

The Mongols went against Russia and enacted a great massacre in the Russian land. They destroyed towns and fortresses and killed people. They besieged Kiev which had been the capital of Russia, and after a long siege they took it and killed the inhabitants of the city. For this reason, when we passed through that land, we found lying in the field countless heads and bones of dead people; for this city had been extremely large and very populous, whereas now it has been reduced to nothing: barely two hundred houses stand there, and those people are held in the harshest slavery.[9]

Having devastated and conquered, the Mongols ruled the eastern Slavs for more than two hundred years. They built their capital of Saray on the lower Volga (see Map 17.3). They forced all the bickering Slavic princes to submit to their rule and to give them tribute and slaves. If the conquered peoples rebelled, the Mongols were quick to punish with death and destruction. Thus, the Mongols unified the eastern Slavs, for the Mongol khan was acknowledged by all as the supreme ruler.

The Mongol unification completely changed the internal political situation. Although the Mongols conquered, they were quite willing to use local princes as their obedient servants and tax collectors. Therefore they did not abolish the title of great prince, bestowing it instead on the prince who served

them best and paid them most handsomely.

Beginning with Alexander Nevsky in 1252, the previously insignificant princes of Moscow became particularly adept at serving the Mongols. They loyally put down popular uprisings and collected the khan's harsh taxes. By way of reward, the princes of Moscow emerged as hereditary great princes. Eventually the Muscovite princes were able to destroy their princely rivals and even to replace the khan as supreme ruler. In this complex process, two princes of Moscow after Alexander Nevsky—Ivan I and Ivan III—were especially noteworthy.

Ivan I (1328–1341) was popularly known as Ivan the Moneybag. A bit like Frederick William of Prussia, he was extremely stingy and built up a large personal fortune. This enabled him to buy more property and to increase his influence by loaning money to less frugal princes to pay their Mongol taxes. Ivan's most serious rival was the prince of Tver, whom the Mongols at one point appointed as great prince.

In 1327 the population of Tver revolted against Mongol oppression, and the prince of Tver joined his people. Ivan immediately went to the Mongol capital of Saray, where he was appointed commander of a large Russian-Mongol army, which then laid waste to Tver and its lands. For this proof of devotion, the Mongols made Ivan the general tax collector for all the Slavic lands they had subjugated and named him great prince. Ivan also convinced the metropolitan of Kiev, the leading churchman of all eastern Slavs, to settle in Moscow; Ivan I thus gained greater prestige, while the church gained a powerful advocate before the khan.

In the next hundred-odd years, in the course of innumerable wars and intrigues, the great princes of Moscow significantly increased their holdings. Then, in the reign of Ivan III (1462–1505), the long process was largely completed. After purchasing Rostov, Ivan conquered and annexed other principalities, of which Novgorod with its lands extending as far as the Baltic Sea was most crucial (see Map 17.3). Thus, more than four hundred years after Iaroslav the Wise had divided the embryonic Kievan state, the princes of Moscow defeated all the rival branches of the house of Ruirik to win complete princely authority.

Another dimension to princely power developed. Not only were the princes of Moscow the *unique* rulers, they were the *absolute* ruler, the autocrat, the *tsar*—the Slavic contraction for "caesar," with all its connotations. This imperious conception of absolute power is expressed in a famous letter from the aging Ivan III to the Holy Roman Emperor Frederick III (1440–1493). Frederick had offered Ivan the title of king in conjunction with the marriage of his daughter to Ivan's nephew. Ivan proudly refused:

We by the grace of God have been sovereigns over our domains from the beginning, from our first forebears, and our right we hold from God, as did our forebears. . . . As in the past we have never needed appointment from anyone, so now do we not desire it.[10]

The Muscovite idea of absolute authority was powerfully reinforced by two developments. First, about 1480 Ivan III stopped acknowledging the khan as his supreme ruler. There is good evidence to suggest that Ivan and his successors saw themselves as khans. Certainly they assimilated the Mongol concept of kingship as the exercise of unrestrained and unpredictable power.

Second, after the fall of Constantinople to the Turks in 1453, the tsars saw themselves as the heirs of both the caesars and Orthodox Christianity, the one true faith. All the other kings of Europe were heretics: only the tsars were rightful and holy rulers. This idea was promoted by Orthodox churchmen, who spoke of "holy Russia" as the "Third Rome." As the monk Pilotheus stated: "Two Romes have fallen, but the third stands, and a fourth there will not be."[11] Ivan's marriage to the daughter of the last Byzantine emperor further enhanced the aura of an eastern imperial inheritance for Moscow. Worthy successor to the mighty khan and the true Christian emperor, the Muscovite tsar was a king above all others.

TSAR AND PEOPLE TO 1689

By 1505 the great prince of Moscow—the tsar—had emerged as the single hereditary ruler of "all the Russias"—all the lands of the eastern Slavs—and he was claiming unrestricted power as his God-given right. In effect, the tsar was demanding the same kind of total authority over all his subjects that the princely descendants of Ruirik had long exercised over their slaves on their own landed estates. This was an extremely radical demand.

While peasants had begun losing their freedom of movement in the fifteenth century, so had the noble boyars begun to lose power and influence. Ivan III pi-

oneered in this regard, as in so many others. When Ivan conquered the principality of Novgorod in the 1480s, he confiscated fully 80 percent of the land, executing the previous owners or resettling them nearer Moscow. He then kept more than half of the confiscated land for himself and distributed the remainder to members of a newly emerging service nobility. The boyars had previously held their land as hereditary private property and been free to serve the prince of their choosing. The new service nobility held the tsar's land on the explicit condition that they serve in the tsar's army. Moreover, Ivan III began to require boyars outside of Novgorod to serve him if they wished to retain their lands. Since there were no competing princes left to turn to, the boyars had to yield.

The rise of the new service nobility accelerated under Ivan IV (1533–1584), the famous Ivan the Terrible. Having ascended the throne at age three, Ivan had suffered insults and neglect at the hands of the haughty boyars after his mother mysteriously died, possibly poisoned, when he was just eight. At age sixteen he suddenly pushed aside his hated boyar advisers. In an awe-inspiring ceremony, complete with gold coins pouring down on his head, he majestically crowned himself and officially took the august title of tsar for the first time.

Selecting the beautiful and kind Anastasia of the popular Romanov family for his wife and queen, the young tsar soon declared war on the remnants of Mongol power. He defeated the faltering khanates of Kazan and Astrakhan between 1552 and 1556, adding vast new territories to Russia. In the course of these wars Ivan virtually abolished the old distinction between hereditary boyar private property and land granted temporarily for service. All nobles, old and new, had to serve the tsar in order to hold any land.

The process of transforming the entire nobility into a service nobility was completed in the second part of Ivan the Terrible's reign. In 1557 Ivan turned westward, and for the next twenty-five years Muscovy waged an exhausting, unsuccessful war primarily with the large Polish-Lithuanian state, which controlled not only Poland but much of the Ukraine in the sixteenth century. Quarreling with the boyars over the war and blaming them for the sudden death of his beloved Anastasia in 1560, the increasingly cruel and demented Ivan turned to strike down all who stood in his way.

Above all, he struck down the ancient Muscovite boyars with a reign of terror. Leading boyars, their relatives, and even their peasants and servants were executed en masse by a special corps of unquestioning servants. Dressed in black and riding black horses, they were forerunners of the modern dictator's secret police. Large estates were confiscated, broken up, and reapportioned to the lower service nobility. The great boyar families were severely reduced. The newer, poorer, more nearly equal service nobility, still less than .5 percent of the total population, was totally dependent on the autocrat.

Ivan also took giant strides toward making all commoners servants of the tsar. His endless wars and demonic purges left much of central Russia depopulated. It grew increasingly difficult for the lower service nobility to squeeze a living for themselves out of the peasants left on their landholdings. As the service nobles demanded more from the remaining peasants, more and more peasants fled toward the wild, recently conquered territories to the east and south. There they formed free groups and outlaw armies known as "Cossacks." The Cossacks maintained a precarious independence beyond the reach of the oppressive landholders and the tsar's hated officials. The solution to this problem was to complete the tying of the peasants to the land, making them serfs perpetually bound to serve the noble landholders, who were bound in turn to serve the tsar.

In the time of Ivan the Terrible, urban traders and artisans were also bound to their towns and jobs, so that the tsar could tax them more heavily. Ivan assumed that the tsar owned Russia's trade and industry, just as he owned all the land. In the course of the sixteenth and seventeenth centuries, the tsars therefore took over the mines and industries and monopolized the country's important commercial activities. The urban classes had no security in their work or property, and even the wealthiest merchants were basically dependent agents of the tsar. If a new commercial activity became profitable, it was often taken over by the tsar and made a royal monopoly. This royal monopolization was in sharp contrast to developments in western Europe, where the capitalist middle classes were gaining strength and security in their private property. The tsar's service obligations checked the growth of the Russian middle classes, just as they led to decline of the boyars, rise of the lower nobility, and the final enserfment of the peasants.

St. Basil's Cathedral in Moscow, with its steeply sloping roofs and proliferation of multicolored onion-shaped domes, was a striking example of powerful Byzantine influences on Russian culture. According to tradition, an enchanted Ivan the Terrible blinded the cathedral's architects to ensure that they would never duplicate their fantastic achievement. *(The New York Public Library)*

Ivan the Terrible's system of autocracy and compulsory service struck foreign observers forcibly. Sigismund Herberstein, a German traveler to Russia, wrote in 1571: "All the people consider themselves to be *kholops,* that is slaves of their Prince." At the same time, Jean Bodin, the French thinker who did so much to develop the modern concept of sovereignty, concluded that Russia's political system was fundamentally different from those of all other European monarchies and comparable only to that of the Turkish empire. In both Turkey and Russia, as in other parts of Asia and Africa, "the prince is become lord of the goods and persons of his subjects . . . governing them as a master of a family does his slaves."[12] The Mongol inheritance weighed heavily on Russia.

As has so often occurred in Russia, the death of an iron-fisted tyrant—in this case, Ivan the Terrible in 1584—ushered in an era of confusion and violent struggles for power. Events were particularly chaotic after Ivan's son Theodore died in 1598 without an heir. The years 1598 to 1613 are aptly called the "Time of Troubles."

The close relatives of the deceased tsar intrigued against and murdered each other, alternately fighting and welcoming the invading Swedes and Poles, who even occupied Moscow. Most serious for the cause of autocracy, there was a great social upheaval as Cossack bands marched northward, rallying peasants and slaughtering nobles and officials. The mass of Cossacks and peasants called for the "true tsar," who would restore their freedom of movement and allow them to farm for whomever they pleased, who would reduce their heavy taxes and lighten the yoke imposed by the landlords.

This social explosion from below, which combined with a belated surge of patriotic opposition to Polish invaders, brought the nobles, big and small, to their senses. In 1613 they elected Ivan's sixteen-year-old grand-nephew, Michael Romanov, the new hereditary tsar. Then they rallied around him in the face of common internal and external threats. Michael's election was a real restoration, and his reign saw the gradual re-establishment of tsarist autocracy. Michael was understandably more kindly disposed toward the supportive nobility than toward the sullen peasants. Thus, while peasants were completely enserfed in 1649, Ivan's heavy military obligations on the nobility were relaxed considerably. In the long reign of Michael's successor, the pious Alexis (1645–1676), this asymmetry of obligations was accentuated. The nobility gained more exemptions from military service, while the peasants were further ground down.

The result was a second round of mass upheaval and protest. In the later seventeenth century, the unity of the Russian Orthodox church was torn apart by a great split. The surface question was the religious reforms introduced in 1652 by the patriarch Nikon, a dogmatic purist who wished to bring "corrupted" Russian practices of worship into line with the Greek Orthodox model. The self-serving church hierarchy quickly went along, but the intensely religious common people resisted. They saw Nikon as the anti-Christ, who was stripping them of the only thing they had—the true religion of "holy Russia."

Great numbers left the church and formed illegal communities of Old Believers, who were hunted down and persecuted. As many as twenty thousand people burned themselves alive, singing the "halleluyah" in their chants three times rather than twice as Nikon had demanded and crossing themselves in the old style, with two rather than three fingers, as they went down in flames. After the great split, the Russian masses were alienated from the established church, which became totally dependent on the state for its authority.

Again the Cossacks revolted against the state, which was doggedly trying to catch up with them on the frontiers and reduce them to serfdom. Under Stenka Razin they moved up the Volga River in 1670 and 1671, attracting a great undisciplined army of peasants, murdering landlords and high church officials, and proclaiming freedom from oppression. This rebellion to overthrow the established order was finally defeated by the government. In response, the thoroughly scared upper classes tightened the screws of serfdom even further. Holding down the peasants, and thereby maintaining the tsar, became almost the principal obligation of the nobility until 1689.

THE REFORMS OF PETER THE GREAT

It is now possible to understand the reforms of Peter the Great (1689–1725) and his kind of monarchial absolutism. Contrary to some historians' assertions, Peter was interested primarily in military power and not in some grandiose westernization plan. A giant for his time, at six feet seven inches, and possessing enormous energy and willpower, Peter was determined to redress the defeats the tsar's armies had occasionally suffered in their wars with Poland and Sweden since the time of Ivan the Terrible.

To be sure, these western foes had never seriously threatened the existence of the tsar's vast kingdom, except perhaps when they had added to the confusion of civil war and domestic social upheaval in the Time of Troubles. Russia had even gained a large mass of the Ukraine from the kingdom of Poland in 1667 (see Map 17.3). And tsarist forces had completed the conquest of the primitive tribes of all Siberia in the seventeenth century. Muscovy, which had been as large

"The Bronze Horseman" This equestrian masterpiece of Peter the Great, finished for Catherine the Great in 1783, dominates the center of St. Petersburg (modern Leningrad). The French sculptor Falconnet has captured the tsar's enormous energy, power, and determination. *(Courtesy of the Courtauld Institute of Art)*

as all the rest of Europe combined in 1600, was three times as large as the rest of Europe in 1689 and by far the largest kingdom on earth. But territorial expansion was the soul of tsardom, and it was natural that Peter would seek further gains. The thirty-six years of his reign knew only one year of peace.

When Peter came to the throne, the heart of his army still consisted of cavalry made up of boyars and service nobility. Foot soldiers played a secondary role, and the whole army served on a part-time basis. The Russian army was lagging behind the professional standing armies being formed in Europe in the seventeenth century. The core of such armies was a highly disciplined infantry—an infantry that fired and refired rifles as it fearlessly advanced, until it charged with bayonets fixed. Such a large permanent army was enormously expensive and could be created only at the cost of great sacrifice. Given the desire to conquer more territory, Peter's military problem was serious.

Peter's solution was, in essence, to tighten up Muscovy's old service system and really make it work. He put the nobility back in harness with a vengeance. Every nobleman, great or small, was once again required to serve in the army or in the civil administration—for life. Since a more modern army and government required skilled technicians and experts, Peter created schools and even universities. One of his most hated reforms required five years of compulsory education away from home for every young nobleman. Peter established an interlocking military-civilian bureaucracy with fourteen ranks, and he decreed that all must start at the bottom and work toward the top. More people of nonnoble origins rose to high positions in the embryonic meritocracy. Peter searched out talented foreigners—twice in his reign he went abroad to study and observe—and placed them in his service. These measures combined to make the army and government more powerful and efficient.

THE RISE OF ABSOLUTISM IN EASTERN EUROPE

1050–1300	Increasing economic development in eastern Europe encourages decline in serfdom
1054	Death of Great Prince Iaroslav the Wise, under whom the Kievan principality reached its height of unity
1054–1237	Kiev is divided into numerous territories ruled by competing princes
1237–1240	Mongol invasion of Russia
1252	Alexander Nevsky, Prince of Moscow, recognizes Mongol overlordship
1327–1328	Suppression of the Tver revolt; Mongol khan recognizes Ivan I as great prince
1400–1650	The nobility reimposes serfdom in eastern Europe
ca 1480	Ivan III rejects Mongol overlordship and adopts the title of tsar
1520–1566	Rule of Suleiman the Magnificent: Ottoman Empire reaches its height
1533–1584	Rule of Tsar Ivan IV (the Terrible): defeat of the khanates of Kazan and Astrakhan; subjugation of the boyar aristocracy
1574	Polish nobles receive the right to inflict the death penalty on their serfs
1598–1613	"Time of Troubles" in Russia
1613	Election of Michael Romanov as tsar: re-establishment of autocracy
1620	Battle of the White Mountain in Bohemia: Ferdinand II initiates Habsburg confiscation of Czech estates
1640–1688	Rule of Frederick William, the Great Elector, who unites Brandenburg, Prussia, and western German holdings intu one state, Brandenburg-Prussia
1649	Tsar Alexis lifts the nine-year limit on the recovery of runaway serfs
1652	Patriarch Nikon's reforms split the Russian Orthodox church
1653	Principle of peasants' "hereditary subjugation" to their lords affirmed in Prussia
1670–1671	Cossack revolt of Stenka Razin in Russia
1683	Siege of Vienna by the Ottoman Turks
1683–1699	Habsburg conquest of Hungary and Transylvania
1689–1725	Rule of Tsar Peter the Great
1700–1721	Great Northern War between Russia and Sweden, resulting in Russian victory and territorial expansion
1701	Elector Frederick III crowned king of Prussia
1703	Foundation of St. Petersburg
	Rebellion of Prince Francis Rakoczy in Hungary
1713	Pragmatic Sanction: Charles VII guarantees Maria Theresa's succession to the Austrian Empire
1713–1740	Rule of King Frederick William I in Prussia

Peter also greatly increased the service requirements of the commoners. He established a regular standing army of more than 200,000 soldiers, made up mainly of peasants commanded by officers from the nobility. In addition, special forces of Cossacks and foreigners numbered more than 100,000. The departure of a drafted peasant boy was celebrated by his family and village almost like a funeral, as indeed it was, since the recruit was drafted for life. The peasantry also served with its taxes, which increased threefold during Peter's reign, as people—"souls"—replaced land as the primary unit of taxation. Serfs were also arbitrarily assigned to work in the growing number of factories and mines. Most of these industrial enterprises were directly or indirectly owned by the state, and they were worked almost exclusively for the military. In general, Russian serfdom became more oppressive under the reforming tsar.

The constant warfare of Peter's reign consumed 80 to 85 percent of all revenues but brought only modest territorial expansion. Yet the Great Northern War with Sweden, which lasted from 1700 to 1721, was crowned in the end by Russian victory. After initial losses, Peter's new war machine crushed the smaller army of Sweden's Charles XII in the Ukraine at Poltava in 1709, one of the most significant battles in Russian history. Sweden never really regained the offensive, and Russia eventually annexed Estonia and much of present-day Latvia (see Map 17.2), lands that had never before been under Russian rule. Russia became the dominant power on the Baltic Sea and very much a European Great Power. If victory or defeat is the ultimate historical criterion, Peter's reforms were a success.

There were other important consequences of Peter's reign. Because of his feverish desire to use modern technology to strengthen the army, many Westerners and Western ideas flowed into Russia for the first time. A new class of educated Russians began to emerge. At the same time, vast numbers of Russians, especially among the poor and weak, hated Peter's massive changes. The split between the enserfed peasantry and the educated nobility thus widened, even though all were caught up in the endless demands of the sovereign.

A new idea of state interest, distinct from the tsar's personal interests, began to take hold. Peter himself fostered this conception of the public interest by claiming time and again to be serving the common good. For the first time a Russian tsar attached explanations to his decrees in an attempt to gain the confidence and enthusiastic support of the populace. Yet, as before, the tsar alone decided what the common good was. Here was a source of future tension between tsar and people.

In sum, Peter built on the service obligations of old Muscovy. His monarchial absolutism was truly the culmination of the long development of a unique Russian civilization. Yet the creation of a more modern army and state introduced much that was new and western to that civilization. This development paved the way for Russia to move much closer to the European mainstream in its thought and institutions during the Enlightenment, especially under that famous administrative and sexual lioness, Catherine the Great.

ABSOLUTISM AND THE BAROQUE

The rise of royal absolutism in eastern Europe had many consequences. Nobles served their powerful rulers in new ways, while the great inferiority of the urban middle classes and the peasants was reconfirmed. Armies became larger and more professional, while taxes rose and authoritarian traditions were strengthened. Nor was this all. Royal absolutism also interacted with baroque culture and art, baroque music and literature. Inspired in part by Louis XIV of France, the great and not-so-great rulers called on the artistic talent of the age to glorify their power and magnificence. This exaltation of despotic rule was particularly striking in the lavish masterpieces of architecture.

BAROQUE ART AND MUSIC

Throughout European history, the cultural tastes of one age have often seemed quite unsatisfactory to the next. So it was with the baroque. The term *baroque* itself may have come from the Portuguese word for an "odd-shaped, imperfect pearl" and was commonly used by late eighteenth-century art critics as an expression of scorn for what they considered an overblown, unbalanced style. The hostility of these critics, who also scorned the Gothic style of medieval cathedrals in favor of a classicism inspired by antiquity and the Renaissance, has long since passed. Specialists agree that the triumphs of the baroque marked one of the high points in the history of Western culture.

The early development of the baroque is complex, but most scholars stress the influence of Rome and the revitalized Catholic church of the later sixteenth century. The papacy and the Jesuits encouraged the growth of an intensely emotional, exuberant art. These patrons wanted artists to go beyond the Renaissance focus on pleasing a small, wealthy cultural elite. They wanted artists to appeal to the senses and thereby touch the souls and kindle the faith of ordinary churchgoers, while proclaiming the power and confidence of the reformed Catholic church. In addition to this underlying religious emotionalism, the baroque drew its sense of drama, motion, and ceaseless striving from the Catholic Reformation. The interior of the famous Jesuit Church of Jesus in Rome

—the Gesù—combined all these characteristics in its lavish, shimmering, wildly active decorations and frescoes.

Taking definite shape in Italy after 1600, the baroque style in the visual arts developed with exceptional vigor in Catholic countries—in Spain and Latin America, Austria, southern Germany, and Poland. Yet baroque art was more than just "Catholic art" in the seventeenth century and the first half of the eighteenth. True, neither Protestant England nor the Netherlands ever came fully under the spell of the baroque, but neither did Catholic France. And Protestants accounted for some of the finest examples of baroque style, especially in music. The baroque style spread partly because its tension and bombast spoke to an agitated age, which was experiencing great violence and controversy in politics and religion.

In painting, the baroque reached maturity early with Peter Paul Rubens (1577–1640), the most outstanding and representative of baroque painters (see color insert IV). Studying in his native Flanders and in Italy, where he was influenced by masters of the High Renaissance such as Michelangelo, Rubens developed his own rich, sensuous, colorful style, which was characterized by animated figures, melodramatic contrasts, and monumental size. Although Rubens excelled in glorifying monarchs such as Queen Mother Marie de' Medici of France, he was also a devout Catholic. Nearly half of his pictures treat Christian subjects. Yet one of Rubens's trademarks was fleshy, sensual nudes, who populate his canvasses as Roman goddesses, water nymphs, and remarkably voluptuous saints and angels.

Rubens was enormously successful. To meet the demand for his work, he established a large studio and hired many assistants to execute his rough sketches and gigantic murals. Sometimes the master artist added only the finishing touches. Rubens's wealth and position—on occasion he was given special diplomatic assignments by the Habsburgs—attest that distinguished artists continued to enjoy the high social status they had won in the Renaissance.

In music, the baroque style reached its culmination almost a century later in the dynamic, soaring lines of the endlessly inventive Johann Sebastian Bach (1685–1750), one of the greatest composers the Western world has ever produced. Organist and choirmaster of several Lutheran churches across Germany, Bach was equally at home writing secular concertos and sublime religious cantatas. Bach's organ music, the greatest ever written, combined the baroque spirit of invention, tension, and emotion in an unforgettable striving toward the infinite. Unlike Rubens, Bach was not fully appreciated in his lifetime, but since the early nineteenth century his reputation has grown steadily.

PALACES AND POWER

As soaring Gothic cathedrals expressed the idealized spirit of the High Middle Ages, so dramatic baroque palaces symbolized the age of absolutist power. By 1700 palace building had become a veritable obsession for the rulers of central and eastern Europe. These baroque palaces were clearly intended to overawe the people with the monarch's strength. The great palaces were also visual declarations of equality with Louis XIV and were therefore modeled after Versailles to a greater or lesser extent. One such palace was Schönbrunn, an enormous Viennese Versailles, begun in 1695 by Emperor Leopold to celebrate Austrian military victories and Habsburg might. Charles XI of Sweden, having reduced the power of the aristocracy, ordered the construction in 1693 of his Royal Palace, which dominates the center of Stockholm to this day. Frederick I of Prussia began his imposing new royal residence in Berlin in 1701, a year after he attained the title of king.

Petty princes also contributed mightily to the palace-building mania. Frederick the Great of Prussia noted that every descendant of a princely family "imagines himself to be something like Louis XIV. He builds his Versailles, has his mistresses, and maintains his army."[13] The not-very-important elector-archbishop of Mainz, the ruling prince of that city, confessed apologetically that "building is a craze which costs much, but every fool likes his own hat."[14] The archbishop of Mainz's own "hat" was an architectural gem, like that of another churchly ruler, the prince-bishop of Würzburg.

In central and eastern Europe, the favorite noble servants of royalty became extremely rich and powerful, and they, too, built grandiose palaces in the capital cities. These palaces were in part an extension of the monarch, for they surpassed the buildings of less-favored nobles and showed all the high road to fame and fortune. Take, for example, the palaces of Prince Eugene of Savoy. A French nobleman by birth and education, Prince Eugene entered the service of Leopold I with the relief of the besieged Vienna in 1683,

Rubens: The Education of Marie One of twenty-one pictures celebrating episodes from the life of Marie de' Medici, the influential widow of France's Henri IV, Rubens's painting is imbued with sensuous vitality. The three muses inspiring the studious young Marie dominate the canvas. *(Louvre/Giraudon/Art Resource)*

The Benedictine Abbey of Melk Rebuilt in the eighteenth century, this masterpiece of the Austrian baroque stands majestically on the heights above the Danube River. *(Archiv/Photo Researchers)*

and he became Austria's most outstanding military hero. It was he who reorganized the Austrian army, smashed the Turks, fought Louis XIV to a standstill, and generally guided the triumph of absolutism in Austria. Rewarded with great wealth by his grateful royal employer, Eugene called on the leading architects of the day, J. B. Fischer von Erlach and Johann Lukas von Hildebrandt, to consecrate his glory in stone and fresco. Fischer built Eugene's Winter (or Town) Palace in Vienna, and he and Hildebrandt collaborated on the prince's Summer Palace on the city's outskirts.

The Summer Palace was actually two enormous buildings, the Lower Belvedere and the Upper Belvedere, completed in 1713 and 1722 respectively and joined by one of the most exquisite gardens in Europe. The Upper Belvedere, Hildebrandt's masterpiece, stood gracefully, even playfully, behind a great

sheet of water. One entered through magnificent iron gates into a hall where sculptured giants crouched as pillars and then moved on to a great staircase of dazzling whiteness and ornamentation. Even today the emotional impact of this building is great: here art and beauty create a sense of immense power and wealth.

Palaces like the Upper Belvedere were magnificent examples of the baroque style. They expressed the baroque delight in bold, sweeping statements, which were intended to provide a dramatic emotional experience. To create this experience, baroque masters dissolved the traditional artistic frontiers: the architect permitted the painter and the artisan to cover his undulating surfaces with wildly colorful paintings, graceful sculptures, and fanciful carvings. Space was used in a highly original way, to blend everything together in a total environment. These techniques shone in all their glory in the churches of southern Germany and in the colossal entrance halls of palaces like that of the prince-bishop of Würzburg (see color insert IV). Artistic achievement and political statement reinforced each other.

ROYAL CITIES

Absolute monarchs and baroque architects were not content with fashioning ostentatious palaces. They remodeled existing capital cities, or even built new ones, to reflect royal magnificence and the centralization of political power. Karlsruhe, founded in 1715 as the capital city of a small German principality, is one extreme example. There, broad, straight avenues radiated out from the palace, so that all roads—like all power—were focused on the ruler. More typically, the monarch's architects added new urban areas alongside the old city; these areas then became the real heart of the expanding capital.

The distinctive features of these new additions were their broad avenues, their imposing government buildings, and their rigorous mathematical layout. Along these major thoroughfares the nobles built elaborate baroque townhouses; stables and servants' quarters were built on the alleys behind. Wide avenues also facilitated the rapid movement of soldiers through the city to quell any disturbance (the king's planners had the needs of the military constantly in mind). Under the arcades along the avenues appeared smart and very expensive shops, the first department stores, with plate-glass windows and fancy displays.

The new avenues brought reckless speed to the European city. Whereas everyone had walked through the narrow, twisting streets of the medieval town, the high and mighty raced down the broad boulevards in their elegant carriages. A social gap opened between the wealthy riders and the gaping, dodging pedestrians. "Mind the carriages!" wrote one eighteenth-century observer in Paris:

Here comes the black-coated physician in his chariot, the dancing master in his coach, the fencing master in his surrey—and the Prince behind six horses at the gallop as if he were in the open country. . . . The threatening wheels of the overbearing rich drive as rapidly as ever over stones stained with the blood of their unhappy victims.[15]

Speeding carriages on broad avenues, an endless parade of power and position: here was the symbol and substance of the baroque city.

THE GROWTH OF ST. PETERSBURG

No city illustrated better than St. Petersburg the close ties among politics, architecture, and urban development in this period. In 1700, when the Great Northern War between Russia and Sweden began, the city did not exist. There was only a small Swedish fortress on one of the water-logged islands at the mouth of the Neva River, where it flows into the Baltic Sea. In 1702 Peter the Great's armies seized this desolate outpost. Within a year the reforming tsar had decided to build a new city there and to make it, rather than ancient Moscow, his capital.

Since the first step was to secure the Baltic coast, military construction was the main concern for the next eight years. A mighty fortress was built on Peter Island, and a port and shipyards were built across the river on the mainland, as a Russian navy came into being. The land was swampy and uninhabited, the climate damp and unpleasant. But Peter cared not at all: for him, the inhospitable northern marshland was a future metropolis, gloriously bearing his name.

After the decisive Russian victory at Poltava in 1709 greatly reduced the threat of Swedish armies, Peter moved into high gear. In one imperious decree after another, he ordered his people to build a city that would equal any in the world. Such a city had to be Western and baroque, just as Peter's army had to be Western and permanent. From such a new city, his "window on Europe," Peter also believed it would be easier to reform the country militarily and administratively.

These general political goals matched Peter's architectural ideas, which had been influenced by his travels in western Europe. First, Peter wanted a comfortable, "modern" city. Modernity meant broad, straight, stone-paved avenues, houses built in a uniform line and not haphazardly set back from the street, large parks, canals for drainage, stone bridges, and street lighting. Second, all building had to conform strictly to detailed architectural regulations set down by the government. Finally, each social group —the nobility, the merchants, the artisans, and so on —was to live in a certain section of town. In short, the city and its population were to conform to a carefully defined urban plan of the baroque type.

St. Petersburg, ca 1760 Rastrelli's remodeled Winter Palace, which housed the royal family until the Russian Revolution of 1917, stands on the left along the Neva River, near the ministries of the tsar's government. The Navy Office with its famous golden spire is in the center. Russia became a naval power and St. Petersburg a great port. *(From G. H. Hamilton,* Art and Architecture in Russia, *Penguin Books, 1954)*

Peter used the traditional but reinforced methods of Russian autocracy to build his modern capital. The creation of St. Petersburg was just one of the heavy obligations he dictatorially imposed on all social groups in Russia. The peasants bore the heaviest burdens. Just as the government drafted peasants for the army, it also drafted twenty-five to forty thousand men each summer to labor in St. Petersburg for three months, without pay. Every ten to fifteen peasant households had to furnish one such worker each summer and then pay a special tax in order to feed that worker in St. Petersburg.

Peasants hated this forced labor in the capital, and each year one-fourth to one-third of those sent risked brutal punishment and ran away. Many peasant construction workers died each summer from hunger, sickness, and accidents. Many also died because peasant villages tended to elect old men or young boys to labor in St. Petersburg, since strong and able-bodied men were desperately needed on the farm in the busy summer months. Thus beautiful St. Peters-

burg was built on the shoveling, carting, and paving of a mass of conscripted serfs.

Peter also drafted more privileged groups to his city, but on a permanent basis. Nobles were summarily ordered to build costly stone houses and palaces in St. Petersburg and to live in them most of the year. The more serfs a noble possessed, the bigger his dwelling had to be. Merchants and artisans were also commanded to settle and build in St. Petersburg. These nobles and merchants were then required to pay for the city's avenues, parks, canals, embankments, pilings, and bridges, all of which were very costly in terms of both money and lives because they were built on a swamp. The building of St. Petersburg was, in truth, an enormous direct tax levied on the wealthy, who in turn forced the peasantry to do most of the work. The only immediate beneficiaries were the foreign architects and urban planners. No wonder so many Russians hated Peter's new city.

Yet the tsar had his way. By the time of his death in 1725, there were at least six thousand houses and numerous impressive government buildings in St. Petersburg. Under the remarkable women who ruled Russia throughout most of the eighteenth century, St. Petersburg blossomed fully as a majestic and well-organized city, at least in its wealthy showpiece sections. Peter's youngest daughter, the quick-witted, sensual beauty, Elizabeth (1741–1762), named as her chief architect Bartolomeo Rastrelli, who had come to Russia from Italy as a boy of fifteen in 1715. Combining Italian and Russian traditions into a unique, wildly colorful St. Petersburg style, Rastrelli built many palaces for the nobility and all the larger government buildings erected during Elizabeth's reign. He also rebuilt the Winter Palace as an enormous, aqua-colored royal residence, now the Hermitage Museum. There Elizabeth established a flashy, luxury-loving, and slightly crude court, which Catherine in turn made truly imperial. All the while St. Petersburg grew rapidly, and its almost 300,000 inhabitants in 1782 made it one of the world's largest cities. Peter and his successors had created out of nothing a magnificent and harmonious royal city, which unmistakably proclaimed the power of Russia's rulers and the creative potential of the absolutist state.

———————

From about 1400 to 1650, social and economic developments in eastern Europe increasingly diverged from those in western Europe. In the east, peasants and townspeople lost precious freedoms, while the nobility increased its power and prestige. It was within this framework of resurgent serfdom and entrenched nobility that Austrian and Prussian monarchs fashioned absolutist states in the seventeenth and early eighteenth centuries. Thus monarchs won absolutist control over standing armies, permanent taxes, and legislative bodies. But they did not question the underlying social and economic relationships. Indeed, they enhanced the privileges of the nobility, which furnished the leading servitors for enlarged armies and growing state bureaucracies.

In Russia, the social and economic trends were similar, but the timing of political absolutism was different. Mongol conquest and rule was a crucial experience, and a harsh, indigenous tsarist autocracy was firmly in place by the reign of Ivan the Terrible in the sixteenth century. More than a century later,

Peter the Great succeeded in tightening up Russia's traditional absolutism and modernizing it by reforming the army, the bureaucracy, and the defense industry. In Russia and throughout eastern Europe, war and the needs of the state in time of war weighed heavily in the triumph of absolutism.

Triumphant absolutism interacted spectacularly with the arts. Baroque art, which had grown out of the Catholic Reformation's desire to move the faithful and exalt the true faith, admirably suited the secular aspirations of eastern rulers. They built grandiose baroque palaces, monumental public squares, and even whole cities to glorify their power and majesty. Thus baroque art attained magnificent heights in eastern Europe, symbolizing the ideal and harmonizing with the reality of imperious royal absolutism.

NOTES

———————

1. Quoted by F. L. Carsten, *The Origins of Prussia,* Clarendon Press, Oxford, 1954, p. 152.
2. Ibid., p. 175.
3. H. Rosenberg, *Bureaucracy, Aristocracy, and Autocracy: The Prussian Experience, 1660–1815,* Beacon Press, Boston, 1966, p. 38.
4. Quoted by R. Ergang, *The Potsdam Führer: Frederick William I, Father of Prussian Militarism,* Octagon Books, New York, 1972, pp. 85, 87.
5. Ibid., pp. 6–7, 43.
6. Quoted by R. A. Dorwart, *The Administrative Reforms of Frederick William I of Prussia,* Harvard University Press, Cambridge, Mass., 1953, p. 226.
7. Quoted by Rosenberg, p. 40.
8. Quoted by R. Pipes, *Russia Under the Old Regime,* Charles Scribner's Sons, New York, 1974, p. 48.
9. Quoted by N. V. Riasanovsky, *A History of Russia,* Oxford University Press, New York, 1963, p. 79.
10. Quoted by I. Grey, *Ivan III and the Unification of Russia,* Collier Books, New York, 1967, p. 39.
11. Quoted by Grey, p. 42.
12. Both quoted by Pipes, pp. 65, 85.
13. Quoted by Ergang, p. 13.
14. Quoted by J. Summerson, in *The Eighteenth Century: Europe in the Age of Enlightenment,* ed. A. Cobban, McGraw-Hill, New York, 1969, p. 80.
15. Quoted by L. Mumford, *The Culture of Cities,* Harcourt Brace Jovanovich, New York, 1938, p. 97.

SUGGESTED READING

All of the books cited in the Notes are highly recommended. F. L. Carsten's *The Origin of Prussia* (1954) is the best study on early Prussian history, and H. Rosenberg, *Bureaucracy, Aristocracy, and Autocracy: The Prussian Experience, 1660–1815* (1966), is a masterful analysis of the social context of Prussian absolutism. In addition to R. Ergang's exciting and critical biography of ramrod Frederick William I, *The Potsdam Führer* (1972), there is G. Ritter, *Frederick the Great* (1968), a more sympathetic study of the talented son by one of Germany's leading conservative historians. G. Craig, *The Politics of the Prussian Army, 1640–1945* (1964), expertly traces the great influence of the military on the Prussian state over three hundred years. R. J. Evans, *The Making of the Habsburg Empire, 1550–1770* (1979), and R. A. Kann, *A History of the Habsburg Empire, 1526–1918* (1974), analyze the development of absolutism in Austria, as does A. Wandruszka, *The House of Habsburg* (1964). J. Stoye, *The Siege of Vienna* (1964), is a fascinating account of the last great Ottoman offensive, which is also treated in the interesting study by P. Coles, *The Ottoman Impact on Europe, 1350–1699* (1968). The Austro-Ottoman conflict is also a theme of L. S. Stavrianos, *The Balkans Since 1453* (1958), and D. McKay's fine biography, *Prince Eugene of Savoy* (1978). A good general account is provided in D. McKay and H. Scott, *The Rise of the Great Powers, 1648–1815* (1983).

On Eastern peasants and serfdom, J. Blum, "The Rise of Serfdom in Eastern Europe," *American Historical Review 62* (July 1957):807–836, is a good point of departure, while R. Mousnier, *Peasant Uprisings in Seventeenth-Century France, Russia, and China* (1970), is an engrossing comparative study. J. Blum, *Lord and Peasant in Russia from the Ninth to the Nineteenth Century* (1961), provides a good look at conditions in rural Russia, and P. Avrich, *Russian Rebels, 1600–1800* (1972), treats some of the violent peasant upheavals those conditions produced. R. Hellie, *Enserfment and Military Change in Muscovy* (1971), is outstanding, as is A. Yanov's provocative *Origins of Autocracy: Ivan the Terrible in Russian History* (1981). In addition to the fine surveys by Pipes and Riasanovsky cited in the Notes, J. Billington, *The Icon and the Axe* (1970), is a stimulating history of early Russian intellectual and cultural developments, such as the great split in the church. M. Raeff, *Origins of the Russian Intelligentsia* (1966), skillfully probes the mind of the Russian nobility in the eighteenth century. B. H. Sumner, *Peter the Great and the Emergence of Russia* (1962), is a fine brief introduction, which may be compared with the brilliant biography by Russia's greatest prerevolutionary historian, V. Klyuchevsky, *Peter the Great* (trans. 1958), and with N. Riasanovsky, *The Image of Peter the Great in Russian History and Thought* (1985). G. Vernadsky and R. Fisher, eds., *A Source Book of Russian History from Early Times to 1917,* 3 vols. (1972), is an invaluable, highly recommended collection of documents and contemporary writings.

Three good books on art and architecture are E. Hempel, *Baroque Art and Architecture in Central Europe* (1965); G. Hamilton, *The Art and Architecture of Russia* (1954); and N. Pevsner, *An Outline of European Architecture,* 6th ed. (1960). Bach, Handel, and other composers are discussed intelligently by M. Bufkozer, *Music in the Baroque Era* (1947).

18

TOWARD A NEW
WORLD-VIEW

OST PEOPLE are not philosophers, but nevertheless they have a basic outlook on life, a more or less coherent world-view. At the risk of oversimplification, one may say that the world-view of medieval and early modern Europe was primarily religious and theological. Not only did Christian or Jewish teachings form the core of people's spiritual and philosophical beliefs, but religious teachings also permeated all the rest of human thought and activity. Political theory relied on the divine right of kings, for example, and activities ranging from marriage and divorce to eating habits and hours of business were regulated by churches and religious doctrines.

In the course of the eighteenth century, this religious and theological world-view of the educated classes of western Europe underwent a fundamental transformation. Many educated people came to see the world primarily in secular and scientific terms. And while few abandoned religious beliefs altogether, many became openly hostile toward established Christianity. The role of churches and religious thinking in earthly affairs and in the pursuit of knowledge was substantially reduced. Among many in the upper and middle classes, a new critical, scientific, and very "modern" world-view took shape. Why did this momentous change occur? How did this new outlook on life affect society and politics? This chapter will focus on these questions.

THE SCIENTIFIC REVOLUTION

The foremost cause of the change in world-view was the scientific revolution. Modern science—precise knowledge of the physical world based on the union of experimental observations with sophisticated mathematics—crystallized in the seventeenth century. Whereas science had been secondary and subordinate in medieval intellectual life, it became independent and even primary for many educated people in the eighteenth century.

The emergence of modern science was a development of tremendous long-term significance. A noted historian has even said that the scientific revolution

of the late sixteenth and seventeenth centuries "outshines everything since the rise of Christianity and reduces the Renaissance and Reformation to the rank of mere episodes, mere internal displacements, within the system of medieval Christendom." The scientific revolution was "the real origin both of the modern world and the modern mentality."[1] This statement is an exaggeration, but not much of one. Of all the great civilizations, only that of the West developed modern science. It was with the scientific revolution that Western society began to acquire its most distinctive traits.

Though historians agree that the scientific revolution was enormously important, they approach it in quite different ways. Some scholars believe that the history of scientific achievement in this period had its own basic "internal" logic and that "nonscientific" factors had quite limited significance. These scholars write brilliant, often highly technical, intellectual studies, but they neglect the broader historical context. Other historians stress "external" economic, social, and religious factors, brushing over the scientific developments themselves. Historians of science now realize that these two approaches need to be brought together, but they are only beginning to do so. It is best, therefore, to examine the milestones on the fateful march toward modern science first and then to search for nonscientific influences along the route.

SCIENTIFIC THOUGHT IN 1500

Since developments in astronomy and physics were at the heart of the scientific revolution, one must begin with the traditional European conception of the universe and movement in it. In the early 1500s, traditional European ideas about the universe were still based primarily on the ideas of Aristotle, the great Greek philosopher of the fourth century B.C. These ideas had gradually been recovered during the Middle Ages and then brought into harmony with Christian doctrines by medieval theologians. According to this revised Aristotelian view, a motionless earth was fixed at the center of the universe. Around it moved ten separate, transparent, crystal spheres. In the first eight spheres were embedded, in turn, the moon, the sun, the five known planets, and the fixed stars. Then followed two spheres added during the Middle Ages to account for slight changes in the positions of the stars over the centuries. Beyond the tenth sphere was heaven, with the throne of God and the souls of the saved. Angels kept the spheres moving in perfect circles.

Aristotle's views, suitably revised by medieval philosophers, also dominated thinking about physics and motion on earth. Aristotle had distinguished sharply between the world of the celestial spheres and that of the earth—the sublunar world. The spheres consisted of a perfect, incorruptible "quintescence," or fifth essence. The sublunar world, however, was made up of four imperfect, changeable elements. The "light" elements—air and fire—naturally moved upward, while the "heavy" elements—water and earth—naturally moved downward. The natural directions of motion did not always prevail, however, for elements were often mixed together and could be affected by an outside force such as a human being. Aristotle and his followers also believed that a uniform force moved an object at a constant speed and that the object would stop as soon as that force was removed.

Aristotle's ideas about astronomy and physics were accepted with minor revisions for two thousand years, and with good reason. First, they offered an understandable, common-sense explanation for what the eye actually saw. Second, Aristotle's science, as interpreted by Christian theologians, fit neatly with Christian doctrines. It established a home for God and a place for Christian souls. It put human beings at the center of the universe and made them the critical link in a "great chain of being" that stretched from the throne of God to the most lowly insect on earth. Thus science was primarily a branch of theology, and it reinforced religious thought. At the same time, medieval "scientists" were already providing closely reasoned explanations of the universe, explanations they felt were worthy of God's perfect creation.

THE COPERNICAN HYPOTHESIS

The desire to explain and thereby glorify God's handiwork led to the first great departure from the medieval system. This departure was the work of the Polish clergyman and astronomer Nicolaus Copernicus (1473–1543). As a young man, Copernicus studied church law and astronomy in various European universities. He saw how professional astronomers were still dependent for their most accurate calculations

Ptolemy's System This 1543 drawing shows how the changing configurations of the planets moving around the earth form the twelve different constellations, or "signs," of the zodiac. The learned astronomer on the right is using his knowledge to predict the future for the king on the left. *(Mary Evans Picture Library)*

Tycho Brahe's Main Observatory Lavishly financed by the king of Denmark, Brahe built his magnificent observatory at Uraniborg between 1576 and 1580. For twenty years he studied the heavens and accumulated a mass of precise but undigested data. *(The British Library)*

on the work of Ptolemy, the last great ancient astronomer, who had lived in Alexandria in the second century A.D. Ptolemy's achievement had been to work out complicated rules to explain the minor irregularities in the movement of the planets. These rules enabled stargazers and astrologers to track the planets with greater precision. Many people then (and now) believed that the changing relationships between planets and stars influenced and even determined an individual's future.

The young Copernicus was uninterested in astrology and felt that Ptolemy's cumbersome and occasionally inaccurate rules detracted from the majesty of a perfect Creator. He preferred an old Greek idea being discussed in Renaissance Italy: that the sun rather than the earth was at the center of the universe. Finishing his university studies and returning to a church position in east Prussia, Copernicus worked on his hypothesis from 1506 to 1530. Never questioning the Aristotelian belief in crystal spheres or the idea that circular motion was most perfect and divine, Copernicus theorized that the stars and planets, including the earth, revolve around a fixed sun. Yet Copernicus was a cautious man. Fearing the ridicule of other astronomers, he did not publish his *On the Revolutions of the Heavenly Spheres* until 1543, the year of his death.

Copernicus's theory had enormous scientific and religious implications, many of which the conservative Copernicus did not anticipate. First, it put the stars at rest, their apparent nightly movement simply a result of the earth's rotation. Thus it destroyed the main reason for believing in crystal spheres capable of moving the stars around the earth. Second, Copernicus's theory suggested a universe of staggering size. If in the course of a year the earth moved around the sun and yet the stars appeared to remain in the same place, then the universe was unthinkably large or even infinite. Finally, by characterizing the earth as just another planet, Copernicus destroyed the basic idea of Aristotelian physics—that the earthly world was quite different from the heavenly one. Where, then, was the realm of perfection? Where was heaven and the throne of God?

The Copernican theory quickly brought sharp attacks from religious leaders, especially Protestants. Hearing of Copernicus's work even before it was published, Martin Luther spoke of him as the "new astrologer who wants to prove that the earth moves and goes round. . . . The fool wants to turn the whole art of astronomy upside down." Luther noted that "as the Holy Scripture tells us, so did Joshua bid the sun stand still and not the earth." Calvin also condemned Copernicus, citing as evidence the first verse of Psalm 93: "The world also is established that it cannot be moved." "Who," asked Calvin, "will venture to place the authority of Copernicus above that of the Holy Spirit?"[2] Catholic reaction was milder at first. The Catholic church had never been hypnotized by literal interpretations of the Bible, and not until 1616 did it officially declare the Copernican theory false.

This slow reaction also reflected the slow progress of Copernicus's theory for many years. Other events were almost as influential in creating doubts about traditional astronomical ideas. In 1572 a new star appeared and shone very brightly for almost two years. The new star, which was actually a distant exploding star, made an enormous impression on people. It seemed to contradict the idea that the heavenly spheres were unchanging and therefore perfect. In 1577 a new comet suddenly moved through the sky, cutting a straight path across the supposedly impenetrable crystal spheres. It was time, as a typical scientific writer put it, for "the radical renovation of astronomy."[3]

FROM TYCHO BRAHE TO GALILEO

One astronomer who agreed was Tycho Brahe (1546–1601). Born into a leading Danish noble family and earmarked for a career in government, Brahe was at an early age tremendously impressed by a partial eclipse of the sun. It seemed to him "something divine that men could know the motions of the stars so accurately that they were able a long time beforehand to predict their places and relative positions."[4] Completing his studies abroad and returning to Denmark, Brahe established himself as Europe's leading astronomer with his detailed observations of the new star of 1572. Aided by generous grants from the king of Denmark, which made him one of the richest men in the country, Brahe built the most sophisticated observatory of his day. For twenty years he meticulously observed the stars and planets with the naked eye. An imposing man who had lost a piece of his nose in a duel and replaced it with a special bridge of gold and silver alloy, a noble who exploited his peasants arrogantly and approached the heavens humbly, Brahe's great contribution was his mass of data. His limited understanding of mathematics prevented him, however, from making much sense out of his data. Part Ptolemaic, part Copernican, he believed that all the planets revolved around the sun and that the entire group of sun and planets revolved in turn around the earth-moon system.

It was left to Brahe's brilliant young assistant, Johannes Kepler (1571–1630), to go much further.

Kepler was a medieval figure in many ways. Coming from a minor German noble family and trained for the Lutheran ministry, he long believed that the universe was built on mystical mathematical relationships and a musical harmony of the heavenly bodies. Working and reworking Brahe's mountain of observations in a staggering sustained effort after the Dane's death, this brilliant mathematician eventually went beyond mystical intuitions.

Kepler formulated three famous laws of planetary motion. First, building on Copernican theory, he demonstrated in 1609 that the orbits of the planets around the sun are elliptical rather than circular. Second, he demonstrated that the planets do not move at a uniform speed in their orbits. Third, in 1619 he showed that the time a planet takes to make its complete orbit is precisely related to its distance from the sun. Kepler's contribution was monumental. Whereas Copernicus had speculated, Kepler proved mathematically the precise relations of a sun-centered (solar) system. His work demolished the old system of Aristotle and Ptolemy, and in his third law he came close to formulating the idea of universal gravitation.

While Kepler was unraveling planetary motion, a young Florentine named Galileo Galilei (1564–1642) was challenging all the old ideas about motion. Like so many early scientists, Galileo was a poor nobleman first marked for a religious career. However, he soon became fascinated by mathematics. A brilliant student, Galileo became a professor of mathematics in 1589 at age twenty-five. He proceeded to examine motion and mechanics in a new way. Indeed, his great achievement was the elaboration and consolidation of the modern experimental method. Rather than speculate about what might or should happen, Galileo conducted controlled experiments to find out what actually *did* happen.

In his famous acceleration experiment, he showed that a uniform force—in this case, gravity—produced a uniform acceleration. Here is how Galileo described his pathbreaking method and conclusion in his *Two New Sciences:*

A piece of wooden moulding . . . was taken; on its edge was cut a channel a little more than one finger in breadth. Having made this groove very straight, smooth and polished, and having lined it with parchment, also as smooth and polished as possible, we rolled along it a

Galileo Explains the two new sciences, astronomy and mechanics, to an eager young visitor to his observatory in this artist's reconstruction. After his trial for heresy by the papal Inquisition, Galileo spent the last eight years of his life under house arrest. *(The Mansell Collection)*

hard, smooth and very round bronze ball. . . . Noting . . . the time required to make the descent . . . we now rolled the ball only one-quarter the length of the channel; and having measured the time of its descent, we found it precisely one-half of the former. . . . In such experiments [over many distances], repeated a full hundred times, we always found that the spaces traversed were to each other as the squares of the times, and that this was true for all inclinations of the plane.[5]

With this and other experiments, Galileo also formulated the law of inertia. That is, rather than rest being the natural state of objects, an object continues in motion forever unless stopped by some external force. Aristotelian physics was in a shambles.

In the tradition of Brahe, Galileo also applied the experimental method to astronomy. His astronomical discoveries had a great impact on scientific development. On hearing of the invention of the telescope in Holland, Galileo made one for himself and trained it on the heavens. He quickly discovered the first four moons of Jupiter, which clearly suggested that Jupiter could not possibly be embedded in any impenetrable crystal sphere. This discovery provided new evidence for the Copernican theory, in which Galileo already believed.

Galileo then pointed his telescope at the moon. He wrote in 1610 in *Siderus Nuncius:*

I feel sure that the moon is not perfectly smooth, free from inequalities, and exactly spherical, as a large school of philosophers considers with regard to the moon and the other heavenly bodies. On the contrary, it is full of inequalities, uneven, full of hollows and protuberances, just like the surface of the earth itself, which is varied. . . . The next object which I have observed is the essence or substance of the Milky Way. By the aid of a telescope anyone may behold this in a manner which so distinctly appeals to the senses that all the disputes which have tormented philosophers through so many ages are exploded by the irrefutable evidence of our eyes, and we are freed from wordy disputes upon the subject. For the galaxy is nothing else but a mass of innumerable stars planted together in clusters. Upon whatever part of it you direct the telescope straightway a vast crowd of stars presents itself to view; many of them are tolerably large and extremely bright, but the number of small ones is quite beyond determination.[6]

Reading these famous lines, one feels that a crucial corner in Western civilization is being turned. The traditional religious and theological world-view, which rested on determining and then accepting the proper established authority, is beginning to give way in certain fields to a critical, "scientific" method. This new method of learning and investigating was the greatest accomplishment of the entire scientific revolution, for it has proved capable of great extension. A historian investigating documents of the past, for example, is not much different from a Galileo studying stars and rolling balls.

Galileo was employed in Florence by the Medici grand dukes of Tuscany, and his work eventually aroused the ire of some theologians. The issue was presented in 1624 to Pope Urban VII, who permitted Galileo to write about different possible systems of the world, as long as he did not presume to judge which one actually existed. After the publication in Italian of his widely read *Dialogue on the Two Chief Systems of the World* in 1632, which too openly lampooned the traditional views of Aristotle and Ptolemy and defended those of Copernicus, Galileo was tried for heresy by the papal Inquisition. Imprisoned and threatened with torture, the aging Galileo recanted, "renouncing and cursing" his Copernican errors. Of minor importance in the development of science, Galileo's trial later became for some writers the perfect symbol of the inevitable conflict between religious belief and scientific knowledge.

NEWTON'S SYNTHESIS

The accomplishments of Kepler, Galileo, and other scientists had taken effect by about 1640. The old astronomy and physics were in ruins, and several fundamental breakthroughs had been made. The new findings had not, however, been fused together in a new synthesis, a single explanatory system that would comprehend motion both on earth and in the skies. That synthesis, which prevailed until the twentieth century, was the work of Isaac Newton (1642–1727).

Newton was born into lower English gentry and attended Cambridge University. A great genius who spectacularly united the experimental and theoretical-mathematical sides of modern science, Newton was also fascinated by alchemy. He sought the elixir of life and a way to change base metals into gold and silver. Not without reason did the twentieth-century economist John Maynard Keynes call Newton the "last of the magicians." Newton was intensely religious. He had a highly suspicious nature, lacked all interest in women and sex, and in 1693 suffered a nervous breakdown from which he later recovered. He was far from being the perfect rationalist so endlessly eulogized by writers in the eighteenth and nineteenth centuries.

Of his intellectual genius and incredible powers of concentration there can be no doubt, however. Arriving at some of his most basic ideas about physics in 1666 at age twenty-four, but unable to prove these theories mathematically, he attained a professorship and studied optics for many years. In 1684 Newton returned to physics for eighteen extraordinarily intensive months. For weeks on end he seldom left his room except to read his lectures. His meals were sent up but he usually forgot to eat them, his mind fastened like a vise on the laws of the universe. Thus did Newton open the third book of his immortal *Mathematical Principles of Natural Philosophy,* published in Latin in 1687 and generally known as the *Principia,* with these lines:

In the preceding books I have laid down the principles of philosophy [that is, science]. . . . These principles are the laws of certain motions, and powers or forces, which chiefly have respect to philosophy. . . . It remains that from the same principles I now demonstrate the frame of the System of the World.

Newton made good his grandiose claim. His towering accomplishment was to integrate in a single explanatory system the astronomy of Copernicus, as corrected by Kepler's laws, with the physics of Galileo and his predecessors. Newton did this by means of a set of mathematical laws that explain motion and mechanics. These laws of dynamics are complex, and it took scientists and engineers two hundred years to work out all their implications. Nevertheless, the key feature of the Newtonian synthesis was the law of universal gravitation. According to this law, every body in the universe attracts every other body in the universe in a precise mathematical relationship, whereby the force of attraction is proportional to the quantity of matter of the objects and inversely proportional to the square of the distance between them. The whole universe—from Kepler's elliptical orbits to Galileo's rolling balls—was unified in one majestic system.

CAUSES OF THE SCIENTIFIC REVOLUTION

With a charming combination of modesty and self-congratulation, Newton once wrote: "If I have seen further [than others], it is by standing on the shoulders of Giants."[7] Surely the path from Copernicus to Newton confirms the "internal" view of the scientific revolution as a product of towering individual genius. The problems of science were inherently exciting, and solution of those problems was its own reward for inquisitive, high-powered minds. Yet there were certainly broader causes as well.

The long-term contribution of medieval intellectual life and medieval universities to the scientific revolution was much more considerable than historians unsympathetic to the Middle Ages once believed. By the thirteenth century, permanent universities with professors and large student bodies had been established in western Europe. The universities were supported by society because they trained the lawyers, doctors, and church leaders society required. By 1300 philosophy had taken its place alongside law, medicine, and theology. Medieval philosophers developed a limited but real independence from theologians and a sense of free inquiry. They nobly pursued a body of knowledge and tried to arrange it meaningfully by means of abstract theories.

Within this framework, science was able to emerge as a minor but distinct branch of philosophy. In the fourteenth and fifteenth centuries, first in Italy and then elsewhere in Europe, leading universities established new professorships of mathematics, astronomy, and physics (natural philosophy) within their faculties of philosophy. The prestige of the new fields was still low among both professors and students. Nevertheless, this pattern of academic science, which grew out of the medieval commitment to philosophy and did not change substantially until the late eighteenth century, undoubtedly promoted scientific development. Rational, critical thinking was applied to scientific problems by a permanent community of scholars. And an outlet existed for the talents of a Galileo or a Newton: all the great pathbreakers either studied or taught at universities.

The Renaissance also stimulated scientific progress. One of the great deficiencies of medieval science was its rather rudimentary mathematics. The recovery of the finest works of Greek mathematics—a by-product of Renaissance humanism's ceaseless search for the knowledge of antiquity—greatly improved European mathematics well into the early seventeenth century. The recovery of more texts also showed that classical mathematicians had had their differences, and Europeans were forced to try to resolve these ancient controversies by means of their own efforts. Finally, the Renaissance pattern of patronage, especially in Italy, was often scientific as well as artistic and humanistic. Various rulers and wealthy businessmen supported scientific investigations, just as the Medicis of Florence supported those of Galileo.

The navigational problems of long sea voyages in the age of overseas expansion were a third factor in the scientific revolution. Ship captains on distant shores needed to be able to chart their positions as accurately as possible, so that reliable maps could be drawn and the risks of international trade reduced. As early as 1484, the king of Portugal appointed a commission of mathematicians to perfect tables to help seamen find their latitude. This resulted in the first European navigation manual.

The problem of fixing longitude was much more difficult. In England, the government and the great capitalistic trading companies turned to science and scientific education in an attempt to solve this pressing practical problem. When the famous Elizabethan financier Sir Thomas Gresham left a large amount of money to establish Gresham College in London, he stipulated that three of the college's seven professors had to concern themselves exclusively with scientific

State Support Governments supported scientific research because they thought it might be useful. Here Louis XIV visits the French Royal Academy of Sciences in 1671 and examines a plan for better military fortifications. The great interest in astronomy, anatomy, and geography is evident. *(Bibliothèque Nationale, Paris)*

subjects. The professor of astronomy was directed to teach courses on the science of navigation. A seventeenth-century popular ballad took note of the new college's calling:

This college will the whole world measure
Which most impossible conclude,
And navigation make a pleasure
By finding out the longitude.[8]

At Gresham College scientists had, for the first time in history, an important, honored role in society. They enjoyed close ties with the top officials of the Royal Navy and with the leading merchants and shipbuilders. Gresham College became the main center of scientific activity in England in the first half of the seventeenth century. The close tie between practical men and scientists also led to the establishment in 1662 of the Royal Society of London, which published scientific papers and sponsored scientific meetings.

Navigational problems were also critical in the development of many new scientific instruments, such as the telescope, the barometer, the thermometer, the pendulum clock, the microscope, and the air pump. Better instruments, which permitted more accurate observations, often led to important new knowledge. Galileo with his telescope was by no means unique.

Better instruments were part of the fourth factor, the development of better ways of obtaining knowledge about the world. Two important thinkers, Francis Bacon (1561–1626) and René Descartes (1596–1650), represented key aspects of this improvement in scientific methodology.

The English politician and writer Francis Bacon was the greatest early propagandist for the new experimental method, as Galileo was its greatest early practitioner. Rejecting the Aristotelian and medieval method of using speculative reasoning to build general theories, Bacon argued that new knowledge had to be pursued through empirical, experimental research. That is, the researcher who wants to learn more about leaves or rocks should not speculate about the subject but rather collect a multitude of specimens and then compare and analyze them. Thus freed from sterile medieval speculation, the facts will speak for themselves, and important general principles will then emerge. Knowledge will increase. Bacon's contribution was to formalize the empirical method, which had already been used by scientists like Brahe and Galileo, into the general theory of inductive reasoning known as *empiricism.*

Bacon claimed that the empirical method would result not only in more knowledge, but in highly practical, useful knowledge. According to Bacon, scientific discoveries like those so avidly sought at Gresham College would bring about much greater control over the physical environment and make people rich and nations powerful. Thus Bacon helped provide a radically new and effective justification for private and public support of scientific inquiry.

The French philosopher René Descartes was a true genius who made his first great discovery in mathematics. As a twenty-three-year-old soldier serving in the Thirty Years' War, he experienced on a single night in 1619 a life-changing intellectual vision. What Descartes saw was that there was a perfect correspondence between geometry and algebra and that geometrical, spatial figures could be expressed as algebraic equations and vice versa. A great step forward in the history of mathematics, Descartes' discovery of analytic geometry provided scientists with an important new tool. Descartes also made contributions to the science of optics, but his greatest achievement was to develop his initial vision into a whole philosophy of knowledge and science.

Like Bacon, Descartes scorned traditional science and had great faith in the powers of the human mind. Yet Descartes was much more systematic and mathematical than Bacon. He decided it was necessary to doubt everything that could reasonably be doubted and then, as in geometry, to use deductive reasoning from self-evident principles to ascertain scientific laws. Descartes' reasoning ultimately reduced all substances to "matter" and "mind"—that is, to the physical and the spiritual. His view of the world as consisting of two fundamental entities is known as *Cartesian dualism.* Descartes was a profoundly original and extremely influential thinker.

It is important to realize that the modern scientific method, which began to crystallize in the late seventeenth century, has combined Bacon's inductive experimentalism and Descartes' deductive, mathematical rationalism. Neither of these extreme approaches was sufficient by itself. Bacon's inability to appreciate the importance of mathematics and his obsession with practical results clearly showed the limitations of antitheoretical empiricism. Likewise, some of

Descartes' positions—he believed, for example, that it was possible to deduce the whole science of medicine from first principles—aptly demonstrated the inadequacy of rigid, dogmatic rationalism. Significantly, Bacon faulted Galileo for his use of abstract formulas, while Descartes criticized the great Italian for being too experimental and insufficiently theoretical. Thus the modern scientific method has typically combined Bacon and Descartes. It has joined precise observations and experimentalism with the search for general laws that may be expressed in rigorously logical, mathematical language.

Finally, there is the question of science and religion. Just as some historians have argued that Protestantism led to the rise of capitalism, others have concluded that Protestantism was a fundamental factor in the rise of modern science. Protestantism, particularly in its Calvinist varieties, supposedly made scientific inquiry a question of individual conscience and not of religious doctrine. The Catholic church, on the other hand, supposedly suppressed scientific theories that conflicted with its teachings and thus discouraged scientific progress.

The truth of the matter is more complicated. *All* religious authorities—Catholic, Protestant, and Jewish—opposed the Copernican system to a greater or lesser extent until about 1630, by which time the scientific revolution was definitely in progress. The Catholic church was initially less hostile than Protestant and Jewish religious leaders. This early Catholic toleration and the scientific interests of Renaissance Italy help account for the undeniable fact that Italian scientists played a crucial role in scientific progress right up to the trial of Galileo in 1633. Thereafter the Counter-Reformation church became more hostile to science, which helps account for the decline of science in Italy (but not in Catholic France) after 1640. At the same time, some Protestant countries became quite "pro-science," especially if the country lacked a strong religious authority capable of imposing religious orthodoxy on scientific questions.

This was the case with England after 1630. English religious conflicts became so intense that it was impossible for the authorities to impose religious unity on anything, including science. It is significant that the forerunners of the Royal Society agreed to discuss only "neutral" scientific questions, so as not to come to blows over closely related religious and political disputes. The work of Bacon's many followers during

René Descartes dismissed the scientific theories of Aristotle and his medieval disciples as outdated dogma. A brilliant philosopher and mathematician, he formulated rules for abstract, deductive reasoning and the search for comprehensive scientific laws. *(Royal Museum of Fine Arts, Copenhagen)*

Cromwell's commonwealth helped solidify the neutrality and independence of science. Bacon advocated the experimental approach precisely because it was open-minded and independent of any preconceived religious or philosophical ideas. Neutral and useful, science became an accepted part of life and developed rapidly in England after about 1640.

SOME CONSEQUENCES OF THE SCIENTIFIC REVOLUTION

The rise of modern science had many consequences, some of which are still unfolding. First, it went hand in hand with the rise of a new and expanding social group—the scientific community. Members of this community were linked together by learned societies, common interests, and shared values. Expansion of knowledge was the primary goal of this community, and scientists' material and psychological rewards depended on their success in this endeavor. Thus science became quite competitive, and even more scientific advance was inevitable.

Second, the scientific revolution introduced not only new knowledge about nature but also a new and revolutionary way of obtaining such knowledge—the modern scientific method. In addition to being both theoretical and experimental, this method was highly critical, and it differed profoundly from the old way of getting knowledge about nature. It refused to base its conclusions on tradition and established sources, on ancient authorities and sacred texts.

The scientific revolution had few consequences for economic life and the living standards of the masses until the late eighteenth century at the very earliest. True, improvements in the techniques of navigation facilitated overseas trade and helped enrich leading merchants. But science had relatively few practical economic applications, and the hopes of the early Baconians were frustrated. The close link between theoretical, or pure, science and applied technology, which we take for granted today, simply did not exist before the nineteenth century. Thus the scientific revolution of the seventeenth century was first and foremost an intellectual revolution. It is not surprising that for more than a hundred years its greatest impact was on how people thought and believed.

THE ENLIGHTENMENT

The scientific revolution was the single most important factor in the creation of the new world-view of the eighteenth-century Enlightenment. This world-view, which has played a large role in shaping the modern mind, was based on a rich mix of ideas, sometimes conflicting, for intellectuals delight in playing with ideas just as athletes delight in playing games. Despite this diversity, three central concepts stand out.

The most important and original idea of the Enlightenment was that the methods of natural science could and should be used to examine and understand all aspects of life. This was what intellectuals meant by *reason,* a favorite word of Enlightenment thinkers. Nothing was to be accepted on faith. Everything was to be submitted to the rational, critical, "scientific" way of thinking. This approach brought the Enlightenment into a head-on conflict with the established churches, which rested their beliefs on the special authority of the Bible and Christian theology. A second important Enlightenment concept was that the scientific method was capable of discovering the laws of human society as well as those of nature. Thus was social science born. Its birth led to the third key idea, that of progress. Armed with the proper method of discovering the laws of human existence, Enlightenment thinkers believed it was at least possible to create better societies and better people. Their belief was strengthened by some genuine improvements in economic and social life during the eighteenth century (see Chapters 19 and 20).

The Enlightenment was therefore profoundly secular. It revived and expanded the Renaissance concentration on worldly explanations. In the course of the eighteenth century, the Enlightenment had a profound impact on the thought and culture of the urban middle and upper classes. It did not have much appeal for the poor and the peasants.

THE EMERGENCE OF THE ENLIGHTENMENT

The Enlightenment did not reach its maturity until about 1750. Yet it was the generation that came of age between the publication of Newton's masterpiece in 1687 and the death of Louis XIV in 1715 that tied the crucial knot between the scientific revolution and a new outlook on life.

Talented writers of that generation popularized hard-to-understand scientific achievements for the educated elite. The most famous and influential popularizer was a versatile French man of letters, Bernard de Fontenelle (1657–1757). Fontenelle practically invented the technique of making highly complicated scientific findings understandable to a broad nonscientific audience. He set out to make science witty and entertaining, as easy to read as a novel. This was a tall order, but Fontenelle largely succeeded.

His most famous work, *Conversations on the Plurality of Worlds* of 1686, begins with two elegant figures walking in the gathering shadows of a large park. One is a woman, a sophisticated aristocrat, and the other is her friend, perhaps even her lover. They gaze at the stars, and their talk turns to a passionate discussion of . . . astronomy! He confides that "each star may well be a different world." She is intrigued by his novel idea: "Teach me about these stars of yours." And he does, gently but persistently stressing how error is giving way to truth. At one point he explains:

There came on the scene a certain German, one Copernicus, who made short work of all those various circles, all those solid skies, which the ancients had pictured to themselves. The former he abolished; the latter he broke in pieces. Fired with the noble zeal of a true astronomer, he took the earth and spun it very far away from the center of the universe, where it had been installed, and in that center he put the sun, which had a far better title to the honor.[9]

Rather than tremble in despair in the face of these revelations, Fontenelle's lady rejoices in the advance of knowledge. Fontenelle thus went beyond entertainment to instruction, suggesting that the human mind was capable of making great progress.

This idea of progress was essentially a new idea of the later seventeenth century. Medieval and Reformation thinkers had been concerned primarily with sin and salvation. The humanists of the Renaissance had emphasized worldly matters, but they had been backward-looking. They had believed it might be possible to equal the magnificent accomplishments of the ancients, but they did not ask for more. Fontenelle and like-minded writers had come to believe that, at least in science and mathematics, their era had gone far *beyond* antiquity. Progress, at least intellectual progress, was clearly possible. During the eighteenth century, this idea would sink deeply into the consciousness of the European elite.

Fontenelle and other literary figures of his generation were also instrumental in bringing science into conflict with religion. Contrary to what is often assumed, many seventeenth-century scientists, both Catholic and Protestant, believed that their work exalted God. They did not draw antireligious implications from their scientific findings. The greatest scientist of them all, Isaac Newton, was a devout if unorthodox Christian who saw all of his studies as directed toward explaining God's message. Newton devoted far more of his time to angels and biblical prophecies than to universal gravitation, and he was convinced that all of his inquiries were equally "scientific."

Fontenelle, on the other hand, was skeptical about absolute truth and cynical about the claims of organized religion. Since such unorthodox views could not be stated openly in Louis XIV's France, Fontenelle made his point through subtle editorializing about science. His depiction of the cautious Copernicus as a self-conscious revolutionary was typical. In his *Eulogies of Scientists,* Fontenelle exploited with endless variations the basic theme of rational, progressive scientists versus prejudiced, reactionary priests. Time and time again Fontenelle's fledgling scientists attended church and studied theology; then, at some crucial moment, each was converted from the obscurity of religion to the clarity of science.

The progressive and antireligious implications that writers like Fontenelle drew from the scientific revolution reflected a very real crisis in European thought at the end of the seventeenth century. This crisis had its roots in several intellectual uncertainties and dissatisfactions, of which the demolition of Aristotelian-medieval science was only one.

A second uncertainty involved the whole question of religious truth. The destructive wars of religion had been fought, in part, because religious freedom was an intolerable idea in the early seventeenth century. Both Catholics and Protestants had believed that religious truth was absolute and therefore worth fighting and dying for. It was also generally believed that a strong state required unity in religious faith. Yet the disastrous results of the many attempts to impose such religious unity, such as Louis XIV's expulsion of the French Huguenots in 1685, led some people to ask if ideological conformity in religious matters was really necessary. Others skeptically asked if religious truth could ever be known with absolute certainty and concluded that it could not.

The most famous of these skeptics was Pierre Bayle (1647–1706), a French Huguenot who took refuge in Holland. A teacher by profession and a crusading journalist by inclination, Bayle critically examined the religious beliefs and persecutions of the past in his *Historical and Critical Dictionary,* published in 1697. Demonstrating that human beliefs had been extremely varied and very often mistaken, Bayle concluded that nothing can ever be known beyond all doubt. In religion as in philosophy, humanity's best hope was open-minded toleration. Bayle's skeptical views were very influential. Many eighteenth-century writers mined his inexhaustible vein of critical skepticism for ammunition in their attacks on superstition and theology. Bayle's four-volume *Dictionary* was found in more private libraries of eighteenth-century France than any other book.

Popularizing Science The frontispiece illustration of Fontenelle's *Conversations on the Plurality of Worlds* invites the reader to share the pleasures of astronomy with an elegant lady and an entertaining teacher. *(University of Illinois)*

The rapidly growing travel literature on non-European lands and cultures was a third cause of uncertainty. In the wake of the great discoveries, Europeans were learning that the peoples of China, India, Africa, and the Americas all had their own very different beliefs and customs. Europeans shaved their faces and let their hair grow. The Turks shaved their heads and let their beards grow. In Europe a man bowed before a woman to show respect. In Siam a man turned his back on a woman when he met her, because it was disrespectful to look directly at her. Countless similar examples discussed in the travel accounts helped change the perspective of educated Europeans. They began to look at truth and morality in relative rather than absolute terms. Anything was possible, and who could say what was right or wrong? As one Frenchman wrote: "There is nothing that opinion, prejudice, custom, hope, and a sense of honor cannot do." Another wrote disapprovingly of religious skeptics who were corrupted "by extensive travel and lose whatever shreds of religion that remained with them. Every day they see a new religion, new customs, new rites."[10]

A fourth cause and manifestation of European intellectual turmoil was John Locke's epoch-making *Essay Concerning Human Understanding.* Published in 1690—the same year Locke published his famous *Second Treatise on Civil Government* (page 530)—Locke's essay brilliantly set forth a new theory about how human beings learn and form their ideas. In doing so, he rejected the prevailing view of Descartes, who had held that all people are born with certain basic ideas and ways of thinking. Locke insisted that all ideas are derived from experience. The human mind is like a blank tablet (*tabula rasa*) at birth, a tablet on which environment writes the individual's understanding and beliefs. Human development is therefore determined by education and social institutions, for good or for evil. Locke's *Essay Concerning Human Understanding* passed through many editions and translations. It was, along with Newton's *Principia,* one of the dominant intellectual inspirations of the Enlightenment.

THE PHILOSOPHES AND THEIR IDEAS

By the death of Louis XIV in 1715, many of the ideas that would soon coalesce into the new world-view had been assembled. Yet Christian Europe was still strongly attached to its traditional beliefs, as witnessed by the powerful revival of religious orthodoxy in the first half of the eighteenth century. By the outbreak of the American Revolution in 1775, however, a large portion of western Europe's educated elite had embraced many of the new ideas. This acceptance was the work of one of history's most influential groups of intellectuals, the *philosophes.* It was the philosophes who proudly and effectively proclaimed that they, at long last, were bringing the light of knowledge to their ignorant fellow creatures in a great Age of Enlightenment.

Philosophe is the French word for "philosopher," and it was in France that the Enlightenment reached its highest development. The French philosophes were indeed philosophers. They asked fundamental philosophical questions about the meaning of life, about God, human nature, good and evil, and cause and effect. But, in the tradition of Bayle and Fontenelle, they were not content with abstract arguments or ivory-tower speculations among a tiny minority of scholars and professors. They wanted to influence and convince a broad audience.

The philosophes were intensely committed to reforming society and humanity, yet they were not free to write as they wished, since it was illegal in France to criticize openly either church or state. Their most radical works had to circulate in France in manuscript form, very much as critical works are passed from hand to hand in unpublished form in dictatorships today. Knowing that direct attacks would probably be banned or burned, the philosophes wrote novels and plays, histories and philosophies, dictionaries and encyclopedias, all filled with satire and double meanings to spread the message.

One of the greatest philosophes, the baron de Montesquieu (1689–1755), brilliantly pioneered this approach in *The Persian Letters,* an extremely influential social satire published in 1721. Montesquieu's work consisted of amusing letters supposedly written by Persian travelers, who see European customs in unique ways and thereby cleverly criticize existing practices and beliefs.

Having gained fame using wit as a weapon against cruelty and superstition, Montesquieu settled down on his family estate to study history and politics. His interest was partly personal for, like many members of the high French nobility, he was dismayed that royal absolutism had triumphed in France under

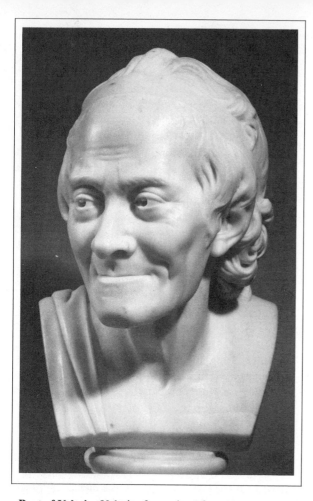

Bust of Voltaire Voltaire first gained fame as a poet and playwright, but today his odes and dramas are remembered largely by scholars. The sparkling short stories he occasionally dashed off still captivate the modern reader, however. *Candide* is the most famous of these witty tales. *(The Fine Arts Museums of San Francisco, Mr. and Mrs. E. John Magnin Gift)*

quieu, because in order to prevent the abuse of power, "it is necessary that by the arrangement of things, power checks power." Admiring greatly the English balance of power among the king, the houses of Parliament, and the independent courts, Montesquieu believed that in France the thirteen high courts —the *parlements*—were front-line defenders of liberty against royal despotism. Clearly no democrat and apprehensive about the uneducated poor, Montesquieu's theory of separation of powers had a great impact on France's wealthy, well-educated elite. The constitutions of the young United States in 1789 and of France in 1791 were based in large part on this theory.

The most famous and in many ways most representative philosophe was François Marie Arouet, who was known by the pen name of Voltaire (1694–1778). In his long career, this son of a comfortable middle-class family wrote over seventy witty volumes, hobnobbed with kings and queens, and died a millionaire because of shrewd business speculations. His early career, however, was turbulent. In 1717 Voltaire was imprisoned for eleven months in the Bastille in Paris for insulting the regent of France. In 1726 a barb from his sharp tongue led a great French nobleman to have him beaten and arrested. This experience made a deep impression on Voltaire. All his life he struggled against legal injustice and class inequalities before the law.

Released from prison after promising to leave the country, Voltaire lived in England for three years. Sharing Montesquieu's enthusiasm for English institutions, Voltaire then wrote various works praising England and popularizing English scientific progress. Newton, he wrote, was history's greatest man, for he had used his genius for the benefit of humanity. "It is," wrote Voltaire, "the man who sways our minds by the prevalence of reason and the native force of truth, not they who reduce mankind to a state of slavery by force and downright violence . . . that claims our reverence and admiration."[11] In the true style of the Enlightenment, Voltaire mixed the glorification of science and reason with an appeal for better people and institutions.

Yet, like almost all of the philosophes, Voltaire was a reformer and not a revolutionary in social and political matters. Returning to France, he was eventually appointed royal historian in 1743, and his *Age of Louis XIV* portrayed Louis as the dignified leader of

Louis XIV. But Montesquieu was also inspired by the example of the physical sciences, and he set out to apply the critical method to the problem of government in *The Spirit of Laws* (1748). The result was a complex comparative study of republics, monarchies, and despotisms—a great pioneering inquiry in the emerging social sciences.

Showing that forms of government were related to history, geography, and customs, Montesquieu focused on the conditions that would promote liberty and prevent tyranny. He argued that despotism could be avoided if political power were divided and shared by a diversity of classes and orders holding unequal rights and privileges. A strong, independent upper class was especially important, according to Montes-

his age. Voltaire also began a long correspondence with Frederick the Great, and he accepted Frederick's flattering invitation to come brighten up the Prussian court in Berlin. The two men later quarreled, but Voltaire always admired Frederick as a free thinker and an enlightened monarch.

Unlike Montesquieu, Voltaire pessimistically concluded that the best one could hope for in the way of government was a good monarch, since human beings "are very rarely worthy to govern themselves." Nor did he believe in social equality in human affairs. The idea of making servants equal to their masters was "absurd and impossible." The only realizable equality, Voltaire thought, was that "by which the citizen only depends on the laws which protect the freedom of the feeble against the ambitions of the strong."[12]

Voltaire's philosophical and religious positions were much more radical. In the tradition of Bayle, his voluminous writings challenged—often indirectly—the Catholic church and Christian theology at almost every point. Though he was considered by many devout Christians to be a shallow blasphemer, Voltaire's religious views were influential and quite typical of the mature Enlightenment. The essay on religion from his widely read *Philosophical Dictionary* sums up many of his criticisms and beliefs:

I meditated last night; I was absorbed in the contemplation of nature; I admired the immensity, the course, the harmony of these infinite globes which the vulgar do not know how to admire.

I admired still more the intelligence which directs these vast forces. I said to myself: "One must be blind not to be dazzled by this spectacle; one must be stupid not to recognize its author; one must be mad not to worship the Supreme Being."

I was deep in these ideas when one of those genii who fill the intermundane spaces came down to me . . . and transported me into a desert all covered with piles of bones. . . . He began with the first pile. "These," he said, "are the twenty-three thousand Jews who danced before a calf, with the twenty-four thousand who were killed while lying with Midianitish women. The number of those massacred for such errors and offences amounts to nearly three hundred thousand.

"In the other piles are the bones of the Christians slaughtered by each other because of metaphysical disputes. . . . "

"What!" I cried, "brothers have treated their brothers like this, and I have the misfortune to be of this brotherhood! . . . Why assemble here all these abominable monuments to barbarism and fanaticism?"

"To instruct you. . . . Follow me now.". . .

I saw a man with a gentle, simple face, who seemed to me to be about thirty-five years old. From afar he looked with compassion upon those piles of whitened bones, through which I had been led to reach the sage's dwelling place. I was astonished to find his feet swollen and bleeding, his hands likewise, his side pierced, and his ribs laid bare by the cut of the lash. "Good God!" I said to him, "is it possible for a just man, a sage, to be in this state? . . . Was it . . . by priests and judges that you were so cruelly assassinated?"

With great courtesy he answered, "Yes."

"And who were these monsters?"

"They were hypocrites."

"Ah! that says everything; I understand by that one word that they would have condemned you to the cruelest punishment. Had you then proved to them, as Socrates did, that the Moon was not a goddess, and that Mercury was not a god?"

"No, it was not a question of planets. My countrymen did not even know what a planet was; they were all arrant ignoramuses. Their superstitions were quite different from those of the Greeks."

"Then you wanted to teach them a new religion?"

"Not at all; I told them simply: 'Love God with all your heart and your neighbor as yourself, for that is the whole of mankind's duty.' Judge yourself if this precept is not as old as the universe; judge yourself if I brought them a new religion." . . .

"Did you not say once that you were come not to bring peace, but a sword?"

"It was a scribe's error; I told them that I brought peace and not a sword. I never wrote anything; what I said can have been changed without evil intention."

"You did not then contribute in any way by your teaching, either badly reported or badly interpreted, to those frightful piles of bones which I saw on my way to consult with you?"

"I have only looked with horror upon those who have made themselves guilty of all these murders."

. . . [Finally] I asked him to tell me in what true religion consisted.

"Have I not already told you? Love God and your neighbor as yourself." . . .

"Well, if that is so, I take you for my only master."[13]

Voltaire was a prodigious worker. This painting shows him dictating to his secretary from the very moment he hops out of bed. *(Bulloz)*

This passage requires careful study, for it suggests many Enlightenment themes of religion and philosophy. As the opening paragraphs show, Voltaire clearly believed in a God. But the God of Voltaire and most philosophes was a distant, deistic God, a great Clockmaker who built an orderly universe and then stepped aside and let it run. Finally, the philosophes hated all forms of religious intolerance. They believed that people had to be wary of dogmatic certainty and religious disputes, which often led to fanaticism and savage, inhuman action. Simple piety and human kindness—the love of God and the golden rule—were religion enough, even Christianity enough, as Voltaire's interpretation of Christ suggests.

The ultimate strength of the philosophes lay, however, in their numbers, dedication, and organization. The philosophes felt keenly that they were engaged in a common undertaking that transcended individuals. Their greatest and most representative intellectual achievement was, quite fittingly, a group effort —the seventeen-volume *Encyclopedia: The Rational Dictionary of the Sciences, the Arts, and the Crafts,* edited by Denis Diderot (1713–1774) and Jean le Rond d'Alembert (1717–1783). Diderot and d'Alembert made a curious pair. Diderot began his career as a hack writer, first attracting attention with a skeptical tract on religion that was quickly burned by the judges of Paris. D'Alembert was one of Europe's leading scientists and mathematicians, the orphaned and illegitimate son of celebrated aristocrats. Moving in different circles and with different interests, the two men set out to find coauthors who would examine the rapidly expanding whole of human knowledge. Even more fundamentally, they set out to teach people how to think critically and objectively about all matters. As Diderot said, he wanted the *Encyclopedia* to "change the general way of thinking."[14]

The editors of the *Encyclopedia* had to conquer innumerable obstacles. After the appearance in 1751 of the first volume, which dealt with such controversial subjects as atheism, the soul, and blind people—all words beginning with *a* in French—the government temporarily banned publication. The pope later placed it on the Index and pronounced excommunication on all who read or bought it. The timid publisher mutilated some of the articles in the last ten volumes without the editors' consent in an attempt to appease the authorities. Yet Diderot's unwavering belief in the importance of his mission held the encyclopedists together for fifteen years, and the enormous work was completed in 1765. Hundreds of thousands of articles by leading scientists and famous writers, skilled workers and progressive priests, treated every aspect of life and knowledge.

Not every article was daring or original, but the overall effect was little short of revolutionary. Science and the industrial arts were exalted, religion and immortality questioned. Intolerance, legal injustice, and out-of-date social institutions were openly criticized. More generally, the writers of the *Encyclopedia* showed that human beings could use the process of reasoning to expand human knowledge. Encyclopedists were convinced that greater knowledge would result in greater human happiness, for knowledge was useful and made possible economic, social, and political progress. The *Encyclopedia* was widely read and extremely influential in France and throughout western Europe as well. It summed up the new world-view of the Enlightenment.

THE LATER ENLIGHTENMENT

After about 1770, the harmonious unity of the philosophes and their thought began to break down. As the new world-view became increasingly accepted by the educated public, some thinkers sought originality by exaggerating certain ideas of the Enlightenment to the exclusion of others. These latter-day philosophes often built rigid, dogmatic systems.

In his *System of Nature* (1770) and other works, the wealthy, aristocratic Baron Paul d'Holbach (1723–1789) argued that human beings were machines completely determined by outside forces. Free will, God, and immortality of the soul were foolish myths. D'Holbach's aggressive atheism and determinism, which were coupled with deep hostility toward Christianity and all other religions, dealt the unity of the Enlightenment movement a severe blow. *Deists* such as Voltaire, who believed in God but not in established churches, were repelled by the inflexible atheism they found in the *System of Nature*. They saw in it the same dogmatic intolerance they had been fighting all their lives.

D'Holbach published his philosophically radical works anonymously to avoid possible prosecution, and in his lifetime he was best known to the public as the generous patron and witty host of writers and intellectuals. At his twice-weekly dinner parties, an inner circle of regulars who knew the baron's secret exchanged ideas with aspiring philosophes and distinguished visitors. One of the most important was the Scottish philosopher David Hume (1711–1776), whose carefully argued skepticism had a powerful long-term influence.

Building on John Locke's teachings on learning (page 585), Hume argued that the human mind is really nothing but a bundle of impressions. These impressions originate only in sense experiences and our habits of joining these experiences together. Since our ideas ultimately reflect only our sense experiences, our reason cannot tell us anything about questions like the origin of the universe or the existence of God, questions that cannot be verified by sense experience (in the form of controlled experiments or mathematics). Paradoxically, Hume's rationalistic inquiry ended up undermining the Enlightenment's faith in the very power of reason itself.

Another French aristocrat, the marquis Marie-Jean de Condorcet (1743–1794), transformed the Enlightenment belief in gradual, hard-won progress into fanciful utopianism. In his *Progress of the Human Mind,* written in 1793 during the French Revolution, Condorcet tracked the nine stages of human progress that had already occurred and predicted that the tenth would bring perfection. Ironically, Condorcet wrote this work while fleeing for his life. Caught and condemned by revolutionary extremists, he preferred death by his own hand to the blade of the guillotine.

figure . 1ère

Canal with Locks The articles on science and the industrial arts in the *Encyclopedia* carried lavish explanatory illustrations. This typical engraving from the section on water and its uses shows advances in canal building and reflects the encyclopedists' faith in technical progress. *(University of Illinois)*

Other thinkers and writers after about 1770 began to attack the Enlightenment's faith in reason, progress, and moderation. The most famous of these was the Swiss Jean-Jacques Rousseau (1712–1778), a brilliant but difficult thinker, an appealing but neurotic individual. Born into a poor family of watchmakers in Geneva, Rousseau went to Paris and was greatly influenced by Diderot and Voltaire. Always extraordinarily sensitive and suspicious, Rousseau came to believe his philosophe friends were plotting against him. In the mid-1750s he broke with them personally and intellectually, living thereafter as a lonely outsider with his uneducated common-law wife and going in his own highly original direction.

Like other Enlightenment thinkers, Rousseau was passionately committed to individual freedom. Un-

like them, however, he attacked rationalism and civilization as destroying rather than liberating the individual. Warm, spontaneous feeling had to complement and correct the cold intellect. Moreover, the individual's basic goodness had to be protected from the cruel refinements of civilization. These ideas greatly influenced the early romantic movement (see Chapter 23), which rebelled against the culture of the Enlightenment in the late eighteenth century.

Applying his heartfelt ideas to children, Rousseau had a powerful impact on the development of modern education. In his famous treatise *Emile* (1762), he argued that education must shield the naturally unspoiled child from the corrupting influences of civilization and too many books. According to Rousseau, children must develop naturally and spontane-

Sugar Cane Mill Enlightenment thinkers were keenly interested in non-European societies and in the comparative study of human development. This engraving depicts an aspect of sugar production with slave labor in the Americas. *(Courtesy of the University of Minnesota Libraries)*

ously, at their own speed and in their own way. It is eloquent testimony to Rousseau's troubled life and complicated personality that he placed all five of his own children in orphanages.

Rousseau also made an important contribution to political theory in the *Social Contract* (1762). His fundamental ideas were the general will and popular sovereignty. According to Rousseau, the general will is sacred and absolute, reflecting the common interests of the people, who have displaced the monarch as the holder of the sovereign power. The general will is not necessarily the will of the majority, however, although minorities have to subordinate themselves to it without question. Little noticed before the French Revolution, Rousseau's concept of the general will appealed greatly to democrats and nationalists after

1789. The concept has also been used since 1789 by many dictators, who have claimed that they, rather than some momentary majority of the voters, represent the general will and thus the true interests of democracy and the sovereign masses.

THE SOCIAL SETTING OF THE ENLIGHTENMENT

The philosophes were splendid talkers as well as effective writers. Indeed, sparkling conversation in private homes spread Enlightenment ideas to Europe's upper-middle class and aristocracy. Paris set the example, and other French cities and European capitals followed. In Paris a number of talented and often rich

Madame Geoffrin's Salon In this stylized group portrait a famous actor reads to a gathering of leading philosophes and aristocrats in 1755. Third from the right presiding over her gathering, is Madame Geoffrin, next to the sleepy ninety-eight-year-old Bernard de Fontenelle. *(Malmaison Chateau/Giraudon/Art Resource)*

women presided over regular social gatherings of the great and near-great in their elegant drawing rooms, or *salons.* There a d'Alembert and a Fontenelle could exchange witty, uncensored observations on literature, science, and philosophy with great aristocrats, wealthy middle-class financiers, high-ranking officials, and noteworthy foreigners. These intellectual salons practiced the equality the philosophes preached. They were open to all men and women with good manners, provided only that they were famous or talented, rich or important. More generally, the philosophes championed greater rights and expanded education for women, arguing that the subordination of females was an unreasonable prejudice and the sign of a barbaric society.

One of the most famous salons was that of Madame Geoffrin, the unofficial godmother of the *Encyclopedia.* Having lost her parents at an early age, the future Madame Geoffrin was married at fifteen by her well-meaning grandmother to a rich and boring businessman of forty-eight. It was the classic marriage of convenience—the poor young girl and the rich old man—and neither side ever pretended that love was a consideration. After dutifully raising her children, Madame Geoffrin sought to break out of her gilded cage as she entered middle age. The very proper businessman's wife became friendly with a neighbor, the marquise de Tencin. In her youth this aristocratic beauty had been rather infamous as the mistress of the regent of France, but she had settled down to run a salon that counted Fontenelle and the philosopher Montesquieu among its regular guests.

When the marquise died in 1749, Madame Geoffrin tactfully transferred these luminaries to her spacious mansion for regular dinners. At first Madame Geoffrin's husband loudly protested the arrival of this horde of "parasites." But his wife's will was much stronger than his, and he soon opened his purse and even appeared at the twice-weekly dinners. "Who was that old man at the end of the table who never said anything?" an innocent newcomer asked one evening. "That," replied Madame Geoffrin without the slightest emotion, "was my husband. He's dead."[15]

When Monsieur Geoffrin's death became official, Madame Geoffrin put the large fortune and a spacious mansion she inherited to good use. She welcomed the encyclopedists—Diderot, d'Alembert, Fontenelle, and a host of others. She gave them generous financial aid and helped to save their enterprise from collapse, especially after the first eight volumes were burned by the authorities in 1759. She also corresponded with the king of Sweden and Catherine the Great of Russia. Madame Geoffrin was, however, her own woman. She remained a practicing Christian and would not tolerate attacks on the church in her house. It was said that distinguished foreigners felt they had not seen Paris unless they had been invited to one of her dinners. The plain and long-neglected Madame Geoffrin managed to become the most renowned hostess of the eighteenth century.

There were many other hostesses, but Madame Geoffrin's greatest rival, Madame du Deffand, was one of the most interesting. While Madame Geoffrin was middle-class, pious, and chaste, Madame du Deffand was a skeptic from the nobility, who lived fast and easy, at least in the early years. Another difference was that women—mostly highly intelligent, worldly members of the nobility—were fully the equal of men in Madame du Deffand's intellectual salon. Forever pursuing fulfillment in love and life, Madame du Deffand was an accomplished and liberated woman. An exceptionally fine letter writer, she carried on a vast correspondence with leading men and women all across Europe. Voltaire was her most enduring friend.

Madame du Deffand's closest female friend was Julie de Lespinasse, a beautiful, talented young woman whom she befriended and made her protégée. The never-acknowledged illegitimate daughter of noble parents, Julie de Lespinasse had a hard youth, but she flowered in Madame du Deffand's drawing room—so much so that she was eventually dismissed by her jealous patroness.

Once again Julie de Lespinasse triumphed. Her friends gave her money so that she could form her own salon. Her highly informal gatherings—she was not rich enough to supply more than tea and cake—attracted the keenest minds in France and Europe. As one philosophe wrote:

She could unite the different types, even the most antagonistic, sustaining the conversation by a well-aimed phrase, animating and guiding it at will. . . . Politics, religion, philosophy, news: nothing was excluded. Her circle met daily from five to nine. There one found men of all ranks in the State, the Church, and the Court, soldiers and foreigners, and the leading writers of the day.[16]

Thus in France the ideas of the Enlightenment thrived in a social setting that graciously united members of the intellectual, economic, and social elites. Never before and never again would social and intellectual life be so closely and so pleasantly joined. In such an atmosphere, the philosophes, the French nobility, and the upper-middle class increasingly influenced one another. Critical thinking became fashionable and flourished alongside hopes for human progress through greater knowledge.

THE EVOLUTION OF ABSOLUTISM

How did the Enlightenment influence political developments? To this important question there is no easy answer. On the one hand, the philosophes were primarily interested in converting people to critical "scientific" thinking and were not particularly concerned with politics. On the other hand, such thinking naturally led to political criticism and interest in political reform. Educated people, who belonged mainly to the nobility and middle class, came to regard political change as both possible and desirable. A further problem is that Enlightenment thinkers had different views on politics. Some, led by the nobleman Montesquieu, argued for curbs on monarchial power in order to promote liberty, and some French judges applied such theories in practical questions.

Until the American Revolution, however, most Enlightenment thinkers believed that political change could best come from above—from the ruler—rather than from below, especially in central and eastern Europe. There were several reasons for this essentially moderate belief. First, royal absolutism was a fact of life, and the kings and queens of Europe's leading states clearly had no intention of giving up their great powers. Therefore the philosophes realistically concluded that a benevolent absolutism offered the best opportunities for improving society. Critical thinking was turning the art of good government into an exact science. It was necessary only to educate and "enlighten" the monarch, who could then make good laws and promote human happiness. Second, philosophes turned toward rulers because rulers seemed to be listening, treating them with respect, and seeking their advice. Finally, although the philosophes did not dwell on this fact, they distrusted the masses. Known simply as "the people" in the eighteenth century, the peasant masses and the urban poor were, according to the philosophes, still enchained by religious superstitions and violent passions. No doubt the people were maturing, but they were still children in need of firm parental guidance.

Encouraged and instructed by the philosophes, several absolutist rulers of the later eighteenth century tried to govern in an "enlightened" manner. Yet, because European monarchs had long been locked in an intense international competition, a more enlightened state often meant in practice a more effective state, a state capable of expanding its territory and defeating its enemies. Moreover, reforms from above had to be grafted onto previous historical developments and existing social structures. Little wonder, then, that the actual programs and accomplishments of these rulers varied greatly. Let us therefore examine the evolution of monarchial absolutism at close range before trying to form any overall judgment regarding the meaning of what historians have often called the "enlightened absolutism" of the later eighteenth century.

THE "GREATS": FREDERICK OF PRUSSIA AND CATHERINE OF RUSSIA

Just as the French culture and absolutism of Louis XIV provided models for European rulers in the late seventeenth century, the Enlightenment teachings of the French philosophes inspired European monarchs in the second half of the eighteenth century. French was the international language of the educated classes, and the education of future kings and queens across Europe lay in the hands of French tutors espousing Enlightenment ideas. France's cultural leadership was reinforced by the fact that it was still the wealthiest and most populous country in Europe. Thus, absolutist monarchs in several west German and Italian states, as well as in Spain and Portugal, proclaimed themselves more enlightened. By far the most influential of the new-style monarchs were Frederick II of Prussia and Catherine II of Russia, both styled "the Great."

FREDERICK THE GREAT. Frederick II (1740–1786), also known as Frederick the Great, built masterfully on his father's work. This was somewhat surprising for, like many children with tyrannical parents, he rebelled against his family's wishes in his early years. Rejecting the crude life of the barracks, Frederick embraced culture and literature, even writing poetry and fine prose in French, a language his father detested. He threw off his father's dour Calvinism and dabbled with atheism. After trying unsuccessfully to run away at age eighteen in 1730, he was virtually imprisoned and even compelled to watch his companion in flight beheaded at his father's command. Yet, like many other rebellious youths, Frederick eventually reached a reconciliation with his father, and by the time he came to the throne ten years later he was determined to use the splendid army his father had left him.

When, therefore, the emperor of Austria, Charles VI, also died in 1740 and his young and beautiful daughter, Maria Theresa, became ruler of the Habsburg dominions, Frederick suddenly and without warning invaded her rich, all-German province of Silesia. This action defied solemn Prussian promises to respect the Pragmatic Sanction, which guaranteed Maria Theresa's succession, but no matter. For Frederick, it was the opportunity of a lifetime to expand the size and power of Prussia. Although Maria Theresa succeeded in dramatically rallying the normally quarrelsome Hungarian nobility, her multinational army was no match for Prussian precision. In 1742, as other greedy powers were falling on her lands in the general European War of the Austrian Succession (1740–1748), she was forced to cede all of Silesia to

Maria Theresa and her husband pose with eleven of their sixteen children at Schön-brunn palace. Joseph, the heir to the throne, stands at the center of the star pattern. Wealthy women often had very large families, in part because they seldom nursed their babies as poor women usually did. *(Kunsthistorisches Museum, Vienna)*

Prussia. In one stroke, Prussia doubled its population to 6 million people. Now Prussia unquestionably towered above all the other German states and stood as a European Great Power.

Successful in 1742, Frederick had to spend much of his reign fighting against great odds to save Prussia from total destruction. Maria Theresa was determined to regain Silesia, and when the ongoing competition between Britain and France for colonial empire brought renewed conflict in 1756, her able chief minister fashioned an aggressive alliance with France and Russia. The aim of the alliance was to conquer Prussia and divide up its territory, just as Frederick II and other monarchs had so recently sought to partition the Austrian Empire. Frederick led his army brilliantly, striking repeatedly at vastly superior forces invading from all sides. At times he believed all was lost, but he fought on with stoic courage. In the end, he was miraculously saved: Peter III came to the Russian throne in 1762 and called off the attack against Frederick, whom he greatly admired.

In the early years of his reign, Frederick II had kept his enthusiasm for Enlightenment culture strictly separated from a brutal concept of international politics. He wrote:

Of all States, from the smallest to the biggest, one can safely say that the fundamental rule of government is the principle of extending their territories. . . . The passions of rulers have no other curb but the limits of their power. Those are the fixed laws of European politics to which every politician submits.[17]

Catherine the Great Intelligent, pleasure-loving, and vain, Catherine succeeded in bringing Russia closer to western Europe than ever before. *(John R. Freeman)*

But the terrible struggle of the Seven Years' War tempered Frederick and brought him to consider how more humane policies for his subjects might also strengthen the state.

Thus Frederick went beyond a superficial commitment to Enlightenment culture for himself and his circle. He tolerantly allowed his subjects to believe as they wished in religious and philosophical matters. He promoted the advancement of knowledge, improving his country's schools and universities.

Second, Frederick tried to improve the lives of his subjects more directly. As he wrote his friend Voltaire, "I must enlighten my people, cultivate their manners and morals, and make them as happy as human beings can be, or as happy as the means at my disposal permit." The legal system and the bureaucracy were Frederick's primary tools. Prussia's laws were simplified, and judges decided cases quickly and impartially. Prussian officials became famous for their hard work and honesty. After the Seven Years' War ended in 1763, Frederick's government also energetically promoted the reconstruction of agriculture and industry in his war-torn country. In all this Frederick set a good example. He worked hard and lived modestly, claiming that he was "only the first servant of the state." Thus Frederick justified monarchy in terms of practical results and said nothing of the divine right of kings.

Frederick's dedication to high-minded principles went only so far, however. He never tried to change Prussia's existing social structure. True, he condemned serfdom in the abstract, but he accepted it in practice and did not even free the serfs on his own estates. He accepted and extended the privileges of the nobility, which he saw as his primary ally in the defense and extension of his realm. It became practically impossible for a middle-class person to gain a top position in the government. The Junker nobility remained the backbone of the army and the entire Prussian state.

CATHERINE THE GREAT. Catherine the Great of Russia (1762–1796) was one of the most remarkable rulers who ever lived, and the philosophes adored her. Catherine was a German princess from Anhalt-Zerbst, a totally insignificant principality sandwiched between Prussia and Saxony. Her father commanded a regiment of the Prussian army, but her mother was related to the Romanovs of Russia, and that proved to be her chance.

Peter the Great had abolished the hereditary succession of tsars so that he could name his successor and thus preserve his policies. This move opened a period of palace intrigue and a rapid turnover of rulers until Peter's youngest daughter Elizabeth came to the Russian throne in 1741. A crude, shrewd woman noted for her hard drinking and hard loving —one of her official lovers was an illiterate shepherd boy—Elizabeth named her nephew Peter heir to the throne and chose Catherine to be his wife in 1744. It was a mismatch from the beginning. The fifteen-year-old Catherine was intelligent and attractive; her husband was stupid and ugly, his face badly scarred by smallpox. Ignored by her childish husband, Catherine carefully studied Russian, endlessly read writers like Bayle and Voltaire, and made friends at court. Soon she knew what she wanted. "I did not care about Peter," she wrote in her *Memoirs,* "but I did care about the crown."[18]

As the old empress Elizabeth approached death, Catherine plotted against her unpopular husband. A dynamic, sensuous woman, Catherine used her sexuality to good political advantage. She selected as her new lover a tall, dashing young officer named Gregory Orlov, who with his four officer brothers commanded considerable support among the soldiers stationed in St. Petersburg. When Peter came to the throne in 1762, his decision to withdraw Russian troops from the coalition against Prussia alienated the army. Nor did Peter III's attempt to gain support from the Russian nobility by freeing it from compulsory state service succeed. At the end of six months, Catherine and the military conspirators deposed Peter III in a palace revolution. Then the Orlov brothers murdered him. The German princess became empress of Russia.

Catherine had drunk deeply at the Enlightenment well. Never questioning the common assumption that absolute monarchy was the best form of government, she set out to rule in an enlightened manner. One of her most enduring goals was to bring the sophisticated culture of western Europe to backward Russia. To do so, she imported Western architects, sculptors, musicians, and intellectuals. She bought masterpieces of Western art in wholesale lots and patronized the philosophes. An enthusiastic letter writer, she corresponded extensively with Voltaire and praised him as the "champion of the human race." When the French government banned the *Encyclopedia,* she offered to publish it in St. Petersburg.

She discussed reform with Diderot in St. Petersburg; and when Diderot needed money, she purchased his library for a small fortune but allowed him to keep it during his lifetime. With these and countless similar actions, Catherine skillfully won a good press for herself and for her country in the West. Moreover, this intellectual ruler, who wrote plays and loved good talk, set the tone for the entire Russian nobility. Peter the Great westernized Russian armies, but it was Catherine who westernized the thinking of the Russian nobility.

Catherine's second goal was domestic reform, and she began her reign with sincere and ambitious projects. Better laws were a major concern. In 1767 she drew up enlightened instructions for the special legislative commission she appointed to prepare a new law code. No new unified code was ever produced, but Catherine did restrict the practice of torture and allowed limited religious toleration. She also tried to improve education and strengthen local government. The philosophes applauded these measures and hoped more would follow.

Such was not the case. In 1773 a simple Cossack soldier named Emelian Pugachev sparked a gigantic uprising of serfs, very much as Stenka Razin had done a century earlier (page 558). Proclaiming himself the true tsar, Pugachev issued "decrees" abolishing serfdom, taxes, and army service. Thousands joined his cause, slaughtering landlords and officials over a vast area of southwestern Russia. Pugachev's untrained hordes eventually proved no match for Catherine's noble-led regular army. Betrayed by his own company, Pugachev was captured and savagely executed.

Pugachev's rebellion was a decisive turning point in Catherine's domestic policy. On coming to the throne she had condemned serfdom in theory, but she was smart enough to realize that any changes would have to be very gradual or else she would quickly follow her departed husband. Pugachev's rebellion put an end to any illusions she might have had about reforming serfdom. The peasants were clearly dangerous, and her empire rested on the support of the nobility. After 1775 Catherine gave the nobles absolute control of their serfs. She extended serfdom into new areas, such as the Ukraine. In 1785 she formalized the nobility's privileged position, freeing them forever from taxes and state service. She also confiscated the lands of the Russian Orthodox church and gave them to favorite officials. Under

Catherine, the Russian nobility attained its most exalted position, and serfdom entered its most oppressive phase.

Catherine's third goal was territorial expansion, and in this respect she was extremely successful. Her armies subjugated the last descendants of the Mongols, the Crimean Tartars, and began the conquest of the Caucasus.

Her greatest coup by far was the partitioning of Poland. Poland showed the dangers of failing to build a strong absolutist state. For decades all important decisions had required the unanimous agreement of every Polish noble, which meant that nothing could ever be done. When between 1768 and 1772 Catherine's armies scored unprecedented victories against the Turks and thereby threatened to disturb the balance of power between Russia and Austria in eastern Europe, Frederick of Prussia obligingly came forward with a deal. He proposed that Turkey be let off easily, and that Prussia, Austria, and Russia each compensate itself by taking a gigantic slice of Polish territory. Catherine jumped at the chance. The first partition of Poland took place in 1772. Two more partitions, in 1793 and 1795, gave all three powers more Polish territory, and the kingdom of Poland simply vanished from the map (see Map 18.1).

Expansion helped Catherine keep the nobility happy, for it provided her vast new lands to give to her faithful servants. Expansion also helped Catherine reward her lovers, of whom twenty-one have been definitely identified. On all these royal favorites she lavished large estates with many serfs, as if to make sure there were no hard feelings when her interest cooled. Until the end this remarkably talented woman—who always believed that, in spite of her domestic setbacks, she was slowly civilizing Russia—kept her zest for life. Fascinated by a new twenty-two-year-old flame when she was a roly-poly grandmother in her sixties, she happily reported her good fortune to a favorite former lover: "I have come back to life like a frozen fly; I am gay and well."[19]

ABSOLUTISM IN FRANCE AND AUSTRIA

The Enlightenment's influence on political developments in France and Austria was complex. In France, the monarchy maintained its absolutist claims, and some philosophes like Voltaire believed that the king was still the best source of needed reform. At the same time, discontented nobles and learned judges drew on thinkers such as Montesquieu for liberal arguments, and they sought with some success to limit the king's power. In Austria, two talented rulers did manage to introduce major reforms, although traditional power politics were more important than Enlightenment teachings.

LOUIS XV OF FRANCE. In building French absolutism, Louis XIV successfully drew on the middle class to curb the political power of the nobility. As long as the Grand Monarch lived, the nobility could only grumble and, like the duke of Saint-Simon in his *Memoirs,* scornfully lament the rise of "the vile bourgeoisie." But when Louis XIV finally died in 1715, to be succeeded by his five-year-old great-grandson, Louis XV (1715–1774), the Sun King's elaborate system of absolutist rule was challenged in a general reaction. Favored by the duke of Orléans, who governed as regent until 1723, the nobility made a strong comeback.

Most importantly, the duke restored to the high court of Paris—the Parlement—the right to "register" and thereby approve the king's decrees. This was a fateful step. The judges of the Parlement of Paris had originally come from the middle class, and their high position reflected the way that Louis XIV (and earlier French monarchs) had chosen to use that class to build the royal bureaucracy so necessary for an absolutist state. By the eighteenth century, however, these "middle-class" judges had risen to become hereditary nobles for, although Louis XIV had curbed the political power of the nobility, he had never challenged its enormous social prestige. Thus, high position in the government continued to bestow the noble status that middle-class officials wanted, either immediately or after three generations of continuous service. The judges of Paris, like many high-ranking officials, actually owned their government jobs and freely passed them on as private property from father to son. By supporting the claim of this well-entrenched and increasingly aristocratic group to register the king's laws, the duke of Orléans sanctioned a counterweight to absolute power.

These implications became clear when the heavy expenses of the War of the Austrian Succession plunged France into financial crisis. In 1748 Louis

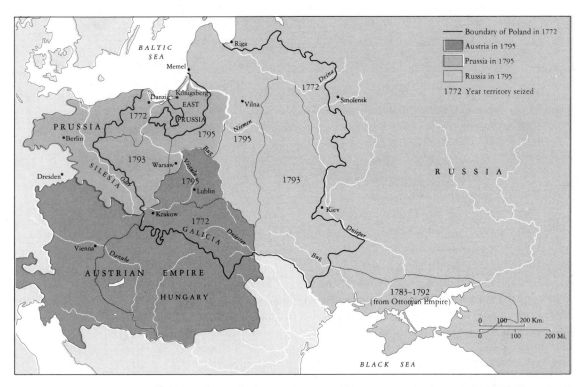

MAP 18.1 The Partition of Poland and Russia's Expansion, 1772–1795 Though all three of the great eastern absolutist states profited from the division of large but weak Poland, Catherine's Russia gained the most.

XV appointed a finance minister who decreed a 5 percent income tax on every individual, regardless of social status. Exemption from most taxation had long been a hallowed privilege of the nobility, and other important groups—the clergy, the large towns, and some wealthy bourgeoisie—had also gained special tax advantages over time. The result was a vigorous protest from many sides, and the Parlement of Paris refused to ratify the tax law. The monarchy retreated; the new tax was dropped.

Following the disastrously expensive Seven Years' War, the conflict re-emerged. The government tried to maintain emergency taxes after the war ended. The Parlement of Paris protested and even challenged the basis of royal authority, claiming that the king's power must necessarily be limited to protect liberty. Once again the government caved in and withdrew the wartime taxes in 1764. Emboldened by its striking victory and widespread support from France's educated elite, the judicial opposition in

Paris and the provinces pressed its demands. In a barrage of pamphlets and legal briefs, it asserted that the king could not levy taxes without the consent of the Parlement of Paris acting as the representative of the entire nation.

Indolent and sensual by nature, more interested in his many mistresses than in affairs of state, Louis XV finally roused himself for a determined defense of his absolutist inheritance. "The magistrates," he angrily told the Parlement of Paris in a famous face-to-face confrontation, "are my officers. . . . In my person only does the sovereign power rest."[20] In 1768 Louis appointed a tough career official named René de Maupeou as chancellor and ordered him to crush the judicial opposition.

Maupeou abolished the Parlement of Paris and exiled its members to isolated backwaters in the provinces. He created a new and docile parlement of royal officials, and he began once again to tax the privileged groups. A few philosophes like Voltaire ap-

THE ENLIGHTENMENT

1686	Fontenelle, *Conversations on the Plurality of Worlds*
1687	Newton, *Principia Mathematica*
1690	Locke, *Essay Concerning Human Understanding* and *Second Treatise on Civil Government*
1697	Bayle, *Historical and Critical Dictionary*
1715–1774	Rule of Louis XV in France
1721	Montesquieu, *The Persian Letters*
1740–1748	War of the Austrian Succession
1740–1780	Rule of Maria Theresa in Austria
1740–1786	Rule of Frederick II (the Great) in Prussia
1742	Austria cedes Silesia to Prussia
1743	Voltaire, *The Age of Louis XIV*
1748	Montesquieu, *The Spirit of the Laws*
	Hume, *An Enquiry Concerning Human Understanding*
1751–1765	Publication of the *Encyclopedia,* edited by Diderot and d'Alembert
1756–1763	Seven Years' War
1762	Rousseau, *The Social Contract*
1762–1796	Rule of Catherine II (the Great) in Russia
1767	Catherine the Great appoints commission to prepare a new law code
1770	D'Holbach, *The System of Nature*
1771	Maupeou, Louis XV's chancellor, abolishes Parlement of Paris
1772	First partition of Poland among Russia, Prussia, and Austria
1774	Ascension of Louis XVI in France: restoration of Parlement of Paris
1780–1790	Rule of Joseph II in Austria
1781	Abolition of serfdom in Austria
1785	Catherine the Great issues the Russian Charter of Nobility, which guarantees the servitude of the peasants
1790–1792	Rule of Leopold II in Austria: serfdom is re-established
1793	Second partition of Poland
	Condorcet, *The Progress of the Human Mind*
1795	Third partition of Poland, which completes its absorption

plauded these measures: the sovereign was using his power to introduce badly needed reforms that had been blocked by a self-serving aristocratic elite. Most philosophes and educated public opinion as a whole sided with the old parlements, however, and there was widespread dissatisfaction with royal despotism. Yet the monarchy's power was still great enough for Maupeou to simply ride over the opposition, and Louis XV would probably have prevailed—if he had lived to a very ripe old age.

But Louis XV died in 1774. The new king, Louis XVI (1774–1792), was a shy twenty-year-old with good intentions. Taking the throne, he is reported to have said: "What I should like most is to be loved."[21] The eager-to-please monarch decided to yield in the face of such strong criticism from so much of France's elite. He dismissed Maupeou and repudiated the strong-willed minister's work. The old Parlement of Paris was reinstated, as enlightened public opinion cheered and hoped for moves toward repre-

ART: A MIRROR OF SOCIETY

Art reveals the interests and values of society and frequently gives intimate and unique glimpses of how people actually lived. In portraits and statues, whether of saints, generals, philosophers, popes, poets, or merchants, it preserves the memory and fame of men and women who shaped society. In paintings, drawings, and carvings, it also shows how people worked, played, relaxed, suffered, and triumphed. Art, therefore, is extremely useful to the historian, especially for periods when written records are scarce. Every work of art and every part of it has meaning and has something of its own to say.

Art also manifests the changes and continuity of European life; as values changed in Europe, so did major artistic themes. Europeans of the sixteenth and seventeenth centuries remained deeply religious and showed a new interest in the world around them; their middle-class attitudes and experiences and their appreciation of agricultural life replaced the activities and interests of a small, wealthy, cultural elite as the subject matter of art. The scenes of everyday life presented in seventeenth-century genre painting were considered interesting in themselves and, as the works following by Siberechts and Van Ostade support, worthy of high art.

Just as Renaissance princes had used art to memorialize themselves, so in the later seventeenth century all the arts were organized to exalt and glorify the Sun King, Louis XIV. As the following painting by Lebrun, *Louis XIV's Visit to the Gobelin Factory,* demonstrates, art served the image of the French absolute monarch as the personification of the French state. Art thereby reflected the political ideal of the times. Known as the court baroque style, it spread throughout France, England, Germany, Italy, and Spain. Politics had triumphed over other expressions of cultural life.

In painting the baroque reached maturity early with Peter Paul Rubens. Court painter to James I and Charles I of England and to Louis XIII and Marie de' Medici of France, Rubens also illustrated many classical and rural subjects. In *Country Fair* (1635), below, against an immense landscape twilight descends, bringing to a close the peasants' day of celebration. (Cliché des Musées Nationaux-Paris.)

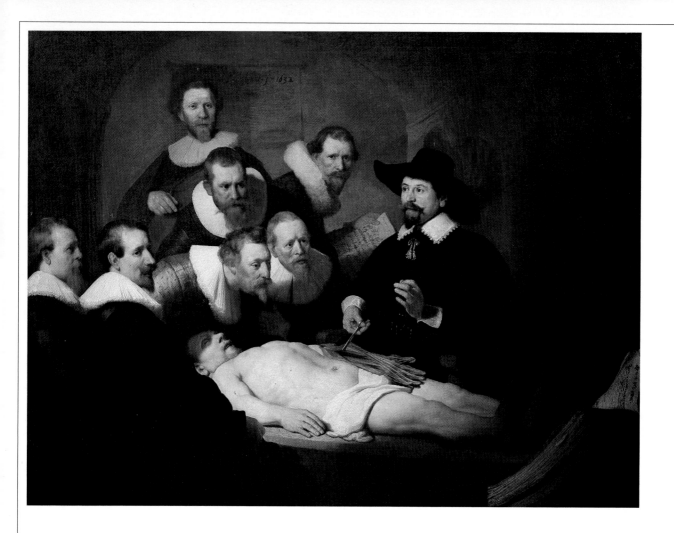

The Anatomy Lesson *(above)* Rembrandt (ca 1632). In the seventeenth century science became modern, in the sense that it was based on direct observation and experimentation. This highly dramatic scene, one of the earliest of Rembrandt (who painted it at age 25), shows his masterful control of light and shade, borne out by the tense atmosphere of the laboratory, and the expressions of deep curiosity on the students' faces. *(Rijksmuseum, Amsterdam)*

A Woman Weighing Gold *(above)* Vermeer (ca 1657). Vermeer painted pictures of middle-class women involved in ordinary activities in the quiet interiors of their homes.
Unrivaled among Dutch masters for his superb control of light, in this painting Vermeer
illuminates the pregnant woman weighing gold on her scales, as Christ in the painting
on the wall weighs the saved and the damned. *(National Gallery of Art, Washington;
Widener Collection)*

Louis XIV's Visit to the Gobelin Factory *(left, above)* Lebrun (1663–1675). World-famous for its superb tapestries, the Gobelin factory was a state-owned industry under Colbert's supervision. It produced everything needed for furnishing the royal palaces. The king (left, holding his hat) gazes around, while the tapestry above celebrates Alexander the Great's triumphs, considered a model of Louis XIV's. *(Haeseler/Art Resource)*

Saint Mark's Square: Looking South-East *(above)* Canaletto (1697–1768). The Venetian Canaletto worked primarily for wealthy aristocrats. His most enthusiastic patrons were English nobles, who became enamoured of his magnificent views of historic Venice while making the continental "grand tour" that topped off the education of the English upper class. This lively scene, showing much a visitor would wish to remember, features Saint Mark's on the left, the Ducal Palace, ships in the lagoon, and market stalls. *(National Gallery, Washington)*

Gersaint's Shop *(left, below)* Watteau. In reaction to the rigid and pompous spirit of Louis XIV's Versailles, Watteau's paintings captured the grace and informality of aristocratic life during the early reign of Louis XV. Here the viewer steps from the cobbles-toned streets directly into Gersaint's salesroom with its busy assistants and rich patrons. A clerk (left) packs away a portrait of Louis XIV, reflecting the end of an era, while the seductive figures in many paintings on the walls suggest the new rococo frivolity. *(Staatliche Museen zu Berlin)*

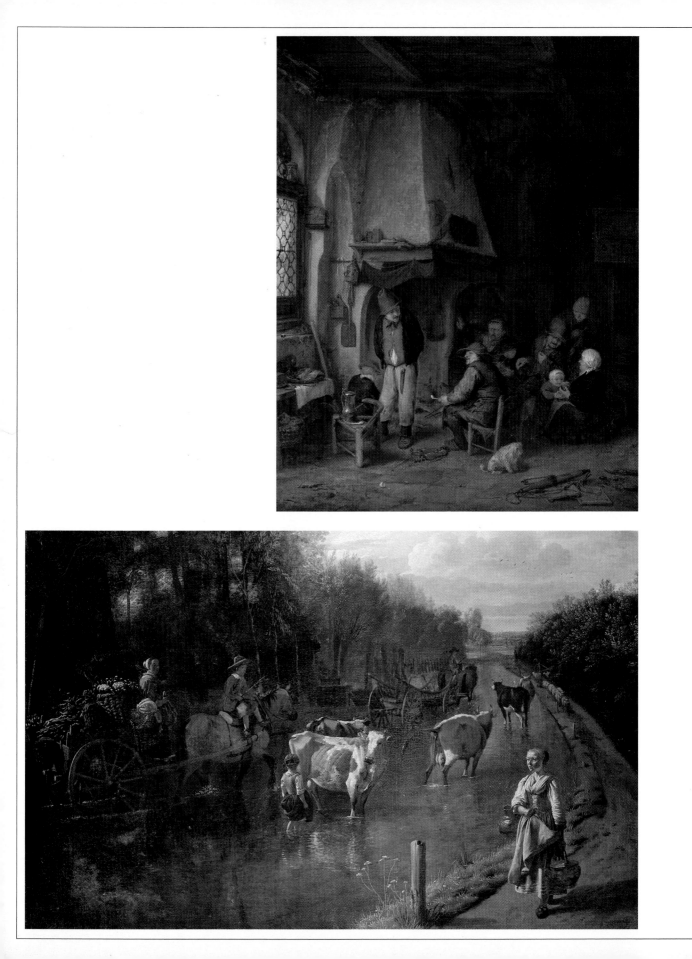

The Ford *(left, below)* Jan Siberechts (1627–1703). Most seventeenth-century artists specialized. The Flemish painter Siberechts concentrated on water-logged landscapes of his native Flanders, which he peopled with peaceful, hardworking peasants absorbed in their daily tasks. Viewing this scene, one feels the slow, difficult movement of beasts and goods along a flooded road, as well as the solid virtues of the little milkmaid and the young peasant woman. *(Niedersächsisches Landesmuseum, Hannover)*

The Spinners *(above)* Diego Velázquez (1599–1660). Spain's master of realism captures women workers in a tapestry workshop and three ladies inspecting a tapestry in the background. Or so people long believed. Modern critics see a mythological weaving competition between the low-born Arachne on the right and the goddess of arts and crafts on the left. (The gutsy Arachne lost and was turned into a spider.) Art historians also have their debates and conflicting interpretations. *(Museo del Prado, Madrid)*

Rustic Interior *(left, above)* Adriaen Van Ostade (1610–1685). Peasant life was also the theme of the Dutch painter Van Ostade. But Van Ostade did not idealize his peasants, who usually lounge in the midst of modest possessions and sport a comical, not-quite-sober air. Pictures of ordinary life, like those of Van Ostade and Siberechts, were small in size, sharply detailed, and well adapted for the living rooms of the Dutch middle class. *(Rijksmuseum, Amsterdam)*

Herrenchiemsee, the Hall of Mirrors by Candlelight *(above)* The splendor of Louis XIV's palace at Versailles long served as a model of royal elegance for European monarchs. Ludwig II of Bavaria built his version of Versailles and its famous Hall of Mirrors on a large lake in southern Germany in the nineteenth century. In the summer season the visitor can still savour the full beauty of Ludwig's Hall of Mirrors at evening concerts by candlelight. *(Werner Neumeister/George Rainbird/Robert Harding)*

sentative government. Such moves were not forth-coming. Instead, a weakened but unrepentant monarchy faced a judicial opposition that claimed to speak for the entire French nation. Increasingly locked in stalemate, the country was drifting toward renewed financial crisis and political upheaval.

THE AUSTRIAN HABSBURGS. Joseph II (1780–1790) was a fascinating individual. For an earlier generation of historians he was the "revolutionary emperor," a tragic hero whose lofty reforms were undone by the landowning nobility he dared to challenge. More recent scholarship has revised this romantic interpretation and stressed how Joseph II continued the state-building work of his mother, the empress Maria Theresa, a remarkable but old-fashioned absolutist.

Maria Theresa's long reign (1740–1780) began with her neighbors, led by Frederick II of Prussia, invading her lands and trying to dismember them (see page 594). Emerging from the long War of the Austrian Succession in 1748 with only the serious loss of Silesia, Maria Theresa and her closest ministers were determined to introduce reforms that would make the state stronger and more efficient. Three aspects were most important in these reforms. First, Maria Theresa introduced measures to bring relations between church and state under government control. Like some medieval rulers, the most devout and very Catholic Maria Theresa aimed at limiting the papacy's political influence in her realm. Second, a whole series of administrative reforms strengthened the central bureaucracy, smoothed out some provincial differences, and revamped the tax system, taxing even the lands of nobles without special exemptions. Finally, the government sought to improve the lot of the agricultural population, cautiously reducing the power of lords over both their hereditary serfs and partially free peasant tenants.

Coregent with his mother from 1765 onward and a strong supporter of change, Joseph II moved forward rapidly when he came to the throne in 1780. He controlled the established Catholic church even more closely, in an attempt to ensure that it produced better citizens. He granted religious toleration and civic rights to Protestants and Jews—a radical innovation that impressed this contemporaries. In even more spectacular peasant reforms, Joseph abolished serfdom in 1781, and in 1789 he decreed that all peasant labor obligations be converted into cash payments.

This ill-conceived measure was violently rejected not only by the nobility but by the peasants it was intended to help, since their primitive barter economy was woefully lacking in money. When a disillusioned Joseph died prematurely at forty-nine, the entire Habsburg Empire was in turmoil. His brother Leopold (1790–1792) was forced to cancel Joseph's radical edicts in order to re-establish order. Peasants lost most of their recent gains, and once again they were required to do forced labor for their lords, as in the 1770s under Maria Theresa.

AN OVERALL EVALUATION

Having examined the evolution of monarchial absolutism in four leading states, it is possible to look for meaningful generalizations and evaluate the overall influence of Enlightenment thought on politics. That thought, it will be remembered, was clustered in two distinct schools: the liberal critique of unregulated monarchy promoted by Montesquieu and the defenders of royal absolutism led by Voltaire.

It is clear that France diverged from its eastern neighbors in its political developments in the eighteenth century. Thus, while neither the French monarchy nor the eastern rulers abandoned the absolutist claims and institutions they had inherited, the monarch's capacity to govern in a truly absolutist manner declined substantially in France, which was not the case in eastern Europe. The immediate cause of this divergence was the political resurgence of the French nobility after 1715 and the growth of judicial opposition, led by the Parlement of Paris. More fundamentally, however, the judicial and aristocratic opposition in France achieved its still rather modest successes because it received major support from educated public opinion, which increasingly made the liberal critique of unregulated royal authority its own. In France, then, the proponents of absolute monarchy were increasingly on the defense, as was the French monarchy itself.

The situation in eastern Europe was different. The liberal critique of absolute monarchy remained an intellectual curiosity, and proponents of reform from above held sway. Moreover, despite their differences, the leading eastern monarchs of the later eighteenth century all claimed that they were acting on the principles of the Enlightenment. The philosophes generally agreed with this assessment and cheered them on. Beginning in the mid-nineteenth century, histo-

Frederick the Great of Prussia Embracing the elegant, intellectual, and international culture of the Enlightenment, Frederick II was also a composer and accomplished musician. This painting shows Frederick playing the flute for family and friends with a chamber orchestra at his favorite retreat, the palace he called *Sans Souci* ("Free from Care"). *(The Mansell Collection)*

rians developed the idea of a common "enlightened despotism" or "enlightened absolutism," and they canonized Frederick, Catherine, and Joseph as its most outstanding examples. More recent research has raised doubts about this old interpretation and has led to a fundamental re-evaluation.

First, there is general agreement that these absolutists, especially Catherine and Frederick, did encourage and spread the cultural values of the Enlightenment. Perhaps this was their greatest achievement. Skeptical in religion and intensely secular in basic orientation, they unabashedly accepted the here and now and sought their happiness in the enjoyment of it. At the same time, they were proud of their intellectual accomplishments and good taste, and they sup-

ported knowledge, education, and the arts. No wonder the philosophes felt the monarchs were kindred spirits.

Historians also agree that the absolutists believed in change from above and tried to enact needed reforms. Yet the results of these efforts brought only very modest improvements, and the life of the peasantry remained very hard in the eighteenth century. Thus some historians have concluded that these monarchs were not really sincere in their reform efforts. Others disagree, arguing that powerful nobilities determined to maintain their privileges blocked the absolutists' genuine commitment to reform. (The old interpretation of Joseph II as the tragic "revolutionary emperor" forms part of this argument.)

The emerging answer to this confusion is that the later eastern absolutists were indeed committed to reform, but that humanitarian objectives were of quite secondary importance. Above all, the absolutists wanted reforms that would strengthen the state and allow them to compete militarily with their neighbors. Modern scholarship has stressed, therefore, how Catherine, Frederick, and Joseph were in many ways simply continuing the state building of their predecessors, reorganizing their armies and expanding their bureaucracies to raise more taxes and troops. The reason for this continuation was simple. The international political struggle was brutal, and the stakes were high. First Austria under Maria Theresa, and then Prussia under Frederick the Great, had to engage in bitter fighting to escape dismemberment, while decentralized Poland was coldly divided and eventually liquidated.

Yet, in their drive for more state power, the later absolutists were also innovators, and the idea of an era of enlightened absolutism retains a certain validity. Sharing the Enlightenment faith in critical thinking and believing that knowledge meant power, these absolutists really were more enlightened because they put their state-building reforms in a new, broader perspective. Above all, they considered how more humane laws and practices could help their populations become more productive and satisfied, and thus able to contribute more substantially to the welfare of the state. It was from this perspective that they introduced many of their most progressive reforms, tolerating religious minorities, simplifying legal codes, and promoting practical education.

The primacy of state as opposed to individual interests—a concept foreign to North Americans long accustomed to easy dominion over a vast continent—also helps to explain some puzzling variations in social policies. For example, Catherine the Great took measures that worsened the peasants' condition because she looked increasingly to the nobility as her natural ally and sought to strengthen it. Frederick the Great basically favored the status quo, limiting only the counter-productive excesses of his trusted nobility against its peasants. On the other hand, Joseph II believed that greater freedom for peasants was the means to strengthen his realm, and he acted accordingly. Each enlightened absolutist sought greater state power, but each believed a different policy would attain that end.

In conclusion, the eastern absolutists of the later eighteenth century combined old-fashioned state building with the culture and critical thinking of the Enlightenment. In doing so, they succeeded in expanding the role of the state in the life of society. Unlike the successors of Louis XIV, they perfected bureaucratic machines that were to prove surprisingly adaptive and capable of enduring into the twentieth century.

This chapter has focused on the complex development of a new world-view in Western civilization. This new view of the world was essentially critical and secular, drawing its inspiration from the Scientific Revolution and crystallizing in the Enlightenment.

The decisive breakthroughs in astronomy and physics in the seventeenth century, which demolished the imposing medieval synthesis of Aristotelian philosophy and Christian theology, had only limited practical consequences despite the expectations of scientific enthusiasts like Bacon. Yet the impact of new scientific knowledge on intellectual life became great. Interpreting scientific findings and Newtonian laws in an antitraditional, antireligious manner, the French philosophes of the Enlightenment extolled the superiority of rational, critical thinking. This new method, they believed, promised not just increased knowledge but even the discovery of the fundamental laws of human society. Although they reached different conclusions when they turned to social and political realities, the philosophes nevertheless succeeded in spreading their radically new world-view. That was a momentous accomplishment.

NOTES

1. H. Butterfield, *The Origins of Modern Science,* Macmillan, New York, 1951, p. viii.
2. Quoted by A. G. R. Smith, *Science and Society in the Sixteenth and Seventeenth Centuries,* Harcourt Brace Jovanovich, New York, 1972, p. 97.

3. Quoted by Butterfield, p. 47.
4. Quoted by Smith, p.100.
5. Ibid., pp. 115–116.
6. Ibid., p. 120.
7. A. R. Hall, *From Galileo to Newton, 1630–1720,* Harper & Row, New York, 1963, p. 290.
8. Quoted by R. K. Merton, *Science, Technology and Society in Seventeenth-Century England,* rev. ed., Harper & Row, New York, 1970, p. 164.
9. Quoted by P. Hazard, *The European Mind, 1680–1715,* Meridian Books, Cleveland, 1963, pp. 304–305.
10. Ibid., pp.11–12.
11. Quoted by L. M. Marsak, ed., *The Enlightenment,* John Wiley & Sons, New York, 1972, p. 56.
12. Quoted by G. L. Mosse et al., eds., *Europe in Review,* Rand McNally, Chicago, 1964, p. 156.
13. M. F. Arouet de Voltaire, *Oeuvres complètes,* Firmin-Didot Frères, Fils et Cie, Paris, 1875, VIII, 188–190.
14. Quoted by P. Gay, "The Unity of the Enlightenment," *History* 3 (1960): 25.
15. Quoted by G. P. Gooch, *Catherine the Great and Other Studies,* Archon Books, Hamden, Conn., 1966, p. 112.
16. Ibid., p. 149.
17. Quoted by L. Krieger, *Kings and Philosophers, 1689–1789,* W. W. Norton, New York, 1970, p. 257.
18. Ibid., p. 15.
19. Ibid., p. 53.
20. Quoted by R. R. Palmer, *The Age of Democratic Revolution,* Princeton University Press, Princeton, N. J., 1959, 1.95–1.96.
21. Quoted by G. Wright, *France in Modern Times,* Rand McNally, Chicago, 1960, p. 42.

SUGGESTED READING

The first three authors cited in the Notes—H. Butterfield, A. G. R. Smith, and A. R. Hall—have written excellent general interpretations of the scientific revolution. Another good study is M. Boas, *The Scientific Renaissance, 1450–1630* (1966), which is especially insightful on the influence of magic on science and on Galileo's trial. T. Kuhn, *The Copernican Revolution* (1957), is the best treatment of the subject; his *The Structure of Scientific Revolutions* (1962) is a challenging, much-discussed attempt to understand major breakthroughs in scientific thought over time. Two stimulating books on the ties between science and society in history are R. Merton, *Science, Technology and Society in Seventeenth-Century England,* rev. ed. (1970), and J. Ben-David, *The Scientist's Role in Society* (1971). E. Andrade, *Sir Isaac Newton* (1958), is a good short biography, which may be compared with F. Manuel, *The Religion of Isaac Newton* (1974).

P. Hazard, *The European Mind, 1680–1715* (1963), is a classic study of the formative years of Enlightenment thought, and his *European Thought in the Eighteenth Century* (1954) is also recommended. A famous, controversial interpretation of the Enlightenment is that of C. Becker, *The Heavenly City of the Eighteenth Century Philosophes* (1932), which maintains that the worldview of medieval Christianity continued to influence the philosophes greatly. Becker's ideas are discussed interestingly in R. O. Rockwood, ed., *Carl Becker's Heavenly City Revisited* (1958). P. Gay has written several major studies on the Enlightenment: *Voltaire's Politics* (1959) and *The Party of Humanity* (1971) are two of the best. I. Wade, *The Structure and Form of the French Enlightenment* (1977), is a recent major synthesis. F. Baumer's *Religion and the Rise of Skepticism* (1969), H. Payne's *The Philosophes and the People* (1976), K. Rogers's *Feminism in Eighteenth-Century England* (1982), and J. B. Bury's old but still exciting *The Idea of Progress* (1932) are stimulating studies of important aspects of Enlightenment thought. Above all, one should read some of the philosophes and let them speak for themselves. Two good anthologies are C. Brinton, ed., *The Portable Age of Reason* (1956), and F. Manuel, ed., *The Enlightenment* (1951). Voltaire's most famous and very amusing novel *Candide* is highly recommended, as are S. Gendzier, ed., *Denis Diderot: The Encyclopedia: Selections* (1967), and A. Wilson's biography, *Diderot* (1972).

In addition to the works mentioned in the Suggested Reading for Chapters 16 and 17, the monarchies of Europe are carefully analyzed in C. Tilly, ed., *The Formation of National States in Western Europe* (1975), and in J. Gagliardo, *Enlightened Despotism* (1967), both of which have useful bibliographies. M. Anderson, *Historians and Eighteenth-Century Europe* (1979), is a valuable introduction to recent scholarship. Other recommended studies on the struggle for power and reform in different countries are: F. Ford, *Robe and Sword* (1953), which discusses the resurgence of the French nobility after the death of Louis XIV; R. Herr, *The Eighteenth-Century Revolution in Spain* (1958), on the impact of Enlightenment thought in Spain; and P. Bernard, *Joseph II* (1968). In addition to I. de Madariaga's masterful *Russia in the Age of Catherine the Great* (1981) and D. Ransel's solid *Politics of Catherinean Russia* (1975), the ambitious reader should look at A. N. Radishchev, *A Journey from St. Petersburg to Moscow* (trans. 1958), a famous 1790 attack on Russian serfdom and an appeal to Catherine the Great to free the serfs, for which Radishchev was exiled to Siberia.

The culture of the time may be approached through A. Cobban, ed., *The Eighteenth Century* (1969), a richly illustrated work with excellent essays, and C. B. Behrens, *The Ancien Régime* (1967). C. Rosen, *The Classical Style: Haydn, Mozart, Beethoven* (1972), brilliantly synthesizes music and society, as did Mozart himself in his great opera *The Marriage of Figaro,* where the count is the buffoon and his servant the hero.

19

THE EXPANSION OF EUROPE IN THE EIGHTEENTH CENTURY

*T*HE WORLD of absolutism and aristocracy, a combination of raw power and elegant refinement, was a world apart from that of ordinary men and women. For the overwhelming majority of the population in the eighteenth century, life remained a struggle with poverty and uncertainty, with the landlord and the tax collector. In 1700 peasants on the land and artisans in their shops lived little better than had their ancestors in the Middle Ages. Only in science and thought, and there only among a few intellectual leaders, had Western society succeeded in going beyond the great achievements of the High Middle Ages, achievements that in turn owed so much to Greece and Rome.

Everyday life was a struggle because European societies, despite their best efforts, still could not produce very much by modern standards. Ordinary people might work like their beasts in the fields, and they often did, but there was seldom enough good food, warm clothing, and decent housing. Life went on; history went on. The wars of religion ravaged Germany in the seventeenth century; Russia rose to become a Great Power; the kingdom of Poland simply disappeared; monarchs and nobles continuously jockeyed for power and wealth. In 1700 or even 1750, the idea of progress—of substantial improvement in the lives of great numbers of people—was still only the dream of a small elite in fashionable salons.

Yet the economic basis of European life was beginning to change. In the course of the eighteenth century, the European economy emerged from the long crisis of the seventeenth century, responded to challenges, and began to expand once again. Some areas were more fortunate than others. The rising Atlantic powers—Holland, France, and above all England—and their colonies led the way. Agriculture and industry, trade and population, began a surge comparable to that of the eleventh- and twelfth-century springtime of European civilization. Only this time, development was not cut short. This time the response to new challenges led toward one of the most influential developments in human history, the Industrial Revolution, considered in Chapter 22. What were the causes of this renewed surge? Why were the fundamental economic underpinnings of European society beginning to change, and what were the dimensions of these changes? How did these changes affect people and their work? These are the questions this chapter will address.

AGRICULTURE AND THE LAND

At the end of the seventeenth century, the economy of Europe was agrarian, as it had been for several hundred years. With the possible exception of Holland, at least 80 percent of the people of all western European countries drew their livelihoods from agriculture. In eastern Europe the percentage was considerably higher.

Men and women lavished their attention on the land, plowing fields and sowing seed, reaping harvests and storing grain. The land repaid these efforts, year after year yielding up the food and most of the raw materials for industry that made life possible. Yet the land was stingy. Even in a rich agricultural region like the Po valley in northern Italy, every bushel of wheat sown yielded on average only five or six bushels of grain at harvest during the seventeenth century. The average French yield in the same period was somewhat less. Such yields were barely more than those attained in fertile, well-watered areas in the thirteenth century or in ancient Greece. By modern standards, output was distressingly low. (For each bushel of wheat seed sown today on fertile land with good rainfall, an American or French farmer can expect roughly forty bushels of produce.) In 1700 European agriculture was much more ancient and medieval than modern.

If the land was stingy, it was also capricious. In most regions of Europe in the sixteenth and seventeenth centuries, harvests were poor, or even failed completely, every eight or nine years. The vast majority of the population who lived off the land might survive a single bad harvest by eating less and drawing on their reserves of grain. But when the land combined with persistent bad weather—too much rain rotting the seed or drought withering the young stalks—the result was catastrophic. Meager grain reserves were soon exhausted, and the price of grain soared. Provisions from other areas with better harvests were hard to obtain.

In such crisis years, which periodically stalked Europe in the seventeenth and even into the eighteenth century, a terrible tightening knot in the belly forced people to tragic substitutes—the "famine foods" of a desperate population. People gathered chestnuts and stripped bark in the forests; they cut dandelions and grass; and they ate these substitutes to escape starvation. In one community in Norway in the early 1700s

Farming the Land Agricultural methods in Europe changed very slowly from the Middle Ages to the early eighteenth century. This realistic picture from Diderot's *Encyclopedia* has striking similarities with agricultural scenes found in medieval manuscripts. *(University of Illinois)*

people were forced to wash dung from the straw in old manure piles in order to bake a pathetic substitute for bread. Even cannibalism occurred in the seventeenth century.

Such unbalanced and inadequate food in famine years made people weak and extremely susceptible to illness and epidemics. Eating material unfit for human consumption, such as bark or grass, resulted in dysentery and intestinal ailments of every kind. Influenza and smallpox preyed with particular savagery on populations weakened by famine. In famine years, the number of deaths soared far above normal. A third of a village's population might disappear in a year or two. The 1690s were as dismal as many of the worst periods of earlier times. One county in Finland, probably typical of the entire country, lost fully 28 percent of its inhabitants in 1696 and 1697. Certain well-studied villages in the Beauvais region of northern France suffered a similar fate. In preindustrial Europe, the harvest was the real king, and the king was seldom generous and often cruel.

To understand why Europeans produced barely enough food in good years and occasionally agonized through years of famine throughout the later seventeenth century, one must follow the plowman, his wife, and his children into the fields to observe their battle for food and life. There the ingenious pattern of farming that Europe had developed in the Middle Ages, a pattern that allowed fairly large numbers of people to survive but could never produce material abundance, was still dominant.

THE OPEN-FIELD SYSTEM

The greatest accomplishment of medieval agriculture was the open-field system of village agriculture developed by European peasants (page 284). That system divided the land to be cultivated by the peasants into a few large fields, which were in turn cut up into long narrow strips. The fields were open, and the strips were not enclosed into small plots by fences or hedges. An individual peasant family—if it were fortunate—held a number of strips scattered throughout the various large fields. The land of those who owned but did not till, primarily the nobility, the clergy, and wealthy townsmen, was also in scattered strips. The peasant community farmed each large field as a community, with each family following the same pattern of plowing, sowing, and harvesting in accordance with tradition and the village leaders.

The ever-present problem was exhaustion of soil. If the community planted wheat year after year in a field, the nitrogen in the soil was soon depleted and crop failure was certain. Since the supply of manure for fertilizer was limited, the only way for the land to recover its life-giving fertility was for a field to lie fallow for a period of time. In the early Middle Ages, a year of fallow was alternated with a year of cropping, so that half the land stood idle in a given year. With time, three-year rotations were introduced, especially on more fertile lands. This system permitted a year of wheat or rye to be followed by a year of oats or beans, and only then by a year of fallow. Even so, only awareness of the tragic consequences of continuous cropping forced undernourished populations to let a third (or half) of their land lie constantly idle, especially when the fallow had to be plowed two or three times a year to keep down the weeds.

Traditional rights reinforced the traditional pattern of farming. In addition to rotating the field crops in a uniform way, villages maintained open meadows for hay and natural pasture. These lands were "common" lands, set aside primarily for the draft horses and oxen so necessary in the fields, but open to the cows and pigs of the village community as well. After the harvest, the people of the village also pastured their animals on the wheat or rye stubble. In many places such pasturing followed a brief period, also established by tradition, for the gleaning of grain. Poor women would go through the fields picking up the few single grains that had fallen to the ground in the course of the harvest. The subject of a great nineteenth-century painting, *The Gleaners* by Jean François Millet, this backbreaking work by hardworking but impoverished women meant quite literally the slender margin of survival for some people in the winter months.

In the age of absolutism and nobility, state and landlord continued to levy heavy taxes and high rents as a matter of course. In so doing they stripped the peasants of much of their meager earnings. The level of exploitation varied. Conditions for the rural population were very different in different areas.

Generally speaking, the peasants of eastern Europe were worst off. As we have seen in Chapter 17, they were still serfs, bound to their lords in hereditary service. Though serfdom in eastern Europe in the eighteenth century had much in common with medieval serfdom in central and western Europe, it was, if anything, harsher and more oppressive. In much of

Millet: The Gleaners Poor French peasant women search for grains and stalks the harvesters (in the background) have missed. The open-field system seen here could still be found in parts of Europe in 1857, when this picture was painted. Millet is known for his great paintings expressing social themes. *(Cliché des Musées Nationaux, Paris)*

eastern Europe there were few limitations on the amount of forced labor the lord could require, and five or six days of unpaid work per week on the lord's land was not uncommon. Well into the nineteenth century, individual Russian serfs and serf families were regularly sold with and without land. Serfdom was often very close to slavery. The only compensating factor in much of eastern Europe was that, as with slavery, differences in well-being among serfs were slight. In Russia, for example, the land available to the serfs for their own crops was divided among them almost equally.

Social conditions were considerably better in western Europe. Peasants were generally free from serfdom. In France and western Germany, they owned land and could pass it on to their children. Yet life in the village was unquestionably hard, and poverty was the great reality for most people. For the Beauvais region of France at the beginning of the eighteenth century, it has been carefully estimated that in good years and bad only a tenth of the peasants could live satisfactorily off the fruits of their landholdings. Owning less than half of the land, the peasants had to pay heavy royal taxes, the church's tithe, and dues to the lord, as well as set aside seed for the next season. Left with only half of their crop for their own use, they had to toil and till for others and seek work far afield in a constant scramble for a meager living. And this was in a country where peasants were comparatively well off. The privileges of the ruling elites weighed heavily on the people of the land.

AGRICULTURAL REVOLUTION

The social conditions of the countryside were well entrenched. The great need was for new farming methods that would enable Europeans to produce more and eat more. The idle fields were the heart of the matter. If peasants could replace the fallow with crops, they could increase the land under cultivation by 50 percent. So remarkable were the possibilities and the results that historians have often spoken of the progressive elimination of the fallow, which occurred slowly throughout Europe from the late seventeenth century on, as an agricultural revolution.

This agricultural revolution, which took longer than historians used to believe, was a great milestone in human development. The famous French scholar Marc Bloch, who gave his life in the resistance to the Nazis in World War Two, summed it up well: "The history of the conquest of the fallow by new crops, a fresh triumph of man over the earth that is just as moving as the great land clearing of the Middle Ages, [is] one of the noblest stories that can be told."[1]

Because grain crops exhaust the soil and make fallowing necessary, the secret to eliminating the fallow lies in alternating grain with certain nitrogen-storing crops. Such crops not only rejuvenate the soil even better than fallowing, but give more produce as well. The most important of these land-reviving crops are peas and beans, root crops such as turnips and potatoes, and clovers and grasses. In the eighteenth century, peas and beans were old standbys; turnips, potatoes, and clover were newcomers to the fields. As time went on, the number of crops that were systematically rotated grew, and farmers developed increas-

Enclosing the Fields This remarkable aerial photograph captures key aspects of the agricultural revolution. Though the long ridges and furrows of the old open-field system still stretch across the whole picture, hedge rows now cut through the long strips to divide the land into several enclosed fields. *(Cambridge University Collection)*

ingly sophisticated patterns of rotation to suit different kinds of soils. For example, farmers in French Flanders near Lille in the late eighteenth century used a ten-year rotation, alternating a number of grain, root, and hay crops on a ten-year schedule. Continuous experimentation resulted in more scientific farming.

Improvements in farming had multiple effects. The new crops made ideal feed for animals. Because peasants and larger farmers had more fodder—hay and root crops—for the winter months, they could build up their small herds of cattle and sheep. More animals meant more meat and better diets for the people. More animals also meant more manure for fertilizer, and therefore more grain for bread and porridge. The vicious cycle in which few animals meant inadequate manure, which meant little grain and less fodder, which led to fewer animals, and so on, could be broken. The cycle became positive: more animals meant more manure, which meant more grain and more fodder, which meant more animals, which meant better diets.

Technical progress had its price, though. The new rotations were scarcely possible within the traditional framework of open fields and common rights. A farmer who wanted to experiment with new methods would have to control the village's pattern of rotation. To wait for the entire village to agree might mean waiting forever. The improving, innovating agriculturalist needed to enclose and consolidate his scattered holdings into a compact, fenced-in field. In doing so, he would also seek to enclose his share of the natural pasture, the common. Yet the common rights were precious to many rural people. Thus when the small landholders and the poor could effectively oppose the enclosure of the open fields, they did so. Only powerful social and political pressures could overcome the traditionalism of rural communities.

The old system of unenclosed open fields and the new system of continuous rotation coexisted in Europe for a very long time. In large parts of central Russia, for example, the old system did not disappear until after the Communist Revolution in 1917. It could also be found in much of France and Germany in the early years of the nineteenth century. Indeed, until the end of the eighteenth century, the promise of the new system was extensively realized only in the Low Countries and in England.

THE LEADERSHIP OF THE LOW COUNTRIES AND ENGLAND

The new methods of the agriculture revolution originated in the Low Countries. The vibrant, dynamic middle-class society of seventeenth-century republican Holland was the most advanced in Europe in many areas of human endeavor. In shipbuilding and navigation, in commerce and banking, in drainage and agriculture, the people of the Low Countries, especially the Dutch, provided models the jealous English and French sought to copy or to cripple.

By the middle of the seventeenth century, intensive farming was well established throughout much of the Low Countries. Enclosed fields, continuous rotation, heavy manuring, and a wide variety of crops—all these innovations were present. Agriculture was highly specialized and commercialized. The same skills that grew turnips produced flax to be spun into linen for clothes and tulip bulbs to lighten the heart with their beauty. The fat cattle of Holland, so beloved by Dutch painters, gave the most milk in Europe. Dutch cheeses were already world renowned.

The reasons for early Dutch leadership in farming were basically threefold. In the first place, since the end of the Middle Ages the Low Countries had been one of the most densely populated areas in Europe. Thus, in order to feed themselves and provide employment, the Dutch were forced at an early date to seek maximum yields from their land and to increase it through the steady draining of marshes and swamps. Even so, they had to import wheat from Poland and eastern Germany.

The pressure of population was connected with the second cause, the growth of towns and cities in the Low Countries. Stimulated by commerce and overseas trade, Amsterdam grew from 30,000 to 200,000 in its golden seventeenth century. The growth of urban population provided Dutch peasants with good markets for all they could produce and allowed each region to specialize in what it did best.

Finally, there was the quality of the people. Oppressed by neither grasping nobles nor war-minded monarchs, the Dutch could develop their potential in a free and capitalistic society. The Low Countries became "the Mecca of foreign agricultural experts who came . . . to see Flemish agriculture with their own eyes, to write about it and to propagate its methods in their home lands."[2]

Harvesting This harvest scene by Pieter Brueghel the Elder (ca 1525–1569) suggests how the relatively prosperous Low Countries might take the lead in agricultural development. Brueghel was a master at depicting everyday peasant life. *(Nelson Gallery–Atkins Museum, Kansas City, Mo., Nelson Fund)*

The English were the best students. Indeed, they were such good students that it is often forgotten that they had teachers at all. Drainage and water control was one subject in which they received instruction. Large parts of seventeenth-century Holland had once been sea and sea marsh, and the efforts of centuries had made the Dutch the world's leaders in the skills of drainage. In the first half of the seventeenth century, Dutch experts made a great contribution to draining the extensive marshes, or fens, of wet and rainy England.

The most famous of these Dutch engineers, Cornelius Vermuyden, directed one large drainage project in Yorkshire and another in Cambridgeshire. The project in Yorkshire was supported by Charles I and financed by a group of Dutch capitalists, who were to receive one-third of all land reclaimed in return for

their investment. Despite local opposition, Vermuyden drained the land by means of a large canal—his so-called Dutch river—and settlers cultivated the new fields in the Dutch fashion. In the Cambridge fens, Vermuyden and his Dutch workers eventually reclaimed forty thousand acres, which were then farmed intensively in the Dutch manner. Although all these efforts were disrupted in the turbulent 1640s by the English civil war, Vermuyden and his countrymen largely succeeded. A swampy wilderness was converted into thousands of acres of some of the best land in England. On such new land, where traditions and common rights were not established, farmers introduced new crops and new rotations fairly easily.

Dutch experience was also important to Viscount Charles Townsend (1674–1738), one of the pioneers of English agricultural improvement. This lord from

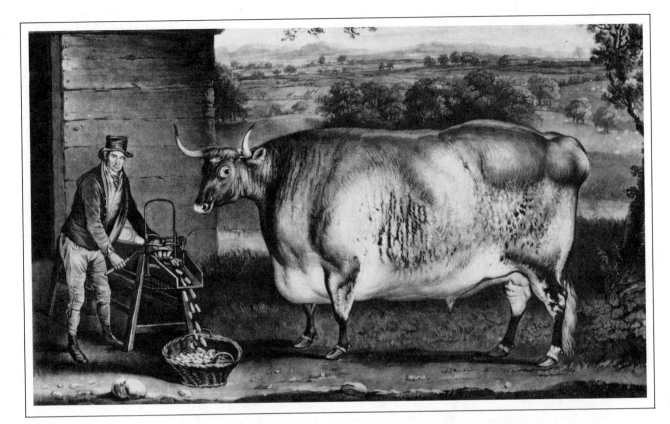

Selective Breeding meant bigger livestock and more meat on English tables. This gigantic champion, one of the new improved shorthorn breed, was known as the New-bus Ox. Such great fat beasts were pictured in the press and praised by poets. *(Institute of Agricultural History and Museum of English Rural Life, University of Reading)*

the upper reaches of the English aristocracy learned about turnips and clover while serving as English ambassador to Holland. In the 1710s, he was using these crops in the sandy soil of his large estates in Norfolk in eastern England, already one of the most innovative agricultural areas in the country. When Lord Charles retired from politics in 1730 and returned to Norfolk, it was said that he spoke of turnips, turnips, and nothing but turnips. This led some wit to nickname his lordship "Turnip" Townsend. But Townsend had the last laugh. Draining extensively, manuring heavily, and sowing crops in regular rotation without fallowing, the farmers who leased Townsend's lands produced larger crops. They and he earned higher incomes. Those who had scoffed reconsidered. By 1740 agricultural improvement in various forms had become something of a craze among the English aristocracy.

Jethro Tull (1674–1741), part crank and part genius, was another important English innovator. A true son of the early Enlightenment, Tull constantly tested accepted ideas about farming in an effort to develop better methods through empirical research. He was especially enthusiastic about using horses for plowing, in preference to slower-moving oxen. He also advocated sowing seed with drilling equipment, rather than scattering it by hand. Drilling distributed seed evenly and at the proper depth. There were also improvements in livestock, inspired in part by the earlier successes of English country gentlemen in breeding ever-faster horses for the races and fox hunts that were their passions. Selective breeding of ordinary livestock was a marked improvement over the old pattern, which has been graphically described as little more than "the haphazard union of nobody's son with everybody's daughter."

By the mid-eighteenth century, English agriculture was in the process of a radical and desirable transformation. The eventual result was that by 1870 English farmers produced 300 percent more food than they had produced in 1700, although the number of people working the land had increased by only 14 percent. This great surge of agricultural production provided food for England's rapidly growing urban population. It was a tremendous achievement.

THE DEBATE OVER ENCLOSURE

To what extent was technical progress a product of social injustice? There are sharp differences of opinion among historians. The oldest and still widely accepted view is that the powerful ruling class, the English landowning aristocracy, enclosed the open fields and divided up the common pasture in such a way that poor people lost their small landholdings and were pushed off the land. The large landowners controlled Parliament, which made the laws. They had Parliament pass hundreds of "enclosure acts," each of which authorized the fencing of open fields in a given district and abolished common rights there. Small farmers who had little land and cottagers who had only common rights could no longer make a living. They lost position and security and had to work for a large landowner for wages or else move to town in search of work. This view, popularized by Karl Marx in the nineteenth century, has remained dear to many historians to this day.

There is some validity to this idea, but more recent studies have shown that the harmful consequences of enclosure in the eighteenth century have often been exaggerated. In the first place, as much as half of English farmland was already enclosed by 1750. A great wave of enclosure of English open fields into sheep pastures had already occurred in the sixteenth and early seventeenth centuries, in order to produce wool for the thriving textile industry. In the later seventeenth and early eighteenth centuries, many open fields were enclosed fairly harmoniously by mutual agreement among all classes of landowners in English villages. Thus parliamentary enclosure, the great bulk of which occurred after 1760 and particularly during the Napoleonic wars early in the nineteenth century, only completed a process that was in full swing. Nor did an army of landless farm laborers appear only in the last years of the eighteenth century.

Much earlier, and certainly by 1700, there were perhaps two landless agricultural workers in England for every self-sufficient farmer. In 1830, after the enclosures were complete, the proportion of landless laborers on the land was not much greater.

Indeed, by 1700 a highly distinctive pattern of landownership existed in England. At one extreme were a few large landowners, at the other a large mass of laborers who held little land and worked for wages. In between stood two other groups: small, self-sufficient farmers who owned their own land and substantial tenant farmers who rented land from the big landowners and hired wage laborers. Yet the small, independent English farmers were already declining in number by 1700, and they continued to do so in the eighteenth century. They could not compete with the profit-minded, market-oriented tenant farmers.

The tenant farmers, many of whom had formerly been independent owners, were the key to mastering the new methods of farming. Well financed by the large landowners, the tenant farmers fenced fields, built drains, and improved the soil with fertilizers. Such improvements actually increased employment opportunities for wage workers in the countryside. So did new methods of farming, for land was farmed more intensively without the fallow, and new crops like turnips required more care and effort. Thus enclosure did not force people off the land by eliminating jobs. By the early nineteenth century, rural poverty was often greatest in those areas of England where the new farming techniques had not been adopted.

THE BEGINNING OF THE POPULATION EXPLOSION

There was another factor that affected the existing order of life and forced economic changes in the eighteenth century. This was the remarkable growth of European population, the beginning of the "population explosion." This population explosion continued in Europe until the twentieth century, by which time it was affecting non-Western areas of the globe. What caused the growth of population, and what did the challenge of more mouths to feed and more hands to employ do to the European economy?

LIMITATIONS ON POPULATION GROWTH

Many commonly held ideas about population in the past are wrong. One such mistaken idea is that people always married young and had large families. A related error is the belief that past societies were so ignorant that they could do nothing to control their numbers and that population was always growing too fast. On the contrary, until 1700 the total population of Europe grew slowly much of the time, and by no means constantly (see Figure 19.1). There were very few occurrences of the frightening increases found in many poor countries today.

In seventeenth-century Europe, births and deaths, fertility and mortality, were in a crude but effective balance. The birthrate—annual births as a proportion of the population—was fairly high, but far lower than it would have been if all women between ages fifteen and forty-five had been having as many children as biologically possible. The death rate in normal years was also high, though somewhat lower than the birthrate. As a result, the population grew modestly in normal years at a rate of perhaps .5 to 1 percent, or enough to double the population in 70 to 140 years. This is, of course, a generalization encompassing many different patterns. In areas like Russia and colonial New England, where there was a great deal of frontier to be settled, the annual rate of increase might well exceed 1 percent. In a country like France, where the land had long been densely settled, the rate of increase might be less than .5 percent.

Although population growth of even 1 percent per year is fairly modest by the standards of many African and Latin American countries today—some of which are growing at about 3 percent annually—it will produce a very large increase over a long period. An annual increase of even 1 percent will result in sixteen times as many people in three hundred years. Such gigantic increases simply did not occur in agrarian Europe before the eighteenth century. In certain abnormal years and tragic periods, many more people died than were born. Total population fell sharply, even catastrophically. A number of years of modest growth would then be necessary to make up for those who had died in an abnormal year. Such savage increases in deaths helped check total numbers and kept the population from growing rapidly for long periods.

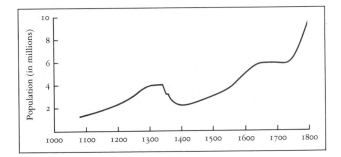

FIGURE 19.1 The Growth of Population in England 1000–1800 England is a good example of both the uneven increase of European population before 1700 and the third great surge of growth, which began in the eighteenth century. *(Source: E. A. Wrigley,* Population and History, *McGraw-Hill, New York, 1969)*

The grim reapers of demographic crisis were famine, epidemic disease, and war. Famine, the inevitable result of poor farming methods and periodic crop failures, was particularly murderous because it was accompanied by disease. With a brutal one-two punch, famine stunned and weakened a population, and disease finished it off. Disease could also ravage independently, even in years of adequate harvests. Bubonic plague returned again and again in Europe for more than three hundred years after the ravages of the Black Death in the fourteenth century. Not until the late 1500s did most countries have as many people as in the early 1300s. Epidemics of dysentery and smallpox also operated independently of famine.

War was another scourge. The indirect effects were more harmful than the organized killing. War spread disease. Soldiers and camp followers passed venereal disease through the countryside to scar and kill. Armies requisitioned scarce food supplies for their own use and disrupted the agricultural cycle. The Thirty Years' War (pages 483–489) witnessed all possible combinations of distress. In the German states, the number of inhabitants declined by more than *two-thirds* in some large areas and by at least one-third almost everywhere else. The Thirty Years' War reduced total German population by no less than 40 percent. But numbers inadequately convey the dimensions of such human tragedy. One needs the vision of the artist. The great sixteenth-century artist, Albrecht Dürer, captured the horror of demographic crisis in his chilling woodcut *The Four Horsemen of*

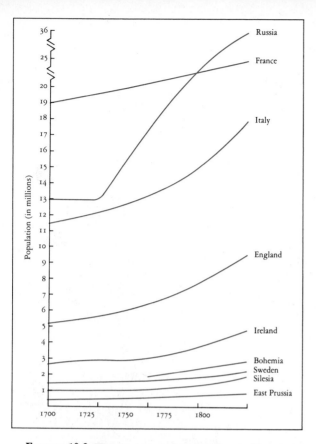

FIGURE 19.2 **The Increase of Population in Europe in the Eighteenth Century** France's large population continued to support French political and intellectual leadership. Russia emerged as Europe's most populous state because natural increase was complemented by growth from territorial expansion.

the *Apocalypse* (page 384). Death, accompanied by his trusty companions War, Famine, and Disease, takes his merciless ride of destruction. The narrow victory of life over death that prevails in normal times is being undone.

THE NEW PATTERN OF THE EIGHTEENTH CENTURY

In the eighteenth century, the population of Europe began to grow markedly. This increase in numbers occurred in all areas of Europe—western and eastern, northern and southern, dynamic and stagnant. Growth was especially dramatic after about 1750, as Figure 19.2 shows.

Although it is certain that Europe's population grew greatly, it is less clear why. Recent painstaking and innovative research in population history has shown that, because population grew everywhere, it

is best to look for general factors and not those limited to individual countries or areas. What, then, caused fewer people to die or, possibly, more babies to be born? In some kinds of families women may have had more babies than before. Yet the basic cause was a decline in mortality—fewer deaths.

The bubonic plague mysteriously disappeared. Following the Black Death in the fourteenth century, plagues had remained a part of the European experience, striking again and again with savage force, particularly in towns. As a German writer of the early sixteenth century noted, "It is remarkable and astonishing that the plague should never wholly cease, but it should appear every year here and there, making its way from one place to another. Having subsided at one time, it returns within a few years by a circuitous route."[3]

As late as 1720, a ship from Syria and the Levant, where plague was ever-present, brought the monstrous disease to Marseilles. In a few weeks, forty thousand of the city's ninety thousand inhabitants died. The epidemic swept southern France, killing one-third, one-half, even three-fourths of those in the larger towns. Once again an awful fear swept across Europe. But the epidemic passed, and that was the last time plague fell on western and central Europe. The final disappearance of plague was due in part to stricter measures of quarantine in Mediterranean ports and along the Austrian border with Turkey. Human carriers of plague were carefully isolated. Chance and plain good luck were more important, however.

It is now understood that bubonic plague is, above all, a disease of rats. More precisely, it is the black rat that spreads major epidemics, for the black rat's flea is the principal carrier of the plague bacillus. After 1600, for reasons unknown, a new rat of Asiatic origin—the brown, or wander, rat—began to drive out and eventually eliminate its black competitor. In the words of a noted authority, "This revolution in the animal kingdom must have gone far to break the lethal link between rat and man."[4] Although the brown rat also contracts the plague, another kind of flea is its main parasite. That flea carries the plague poorly and, for good measure, has little taste for human blood.

Advances in medical knowledge did not contribute much to reducing the death rate in the eighteenth century. The most important advance in preventive medicine in this period was inoculation against smallpox. Yet this great improvement was long confined mainly to England and probably did little to reduce deaths throughout Europe until the latter part of the century. Improvements in the water supply and sewerage promoted somewhat better public health and helped reduce such diseases as typhoid and typhus in some urban areas of western Europe. Yet those early public-health measures had only limited general significance. In fact, changes in the rat population helped much more than did doctors and medical science in the eighteenth century.

Human beings were more successful in their efforts to safeguard the supply of food and protect against famine. The eighteenth century was a time of considerable canal and road building in western Europe. These advances in transportation, which were among the more positive aspects of strong absolutist states, lessened the impact of local crop failure and famine. Emergency supplies could be brought in. The age-old spectacle of localized starvation became less frequent. Wars became more gentlemanly and less destructive than in the seventeenth century and spread fewer epidemics. New foods, particularly the potato, were introduced. Potatoes served as an important alternative source of vitamins A and C for the poor, especially when the grain crops were skimpy or failed. In short, population grew in the eighteenth century primarily because years of abnormal death rates were less catastrophic. Famines, epidemics, and wars continued to occur, but their severity moderated.

The growth of population in the eighteenth century cannot be interpreted as a sign of human progress, however. Plague faded from memory, transport improved, people learned to eat potatoes; yet for the common people, life was still a great struggle. Indeed, it was often more of a struggle than ever, for in many areas increasing numbers led to overpopulation. A serious imbalance between the number of people and the economic opportunities available to them developed. There was only so much land available, and tradition slowed the adoption of better farming methods. Therefore agriculture could not provide enough work for the rapidly growing labor force. Everyone might work steadily during planting and harvesting, when many hands were needed, but at other times rural people were often unemployed or underemployed.

Doctor in Protective Clothing Most doctors believed, incorrectly, that poisonous smells carried the plague. This doctor has placed strong-smelling salts in his "beak" to protect himself against deadly plague vapors. *(Germanisches Nationalmuseum, Nuremberg)*

Women Working This mother and her daughters may well be knitting, lace-making, and spinning for some merchant capitalist. The close ties between cottage industry and agriculture are well illustrated in this summer scene. *(University of Illinois, Champaign)*

Growing numbers increased the challenge of poverty, especially the severe poverty of the rural poor. People in the countryside had to look for new ways to make a living. Even if work outside of farming paid poorly, small wages were better than none. Thus in the eighteenth century, growing numbers of people and acute poverty were even more influential than new farming methods as forces for profound change in agrarian Europe.

THE GROWTH OF COTTAGE INDUSTRY

The growth of population contributed to the development of industry in rural areas. The poor in the countryside were eager to supplement their earnings from agriculture with other types of work, and capitalists from the city were eager to employ them, often at lower wages than urban workers commanded. Manufacturing with hand tools in peasant cottages grew markedly in the eighteenth century. Rural industry became a crucial feature of the European economy.

To be sure, peasant communities had always made some clothing, processed some food, and constructed some housing for their own use. But in the High Middle Ages, peasants did not produce manufactured goods on a large scale for sale in a market; they were not handicraft workers as well as farmers and field laborers. Industry in the Middle Ages was dominated and organized by urban craft guilds and urban merchants, who jealously regulated handicraft production and sought to maintain it as an urban monopoly. By the eighteenth century, however, the pressures of rural poverty and the need for employment in the countryside had proved too great, and a new system

The Linen Industry in Ireland Many steps went into making textiles. Here the women are beating away the coarse woody part of the flax plant so that the man can draw the soft part through a series of combs. The fine flax fibers will then be spun into thread and woven into cloth by this family enterprise. *(The Mansell Collection)*

was expanding lustily. The new system had many names. Sometimes referred to as "cottage industry" or "domestic industry," it has often been called the "putting-out system."

THE PUTTING-OUT SYSTEM

The two main participants in the putting-out system were the merchant-capitalist and the rural worker. The merchant loaned or "put out" raw materials—raw wool, for example—to several cottage workers. Those workers processed the raw material in their own homes, spinning and weaving the wool into cloth in this case, and returned the cloth to the merchant. The merchant paid the outworkers for their work by the piece and proceeded to sell the finished product. There were endless variations on this basic relationship. Sometimes rural workers would buy their own materials and work as independent pro-

ducers before they sold to the merchant. The relative importance of earnings from the land and from industry varied greatly for handicraft workers. In all cases, however, the putting-out system was a kind of capitalism. Merchants needed large amounts of capital, which they held in the form of goods being worked up and sold in distant markets. They sought to make profits and increase their capital in their businesses.

The putting-out system was not perfect, but it had definite advantages. It increased employment in the countryside and provided the poor with additional income. Since production in the countryside was unregulated, workers and merchants could change procedures and experiment as they saw fit. Because they did not need to meet rigid guild standards, which maintained quality but discouraged the development of new methods, cottage industry became capable of producing many kinds of goods. Textiles, all

Rural Industry in Action This French engraving suggests just how many things could be made in the countryside with simple hand tools. These men are making inexpensive but long-lasting wooden shoes, which were widely worn by the poor. *(University of Illinois)*

manner of knives, forks, and housewares, buttons and gloves, clocks and musical instruments could be produced quite satisfactorily in the countryside. Luxury goods for the rich, such as exquisite tapestries and fine porcelain, demanded special training, close supervision, and centralized workshops. Yet such goods were as exceptional as those who used them. The skills of rural industry were sufficient for everyday articles.

Rural manufacturing did not spread across Europe at an even rate. It appeared first in England and developed most successfully there, particularly for the spinning and weaving of woolen cloth. By 1500 half of England's textiles were being produced in the countryside. By 1700 English industry was generally more rural than urban and heavily reliant on the putting-out system. Continental countries developed rural industry more slowly.

In France at the time of Louis XIV, Colbert had revived the urban guilds and used them as a means to control the cities and collect taxes (page 515). But the pressure of rural poverty proved too great. In 1762 the special privileges of urban manufacturing were

abolished in France, and the already-developing rural industries were given free rein from then on. The royal government in France had come to believe that the best way to help the poor peasants was to encourage the growth of cottage manufacturing. Thus in France, as in Germany and other areas, the later part of the eighteenth century witnessed a remarkable expansion of rural industry in certain densely populated regions. The pattern established in England was spreading to the Continent.

THE TEXTILE INDUSTRY

Throughout most of history, until at least the nineteenth century, the industry that has employed the most people has been textiles. The making of linen, woolen, and eventually cotton cloth was the typical activity of cottage workers engaged in the putting-out system. A look inside the cottage of the English rural textile worker illustrates a way of life as well as an economic system.

The rural worker lived in a small cottage, with tiny windows and little space. Indeed, the worker's cottage was often a single room that served as workshop, kitchen, and bedroom. There were only a few pieces of furniture, of which the weaver's loom was by far the largest and most important. That loom had changed somewhat in the early eighteenth century, when John Kay's invention of the flying shuttle enabled the weaver to throw the shuttle back and forth between the threads with one hand. Aside from that improvement, however, the loom was as it had been for much of history. In the cottage there were also spinning wheels, tubs for dyeing cloth and washing raw wool, and carding pieces to comb and prepare the raw material.

These different pieces of equipment were necessary because cottage industry was first and foremost a family enterprise. All the members of the family helped in the work, so that "every person from seven to eighty (who retained their sight and who could move their hands) could earn their bread," as one eighteenth-century English observer put it.[5] While the women and children prepared the raw material and spun the thread, the man of the house wove the cloth. There was work for everyone, even the youngest. After the dirt was beaten out of the raw cotton, it had to be thoroughly cleaned with strong soap in a tub, where tiny feet took the place of the agitator in a washing machine. George Crompton, the son of Samuel Crompton, who in 1784 invented the mule for cotton spinning, recalled that "soon after I was able to walk I was employed in the cotton manufacture.... My mother tucked up my petticoats about my waist, and put me into the tub to tread upon the cotton at the bottom."[6] Slightly older children and aged relatives carded and combed the cotton or wool, so that the woman and the older daughter she had taught could spin it into thread. Each member had a task. The very young and very old worked in the family unit as a matter of course.

There was always a serious imbalance in this family enterprise: the work of four or five spinners was needed to keep one weaver steadily employed. Therefore, the wife and the husband had constantly to try to find more thread and more spinners. Widows and unmarried women—those "spinsters" who spun for their living—were recruited by the wife. Or perhaps the weaver's son went off on horseback to seek thread. The need for more thread might even lead the weaver and his wife to become small capitalist employers. At the end of the week, when they received the raw wool or cotton from the merchant-manufacturer, they would put out some of this raw material to other cottages. The following week they would return to pick up the thread and pay for the spinning—spinning that would help keep the weaver busy for a week until the merchant came for the finished cloth.

Relations between workers and employers were not always harmonious. In fact, there was continuous conflict. An English popular song written about 1700, called "The Clothier's Delight, or the Rich Men's Joy and The Poor Men's Sorrow," has the merchant boasting of his countless tricks used to "beat down wages":

We heapeth up riches and treasure great store
Which we get by griping and grinding the poor.
And this is a way for to fill up our purse
Although we do get it with many a curse.[7]

There were constant disputes over weights of materials and the quality of the cloth. Merchants accused workers of stealing raw materials, and weavers complained that merchants delivered underweight bales. Both were right; each tried to cheat the other, even if only in self-defense.

There was another problem, at least from the merchant-capitalist's point of view. Rural labor was cheap, scattered, and poorly organized. For these reasons it was hard to control. Cottage workers tended to work in spurts. After they got paid on Saturday afternoon, the men in particular tended to drink and carouse for two or three days. Indeed, Monday was called "holy Monday" because inactivity was so religiously observed. By the end of the week the weaver was probably working feverishly to make his quota. But if he did not succeed, there was little the merchant could do. When times were good and the merchant could easily sell everything produced, the weaver and his family did fairly well and were particularly inclined to loaf, to the dismay of the capitalist. Thus, in spite of its virtues, the putting-out system in the textile industry had definite shortcomings. There was an imbalance between spinning and weaving. Labor relations were often poor, and the merchant was unable to control the quality of the cloth or the schedule of the workers. The merchant-capitalist's search for more efficient methods of production became intense.

BUILDING THE ATLANTIC ECONOMY

In addition to agricultural improvement, population pressure, and expanding cottage industry, the expansion of Europe in the eighteenth century was characterized by the growth of world trade. Spain and Portugal revitalized their empires and began drawing more wealth from renewed development. Yet once again, the countries of northwestern Europe—the Netherlands, France, and above all Great Britain—benefited most. Great Britain (formed in 1707 by the union of England and Scotland in a single kingdom), gradually became the leading maritime power. In the eighteenth century, British ships and merchants succeeded in dominating long-distance trade, particularly the fast-growing intercontinental trade across the Atlantic Ocean. The British played the critical role in building a fairly unified Atlantic economy, which offered remarkable opportunities for them and their colonists.

MERCANTILISM AND COLONIAL WARS

Britain's commercial leadership in the eighteenth century had its origins in the mercantilism of the seventeenth century (page 515). European mercantilism was a system of economic regulations aimed at increasing the power of the state. As practiced by a leading advocate like Colbert under Louis XIV, mercantilism aimed particularly at creating a favorable balance of foreign trade in order to increase a country's stock of gold. A country's gold holdings served as an all-important treasure chest, to be opened periodically to pay for war in a violent age.

Early English mercantilists shared these views. As Thomas Mun, a leading merchant and early mercantilist, wrote in *England's Treasure by Foreign Trade* (1630, published 1664): "The ordinary means therefore to increase our wealth and treasure is by foreign trade wherein we must observe this rule; to sell more to strangers yearly than we consume of theirs in value." What distinguished English mercantilism was the unusual idea that governmental economic regulations could and should serve the private interest of individuals and groups as well as the public

needs of the state. As Josiah Child, a very wealthy brewer and director of the East India Company, put it, in the ideal economy "Profit and Power ought jointly to be considered."[8]

In France and other continental countries, by contrast, seventeenth-century mercantilists generally put the needs of the state far above those of businessmen and workers. And they seldom saw a possible union of public and private interests for a common good.

The result of the English desire to increase both its military power and private wealth was the mercantile system of the Navigation Acts. Oliver Cromwell established the first of these laws in 1651, and the restored monarchy of Charles II extended them further in 1660 and 1663; the Navigation Acts of the seventeenth century were not seriously modified until 1786. The acts required that most goods imported from Europe into England and Scotland be carried on British-owned ships with British crews or on ships of the country producing the article. Moreover, these laws gave British merchants and shipowners a virtual monopoly on trade with the colonies. The colonists were required to ship their products—sugar, tobacco, and cotton—on British ships and to buy almost all of their European goods from the mother country. It was believed that these economic regulations would provide British merchants and workers with profits and employment, and colonial plantation owners and farmers with a guaranteed market for their products. And the state would develop a shipping industry with a large number of tough, experienced deepwater seamen, who could be drafted when necessary into the Royal Navy to protect the island nation.

The Navigation Acts were a form of economic warfare. Their initial target was the Dutch, who were far ahead of the English in shipping and foreign trade in the mid-seventeenth century. The Navigation Acts, in conjunction with three Anglo-Dutch wars between 1652 and 1674, did seriously damage Dutch shipping and commerce. The thriving Dutch colony of New Amsterdam was seized in 1664 and rechristened "New York." By the later seventeenth century, when the Dutch and the English became allies to stop the expansion of France's Louis XIV, the Netherlands was falling behind England in shipping, trade, and colonies.

As the Netherlands followed Spain into relative decline, France stood clearly as England's most serious

1651–1663	British Navigation Acts create the mercantile system, which is not seriously modified until 1786
1652–1674	Three Anglo-Dutch wars damage Dutch shipping and commerce
1664	New Amsterdam is seized and renamed New York
1701–1714	War of the Spanish Succession
1713	Peace of Utrecht: Britain wins parts of Canada from France and control of the western African slave trade from Spain
1740–1748	War of the Austrian Succession, resulting in no change in territorial holdings in North America
1756–1763	Seven Years' War (known in North America as the French and Indian War), a decisive victory for Britain
1763	Treaty of Paris: Britain receives all French territory on the North American mainland and achieves dominance in India

rival in the competition for overseas empire. Rich in natural resources and endowed with a population three or four times that of England, continental Europe's leading military power was already building a powerful fleet and a worldwide system of rigidly monopolized colonial trade. And France, aware that Great Britain coveted large parts of Spain's American empire, was determined to revitalize its Spanish ally. Thus from 1701 to 1763, Britain and France were locked in a series of wars to decide, in part, which nation would become the leading maritime power and claim a lion's share of the profits of Europe's overseas expansion (see Map 19.1).

The first round was the War of the Spanish Succession (page 520), which started when Louis XIV declared his willingness to accept the Spanish crown willed to his grandson. Besides upsetting the continental balance of power, a union of France and Spain threatened to destroy the British colonies in North America. The thin ribbon of British settlements along the Atlantic seaboard from Massachusetts to the Carolinas would be surrounded by a great arc of Franco-Spanish power stretching south and west from French Canada to Florida and the Gulf of Mexico (see Map 19.1). Defeated by a great coalition of states after twelve years of fighting, Louis XIV was forced in the Peace of Utrecht (1713) to cede Newfoundland, Nova Scotia, and the Hudson Bay territory to Britain. Spain was compelled to give Britain

control of the lucrative West African slave trade—the so-called *asiento*—and to let Britain send one ship of merchandise into the Spanish colonies annually, through Porto Bello on the Isthmus of Panama.

France was still a mighty competitor. In 1740 the War of the Austrian Succession (1740–1748), which started when Frederick the Great of Prussia seized Silesia from Austria's Maria Theresa (page 594), became a world war, including Anglo-French conflicts in India and North America. Indeed, it was the seizure of French territory in Canada by New England colonists that led France to sue for peace in 1748 and to accept a return to the territorial situation existing in North America at the beginning of the war. France's Bourbon ally, Spain, defended itself surprisingly well, and Spain's empire remained intact.

This inclusive stand-off helped set the stage for the Seven Years' War (1756–1763). In central Europe, Austria's Maria Theresa sought to win back Silesia and crush Prussia, thereby re-establishing the Habsburgs' traditional leadership in German affairs. She almost succeeded (see page 595), skillfully winning both France—the Habsburgs' long-standing enemy —and Russia to her cause. Yet the Prussian state survived, saved by its army and the sudden decision of Russia to withdraw from the war in 1762.

Outside of Europe, the Seven Years' War was the decisive round in the Franco-British competition for colonial empire (see Map 19.2). Led by William Pitt,

The East India Dock, London This painting by Samuel Scott captures the spirit and excitement of British maritime expansion. Great sailing ships line the quay, bringing profit and romance from far-off India. London grew in population from 350,000 in 1650 to 900,000 in 1800, when it was twice as big as Paris, its nearest rival. *(Victoria & Albert Museum, London)*

MAP 19.1 The Economy of the Atlantic Basin in 1701 The growth of trade encouraged both economic development and military conflict in the Atlantic Basin.

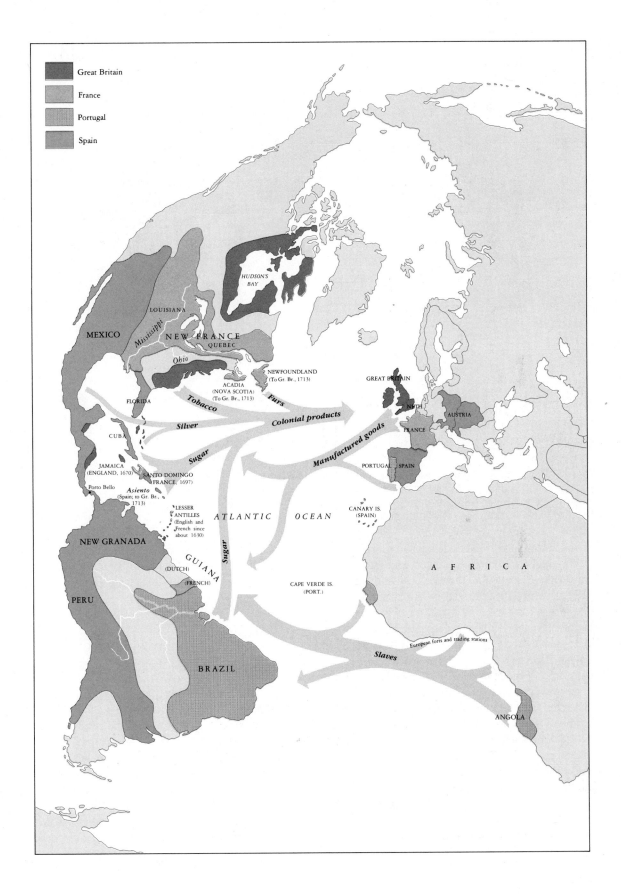

Great Britain

France

Portugal

Spain

HUDSON'S
BAY

LOUISIANA

MEXICO   NEW FRANCE
                      QUEBEC

Mississippi   *Ohio*

                                        NEWFOUNDLAND
                                        (To Gr. Br., 1713)   GREAT BRITAIN
                         ACADIA
                         (NOVA SCOTIA)   NETH.
FLORIDA   (To Gr. Br., 1713)   *Furs*   AUSTRIA
                         Tobacco   FRANCE
                                            Colonial products
                 Silver   *Manufactured goods*
CUBA   PORTUGAL SPAIN
JAMAICA
(ENGLAND, 1670)   *Sugar*
                 SANTO DOMINGO
Porto Bello   FRANCE, 1697)
        Asiento   CANARY IS.
        (Spain; to Gr. Br.,   (SPAIN)
           1713)   LESSER
                       ANTILLES   ATLANTIC   OCEAN
NEW GRANADA   (English and
                       French since
                       about 1630)   A F R I C A
PERU   GUIANA   *Sugar*
                (DUTCH)
                (FRENCH)   CAPE VERDE IS.
                                                (PORT.)

                                                            European forts and trading stations

                    BRAZIL   *Slaves*

                                                                              ANGOLA

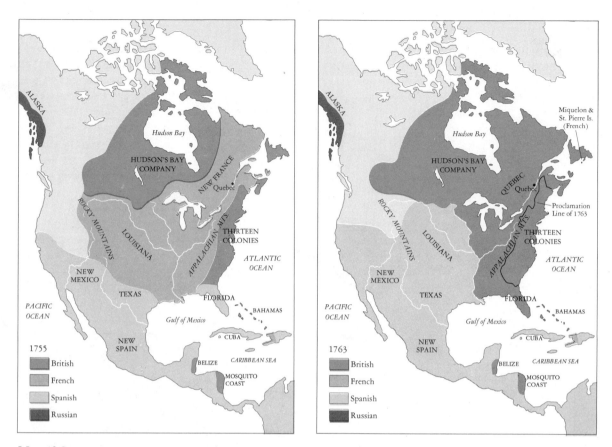

MAP 19.2 European Claims in North America Before and After the Seven Years' War (1756–1763) France lost its vast claims in North America, though the British government then prohibited colonists from settling west of a line drawn in 1763. The British wanted to avoid costly wars with Indians living in the newly conquered territory.

whose grandfather had made a fortune as a trader in India, the British concentrated on using superior sea power to destroy the French fleet and choke off French commerce around the world. Capturing Quebec in 1759 and winning a great naval victory at Quiberon Bay, the British also strangled France's valuable sugar trade with its Caribbean islands and smashed French forts in India. After Spain entered the war on France's side in 1761, the surging British temporarily occupied Havana in Cuba and Manila in the Philippines. With the Treaty of Paris (1763), France lost all its possessions on the mainland of North America. French Canada as well as French territory east of the Mississippi River passed to Britain, and France ceded Louisiana to Spain as compensation for Spain's loss of Florida to Britain. France also gave up most of its holdings in India, opening the way to British dominance on the subcontinent. By 1763 British naval power, built in large part on the rapid growth of the British shipping industry after the pas-

sage of the Navigation Acts, had triumphed decisively. Britain had realized its goal of monopolizing a vast trading and colonial empire for its exclusive benefit.

LAND AND WEALTH IN NORTH AMERICA

Of all Britain's colonies, those on the North American mainland proved most valuable in the long run. The settlements along the Atlantic coast provided an important outlet for surplus population, so that migration abroad limited poverty in England, Scotland, and northern Ireland. The settlers also benefited. In the mainland colonies, they had privileged access to virtually free and unlimited land. The availability of farms was a precious asset in preindustrial Europe, where agriculture was the main source of income and prestige.

The possibility of having one's own farm was particularly attractive to ordinary men and women from the British Isles. Land in England was already highly concentrated in the hands of the nobility and gentry in 1700 and became more so with agricultural improvement in the eighteenth century. White settlers who came to the colonies as free men and women, or as indentured servants pledged to work seven years for their passage, or as prisoners and convicts, could obtain their own farms on easy terms as soon as they had their personal freedom. Many poor white farmers also came to the mainland from the British West Indies, crowded out of those islands by the growth of big sugar plantations using black slave labor. To be sure, life in the mainland colonies was hard, especially on the frontier. Yet the settlers succeeded in paying little or no rent to grasping landlords, and taxes were very low. Unlike the great majority of European peasants, who had to accept high rents and taxes as part of the order of things, American farmers could keep most of what they managed to produce.

The availability of land made labor expensive in the colonies. This basic fact, rather than any repressive aspects of the Navigation Acts, limited the growth of industry in the colonies. As the Governor of New York put it in 1767:

The price of labor is so great in this part of the world that it will always prove the greatest obstacle to any manufacturers attempting to set up here, and the genius of the people in a country where everyone can have land to work upon leads them so naturally into agriculture that it prevails over every other occupation.[9]

The advantage for colonists was in farming, and farm they did.

Cheap land and scarce labor were also critical factors in the growth of slavery in the southern colonies. By 1700 British indentured servants were carefully avoiding the Virginia lowlands, where black slavery was spreading, and by 1730 the large plantations there had gone over completely to black slaves. Slave labor permitted an astonishing tenfold increase in tobacco production between 1700 and 1774 and created a wealthy aristocratic planter class in Maryland and Virginia.

In the course of the eighteenth century, the farmers of New England and, particularly, the middle colonies of Pennsylvania and New Jersey began to produce more food than they needed. They exported ever more foodstuffs, primarily to the West Indies. There the owners of the sugar plantations came to depend on the mainland colonies for grain and dried fish to feed their slaves. The plantation owners, whether they grew tobacco in Virginia and Maryland or sugar in the West Indies, had the exclusive privilege of supplying the British Isles with their products. Englishmen could not buy cheaper sugar from Brazil, nor were they allowed to grow tobacco in the home islands. Thus the colonists, too, had their place in the protective mercantile system of the Navigation Acts. The American shipping industry grew rapidly in the eighteenth century, for example, because colonial shippers enjoyed the same advantages as their fellow British citizens in the mother country.

The abundance of almost-free land resulted in a rapid increase in the colonial population in the eighteenth century. In a mere three-quarters of a century after 1700, the white population of the mainland colonies multiplied a staggering ten times, as immigrants arrived and colonial couples raised large families. In 1774, 2.2 million whites and 330,000 blacks inhabited what would soon become the independent United States.

Rapid population growth did not reduce the settlers to poverty. On the contrary, agricultural development resulted in fairly high standards of living, in eighteenth-century terms, for mainland colonists. There was also an unusual degree of economic equality, by European standards. Few people were extremely rich and few were extremely poor. Most remarkable of all, on eve of the American Revolution, the *average* white man or woman in the mainland British colonies probably had the highest income and standard of living in the world. It has been estimated that between 1715 and 1775 the real income of the average American was increasing about 1 percent per year per person, almost two-thirds as fast as it increased with massive industrialization between 1840 and 1959. When one considers that between 1775 and 1840 Americans experienced no improvement in their standard of living, it is clear just how much the colonists benefited from hard work and the mercantile system created by the Navigation Acts.[10]

THE GROWTH OF FOREIGN TRADE

England also profited greatly from the mercantile system. Above all, the rapidly growing and increasingly wealthy agricultural populations of the main-

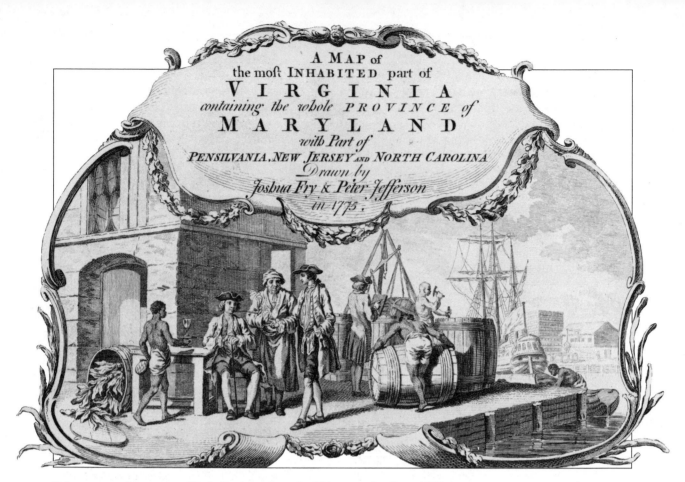

Tobacco was a key commodity in the Atlantic trade. This engraving from 1775 shows a merchant and his slaves preparing a cargo for sail. *(The British Library)*

land colonies provided an expanding market for English manufactured goods. This situation was extremely fortunate, for England in the eighteenth century was gradually losing, or only slowly expanding, its sales to many of its traditional European markets. However, rising demand for manufactured goods from North America, as well as from the West Indies, Africa, and Latin America, allowed English cottage industry to continue to grow and diversify. Merchant-capitalists and manufacturers found new and exciting opportunities for profit and wealth.

Since the late Middle Ages, England had relied very heavily on the sale of woolen cloth in foreign markets. Indeed, as late as 1700, woolen cloth was the only important manufactured good exported from England, and fully 90 percent of it was sold to Europeans. In the course of the eighteenth century, the states of continental Europe were trying to develop their own cottage textile industries in an effort to deal with rural poverty and overpopulation. Like England earlier, these states adopted protectionist, mercantilist policies. They tried by means of tariffs and other measures to exclude competing goods from abroad, whether English woolens or the cheap but beautiful cotton calicos the English East India Company brought from India and sold in Europe.

France had already closed its markets to the English in the seventeenth century. In the eighteenth century, German states purchased much less woolen cloth from England and encouraged cottage production of coarse, cheap linens, which became a feared competitor in all of central and southern Europe. By 1773 England was selling only about two-thirds as much woolen cloth to northern and western Europe as it had in 1700. The decline of sales to the Continent meant that the English economy badly needed new markets and new products in order to develop and prosper.

Protected colonial markets came to the rescue. More than offsetting stagnating trade with Europe, they provided a great stimulus for many branches of English manufacturing. The markets of the Atlantic economy led the way, as may be seen in Figure 19.3. English exports of manufactured goods to continental Europe increased very modestly, from roughly £2.9 million in 1700 to only £3.3 million in 1773. Meanwhile, sales of manufactured products to the Atlantic economy—primarily the mainland colonies of North America and the West Indian sugar islands, with an important assist from West Africa and Latin America—soared from £500,000 to £3.9 million. Sales to other "colonies"—Ireland and India—also rose substantially in the eighteenth century.

English exports became much more balanced and diversified. To America and Africa went large quantities of metal items—axes to frontiersmen, firearms, chains for slaveowners. There were also clocks and coaches, buttons and saddles, china and furniture, musical instruments and scientific equipment, and a host of other things. By 1750 half the nails made in England were going to the colonies. Foreign trade became the bread and butter of some industries.

Thus the mercantile system formed in the seventeenth century to attack the Dutch and to win power and profit for England continued to shape trade in the eighteenth century. The English concentrated in their hands much of the demand for manufactured goods from the growing Atlantic economy. The pressure of demand from three continents on the cottage industry of one medium-sized country heightened the efforts of English merchant-capitalists to find new and improved ways to produce more goods. By the 1770s England stood on the threshold of the radical industrial change to be described in Chapter 22.

REVIVAL IN COLONIAL LATIN AMERICA

When the last Spanish Habsburg, the feeble-minded Charles II, died in 1700 (page 520), Spain was "little less cadaverous than its defunct master."[11] Its vast empire lay before Europe awaiting dismemberment. Yet, in one of those striking reversals with which history is replete, Spain revived. The empire held together and even prospered, while a European-oriented landowning aristocracy enhanced its position in colonial society.

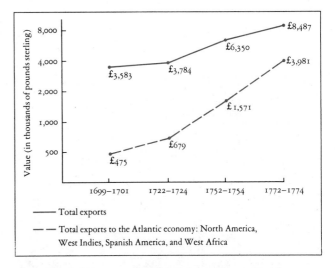

FIGURE 19.3 Exports of English Manufactured Goods 1700–1774 While trade between England and Europe stagnated after 1700, English exports to Africa and the Americas boomed and greatly stimulated English economic development. (Source: R. Davis, "English Foreign Trade, 1700–1774," Economic History Review, 2d series, 15 [1962]; 302–303)

Spain recovered in part because of better leadership. Louis XIV's grandson, who took the throne as Philip V (1700–1746), brought new men and fresh ideas with him from France and rallied the Spanish people to his Bourbon dynasty in the long War of the Spanish Succession. When peace was restored, a series of reforming ministers reasserted royal authority, overhauling state finances and strengthening defense. To protect the colonies, they restored Spain's navy to a respectable third place in Europe behind Great Britain and France. Philip's ministers also promoted the economy with vigorous measures that included a gradual relaxation of the state monopoly on colonial trade. The able Charles III (1759–1788), a truly enlightened monarch, further extended economic and administrative reform.

Revitalization in Madrid had positive results in the colonies. The colonies succeeded in defending themselves from numerous British attacks and even increased in size. Spain received Louisiana from France in 1763, and missionaries and ranchers extended Spanish influence all the way to northern California.

Political success was matched by economic improvement. After declining markedly in the seven-

Porto Bello Located on the isthmus of Panama, little Porto Bello was a major port in Spanish America. When ships arrived for 40-day trade fairs, it bustled with the energy of merchants, slaves, and soldiers. *(Pierpont Morgan Library)*

teenth century, silver mining recovered in Mexico and Peru. Output quadrupled between 1700 and 1800, when Spanish America accounted for half of world silver production. Ever a risky long shot at sudden riches, silver mining encouraged a gambler's attitude toward wealth and work. The big profits of the lucky usually went into land. Silver mining also encouraged food production for large mining camps and gave the *creoles*—people of Spanish blood born in America—the means to purchase more and more European luxuries and manufactured goods. A class of wealthy merchants arose to handle this flourishing trade, which often relied on smuggled goods from Great Britain. As in British North America, industry remained weak, although workshops employing forced Indian labor were occupied with fashioning Mexican and Peruvian wool into coarse fabrics for purchase by the Latin American masses. Spain's col-

onies were an important element of the Atlantic economy.

Economic development strengthened the creole elite, which came to rival the top government officials dispatched from Spain. As in most preindustrial societies, land was the main source of wealth. In contrast to their British American counterparts but like Spanish and eastern European aristocracy, creole estate owners controlled much of the land. Small independent farmers were rare.

The Spanish crown had given large holdings to the conquering pioneers and their followers, and beginning in the late sixteenth century many large tracts of state land were sold to favored settlers. Thus, though the crown decreed that Indian communities were to retain the use of their tribal lands, a class of large landholders grew up in sparsely settled regions and in the midst of the defeated Indian populations.

The Spanish settlers strove to become a genuine European aristocracy, and they largely succeeded. As good aristocrats, they believed that work in the fields was the proper occupation of a depressed, impoverished peasantry. The defenseless Indians suited their needs. As the Indian population recovered in numbers, slavery and periodic forced labor gave way to widespread debt peonage from 1600 on. Under this system, a planter or rancher would keep his christianized, increasingly hispanicized Indians in perpetual debt bondage by periodically advancing food, shelter, and a little money. Debt peonage subjugated the Indians and was a form of agricultural serfdom.

The landowning class practiced *primogeniture,* passing everything from eldest son to eldest son to prevent fragmentation of land and influence. Also like European nobles, wealthy creoles built ornate townhouses, contributing to the development of a lavish colonial baroque style that may still be seen in Lima, Peru, and Mexico City. The creole elite followed European cultural and intellectual trends. Enlightenment ideas spread to colonial salons and universities, encouraging a questioning attitude and preparing the way for the creoles' rise to political power with the independence movements of the early nineteenth century.

There were also creoles of modest means, especially in the cities, since estate agriculture discouraged small white farmers. (Chile was an exception: since it had few docile Indians to exploit, white settlers had to work their small farms to survive.) The large middle group in Spanish colonies consisted of racially mixed *mestizos,* the offspring of Spanish men and Indian women. The most talented mestizos realistically aspired to join the creoles, for enough wealth and power could make one white. This ambition siphoned off the most energetic mestizos and lessened the build-up of any lower-class discontent. Thus, by the end of the colonial era roughly 20 percent of the population was classified as white and about 30 percent as mestizo. Pure-blooded Indians accounted for most of the remainder, for only on the sugar plantations of Cuba and Puerto Rico did black slavery ever take firm root in Spanish America.

The situation was quite the opposite in Portuguese Brazil. As in the West Indies, enormous numbers of blacks were brought in chains to work the sugar plantations. About half the population of Brazil was of African origin in the early nineteenth century. Even more than in the Spanish territories, the people of Brazil intermingled sexually and culturally. In contrast to North America, where racial lines were hard and fast, at least in theory, colonial Brazil made a virtue of miscegenation, and the population grew to include every color in the racial rainbow.

While some European intellectual elites were developing a new view of the world in the eighteenth century, Europe as a whole was experiencing a gradual but far-reaching expansion. As agriculture showed signs of modest improvement across the continent, first the Low Countries and then England succeeded in launching the epoch-making Agricultural Revolution. Plague disappeared, and the populations of all countries grew significantly, encouraging the progress of cottage industry and merchant capitalism.

Europeans also continued their overseas expansion, fighting for empire and profit and consolidating their hold on the Americas in particular. A revived Spain and its Latin American colonies participated fully in this expansion. As in agriculture and cottage industry, however, England and its empire proved most successful. The English concentrated much of the growing Atlantic trade in their hands, which challenged and enriched English industry and intensified the search for new methods of production. Thus, by the 1770s, England was on the verge of an economic breakthrough fully as significant as the great political upheaval destined to develop shortly in neighboring France.

NOTES

1. M. Bloch, *Les caractères originaux de l'histoire rurale française,* Librarie Armand Colin, Paris, 1960, 1.244–245.
2. B.H. Slicher van Bath, *The Agrarian History of Western Europe, A.D. 500–1850,* St. Martin's Press, New York, 1963, p. 240.

3. Quoted in E. E. Rich and C. H. Wilson, eds., *Cambridge Economic History of Europe,* Cambridge University Press, Cambridge, Eng., 4.74.
4. Ibid., p. 85.
5. Quoted by I. Pinchbeck, *Women Workers and the Industrial Revolution, 1750–1850,* F. S. Crofts, New York, 1930, p. 113.
6. Quoted by S. Chapman, *The Lancashire Cotton Industry,* Manchester University Press, Manchester, Eng., 1903, p. 13.
7. Quoted by P. Mantoux, *The Industrial Revolution in the Eighteenth Century,* Harper & Row, New York, 1961, p. 75.
8. Quoted by C. Wilson, *England's Apprenticeship, 1603–1763,* Longmans, Green, London, 1965, p. 169.
9. Quoted by D. Dillard, *Economic Development of the North Atlantic Community,* Prentice-Hall, Englewood Cliffs, N.J., 1967, p. 192.
10. G. Taylor, "America's Growth Before 1840," *Journal of Economic History* 24 (December 1970):427–444.
11. J. Rippy, *Latin America: A Modern History,* rev. ed., University of Michigan Press, Ann Arbor, 1968, p. 97.

SUGGESTED READING

B. H. Slicher van Bath, *The Agrarian History of Western Europe, A.D. 500–1850* (1963), is a wide-ranging general introduction to the gradual transformation of European agriculture, as is M. Bloch's great classic, cited in the Notes, which has been translated as *French Rural History* (1966). J. Blum, *The End of the Old Order in Rural Europe* (1978), is an impressive comparative study. J. de Vries, *The Dutch Rural Economy in the Golden Age, 1500–1700* (1974), skillfully examines the causes of early Dutch leadership in farming, and E. L. Jones, *Agriculture and Economic Growth in England, 1650–1815* (1967), shows the importance of the agricultural revolution for England. Two recommended and complementary studies on landowning nobilities are R. Forster, *The Nobility of Toulouse in the Eighteenth Century* (1960), and G. E. Mingay, *English Landed Society in the Eighteenth Century* (1963). A. Goodwin, ed., *The European Nobility in the Eighteenth Century* (1967), is an exciting group of essays on aristocrats in different countries. R. and E. Forster, eds., *European Society in the Eighteenth Century* (1969), assembles a rich collection of contemporary writing on a variety of economic and social topics. E. Le Roy Ladurie, *The Peasants of Languedoc* (1976), a brilliant and challenging study of rural life in southern France for several centuries, complements J. Goody et al., eds., *Family and Inheritance: Rural Society in Western Europe, 1200–1800* (1976). Life in small-town preindustrial France comes alive in P. Higonnet, *Pont-de-Montvert: Social Structure and Politics in a French Village, 1700–1914* (1971), while O. Hufton deals vividly and sympathetically with rural migration, work, women, and much more in *The Poor in Eighteenth-Century France* (1974). P. Mantoux, *The Industrial Revolution in the Eighteenth Century* (1928), and D. Landes, *The Unbound Prometheus* (1969), provide excellent discussions of the development of cottage industry.

Two excellent multivolume series, *The Cambridge Economic History of Europe,* and C. Cipolla, ed., *The Fontana Economic History of Europe,* cover the sweep of economic developments from the Middle Ages to the present and have extensive bibliographies. F. Braudel, *Civilization and Capitalism, 15th–18th Century* (1981–1984), is a monumental and highly recommended three-volume synthesis. In the area of trade and colonial competition, V. Barbour, *Capitalism in Amsterdam* (1963), and C. R. Boxer, *The Dutch Seaborne Empire* (1970), are very interesting on Holland. C. Wilson, *Profit and Power: A Study of England and the Dutch Wars* (1957), is exciting scholarship, as are W. Dorn, *The Competition for Empire, 1740–1763* (1963), D. K. Fieldhouse, *The Colonial Empires* (1971), and R. Davies, *The Rise of Atlantic Economies* (1973). R. Pares, *Yankees and Creoles* (1956), is a short, lively work on trade between the mainland colonies and the West Indies. E. Williams, *Capitalism and Slavery* (1966), provocatively argues that slavery provided the wealth necessary for England's industrial development. Another exciting work is J. Nef, *War and Human Progress* (1968), which examines the impact of war on economic and industrial development in European history between about 1500 and 1800 and may be compared with M. Gutmann, *War and Rural Life in the Early Modern Low Countries* (1980). J. Fagg's *Latin America* (1969) provides a good introduction to the colonial period and has a useful bibliography, while C. Haring, *The Spanish in America* (1947), is a fundamental modern study.

Three very fine books on the growth of population are C. Cipolla's short and lively *The Economic History of World Population* (1962); E. A. Wrigley's more demanding *Population and History* (1969); and T. McKeown's scholarly *The Modern Rise of Population* (1977). W. McNeill, *Plagues and Peoples* (1976), is also noteworthy. In addition to works on England cited in the Suggested Reading for Chapter 22, D. George, *England in Transition* (1953), and C. Wilson, *England's Apprenticeship, 1603–1763* (1965), are highly recommended. The greatest novel of eighteenth-century English society is Henry Fielding's unforgettable *Tom Jones,* although Jane Austen's novels about country society, *Emma* and *Pride and Prejudice,* are not far behind.

20

THE LIFE OF THE
PEOPLE

*T*HE DISCUSSION of agriculture and industry in the last chapter showed the ordinary man and woman at work, straining to make ends meet and earn a living. Yet work is only part of human experience. What about the rest? What about such basic things as marriage and childhood, food and diet, health and religion? How, in short, did the peasant masses and urban poor really live in western Europe before the age of revolution at the end of the eighteenth century? This is the simple but profound question that the economic and social developments naturally raise.

MARRIAGE AND THE FAMILY

The basic unit of social organization is the family. It is within the structure of the family that human beings love, mate, and reproduce themselves. It is primarily the family that teaches the child, imparting values and customs that condition an individual's behavior for a lifetime. The family is also an institution woven into the web of history. It evolves and changes, assuming different forms in different times and places.

EXTENDED AND NUCLEAR FAMILIES

In many traditional Asian and African societies, the typical family has often been an extended family. A newly married couple, instead of establishing their own home, will go to live with either the bride's or the groom's family. The couple raises their children while living under the same roof with their own brothers and sisters, who may also be married. The family is a big, three- or four-generation clan, headed by a patriarch or perhaps a matriarch, and encompassing everyone from the youngest infant to the oldest grandparent.

Extended families, it is often said, provide security for adults and children in traditional agrarian peasant economies. Everyone has a place within the extended family, from cradle to grave. Sociologists frequently assume that the extended family gives way to the conjugal, or nuclear, family with the advent of industrialization and urbanization. Couples establish their own households and their own family identities when they marry. They live with the children they raise, apart from their parents. Something like this is indeed happening in much of Asia and Africa today. And since Europe was once agrarian and preindustrial, it has often been believed that the extended family must also have prevailed in Europe before being destroyed by the Industrial Revolution.

In fact, the situation was quite different in western and central Europe. By 1700 the extended three-generational family was a great rarity. Indeed, the extended family may never have been common in Europe, although it is hard to know about the Middle Ages because fewer records survive. When young European couples married, they normally established their own households and lived apart from their parents. When a three-generation household came into existence, it was usually a parent who moved in with a married child, rather than a newly married couple moving in with either set of parents. The married couple, and the children that were sure to follow, were on their own from the beginning.

Perhaps because European couples set up separate households when they married, people did not marry young in the seventeenth and early eighteenth centuries. Indeed, the average person, who was neither rich nor aristocratic, married surprisingly late, many years after reaching adulthood and many more after beginning to work. In one well-studied, typical English village, both men and women married for the first time at an average age of twenty-seven or older in the seventeenth and eighteenth centuries. A similar pattern existed in early eighteenth-century France. Moreover, a substantial portion of men and women never married at all.

Between two-fifths and three-fifths of European women capable of bearing children—that is, women between fifteen and forty-four—were unmarried at any given time. The contrast with traditional non-Western societies is once again striking. In those societies, the pattern has very often been almost universal and very early marriage. The union of a teenage bride and her teenage groom has been the general rule.

The custom of late marriage and nuclear family was a distinctive characteristic of European society. The consequences have been tremendous, though still only partially explored. It seems likely that the dynamism and creativity that have characterized European society were due in large part to the pattern of marriage and family. This pattern fostered and required self-reliance and independence. In preindustrial western Europe in the sixteenth through eighteenth centuries, marriage normally joined a mature man and a mature woman—two adults who had already experienced a great deal of life and could transmit self-reliance and real skills to the next generation.

Why was marriage delayed? The main reason was that couples normally could not marry until they could support themselves economically. The land was the main source of income. The peasant son often needed to wait until his father's death to inherit the family farm and marry his sweetheart. Similarly, the peasant daughter and her family needed to accumulate a small dowry to help her boyfriend buy land or build a house.

There were also laws and regulations to temper impetuous love and physical attraction. In some areas, couples needed the legal permission or tacit approval of the local lord or landowner in order to marry. In Austria and Germany, there were legal restrictions on marriage, and well into the nineteenth century poor couples had particular difficulty securing the approval of local officials. These officials believed that freedom to marry for the lower classes would mean more paupers, more abandoned children, and more money for welfare. Thus prudence, custom, and law combined to postpone the march to the altar. This pattern helped society maintain some kind of balance between the number of people and the available economic resources.

Work Away from Home

Many young people worked within their families until they could start their own households. Boys plowed and wove; girls spun and tended the cows. Many others left home to work elsewhere. In the towns, a lad might be apprenticed to a craftsman for seven or fourteen years to learn a trade. During that time he would not be permitted to marry. In most trades he earned little and worked hard, but if he were lucky he might eventually be admitted to a guild and establish his economic independence. More often, the young man would drift from one tough job to another: hired hand for a small farmer, laborer on a new road, carrier of water in a nearby town. He was always subject to economic fluctuations, and unemployment was a constant threat.

The Chimney Sweep Some boys and girls found work as chimney sweeps, especially if they were small. Climbing up into chimneys was dirty, dangerous work. Hot stones could set the sweep's clothing on fire. Such work for youngsters eventually died out in the nineteenth century, the period of this drawing. *(Photo: Caroline Buckler)*

Girls also left their families to work, at an early age and in large numbers. The range of opportunities open to them was more limited, however. Service in another family's household was by far the most common job. Even middle-class families often sent their daughters into service and hired others as servants in return. Thus, a few years away from home as a servant was a normal part of growing up. If all went well, the girl (or boy) would work hard and save some money for parents and marriage. At the least, there would be one less mouth to feed at home.

The legions of young servant girls worked hard but had little real independence. Sometimes the employer paid the girl's wages directly to her parents. Constantly under the eye of her mistress, her tasks were many—cleaning, shopping, cooking, caring for the baby—and often endless, for there were no laws to limit her exploitation. Few girls were so brutalized that they snapped under the strain, like the Russian servant girl Varka in Chekhov's chilling story, "Sleepy," who, driven beyond exhaustion, finally quieted her mistress's screaming child by strangling it in its cradle. But court records are full of complaints by servant girls of physical mistreatment by their mistresses. There were many others like the fifteen-year-old English girl in the early eighteenth century who told the judge that her mistress had not only called her "very opprobrious names, as Bitch, Whore and the like," but also "beat her without provocation and beyond measure."[1]

There was also the pressure of seducers and sexual attack. In theory, domestic service offered protection and security for a young girl leaving home. The girl had food, lodging, and a new family. She did not drift in a strange and often dangerous environment. But in practice, she was often the easy prey of a lecherous master, or his sons, or his friends. Indeed, "the evidence suggests that in all European countries, from Britain to Russia, the upper classes felt perfectly free to exploit sexually girls who were at their mercy."[2] If the girl became pregnant, she was quickly fired and thrown out in disgrace to make her own way. Prostitution and petty thievery were often the harsh alternatives that lay ahead. "What are we?" exclaimed a bitter Parisian prostitute. "Most of us are unfortunate women, without origins, without education, servants and maids for the most part."[3]

PREMARITAL SEX AND BIRTH-CONTROL PRACTICES

Did the plight of some ex-servant girls mean that late marriage in preindustrial Europe went hand in hand with premarital sex and many illegitimate children? For most of western and central Europe, until at least 1750, the answer seems to be no. English parish registers, in which the clergy recorded the births and deaths of the population, seldom list more than one bastard out of every twenty children baptized. Some French parishes in the seventeenth century had extraordinarily low rates of illegitimacy, with less than 1 percent of the babies born out of wedlock. Illegitimate babies were apparently a rarity, at least as far as the official church records are concerned.

At the same time, premarital sex was clearly commonplace. In one well-studied English village, one-third of all first children were conceived before the couple was married, and many were born within three months of the marriage ceremony. No doubt many of these couples were already betrothed, or at least "going steady," before they entered into an intimate relationship. But the very low rates of illegitimate birth also reflect the powerful social controls of the traditional village, particularly the open-field village with its pattern of cooperation and common action. Irate parents and village elders, indignant priests and authoritative landlords, all combined to pressure any young people who wavered about marriage in the face of unexpected pregnancy. These controls meant that premarital sex was not entered into lightly. In the countryside it was generally limited to those contemplating marriage.

Once a woman was married, she generally had several children. This does not mean that birth control within marriage was unknown in western and central Europe before the nineteenth century. But it was primitive and quite undependable. The most common method was *coitus interruptus*—withdrawal by the male before ejaculation. The French, who were apparently early leaders in contraception, were using this method extensively to limit family size by the end of the eighteenth century. Withdrawal as a method of birth control was in keeping with the European pattern of nuclear family, in which the father bore the direct responsibility of supporting his children. Withdrawal—a male technique—was one way to meet that responsibility.

Mechanical and other means of contraception were not unknown in the eighteenth century, but they appear to have been used mainly by certain sectors of the urban population. The "fast set" of London used the "sheath" regularly, although primarily to protect against venereal disease, not pregnancy. Prostitutes used various contraceptive techniques to prevent pregnancy, and such information was probably available to anyone who really sought it. The second part of an indictment for adultery against a late-sixteenth-century English vicar charged that the wayward minister was "also an instructor of young folks [in] how to commit the sin of adultery or fornication and not to beget or bring forth children."[4]

New Patterns of Marriage and Illegitimacy

In the second half of the eighteenth century, the pattern of late marriage and few illegitimate children began to break down. It is hard to say why. Certainly, changes in the economy had a gradual but profound impact. The growth of cottage industry created new opportunities for earning a living, opportunities not tied to limited and hard-to-get land. Because a scrap of ground for a garden and a cottage for the loom and spinning wheel could be quite enough for a modest living, young people had greater independence and did not need to wait for a good-sized farm. A contemporary observer of an area of rapidly growing cottage industry in Switzerland at the end of the eighteenth century described these changes: "The increased and sure income offered by the combination of cottage manufacture with farming hastened and multiplied marriages and encouraged the division of landholdings, while enhancing their value; it also promoted the expansion and embellishment of houses and villages."[5]

As a result, cottage workers married not only earlier but for different reasons. Nothing could be so businesslike, so calculating, as a peasant marriage which was often dictated by the needs of the couple's families. After 1750, however, courtship became more extensive and freer as cottage industry grew. It was easier to yield to the attraction of the opposite sex and fall in love. The older generation was often shocked by the lack of responsibility they saw in the early marriages of the poor, the union of "people with only two spinning wheels and not even a bed." But the laws and regulations they imposed, especially in Germany, were often disregarded. Unions based on love rather than on economic considerations were increasingly the pattern for cottage workers. Factory workers, numbers of whom first began to appear in England after about 1780, followed the path blazed by cottage workers.

Changes in the timing and motivation of marriage went hand in hand with a rapid increase in illegitimate births between about 1750 and 1850. Some historians even speak of an "illegitimacy explosion." In Frankfurt, Germany, for example, only about 2 percent of all births were illegitimate in the early 1700s. This figure rose to 5 percent in about 1760, to about 10 percent in 1800, and peaked at about 25 percent around 1850. In Bordeaux, France, illegitimate births rose steadily until by 1840 one out of every three babies was born out of wedlock. Small towns and villages less frequently experienced such startlingly high illegitimacy rates, but increases from a range of 1 to 3 percent initially to 10 to 20 percent between 1750 and 1850 were commonplace. A profound sexual and cultural transformation was taking place. Fewer girls were abstaining from premarital intercourse, and fewer boys were marrying the girls they got pregnant.

It is hard to know exactly why this change occurred and what it meant. The old idea of a safe, late, economically secure marriage did not reflect economic and social realities. The growing freedom of thought in the turbulent years beginning with the French Revolution in 1789 influenced sexual and marital behavior. And illegitimate births, particularly in Germany, were also the result of open rebellion against class laws limiting the right of the poor to marry. Unable to show a solid financial position and thereby obtain a marriage license, couples asserted their independence and lived together anyway. Children were the natural and desired result of "true love" and greater freedom. Eventually, when the stuffy, old-fashioned propertied classes gave in and repealed their laws against "imprudent marriage," poor couples once again went to the altar, often accompanied by their children, and the number of illegitimate children declined.

More fundamentally, the need to seek work outside farming and the village made young people more mobile. Mobility in turn encouraged new sexual and marital relationships, which were less subject to parental pressure and village tradition. As in the case of young servant girls who became pregnant and were then forced to fend for themselves, some of these relationships promoted loose living or prostitution. This resulted in more illegitimate births and strengthened an urban subculture of habitual illegitimacy.

Early Sexual Emancipation?

It has been suggested that the increase in illegitimate births represented a stage in the emancipation of women. According to this view, new economic opportunities outside the home, in the city and later in the factory, revolutionized women's attitudes about themselves. Young working women became indivi-

The Face of Poverty This woman and her children live by begging. Families of beggars roamed all over eighteenth-century Europe, and many observers feared they were increasing in numbers. *(John Freeman/Fotomas Index)*

dualistic and rebelled against old restrictions like late marriage. They sought fulfillment in the pleasure of sexuality. Since there was little birth control, freer sex for single women meant more illegitimate babies.

No doubt single working women in towns and cities were of necessity more independent and self-reliant. Yet, until at least the late nineteenth century, it seems unlikely that such young women were motivated primarily by visions of emancipation and sexual liberation. Most women were servants or textile workers. These jobs paid poorly, and the possibility of a truly independent "liberated" life was correspondingly limited. Most women in the city probably looked to marriage and family life as an escape from hard, poorly paid work and as the foundation of a satisfying life.

Hopes and promises of marriage from men of the working girl's own class led naturally enough to sex.[6]

In one medium-sized French city in 1787 to 1788, the great majority of unwed mothers stated that sexual intimacy had followed promises of marriage. Many soldiers, day laborers, and male servants were no doubt sincere in their proposals. But their lives were insecure, and many hesitated to take on the heavy economic burdens of wife and child. Nor were their backbones any longer stiffened by the traditional pressures of the village.

In a growing number of cases, therefore, the intended marriage did not take place. The romantic yet practical dreams and aspirations of many young working women and men were frustrated by low wages, inequality, and changing economic and social conditions. Old patterns of marriage and family were breaking down among the common people. Only in the late nineteenth century would more stable patterns reappear.

WOMEN AND CHILDREN

In the traditional framework of preindustrial Europe, women married late but then began bearing children rapidly. If a woman married before she was thirty, and if both she and her husband lived to forty-five, the chances were roughly one in two that she would give birth to six or more children. The newborn child entered a dangerous world. Infant mortality was high. One in five was sure to die and one in three was quite likely to, in the poorer areas. Newborn children were very likely to catch infectious diseases of the stomach and chest, which were not understood. Thus little could be done for an ailing child, even in rich families. Childhood itself was dangerous. Parents in preindustrial Europe could count themselves fortunate if half their children lived to adulthood.

CHILD CARE AND NURSING

Women of the lower classes generally breast-fed their infants, and for much longer periods than is customary today. Breast-feeding decreases the likelihood of pregnancy for the average woman by delaying the resumption of ovulation. Although women may have been only vaguely aware of the link between nursing and not getting pregnant, they were spacing their children—from two to three years apart—and limiting their fertility by nursing their babies. If a newborn baby died, nursing stopped and a new life could be created. Nursing also saved lives: the breast-fed infant was more likely to survive on its mother's milk than on any artificial foods. In many areas of Russia, where common practice was to give a new child a sweetened (and germ-laden) rag to suck on for its subsistence, half the babies did not survive the first year.

In contrast to the laboring poor, the women of the aristocracy and upper-middle class seldom nursed their own children. The upper-class woman felt that breast-feeding was crude, common, and well beneath her dignity. Instead she hired a wet nurse to suckle her child. The urban mother of more modest means —the wife of a shopkeeper or artisan—also commonly used a wet nurse, sending her baby to some poor woman in the country as soon as possible.

Wet-nursing was a very widespread and flourishing business in the eighteenth century, a dismal business within the framework of the putting-out system. The traffic was in babies rather than in wool and cloth, and two or three years often passed before the wet-nurse worker finished her task. The great French historian Jules Michelet described with compassion the plight of the wet nurse, who was still going to the homes of the rich in early nineteenth-century France:

People do not know how much these poor women are exploited and abused, first by the vehicles which transport them (often barely out of their confinement), and afterward by the employment offices which place them. Taken as nurses on the spot, they must send their own child away, and consequently it often dies. They have no contact with the family that hires them, and they may be dismissed at the first caprice of the mother or doctor. If the change of air and place should dry up their milk, they are discharged without any compensation. If they stay here [in the city] they pick up the habits of the easy life, and they suffer enormously when they are forced to return to their life of [rural] poverty. A good number become servants in order to stay in the town. They never rejoin their husbands, and the family is broken.[7]

Other observers noted the flaws of wet nursing. It was a common belief that a nurse passed her bad traits to the baby with her milk. When a child turned out poorly, it was assumed that "the nurse changed it." Many observers charged that nurses were often negligent and greedy. They claimed that there were large numbers of "killing nurses" with whom no child every survived. The nurse let the child die quickly, so that she could take another child and another fee. No matter how the adults fared in the wet-nurse business, the child was a certain loser.

FOUNDLINGS AND INFANTICIDE

In the ancient world and in Asian societies it was not uncommon to allow or force newborn babies, particularly girl babies, to die when there were too many mouths to feed. To its great and eternal credit, the early medieval church, strongly influenced by Jewish law, denounced infanticide as a pagan practice and insisted that every human life was sacred. The willful destruction of newborn children became a crime punishable by death. And yet, as the reference to "killing nurses" suggests, direct and indirect methods of eliminating unwanted babies did not disap-

Abandoned Children At this Italian foundlings' home a frightened, secretive mother could discreetly deposit her baby. *(The Bettmann Archive)*

pear. There were, for example, many cases of "overlaying"—parents rolling over and suffocating the child placed between them in their bed. Such parents claimed they were drunk and had acted unintentionally. In Austria in 1784, suspicious authorities made it illegal for parents to take children under five into bed with them. Severe poverty on the one hand and increasing illegitimacy on the other conspired to force the very poor to thin their own ranks.

The young girl—very likely a servant—who could not provide for her child had few choices. If she would not stoop to abortion or the services of a killing nurse, she could bundle up her baby and leave it on the doorstep of a church. In the late seventeenth century, Saint Vincent de Paul was so distressed by the number of babies brought to the steps of Notre Dame in Paris that he established a home for foundlings. Others followed his example. In England the government acted on a petition calling for a foundling hospital "to prevent the frequent murders of poor, miserable infants at birth" and "to suppress the inhuman custom of exposing newborn children to perish in the streets."

In much of Europe in the eighteenth century, foundling homes became a favorite charity of the rich and powerful. Great sums were spent on them. The foundling home in St. Petersburg, perhaps the most elaborate and lavish of its kind, occupied the former palaces of two members of the high nobility. In the early nineteenth century it had 25,000 children in its care and was receiving 5,000 new babies a year. At their best, the foundling homes of the eighteenth century were a good example of Christian charity and social concern in an age of great poverty and inequality.

Yet the foundling home was no panacea. By the 1770s one-third of all babies born in Paris were immediately abandoned to the foundling home by their mothers. Fully a third of all those foundlings were abandoned by married couples, a powerful commentary on the standard of living among the working poor, for whom an additional mouth to feed often meant tragedy. In London competition for space in the foundling home soon became so great that it led "to the disgraceful scene of women scrambling and fighting to get to the door, that they might be of the fortunate few to reap the benefit of the Asylum."[8]

Furthermore, great numbers of babies entered, but few left. Even in the best of these homes half the babies normally died within a year. In the worst, fully 90 percent did not survive. They succumbed to long journeys over rough roads, the intentional and unintentional neglect of their wet nurses, and the customary childhood illnesses. So great was the carnage that some contemporaries called the foundling hospitals "legalized infanticide."

Certainly some parents and officials looked on the hospitals as a dump for unwanted babies. In the early 1760s, when the London Foundling Hospital was obliged to accept all babies offered, it was deluged with babies from the countryside. Many parish officers placed with the foundling home the abandoned children in their care, just as others apprenticed five-year-old children to work in factories. Both practices reduced the cost of welfare at the local level. Throughout the eighteenth century, millions of children of the poor continued to exit after the briefest of appearances on the earthly stage. True, they died after being properly baptized, an important consideration in still-Christian Europe. Yet those who dream of an idyllic past would do well to ponder the foundling's fate.

ATTITUDES TOWARD CHILDREN

What were the more typical circumstances of children's lives? Did the treatment of foundlings reflect the attitudes of normal parents? Harsh as it may sound, the young child was very often of little concern to its parents and to society in the eighteenth century. This indifference toward children was found in all classes; rich children were by no means exempt. The practice of using wet nurses, who were casually selected and often negligent, is one example of how even the rich and the prosperous put the child out of sight and out of mind. One French moralist, writing in 1756 about how to improve humanity, observed that "one blushes to think of loving one's children." It has been said that the English gentleman of the period "had more interest in the diseases of his horses than of his children."[9]

Parents believed that the world of the child was an uninteresting one. When parents did stop to notice their offspring, they often treated them as dolls or playthings—little puppies to fondle and cuddle in a moment of relaxation. The psychological distance between parent and child remained vast.

Much of the indifference was due to the terrible frequency, the terrible banality, of death among children of all classes. Parents simply could not afford to become too emotionally involved with their children, who were so unlikely to survive. The great eighteenth-century English historian Edward Gibbon (1737–1794) wrote that "the death of a new born child before that of its parents may seem unnatural but it is a strictly probable event, since of any given number the greater part are extinguished before the ninth year, before they possess the faculties of the mind and the body." Gibbon's father named all his boys Edward, hoping that at least one of them would survive to carry his name. His prudence was not misplaced. Edward the future historian and eldest survived. Five brothers and sisters who followed him all died in infancy.

Doctors were seldom interested in the care of children. One contemporary observer quoted a famous doctor as saying that "he never wished to be called to a young child because he was really at a loss to know what to offer for it." There were "physicians of note who make no scruple to assert that there is nothing to be done for children when they are ill." Children were caught in a vicious circle: they were neglected because they were very likely to die and they were likely to die because they were neglected.

Indifference toward children often shaded off into brutality. When parents and other adults did turn toward children, it was normally to discipline and control them. The novelist Daniel Defoe (1660?–1731), always delighted when he saw very young children working hard in cottage industry, coined the axiom "Spare the rod and spoil the child." He meant it. So did Susannah Wesley, mother of John Wesley (1703–1791), the founder of Methodism. According to her, the first task of a parent toward her children was "to conquer the will, and bring them to an obedient temper." She reported that her babies were "taught to fear the rod, and to cry softly; by which means they escaped the abundance of correction they might otherwise have had, and that most odious noise of the crying of children was rarely heard in the house."[10]

It was hardly surprising that, when English parish officials dumped their paupers into the first factories late in the eighteenth century, the children were beaten and brutalized (see page 703). That was part of the childrearing pattern—widespread indifference on the one hand and strict physical discipline on the other—that prevailed through most of the eighteenth century.

Late in the century, this pattern came under attack. Critics like Jean-Jacques Rousseau called for greater love, tenderness, and understanding toward children. In addition to supporting foundling homes to discourage infanticide and urging wealthy women to nurse their own babies, these new voices ridiculed the practice of swaddling. Wrapping youngsters in tight-fitting clothes and blankets was generally believed to form babies properly by "straightening them out."

The Five Senses Published in 1774, J. B. Basedow's *Elementary Reader* helped spread new attitudes toward child development and education. Drawing heavily upon the theories of Locke and Rousseau, the German educator advocated nature study and contact with everyday life. In this illustration for Basedow's reader, gentle teachers allow uncorrupted children to learn about the five senses through direct experience. *(Photo: Caroline Buckler)*

By the end of the century, small children were often dressed in simpler, more comfortable clothing, allowing much greater freedom of movement. More parents expressed a delight in the love and intimacy of the child and found real pleasure in raising their offspring. These changes were part of the general growth of humanitarianism and optimism about human potential that characterized the eighteenth-century Enlightenment.

SCHOOLS AND EDUCATION

The role of formal education outside the home, in those special institutions called schools, was growing more important. The aristocracy and the rich had led the way in the sixteenth century with special colleges, often run by the Jesuits. But "little schools," charged with elementary education of the children of the masses, did not appear until the seventeenth century.

Unlike medieval schools, which mingled all age groups, the little schools specialized in boys and girls from seven to twelve, who were instructed in basic literacy and religion.

Although large numbers of common people got no education at all in the eighteenth century, the beginnings of popular education were recognizable. France made a start in 1682 with the establishment of Christian schools, which taught the catechism and prayers as well as reading and writing. The Church of England and the dissenting congregations established "charity schools" to instruct the children of the poor. As early as 1717, Prussia made attendance at elementary schools compulsory. Inspired by the old Protestant idea that every believer should be able to read and study the Bible in the quest for personal salvation and by the new idea of a population capable of effectively serving the state, Prussia led the way in the development of universal education. Religious motives were also extremely important elsewhere. From the middle of the seventeenth century, Presbyterian Scotland was convinced that the path to salvation lay in careful study of the Scriptures, and this belief led to an effective network of parish schools for rich and poor alike. The Enlightenment commitment to greater knowledge through critical thinking reinforced interest in education in the eighteenth century.

The result of these efforts was a remarkable growth of basic literacy between 1600 and 1800, especially after 1700. Whereas in 1600 only one male in six was barely literate in France and Scotland, and one in four in England, by 1800 almost 90 percent of the Scottish male population was literate. At the same time, two out of three males were literate in France, and in advanced areas such as Normandy, literacy approached 90 percent (see Map 20.1). More than half of English males were literate by 1800. In all three countries the bulk of the jump occurred in the eighteenth century. Women were also increasingly literate, although they probably lagged behind men somewhat in most countries. Some elementary education was becoming a reality for European peoples, and schools were of growing significance in everyday life.

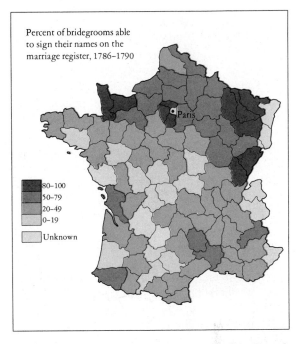

Percent of bridegrooms able to sign their names on the marriage register, 1786–1790

80–100
50–79
20–49
0–19
Unknown

MAP 20.1 Literacy in France on the Eve of the French Revolution Literacy rates varied widely between and within states in eighteenth-century Europe. Northern France was clearly ahead of southern France.

THE EUROPEAN'S FOOD

Plague and starvation, which recurred often in the seventeenth century, gradually disappeared in the eighteenth century. This phenomenon probably accounts in large part for the rapid growth in the total number of Europeans and for their longer lives. The increase in the average life span, allowing for regional variations, was remarkable. In 1700 the average European could expect at birth to live only twenty-five years. A century later, a newborn European could expect to live fully ten years longer, to age thirty-five. The doubling of the adult life span meant that there was more time to produce and create, and more reason for parents to stress learning and preparation for adulthood.

People also lived longer because ordinary years were progressively less deadly. People ate better and somewhat more wisely. Doctors and hospitals probably saved a few more lives than they had in the past. How and why did health and life expectancy im-

prove, and how much did they improve? And what were the differences between rich and poor? To answer these questions, it is necessary first to follow the eighteenth-century family to the table and then to see what contribution doctors made.

DIETS AND NUTRITION

Although the accomplishments of doctors and hospitals are constantly in the limelight today, the greater, if less spectacular, part of medicine is preventive medicine. The great breakthrough of the second half of the nineteenth century was the development of public health techniques—proper sanitation and mass vaccinations—to prevent outbreaks of communicable diseases. Even before the nineteenth century, when medical knowledge was slight and doctors were of limited value, prevention was the key to longer life. Good clothing, warm dry housing, and plentiful food make for healthier populations, much more capable of battling off disease. Clothing and housing for the masses probably improved only modestly in the eighteenth century, but the new agricultural methods and increased agricultural output had a beneficial effect. The average European ate more and better food and was healthier as a result in 1800 than in 1700. This pattern is apparent if we look at the fare of the laboring poor.

At the beginning of the eighteenth century, ordinary men and women depended on grain as fully as they had in the past. Bread was quite literally the staff of life. Peasants in the Beauvais region of France ate two pounds of bread a day, washing it down with water, green wine, beer, or a little skimmed milk. Their dark bread was made from a mixture of rough-ground wheat and rye—the standard flour of the poor. The poor also ate grains in soup and gruel. In rocky northern Scotland, for example, people depended on oatmeal, which they often ate half-cooked so it would swell in their stomachs and make them feel full. No wonder, then, that the supply of grain and the price of bread were always critical questions for most of the population.

The poor, rural and urban, also ate a fair quantity of vegetables. Indeed, vegetables were considered "poor people's food." Peas and beans were probably the most common; grown as field crops in much of Europe since the Middle Ages, they were eaten fresh in late spring and summer. Dried, they became the basic ingredients in the soups and stews of the long winter months. In most regions, other vegetables appeared in season on the tables of the poor, primarily cabbages, carrots, and wild greens. Fruit was uncommon and limited to the summer months.

The European poor loved meat and eggs, but even in England—the wealthiest country in Europe in 1700—they seldom ate their fill. Meat was too expensive. When the poor did eat meat—on a religious holiday or at a wedding or other special occasion—it was most likely lamb or mutton. Sheep could survive on rocky soils and did not compete directly with humans for the slender resources of grain.

Milk was rarely drunk. It was widely believed that milk caused sore eyes, headaches, and a variety of ills, except among the very young and very old. Milk was used primarily to make cheese and butter, which the poor liked but could afford only occasionally. Medical and popular opinion considered whey, the watery liquid left after milk was churned, "an excellent temperate drink."

The diet of the rich—aristocrats, officials, and the comfortable bourgeoisie—was traditionally quite different from that of the poor. The men and women of the upper classes were rapacious carnivores, and a person's standard of living and economic well-being were often judged by the amount of meat eaten. A truly elegant dinner among the great and powerful consisted of one rich meat after another—a chicken pie, a leg of lamb, a grilled steak, for example. Three separate meat courses might be followed by three fish courses, laced with piquant sauces and complemented with sweets, cheeses, and nuts of all kinds. Fruits and vegetables were not often found on the tables of the rich. The long-standing dominance of meat and fish in the diet of the upper classes continued throughout the eighteenth century. There was extravagant living, and undoubtedly great overeating and gluttony, not only among the aristocracy but also among the prosperous professional classes.

There was also an enormous amount of overdrinking among the rich. The English squire, for example, who loved to ride with his hounds, loved drink with a similar passion. He became famous as the "four-bottle man." With his dinner he drank red wine from France or white wine from the Rhineland, and with his desert he took sweet but strong port or Maderia from Portugal. Sometimes he ended the evening under the table in a drunken stupor, but very often he did not. The wine and the meat were consumed together in long hours of sustained excess, permitting the gentleman and his guests to drink enormous quantities.

The diet of small traders, master craftsmen, minor bureaucrats—the people of the towns and cities—was probably less monotonous than that of the peasantry. The markets, stocked by market gardens in the outskirts, provided a substantial variety of meats, vegetables, and fruits, although bread and beans still formed the bulk of the poor family's diet.

There were also regional dietary differences in 1700. Generally speaking, northern, Atlantic Europe ate better than southern, Mediterranean Europe. The poor of England probably ate best of all. Contemporaries on both sides of the Channel often contrasted the Englishman's consumption of meat with the French peasant's greater dependence on bread and vegetables. The Dutch were also considerably better fed than the average European, in large part because of their advanced agriculture and diversified gardens.

Le Nain: Peasant Family A little wine and a great deal of dark bread: the traditional food of the poor French peasantry accentuates the poetic dignity of this masterpiece, painted about 1640 by Louis Le Nain. *(Cliché des Musées Nationaux, Paris)*

THE IMPACT OF DIET ON HEALTH

How were the poor and the rich served by their quite different diets? Good nutrition depends on a balanced supply of food as well as on an adequate number of calories. Modern research has shown that the chief determinant of nutritional balance is the relationship between carbohydrates (sugar and starch) and proteins. A diet consisting primarily of carbohydrates is seriously incomplete.

At first glance, the diet of the laboring poor, relying as it did on carbohydrates, seems unsatisfactory. Even when a peasant got his daily two or three pounds of bread, his supply of protein and essential vitamins would seem too low. A closer look reveals a brighter picture. Most bread was "brown" or "black," made from wheat or rye. The flour of the eighteenth century was a whole-meal flour, produced by stone grinding. It contained most of the bran—the ground-up husk—and the all-important wheat germ. The bran and germ contain higher proportions of some minerals, vitamins, and good-quality proteins than does the rest of the grain. Only when they are removed does bread become a foodstuff providing relatively more starch and less of the essential nutrients.

In addition, the field peas and beans eaten by poor people since Carolingian days contained protein that complemented the proteins in whole-meal bread.

The proteins in whey, cheese, and eggs, which the poor ate at least occasionally, also supplemented the value of the protein in the bread and vegetables. Indeed, a leading authority concludes that if a pint of milk and some cheese and whey were eaten each day, the balance of the poor people's diet "was excellent, far better indeed than in many of our modern diets."[11]

The basic bread-and-vegetables diet of the poor *in normal times* was satisfactory. It protected effectively against most of the disorders associated with a deficiency of the vitamin B complex, for example. The lack of sugar meant that teeth were not so plagued by cavities. Constipation was almost unknown to peasants and laborers living on coarse cereal breads, which provided the roughage modern diets lack. The common diet of the poor also generally warded off anemia, although anemia among infants was not uncommon.

The key dietary problem was probably getting enough green vegetables (or milk), particularly in the late winter and early spring, to ensure adequate supplies of vitamins A and C. A severe deficiency of vitamin C produces scurvy, a disease that leads to rotting gums, swelling of the limbs, and great weakness. Before the season's first vegetables, many people had used up their bodily reserves of vitamin C and were suffering from mild cases of scurvy. Sailors on long voyages suffered most. By the end of the sixteenth century the exceptional antiscurvy properties of lemons and limes led to the practice of supplying some crews with a daily ration of lemon juice, which had highly beneficial effects. "Scurvy grass"—a kind of watercress—also guarded against scurvy, and this disease was increasingly controlled on even the longest voyages.

The practice of gorging on meat, sweets, and spirits caused the rich their own nutritional problems. They, too, were very often deficient in vitamins A and C because of their great disdain for fresh vegetables. Gout was a common affliction of the overfed and underexercised rich. No wonder they were often caricatured dragging their flabby limbs and bulging bellies to the table, to stuff their swollen cheeks and poison their livers. People of moderate means, who could afford some meat and dairy products with fair regularity but who had not abandoned the bread and vegetables of the poor, were probably best off from a nutritional standpoint.

NEW FOODS AND NEW KNOWLEDGE

In nutrition and food consumption, Europe in the early eighteenth century had not gone beyond its medieval accomplishments. This situation began to change markedly as the century progressed. Although the introduction of new methods of farming was confined largely to the Low Countries and England, a new food—the potato—came to the aid of the poor everywhere.

Introduced into Europe from the Americas, along with corn, squash, tomatoes, chocolate, and many other useful plants, the humble potato is an excellent food. It contains a good supply of carbohydrates and calories, and is rich in vitamins A and C, especially if the skin is eaten and it is not overcooked. The lack of green vegetables that could lead to scurvy was one of the biggest deficiencies in the poor person's winter and early spring diet. The potato, which gave a much higher caloric yield than grain for a given piece of land, provided the needed vitamins and supplemented the bread-based diet. Doctors, increasingly aware of the dietary benefits of potatoes, prescribed them for the general public and in institutions such as schools and prisons.

For some poor people, especially desperately poor peasants who needed to get every possible calorie from a tiny plot of land, the potato replaced grain as the primary food in the eighteenth century. This happened first in Ireland, where in the seventeenth century Irish rebellion had led to English repression and the perfection of a system of exploitation worthy of the most savage Eastern tyrant. The foreign (and Protestant) English landlords took the best land, forcing large numbers of poor (and Catholic) peasants to live off tiny scraps of rented ground. By 1700 the poor in Ireland lived almost exclusively on the bountiful fruits of the potato plot.

Elsewhere in Europe, the potato took hold more slowly. Potatoes were first fed to pigs and livestock, and there was considerable debate over whether they were fit for humans. In Germany the severe famines caused by the Seven Years' War (page 595) settled the matter: potatoes were edible and not "famine food." By the end of the century, the potato was an important dietary supplement in much of Europe.

There was also a general growth of market gardening and a greater variety of vegetables in towns and cities. Potatoes, cabbages, peas, beans, radishes, spinach, asparagus, lettuce, parsnips, carrots, and other vegetables were sold in central markets and streets. In the course of the eighteenth century, the large towns and cities of maritime Europe began to receive semitropical fruit, such as oranges, lemons, and limes, from Portugal and the West Indies, although they were not cheap.

The growing variety of food was matched by some improvement in knowledge about diet and nutrition. For the poor, such improvement was limited primarily to the insight that the potato and other root crops improved health in the winter and helped to prevent scurvy. The rich began to be aware of the harmful effects of their meat-laden, wine-drowned meals.

The waning influence of Galen's medical teachings was another aspect of progress. Galen's Roman synthesis of ancient medical doctrines held that the four basic elements—air, fire, water, and earth—combine to produce in each person a complexion and a corresponding temperament. Foods were grouped into four categories appropriate for each complexion. Galen's notions dominated the dietary thinking of the seventeenth-century medical profession: "Galen said that the flesh of a hare preventeth fatness, causeth sleep and cleanseth the blood," and so on for a thousand things. For instance, vegetables were seen as "windy" and tending to cause fevers, and fruits as dangerous except in very small amounts.

The growth of scientific experimentation in the seventeenth century led to a generally beneficial questioning of the old views. Haphazardly, by trial and error, and influenced by advances in chemistry, saner ideas developed. Experiments with salts led to the belief that foods were by nature either acid (all fruits and most vegetables) or alkaline (all meats). Doctors and early nutritionists came to believe that one key to good health was a *balance* of the two types.

Not all changes in the eighteenth century were for the better, however. Bread began to change, most noticeably in England. Rising incomes and new tastes led to a shift from whole-meal black or brown bread to white bread made from finely ground and sifted flour. On the Continent, such white bread was generally limited to the well-to-do. To the extent that the preferred wheaten flour was stone-ground and sifted for coarse particles only, white bread remained satisfactory. But the desire for "bread as white as snow" was already leading to a decline in nutritional value.

The coarser bran, which is necessary for roughage,

and at least some of the germ, which darkened the bread but contained the grain's nutrients, were already being sifted out to some extent. Bakers in English cities added the chemical alum to their white loaves to make them smoother, whiter, and larger. In the nineteenth century, "improvements" in milling were to lead to the removal of almost all the bran and germ from the flour, leaving it perfectly white and greatly reduced in nutritional value. The only saving grace in the sad deterioration of bread was that people began to eat and therefore to depend on it less.

Another sign of nutritional decline was the growing consumption of sweets in general and sugar in particular. Initially a luxury, sugar dropped rapidly in price, as slave-based production increased in the Americas, and it was much more widely used in the eighteenth century. This development probably led to an increase in cavities and to other ailments as well. Overconsumption of refined sugar can produce, paradoxically, low blood sugar (hypoglycemia) and, for some individuals at least, a variety of physical and mental ailments. Of course the greater or lesser poverty of the laboring poor saved most of them from the problems of the rich and well-to-do.

MEDICAL SCIENCE AND THE SICK

Advances in medical science played a very small part in improving the health and lengthening the lives of people in the eighteenth century. Such seventeenth-century advances as William Harvey's discovery of the circulation of blood were not soon translated into better treatment. The sick had to await the medical revolution of the later nineteenth century for much help from doctors.

Yet developments in medicine reflected the general thrust of the Enlightenment. The prevailing focus on discovering the laws of nature and on human problems, rather than on God and the heavens, gave rise to a great deal of research and experimentation. The century saw a remarkable rise in the number of doctors, and a high value was placed on their services. Thus when the great breakthroughs in knowledge came in the nineteenth century, they could be rapidly diffused and applied. Eighteenth-century medicine, in short, gave promise of a better human existence, but most of the realization lay far in the future.

Knives for Bloodletting In the eighteenth century doctors continued to use these diabolical instruments to treat almost every illness, with disastrous results. *(Courtesy, World Heritage Museum. Photo: Caroline Buckler)*

THE MEDICAL PROFESSIONALS

Care of the sick was the domain of several competing groups—faith healers, apothecaries, surgeons, and physicians. Since the great majority of common ailments have a tendency to cure themselves, each group could point to successes and win adherents. When the doctor's treatment made the patient worse, as it often did, the original medical problem could always be blamed.

Faith healers, who had been one of the most important kinds of physicians in medieval Europe, remained active. They and their patients believed that demons and evil spirits caused disease by lodging in people and that the proper treatment was to exorcise or drive out the offending devil. This demonic view of disease was strongest in the countryside, as was faith in the healing power of religious relics, prayer, and the laying on of hands. Faith healing was particularly effective in the treatment of mental disorders like hysteria and depression, where the link between attitude and illness is most direct.

Apothecaries, or pharmacists, sold a vast number of herbs, drugs, and patent medicines for every conceivable "temperament and distemper." Early pharmacists were seldom regulated, and they frequently diagnosed as freely as the doctors whose prescriptions they filled. Their prescriptions were incredibly complex—a hundred or more drugs might be included in a single prescription—and often very expensive. Some of the drugs undoubtedly worked: strong laxatives were given to the rich for their constipated bowels. Indeed the medical profession continued to believe that regular "purging" of the bowels was essential for good health and the treatment of illness. Much purging was harmful, however, and only bloodletting for the treatment of disease was more effective in speeding patients to their graves.

Drugs were prescribed and concocted in a helter-skelter way. With so many different drugs being combined, it was impossible to isolate cause and effect. Nor was there any standardization. A complicated prescription filled by ten different pharmacists would result in ten different preparations with different medical properties.

Surgeons competed vigorously with barbers and "bone benders," the forerunners of chiropractors. The eighteenth-century surgeon (and patient) labored in the face of incredible difficulties. Almost all operations were performed without any painkiller, for anesthesia was believed too dangerous. The terrible screams of people whose limbs were being sawed off shattered hospitals and battlefields. Such operations were common, because a surgeon faced with an extensive wound sought to obtain a plain surface that he could cauterize with fire. Thus, if a person broke an arm or a leg and the bone stuck out, off came the limb. Many patients died from the agony and shock of such operations.

Surgery was also performed in the midst of filth and dirt. There simply was no knowledge of bacteriology and the nature of infection. The simplest wound treated by a surgeon festered, often fatally. In fact, surgeons encouraged wounds to fester in the belief—a remnant of Galen's theory—that the pus was beneficially removing the base portions of the body.

Physicians, the fourth major group, were trained like surgeons. They were apprenticed in their teens to a practicing physician for several years of on-the-job training. This training was then rounded out with hospital work or some university courses. To their credit, physicians in the eighteenth century were increasingly willing to experiment with new methods, but the hand of Galen lay heavily on them. Bloodletting was still considered a medical cure-all. It was the way "bad blood," the cause of illness, was removed and the balance of humors necessary for good health restored.

According to a physician practicing medicine in Philadelphia in 1799, "No operation of surgery is so frequently necessary as bleeding. . . . But though practiced by midwives, gardeners, blacksmiths, etc., very few know when it is proper." The good doctor went on to explain that bleeding was proper at the onset of all inflammatory fevers, in all inflammations, and for "asthma, sciatic pains, coughs, headaches, rheumatisms, the apoplexy, epilepsy, and bloody fluxes."[12] It was also necessary after all falls, blows, and bruises.

Physicians, like apothecaries, laid great stress on purging. They also generally believed that disease was caused by bad odors, and for this reason they carried canes whose heads contained ammonia salts. As they made their rounds in the filthy, stinking hospitals, physicians held their canes to their noses to protect themselves from illness.

While ordinary physicians were bleeding, apothecaries purging, surgeons sawing, and faith healers praying, the leading medical thinkers were attempting to pull together and assimilate all the information and misinformation they had been accumulating. The attempt was ambitious: to systematize medicine around simple, basic principles, as Newton had done in physics. But the schools of thought resulting from such speculation and theorizing did little to improve medical care. Proponents of *animism* explained life and disease in terms of *anima,* the "sensitive soul," which they believed was present throughout the body and prevented its decay and self-destruction. Another school, *vitalism,* stressed "the vital principle," which inhabited all parts of the body. Vitalists tried to classify diseases systematically.

More interesting was the *homeopathic* system of Samuel Hahnemann of Leipzig. Hahnemann believed that very small doses of drugs that produce certain symptoms in a healthy person will cure a sick person with those symptoms. This theory was probably preferable to most eighteenth-century treatments, in that it was a harmless alternative to the extravagant and often fatal practices of bleeding, purging, drug taking, and induced vomiting. The patient gained confidence, and the body had at least a fighting chance of recovering.

HOSPITALS

Hospitals were terrible throughout most of the eighteenth century. There was no isolation of patients. Operations were performed in the patient's bed. The nurses were old, ignorant, greedy, and often drunk women. Fresh air was considered harmful, and infections of every kind were rampant. Diderot's article in the *Encyclopedia* on the Hôtel-Dieu in Paris, the "richest and most terrifying of all French hospitals," vividly describes normal conditions of the 1770s:

Imagine a long series of communicating wards filled with sufferers of every kind of disease who are sometimes packed three, four, five or even six into a bed, the living alongside the dead and dying, the air polluted by this mass of unhealthy bodies, passing pestilential germs of their afflictions from one to the other, and the spectacle of suffering and agony on every hand. That is the Hôtel-Dieu.

The result is that many of these poor wretches come out with diseases they did not have when they went in, and often pass them on to the people they go back to live

Hospital Life Patients crowded into hospitals like this one in Hamburg in 1746 had little chance of recovery. A priest by the window administers last rites, while in the center a surgeon coolly saws off the leg of a man who has received no anesthesia. *(Germanisches Nationalmuseum, Nuremberg)*

with. Others are half-cured and spend the rest of their days in an invalidism as hard to bear as the illness itself; and the rest perish, except for the fortunate few whose strong constitutions enable them to survive.[13]

No wonder the poor of Paris hated hospitals and often saw confinement there as a plot to kill paupers.

In the last years of the century, the humanitarian concern already reflected in Diderot's description of the Hôtel-Dieu led to a movement for hospital reform through western Europe. Efforts were made to improve ventilation and eliminate filth, on the grounds that bad air caused disease. The theory was wrong, but the results were beneficial, since the spread of infection was somewhat reduced.

MENTAL ILLNESS

Mental hospitals, too, were incredibly savage institutions. The customary treatment for mental illness was bleeding and cold water, administered more to maintain discipline than to effect a cure. Violent persons were chained to the wall and forgotten. A breakthrough of sorts occurred in the 1790s, when William Tuke founded the first humane sanatorium in England. In Paris an innovative warden, Philippe Pinel, took the chains off the mentally disturbed in 1793 and tried to treat them as patients rather than as prisoners.

In the eighteenth century, there were all sorts of wildly erroneous ideas about mental illness. One was that moonlight caused madness, a belief reflected in the word *lunatic*—someone harmed by lunar light. Another mid-eighteenth-century theory, which lasted until at least 1914, was that masturbation caused madness, not to mention acne, epilepsy, and premature ejaculation. Thus parents, religious institutions, and schools waged relentless war on masturbation by males, although they were curiously uninterested in female masturbation. In the nineteenth century, this misguided idea was to reach its greatest height, resulting in increasingly drastic medical treatment. Doctors ordered their "patients" to wear mittens, fitted them with wooden braces between the knees, or simply tied them up in straitjackets.

MEDICAL EXPERIMENTS AND RESEARCH

In the second half of the eighteenth century, medicine in general turned in a more practical and experimental direction. Some of the experimentation was creative quackery involving the recently discovered phenomenon of electricity. One magnificent quack, James Graham of London, opened a great hall filled with the walking sticks, crutches, eyeglasses, and ear trumpets of supposedly cured patients, which he kept as symbols of his victory over disease. Great glass globes and the rich perfumes of burning incense awaited all who entered. Graham's principal treatment involved his Celestial Bed, which was lavishly decorated with magnets and electrical devices. Graham claimed that by sleeping in it youths would keep their good looks, their elders would be rejuvenated, and couples would have beautiful, healthy children. The fee for a single night in the Medico-Magnetico-Musico-Electrical Bed was £100—a great sum of money.

The rich could buy expensive treatments, but the prevalence of quacks and the general lack of knowledge meant they often got little for their money. Because so many treatments were harmful, the poor were probably much less deprived by their almost total lack of access to medical care than one might think.

Renewed experimentation and the intensified search for solutions to human problems also led to some real, if still modest, advances in medicine after 1750. The eighteenth century's greatest medical triumph was the conquest of smallpox.

With the progressive decline of bubonic plague, smallpox became the most terrible of the infectious diseases. In the words of the historian Thomas Macaulay, "smallpox was always present, filling the churchyard with corpses, tormenting with constant fears all whom it had not stricken." In the seventeenth century, one in every four deaths in the British Isles was due to smallpox, and it is estimated that 60 million Europeans died of it in the eighteenth century. Fully 80 percent of the population was stricken at some point in life, and 25 percent of the total population was left permanently scarred. If ever a human problem cried out for solution, it was smallpox.

The first step in the conquest of this killer came in the early eighteenth century. An English aristocrat whose great beauty had been marred by the pox, Lady Mary Wortley Montague, learned about the practice of inoculation in the Ottoman Empire while her husband was serving as British ambassador there. She had her own son successfully inoculated in Constantinople and was instrumental in spreading the practice in England after her return in 1722.

Inoculation against smallpox had long been practiced in the Middle East. The skin was deliberately broken, and a small amount of matter taken from the pustule of a smallpox victim was applied. The person thus contracted a mild case of smallpox that gave lasting protection against further attack. Inoculation was risky, however, and about one person in fifty died from it. In addition, people who had been inoculated were just as infectious as those who had caught the disease by chance. Inoculated people thus spread the disease, and the practice of inoculation against smallpox was widely condemned in the 1730s.

Success in overcoming this problem in British colonies led the British College of Physicians in 1754 to strongly advocate inoculation. Moreover, a successful search for cheaper methods led to something approaching mass inoculation in England in the 1760s. One specialist treated seventeen thousand patients and only five died. Both the danger and the cost had been reduced, and deadly smallpox struck all classes less frequently. On the Continent, the well-to-do were also inoculated, beginning with royal families like those of Maria Theresa and Catherine the Great. The practice then spread to the middle classes. Smallpox inoculation played some part in the decline of the death rate at the end of the century and the increase in population.

The Fight Against Smallpox This Russian illustration dramatically urges parents to inoculate their children against smallpox. The good father's healthy youngsters flee from their ugly and infected playmates, who hold their callous father responsible for their shameful fate. *(Yale Medical Library)*

The final breakthrough against smallpox came at the end of the century. Edward Jenner (1749–1823), a talented country doctor, noted that in the English countryside there was a long-standing belief that dairy maids who had contracted cowpox did not get smallpox. Cowpox produces sores on the cow's udder and on the hands of the milker. The sores resemble those of smallpox, but the disease is mild and not contagious.

For eighteen years Jenner practiced a kind of Baconian science, carefully collecting data on protection against smallpox by cowpox. Finally, in 1796 he performed his first vaccination on a young boy, using matter taken from a milkmaid with cowpox. Two months later he inoculated the boy with smallpox pus, but the disease did not take. In the next two years, twenty-three successful vaccinations were performed, and in 1798 Jenner published his findings. There was some skepticism and hostility, but after Austrian medical authorities replicated Jenner's re-

sults, the new method of treatment spread rapidly. Smallpox soon declined to the point of disappearance in Europe and then throughout the world. Jenner eventually received prizes of £30,000 from the British government for his great discovery, a fitting recompense for a man who gave an enormous gift to humanity and helped lay the foundation for the science of immunology in the nineteenth century.

RELIGION AND CHRISTIAN CHURCHES

Though the critical spirit of the Enlightenment spread among the educated elite in the eighteenth century, the great mass of ordinary men and women remained firmly committed to the Christian religion, especially in rural areas. Religion offered answers to life's mysteries and gave comfort and courage in the face of sorrow and fear. Religion also remained strong because it was usually embedded in local traditions and everyday social experience.

Yet the popular religion of village Europe was everywhere enmeshed in a larger world of church hierarchies and state power. These powerful outside forces sought to regulate religious life at the local level. These efforts created tensions that helped set the scene for a vigorous religious revival in Germany and England.

THE INSTITUTIONAL CHURCH

As in the Middle Ages (Chapters 9–12), the local parish church remained the basic religious unit all across Europe. Still largely coinciding with the agricultural village, the parish fulfilled many needs. The parish church was the focal point of religious devotion, which went far beyond sermons and Holy Communion. The parish church organized colorful processions and pilgrimages to local shrines. Even in Protestant countries, where such activities were severely restricted, congregations gossiped and swapped stories after services, and neighbors came together in church for baptisms, marriages, funerals, and special events. Thus the parish church was woven into the very fabric of community life.

Moreover, the local church had important administrative tasks. Priests and parsons were truly the bookkeepers of agrarian Europe, and it is because parish registers were so complete that historians have learned so much about population and family life. Parishes also normally distributed charity to the destitute, looked after orphans, and provided whatever primary education was available.

The many tasks of the local church were usually the responsibility of a resident priest or pastor, a full-time professional working with assistants and lay volunteers. Moreover, all clerics—whether Catholic, Protestant, or Orthodox—shared the fate of middlemen in a complicated institutional system. Charged most often with ministering to poor peasants, the priest or parson was the last link in a powerful church-state hierarchy that was everywhere determined to control religion down to the grassroots. However, the regulatory framework of belief, which went back at least to the fourth century, when Christianity became the official religion of the Roman Empire, had undergone important changes since 1500.

The Protestant Reformation had begun as a culmination of medieval religiosity and a desire to purify Christian belief. Martin Luther, the greatest of the reformers (pages 431–438), preached that all men and women were saved from their sins and God's damnation only by personal faith in Jesus Christ. The individual could reach God directly, without need of priestly intermediaries. This was the revolutionary meaning of Luther's "priesthood of all believers," which broke forever the monopoly of the priestly class over medieval Europe's most priceless treasury —eternal salvation.

As the Reformation gathered force, with peasant upheaval and doctrinal competition, Luther turned more conservative. The monkish professor called on willing German princes to put themselves at the head of official churches in their territories. Other monarchs in northern Europe followed suit. Protestant authorities, with generous assistance from state-certified theologians like Luther, then proceeded to regulate their "territorial churches" strictly, selecting personnel and imposing detailed rules. They joined with Catholics to crush the Anabaptists (pages 445–447), who with their belief in freedom of conscience and separation of church and state had become the real revolutionaries. Thus the Reformation, initially so radical in its rejection of Rome and its stress on individual religious experience, eventually resulted in a bureaucratization of the church and local religious life in Protestant Europe.

The Reformation era also increased the practical power of Catholic rulers over "their" churches, but it was only in the eighteenth century that some Catholic monarchs began to impose striking reforms. These reforms, which had their counterparts in Orthodox Russia, had a very "Protestant" aspect. They increased state control over the Catholic church, making it less subject to papal influence.

Spain provides a graphic illustration of changing church-state relations in Catholic lands. A deeply Catholic country with devout rulers, Spain nevertheless took firm control of ecclesiastical appointments. Papal proclamations could not even be read in Spanish churches without prior approval from the government. Spain also asserted state control over the Spanish Inquisition (pages 420–421), which had been ruthlessly pursuing heresy as an independent agency under Rome's direction for two hundred years. In sum, Spain went far toward creating a "national" Catholic church, as France had done earlier.

A more striking indication of state power and papal weakness was the fate of the Society of Jesus. As the most successful of the Catholic Reformation's new religious orders (see page 457), the well-educated Jesuits were extraordinary teachers, missionaries, and agents of the papacy. In many Catholic countries, the Jesuits exercised tremendous political influence, since individual members held high government positions, and Jesuit colleges formed the minds of Europe's Catholic nobility. Yet, by playing politics so effectively, the Jesuits eventually raised a broad coalition of enemies, which destroyed their order. Especially bitter controversies over the Jesuits rocked the entire Catholic hierarchy in France. Following the earlier example of Portugal, the French king ordered the Jesuits out of France in 1763 and confiscated their property. France and Spain then pressured Rome to dissolve the Jesuits completely. In 1773 a reluctant pope caved in, although the order was revived after the French Revolution.

Some Catholic rulers also turned their reforming efforts on monasteries and convents, believing that the large monastic clergy should make a more practical contribution to social and religious life. Austria, a leader in controlling the church (see page 601), showed how far the process could go. Whereas Maria Theresa sharply restricted entry into "unproductive" orders, Joseph II recalled the radical initiatives of the Protestant Reformation. In his *Edict on Idle Institutions,* Joseph abolished contemplative orders, henceforth permitting only orders engaged in teaching, nursing, or other practical work. The number of monks plunged from 65,000 to 27,000. The state also expropriated the dissolved monasteries and used their great wealth for charitable purposes and higher salaries for ordinary priests.

CATHOLIC PIETY

Catholic territorial churches also sought to purify religious practice somewhat. As might be expected, Joseph II went the furthest. Above all, he and his agents sought to root out what they considered to be idolatry and superstition. Yet pious peasants saw only an incomprehensible attack on the true faith and drew back in anger. Joseph's sledgehammer approach and the resulting reaction dramatized an underlying tension between Christian reform and popular piety after the Reformation.

Protestant reformers had taken very seriously the commandment that "Thou shalt not make any graven image" (Exodus 20:4), and their radical reforms had reordered church interiors. Relics and crucifixes had been permanently removed from crypt and altar, while stained-glass windows had been smashed and walls and murals covered with whitewash. Processions and pilgrimages, saints and shrines—all such nonessentials had been rigorously suppressed in the attempt to recapture the vital core of the Christian religion. Such revolutionary changes had often troubled ordinary churchgoers, but by the late seventeenth century, these reforms had been thoroughly routinized by official Protestant churches.

The situation was quite different in Catholic Europe around 1700. First of all, the visual contrast was striking; baroque art (pages 561–562) had lavished rich and emotionally exhilarating figures and images on Catholic churches, just as Protestants had removed theirs. From almost every indication, people in Catholic Europe remained intensely religious. More than 95 percent of the population probably attended church for Easter Communion, the climax of the Catholic church year.

Much of the tremendous popular strength of religion in Catholic countries reflected that its practice went far beyond Sunday churchgoing and was an important part of community life. Thus, although Catholics reluctantly confessed their sins to the priest, they enthusiastically joined together in public processions to celebrate the passage of the liturgical year. In addition to the great processional days— such as Palm Sunday, the joyful re-enactment of Jesus' triumphal entry into Jerusalem, or Rogations, with its chanted supplications and penances three days before the bodily ascent of Jesus into heaven on Ascension Day—each parish had its own local processions. Led by its priest, a congregation might march around the village, or across the countryside to a local shrine or chapel. There were endless variations. In the southern French Alps, the people looked forward especially to "high-mountain" processions in late spring. Parishes came together from miles around on some high mountain. There the assembled priests asked God to bless the people with healthy flocks and pure waters, and then all joined together in an enormous picnic. Before each procession, the priest explained its religious significance to kindle group piety. But processions were also folklore

Planting Crosses After their small wooden crosses were blessed at church as part of the Rogation Days' ceremonies in early May, French peasants traditionally placed them in their fields and pastures to protect the seed and help the crops grow. This illustration shows the custom being performed in the French Alps in the mid-nineteenth century, when it was still widely practiced. *(L' Illustration/Library of Congress)*

and tradition, an escape from work and a form of recreation. A holiday atmosphere sometimes reigned on longer processions, with drinking and dancing and couples disappearing into the woods.

Devout Catholics held many religious beliefs that were marginal to the Christian faith, often of obscure or even pagan origin. On the feast of Saint Anthony, priests were expected to bless salt and bread for farm animals to protect them from disease. One saint's relics could help cure a child of fear, and there were healing springs for many ailments. The ordinary person combined a strong Christian faith with a wealth of time-honored superstitions.

Parish priests and Catholic hierarchies were frequently troubled by the limitations of their parishioners' Christian understanding. One parish priest in France, who kept an invaluable daily diary, lamented that his parishioners were "more superstitious than devout . . . and sometimes appear as baptized idolators."[14]

Many parish priests in France, often acting on instructions from their bishops, made an effort to purify popular religious culture. For example, one priest tried to abolish pilgrimages to a local sacred spring of Our Lady, reputed to revive dead babies long enough for a proper baptism. French priests denounced par-

"Clipping the Church" The ancient English ceremony of dancing around the church once each year on the night before Lent undoubtedly had pre-Christian origins, for its purpose was to create a magical protective chain against evil spirits and the devil. The Protestant reformers did their best to stamp them out, but such "pagan practices" sometimes lingered on. *(Somerset Archaeological and Natural History Society)*

ticularly the "various remnants of paganism" found in popular bonfire ceremonies during Lent, in which young men, "yelling and screaming like madmen," tried to jump over the bonfires in order to help the crops grow and protect themselves from illness. One priest saw rational Christians turning back into pagan animals—"the triumph of Hell and the shame of Christianity."[15]

Yet, whereas Protestant reformers had already used the power of the territorial state to crush such practices, Catholic church leaders generally pro-ceeded cautiously in the eighteenth century. They knew that old beliefs—such as the belief common throughout Europe that the priest's energetic ringing of churchbells and his recitation of ritual prayers would protect the village from hail and thunder-storms—were an integral part of the people's reli-gion. Thus Catholic priests and hierarchies generally preferred a compromise between theological purity and the people's piety, realizing perhaps that the line between divine truth and mere superstition is not easily drawn.

PROTESTANT REVIVAL

By the late seventeenth century, official Protestant churches had completed their vast reforms and had generally settled into a smug complacency. In the Reformation heartland, one concerned German minister wrote that the Lutheran church "had become paralyzed in forms of dead doctrinal conformity" and badly needed a return to its original inspiration.[16] This voice was one of many that would prepare and then guide a powerful Protestant revival, a revival largely successful because it answered the intense but increasingly unsatisfied needs of common people.

The Protestant revival began in Germany. It was known as "Pietism," and three aspects helped explain its powerful appeal. First, Pietism called for warm emotional religion that everyone could experience. Enthusiasm—in prayer, in worship, in preaching, in life itself—was the key concept. "Just as a drunkard becomes full of wine, so must the congregation become filled with spirit," declared one exuberant writer. Another said simply, "The heart must burn."[17]

Second, Pietism reasserted the earlier radical stress on the "priesthood of all believers," thereby reducing the large gulf between the official clergy and the Lutheran laity, which had continued to exist after the Reformation. Bible reading and study were enthusiastically extended to all classes, which provided a powerful spur for popular education as well as individual religious development. Finally, Pietists believed in the practical power of Christian rebirth in everyday affairs. Reborn Christians were expected to lead good, moral lives and come from all walks of life.

Pietism had a major impact on John Wesley (1703–1791), who served as the catalyst for popular religious revival in England. Wesley came from a long line of ministers, and when he went to Oxford University to prepare for the clergy, he mapped a fanatically earnest "scheme of religion." Like some students during final exam period, he organized every waking moment. After becoming a teaching fellow at Oxford, he organized a Holy Club for similarly minded students, who were soon known contemptuously as "Methodists" because they were so methodical in their devotion. Yet, like the young Luther, Wesley remained intensely troubled about his own salvation, even after his ordination as an Anglican priest in 1728.

Wesley's anxieties related to grave problems of the faith in England. The Church of England was shamelessly used by the government to provide favorites with high-paying jobs and sinecures. Building of churches practically stopped while the population grew, and in many parishes there was a grave shortage of pews. Services and sermons had settled into an uninspiring routine. That the properly purified religion had been separated from local customs and social life was symbolized by church doors that were customarily locked on weekdays. Moreover, the skepticism of the Enlightenment was making inroads among the educated classes, and deism was becoming popular. Some bishops and church leaders acted as if they believed that doctrines like the Virgin Birth or the Ascension were little more than particularly elegant superstitions.

Living in an atmosphere of religious decline and uncertainty, Wesley became profoundly troubled by his lack of faith in his own salvation. Yet spiritual counseling from a sympathetic Pietist minister from Germany prepared Wesley for a mystical, emotional "conversion" in 1738. He described this critical turning point in his *Journal*:

In the evening I went to a [Christian] society in Aldersgate Street where one was reading Luther's preface to the Epistle to the Romans. *About a quarter before nine, while he was describing the change which God works in the heart through faith in Christ, I felt my heart strangely warmed. I felt I did trust in Christ, Christ alone for salvation; and an assurance was given me that he had taken away* my *sins, even* mine, *and saved* me *from the law of sin and death.*[18]

Wesley's emotional experience resolved his intellectual doubts. Moreover, he was convinced that any person, no matter how poor or simple, might have a similar heartfelt conversion and gain the same blessed assurance.

Wesley took the good news to the people. Since existing churches were often overcrowded and the church-state establishment was hostile, Wesley preached in open fields. People came in large numbers. Of critical importance, Wesley expanded on earlier Dutch theologians' views and emphatically rejected Calvinist predestination—the doctrine of salvation granted only to a select few (see page 444). Rather, *all* men and women who earnestly sought salvation might be saved. It was a message of hope and joy, of free will and universal salvation.

Open-Air Preaching In an age when the established Church of England lacked religious fervor and smugly reinforced the social status quo, John Wesley carried the good news of salvation to common people across the land. Wesley sparked a powerful religious revival, symbolized by the joyful bell-ringing of the town crier in this illustration. *(Historical Pictures Service)*

Traveling some 225,000 miles by horseback and preaching more than 40,000 sermons in fifty years, Wesley's ministry won converts, formed Methodist cells, and eventually resulted in a new denomination. Evangelicals in the Church of England and the old dissenting groups also followed Wesley's example, giving impetus to an even broader awakening among the lower classes. That result showed that in England, as throughout Europe despite different churches and different practices, religion remained a vital force in the lives of the people.

In recent years, imaginative research has greatly increased the specialist's understanding of ordinary life and social patterns in the past. The human experience, as recounted by historians, has become richer and more meaningful, and many mistaken ideas

have fallen. This has been particularly true of eighteenth-century, preindustrial Europe. The intimacies of family life, the contours of women's history and of childhood, and vital problems of medicine and religion are emerging from obscurity. Nor is this all. A deeper, truer understanding of the life of common people can shed light on the great economic and political developments of long-standing concern, to be seen in the next chapter.

NOTES

1. Quoted by J. M. Beattie, "The Criminality of Women in Eighteenth-Century England," *Journal of Social History* 8 (Summer 1975): 86.
2. W. L. Langer, "Infanticide: A Historical Survey," *History of Childhood Quarterly* 1 (Winter 1974): 357.

3. Quoted in R. Cobb, *The Police and the People: French Popular Protest, 1789–1820,* Clarendon Press, Oxford, Eng., 1970, p. 238.

4. Quoted by E. A. Wrigley, *Population and History,* McGraw-Hill, New York, 1969, p. 127.

5. Quoted in D. S. Landes, ed., *The Rise of Capitalism,* Macmillan, New York, 1966, pp. 56–57.

6. See L. A. Tilly, J. W. Scott, and M. Cohen, "Women's Work and European Fertility Patterns," *Journal of Interdisciplinary History* 6 (Winter 1976): 447–476.

7. J. Michelet, *The People,* trans. with an introduction by J. P. McKay, University of Illinois Press, Urbana, 1973 (original publication, 1846), pp. 38–39.

8. J. Brownlow, *The History and Design of the Foundling Hospital,* London, 1868, p. 7.

9. Quoted by B. W. Lorence, "Parents and Children in Eighteenth-Century Europe," *History of Childhood Quarterly* 2 (Summer 1974): 1–2.

10. Ibid., pp. 13, 16.

11. J. C. Drummond and A. Wilbraham, *The Englishman's Food: A History of Five Centuries of English Diet,* 2nd ed., Jonathan Cape, London, 1958, p. 75.

12. Quoted by L. S. King, *The Medical World of the Eighteenth Century,* University of Chicago Press, Chicago, 1958, p. 320.

13. Quoted by R. Sand, *The Advance to Social Medicine,* Staples Press, London, 1952, pp. 86–87.

14. Quoted by I. Woloch, *Eighteenth-Century Europe: Tradition and Progress, 1715–1789,* W. W. Norton, New York, 1982, p. 292.

15. Quoted by T. Tackett, *Priest and Parish in Eighteenth-Century France,* Princeton University Press, Princeton, N. J., 1977, p. 214.

16. Quoted by K. Pinson, *Pietism as a Factor in the Rise of German Nationalism,* Columbia University Press, New York, 1934, p. 13.

17. Ibid., pp. 43–44.

18. Quoted by S. Andrews, *Methodism and Society,* Longmans, Green, London, 1970, p. 327.

SUGGESTED READING

Though often ignored in many general histories of the Western world, social topics of the kind considered in this chapter flourish in specialized journals today. The articles cited in the Notes are typical of the exciting work being done, and the reader is strongly advised to take time to look through recent volumes of some leading journals: *Journal of Social History, Past and Present, History of Childhood Quarterly,* and *Journal of Interdisciplinary History.* In addition, the number of book-length studies has begun to expand rapidly.

Among general introductions to the history of the family, women, and children, E. A. Wrigley, *Population and History* (1969), is excellent. P. Laslett, *The World We Have Lost* (1965), is an exciting, pioneering investigation of England before the Industrial Revolution, though some of his conclusions have been weakened by further research. L. Stone, *The Family, Sex and Marriage in England, 1500–1800* (1977), is a brilliant general interpretation, and L. Tilly and J. Scott, *Women, Work and Family* (1978), is excellent. P. Ariès, *Centuries of Childhood: A Social History of Family Life* (1962), is another stimulating study. E. Shorter, *The Making of the Modern Family* (1975), is an all-too-lively and rather controversial interpretation. All four works are highly recommended. T. Rabb and R. I. Rothberg, eds., *The Family in History* (1973), is a good collection of articles dealing with both Europe and the United States. A. MacFarlane, *The Family Life of Ralph Josselin* (1970), is a brilliant re-creation of the intimate family circle of a seventeenth-century English clergyman who kept a detailed diary; MacFarlane's *Origins of English Individualism: The Family, Property and Social Transition* (1978) is a major work. I. Pinchbeck and M. Hewitt, *Children in English Society* (1973), is a good introduction. E. Flexner has written a fine biography on the early feminist Mary Wollstonecraft (1972). Various aspects of sexual relationships are treated imaginatively by M. Foucault, *The History of Sexuality* (1981), and R. Wheaton and T. Hareven, eds., *Family and Sexuality in French History* (1980).

J. Burnett, *A History of the Cost of Living* (1969), has a great deal of interesting information about what people spent their money on in the past and complements the fascinating work of J. C. Drummond and A. Wilbraham, *The Englishman's Food: A History of Five Centuries of English Diet* (1958). J. Knyveton, *Diary of a Surgeon in the Year 1751–1752* (1937), gives a contemporary's unforgettable picture of both eighteenth-century medicine and social customs. Good introductions to the evolution of medical practices are B. Ingles, *History of Medicine* (1965); O. Bettmann, *A Pictorial History of Medicine* (1956); and H. Haggard's old but interesting *Devils, Drugs, and Doctors* (1929). W. Boyd, *History of Western Education* (1966), is a standard survey, which may be usefully supplemented by an important article by L. Stone, "Literacy and Education in England, 1640–1900," *Past and Present* 42 (February 1969): 69–139. M. D. George, *London Life in the Eighteenth Century* (1965), is a delightfully written book, while L. Chevalier, *Labouring Classes and Dangerous Classes* (1973), is a keen analysis of the poor people of Paris in a slightly later period. G. Rudé, *The Crowd in History, 1730–1848* (1964), is an innovative effort to see politics and popular protest from below. An important series edited by R. Forster and O. Ranuum considers neglected social questions such as diet, abandoned children, and deviants, as does P. Burke's excellent study, *Popular Culture in Early Modern Europe* (1978).

Good works on religious life include J. Delumeau, *Catholicism Between Luther and Voltaire: A New View of the Counter-Reformation* (1977); T. Tackett, *Priest and Parish in Eighteenth-Century France* (1977); B. Semmel, *The Methodist Revolution* (1973); and J. Bettey, *Church and Community: The Parish Church in English Life* (1979).

THE REVOLUTION IN POLITICS, 1775–1815

HE LAST YEARS of the eighteenth century were a time of great upheaval. A series of revolutions and revolutionary wars challenged the old order of kings and aristocrats. The ideas of freedom and equality, ideas that have not stopped shaping the world since that era, flourished and spread. The revolution began in North America in 1775. Then in 1789 France, the most influential country in Europe, became the leading revolutionary nation. It established first a constitutional monarchy, then a radical republic, and finally a new empire under Napoleon. The armies of France also joined forces with patriots and radicals abroad in an effort to establish new governments based on new principles throughout much of Europe. The world of modern domestic and international politics was born.

What caused this era of revolution? What were the ideas and objectives of the men and women who rose up violently to undo the established system? What were the gains and losses for privileged groups and for ordinary people in a generation of war and upheaval? These are the questions on which this chapter's examination of the French and American revolutions will be based.

LIBERTY AND EQUALITY

Two ideas fueled the revolutionary period in both America and Europe: liberty and equality. What did eighteenth-century politicians and other people mean by liberty and equality, and why were those ideas so radical and revolutionary in their day?

The call for liberty was first of all a call for individual human rights. Even the most enlightened monarchs customarily claimed that it was their duty to regulate what people wrote and believed. Liberals of the revolutionary era protested such controls from on high. They demanded freedom to worship according to the dictates of their consciences instead of according to the politics of their prince. They demanded the end of censorship and the right to express their beliefs freely in print and at public meetings. They demanded freedom from arbitrary laws and from judges who simply obeyed orders from the government.

These demands for basic personal freedoms, which were incorporated into the American Bill of Rights and other liberal constitutions, were very far-reaching. Indeed, eighteenth-century revolutionaries demanded more freedom than most governments today believe it is desirable to grant. The Declaration of the Rights of Man, issued at the beginning of the French Revolution, proclaimed, "Liberty consists in being able to do anything that does not harm another person." A citizen's rights had, therefore, "no limits except those which assure to the other members of society the enjoyment of these same rights." Liberals called for the freedom of the individual to develop and to create to the fullest possible extent. In the context of the aristocratic and monarchial forms of government that then dominated Europe, this was a truly radical idea.

The call for liberty was also a call for a new kind of government. The revolutionary liberals believed that the people were sovereign—that is, that the people alone had the authority to make laws limiting the individual's freedom of action. In practice, this system of government meant choosing legislators who represented the people and who were accountable to them. Moreover, liberals of the revolutionary era believed that every people—every ethnic group—had this right of self-determination and thus the right to form a free nation.

By equality, eighteenth-century liberals meant that all citizens were to have identical rights and civil liberties. Above all, the nobility had no right to special privileges based on the accident of birth.

Liberals did not define equality as meaning that everyone should be equal economically. Quite the contrary. As Thomas Jefferson wrote in an early draft of the American Declaration of Independence, before changing "property" to the more noble-sounding "happiness," everyone was equal in "the pursuit of property." Jefferson and other liberals certainly did not expect equal success in that pursuit. Great differences in wealth and income between rich and poor were perfectly acceptable to liberals. The essential point was that everyone should legally have an equal chance. French liberals and revolutionaries said they wanted "careers opened to talent." They wanted employment in government, in business, and in the professions to be based on ability, not on family background or legal status.

Equality of opportunity was a very revolutionary idea in eighteenth-century Europe. Legal inequality between classes and groups was the rule, not the exception. Society was still legally divided into groups with special privileges, such as the nobility and the clergy, and groups with special burdens, like the peasantry. In many countries, various middle-class groups—professionals, businessmen, townspeople, and craftsmen—enjoyed privileges that allowed them to monopolize all sorts of economic activity. It was this kind of economic inequality, an inequality based on artificial legal distinctions, against which liberals protested.

THE ROOTS OF LIBERALISM

The ideas of liberty and equality—the central ideas of classical liberalism—have deep roots in Western history. The ancient Greeks and the Judeo-Christian tradition had affirmed for hundreds of years the sanctity and value of the individual human being. The Judeo-Christian tradition, reinforced by the Reformation, had long stressed personal responsibility on the part of both common folk and exalted rulers, thereby promoting the self-discipline without which liberty becomes anarchy. The hounded and persecuted Protestant radicals of the later sixteenth century had died for the revolutionary idea that individuals were entitled to their own religious beliefs.

Although the liberal creed had roots deep in the Western tradition, classical liberalism first crystallized at the end of the seventeenth century and during the Enlightenment of the eighteenth century.

Liberal ideas reflected the Enlightenment's stress on human dignity and human happiness on earth. Liberals shared the Enlightenment's general faith in science, rationality, and progress: the adoption of liberal principles meant better government and a better society for all. Almost all the writers of the Enlightenment were passionately committed to greater personal liberty. They preached religious toleration, freedom of press and speech, and fair and equal treatment before the law.

Certain English and French thinkers were mainly responsible for joining the Enlightenment's concern for personal freedom and legal equality to a theoretical justification of liberal self-government. The two most important were John Locke and the baron de Montesquieu, considered earlier. Locke (page 530) maintained that England's long political tradition rested on "the rights of Englishmen" and on representative government through Parliament. Locke ad-

The Marquis de Lafayette was the most famous great noble to embrace the liberal revolution. Shown here directing a battle in the American Revolution, he returned to champion liberty and equality in France. For admirers he was the "hero of two worlds." *(Anne S. K. Brown Military Collection, John Hay Library, Brown University)*

mired especially the great Whig noblemen who had made the bloodless revolution of 1688, and he argued that a government that oversteps its proper functions —protecting the natural rights of life, liberty, and private property—becomes a tyranny. Montesquieu (page 585) was also inspired by English constitutional history. He, too, believed that powerful "intermediary groups"—such as the judicial nobility of which he was a proud member—offered the best defense of liberty against despotism.

THE ATTRACTION OF LIBERALISM

The belief that representative institutions could defend their liberty and interests appealed powerfully to ambitious and educated bourgeois. Yet it is important to realize that liberal ideas about individual rights and political freedom also appealed to much of the aristocracy, at least in western Europe and as formulated by Montesquieu. Representative government did not mean democracy, which liberal thinkers tended to equate with mob rule. Rather, they envisioned voting for representatives as being restricted to those who owned property, those with "a stake in society." England had shown the way. After 1688 it had combined a parliamentary system and considerable individual liberty with a restricted franchise and unquestionable aristocratic pre-eminence. In the course of the eighteenth century, many leading French nobles, led by a judicial nobility inspired by Montesquieu, as shown in Chapter 18, were increasingly eager to follow the English example.

Eighteenth-century liberalism, then, appealed not only to the middle class, but also to some aristocrats. It found broad support among the educated elite and the substantial classes in western Europe. What it lacked from the beginning was strong mass support. For comfortable liberals, the really important questions were theoretical and political. They had no need to worry about their stomachs and the price of bread. For the much more numerous laboring poor, the great questions were immediate and economic. Getting enough to eat was the crucial challenge. These differences in outlook and well-being were to lead to many misunderstandings and disappointments for both groups in the revolutionary era.

THE AMERICAN REVOLUTION, 1775–1789

The era of liberal revolution began in the New World. The thirteen mainland colonies of British North America revolted against their mother country and then succeeded in establishing a new unified government.

Americans have long debated the meaning of their revolution. Some have even questioned whether or not it was a real revolution, as opposed to a war for independence. According to some scholars, the Revolution was conservative and defensive in that its demands were for the traditional liberties of Englishmen; Americans were united against the British, but otherwise they were a satisfied people, not torn by internal conflict. Other scholars have argued that, on

the contrary, the American Revolution was quite radical. It split families between patriots and Loyalists and divided the country. It achieved goals that were fully as advanced as those obtained by the French in their great revolution a few years later.

How does one reconcile these positions? Both contain large elements of truth. The American revolutionaries did believe they were demanding only the traditional rights of English men and women. But those traditional rights were liberal rights, and in the American context they had very strong democratic and popular overtones. Thus the American Revolution was fought in the name of established ideals that were still quite radical in the context of the times. And in founding a government firmly based on liberal principles, the Americans set an example that had a forceful impact on Europe and speeded up political development there.

THE ORIGINS OF THE REVOLUTION

The American Revolution had its immediate origins in a squabble over increased taxes. The British government had fought and decisively won the Seven Years' War (page 625) on the strength of its professional army and navy. The American colonists had furnished little real aid. The high cost of the war to the British, however, had led to a doubling of the British national debt. Anticipating further expense defending its recently conquered western lands from Indian uprisings like that of Pontiac, the British government in London set about reorganizing the empire with a series of bold, largely unprecedented measures. Breaking with tradition, the British decided to maintain a large army in North America after peace was restored in 1763. Moreover, they sought to exercise strict control over their newly conquered western lands and to tax the colonies directly. In 1765 the government pushed through Parliament the Stamp Act, which levied taxes on a long list of commercial and legal documents, diplomas, pamphlets, newspapers, almanacs, dice, and playing cards. A stamp glued to each article indicated the tax had been paid.

The effort to increase taxes as part of tightening up the empire seemed perfectly reasonable to the British. Heavier stamp taxes had been collected in Great Britain for two generations, and Americans were being asked only to pay a share of their own defense.

The Boston Tea Party This contemporary illustration shows men disguised as Indians dumping East India Company tea into Boston's harbor. The enthusiastic crowd cheering from the wharf indicates widespread popular support. *(Library of Congress)*

Moreover, Americans had been paying only very low local taxes. The Stamp Act would have doubled taxes to about two shillings per person. No other people in the world (except the Poles) paid so little. The British, meanwhile, paid the world's highest taxes in about 1765—twenty-six shillings per person. It is not surprising that taxes per person in the newly independent American nation were much higher in 1785 than in 1765, when the British no longer subsidized American defense. The colonists protested the Stamp Act vigorously and violently, however, and after rioting and boycotts against British goods, Parliament reluctantly repealed the new tax.

As the fury of the Stamp Act controversy revealed, much more was involved than taxes. The key question was political. To what extent could the home government refashion the empire and reassert its power while limiting the authority of colonial legislatures and their elected representatives? Accordingly, who should represent the colonies, and who had the right to make laws for Americans? While a troubled majority of Americans searched hard for a compromise, some radicals began to proclaim that "taxation without representation is tyranny." The British government replied that Americans were represented in Parliament, albeit indirectly (like most Englishmen themselves), and that the absolute supremacy of Parliament throughout the empire could not be questioned. Many Americans felt otherwise. As John Adams put it, "A Parliament of Great Britain can have no more rights to tax the colonies than a Parliament of Paris." Thus imperial reorganization and Parliamentary supremacy came to appear as grave threats to Americans' existing liberties and time-honored institutions.

Americans had long exercised a great deal of independence and gone their own way. In British North America, unlike England and Europe, no powerful established church existed, and personal freedom in questions of religion was taken for granted. The colonial assemblies made the important laws, which were seldom overturned by the home government. The right to vote was much more widespread than in England. In many parts of colonial Massachusetts, for example, as many as 95 percent of the adult males could vote.

Moreover, greater political equality was matched by greater social and economic equality. Neither a hereditary nobility nor a hereditary serf population existed, although the slavery of the Americas consigned blacks to a legally oppressed caste. Independent farmers were the largest group in the country and set much of its tone. In short, the colonial experience had slowly formed a people who felt themselves separate and distinct from the home country. The controversies over taxation intensified those feelings of distinctiveness and separation and brought them to the fore.

In 1773 the dispute over taxes and representation flared up again. The British government had permitted the financially hard-pressed East India Company to ship its tea from China directly to its agents in the colonies, rather than through London middlemen, who then sold to independent merchants in the colonies. Thus the company secured a vital monopoly on the tea trade, and colonial merchants were suddenly excluded from a highly profitable business. The colonists were quick to protest.

In Boston, men disguised as Indians had a rowdy "tea party" and threw the company's tea into the harbor. This led to extreme measures. The so-called Coercive Acts closed the port of Boston, curtailed local elections and town meetings, and greatly expanded the royal governor's power. County conventions in Massachusetts protested vehemently and urged that the acts be "rejected as the attempts of a wicked administration to enslave America." Other colonial assemblies joined in the denunciations. In September 1774, the First Continental Congress met in Philadelphia, where the more radical members argued successfully against concessions to the crown. Compromise was also rejected by the British parliament, and in April 1775 fighting began at Lexington and Concord.

INDEPENDENCE

The fighting spread, and the colonists moved slowly but inevitably toward open rebellion and a declaration of independence. The uncompromising attitude of the British government and its use of German mercenaries went a long way toward dissolving long-standing loyalties to the home country and rivalries among the separate colonies. *Common Sense* (1775), a brilliant attack by the recently arrived English radical Thomas Paine (1737–1809), also mobilized public opinion in favor of independence. A runaway best seller with sales of 120,000 copies in a few months,

Paine's tract ridiculed the idea of a small island ruling a great continent. In his call for freedom and republican government, Paine expressed Americans' growing sense of separateness and moral superiority.

On July 4, 1776, the Second Continental Congress adopted the Declaration of Independence. Written by Thomas Jefferson, the Declaration of Independence boldly listed the tyrannical acts committed by George III (1760–1820) and confidently proclaimed the natural rights of man and the sovereignty of the American states. Sometimes called the world's greatest political editorial, the Declaration of Independence in effect universalized the traditional rights of Englishmen and made them the rights of all mankind. It stated that "all men are created equal . . . they are endowed by their Creator with certain unalienable rights . . . among these are life, liberty, and the pursuit of happiness." No other American political document has ever caused such excitement, both at home and abroad.

Many American families remained loyal to Britain; many others divided bitterly. After the Declaration of Independence, the conflict often took the form of a civil war pitting patriot against Loyalist. The Loyalists tended to be wealthy and politically moderate. Many patriots, too, were wealthy—individuals such as John Hancock and George Washington—but willingly allied themselves with farmers and artisans in a broad coalition. This coalition harassed the Loyalists and confiscated their property to help pay for the American war effort. The broad social base of the revolutionaries tended to make the liberal revolution democratic. State governments extended the right to vote to many more people in the course of the war and re-established themselves as republics.

On the international scene, the French sympathized with the rebels from the beginning. They wanted revenge for the humiliating defeats of the Seven Years' War. Officially neutral until 1776, they supplied the great bulk of guns and gunpowder used by the American revolutionaries, very much as neutral great powers supply weapons for "wars of national liberation" today. By 1777 French volunteers were arriving in Virginia, and a dashing young nobleman, the marquis de Lafayette (1757–1834), quickly became one of Washington's most trusted generals. In 1778 the French government offered the Americans a formal alliance, and in 1779 and 1780 the Spanish and Dutch declared war on Britain. Catherine the

The Signing of the Declaration, July 4, 1776 John Trumbull's famous painting shows the dignity and determination of America's revolutionary leaders. An extraordinarily talented group, they succeeded in rallying popular support without losing power to more radical forces in the process. *(Yale University Art Gallery)*

Great of Russia helped organize a League of Armed Neutrality in order to protect neutral shipping rights, which Britain refused to recognize.

Thus, by 1780 Great Britain was engaged in an imperial war against most of Europe as well as the thirteen colonies. In these circumstances, and in the face of severe reverses in India, the West Indies, and at Yorktown in Virginia, a new British government decided to cut its losses. American negotiators in Paris were receptive. They feared that France wanted a treaty that would bottle up the new United States east of the Alleghenies and give British holdings west of the Alleghenies to France's ally, Spain. Thus the American negotiators ditched the French and accepted the extraordinarily favorable terms Britain offered.

By the Treaty of Paris of 1783, Britain recognized the independence of the thirteen colonies and ceded all its territory between the Appalachians and the Mississippi River to the Americans. Out of the bitter rivalries of the Old World, the Americans snatched dominion over half a continent.

FRAMING THE CONSTITUTION

The liberal program of the American Revolution was consolidated by the federal Constitution, the Bill of Rights, and the creation of a national republic. Assembling in Philadelphia in the summer of 1787, the delegates to the Constitutional Convention were determined to end the period of economic depression, social uncertainty, and very weak central government that had followed independence. The delegates decided, therefore, to grant the federal, or central, government important powers: regulation of domestic and foreign trade, the right to levy taxes, and the means to enforce its laws.

Strong rule was placed squarely in the context of representative self-government. Senators and congressmen would be the lawmaking delegates of the voters, and the president of the republic would be an elected official. The central government was to operate in Montesquieu's framework of checks and balances. The executive, legislative, and judicial branches would systematically balance each other.

Benjamin Franklin in France Franklin served with distinction as American ambassador to France during the War for Independence. Shown signing the crucial Treaties of Commerce and Alliance in 1778, Franklin was lionized by the French as scientist and sage. *(Brown Brothers)*

The power of the federal government would in turn be checked by the powers of the individual states.

When the results of the secret deliberation of the Constitutional Convention were presented to the states for ratification, a great public debate began. The opponents of the proposed constitution—the Anti-Federalists—charged that the framers of the new document had taken too much power from the individual states and made the federal government too strong. Moreover, many Anti-Federalists feared for the personal liberties and individual freedoms for which they had just fought. In order to overcome these objections, the Federalists solemnly promised to spell out these basic freedoms as soon as the new constitution was adopted. The result was the first ten amendments to the Constitution, which the first Congress passed shortly after it met in New York in March 1789. These amendments formed an effective bill of rights to safeguard the individual. Most of them—trial by jury, due process of law, right to assemble, freedom from unreasonable search—had their origins in English law and the English Bill of Rights of 1689. Others—the freedoms of speech, the press, and religion—reflected natural-law theory and the American experience.

The American Constitution and the Bill of Rights exemplified the great strengths and the limits of what came to be called "classical liberalism." Liberty meant individual freedoms and political safeguards. Liberty also meant representative government but did not necessarily mean democracy with its principle of one man, one vote.

Equality—slaves excepted—meant equality before the law, not equality of political participation or economic well-being. Indeed, economic inequality was resolutely defended by the elite who framed the Constitution. The right to own property was guaranteed by the Fifth Amendment, and if the government took private property, the owner was to receive "just compensation." The radicalism of liberal revolution in America was primarily legal and political, not economic or social.

THE REVOLUTION'S IMPACT ON EUROPE

Hundreds of books, pamphlets, and articles analyzed and romanticized the American upheaval. Thoughtful Europeans noted, first of all, its enormous long-term implications for international politics. A secret report by the Venetian ambassador to Paris in 1783 stated what many felt: "If only the union of the Provinces is preserved, it is reasonable to expect that, with the favorable effects of time, and of European arts and sciences, it will become the most formidable power in the world."[1] More generally, American independence fired the imaginations of those few aristocrats who were uneasy with their privileges and of those commoners who yearned for greater equality. Many Europeans believed that the world was advancing now and that America was leading the way. As one French writer put it in 1789: "This vast continent which the seas surround will soon change Europe and the universe."

Europeans who dreamed of a new era were fascinated by the political lessons of the American Revolution. The Americans had begun with a revolutionary defense against tyrannical oppression, and they had been victorious. They had then shown how rational beings could assemble together to exercise sovereignty and write a permanent constitution—a new social contract. All this gave greater reality to the concepts of individual liberty and representative government. It reinforced one of the primary ideas of the Enlightenment, the idea that a better world was possible.

THE FRENCH REVOLUTION, 1789–1791

No country felt the consequences of the American Revolution more directly than France. Hundreds of French officers served in America and were inspired by the experience. The most famous of these, the young and impressionable marquis de Lafayette left home as a great aristocrat determined only to fight France's traditional foe, England. He returned with a love of liberty and firm republican convictions. French intellectuals and publicists engaged in passionate analysis of the federal Constitution, as well as the constitutions of the various states of the new United States. The American Revolution undeniably hastened upheaval in France.

Yet the French Revolution did not mirror the American example. It was more violent and more complex, more influential and more controversial, more loved and more hated. For Europeans and most of the rest of the world, it was *the* great revolution of the eighteenth century, the revolution that opened the modern era in politics.

THE BREAKDOWN OF THE OLD ORDER

Like the American Revolution, the French Revolution had its immediate origins in the financial difficulties of the government. As we noted in Chapter 18, the efforts of Louis XV's ministers to raise taxes had been thwarted by the Parlement of Paris, strengthened in its opposition by widespread popular support. When renewed efforts to reform the tax system met a similar fate in 1776, the government was

Louis XVI Idealized in this stunning portrait by Duplessis as a majestic, self-confident ruler and worthy heir of Louis XIV, Louis XVI was actually shy, indecisive, and somewhat stupid. *(Château de Versailles/Giraudon/Art Resource)*

forced to finance all of its enormous expenditures during the American war with borrowed money. The national debt and the annual budget deficit soared. By the 1780s fully half of France's annual budget went for ever-increasing interest payments on the ever-increasing debt. Another quarter went to maintain the military, while 6 percent was absorbed by the costly and extravagant king and his court at Versailles. Less than one-fifth of the entire national budget was available for the productive functions of the state, such as transportation and general administration. It was an impossible financial situation.

One way out would have been for the government to declare partial bankruptcy, forcing its creditors to accept greatly reduced payments on the debt. The powerful Spanish monarchy had regularly repudiated large portions of its debt in earlier times, and France had done likewise, after an attempt to establish a French national bank ended in financial disaster in 1720. Yet by the 1780s the French debt was held by an army of aristocratic and bourgeois creditors, and the French monarchy, though absolute in theory, had become far too weak for such a drastic and unpopular action.

Nor could the king and his ministers, unlike modern governments, print money and create inflation to cover their deficits. Unlike England and Holland, which had far larger national debts relative to their populations, France had no central bank, no paper currency, and no means of creating credit. French money was good gold coin. Therefore, when a depressed economy and a lack of public confidence made it increasingly difficult for the government to obtain new gold loans in 1786, it had no alternative but to try to increase taxes. And since France's tax system was unfair and out of date, increased revenues were possible only through fundamental reforms. Such reforms would affect all groups in France's complex and fragmented society and opened a Pandora's box of social and political demands.

LEGAL ORDERS AND SOCIAL REALITIES

As in the Middle Ages, France's 25 million inhabitants were still legally divided into three orders or "estates"—the clergy, the nobility, and everyone else. As the nation's first estate, the clergy numbered about 100,000 and had important privileges. It owned about 10 percent of the land and paid only a "voluntary gift" to the government every five years. Moreoever, the church levied a tax (the tithe) on landowners, which averaged somewhat less than 10 percent. Much of the church's income was actually drained away from local parishes by political appointees and worldly aristocrats at the top of the church hierarchy, to the intense dissatisfaction of the poor parish priests.

The second legally defined estate consisted of some 400,000 noblemen and noblewomen—the descendants of "those who fought" in the Middle Ages. The nobles owned outright about 25 percent of the land in France, and they, too, were taxed very lightly. Moreover, nobles continued to enjoy certain manorial rights, or privileges of lordship, that dated back to medieval times and allowed them to tax the peasantry for their own profit. This was done by means of exclusive rights to hunt and fish, village monopolies on baking bread and pressing grapes for wine, fees for justice, and a host of other "useful privileges." In addition, nobles had "honorific privileges," such as the right to precedence on public occasions and the right to wear a sword. These rights conspicuously proclaimed the nobility's legal superiority and exalted social position.

Everyone else was a commoner, a member of the third estate. A few commoners were rich merchants or highly successful doctors and lawyers. Many more were urban artisans and unskilled day laborers. The vast majority of the third estate consisted of the peasants and agricultural workers in the countryside. Thus the third estate was a conglomeration of vastly different social groups, united only by their shared legal status as distinct from the privileged nobility and clergy.

In discussing the long-term origins of the French Revolution, historians have long focused on growing tensions between the nobility and the comfortable members of the third estate, usually known as the bourgeoisie, or middle class. A dominant historical interpretation has held sway for at least two generations. According to this interpretation, the bourgeoisie was basically united by economic position and class interest. Aided by the general economic expansion discussed in Chapter 19, the middle class grew rapidly in the eighteenth century, tripling to about 2.3 million persons, or about 8 percent of France's population. Increasing in size, wealth, culture, and self-confidence, this rising bourgeoisie became progressively exasperated by archaic "feudal" laws restraining the economy and by the growing pretensions of a reactionary nobility, which was closing ranks against middle-class needs and aspirations. As a result, the French bourgeoisie eventually rose up to lead the entire third estate in a great social revolution, a revolution that destroyed feudal privileges and established a capitalist order based on individualism and a market economy.

In recent years, a flood of new research has challenged these accepted views, and once again the French Revolution is a subject of heated scholarly debate. Above all, revisionist historians have questioned the existence of a growing social conflict between a progressive capitalistic bourgeoisie and a reactionary feudal nobility in eighteenth-century France. Instead, these historians see both bourgeoisie and nobility as being riddled with internal rivalries and highly fragmented. The great nobility, for example, was profoundly separated from the lesser nobility by differences in wealth, education, and worldview. Differences within the bourgeoisie—between wealthy financiers and local lawyers, for example—were no less profound. Rather than standing as unified blocs against each other, nobility and bourgeoisie formed two parallel social ladders, increasingly linked together at the top by wealth, marriage, and Enlightenment culture.

Revolutionist historians stress in particular three developments in their reinterpretation. First, the nobility remained a fluid and relatively open order. Throughout the eighteenth century, substantial numbers of successful commoners continued to seek and obtain noble status through government service and purchase of expensive positions conferring nobility. Thus the nobility of the robe continued to attract the wealthiest members of the middle class and to permit social mobility. Second, key sections of the nobility were no less liberal than the middle class, which until revolution actually began, generally supported the judicial opposition led by the Parlement of Paris. Finally, the nobility and the bourgeoisie were not really at odds in the economic sphere. Both looked to investment in land and government service as their preferred activities, and the ideal of the merchant capitalist was to gain enough wealth to retire from trade, purchase estates, and live nobly as a large landowner. At the same time, wealthy nobles often acted as aggressive capitalists, investing especially in mining, metallurgy, and foreign trade.

The revisionists have clearly shaken the belief that the bourgeoisie and the nobility were inevitably locked in growing conflict before the revolution. But in stressing the similarities between the two groups, especially at the top, they have also reinforced the view, long maintained by historians, that the Old Regime had ceased to correspond with social reality by the 1780s. Legally, society was still based on rigid orders inherited from the Middle Ages. In reality, France had already moved far toward being a society based on wealth and economic achievement, where an emerging elite that included both aristocratic and bourgeois notables was frustrated by a bureaucratic monarchy that had long claimed the right to absolute power.

THE FORMATION OF THE NATIONAL ASSEMBLY

The Revolution was under way by 1787, though no one could have realized what was to follow. Spurred by a depressed economy and falling tax receipts, Louis XVI's minister of finance revived old proposals to impose a general tax on all landed property, as well as provincial assemblies to help administer the tax, and he convinced the king to call an Assembly of Notables to gain support for the idea. The assembled notables, who were mainly important noblemen and high-ranking clergy, were not in favor of it. In return for their support, they demanded that control over all government spending be given to the provincial assemblies, which they expected to control. When the government refused, the nobles responded that such sweeping tax changes required the approval of the Estates General, the representative body of all three estates, which had not met since 1614.

Facing imminent bankruptcy, the king tried to reassert his authority. He dismissed the nobles and established new taxes by decree. In stirring language, the Parlement of Paris promptly declared the royal initiative null and void. The Parlement went so far as to specify some of the "fundamental laws" against which no king could transgress, such as national consent to taxation and freedom from arbitrary arrest and imprisonment. When the king tried to exile the judges, a tremendous wave of protest swept the country. Frightened investors also refused to advance more loans to the state. Finally, in July 1788, a beaten Louis XVI called for a spring session of the Estates General. Absolute monarchy was collapsing.

What would replace it? Throughout the unprecedented election campaign of 1788 and 1789, that question excited France. All across the country, clergy, nobles, and commoners came together in their respective orders to draft petitions for change and to elect their respective delegates to the Estates General. The local assemblies of the clergy showed considerable dissatisfaction with the church hierarchy, and two-thirds of the delegates were chosen from the poorer parish priests, who were commoners by birth. The nobles, already badly split by wealth and education, remained politically divided. A conservative majority was drawn from the poorer and more numerous provincial nobility, but fully a third of the nobility's representatives were liberals committed to major changes.

As for the third estate, there was great popular participation in the elections. Almost all male commoners twenty-five years or older had the right to vote. However, voting required two stages, which meant that most of the representatives finally selected by the third estate were well-educated, prosperous members of the middle class. Most of them were not businessmen, but lawyers and government officials. Social status and prestige were matters of particular concern to this economic elite. There were no delegates from the great mass of laboring poor— the peasants and the artisans.

The petitions for change from the three estates showed a surprising degree of agreement on most issues, as recent research has clearly revealed. There was general agreement that royal absolutism should give way to constitutional monarchy, in which laws and taxes would require the consent of an Estates General meeting regularly. All agreed that, in the future, individual liberties must be guaranteed by law and that the position of the parish clergy had to be improved. It was generally acknowledged that economic development required reforms, such as the abolition of internal trade barriers. The striking similarities in the grievance petitions of the clergy, nobility, and third estate reflected the broad commitment of France's elite to liberalism.

Yet an increasingly bitter quarrel undermined this consensus during the intense election campaign: *how* would the Estates General vote, and precisely *who* would lead in the political reorganization that was generally desired? The Estates General of 1614 had sat as three separate houses. Any action had required the agreement of at least two branches, a requirement that had guaranteed control by the privileged orders —the nobility and the clergy. Immediately after its victory over the king, the aristocratic Parlement of Paris ruled that the Estates General should once again sit separately, mainly out of respect for tradition but partly to enhance the nobility's political position. The ruling was quickly denounced by certain middle-class intellectuals and some liberal nobles. They demanded instead a single assembly dominated by representatives of the third estate, to ensure fundamental reforms. Reflecting a growing hostility toward aristocratic aspirations, the abbé Sieyès argued in 1789 in his famous pamphlet, *What Is the Third Estate?*, that the nobility was a tiny, overprivileged minority and that the neglected third estate constituted the true strength of the French nation. When the government agreed that the third estate should have as many delegates as the clergy and the nobility combined, but then rendered its act meaningless by upholding voting by separate order, middle-class leaders saw fresh evidence of an aristocratic conspiracy.

In May 1789, the twelve hundred delegates of the three estates paraded in medieval pageantry through the streets of Versailles to an opening session resplen-dent with feudal magnificence. The estates were almost immediately deadlocked. Delegates of the third estate refused to transact any business until the king ordered the clergy and nobility to sit with them in a single body. Finally, after a six-week war of nerves, a few parish priests began to go over to the third estate, which on June 17 voted to call itself the National Assembly. On June 20, excluded from their hall because of "repairs," the delegates of the third estate moved to a large indoor tennis court. There they swore the famous Oath of the Tennis Court, pledging never to disband until they had written a new constitution.

The king's actions were then somewhat contradictory. On June 23, he made a conciliatory speech to a joint session, urging reforms, and then ordered the three estates to meet together. At the same time, he apparently followed the advice of relatives and court nobles, who urged the king to dissolve the Estates General by force. The king called an army of eighteen thousand troops toward Versailles, and on July 11 he dismissed his finance minister and his other more liberal ministers. Faced with growing opposition since 1787, Louis XVI had resigned himself to bankruptcy. Now he sought to reassert his divine and historic right to rule. The middle-class delegates had done their best, but they were resigned to being disbanded at bayonet point. One third-estate delegate reassured a worried colleague: "You won't hang—you'll only have to go back home."[2]

THE REVOLT OF THE POOR AND THE OPPRESSED

While the third estate pressed for symbolic equality with the nobility and clergy in a single legislative body at Versailles, economic hardship gripped the masses of France in a tightening vise. Grain was the basis of the diet of ordinary people, and in 1788 the harvest had been extremely poor. The price of bread, which had been rising gradually since 1785, began to soar. By July 1789, the price of bread in the provinces climbed as high as eight sous per pound. In Paris, where bread was subsidized by the government in an attempt to prevent popular unrest, the price rose to four sous. The poor could scarcely afford to pay two sous per pound, for even at that price a laborer with a wife and three children had to spend half of his wages to buy the family's bread.

Harvest failure and high bread prices unleashed a classic economic depression of the preindustrial age.

Storming the Bastille This contemporary drawing conveys the fury and determination of the revolutionary crowd on July 14, 1789. This successful popular action had enormous symbolic significance, and July 14 has long been France's most important national holiday. *(Photo: Flammarion)*

With food so expensive and with so much uncertainty, the demand for manufactured goods collapsed. Thousands of artisans and small traders were thrown out of work. By the end of 1789, almost half of the French people would be in need of relief. One person in eight was a pauper, living in extreme want. In Paris the situation was desperate in July 1789: perhaps 150,000 of the city's 600,000 people were without work.

Against this background of dire poverty and excited by the political crisis, the people of Paris entered decisively onto the revolutionary stage. They believed in a general, though ill-defined, way that the economic distress had human causes. They believed that they should have steady work and enough bread to survive. Specifically, they feared that the dismissal

of the king's moderate finance minister would throw them at the mercy of aristocratic landowners and grain speculators. Stories like that quoting the wealthy financier Joseph François Foulon as saying that the poor "should eat grass, like my horses," and rumors that the king's troops would sack the city began to fill the air. Angry crowds formed and passionate voices urged action. On July 13, the people began to seize arms for the defense of the city, and on July 14, several hundred of the most determined people marched to the Bastille to search for gunpowder.

An old medieval fortress with walls ten feet thick and eight great towers each a hundred feet high, the Bastille had long been used as a prison. It was guarded by eighty retired soldiers and thirty Swiss guards. The governor of the fortress-prison refused to hand over the powder, panicked, and ordered his men to fire, killing ninety-eight people attempting to enter. Cannon were brought to batter the main gate, and fighting continued until the governor of the prison surrendered. While he was being taken under guard to city hall, a band of men broke through and hacked him to death. His head and that of the mayor of Paris, who had been slow to give the crowd arms, were stuck on pikes and paraded through the streets. The next day, a committee of citizens appointed the marquis de Lafayette commander of the city's armed forces. Paris was lost to the king, who was forced to recall the finance minister and to disperse his troops. The uprising had saved the National Assembly.

As the delegates resumed their long-winded and inconclusive debates at Versailles, the people in the countryside sent them a radical and unmistakable message. All across France, peasants began to rise in spontaneous, violent, and effective insurrection against their lords, ransacking manor houses and burning feudal documents that recorded the peasants' obligations. Neither middle-class landowners, who often owned manors and village monopolies, nor the larger, more prosperous farmers were spared. In some areas, the nobles and bourgeoisie combined forces and organized patrols to protect their property. Yet the peasant insurrection went on. Recent enclosures were undone, old common lands were reoccupied, and the forests were seized. Taxes went unpaid. Fear of vagabonds and outlaws—the so-called Great Fear—seized the countryside and fanned the flames of rebellion. The long-suffering peasants were doing their best to free themselves from aristocratic privilege and exploitation.

Faced with chaos, yet fearful of calling on the king to restore order, some liberal nobles and middle-class delegates at Versailles responded to peasant demands with a surprise maneuver on the night of August 4, 1789. The duke of Aiguillon, one of France's greatest noble landowners, declared that

in several provinces the whole people forms a kind of league for the destruction of the manor houses, the ravaging of the lands, and especially for the seizure of the archives where the title deeds to feudal properties are kept. It seeks to throw off at last a yoke that has for many centuries weighted it down.[3]

He urged equality in taxation and the elimination of feudal dues. In the end, all the old exactions were abolished, generally without compensation: serfdom where it still existed, exclusive hunting rights for nobles, fees for justice, village monopolies, the right to make peasants work on the roads, and a host of other dues. Though a clarifying law passed a week later was less generous, the peasants ignored the "fine print." They never paid feudal dues again. Thus the French peasantry, which already owned about 30 percent of all the land, quickly achieved a great and unprecedented victory. Henceforth, the French peasants would seek mainly to consolidate their triumph. As the Great Fear subsided, they became a force for order and stability.

A LIMITED MONARCHY

The National Assembly moved forward. On August 27, 1789, it issued the Declaration of the Rights of Man. This great liberal document had a very American flavor, and Lafayette even discussed his draft in detail with the American ambassador in Paris, Thomas Jefferson, the author of the American Declaration of Independence. According to the French declaration, "men are born and remain free and equal in rights." Mankind's natural rights are "liberty, property, security, and resistance to oppression." Also, "every man is presumed innocent until he is proven guilty." As for law, "it is an expression of the general will; all citizens have the right to concur personally or through their representatives in its formation. . . . Free expression of thoughts and opinions is one of the most precious rights of mankind: every citizen may therefore speak, write, and publish freely." In short, this clarion call of the liberal revolutionary ideal guaranteed equality before the law, representative government for a sovereign people, and individual freedom. This revolutionary credo, only two pages long, was propagandized throughout France and Europe and around the world.

Moving beyond general principles to draft a constitution proved difficult. The questions of how much power the king should retain and whether he could permanently veto legislation led to another deadlock. Once again the decisive answer came from the poor, in this instance the poor women of Paris.

To understand what happened, one must remember that the work and wages of women and children were essential in the family economy of the laboring poor. In Paris great numbers of women worked, particularly within the putting-out system in the garment industry—making lace, fancy dresses, embroidery, ribbons, bonnets, corsets, and so on. Many of these goods were beautiful luxury items, destined for an aristocratic and international clientele.[4] Immediately after the fall of the Bastille, many of France's great court nobles began to leave Versailles for foreign lands, so that a plummeting demand for luxuries intensified the general economic crisis. International markets also declined, and the church was no longer able to give its traditional grants of food and money to the poor. Unemployment and hunger increased further, and the result was another popular explosion.

On October 5, some seven thousand desperate women marched the twelve miles from Paris to Versailles to demand action. A middle-class deputy looking out from the assembly saw "multitudes arriving from Paris including fishwives and bullies from the market, and these people wanted nothing but bread." This great crowd invaded the assembly, "armed with scythes, sticks and pikes." One coarse, tough old woman directing a large group of younger women defiantly shouted into the debate: "Who's that talking down there? Make the chatterbox shut up. That's not the point: the point is that we want bread."[5] Hers was the genuine voice of the people, essential to any understanding of the French Revolution.

The women invaded the royal apartments, slaughtered some of the royal bodyguards, and furiously searched for the despised queen, Marie Antoinette. "We are going to cut off her head, tear out her heart, fry her liver, and that won't be the end of it," they shouted, surging through the palace in a frenzy. It seems likely that only the intervention of Lafayette and the National Guard saved the royal family. But the only way to calm the disorder was for the king to go and live in Paris, as the crowd demanded.

The next day, the king, the queen, and their son left for Paris in the midst of a strange procession. The heads of two aristocrats, stuck on pikes, led the way. They were followed by the remaining members of the royal bodyguard, unarmed and surrounded and mocked by fierce men holding sabers and pikes. A mixed and victorious multitude surrounded the king's carriage, hurling crude insults at the queen. There was drinking and eating among the women. "We are bringing the baker, the baker's wife, and the baker's boy," they joyfully sang. The National Assembly followed the king to Paris. Reflecting the more radical environment, it adopted a constitution that gave the virtually imprisoned "baker" only a temporary veto in the lawmaking process. And, for a time, he and the government made sure that the masses of Paris did not lack bread.

"To Versailles" This print is one of many commemorating the women's march on Versailles. Notice on the left that the fashionable lady from the well-to-do is a most reluctant revolutionary. *(Photo: Flammarion)*

The next two years until September 1791 saw the consolidation of the liberal Revolution. Under middle-class leadership, the National Assembly abolished the French nobility as a legal order and pushed forward with the creation of a constitutional monarchy, which Louis XVI reluctantly agreed to accept in July 1790. In the final constitution, the king remained the head of state, but all lawmaking power was placed in the hands of the National Assembly, elected by the economic upper half of French males. Eighty-three departments of approximately equal size replaced the complicated old patchwork of provinces with their many historic differences. The jumble of weights and measures that varied from province to province was reformed, leading to the introduction of the simple, rational metric system in 1793. The National Assembly promoted economic freedom. Monopolies, guilds, and workers' combinations were prohibited, and barriers to trade within France were abolished in the name of economic liberty. Thus the National Assembly applied the critical spirit of the Enlightenment to reform France's laws and institutions completely.

The assembly also threatened nobles who had emigrated from France with the loss of their lands. It nationalized the property of the church and abolished the monasteries as useless relics of a distant past. The government used all former church property as collateral to guarantee a new paper currency, the *assignats,* and then sold these properties in an attempt to put the state's finances on a solid footing. Although the church's land was sold in large blocks, a procedure that favored nimble speculators and the rich, peasants eventually purchased much of it as it was subdivided. These purchases strengthened their attachment to the revolutionary state.

The most unfortunate aspect of the reorganization of France was that it brought the new government into conflict with the Catholic church. Many middle-class delegates to the National Assembly, imbued with the rationalism and skepticism of the eighteenth-century philosophes, harbored a deep distrust of popular piety and "superstitious religion." They were interested in the church only to the extent that they could seize its land and use the church to strengthen the new state. Thus they established a national church, with priests chosen by voters. In the face of resistance, the National Assembly required the clergy to take a loyalty oath to the new government. The clergy became just so many more em-

ployees of the state. The pope formally condemned this attempt to subjugate the church. Against such a backdrop, it is not surprising that only half the priests of France took the oath of allegiance. The result was a deep division within both the country and the clergy itself on the religious question, and confusion and hostility among French Catholics were pervasive. The attempted reorganization of the Catholic church was the revolutionary government's first important failure.

WORLD WAR AND REPUBLICAN FRANCE, 1791–1799

When Louis XVI accepted the final version of the completed constitution in September 1791, a young and still obscure provincial lawyer and member of the National Assembly named Maximilien Robespierre (1758–1794) evaluated the work of two years and concluded, "The Revolution is over." Robespierre was both right and wrong. He was right in the sense that the most constructive and lasting reforms were in place. Nothing substantial in the way of liberty and equality would be gained in the next generation, though much would be lost. He was wrong in the sense that a much more radical stage lay ahead. New heroes and new ideologies were to emerge in revolutionary wars and international conflict.

THE BEGINNING OF WAR

The outbreak and progress of revolution in France produced great excitement and a sharp division of opinion in Europe and the United States. Liberals and radicals such as the English scientist Joseph Priestly (1733–1804) and the American patriot Thomas Paine saw a mighty triumph of liberty over despotism. Conservative spirits like Edmund Burke (1729–1797) were deeply troubled. In 1790 Burke published *Reflections on the Revolution in France,* one of the great intellectual defenses of European conservatism. He defended inherited privileges in general and those of the English monarchy and aristocracy in particular. He predicted that unlimited reform would lead only to chaos and renewed tyranny. By 1791 fear was growing outside France that the great hopes raised by the Revolution might be tragically dashed. The moderate German writer Friedrich von Gentz was apprehensive that, if moderate and intelligent revolution failed in France, all the old evils would be ten times worse: "It would be felt that men could be happy only as slaves, and every tyrant, great or small, would use this confession to seek revenge for the fright that the awakening of the French nation had given him."[6]

The kings and nobles of Europe, who had at first welcomed the revolution in France as weakening a competing power, began to feel threatened themselves. At their courts they listened to the diatribes of great court nobles who had fled France and were urging intervention in France's affairs. When Louis XVI and Marie Antoinette were arrested and returned to Paris after trying unsuccessfully to slip out of France in June 1791, the monarchs of Austria and Prussia issued the Declaration of Pillnitz. This carefully worded statement declared their willingness to intervene in France, but only with the unanimous agreement of all the Great Powers, which they did not expect to receive. Austria and Prussia expected their threat to have a sobering effect on revolutionary France without causing war.

The crowned heads of Europe misjudged the revolutionary spirit in France. When the National Assembly had disbanded, it had sought popular support by decreeing that none of its members would be eligible for election to the new Legislative Assembly. This meant that, when the new representative body was duly elected and convened in October 1791, it had a different character. The great majority were still prosperous, well-educated, and middle class, but they were younger and less cautious than their predecessors. Loosely allied as "Jacobins," so named after their political club, the new representatives to the Assembly were passionately committed to liberal revolution.

The Jacobins increasingly lumped "useless aristocrats" and "despotic monarchs" together, and they easily whipped themselves into a patriotic fury with bombastic oratory. So the courts of Europe were attempting to incite a war of kings against France; well then, "we will incite a war of people against kings. . . . Ten million Frenchmen, kindled by the fire of liberty, armed with the sword, with reason, with eloquence would be able to change the face of the world and make the tyrants tremble on their thrones."[7] Only Robespierre and a very few others argued that people do not welcome liberation at the point of a gun. Such

warnings were brushed aside. France would "rise to the full height of her mission," as one deputy urged. In April 1792, France declared war on Francis II, archduke of Austria and king of Hungary and Bohemia.

France's crusade against tyranny went poorly at first. Prussia joined Austria in the Austrian Netherlands (present-day Belgium), and French forces broke and fled at their first encounter with armies of this First Coalition. The road to Paris lay open, and it is possible that only conflict between the eastern monarchs over the division of Poland saved France from defeat.

Military reversals and Austro-Prussian threats caused a wave of patriotic fervor to sweep France. The Legislative Assembly declared the country in danger. Volunteer armies from the provinces streamed through Paris, fraternizing with the people and singing patriotic songs like the stirring *Marseillaise,* later the French national anthem.

In this supercharged wartime atmosphere, rumors of treason by the king and queen spread in Paris. Once again, as in the storming of the Bastille, the common people of Paris acted decisively. On August 10, 1792, a revolutionary crowd attacked the royal palace at the Tuileries, capturing it after heavy fighting with the Swiss Guards. The king and his family fled for their lives to the nearby Legislative Assembly, which suspended the king from all his functions, imprisoned him, and called for a new National Convention to be elected by universal male suffrage. Monarchy in France was on its deathbed, mortally wounded by war and popular revolt.

THE SECOND REVOLUTION

The fall of the monarchy marked a rapid radicalization of the Revolution, which historians often call the "second revolution." Louis's imprisonment was followed by the September Massacres, which sullied the Revolution in the eyes of most of its remaining foreign supporters. Wild stories seized the city that imprisoned counter-revolutionary aristocrats and priests were plotting with the allied invaders. As a result, angry crowds invaded the prisons of Paris and summarily slaughtered half the men and women they found. In late September 1792, the new, popularly elected National Convention proclaimed France a republic. The republic adopted a new revolutionary calendar, and citizens were expected to address each other with the friendly "thou" of the people, rather than with the formal "you" of the rich and powerful.

All of the members of the National Convention were Jacobins and republicans, and the great majority continued to come from the well-educated middle class. But the convention was increasingly divided into two well-defined, bitterly competitive groups—the Girondists and the Mountain, so called because its members, led by Danton and Robespierre, sat on the uppermost left-hand benches of the assembly hall. Many indecisive members seated in the "Plain" below floated back and forth between the rival factions.

The division was clearly apparent after the National Convention overwhelmingly convicted Louis XVI of treason. By a single vote, 361 of the 720 members of the convention then unconditionally sentenced him to death in January 1793. Louis died with tranquil dignity on the newly invented guillotine. One of his last statements was, "I am innocent and shall die without fear. I would that my death might bring happiness to the French, and ward off the dangers which I foresee."[8]

Both the Girondists and the Mountain were determined to continue the "war against tyranny." The Prussians had been stopped at the indecisive battle of Valmy on September 20, 1792, one day before the republic was proclaimed. Republican armies then successfully invaded Savoy and captured Nice. A second army corps invaded the German Rhineland and took the city of Frankfurt. To the north, the revolutionary armies won their first major battle at Jemappes and occupied the entire Austrian Netherlands by November 1792. Everywhere they went, French armies of occupation chased the princes, "abolished feudalism," and found support among some peasants and middle-class people.

But the French armies also lived off the land, requisitioning food and supplies and plundering local treasures. The liberators looked increasingly like foreign invaders. International tensions mounted. In February 1793, the National Convention, at war with Austria and Prussia, declared war on Britain, Holland, and Spain as well. Republican France was now at war with almost all of Europe, a great war that would last almost without interruption until 1815.

As the forces of the First Coalition drove the French from the Austrian Netherlands, peasants in western France revolted against being drafted into the army. They were supported and encouraged in their resistance by devout Catholics, royalists, and foreign agents.

In Paris, the quarrelsome National Convention found itself locked in a life-and-death political struggle between the Girondists and the Mountain. The two groups were in general agreement on questions of policy. Sincere republicans, they hated privilege and wanted to temper economic liberalism with social concern. Yet personal hatreds ran deep. The Girondists feared a bloody dictatorship by the Mountain, and the Mountain was no less convinced that the more moderate Girondists would turn to conservatives and even royalists in order to retain power. With the middle-class delegates so bitterly divided, the laboring poor of Paris emerged as the decisive political factor.

The great mass of the Parisian laboring poor always constituted—along with the peasantry in the summer of 1789—the elemental force that drove the Revolution forward. It was the artisans, shopkeepers, and day laborers who had stormed the Bastille, marched on Versailles, driven the king from the Tuileries, and carried out the September Massacres. The petty traders and laboring poor were often known as the *sans-culottes,* "without breeches," because they wore trousers instead of the knee breeches of the aristocracy and the solid middle class. The immediate interests of the sans-culottes were mainly economic, and in the spring of 1793, the economic situation was as bad as the military situation. Rapid inflation, unemployment, and food shortages were again weighing heavily on the poor.

Moreover, by the spring of 1793, the sans-culottes were keenly interested in politics. Encouraged by the so-called angry men, such as the passionate young ex-priest and journalist Jacques Roux, the sans-culottes were demanding radical political action to guarantee them their daily bread. At first the Mountain joined the Girondists in violently rejecting these demands. But in the face of military defeat, peasant revolt, and hatred of the Girondists, the Mountain and especially Robespierre became more sympathetic. The Mountain joined with sans-culottes activists in the city government to engineer a popular uprising, which forced the convention to arrest thirty-one Girondist deputies for treason on June 2. All power passed to the Mountain.

Robespierre and others from the Mountain joined the recently formed Committee of Public Safety, to which the convention had given dictatorial power to

The Reign of Terror A man, woman, and child accused of political crimes are brought before a special revolutionary committee for trial. The Terror's iron dictatorship crushed individual rights as well as treason and opposition. *(Photo: Flammarion)*

deal with the national emergency. These developments in Paris triggered revolt in leading provincial cities, such as Lyons and Marseilles, where moderates denounced Paris and demanded a decentralized government. The peasant revolt spread and the republic's armies were driven back on all fronts. By July 1793, only the areas around Paris and on the eastern frontier were firmly controlled by the central government. Defeat appeared imminent.

TOTAL WAR AND THE TERROR

A year later, in July 1794, the Austrian Netherlands and the Rhineland were once again in the hands of conquering French armies, and the First Coalition was falling apart. This remarkable change of fortune was due to the revolutionary government's success in harnessing, for perhaps the first time in history, the explosive forces of a planned economy, revolutionary terror, and modern nationalism in a total war effort.

Robespierre and the Committee of Public Safety advanced with implacable resolution on several fronts in 1793 and 1794. In an effort to save revolutionary France, they collaborated with the fiercely patriotic and democratic sans-culottes. They established, as best they could, a planned economy with egalitarian social overtones. Rather than let supply and demand determine prices, the government decreed the maximum allowable prices, fixed in paper assignats, for a host of key products. Though the state was too weak to enforce all its price regulations, it did fix the price of bread in Paris at levels the poor could afford. Rationing and ration cards were introduced to make sure that the limited supplies of bread were shared fairly. Quality was also controlled. Bakers were permitted to make only the "bread of equality" —a brown bread made of a mixture of all available flours. White bread and pastries were outlawed as frivolous luxuries. The poor of Paris may not have eaten well, but they ate.

They also worked, mainly to produce arms and munitions for the war effort. Craftsmen and small manufacturers were told what to produce and when to deliver. The government nationalized many small

THE FRENCH REVOLUTION

May 5, 1789	Estates General convene at Versailles
June 17, 1789	Third Estate declares itself the National Assembly
June 20, 1789	Oath of the Tennis Court
July 14, 1789	Storming of the Bastille
July–August 1789	The Great Fear in the countryside
August 4, 1789	National Assembly abolishes feudal privileges
August 27, 1789	National Assembly issues Declaration of the Rights of Man
October 5, 1789	Parisian women march on Versailles and force royal family to return to Paris
November 1789	National Assembly confiscates church lands
July 1790	Civil Constitution of the Clergy establishes a national church
	Louis XVI reluctantly agrees to accept a constitutional monarchy
June 1791	Arrest of the royal family while attempting to flee France
August 1791	Declaration of Pillnitz by Austria and Prussia
April 1792	France declares war on Austria
August 1792	Parisian mob attacks palace and takes Louis XVI prisoner
September 1792	September Massacres
	National Convention declares France a republic and abolishes monarchy
January 1793	Execution of Louis XVI
February 1793	France declares war on Britain, Holland, and Spain
	Revolts in provincial cities
March 1793	Bitter struggle in the National Convention between Girondists and the Mountain
April–June 1793	Robespierre and the Mountain organize the Committee of Public Safety and arrest Girondist leaders
September 1793	Price controls to aid the sans-culottes and mobilize war effort
1793–1794	Reign of Terror in Paris and the provinces
Spring 1794	French armies victorious on all fronts
July 1794	Execution of Robespierre
	Thermidorean Reaction begins
1795–1799	The Directory
1795	End of economic controls and suppression of the sans-culottes
1797	Napolean defeats Austrian armies in Italy and returns triumphant to Paris
1798	Austria, Great Britain, and Russia form the Second Coalition against France
1799	Napoleon overthrows the Directory and seizes power

workshops and requisitioned raw materials and grain from the peasants. Sometimes planning and control did not go beyond orders to meet the latest emergency: "Ten thousand soldiers lack shoes. You will take the shoes of all the aristocrats in Strasbourg and deliver them ready for transport to headquarters at 10 A.M. tomorrow." Failures to control and coordinate were failures of means and not of desire: seldom if ever before had a government attempted to manage an economy so thoroughly. The second revolution and the ascendancy of the sans-culottes had produced an embryonic emergency socialism, which was to have great influence on the subsequent development of socialist ideology.

While radical economic measures supplied the poor with bread and the armies with weapons, a Reign of Terror (1793–1794) was solidifying the home front. Special revolutionary courts, responsible only to Robespierre's Committee of Public Safety, tried rebels and "enemies of the nation" for political crimes. Drawing on popular, sans-culottes support centered in the local Jacobin clubs, these local courts ignored normal legal procedures and judged severely. Some 40,000 French men and women were executed or died in prison. Another 300,000 suspects crowded the prisons and often brushed close to death in a revolutionary court.

Robespierre's Reign of Terror was one of the most controversial phases of the French Revolution. Most historians now believe that the Terror was not directed against any single class. Rather, it was a political weapon directed impartially against all who might oppose the revolutionary government. For many Europeans of the time, however, the Reign of Terror represented a terrifying perversion of the generous ideals of 1789. It strengthened the belief that France had foolishly replaced a weak king with a bloody dictatorship.

The third and perhaps most decisive element in the French republic's victory over the First Coalition was its ability to continue drawing on the explosive power of patriotic dedication to a national state and a national mission. This is the essence of modern nationalism. With a common language and a common tradition, newly reinforced by the ideas of popular sovereignty and democracy, the French people were stirred by a common loyalty. The shared danger of foreign foes and internal rebels unified all classes in a heroic defense of the nation.

In such circumstances, war was no longer the gentlemanly game of the eighteenth century, but a life-and-death struggle between good and evil. Everyone had to participate in the national effort. According to a famous decree of August 23, 1793:

The young men shall go to battle and the married men shall forge arms. The women shall make tents and clothes, and shall serve in the hospitals; children shall tear rags into lint. The old men will be guided to the public places of the cities to kindle the courage of the young warriors and to preach the unity of the Republic and the hatred of kings.

Like the wars of religion, war in 1793 was a crusade; this war, though, was fought for a secular rather than a religious ideology.

As all unmarried young men were subject to the draft, the French armed forces swelled to 1 million men in fourteen armies. A force of this size was unprecedented in the history of European warfare. The soldiers were led by young, impetuous generals, who had often risen rapidly from the ranks and personified the opportunities the Revolution seemed to offer gifted sons of the people. These generals used mass attacks at bayonet point by their highly motivated forces to overwhelm the enemy. By the spring of 1794, French armies were victorious on all fronts. The republic was saved.

THE THERMIDORIAN REACTION AND THE DIRECTORY, 1794–1799

The success of the French armies led Robespierre and the Committee of Public Safety to relax the emergency economic controls, but they extended the political Reign of Terror. Their lofty goal was increasingly an ideal democratic republic, where justice would reign and there would be neither rich nor poor. Their lowly means were unrestrained despotism and the guillotine, which struck down any who might seriously question the new order. In March 1794, to the horror of many sans-culottes, Robespierre's Terror wiped out many of the "angry men," led by the radical social democrat Jacques Hébert. Two weeks later, several of Robespierre's long-standing collaborators, led by the famous orator Danton, marched up the steps to the guillotine. Knowing that they might be next, a strange assortment of radicals and moderates in the convention organized a conspiracy. They howled down Robespierre when he tried to speak to the National Convention on 9 Thermidor (July 27, 1794). On the following day, it was Robespierre's turn to be shaved by the revolutionary razor.

As Robespierre's closest supporters followed their leader, France unexpectedly experienced a thorough reaction to the despotism of the Reign of Terror. In a general way, this "Thermidorian reaction" recalled the early days of the Revolution. The respectable middle-class lawyers and professionals who had led the liberal Revolution of 1789 reasserted their authority. Drawing support from their own class, the provincial cities, and the better-off peasants, the National Convention abolished many economic controls, printed more paper currency, and let prices rise sharply. It severely restricted the local political organizations where the sans-culottes had their strength. And all the while, the wealthy bankers and newly rich speculators celebrated the sudden end of the Terror with an orgy of self-indulgence and ostentatious luxury.

The collapse of economic controls, coupled with runaway inflation, hit the working poor very hard. The gaudy extravagance of the rich wounded their pride. The sans-culottes accepted private property, but they believed passionately in small business and the right of all to earn a decent living. Increasingly disorganized after Robespierre purged their radical spokesmen, the common people of Paris finally revolted against the emerging new order in early 1795. The Convention quickly used the army to suppress these insurrections. For the first time since the fall of the Bastille, bread riots and uprisings by Parisians living on the edge of starvation were effectively put down by a government that made no concessions to the poor.

In the face of all these catastrophes, the revolutionary fervor of the laboring poor finally subsided. As far as politics was concerned, their interest and influence would remain very limited until 1830. There arose, especially from the women, a great cry for peace and a turning toward religion. As the government looked the other way, the women brought back the Catholic church and the worship of God. In one French town, women fought with each other over which of their children should be baptized first. After six tumultuous years, the women of the poor concluded that the Revolution was a failure.

As for the middle-class members of the National Convention, they wrote yet another constitution, which they believed would guarantee their economic position and political supremacy. The mass of the population could vote only for electors, who would be men of means. Electors then elected the members of a reorganized legislative assembly, as well as key officials throughout France. The assembly also chose the five-man executive—the Directory.

The men of the Directory continued to support French military expansion abroad. War was no longer so much a crusade as a means to meet the ever-present, ever-unsolved economic problem. Large, victorious French armies reduced unemployment at home, and they were able to live off the territories they conquered and plundered.

The unprincipled action of the Directory reinforced widespread disgust with war and starvation. This general dissatisfaction revealed itself clearly in the national elections of 1797, which returned a large number of conservative and even monarchist deputies who favored peace at almost any price. Fearing for their skins, the members of the Directory used the army to nullify the elections and began to govern dictatorially. Two years later, Napoleon Bonaparte ended the Directory in a coup d'état and substituted a strong dictatorship for a weak one. Truly, the Revolution was over.

THE NAPOLEONIC ERA, 1799–1815

For almost fifteen years, from 1799 to 1814, France was in the hands of a keen-minded military dictator of exceptional ability. One of history's most fascinating leaders, Napoleon Bonaparte realized the need to put an end to civil strife in France, in order to create unity and consolidate his rule. And he did. But Napoleon saw himself as a man of destiny, and the glory of war and the dream of universal empire proved irresistible. For years he spiraled from victory to victory; but in the end he was destroyed by a mighty coalition united in fear of his restless ambition.

NAPOLEON'S RULE OF FRANCE

In 1799, when he seized power, young General Napoleon Bonaparte was a national hero. Born in Corsica into an impoverished noble family in 1769, Napoleon left home to become a lieutenant in the French artillery in 1785. After a brief and unsuccessful adventure fighting for Corsican independence in 1789, he returned to France as a French patriot and a dedicated revolutionary. Rising rapidly in the new army, Napoleon was placed in command of French forces in Italy and won brilliant victories there in 1796 and 1797. His next campaign, in Egypt, was a failure, but Napoleon made his way back to France before the fiasco was generally known. His reputation remained intact.

Napoleon soon learned that some prominent members of the legislative assembly were plotting against the Directory. The dissatisfaction of these plotters stemmed not so much from the fact that the Directory was a dictatorship, as from the fact that it was a weak dictatorship. Ten years of upheaval and uncertainty had made firm rule much more appealing than liberty and popular politics to these disillusioned revolutionaries. The abbé Sieyès personified this evolution in thinking. In 1789 he had written in his famous pamphlet, *What Is the Third Estate?,* that the nobility was grossly overprivileged and that the entire people should rule the French nation. Now Sieyès's motto was "confidence from below, authority from above."

Like the other members of his group, Sieyès wanted a strong military ruler. The flamboyant thirty-year-old Napoleon was ideal. Thus the conspirators and Napoleon organized a takeover. On November 9, 1799, they ousted the Directors, and the following day soldiers disbanded the assembly at bayonet point. Napoleon was named first consul of the republic, and a new constitution consolidating his position was overwhelmingly approved in a plebiscite in December 1799. Republican appearances were maintained, but Napoleon was already the real ruler of France.

The essence of Napoleon's domestic policy was to use his great and highly personal powers to maintain order and put an end to civil strife. He did so by working out unwritten agreements with powerful groups in France, whereby these groups received favors in return for loyal service. Napoleon's bargain with the solid middle class was codified in the famous Civil Code of 1804, which reasserted two of the fundamental principles of the liberal and essentially moderate revolution of 1789: equality of all citizens before the law and absolute security of wealth and private property. Napoleon and the leading bankers of Paris established a privately owned Bank of France, which loyally served the interests of both the state and the financial oligarchy. Napoleon's defense of the economic status quo also appealed to the peasants, who had bought some of the lands confiscated from the church and nobility. Thus Napoleon reconfirmed the gains of the peasantry and reassured the middle class, which had already lost a large number of its revolutionary illusions in the face of social upheaval.

Napoleon Crossing the Alps: David Bold and commanding, with flowing cape and surging stallion, the daring young Napoleon Bonaparte leads his army across the Alps from Italy to battle the Austrians in 1797. This painting by the great Jacques-Louis David (1748–1825) is a stirring glorification of Napoleon, a brilliant exercise in mythmaking. *(The Granger Collection)*

At the same time, Napoleon accepted and strengthened the position of the French bureaucracy. Building on the solid foundations that revolutionary governments had inherited from the Old Regime, he perfected a thoroughly centralized state. A network of prefects, subprefects, and centrally appointed mayors depended on Napoleon and served him well. Nor were members of the old nobility slighted. In 1800 and again in 1802 Napoleon granted amnesty to a hundred thousand émigrés on the condition that they return to France and take a loyalty oath. Members of this returning elite soon ably occupied many high posts in the expanding centralized state. Only a thousand diehard monarchists were exempted and remained abroad. Napoleon also created a new imperial nobility in order to reward his most talented generals and officials.

The Napoleonic Era

November 1799	Napoleon overthrows the Directory
December 1799	French voters overwhelmingly approve Napoleon's new constitution
1800	Napoleon founds the Bank of France
1801	France defeats Austria and acquires Italian and German territories in the Treaty of Lunéville
	Napoleon signs a concordat with the pope
1802	Treaty of Amiens with Britain
March 1804	Execution of the Duke of Engheim
December 1804	Napoleon crowns himself emperor
October 1805	Battle of Trafalgar: Britain defeats the French and Spanish fleets
December 1805	Battle of Austerlitz: Napoleon defeats Austria and Prussia
1807	Treaties of Tilsit: Napoleon redraws the map of Europe
1810	Height of the Grand Empire
June 1812	Napoleon invades Russia with 600,000 men
Winter 1812	Disastrous retreat from Russia
March 1814	Russia, Prussia, Austria, and Britain form the Quadruple Alliance to defeat France
April 1814	Napoleon abdicates and is exiled to Elba
February–June 1815	Napoleon escapes from Elba and rules France until suffering defeat at Battle of Waterloo

Napoleon's great skill in gaining support from important and potentially hostile groups is illustrated by his treatment of the Catholic church in France. In 1800 the French clergy was still divided into two groups: those who had taken an oath of allegiance to the revolutionary government and those in exile or hiding who had refused to do so. Personally uninterested in religion, Napoleon wanted to heal the religious division so that a united Catholic church in France could serve as a bulwark of order and social peace. After long and arduous negotiations, Napoleon and Pope Pius VII (1800–1823) signed the Concordat of 1801. The pope gained for French Catholics the precious right to practice their religion freely, but Napoleon gained the most politically. His government now nominated bishops, paid the clergy, and exerted great influence over the church in France.

The domestic reforms of Napoleon's early years were his greatest achievement. Much of his legal and administrative reorganization has survived in France to this day. More generally, Napoleon's domestic initiatives gave the great majority of French people a welcome sense of order and stability. And when Napoleon added the glory of military victory, he rekindled a spirit of national unity that would elude France throughout most of the nineteenth century.

Order and unity had their price: Napoleon's authoritarian rule. Free speech and freedom of the press —fundamental rights of the liberal revolution, enshrined in the Declaration of the Rights of Man— were continually violated. Napoleon constantly reduced the number of newspapers in Paris. By 1811 only four were left, and they were little more than organs of government propaganda. The occasional elections were a farce. Later laws prescribed harsh penalties for political offenses.

These changes in the law were part of the creation of a police state in France. Since Napoleon was usually busy making war, this task was largely left to Joseph Fouché, an unscrupulous opportunist who had earned a reputation for brutality during the Reign of Terror. As minister of police, Fouché organized a ruthlessly efficient spy system, which kept thousands of citizens under continuous police surveillance. People suspected of subversive activities were arbitrarily detained, placed under house arrest, or even consigned to insane asylums. After 1810 political suspects were held in state prisons, as they had been during the Terror. There were about 2,500 such political prisoners in 1814.

Napoleon on Campaign This picture of the bloody Battle of Bordino in Russia in 1812 captures important features of Napoleonic warfare. While cannon boomed, infantry fired, and cavalry charged, commanders watching from on high directed their forces like pieces on a chessboard. *(Brown Brothers)*

NAPOLEON'S WARS AND FOREIGN POLICY

Napoleon was above all a military man, and a great one. After coming to power in 1799, he sent peace feelers to Austria and Great Britain, the two remaining members of the Second Coalition, which had been formed against France in 1798. When these overtures were rejected, French armies led by Napoleon decisively defeated the Austrians. In the Treaty of Lunéville (1801) Austria accepted the loss of its Italian possessions, and German territory on the west bank of the Rhine was incorporated into France. Once more, as in 1797, the British were alone, and war-weary, like the French.

Still seeking to consolidate his regime domestically, Napoleon concluded the Treaty of Amiens with Great Britain in 1802. Britain agreed to return Trinidad and the Caribbean islands, which it had seized from France in 1793. The treaty said very little about Europe, though. France remained in control of Holland, the Austrian Netherlands, the west bank of the Rhine, and most of the Italian peninsula. Napoleon was free to reshape the German states as he wished. To the dismay of British businessmen, the Treaty of Amiens did not provide for expansion of the commerce between Britain and the Continent. It was clearly a diplomatic triumph for Napoleon, and peace with honor and profit increased his popularity at home.

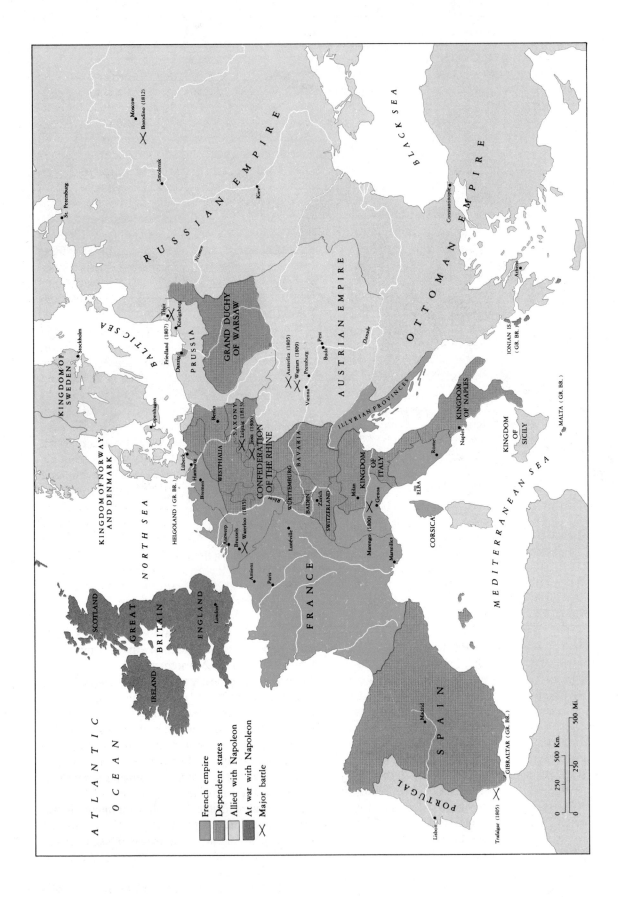

In 1802 Napoleon was secure but unsatisfied. Ever a romantic gambler as well as a brilliant administrator, he could not contain his power drive. Aggressively redrawing the map of Germany so as to weaken Austria and attract the secondary states of southwestern Germany toward France, Napoleon was also mainly responsible for renewed war with Great Britain. Regarding war with Britain as inevitable, he threatened British interests in the eastern Mediterranean and tried to restrict British trade with all of Europe. Britain had technically violated the Treaty of Amiens by failing to evacuate the island of Malta, but it was Napoleon's decision to renew war in May 1803. He concentrated his armies in the French ports on the Channel in the fall of 1803 and began making preparations to invade England. Yet Great Britain remained mistress of the seas. When Napoleon tried to bring his Mediterranean fleet around Gibraltar to northern France, a combined French and Spanish fleet was, after a series of mishaps, virtually annihilated by Lord Nelson at the Battle of Trafalgar on 21 October 1805. Invasion of England was henceforth impossible. Renewed fighting had its advantages, however, for the first consul used the wartime atmosphere to have himself proclaimed emperor in late 1804.

Austria, Russia, and Sweden joined with Britain to form the Third Coalition against France shortly before the Battle of Trafalgar. Actions like Napoleon's assumption of the Italian crown had convinced both Alexander I of Russia and Francis II of Austria that Napoleon was a threat to their interests and to the European balance of power. Yet the Austrians and the Russians were no match for Napoleon, who scored a brilliant victory over them at the Battle of Austerlitz in December 1805. Alexander I decided to pull back, and Austria accepted large territorial losses in return for peace as the Third Coalition collapsed.

Victorious at Austerlitz, Napoleon proceeded to reorganize the German states to his liking. In 1806 he abolished many of the tiny German states, as well as the ancient Holy Roman Empire, whose emperor had traditionally been the ruler of Austria. Napoleon

established by decree a German Confederation of the Rhine, a union of fifteen German states minus Austria, Prussia, and Saxony. Naming himself "protector" of the confederation, Napoleon firmly controlled western Germany.

Napoleon's intervention in German affairs alarmed the Prussians, who had been at peace with France for more than a decade. Expecting help from his ally Russia, Frederick William III of Prussia mobilized his armies. Napoleon attacked and won two more brilliant victories in October 1806 at Jena and Auerstädt, where the Prussians were outnumbered two to one. The war with Prussia and Russia continued into the following spring, and after Napoleon's larger armies won another victory, Alexander decided to seek peace.

For several days in June 1807, the young tsar and the French emperor negotiated face to face on a raft anchored in the middle of the Niemen River. All the while, the helpless Frederick William rode back and forth on the shore, anxiously awaiting the results. As the German poet Heinrich Heine said later, Napoleon had but to whistle and Prussia would have ceased to exist. In the subsequent treaties of Tilsit, Prussia lost half of its population, while Russia accepted Napoleon's reorganization of western and central Europe. Alexander also promised to enforce Napoleon's recently decreed economic blockade against British goods and to declare war on Britain if Napoleon could not make peace on favorable terms with his island enemy.

After the victory of Austerlitz and even more after the treaties of Tilsit, Napoleon saw himself as the emperor of Europe and not just of France. The so-called Grand Empire he built had three parts. The core was an ever-expanding France, which by 1810 included Belgium, Holland, parts of northern Italy, and much German territory on the east bank of the Rhine. Beyond French borders Napoleon established a number of dependent satellite kingdoms, on the thrones of which he placed (and replaced) the members of his large family. Third, there were the independent but allied states of Austria, Prussia, and Russia. Both satellites and allies were expected after 1806 to support Napoleon's continental system and thus to cease all trade with Britain.

The impact of the Grand Empire on the peoples of Europe was considerable. In the areas incorporated into France and in the satellites (see Map 21.1), Napoleon introduced many French laws, abolishing

MAP 21.1 Napoleonic Europe in 1810

feudal dues and serfdom where French revolutionary armies had not already done so. Some of the peasants and middle class benefited from these reforms. Yet while he extended progressive measures to his cosmopolitan empire, Napoleon had to put the prosperity and special interests of France first in order to safeguard his power base. Levying heavy taxes in money and men for his armies, Napoleon came to be regarded more as a conquering tyrant than as an enlightened liberator.

The first great revolt occurred in Spain. In 1808 a coalition of Catholics, monarchists, and patriots rebelled against Napoleon's attempts to make Spain a French satellite with a Bonaparte as its king. French armies occupied Madrid, but the foes of Napoleon fled to the hills and waged uncompromising guerrilla warfare. Spain was a clear warning. Resistance to French imperialism was growing.

Yet Napoleon pushed on, determined to hold his complex and far-flung empire together. In 1810, when the Grand Empire was at its height, Britain still remained at war with France, helping the guerrillas in Spain and Portugal (see Map 21.1). The continental system, organized to exclude British goods from the Continent and force that "nation of shopkeepers" to its knees, was a failure. Instead, it was France that suffered from Britain's counter-blockade, which created hard times for French artisans and the middle class. Perhaps looking for a scapegoat, Napoleon turned on Alexander I of Russia, who had been fully supporting Napoleon's war of prohibitions against British goods.

Napoleon's invasion of Russia began in June 1812 with a force that eventually numbered 600,000, probably the largest force yet assembled in a single army. Only one-third of this force was French, however; nationals of all the satellites and allies were drafted into the operation. Originally planning to winter in the Russian city of Smolensk if Alexander did not sue for peace, Napoleon reached Smolensk and recklessly pressed on. The great battle of Borodino that followed was a draw, and the Russians retreated in good order. Alexander ordered the evacuation of Moscow, which then burned, and refused to negotiate. Finally, after five weeks in the burned-out city, Napoleon ordered a retreat. That retreat was one of the great military disasters in history. The Russian army and the Russian winter cut Napoleon's army to pieces. Only 30,000 men returned to their homelands.

Leaving his troops to their fate, Napoleon raced to Paris to raise yet another army. Possibly he might still have saved his throne if he had been willing to accept a France reduced to its historical size—the proposal offered by Austria's foreign minister Metternich. But Napoleon refused. Austria and Prussia deserted Napoleon and joined Russia and Great Britain in the Fourth Coalition. All across Europe, patriots called for a "war of liberation" against Napoleon's oppression, and the well-disciplined regular armies of Napoleon's enemies closed in for the kill. This time the coalition held together, cemented by the Treaty of Chaumont, which created a Quadruple Alliance to last for twenty years. Less than a month later, on April 4, 1814, a defeated, abandoned Napoleon abdicated his throne. After this unconditional abdication, the victorious allies granted Napoleon the island of Elba off the coast of Italy as his own tiny state. Napoleon was even allowed to keep his imperial title, and France was required to pay him a large yearly income of 2 million francs.

The allies also agreed to the restoration of the Bourbon dynasty, in part because demonstrations led by a few dedicated French monarchists indicated some support among the French people for that course of action. The new monarch, Louis XVIII (1814–1824), tried to consolidate that support by issuing the Constitutional Charter, which accepted many of France's revolutionary changes and guaranteed civil liberties. Indeed, the Charter gave France a constitutional monarchy roughly similar to that established in 1791, although far fewer people had the right to vote for representatives to the resurrected Chamber of Deputies. Moreover, after Louis XVIII stated firmly that his government would not pay any war reparations, France was treated leniently by the allies, who agreed to meet in Vienna to work out a general peace settlement.

Yet Louis XVIII—old, ugly, and crippled by gout —totally lacked the glory and magic of Napoleon. Hearing of political unrest in France and diplomatic tensions in Vienna, Napoleon staged a daring escape from Elba in February 1815. Landing in France, he issued appeals for support and marched on Paris with a small band. French officers and soldiers who had fought so long for their emperor responded to the call. Louis XVIII fled, and once more Napoleon took command. But Napoleon's gamble was a desperate long shot, for the allies were united against him. At the end of a frantic period known as the Hundred

Days, they crushed his forces at Waterloo on June 18, 1815, and imprisoned him on the rocky island of St. Helena, far off the western coast of Africa. Old Louis XVIII returned again—this time "in the baggage of the allies," as his detractors scornfully put it—and recommenced his reign. The allies now dealt rather harshly with the apparently incorrigible French (see Chapter 23). And Napoleon, doomed to suffer crude insults at the hands of sadistic English jailers on distant St. Helena, could take revenge only by writing his memoirs, skillfully nurturing the myth that he had been Europe's revolutionary liberator, a romantic hero whose lofty work had been undone by oppressive reactionaries. An era had ended.

———————

The revolution that began in America and spread to France was a liberal revolution. Inspired by English history and some of the teachings of the Enlightenment, revolutionaries on both sides of the Atlantic sought to establish civil liberties and equality before the law within the framework of representative government. Success in America was subsequently matched by success in France. There liberal nobles and an increasingly class-conscious middle class overwhelmed declining monarchial absolutism and feudal privilege, thanks to the common people—the sans-culottes and the peasants. The government and society established by the Declaration of the Rights of Man and the French constitution of 1791 were remarkably similar to those created in America by the federal Constitution and the Bill of Rights. Thus the new political system, based on electoral competition and civil equality, came into approximate harmony with France's evolving social structure, which had become increasingly based on wealth and achievement rather than on tradition and legal privileges.

Yet the Revolution in France did not end with the liberal victory of 1789 to 1791. As Robespierre led the determined country in a total war effort against foreign foes, French revolutionaries became more democratic, radical, and violent. Their effort succeeded, but at the price of dictatorship—first by Robespierre himself and then by the Directory and Napoleon. Some historians blame the excesses of the French revolutionaries for the emergence of dictatorship, while others hold the conservative monarchs of

Europe responsible. In any case, historians have often concluded that the French Revolution ended in failure.

This conclusion is highly debatable, though. After the fall of Robespierre, the solid middle class, with its liberal philosophy and Enlightenment world-view, reasserted itself. Under the Directory, it salvaged a good portion of the social and political gains that it and the peasantry had made between 1789 and 1791. In so doing, the middle-class leaders repudiated the radical social and economic measures associated with Robespierre, but they never re-established the old pattern of separate legal orders and absolute monarchy. Napoleon built on the policies of the Directory. With considerable success he sought to add the support of the old nobility and the church to that of the middle class and the peasantry. And though Napoleon sharply curtailed thought and speech, he effectively promoted the reconciliation of old and new, of centralized government and careers open to talent, of noble and bourgeois in a restructured property-owning elite. Little wonder, then, that Louis XVIII had no choice but to accept a French society solidly based on wealth and achievement. In granting representative government and civil liberties to facilitate his restoration to the throne in 1814, Louis XVIII submitted to the rest of the liberal triumph of 1789 to 1791. The core of the French Revolution had survived a generation of war and dictatorship. Old Europe would never be the same.

NOTES

———————

1. Quoted by R. R. Palmer, *The Age of the Democratic Revolution,* Princeton University Press, Princeton, N.J., 1959, 1.239.
2. G. Lefebvre, *The Coming of the French Revolution,* Vintage Books, New York, 1947, p. 81.
3. P. H. Beik, ed., *The French Revolution,* Walker, New York, 1970, p. 89.
4. O. Hufton, "Women in Revolution," *Past and Present* 53 (November 1971): 91–95.
5. G. Pernoud and S. Flaisser, eds., *The French Revolution,* Fawcett Publications, Greenwich, Conn., 1960, p. 61.
6. L. Gershoy, *The Era of the French Revolution, 1789–1799,* Van Nostrand, New York, 1957, p. 135.
7. Ibid., p. 150.
8. Pernoud and Flaisser, pp. 193–194.

SUGGESTED READING

In addition to the fascinating eyewitness reports on the French Revolution in P. Beik, *The French Revolution,* and G. Pernoud and S. Flaisser, eds., *The French Revolution* (1960), A. Young's *Travels in France During the Years 1787, 1788 and 1789* (1969) offers an engrossing contemporary description of France and Paris on the eve of revolution. Edmund Burke, *Reflections on the Revolution in France,* first published in 1790, is the classic conservative indictment. The intense passions the French Revolution has generated may be seen in the nineteenth-century French historians, notably the enthusiastic Jules Michelet, *History of the French Revolution;* the hostile Hippolyte Taine; and the judicious Alexis de Tocqueville, whose masterpiece, *The Old Regime and the French Revolution,* was first published in 1856. Important recent general studies on the entire period are R. R. Palmer, *The Age of the Democratic Revolution* (1959, 1964), which paints a comparative international picture; E. J. Hobsbawm, *The Age of Revolution, 1789–1848* (1962); C. Breunig, *The Age of Revolution and Reaction, 1789–1850* (1970); O. Connelly, *French Revolution—Napoleonic Era* (1979); and L. Dehio, *The Precarious Balance: Four Centuries of the European Power Struggle* (1962). C. Brinton's older but delightfully written *A Decade of Revolution, 1789–1799* (1934) complements his stimulating *Anatomy of Revolution* (1938, 1965), an ambitious comparative approach to revolution in England, America, France, and Russia. A Cobban, *The Social Interpretation of the French Revolution* (1964), and F. Furet, *Interpreting the French Revolution* (1981), are exciting reassessments of many wellworn ideas, to be compared with W. Doyle, *Origins of the French Revolution* (1981); G. Lefebvre, *The Coming of the French Revolution* (1947); and N. Hampson, *A Social History of the French Revolution* (1963). G. Rudé makes the men and women of the great days of upheaval come alive in his *The Crowd in the French Revolution* (1959). R. R. Palmer studies sympathetically the leaders of the Terror in *Twelve Who Ruled* (1941). Two other particularly interesting, detailed works are C. L. R. James, *The Black Jacobins* (1938, 1980), on black slave revolt in Haiti, and J. C. Herold, *Mistress to an Age* (1955), on the remarkable Madame de Staël. On revolution in America, E. Morgan, *The Birth of the Republic, 1763–89* (1956), and B. Bailyn, *The Ideological Origins of the American Revolution* (1967), are noteworthy. Three important recent studies on aspects of revolutionary France are D. Jordan's vivid *The King's Trial: Louis XVI vs. the French Revolution* (1979); W. Sewell, Jr.'s imaginative *Work and Revolution in France: The Language of Labor from the Old Regime to 1848* (1980); and P. Higonnet, *Class, Ideology and the Rights of Nobles During the French Revolution* (1981).

P. Geyl, *Napoleon, For and Against* (1949), is a delightful discussion of changing historical interpretations of Napoleon, which may be compared with a more recent treatment by R. Jones, *Napoleon: Man and Myth* (1977). Good biographies are J. M. Thompson, *Napoleon Bonaparte: His Rise and Fall* (1952); F. H. M. Markham, *Napoleon* (1964); and V. Cronin, *Napoleon Bonaparte* (1972). L. Bergeron, *France Under Napoleon* (1981), is an important synthesis. Wonderful novels inspired by this period include Raphael Sabatini's *Scaramouche,* a swashbuckler of revolutionary intrigue with accurate historical details; Charles Dickens's classic *Tale of Two Cities;* and Leo Tolstoy's monumental saga of Napoleon's invasion of Russia (and much more), *War and Peace.*

CHAPTER OPENER CREDITS

NOTES ON THE ILLUSTRATIONS

CHAPTER OPENER CREDITS

Title page: Anderson/Art Resource
Chapter 12: Bibliothèque Royale Albert I, Brussels
Chapter 13: Anderson/Art Resource
Chapter 14: Courtesy, Museum of Fine Arts, Boston
Chapter 15: The Mansell Collection
Chapter 16: Rijksmuseum, Amsterdam
Chapter 17: Historical Pictures Service, Chicago
Chapter 18: The Fotomas Index
Chapter 19: Louvre/Cliché des Musées Nationaux
Chapter 20: Library of Congress
Chapter 21: Versailles/Cliché des Musées Nationaux

NOTES ON THE ILLUSTRATIONS

Page 354 Burying victims of the Black Death. MS. 13076–77, fol. 24v.

Page 359 *St. Sebastian Interceding for the Plague-Stricken* by Josse Lieferinxe, in the Walters Art Gallery, Baltimore, Maryland.

Page 366 Episode from the battle of Crècy from Froissart's *Chronicle,* as reproduced in *Larousse Ancient and Medieval History,* p. 363.

Page 378 Misericord from Henry VII's Chapel, Westminster Abbey.

Page 381 From Froissart's *Chronicles,* MS. Fr. 2643, fol. 125, in Bibliothèque Nationale, Paris.

Page 384 *The Four Horsemen of the Apocalypse,* woodcut ca 1498, by Albrecht Dürer, German painter and engraver (1471–1528) regarded as leader of the German Renaissance school of painting.

Page 388 Andrea Mantegna (Italian, 1431–1506), *Lodovico Gonzaga, Marquess of Mantua, with Members of His Family* (Mantua).

Page 391 Sixteenth-century woodcut.

Page 394 The Palazzo Vecchio (1298–1314) is attributed to Arnolfo di Cambio and was built as the seat of the Signoria, the government of the Florentine republic. This fortified palace is also known as the Palazzo della Signoria.

Page 396 Michelangelo Buonarroti (1475–1564) was a Florentine painter, sculptor, and architect of the High Renaissance. Michelangelo's tomb of the

Medicis was executed between 1524 and 1533, and includes the sculpted marble figures of Lorenzo and Giuliano de Medici as well as four allegorical figures: *Dawn* and *Evening* on one side of the tomb and *Night* and *Day* on the other.

Page 399 The original painting can be found in the Uffizi Gallery, Florence.

Page 401 Hans Memling (real name Mimmelinghe, also spelled Memline and Hemmelinck), active ca 1465–d. 1494. Tommaso Portinari (ca 1432–1501), tempera and oil on wood, 17⅜ inches high by 13¼ wide. Maria Portinari (b. 1456), tempera and oil on wood, 17⅜ inches high by 13⅜ wide.

Page 402 Terracotta, School of Luca della Robbia, late fifteenth century. Della Robbia's invention of the process of making polychrome-glazed terracottas led contemporaries to consider him one of the great artistic innovators. The warm humanity of this roundel (circular panel) is characteristic of della Robia's art.

Page 406 Engraving by Johannes Stradanus (J. van der Straet), Belgian painter (1523–1605).

Page 408 Tiziano Vecellio, Italian painter, 1477–1576.

Page 410 *The Adoration,* 1507, by Hans Baldung (also called Hans Grien or Grün), German painter, engraver, and designer of woodcuts and glass painting (1476?–1545).

Page 415 Hieronymus Bosch (Hieronymus van Aeken), Dutch painter (ca 1450–1516).

Page 417 Fifteenth-century miniature from *Ethique d'Aristotle,* ms. I.2, fol. 145, in Bibliothèque Municipale, Rouen, France.

Page 426 Workshop of Pieter Bruegel the Elder (Flemish, active ca 1551, died 1569), *Combat between Carnival and Lent.* Oil on panel, 14⅜ by 25 inches. Seth K. Sweetser Fund, Abbot Lawrence Fund, Ernest Wadsworth Longfellow Fund, Warren Collection, and Juliana Cheney Edwards Collection.

Page 430 Lucas Cranach the Elder (1472–1553), German painter, etcher, and woodcutter; *Passional Christi und Antichristi,* Wittenburg, 1521.

Page 433 *The Small Crucifixion,* ca 1510, by Matthias Grünewald, German painter (ca 1465–1528). Wood.

Page 435 Attributed to Hieronymus (Dutch, ca 1450–1516), *Christ Before Pilate.* Oil on panel, 80 by 100 cm.

Page 440 Titian (Tiziano Vecellio, 1487–1577), *Charles V.* Oil on canvas, 10 feet 10½ inches by 9 feet 2 inches. Museo del Prado, Madrid. Titian was the greatest Venetian painter of his time. The colors of this portrait of Charles V are dark, perhaps emphasized by restorations that followed the damage incurred in a fire in the eighteenth century.

Page 450 *Sir Thomas More* (1478–1535), painted in 1527 by Hans Holbein the Younger, German painter (1497?–1543) and court painter to Henry VIII.

Page 458 Engraving by Charles van Mallery.

Page 462 Engagement near Isle of Wight in 1588, Spanish fleet on the left, English on the right. In left foreground a Spanish vessel is exchanging shots with two English vessels to the right.

Page 465 World chart by Vesconte Maggiolo, 1511, showing North America as a promontory of Asia.

Page 476 Long the site of a royal residence and hunting lodge, Fontainebleau was expanded and transformed by Francis I in 1530–1540. Il Rosso (Giovanni Battista de'Rossi, 1494–1540), Florentine painter; Francesco Primaticcio, Italian painter and architect (1504–1570); and Sebastiano Serlio, Italian architect and writer on art (1475–1554) were called by Francis I from Italy to build and decorate the palace. The gallery of Francis I set a fashion in decoration imitated throughout Europe.

Page 480 *Iconoclasts in The Netherlands,* 1583.

Page 483 *Crescent Formation of the Spanish Armada,* 1588.

Page 488 David Vinckboons (Flemish-Dutch, 1576/1578–1632), *The Peasant's Misfortune.* Panel, 26.5 by 42 cm.

Page 490 *Mars and Venus United by Love,* ca 1580, by Paolo Veronese, Italian painter (1528–1588).

Page 499 From George Turberville, *The Noble Art of Venery* (1575).

Page 500 *Portrait of Juan de Paraja,* ca 1650, by Diego Rodriguez de Silva y Velasquez, Spanish painter (1599–1660). Oil.

Page 506 A senior merchant of the Dutch East India Company pointing out the Company's ships in Batavia Bay (now Djakarta, Indonesia) to his wife. Detail of a painting by Albert Cuyp (1620–1691).

Page 511 Philippe de Champaigne (French, 1602–1674), *Cardinal Richelieu Swearing the Order of the Holy Ghost.*

Page 512 Antoine Coysevox (French, 1640–1730), *Louis XIV,* statue for the Town Hall of Paris, 1687–1689.

Page 517 "The Noble Is the Spider," from Jacques Lagniet, *Receuil des Proverbes,* 1657–1663.

Page 518 *The Rape of the Sabine Women,* ca 1636–1637, by Nicolas Poussin, French painter (1594–1665). Oil on canvas.

eighteenth-century fact of life. Hangings were popular events, a spectator sport for rowdy crowds.

Page 644   After an engraving by R. Lehman of the foundlings' home called La Rota.

Page 649   *Famille de paysans,* ca 1640, by Louis Le Nain, French painter (1593–1648). The original can be seen in the Louvre.

Page 656   "The Remarkable Effects of Vaccination," an anonymous nineteenth-century Russian cartoon in the Clements C. Fry Collection of Medical Prints and Drawings, Yale Medical Library.

Page 659   From *L'Illustration,* 1855 (1ᵉ sem.), p. 309.

Page 660   W. W. Wheatley, "Dancing around the Church," 1848.

Page 666   Jacques-Louis David (French, 1748–1825), *The Tennis Court Oath.*

Page 673   John Trumbull (American, 1756–1843), *The Declaration of Independence.* Oil on canvas, 1786. 21⅛ by 31⅛ inches.

Page 674   E. C. Mills, *Signing of Treaty of Amity and Commerce and of Alliance between France and the United States.*

Page 675   This portrait of Louis XVI by Joseph-Siffrein Duplessis, French portrait painter (1725–1802), hangs in the Marie Antoinette Gallery at Versailles.

Page 679   This drawing by Persin de Prieur, "Premier assaut contre La Bastille," can be seen in the Musée Carnavalet, Paris.

Page 685   "Un Comité révolutionnaire sous la Terreur," after Alexandre Évariste Fragonard, French historical painter (1780–1850).

Page 689   Jacques-Louis David (French, 1748–1825), *Napoleon Crossing the Alps.* Oil on canvas.

INDEX

Henry the Navigator, prince of Portugal, 466, 470, 495
Herberstein, Sigismund, 557
Heresy, 368, 453, 457–458
Hermandades, 419
Hermitage Museum, 567
Hildebrandt, Johann Lukas von, 564
Hispaniola, 469
Historical and Critical Dictionary (Bayle), 583
History: Valla as historian, 397; Shakespeare's history plays, 498
Hobbes, Thomas, 528, 535
Hohenzollern family, 377, 547–552
Holland: 430, 470; leads southern Netherlands to independence from Spain, 481; Louis XIV invades, 519; 1702 war with France, 520; in 17th century, 532–535; agriculture in 17th, 18th century, 613–615; Anglo–Dutch wars, 624. *See also* Netherlands
Holy Roman Empire: lack of centralization, 415; rise of Habsburg dynasty, 439–440; Protestant Reformation's effect on, 436–443; and Thirty Years' War, 483–489; 1702 war with France, 520; abolished by Napoleon, 693
Homeopathy, 653
Horace, 497
Hospitals, 653–654
Hôtel–Dieu, 653–654
House of Commons. *See* Parliament, English
House of Lords. *See* Parliament, English
Huguenots, 445, 478, 510, 516–517
Humanism: secular, 396–397, 407, 428; Christian, 409–415, 428, 429, 443
Hume, David, 589
Hundred Years' War, 361–371
Hungary: Habsburgs conquer from Ottoman Turks, 545–547; nobles thwart Habsburgs, 547
Hus, John, 375, 376 (illus.), 434
Hutton, Ulrich von, 438

Iaroslav the Wise, prince of Kiev, 553
Iberian Peninsula: pluralism of, 419; united by Isabella and Ferdinand, 419–421
Illegitimacy: in 17th, 18th centuries, 638, 640–641
Imitation of Christ, The (Thomas à Kempis), 430–431
Immigration. *See* Migrations
Immunology, 656
Imperialism, Western: 18th–century competition for colonies, 624–629
Inca empire, 469
Income: in 18th–century colonies, 629
Indentured servants, 629
Index of Prohibited Books, 458, 589
India, 631; and Seven Years' War, 628
Indians, American, 469, 470; enslavement of, 495–496; and Spanish settlers, 632–633
Individualism, in the Renaissance, 395
Indulgences, 432–434, 456
Industry: cottage, 620–623, 641; growth limited in colonies, 629
Infanticide: in Renaissance, 407–408; in 18th century, 643–644

Infantry, 559
Infants: nursing of, in 18th century, 643. *See also* Children
Inflation: in 16th–century Spain, 473; after Thirty Years' War, 489
Innocent III, pope, 407
Innocent VIII, pope, 429
Inquisition: Spanish, 420–421, 445; Roman, 458
Institutes of the Christian Religion, The (Calvin), 445–446, 476
Ireland, 419; and Church of Ireland, 452; uprising in 1641, 527; under Cromwell, 529; potatoes as staple food, 650
Isabella, queen of England, 361–363
Isabella, queen of Spain, 416, 418, 419–421, 470
Isabella of Este, duchess of Mantua, 409
Islam: in Spain, 419, 420; racial views of, 496; and Ottoman Turks, 545–546
Israel. *See* Jews
Italy: origins of Renaissance in, 390–394; city–states, 392–394, 416; Napoleon's victories in, 688, 693. *See also* Rome
Ivan I, Russian prince, 555
Ivan III, prince of Moscow, 555–556
Ivan IV (The Terrible), tsar of Russia, 556–558

Jacobins, 683, 684
Jacquerie, 381
Jamaica, 469
James I, king of England, 498, 501, 525
James II, king of England, 530
James V, king of Scotland, 452
Janissary corps, 546
Jefferson, Thomas, 668, 672, 680
Jena, battle of, 693
Jenner, Edward, 656
Jesuits, 457, 483, 485, 658
Jesus Christ. *See* Christianity
Jews: attacked in 14th century, 383; in Spain, 420; welcomed by Cromwell, 529
Jiménez, Francisco, 420, 430
Joan of Arc, 367–368
Joanna of Castile, 421, 439
John, archbishop of Plano Carpini, 555
John II, king of Portugal, 466
John of Salisbury, 396
John of Spoleto, 374
Joliet, Louis, 516
Joseph II, king of Austria, 601, 603, 658
Judeo–Christian tradition, 669
Julius II, pope, 398, 429, 431, 448, 453
Junkers, 549, 552, 596
Justices of the peace, in Tudor England, 419
Justinian Code, 416

Karlsruhe, 565
Kay, John, 623
Kazan, 556
Kepler, Johannes, 575
Keynes, John Maynard, 577
Khaldun, Ibn, 496

Maryland, planter class in, 629
Mary of Burgundy, 439
Masaccio, 400
Massachusetts, beginnings of American Revolution in, 672
Masturbation, 655
Mathematical Principles of Natural Philosophy (Newton), 577
Mathematics: and scientific progress, 578; Descartes and, 580
Matthias, Holy Roman emperor, 484
Maupéou, René de, 599–600
Maximilian I, Holy Roman emperor, 394, 439
Mazarin, Jules, 511, 513
Mechanics: Galileo's, 575–577; Newton's, 577–578
Mecklenburg, dukes of, 486
Medici, Catherine de', queen of France, 475, 478
Medici, Cosimo de', 392, 400, 403
Medici, Giovanni de'. *See* Leo X, pope
Medici, Lorenzo de', 392, 394, 399
Medici, Marie de', 509
Medici, Piero de', 394
Medici family, 392, 404, 430, 577
Medicine: and bubonic plague, 358–360; in 18th century, 618–619, 651–656; early views of, on nutrition, 650–651
Medieval. *See* Middle Ages
Melton, William, 447
Memling, Hans, 414
Mendicants, 380
Mental illness, 654–655
Mercantilism: as principle of Spanish colonial government, 473–474; as practiced by Colbert, 515, 624; as colonial economic warfare, 624–633
Merchants: in cottage industry, 621–623
Merici, Angela, 457
Metternich, Klemens von: and Napoleon, 694
Mexico, 469, 631–632
Michael, tsar of Russia, 558
Michelangelo, 398, 400, 403
Michelet, Jules, 368, 643
Middle Ages, 385; late, crises in, 356–385; rejected by Renaissance, 395
Middle class: and Calvinist values, 526; rise of in France, 676–677
Middleton, Sir Gilbert de, 380
Migrations: of Europeans in 16th century, 464–474, 493–496
Milan: prosperity in 14th century, 390; as city-state, 392, 393; Spain in, 474
Militarism: of Prussians, 550–552
Millet, Jean François, 610, 611 (illus.)
Mirandola, Pico della, 397, 409
Mohammed II, sultan, 466, 472
Mohammedanism. *See* Islam
Molière, 519
Mona Lisa, 475
Monarchy: in Renaissance, 416, 420–421; Charles V's ideas of, 439–441; absolutist, 508–515, 519–524, 525; development of modern English, 529–532; influence on east, west systems, 542–543; Enlightenment's influence on, 596–603; fall of in France, 675–688
Monasteries: dissolved by Henry VIII, 450; in Spain, 522; contemplative orders abolished, 658

Mongols, 553–555
Monks. *See* Monasteries
Montague, Lady Mary Wortley, 655
Montaigne, Michel de, 497–498
Montesquieu, Charles-Louis, Baron de, 585–586, 593, 669
Montezuma, 469
More, Thomas, 411–412, 428, 450
Mortality rates, 360
Moscow, 694; rise of, 555–556
Motherhood. *See* Family; Women
Motion: Galileo's work on, 555–557; Newton's synthesis of findings on, 577–578
Mountain (French political group), 684
Mun, Thomas, 624
Münster, 446, 486
Music: French classicism, 519; baroque, 562
Muslims: racial views of, 496
Mussolini, Benito: defeats in Greece, 451

Naples: as city-state, 392; Spain in, 474
Napoleon I, emperor of France: rules France, 688–690; England defeats at Trafalgar, 693; becomes emperor, 693; European wars of, 693–695; defeated in Russia, abdicates, 694; defeated at Waterloo, 695
National Assembly of France, 677–678, 680, 681, 682
National Convention (France), 684, 687–688
Nationalism: stimulated by Hundred Years' War, 371; concept of state in Renaissance, 415–416; in Hungary, 547; and French Republic's success in war, 687
Nations. *See* Nationalism; State
Navarre, 419, 420
Navigation: 15th-century developments in, 473; and scientific revolution, 578–580; instruments developed for, 580
Navigation Acts, 624, 629
Nelson, Horatio, 693
Netherlands, 469–470; ravaged by Louis XI, 416; under Charles V, 478–479; revolt of, 479–482; United Provinces formed, 481, 488; in 17th century, 532–535; new farming system in, 613–614; and competition for colonial empire, 624–625; declares war on Britain, 672; and First Coalition against France, 683, 684, 685, 687; as part of Grand Empire, 693
Newfoundland, discovery of, 470
New Jersey, 629
New Testament. *See* Bible
Newton, Isaac, 577–578, 583, 585, 586
Nicholas V, pope, 395
Nijmegen, Treaty of, 519
Nikon, Patriarch, 558
Ninety-five Theses, Luther, 432–434
Nitrogen-storing crops, 612
Nobility: fur-collar crime by, 380; urban, formation of, 391; absolute monarchy's mastery over, 508; Richelieu breaks power of, 510; Louis XIV humbles, 513, 514, 515; in eastern Europe, 540–543; of Hungary, thwarts Habsburgs, 546; Prussian, 547, 549; Russian, 553, 556–559; under Catherine the Great, 597–598; under Louis XV, 598; and French Revolution, 676–678
Nördlingen, battle of, 486